Pioneering Sustainable Innovations in Renewable Energy Technologies

Kamal Hirech
Mohammed First University, Morocco

IGI Global
Scientific Publishing
Publishing Tomorrow's Research Today

Vice President of Editorial	Melissa Wagner
Managing Editor of Acquisitions	Mikaela Felty
Managing Editor of Book Development	Jocelynn Hessler
Production Manager	Mike Brehm
Cover Design	Phillip Shickler

Published in the United States of America by
IGI Global Scientific Publishing
701 East Chocolate Avenue
Hershey, PA, 17033, USA
Tel: 717-533-8845
Fax: 717-533-7115
Website: https://www.igi-global.com E-mail: cust@igi-global.com

Library of Congress Cataloging-in-Publication Data

LCCN: 2024060461 (CIP Data Pending)
ISBN13: 9798369399248
EISBN13: 9798369399262

British Cataloguing in Publication Data
A Cataloguing in Publication record for this book is available from the British Library.

Table of Contents

Detailed Table of Contents

In this article, we have improved the control of the multilevel inverter for a photovoltaic system connected to the grid, in order to minimize the total harmonic distortion of current. This topology being considered as a new converter topology applied to photovoltaic systems and has the advantage of very high-energy efficiency, but has the disadvantage linked to harmonics injected into the grid, which causes switching of states if the number of levels increases, which increases the rate of harmonic distortion. An improved control based on sinusoidal control (SPWM) and space vector control (SVPWM) was proposed to improve the control of multilevel inverters. The evaluation of our results obtained is based on the value of the total harmonic distortion(THD). This control strategy eliminates current harmonics and improves the quality of energy supplied to the grid by the photovoltaic system, and It can be seen that among all the control methods, the THD is minimum at the output of different multilevel inverter topologies

India is divided into many areas, each with its own distinct characteristics such as sunshine exposure, energy consumption habits, and physical factors. The goal is to create specialized solar power plans that meet the individual needs of each place. This includes assessing parameters such as solar irradiation, energy demand, and available land or infrastructure for solar panel installation. By tailoring methods to the circumstances of each region, it guarantees that solar energy resources are deployed optimally, easing India's transition to sustainable and dependable energy sources. The study will conduct a comparative examination of solar power generation in five different Indian regions. This comparison considers a

irradiance conditions. The simulation evaluates the boost converter's input and output characteristics, including voltage, current, and power, using two irradiance profiles: a smooth Gaussian profile representing a clear day and abrupt irradiance changes simulating a non-clear day. Results demonstrate the converter's efficiency and adaptability, providing insights for practical applications in renewable energy systems.

Solar irradiance forecasting plays a crucial role in renewable energy, weather prediction, and climate modeling. Accurate forecasts are essential for optimizing solar power efficiency, grid integration, and energy planning. Traditional forecasting methods, based on physical and statistical models, struggle to capture the complex, nonlinear relationships inherent in solar irradiance. To address these challenges, this chapter presents a comparative analysis of advanced machine learning (ML) and deep learning (DL) models. Techniques like CNNs, RNNs, and hybrid models have demonstrated strong capabilities in extracting temporal and spatial patterns from Solar data. The integration of K-means++ clustering with DL frameworks further enhances model robustness, generalization, and interpretability. This chapter evaluates hybrid models, such as Temporal CNN-LSTM, CNN-GRU using metrics based on Solcast data. Results reveal that the TCNN-GRU model outperforms other state-of-the-art approaches, underscoring the value of clustering-enhanced DL frameworks for accurate solar irradiance forecasting.

Green energy systems offer a promising solution to meet the increasing energy demands. In this paper, we introduce a supervisory approach for intelligent household power control. This approach integrates Plug-in Electric Vehicles (PEVs) with Lithium-ion battery packs and wind energy systems, linking them to the smart grid. The key objective is to investigate a power management strategy that utilizes a Wind turbine Generator system (WG) as a renewable energy source and PEVs not only as energy storage but also as an emergency power supply during peak demand periods. By smoothing the household power curve, this approach improves grid stability by filling valleys during off-peak times and shaving peaks during rush periods. The strategy adjusts power demand based on PEV storage levels and WG power generation, ensuring optimal utilization. Simulation results across nine different residential load scenarios demonstrate the effectiveness of this strategy in achieving a balance between demand and supply, fostering harmony within the system.

This paper, is dedicated to a Brackish Water Reverse Osmosis (BWRO) pumping/desalination unit systemic modeling. The studied system is likely to operate in remote off-grid communities lacking access to freshwater. Without battery storage, the unit is powered by hybrid PV/Wind-turbine generator to provide freshwater for domestic and agricultural uses. The investigated system includes three concurrent processes: pumping, desalination and irrigation putting ahead different disciplinary notably hydraulic, electric, mechanical and thermal domains. Therefore, based on bond graph technique, a multidisciplinary approach is adopted for such a multi-physical system modeling. The established model is assessed using BWRO test bench through an experimental characterization of the studied processes. The experimental results demonstrated the significant accuracy of the bond graph models obtained from simulation drawn with a dedicated software tool.

Photovoltaic systems, crucial for sustainable energy generation, often suffer from various material and electrical defects that can significantly hinder their performance and longevity. Material defects such as microcracks, delamination, and degradation of encapsulants affect the mechanical integrity of PV modules, leading to efficiency losses and increased failure rates. Electrical defects, including hotspots, mismatched cells, and connection issues, further exacerbate these problems, reducing the overall power output and system reliability. These defects not only compromise the efficiency of energy conversion but also have broader environmental implications. Increased failure rates necessitate more frequent replacements and disposal of PV modules, contributing to electronic waste and resource consumption. Furthermore, the energy payback time of PV systems is extended, delaying the environmental benefits of transitioning to renewable energy sources. Understanding these defects' nature and impact is vital for improving PV durability and maximizing environmental contributions.

This chapter focuses on the optimal control of a wind energy system based on DFIG by highlighting two optimal control approaches: Linear Quadratic Regulator based on a genetic algorithm, and robust LQMinMax regulator. The study aims to enhance the system's dynamic response, stability, and robustness under disturbances in stator voltages and wind speed. The LQR regulator minimizes errors in active/

reactive powers and DC-link voltage, with the genetic algorithm ensuring optimal tuning of the Q and R weighting matrices. The LQMinMax regulator minimizes errors and rejects disturbances in the worst-case scenarios. A linearized state-space model of the system, including the generator, wind turbine, grid-side filter, and DC-link, is developed in the d-q rotating frame. The dynamic performances of the proposed controllers are compared to PI controller and standard LQR regulator. Simulation results demonstrate that proposed controllers significantly improve time-domain specifications, stability, and robustness, with eigenvalue and participation factor analyses confirming these enhancements.

This study investigates the optical absorption and thermal conductivity properties of GaAsPN/GaP samples, fabricated under various growth conditions and post-growth annealing treatments. The results show that optical absorption is strongly influenced by arsenic content, with GaAsPN alloys containing 10% arsenic and 2% nitrogen offering optimal characteristics for tandem solar cells on silicon, due to a perfect lattice match and absorption in the target energy range of 1.7-1.8 eV. Thermal annealing significantly improves both optical absorption and thermal conductivity, reaching 38,000 cm^{-1} and 4 W/mK, respectively. These properties were characterized using Photothermal Deflection Spectroscopy (PDS), a crucial technique for assessing thermal conductivity. These findings highlight the importance of GaAsPN materials for developing high-performance multijunction solar cells, where efficient thermal management and a broad absorption range are key to maximizing photovoltaic conversion efficiency.

The present study focuses on how the variables like energy consumption, domestic material consumption, access to clean fuels for cooking and trade openness affect the economic growth and other explanatory variables like renewable energy use and access to clean fuels for cooking affect the Carbon dioxide emissions in the short run and long run of South Asian Economies. So, the study made an attempt to evaluate the impact of above-mentioned variables on economic growth and carbon dioxide emission using the Panel Autoregressive Distributed Lag (ARDL) method. All the preliminary tests including unit root assessments, cross-sectional dependence evaluation, Kao-co-integration etc. were used before applying the ARDL panel model. There is a positive impact of energy consumption, access to clean fuels along with the trade openness (measured by trade¬ as % of GDP) on economic growth. However, the increased domestic material consumption has a negative impact on GDP per capita. Addition to this, the renewable energy consumption is significant and negatively impacts the emissions.

The integration of renewable energy into sustainable development strategies is a global challenge. Teaching these concepts requires innovative pedagogical approaches that combine educational effectiveness with technical complexity. This paper explores the use of machine learning (ML) systems to enhance the teaching of renewable energy and sustainable development. We propose a pedagogical architecture using adaptive learning algorithms to analyze student performance and interactions, personalizing learning paths. The system adjusts teaching modules in real-time, providing individualized support. We also implement predictive models simulating the impact of renewable energy on ecosystems and infrastructure through interactive simulations, helping students understand sustainability challenges. Preliminary results show improved knowledge retention and better understanding of energy transition concepts. We discuss the pedagogical implications of ML systems in higher education and suggest directions for their large-scale implementation

In this work, natural convection in a grooved rectangular cavity with aspect ratio R =2 is analyzed using a hybrid numerical approach combining the lattice Boltzmann method and the finite-difference technique. The introduction of two grooves in the cavity geometry aims to improve heat transfer passively, using air as the fluid with a Prandtl number of 0.71. The results show a significant improvement in heat transfer efficiency, with a 76% gain in Nusselt number at a Rayleigh number (Ra) set at 10^3 and 43% at Ra=10^5, compared with the conventional Rayleigh-Bénard configuration. These improvements are attributed to modified flow structures and deformed isotherms near the grooves, favoring complex fluid dynamics and improved heat exchange. These results suggest that grooved geometries can be used effectively to enhance heat transfer in natural convection, making them suitable for a variety of industrial applications.

This paper presents a numerical study of magnetohydrodynamic (MHD) natural convection in a three-dimensional L-shaped cavity using the lattice Boltzmann approach coupled with the finite difference method. The analysis focuses on the effect of a uniform vertical magnetic field on heat transfer, isotherms, and velocity field. The numerical code was validated by comparing our results with data available in the literature. The simulations revealed that, in the absence of a magnetic field, a significant improvement in heat transfer of 93.27% on the left vertical hot wall and 225.65% on the bottom hot wall was observed when the Rayleigh number (Ra) increased from 10^3 to 10^5. In the presence of the magnetic field, heat transfer decreased by 47.28% on the left vertical wall and 66.58% on the bottom wall, when the Hartmann number was increased from zero to 200 for Ra fixed at 10^5.

Conventional extrarenal technologies are often energy-hungry and responsible for substantial greenhouse gas emissions. In response to this problematic, the Moroccan Wearable Artificial Kidney (MorWAK) project proposes an alternative, more sustainable solution in the context of green dialysis. MorWAK differs radically from existing extrarenal technologies, which consume a huge amount of energy to ensure the filtration processes (the consumption of a hemodialysis generator per 4-hour session is around 3 kWh, and that of the water treatment unit to produce purified water is 150 kW/d). On the other hand, the MorWAK operates without being connected to a water source, consuming very little energy (10 W). This contributes to eco-responsible energy management, which has a direct impact on reducing greenhouse gas emissions. MorWAK thus represents a breakthrough innovation in improving the sustainability of healthcare technologies, offering a more eco-friendly alternative to current energy challenges

This research aims to study flow control for reducing aerodynamic drag and cooling a rectangular electronic component using an inclined partition. Unlike existing studies focusing on horizontal or vertical plates, this work numerically investigates the impact of an inclined partition on flow and temperature fields using the lattice Boltzmann method with multiple relaxation times. Results show that increasing the Reynolds number enhances heat exchange and reduces the drag coefficient, with a maximum Cd reduction of 125.41% at Re = 250. Additionally, at Re = 150, positioning the partition at g = 2d achieves a Cd reduction of 130.97%. For effective cooling of electronic components, placing a sloping partition at g = 2d before the component yields significant energy savings.

In this research, a numerical study of airflow control coupled to heat transfer around a heated square cylinder is carried out at a fixed Reynolds number Re= 150. The effect of the position and length of two partitions implemented downstream of the square obstacle is examined. The numerical simulations are performed using the lattice Boltzmann method with multiple relaxation times. The results obtained are presented in the form of graphs of the isotherms, the temporal and mean drag coefficients and as a function of the mean Nusselt number. These results show that the successive inclusion of control partitions has a significant impact on the increase of the reduction percentage of drag coefficient that reaches 30.74% in a critical case. In a parallel implementation of the two partitions, this reduction percentage reaches

17.26%, with an enhancement of over 10% in heat exchange between the air and the heated obstacle's top, bottom, and rear surfaces compared to the uncontrolled case.

Today's manufacturing industry consumes a substantial amount of energy, so controlling energy usage and improving energy efficiency in Industry 4.0 are key priorities, both for the environment and for the competitiveness of our industry. At the same time, manufacturing industries, particularly energy-intensive ones, are placing increasing emphasis on reducing energy consumption. In this respect, energy-saving programming has proved effective in reducing energy consumption. In this context, this article focuses on the Industry 4.0 issues of coordinated scheduling of production and preventive maintenance activities over fixed and flexible periods, with a view to minimizing total energy consumption. In our study, we examine both types of unavailability constraints. We propose two models based on mixed-integer binary programming to determine optimum schedules, respectively. We evaluate the effectiveness of these two types of programming in mixed integer binary form using exact data separation B&B method and seamless evaluation using the CPLEX software.

This study aims to forecast Ghana's carbon dioxide (CO2) emissions from 2015 to 2027, using the Autoregressive Integrated Moving Average (ARIMA) model to examine time-series dynamics of emissions. The purpose is to shed more light on the future profile of emissions and underscore the value of sustainable development practices in the country. This study uses time-series data of CO2 emissions from Ghana for the period 2015 to 2016. The ARIMA model is used in the analysis and prediction of the data to determine the future trend in carbon emissions. This model accounts for the historical patterns and trends in emissions, ensuring a robust and reliable forecast. An upward trend in the Ghanaian CO2 emissions emerges, with a projected rise to 3.10 kilotonnes (kt) of CO2 per annum to 2027. Results point to the rising challenges of emission management and demand effective intervention to check on ecological impacts. This trend spells out the need for the needful strategies in Ghana towards attaining carbon footprint reduction and fostering sustainability.

A polyhouse, known as a greenhouse, is a building constructed for the controlled growing of plants. It is built of translucent materials such as glass or plastic and is intended to offer a regulated temperature for plants, shielding them from harsh weather but allowing sunlight to pass through. The chapter considers cost, practicability, sensors values, emerging IoT methods, employing arduino microcontroller with remote control, and building a smart greenhouse monitoring system with performance monitoring. This technology will assist farmers in avoiding physical visits to the field while increasing productivity by maintaining precise parameters such as CO2, soil moisture, temperature, and light in the greenhouse via IoT.

This study assessed biomass briquettes as a sustainable energy alternative in Bule Hora town, Ethiopia, addressing waste and traditional biomass reliance. Feedstocks (coffee husks, sawdust, maize cobs, and waste paper) were carbonized and compacted. Briquette properties were analyzed using ASTM standards, with ANOVA, regression, and RSM applied for optimization. Maize cobs had the highest calorific value (1820.54±0.37 kcal/kg). Carbonized briquettes excelled in density (0.61 g/cm^3) and calorific value (4670.92 kcal/kg). Compression pressure, binder content, and carbonization temperature significantly impacted performance (p < 0.01). Optimal conditions: 12 MPa pressure, 11% binder, 450°C temperature, and 16% moisture. Surveys showed 70% willingness to adopt briquettes for reduced pollution and cost savings, despite awareness and market challenges. Biomass briquetting offers a solution to deforestation, energy security, and waste management, recommending scaled production, optimized processes, and awareness initiatives.

Environmental degradation remains a critical global challenge, driving many countries to prioritize financial development as a key enabler of renewable energy adoption. Additionally, urbanization presents opportunities for expanding the use of sustainable energy sources. In this context, the present study examines the impact of financial development, renewable energy consumption, green investment, urbanization, and economic growth on environmental quality in Tunisia from 1990 to 2020. Utilizing both linear and non-linear autoregressive distributed lag (ARDL and NARDL) models, the study assesses short-term and long-term dynamics while also employing the Granger causality test to explore causal relationships. The ARDL and NARDL bounds tests confirm the existence of a long-term equilibrium relationship among the variables. The findings indicate that economic growth significantly contributes to rising carbon emissions in both the short and long run. The non-linear results further reveal that while a negative energy shock improves environmental quality over the long term, a positive energy shock enhances environmental quality in the short term. This underscores the significant role of renewable energy consumption in reducing CO_2 emissions in Tunisia. However, green investment appears to have only a marginal impact on environmental quality. To effectively combat climate change and foster sustainable economic growth, Tunisia must implement an environmentally friendly development strategy and further investigate the interplay between technological innovation and renewable energy. Accordingly, policymakers should actively promote and incentivize the adoption of innovative clean technologies in renewable energy production.

Preface

The transition to a sustainable energy future is one of the most pressing challenges of our time. As climate change accelerates and energy demands grow, the need for innovative solutions in renewable energy technologies has never been more urgent. The advancements in this field are not only transforming global energy systems but also driving economic growth, improving energy security, and fostering environmental sustainability. This book, *Pioneering Sustainable Innovations in Renewable Energy Technologies*, serves as a platform to explore the latest breakthroughs and their implications for a cleaner, more resilient future.

The chapters in this volume present a diverse range of research, from emerging photovoltaic materials and wind energy innovations to smart grid integration and AI-driven energy management. By bringing together leading researchers, industry experts, and policymakers, this book provides a comprehensive perspective on how technological advancements are shaping the renewable energy landscape. Through rigorous analysis and case studies, the contributors examine the practical applications and policy frameworks needed to scale these innovations effectively.

A key objective of this work is to bridge the gap between academic research and real-world implementation. The insights presented here are intended to support decision-makers in developing policies that encourage sustainable energy adoption while also providing researchers and industry professionals with the technical knowledge needed to drive progress. Additionally, by addressing topics such as financing renewable energy, off-grid solutions, and the socioeconomic impacts of clean energy transitions, this book highlights the broader implications of renewable energy beyond technological development.

This volume is intended for a wide audience, including researchers, engineers, educators, industry leaders, and policymakers. Whether you are exploring cutting-edge innovations, seeking practical solutions for integrating renewable energy into existing infrastructures, or examining the role of policy and regulation, this book offers valuable insights that will deepen your understanding of the field.

As editor, I extend my gratitude to the esteemed contributors whose expertise and dedication have made this volume possible. Their work underscores the transformative potential of renewable energy technologies in building a sustainable world. It is my hope that this book will inspire further research, collaboration, and action towards a cleaner and more sustainable energy future.

Chapter 1 presents an improved control strategy for multilevel photovoltaic (PV) inverters to minimize total harmonic distortion (THD). By combining sinusoidal pulse-width modulation (SPWM) and space vector control (SVPWM), this approach enhances energy quality and eliminates current harmonics in grid-connected PV systems. The findings confirm that the proposed control method significantly reduces THD across different multilevel inverter topologies, improving power efficiency and reliability.

Chapter 2 explores solar energy potential across different Indian regions using the System Advisor Model (SAM) software. By analyzing solar irradiation, energy demand, and economic feasibility, the study provides a comparative evaluation of solar power generation in five key regions. The assessment considers parameters such as levelized cost of energy, annual AC energy production, and performance ratio, ensuring optimal solar resource utilization tailored to each region's needs.

Chapter 3 delves into Maximum Power Point Tracking (MPPT) control strategies for wind turbines, aiming to maximize energy extraction. Through aerodynamic modeling and control implementation, including indirect speed control, the study demonstrates enhanced wind energy conversion efficiency. Simulations validate system stability, optimizing rotor speed and ensuring balanced sinusoidal stator currents, thus underscoring the importance of advanced control algorithms in wind farm operations.

Chapter 4 introduces an optimized transformer-less PV inverter design for grid operation utilizing super junction MOSFETs and SiC diodes. Addressing challenges related to switching losses and leakage currents, the proposed inverter achieves high efficiency (98.5%) by maintaining a stable common-mode voltage and minimizing waveform distortions. Experimental validation highlights its potential for next-generation grid-connected PV systems.

Chapter 5 presents an off-grid PV solar design system integrated with an IoT-driven software-defined network (SDN). By leveraging artificial neural networks (ANN) for solar radiation prediction and SDN for real-time energy distribution management, the system improves grid performance through enhanced load balancing, energy optimization, and demand response. The findings emphasize the role of automation and IoT in future renewable energy networks.

Chapter 6 examines PV energy harvesting using boost converters under dynamic irradiance conditions. Through MATLAB/Simulink simulations, the study evaluates the performance of MPPT algorithms in adapting to fluctuating sunlight conditions. Results demonstrate the effectiveness of the boost converter in maintaining efficient energy transfer, offering valuable insights for renewable energy applications.

Chapter 7 investigates solar irradiance forecasting using K-means++ clustering combined with hybrid deep learning models. By integrating convolutional neural networks (CNNs) and recurrent architectures (GRU, LSTM), the study enhances predictive accuracy. Performance analysis using Solcast data reveals that the TCNN-GRU model outperforms conventional approaches, highlighting the significance of clustering-enhanced deep learning frameworks in solar forecasting.

Chapter 8 proposes a power management strategy for wind energy conversion systems and plug-in electric vehicles (PEVs) in residential applications. By optimizing power flow between wind turbines, smart grids, and PEVs, the study enhances grid stability. Simulation results validate the strategy's ability to balance energy demand and supply, mitigating peak loads and improving overall system efficiency.

Chapter 9 explores bond graph modeling for an off-grid brackish water pumping and desalination unit powered by hybrid PV/wind energy. Through a multidisciplinary approach encompassing hydraulic, electrical, mechanical, and thermal domains, the study presents a systemic model validated against experimental test bench results. The findings demonstrate the feasibility of using renewable energy for freshwater production and agricultural irrigation in remote communities.

Chapter 10 addresses material and electrical defects in photovoltaic systems and their environmental implications. By analyzing common failures such as microcracks, hotspots, and encapsulant degradation, the study highlights the impact of these defects on system longevity and electronic waste. Understanding these challenges is crucial for improving PV durability and maximizing the environmental benefits of solar energy.

Chapter 11 presents optimal control strategies for wind energy systems based on doubly-fed induction generators (DFIG). By comparing Linear Quadratic Regulator (LQR) methods with genetic algorithm optimization and robust LQMinMax control, the study enhances system stability and dynamic response under varying wind conditions. Simulation results confirm the superiority of the proposed controllers over conventional PI regulators, ensuring reliable wind energy integration.

Chapter 12 investigates the opto-thermal properties of GaAsPN materials for tandem solar cells on silicon. Through photothermal deflection spectroscopy (PDS), the study evaluates the impact of arsenic and nitrogen content on optical absorption and thermal conductivity. Findings indicate that optimized GaAsPN compositions enhance photovoltaic performance, making them suitable for high-efficiency multijunction solar cells.

Chapter 13 explores the intricate relationships between energy consumption, trade openness, and domestic material consumption on economic growth, as well as the impact of renewable energy and clean fuels on carbon dioxide emissions in South Asian economies. Employing the Panel Autoregressive Distributed Lag (ARDL) method, the study rigorously evaluates these connections through various econometric tests. The findings indicate that energy consumption and trade openness positively contribute to GDP, while excessive material consumption negatively affects economic performance. Additionally, renewable energy consumption significantly reduces emissions, reinforcing the importance of sustainable energy policies in fostering economic growth without exacerbating environmental degradation.

As global energy transition challenges intensify, chapter 14 presents an innovative approach to education through machine learning (ML) systems. By integrating adaptive learning algorithms, this study proposes a pedagogical model that personalizes learning experiences, dynamically adjusting teaching modules based on student interactions and performance. Additionally, predictive models are implemented to simulate renewable energy impacts on ecosystems and infrastructure. Preliminary results demonstrate improved knowledge retention and a deeper understanding of sustainability issues, emphasizing the potential of ML-driven education to enhance the effectiveness of teaching renewable energy and sustainable development concepts.

Chapter 15 delves into the enhancement of natural convection heat transfer within a grooved rectangular cavity using a hybrid numerical approach. By integrating the lattice Boltzmann method with finite-difference techniques, the study demonstrates that modified cavity geometries significantly improve heat transfer efficiency. The research finds that introducing grooves results in substantial gains in Nusselt numbers, with efficiency improvements varying by Rayleigh numbers. These findings suggest that optimized grooved geometries can serve as passive solutions for improving thermal management in industrial applications.

Chapter 16 presents a numerical investigation into magnetohydrodynamic (MHD) convection heat transfer in an L-shaped cavity using the lattice Boltzmann method. The study examines the influence of a uniform magnetic field on thermal flow dynamics, revealing that while increasing Rayleigh numbers enhance heat transfer significantly, applying a magnetic field dampens this effect. These results underscore the complex interplay between magnetic fields and convective heat transfer, providing insights into their applications in engineering and industrial thermal management.

Addressing the high energy demands of conventional dialysis technologies, chapter 17 introduces the Moroccan Wearable Artificial Kidney (MorWAK) project. By operating with minimal energy requirements and eliminating the need for a continuous water source, MorWAK presents a groundbreaking advancement in sustainable dialysis solutions. This innovation contributes to reducing greenhouse gas emissions while maintaining effective filtration processes, offering a viable alternative for improving the sustainability of healthcare technologies.

Chapter 18 investigates aerodynamic drag reduction and cooling efficiency of a rectangular electronic component through flow control techniques. Using the lattice Boltzmann method, the study finds that employing an inclined partition significantly reduces drag coefficients and enhances heat dissipation at various Reynolds numbers. The results suggest that strategic positioning of partitions can optimize energy efficiency in electronic cooling systems, making this a valuable contribution to the field of thermal management in electronics.

Focusing on airflow control and heat transfer efficiency, chapter 19 explores the impact of dual partitions on fluid dynamics around a heated square cylinder. Through numerical simulations, the findings indicate that the inclusion of partitions significantly reduces drag coefficients while improving heat exchange. These insights offer potential applications in energy-efficient thermal regulation and aerodynamic design.

In light of increasing energy demands in manufacturing, chapter 20 examines energy-efficient scheduling strategies in Industry 4.0. The study introduces two mixed-integer binary programming models to optimize production and maintenance scheduling, reducing overall energy consumption. Using the CPLEX software for evaluation, the findings highlight the potential of intelligent scheduling to enhance industrial sustainability and operational efficiency.

Chapter 21 employs the ARIMA model to forecast Ghana's carbon emissions trajectory from 2015 to 2027. The analysis reveals an upward trend in CO2 emissions, emphasizing the need for proactive sustainable development policies. The study underscores the importance of emission management strategies to mitigate ecological impacts and promote long-term environmental sustainability in Ghana.

With agriculture increasingly reliant on technological advancements, chapter 22 explores the implementation of IoT-based smart polyhouses. Utilizing Arduino microcontrollers and sensor-based monitoring, the study demonstrates how automation can optimize greenhouse conditions while reducing manual labor. This innovation offers promising prospects for improving agricultural efficiency and sustainability.

Addressing energy security and waste management challenges, chapter 23 evaluates biomass briquettes as a sustainable energy alternative in Ethiopia. By analyzing the calorific value, density, and compressibility of various feedstocks, the findings demonstrate the viability of biomass briquettes in reducing reliance on traditional fuels. The research advocates for scaled-up production and market awareness initiatives to enhance biomass adoption.

Chapter 24 investigates the relationship between financial development, urbanization, and environmental sustainability in Tunisia using the ARDL model. The findings reveal that while economic growth increases carbon emissions, renewable energy consumption and urbanization contribute to emissions reduction. These results highlight the need for policies that balance economic growth with environmental sustainability through strategic green investments and renewable energy adoption.

The research and insights presented in this book illustrate the transformative power of renewable energy technologies in addressing the global energy crisis. From innovative materials and energy-efficient systems to policy frameworks and socioeconomic impacts, this volume offers a comprehensive exploration of the many dimensions of renewable energy development. The contributions of the authors provide valuable perspectives on both the challenges and opportunities that lie ahead in the transition to sustainable energy.

It is my hope that this book will serve as an essential resource for researchers, educators, industry professionals, and policymakers committed to advancing renewable energy technologies. By fostering interdisciplinary collaboration and knowledge exchange, we can collectively drive the innovation needed to secure a cleaner, more sustainable future for generations to come.

Chapter 1
A Multilevel PV Inverter Control With Minimal Total Harmonic Distortion (THD)

Lamreoua Abdelhak
https://orcid.org/0000-0002-4015-0957
Independent Researcher, Morocco

Benslimane Anas
National School of Applied Sciences, University of Mohammed I, Oujda, Morocco

Bouchnaif Jamal
https://orcid.org/0000-0003-1653-4608
Higher School of Technology, University of Mohammed I, Oujda, Morocco

Hirech Kamal
Higher School of Education and Training, University of Mohammed I, Oujda, Morocco

El Ouariachi Mostafa
Higher School of Technology, University of Mohammed I, Oujda, Morocco

ABSTRACT

In this article, we have improved the control of the multilevel inverter for a photovoltaic system connected to the grid, in order to minimize the total harmonic distortion of current. This topology being considered as a new converter topology applied to photovoltaic systems and has the advantage of very high-energy efficiency, but has the disadvantage linked to harmonics injected into the grid, which causes switching of states if the number of levels increases, which increases the rate of harmonic distortion. An improved control based on sinusoidal control (SPWM) and space vector control (SVPWM) was proposed to improve the control of multilevel inverters. The evaluation of our results obtained is based on the value of the total harmonic distortion(THD). This control strategy eliminates current harmonics and improves the quality of energy supplied to the grid by the photovoltaic system, and It can be seen that among all the control methods, the THD is minimum at the output of different multilevel inverter topologies

DOI: 10.4018/979-8-3693-9924-8.ch001

INTRODUCTION

Over the past few years, power electronic converters, especially DC-AC and AC-DC converter types, have been able to find their place in the industrial power market and are today used in various applications such as such as renewable energy systems, electric drives, electric vehicles, and so on (Lamreoua et al., 2021; Azeem, et al., 2019). Due to this great attention of the industry towards power electronic converters, many researches have comprehensibly studied different aspects of them, including the topology structure, the method of control and the future of applications (Lakshmi et al., 2013). Multilevel topologies have emerged to overcome the drawbacks that prevent the use of power converters in higher voltage and current applications (Lamreoua et al., 2022). Various multilevel converter topologies have been introduced, each of which has an important advantage that makes them more suitable for particular applications (Kavali & Mittal, 2014). While the main multilevel topologies focus on the use of many semiconductor devices and main source DC circuits, later, many efforts have been made to optimize conventional multilevel structures or to design a new optimized structure (Natarajan & Rao, 2020; Khan, et al., 2021). The optimization was a compromise between the number of output voltages, the number of active and passive devices, and the DC sources.

Generally, the switching techniques of multilevel converters are classified into high and low switching frequency methods. Since the multilevel converter includes a considerable number of semiconductor devices, they can generate notable power losses, particularly in high power applications, if high switching frequency modulation techniques are used (Khalid et al., 2021; Lamreoua, et al., 2022). On the other hand, the low frequency modulation technique should be optimized in an adaptable manner depending on the targeted application in order to improve the system efficiency and converter performance (Lamreoua et al., 2022). In this article, a study and improvement of the control of multilevel inverters and their modulation techniques. An improved control based on sinusoidal control (SPWM) and space vector control (SVPWM) was proposed to improve the control of multilevel inverters. This control is composed of two regulation loops the first allows to regulate the active and reactive currents injected into the grid (Odeh, et al., 2021). The PR controllers are implemented to adjust the grid currents in the synchronous frame (Althobaiti, et al., 2016). To generate the reference current and maintain synchronism between the inverter and the grid, a Phase-locked loop technique (PLL) can be used. The main advantage and objective of this method is to effectively compensate the harmonic current content of the grid current. Multilevel converters provide higher quality waveforms, reducing current and voltage ripple after reducing losses caused by high frequency harmonics. Less expensive and less bulky passive components are used, which is to reduce manufacturing costs, gain in efficiency, reliability and energy quality. The main objective is to address, in terms of cost, efficiency, power management and power quality, the problems found with classic three phase photovoltaic inverter. After comparing the results of the two system, we notice that the current THDi of the interlaced inverter is lower than that obtained with the classical inverter. The effectiveness of these techniques proposed in this article is demonstrated by the simulation results using the MATLAB / SIMULINK environment.

DESCRIPTION OF OUR RESEARCH SYSTEM

The shape of the proposed three-phase photovoltaic inverter is proven in Figure 1. It includes a PV generator as a energy source, an MPPT block to make certain the tracking of the MPPT, a transistor bridge which include a current converter provided through the PV generator and on the cease a clear out to put off harmonics at excessive frequencies.

Figure 1. Diagrams of the proposed photovoltaic inverter with the grid

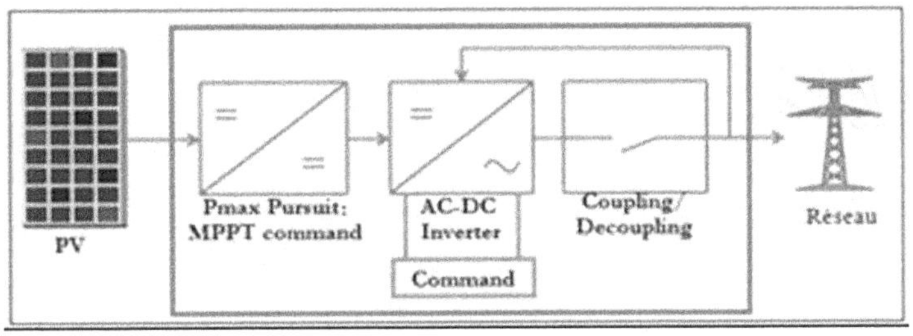

THREE-PHASE MULTI-LEVEL INVERTER STRUCTURE.

1. Cascading H-Bridge structure

Cascading H-Bridge seven-level inverter, are obtained by connecting two or more single-phase HBridge inverters, each consisting of four switches, in series. Which gives multiple voltage levels at the inverter. Thus, to obtain a seven-level voltage, this inverter consists of 3 H-bridges in series, i.e. 12 switches and a maximum output voltage of 3Vdc. Figure 2 shows the CHB seven-level for one phase.

Figure 2. Seven-level CHB inverter (one phase)

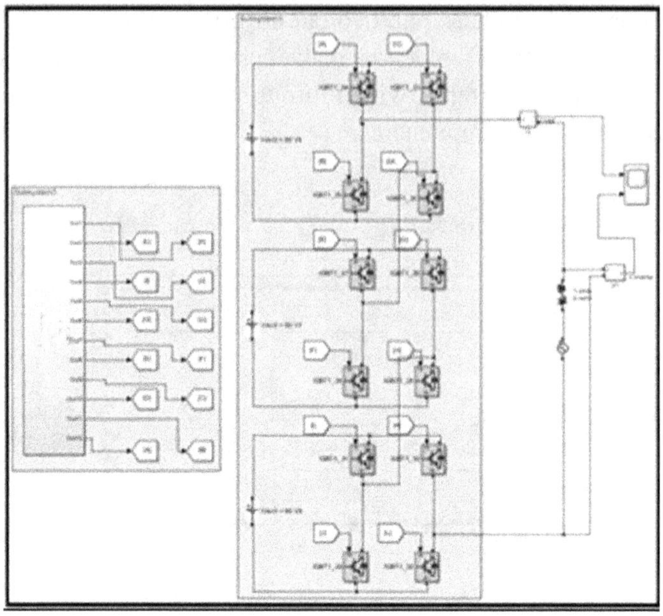

Figure 3. Multi-level Interlaced inverter (one phase)

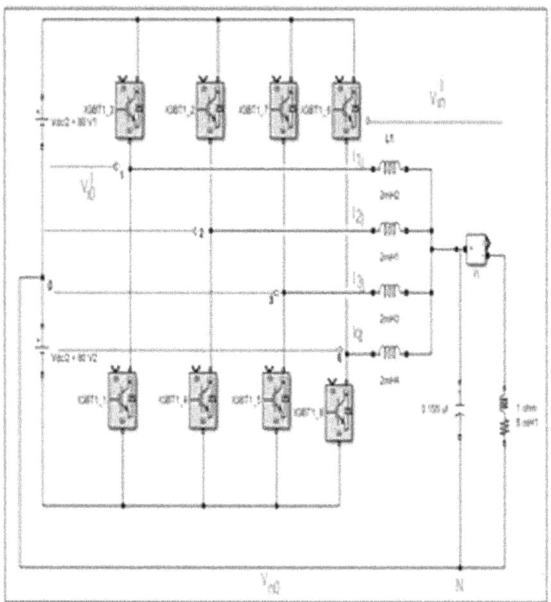

2. Interlaced inverter structure

Figure 3 shows the proposed architecture of the system *PV* connected to the power grid using an interlaced inverter (q =4). Our system is consist of a voltage source *Vdc*, an interlaced inverter, a filter and the grid. In our case study, The $q = 4$ inductors $L1$ represent the filter inductances on the inverter side. From the figure 3, we will note that the quantities, $Vi0j$ is the voltage vector of cells, $Vinj$ is the voltage vector on the inverter side, $Vj2$ is the voltage vector of the grid, iij is the vector of the current flowing through the cells of switching of the inverter. q the number of cells, ijc is the current through the capacitor; $ij1$ and $ij2$ are the output currents on the inverter and grid. The subscript j denotes the number of the phase (j = 1, 2, 3) while i denotes the row of interlaced cells of a phase j (i = 1, 2...q). The inverter is controlled by open loop and closed-loop pulse width modulation (MLI). The diagram of the control command of the inverter is shown in Figure. 3. The control of the three arms is obtained by comparing a modulating sine wave ($Vj*$; j = 1,2, 3) of frequency f = 50Hz with q high frequency triangular carrier waves (fsw = 20 kHz). The q carriers are offset by Tsw/q (Tsw being the switching period) and the three-reference sinusoidal voltages $Vj*$ are phase-shifted by $2\pi/3$. The two switches of the same switching cell are complementary $Ki = \overline{Ki}'$ (i = 1, 2, 3...).

In this part, we add a balancing loop to the previous command. This control contains two current regulation loops: an external loop for controlling the output current and a loop internal for cell current balancing (Figure 5).

3. Seven-level packed U-Cell inverter structure:

The three-Phase seven-level MPUC inverter is shown in Figure 4. The PUC type multilevel inverter delivers high range of power quality using a small number of active and passive components. This topology comprises a set of arms with two switches and a capacitor. If we have a converter of n arms, the latter will be composed of 2n switches and n-1 with 2n-1 levels. Thus, our seven-level converter composed of six switches and two capacitors per phase, on the other hand that of a cascade multi-level inverter consists of 12 switches per phase. It can be noticed that the PUC inverter are only two redundant switching states for the zero-voltage vector and the remaining six states give a unique voltage level at the output. Depending on the voltage and current polarities, the cell capacitors are charged, discharged or bypassed. The PUC-cell inverter has the most minimum number of components compared to the various other structures of multilevel converters.

Figure 4. Three-phase seven-level packed U-Cell Inverter

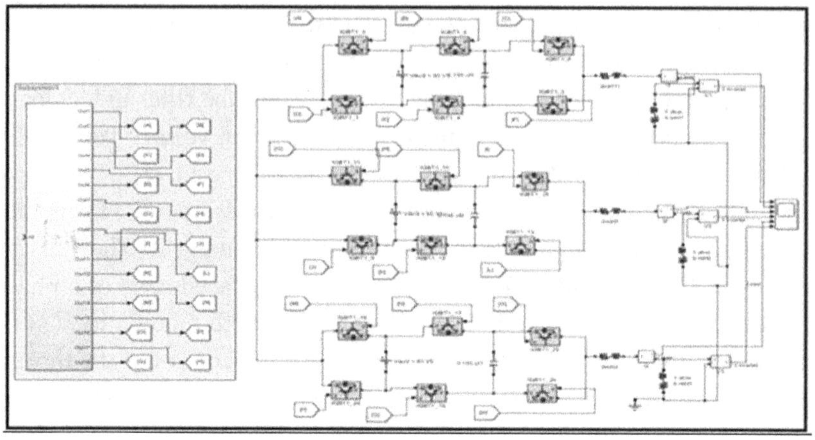

CONTROL METHOD:

- *The PQ current control unit (Figure 5).*

In this part, we directly control the output current into the load, being identical to the sum of the currents of the cells. The current loop generates the reference voltages $Ud*$ and $Uq*$ which are transformed in reverse Park to obtain three sinusoidal voltages in the reference abc.

The transfer function of our system is given by the matrix equation in coordinate's dq:

$$G_{PI}^{(dq)}(s) = \begin{bmatrix} K_p + \frac{K_i}{s} & 0 \\ 0 & K_p + \frac{K_i}{s} \end{bmatrix} \tag{3}$$

- *PR Regulator.*

The Proportional Resonant Regulator for the current is given by the following expression: $G_{PR}(s) = K_P + K_I \frac{S}{S^2 + 0^2}$ (4)

Wherever KP and KI succinctly represent the proportional gain and the integral gain.

Figure 5. Block illustration of the proposed control system

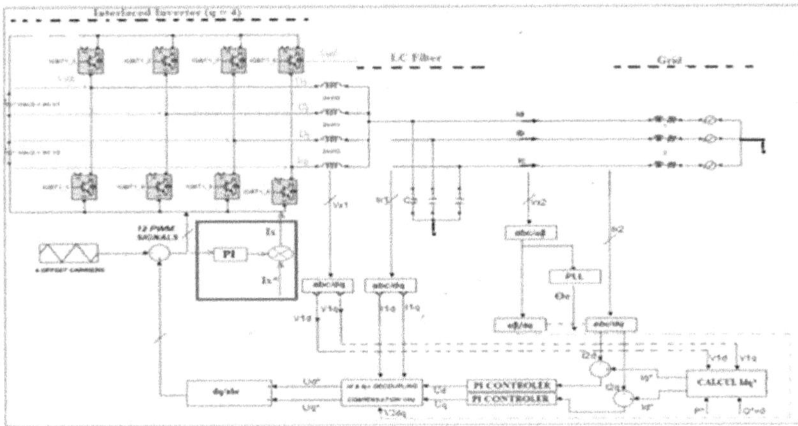

■ *Transport Delay Based PLL structure.*

The principle of PLL is to use a PI controller to reduce the phase difference between the line phase angle and the voltage output phase angle to zero (Figure 6).

Figure 6. Phase locked loop algorithm

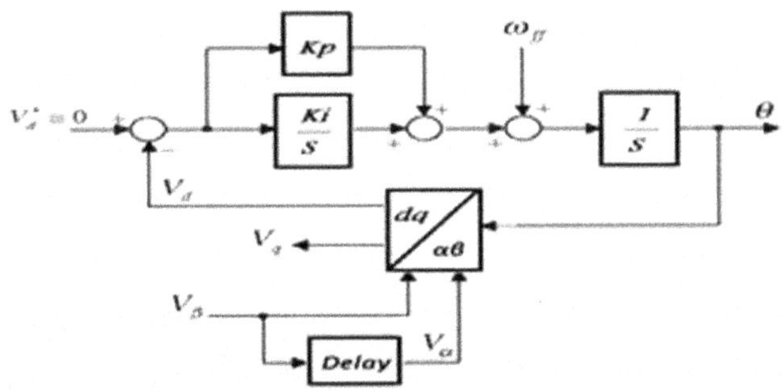

PARAMETERS

Table 1. Simulation Parameters

PV Power	500W
Vd	80V
Imin	1A
fg	50Hz
fpwm	10KHz

SIMULATION RESULTS

After recalling the operating principle of the SPWM and SVPWM command, we will model it under the environment of the MATLAB / SIMULINK software

Figure 7. Simulation Results

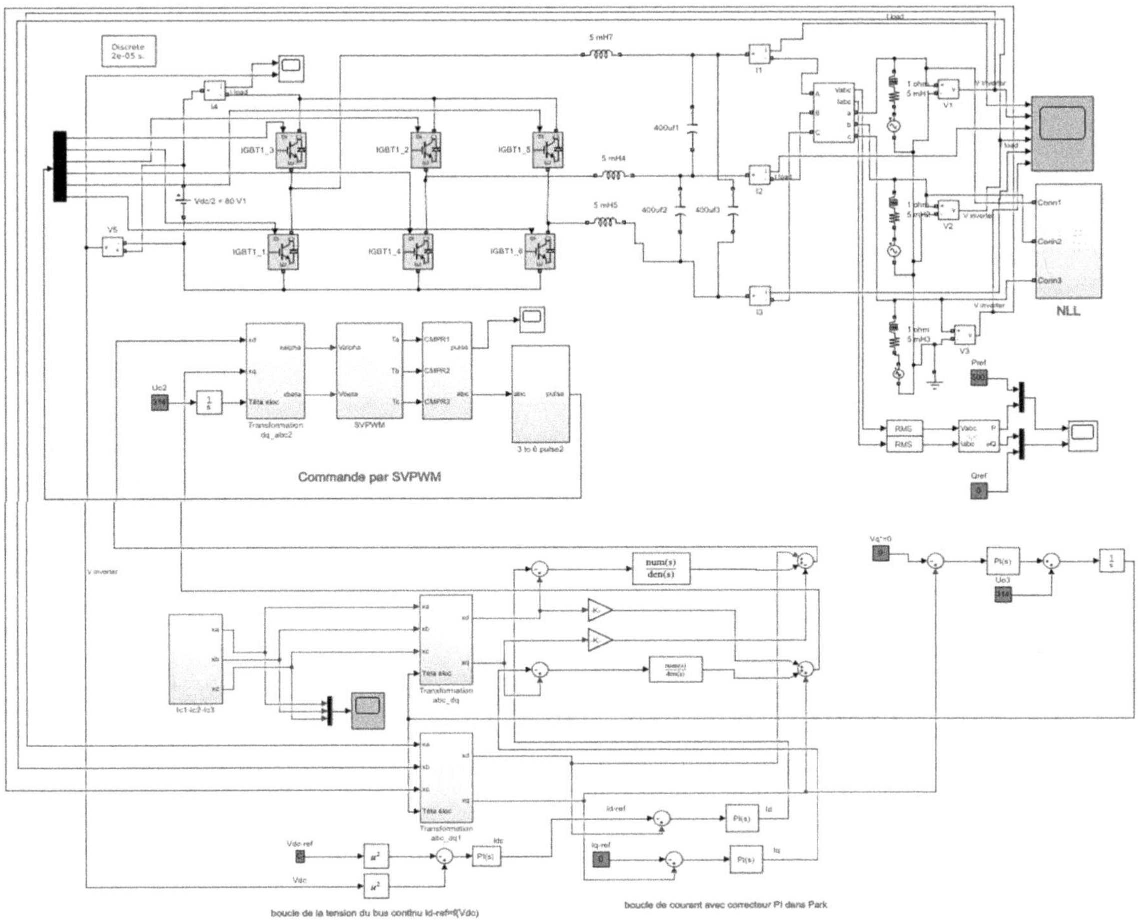

CASCADING H-BRIDGE SEVEN-LEVEL INVERTER

- *Sinusoidal Pulse with modulation control (SPWM):*

Figure 8. FFT spectrum current and voltage by the SPWM control

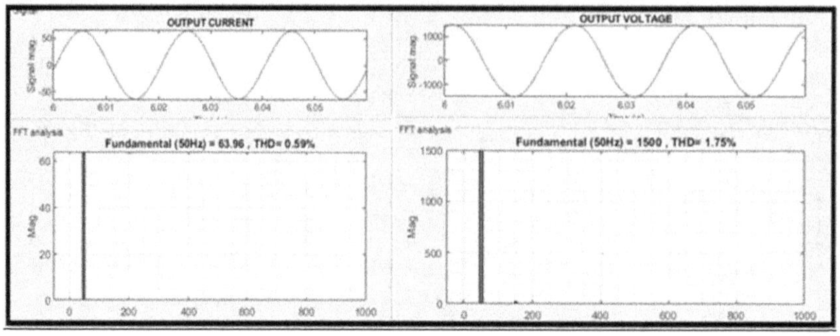

- *Space vector pulse with modulation control (SVPWM):*

Figure 9. FFT spectrum current and voltage by the SVPWM control

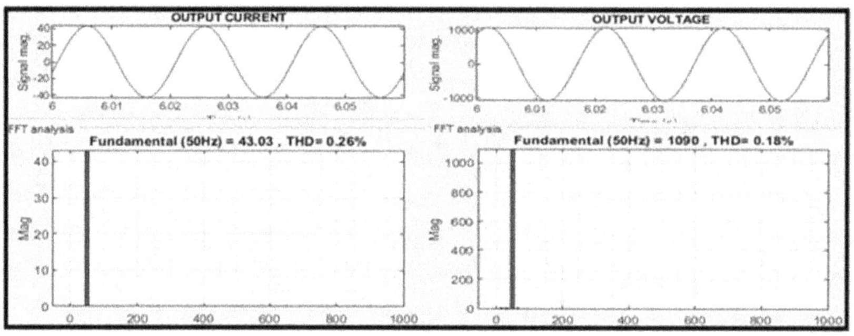

By applying the same control strategy presented previously, the PWM method is used to generate voltage signals at the output of the inverter. This control method is based on the indirect control of active and reactive currents injected into the grid.

With the SPWM control strategy and from the FFT analysis shown in figure 7, it was possible to see that the total THD of this topology is better and equal to THDi=0.59% and THDv=1.75% for the current and the voltage respectively. With this SPWM control strategy, the THD of the seven levels topology is less than that of the value obtained by an open-loop control, et all the harmonics of high frequencies, disappeared after application of the sinusoidal command, therefore the SPWM control strategy allows us to obtain a purely sinusoidal signal in the grid.

In the second figure 8, the total harmonic distortion observed for the space vector control is THDi=0.26% and THDv=0.18%. The all harmonics are eliminated, using the SVPWM control, providing a reduction in harmonics present at the output of the multi-level inverter. The benefits of this control were confirmed by the reduced harmonic distortion, as well as the quality of the current and voltage signal at the output of this inverter into the grid.

SEVEN-LEVEL THREE-PHASE INVERTER (INTERLACED INVERTER Q = 4):

Figure 10. Grid current and voltage FFT spectrums before filtering

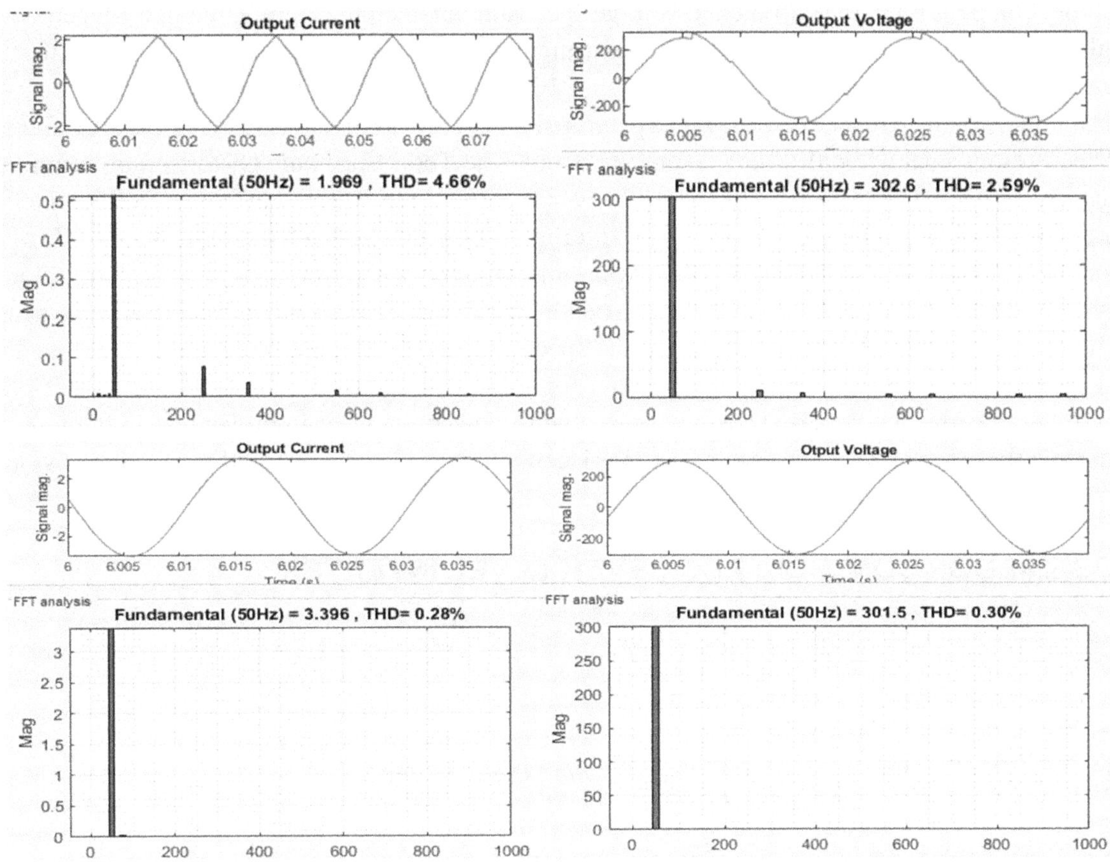

Figure 11. Grid current and voltage FFT spectrums after filtering (a) by the SPWM control ;(b) by the SVPWM control

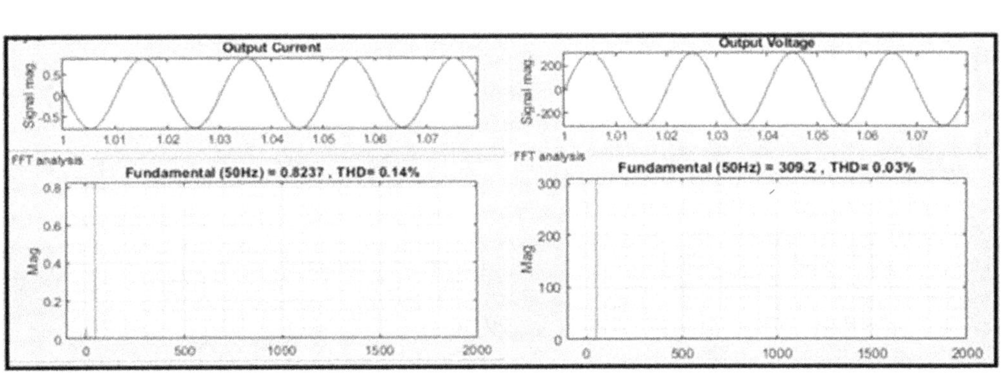

For the seven-level interleaved inverter, each phase of the inverter is composed of q=4 arms. The control of the three arms is obtained by comparing a modulating sine wave with q triangular carrier waves. In this part, we add a balancing loop to the previous command. This control contains two current regulation loops: an external loop for controlling the output current and a loop internal for cell current balancing.

Before compensation, the current and voltage spectrum obtained in figure.9, the one-shaped output signals are not purely sinusoidal because of the harmonics of currents and voltage. After compensation, the waveforms of the current and voltage in figure 10 (a) and (b), are almost sinusoidal and identical. We can notice the difference between FFT spectrums before and after filtering, such as, after filtering is perfectly sinusoidal while that before filtering is not. The Toral Harmonic Distortion of current and voltage are respectively THDi = 0.28% and THDv = 0.30% for the 50 first harmonics, unlike the indicators of THD obtained by the system before filtering (4.66% and 2.59%), the indicators THD of the system using an interlace inverter after filtering are much better. This shows that the THD indicators of the system using an interlaced inverter are much better than those obtained with the conventional inverter

Therefore, by comparing the results from the spectral analysis, it could be seen that the total THD of the multi-level three-phase interlaced inverter topology is lower than in the two-level three-phase photovoltaic inverter typologies. Thus, we were able to ob-serve a clear improvement in the transient regime and the waveforms in the case of the interlaced inverter system compared to the classic system.

CLOSED-LOOP CONTROL: PUC SEVEN-LEVEL INVERTER

- *Sinusoidal Pulse with modulation control:*

Figure 12. FFT spectrum current and voltage by the SPWM control

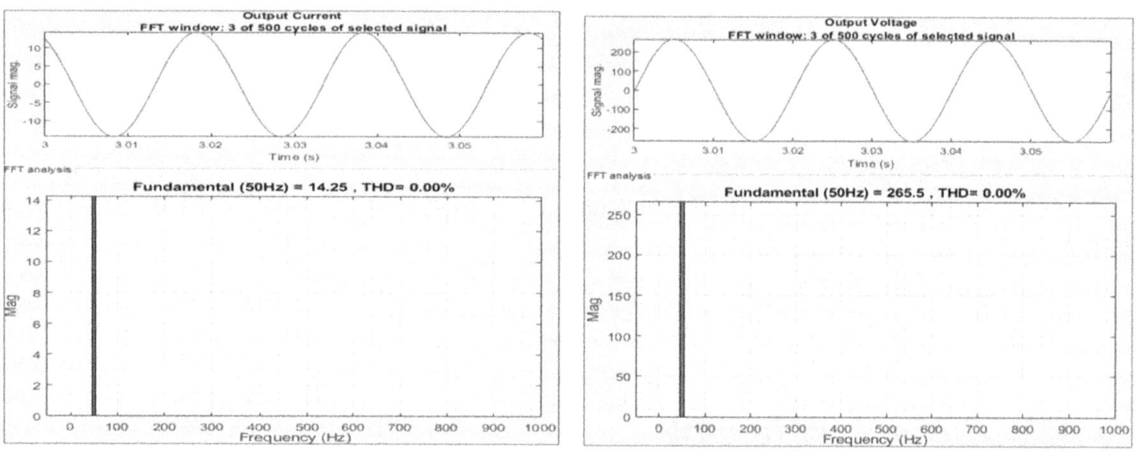

- *Space vector pulse with modulation control (SVPWM):*

Figure 13. FFT spectrum current and voltage by the SVPWM control

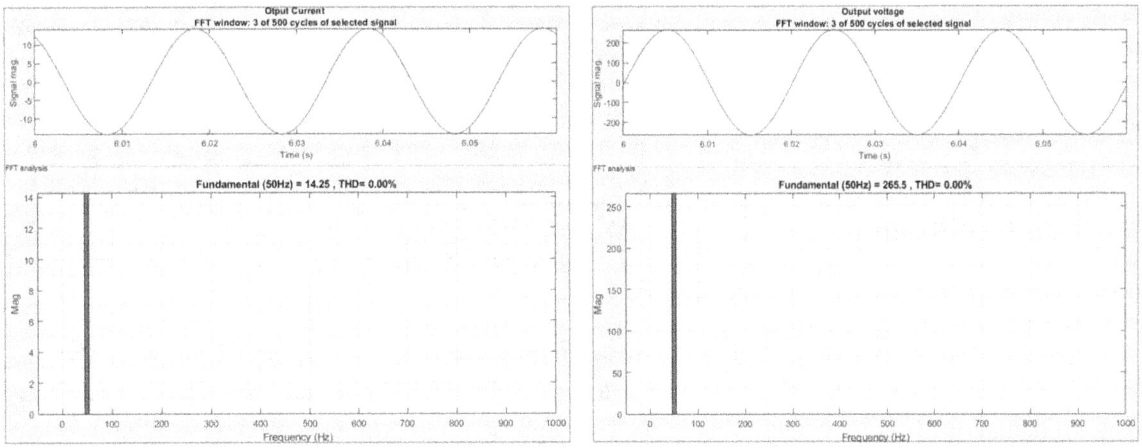

To generate the control reference voltage of the multi-level PUC inverter, this voltage is a sinusoidal waveform based on pulse control with modulation; This reference voltage can be subdivided into six positive and negative intervals after comparison with six triangular signals. Our control method allows indirect control of the active and reactive current delivered in the grid.

An SPWM control simulation of a three-phaśe seven-level PUC inverter is evaluated for nonlinear overload and the above spectrum is shown the figure 11. The results obtained in closed loop make it possible to produce a purely sinusoidal current and voltage, is illustrated in the figure 11 with a THD of 0% for the current and the voltage successively. The load voltages are equal to the reference value, and in phase with the current, with a zero THD (<<<5%).

After application of Vector Pulse Width Modulation (SVPWM), the results obtained by this method is identical to Sinusoidal Control (SPWM), with one difference that, Spatial Vector PWM is a digital modulation method that has previously been used to generate levels of voltage 2N+1. However, not suitable for high power applications, due to the complexity of vector calculations and associated times. It requires two reference voltages for the upper and lower arms. The waveforms and the THD spectrum for the SVPWM control strategy are shown in figure 12. That the current and voltage waveform are purely sinusoidal after applying the SVPWM command. The THD of current and voltage at seven levels is measured at 0.0% respectively. From the simulation results with two control strategies, it is concluded that with the multi-level PUC inverter, the harmonic distortion is less after applying SPWM and SVPWM control. In addition to this, it also has a reduced number of components, which has made it possible to completely improve the conduction losses and the current losses injected into the grid, and finally to have the highest output voltage;

Table 2. Comparison of THD for different controls of the three-phase Seven-Level inverter.

Structure	Control strategy	THDi	THDv
Three-Phase seven level interlaced inverter	Closed-loop: SPWM Control	**0.28%**	**0.30%**
	Closed-loop: SVPWM Control	**0.14%**	**0.03%**
Three-Phase Seven-level CHB Inverter	Closed-loop: SPWM Control	**0.59%**	**1.75%**
	Closed-loop: SVPWM Control	**0.26%**	**0.18%**
Three-Phase Seven-level PUC Inverter	Open-loop seven-level inverter Control	**16.48%**	**16.48%**
	Closed-loop: SPWM Control	**0.0%**	**0.0%**
	Closed-loop: SVPWM Control	**0.0%**	**0.0%**

From the simulation results with two-control strategies table 2, it is concluded that with the multi-level PUC inverter has less harmonic distortion after applied of SPWM and SVPWM control the grid have been completely improved, and finally has the highest output voltage compared to the interleaved and CHB topology. In addition to this, it also has a reduced number of components, so lower conduction losses and the losses of current injected into the grid, and finally to have the highest output voltage;

FUTURE RESEARCH DIRECTIONS

The focus of future research will be mostly on three main interrelated areas, namely:

- Application and experimental validation of the approaches proposed for multilevel inverters, using higher power PV sources and injecting the latter into the Moroccan national electricity grid;
- Exploration of new control strategies and study of the impacts of the integration of these renewable photovoltaic energy production sources into the grid;
- Development and design of new control techniques based on hysteresis and artificial intelligence.

CONCLUSION

This article; is a contribution towards the improvement the control of the three-phase Multi-level photovoltaic inverter, with a new control strategy, by the sinusoidal (SPWM) and space vector (SVPWM) control for a nonlinear load. The comparison of the results of two controls applied to the three-phase photovoltaic inverter at seven levels shows an effective reduction of the THD (<<5%) by the vector control (SVPWM) compared to that of the sinusoidal control (SPWM), with better quality of the current and the voltage injected into the grid by the two control strategies. And our findings suggest that the PUC topology offers several advantages compared to the interleaved and CHB topology, such as the low THD, the reduced number of components, and the highest output voltage which makes it possible to obtain better performance and a low manufacturing cost. The proposed control scheme presents a simple control structure with satisfactory operating performances that are comparable with those reported with control structures for two level topologies we result that this control method respond, in terms of cost and energy efficiency, to the problems encountered with the photovoltaic inverter, and effectively compensates the harmonics current injected in the grid. . The effectiveness of these techniques proposed in this article is demonstrated by the simulation results using the MATLAB / SIMULINK environment. After validation of the experimental results for a two-level three-phase system, the implementation of these simulation results for multi-level three-phase inverters in a complete system is the subject of the next step, which allowed us to verify the reliability of these results and achieve the objectives of this regulation.

RREFERENCES

Althobaiti, A., Armstrong, M. and Elgendy. M. A. (2016) '' Current Control of Three-phase Grid-connected PV Inverters using Adaptive PR Controller'', *IEEE Trans* 978-1-4673-9768, 2016.

Azeem, A., Ansari, M. K., Tariq, M., Sarwar, A., & Ashraf, I. (2019). Design and Modeling of Solar Photovoltaic System Using Seven-Level Packed U-Cell (PUC) Multilevel Inverter and Zeta Converter for Off-Grid Application in India. *Electrica, 19*(2), 101–112. DOI: 10.26650/electrica.2019.18053

Kavali, J., Mittal, A., (2014). Comparative study of various Cascaded H-Bridge Multilevel Inverter Topologies. *Int. J. Electr. Electron. Eng. Res. (IJEEER)* 4 (3), 51–62 (ISSN (P): 2250-155X; ISSN(E): 2278-943X © TJPRC Pvt., Ltd.).

Khalid, H. A, Al-Emadi, N.A., BenBrahim, L., Gastli, A., Cecati, C.(2021) '' A novel model predictive control with an integrated SOC and floating DC-link voltage balancing for 3-phase 7-level PUC converter-based MV BESS', International Journal of Electrical Power and Energy Systems, , Article 106895, 24 March (2021).DOI: 10.1016/j.ijepes.2021.106895

Khan, S. A., Guo, Y., Khan, M. N. H., Siwakoti, Y., Zhu, J., & Blaabjerg, F. (2020, November). A novel single source three phase seven-level inverter topology for grid-tied photovoltaic application. In 2020 IEEE 9th International Power Electronics and Motion Control Conference (IPEMC2020-ECCE Asia) (pp. 372-376). IEEE.

Lakshmi, T. V. V. S., George, N., Umashankar, S., & Kothari, D. P. (2013, February). Cascaded seven level inverter with reduced number of switches using level shifting PWM technique. In *2013 International Conference on Power, Energy and Control (ICPEC)* (pp. 676-680). IEEE.

Lamreoua, A., Benslimane, A., Bouchnaif, J., & El Ouariachi, M. (2021) "An Improved Sinusoidal (PWM) and Vector (SVPWM) Current Control for a Three-Phase Photovoltaic Inverter Connected to a Non-linear Load" *Proceedings of the 2nd International Conference on Electronic Engineering and Renewable Energy Systems. ICEERE 2020.* , Print ISBN: 978-981-15-6258-7, Online ISBN: 978-981-15-6259-4.DOI: 10.1007/978-981-15-6259-4_51

Lamreoua, A., Benslimane, A., Boucnaif, J., Hirech, K. A., Adaliou, A., & Elouariachi, M. (2022) Current harmonic reduction for grid-connected photovoltaic system (PV) based on improved control of three-phase seven-level PUC inverter, *Elsevier journal (Materiels todady: Proceedings)*, (ISSN 2214-7853) ", Volume 72, Part 7, p.p 3536-3543 2023. DOI: <ALIGNMENT.qj></ALIGNMENT>10.1016/j.matpr.2022

Lamreoua, A., Benslimane, A., & El Ouariachi, M. (2022). ''An improved sinusoidal (SPWM) and Space Vector (SVPWM) Current Control for Multi-Levels Three-phase Photovoltaic Inverter Connected to a Non-Linear Load'', *proceedings of the International Conference on Digital Technologies and Applications (ICDTA 22). Digital Technologies and Applications*, DOI: , (2022) pp 672–683DOI: 10.1007/978-3-031-01942-5_67

Lee, J. H., Lee, J. S., Moon, H. C., & Lee, K. B. (2018). An improved finite set model predictive control based on discrete space vector modulation methods for grid-connected three-level voltage source inverter. *IEEE Journal of Emerging and Selected Topics in Power Electronics, 6*(4), 1744–1760. DOI: 10.1109/JESTPE.2018.2830783

Natarajan, G. V. V., & Rao, G. R. (2020) '' Three phase PUC5 inverter fed induction motor for renewable energy application'', International Journal of Power Electronics and Drive System (IJPEDS), ISSN: 2088-8694, DOI: , Vol. 11, No. 1, March (2020) 1-9.DOI: 10.11591/ijpeds.v11.i1.pp1-9

Odeh, C. I., Kondratenko, D., Lewicki, A., Morawiec, M., Ja̧derko, A., & Baran, J. (2021). Pulse-Width Modulation Template for Five-Level Switch-Clamped H-Bridge-Based Cascaded Multilevel Inverter. *Energies*, *14*(22), 7726. DOI: 10.3390/en14227726

Chapter 2
Assessing Solar Energy Potential for Green Energy Production Across Indian Regions Using SAM Software

Karuppiah Natarajan
https://orcid.org/0000-0003-3325-566X
Vardhaman College of Engineering, India

S. Tamilselvi
https://orcid.org/0000-0002-4167-9115
Sri Sivasubramaniya Nadar College of Engineering, India

Patil Mounica
https://orcid.org/0000-0002-8146-0572
Vardhaman College of Engineering, India

Murugaperumal Krishnamoorthy
https://orcid.org/0000-0002-0203-0700
Vardhaman College of Engineering, India

Hariharan Shanmugasundaram
https://orcid.org/0000-0001-9686-4329
Vardhaman College of Engineering, India

K. ShivaBhargavi
https://orcid.org/0009-0004-8091-145X
Vardhaman College of Engineering, India

ABSTRACT

India is divided into many areas, each with its own distinct characteristics such as sunshine exposure, energy consumption habits, and physical factors. The goal is to create specialized solar power plans that meet the individual needs of each place. This includes assessing parameters such as solar irradiation, energy demand, and available land or infrastructure for solar panel installation. By tailoring methods to the circumstances of each region, it guarantees that solar energy resources are deployed optimally, easing India's transition to sustainable and dependable energy sources. The study will conduct a comparative examination of solar power generation in five different Indian regions. This comparison considers a variety of measures, including the levelized cost of energy, yearly AC energy production, performance ratio, and simple payback period. The goal is to assess the effectiveness and efficiency of solar energy installations in these various locales.

DOI: 10.4018/979-8-3693-9924-8.ch002

1. INTRODUCTION

India, the third-largest energy consumer globally, has set ambitious sustainability goals, aiming to cut carbon intensity by 45% by 2030, generate 50% of electricity from renewables, and achieve net-zero emissions by 2070. Renewable energy sources like solar, wind, hydro, and geothermal power offer a sustainable alternative to fossil fuels, producing minimal greenhouse gas emissions. Transitioning to renewables is essential to combat climate change, reduce carbon footprints, enhance energy security, and drive economic growth. As fossil fuel reserves dwindle, investing in clean energy ensures a more resilient and environmentally friendly future.

Among renewables, solar energy stands out for its abundance, accessibility, and versatility. The sun provides an immense, inexhaustible energy supply that can be harnessed globally, from urban centers to remote locations (Krishnamoorthy et. al., 2020). Advances in photovoltaic (PV) technology have made solar power more efficient and cost-effective, enabling widespread adoption in residential, commercial, and utility-scale applications. Solar panels can be installed on rooftops, open fields, deserts, or integrated into buildings, facilitating decentralized power generation and expanding energy access (Rus-Casas, C., et. al, 2024). By prioritizing solar and other renewable sources, India can move towards a cleaner, more sustainable energy future.

India, as a developing nation with a vast population, has a growing demand for electricity. However, the existing power infrastructure is insufficient to meet energy needs across the entire country. To address this challenge, the Indian government is promoting the installation of rooftop solar panels by offering subsidies to encourage adoption. This project focuses on evaluating and comparing solar energy generation across different regions in India using the System Advisor Model (SAM). The study serves as a case analysis to understand the effectiveness of solar power generation in diverse climatic and geographical conditions.

SAM is a free techno-economic software tool designed to assist professionals in the renewable energy sector, including project managers, engineers, policy analysts, technology developers, and researchers (Jia Liu, et. al, 2020). It can simulate various renewable energy systems, including photovoltaic (PV) setups, ranging from small residential installations to large-scale utility projects (Yusheng Zhang, et. al, 2019). To model a renewable energy project in SAM, users must select a performance model and a financial model while assigning values to key input parameters. These include project location, system components, installation and operational costs, as well as financial incentives. By leveraging SAM, stakeholders can make informed decisions about the feasibility and efficiency of solar energy projects in different regions of India (Abdin, Z., and Mérida, W., 2019).

Figure 1. Proposed system feasibility analysis methodology

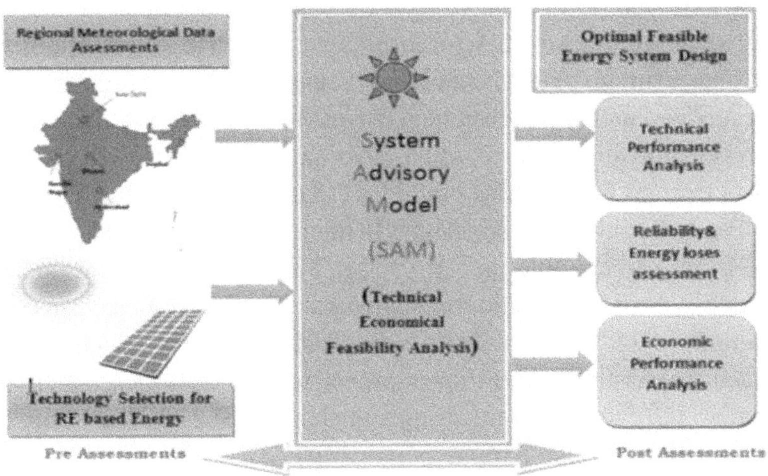

The objectives of this research are assess and compare solar energy potential across diverse Indian regions using SAM software, analyze the impact of technological advancements on energy production and economic feasibility and provide data-driven insights for policymakers, guiding strategic decisions to optimize solar energy utilization and foster sustainable development.

1.1. Recent Work Related to HRE System Design

The increasing focus on renewable energy has led to numerous studies exploring different configurations of renewable energy systems (RES). These configurations include both single-source and hybrid systems. Examples of such setups include photovoltaic (PV) panels integrated with battery storage, wind turbine systems, or hybrid models combining PV panels with wind turbines and batteries (Hassan Zahboune et. al., 2016). Additionally, more complex hybrid systems have been developed, such as PV and wind turbines coupled with micro-Pumped Hydro Energy Storage, wind energy integrated with hydrogen storage, and PV/wind systems combined with split diesel generators and batteries (Chennaif, M., Zahboune, H., and Elhafyani, M., 2020). Other advanced configurations include PV/wind/concentrated solar power (CSP) with battery and thermal energy storage, as well as PV/CSP hybrid plants. These diverse setups aim to enhance efficiency, reliability, and sustainability in renewable energy generation (Aniello, G., Shamon, H., and Kuckshinrichs, W., 2021).

Despite extensive research on renewable energy systems (RES), optimizing their size remains a crucial area of study, focusing on techno-economic analysis and parametric evaluations. Various studies have developed optimization algorithms to minimize intermittency, ensure continuous power supply, and reduce energy production costs (Loukatou, A., Johnson, P., and Howell, S., 2019). Different approaches have been explored, including Mixed Integer Linear Programming for Hybrid Renewable Energy Systems (HRES), Genetic Algorithms (GA) to lower Life Cycle Cost (LCC) and CO_2 emissions, and adaptive differential mutation evolution strategies for standalone solar systems. Additionally, hybrid methods combining chaotic search, harmonic search, and simulated annealing have been applied to optimize solar-wind-hydrogen hybrid systems. These techniques enhance the efficiency, reliability, and cost-effectiveness of renewable energy solutions (Papadopoulos et. al., 2019).

Naaim et al. (2024) explored the optimal sizing of a standalone hybrid system designed to meet both electrical and thermal needs while accounting for surplus energy and heat loss. Another study proposed an optimization method to determine the ideal battery size for standalone microgrids (Ekren, O., and Ekren, B.Y., 2010). Additionally, research has been conducted on the sizing of grid-connected hybrid renewable energy systems for residential applications, while other studies have focused on the techno-economic optimization of hybrid systems for high-rise buildings (shabani et. al, 2020). Firtina-Ertis et al. (2020) utilized the Strength Pareto Evolutionary Algorithm to minimize the levelized cost of energy and carbon dioxide life cycle emissions in renewable energy systems. Further advancements include a novel method for designing large-scale hydro-photovoltaic complementary systems and a techno-economic assessment of hybrid energy solutions for off-grid power generation and hydrogen production. Ogunjuyigbe, A.S.O et al. (2016) introduced a heuristic optimization approach using stochastic gradient search to minimize the overall cost of renewable energy systems, while Madziga et al. (2018) examined the optimal scale of hybrid renewable energy systems in Africa.

2. PROPOSED METHODOLOGY USING SAM SOFTWARE

The methodology outlines the step-by-step process for conducting the project. When using SAM software to model a photovoltaic (PV) system, the first step involves extracting weather data for the selected regions.

Figure 2. Process flow chart for optimal HRE system selection

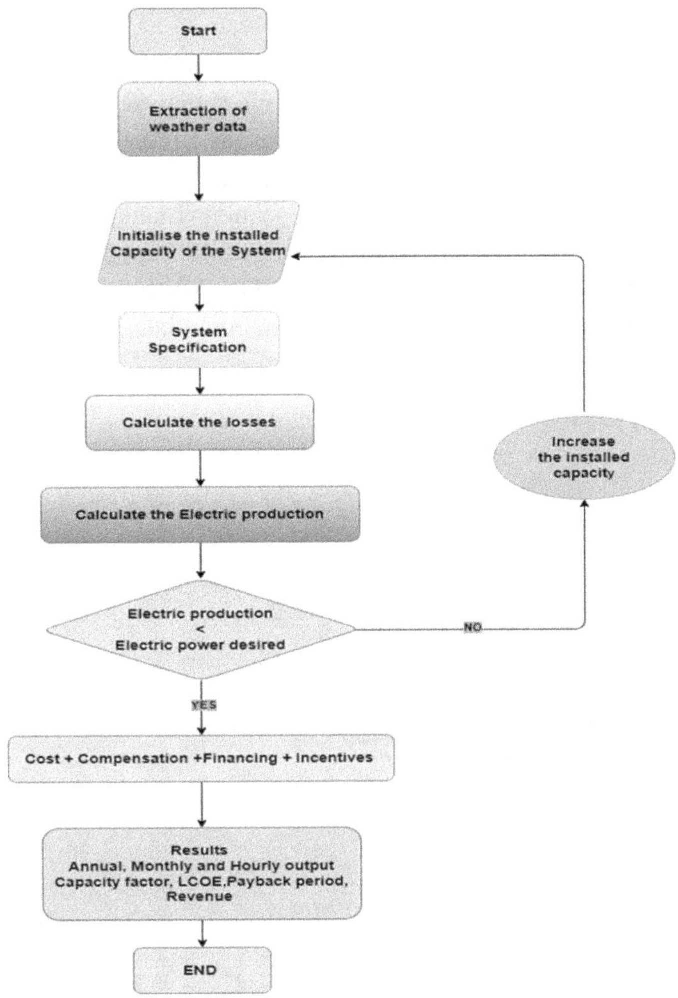

SAM enables both performance and financial modeling, requiring the initial determination of the power plant's capacity (Liu, M. et. al, 2023). Once the system's installed capacity is set, the next step is to define its characteristics, including module type, inverter specifications, and system design parameters. For the financial model, factors such as installation costs, operational expenses, and electricity rates are specified. The system's losses and electricity generation are then calculated based on the input values (Dounia Ziyati et. al, 2021). If the generated electricity is insufficient to meet the required power output, the installed capacity is adjusted, and system specifications are updated accordingly.

This process is repeated until the generated power meets the desired level. Once the performance criteria are satisfied, SAM proceeds with the financial analysis and provides results, including annual, monthly, and hourly energy output, capacity factor, levelized cost of energy (LCOE), and payback period.

3. CASE STUDY: COMPARATIVE ANALYSIS FOR HRE SYSTEM IMPLEMENTATION

Modeling a photovoltaic (PV) system requires accurate resource data, which necessitates downloading weather files for the selected region from NREL. This project analyzes solar energy generation across five different locations in India: Hyderabad, Delhi, Gandhinagar, Bhopal, and Imphal. Therefore, weather data for these regions is obtained and added to the software library. To download the weather files, it is essential to know the latitude and longitude of each location (Lian, J et. al., 2019). The selected regions represent different geographical zones of India, with Delhi from the north, Imphal from the east, Gandhinagar from the west, Hyderabad from the south, and Bhopal from the central part of the country. This selection ensures a comprehensive comparison of solar potential across diverse climatic conditions.

Figure 3. Load assessment at the selective site of the Indian regions

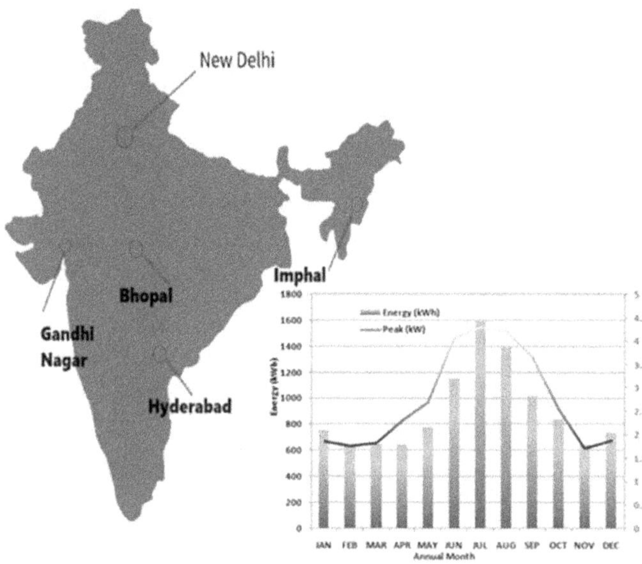

3.1 RE Resource Assessment

The table presents the geographical coordinates (latitude and longitude) along with radiation data, indicating the average solar energy received at the Earth's surface per day. The selected regions—Delhi, Hyderabad, Bhopal, Gandhinagar, and Imphal—represent different parts of India and were chosen based on variations in solar irradiance across different locations. Since the project focuses on comparing solar potential, regions from all parts of India, including the north, south, east, west, and central areas, were considered. A higher radiation value signifies greater potential for solar energy generation, making these locations suitable for analysis.

Table 1. Comparison of Geographical Coordinates and Solar Resource Potential

AREA	Latitude	Longitude	Available Radiation (KWh/m²/day)
Delhi	28.58	77.45	5.47
Hyderabad	17.38	78.49	5.4
Bhopal	23.26	77.41	5.39
Gandhi Nagar	23.21	72.62	5.2
Imphal	24.82	93.93	4.83

3.2 Selection of Solar PV Module

A module is an assembly of interconnected solar cells and is used to evaluate photovoltaic performance in SAM software. The module model estimates the DC electrical output of a single unit based on its design specifications and the incident solar radiation (plane-of-array irradiance) derived from weather data. The figure shows characteristic graph of module .The Selected module is Sun Power SPR-X21-335.

Figure 4. PV Module characteristics

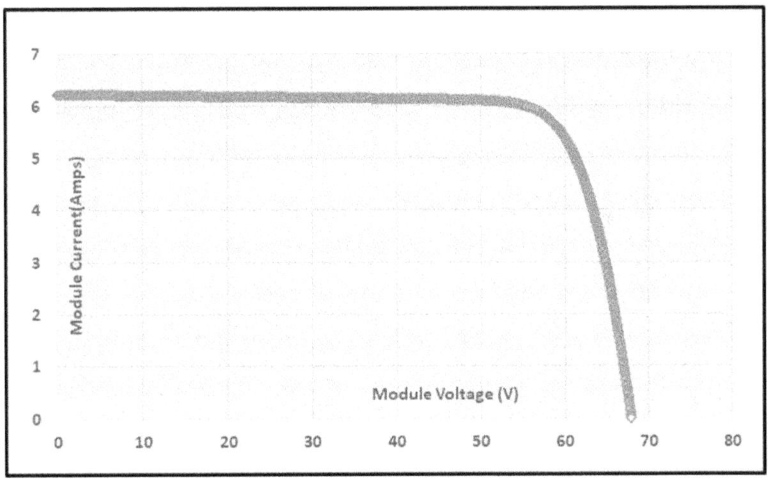

Under reference conditions, the module operates with a total irradiance of 1000 W/m² and a cell temperature of 25°C. It is constructed using Mono-c-Si material and has a surface area of 1.631 m².

Table 2. Module specifications

Nominal efficiency	20.55%
Maximum power(Pmp)	335.205 W_{dc}
Max power Voltage(Vmp)	57.3 Vdc
Max power current(Imp)	5.8 A_{dc}

continued on following page

Table 2. Continued

Open circuit voltage(Voc)	67.9 V_{dc}
Short circuit current(Isc)	6.2 A_{dc}

The table presents the module's specifications, including a nominal efficiency of 20.55%. The maximum power (Pmp) is 335.205 W_{dc}, with a corresponding maximum power voltage (Vmp) of 57.3 V_{dc}. The maximum power current (Imp) is 5.8 A_{dc}, while the open-circuit current (Isc) is 6.2 A_{dc}. Additionally, the open-circuit voltage (Voc) is 67.9 Vdc.

3.3 Inverter Module Selection

An inverter is a device that converts DC voltage into AC voltage. In SAM, a photovoltaic system can be modeled using only a single type of inverter (Carlo, J. et. al, 2020). The selection of an inverter is based on various characteristics and parameters, including maximum AC power, maximum DC power, maximum DC voltage, and DC current. The selected inverter is SolarEdge Technologies Ltd: SE6000x[208V].

Figure 5. Inverter Characteristics

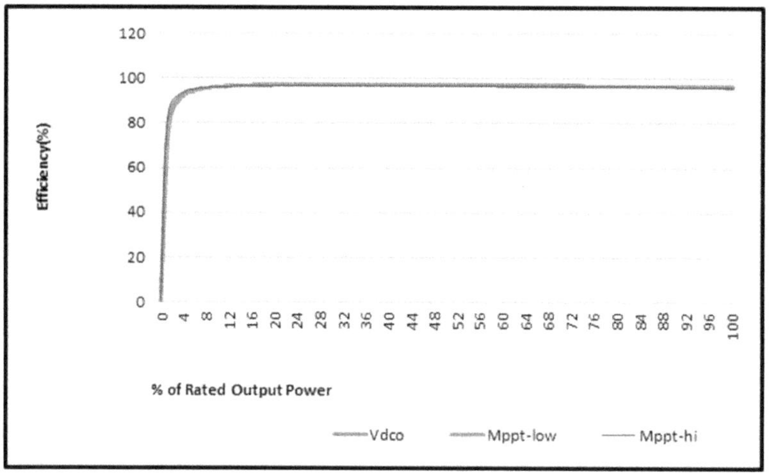

The figure displays a graph illustrating the characteristics of an inverter, with Rated Output Power on the X-axis and Efficiency on the Y-axis. The red line represents Mppt-Low, while the blue line indicates Vdco.

Table 3. Inverter datasheet parameters

Maximum AC Power	5330 Wac
Maximum DC Power	5530.11 Wdc
Power use during operation	11.9749 Wdc
Power use at night	5.22 Wac
Nominal AC voltage	208 Vac
Maximum DC Voltage	500 Vdc
Maximum DC Current	13.4881 Adc
Minimum MPPT DC Voltage	320 Vdc
Nominal DC Voltage	410 Vdc
Maximum MPPT DC Voltage	500 Vdc

The table outlines the key specifications of the inverter. The maximum AC power output is 5330 watts (Wac), while the maximum DC power output reaches 5530.11 watts (Wdc). During operation, the system consumes 11.9749 watts, indicating the energy required for its functionality. At night, when PV panels are inactive, power consumption is 5.22 watts (Wac). The table also includes voltage and current specifications, such as a nominal AC voltage of 208 volts (Voc), a maximum DC voltage of 500 volts (Vdc), and a maximum DC current of 13.4881 amperes (Adc). Additionally, it provides details on voltage regulation, with a minimum MPPT (Maximum Power Point Tracking) DC voltage of 320 volts (Vdc), a nominal DC voltage of 410 volts (Vdc), and a maximum MPPT DC voltage of 500 volts (Vdc).

3.4 HRE System Design

System design variables are essential for sizing the PV system, including the number of inverters for AC sizing, the number of modules per string in the subarray, and the number of strings connected in parallel. The DC to AC ratio is 1.13. The system's nameplate DC capacity is 6.034 kWdc, while the total inverter capacity is 5.530 kWdc. The total AC capacity is 5.330 kWac, and the overall module area is 29.358 m^2.

Table 4. Tracking & Orientation

Axis	*fixed*
Tilt (180)	20
Azimuth (deg)	180
Ground coverage ratio (GCR)	0.3
Tracker rotation limit (deg)	45

Figure 6. Orientation

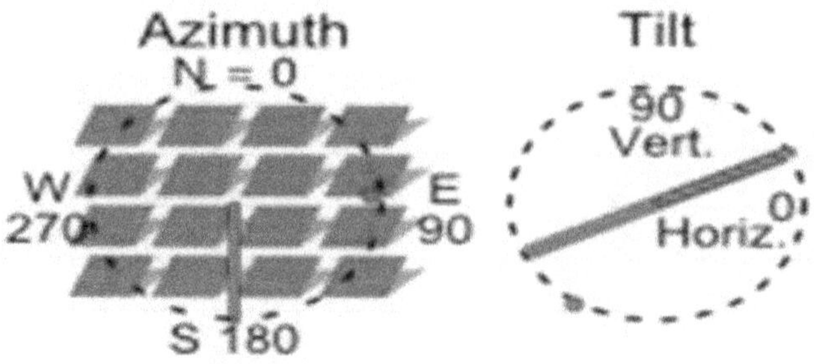

The table provides detail specifications of solar photovoltaic system. The solar panels are stationary, shown by axis is fixed. The tilt angle is set to 20 degrees. Azimuth angle specified as 180 degrees that means panels are facing directly south. The ground coverage ratio is the total area covered by the solar panels to the total ground and it is 0.3. Tracker rotation limit is 45 degrees, rotate to adjust the orientation of the panel.

Table 5. DC sizing and Configuration

Electrical Configuration	Subarray1
Module per string in subarray	6
Strings in parallel in subarray	3
Number of modules in subarray	18
String Voc at reference condition (V)	407.4
String Vmp at reference condition (V)	343.8

The table presents the electrical configuration of the subarrays. This project is designed with a single subarray. In Subarray 1, each string consists of 6 modules connected in series, meaning six solar panels are linked end-to-end to form one string. Additionally, there are three parallel sets of these 6-module strings, resulting in a total of 18 modules within Subarray 1. At reference conditions, the open-circuit voltage of the string is 407.4 volts, while the maximum power point voltage (Vmp) is 343.3 volts. This Vmp value represents the optimal operating voltage when the string generates maximum power.

3.5 Impact of Shading and System Layout

Shading and snow losses refer to the reduction in solar radiation reaching photovoltaic modules due to obstructions. Shading can be classified into external shading and self-shading. External shading occurs when nearby objects such as trees, buildings, or roof structures cast shadows on the solar panels (Ridha H.M. et. al, 2020). Self-shading happens when modules in adjacent rows within the array block cast shadows on each other. Snow losses occur when accumulated snow covers the modules in a subarray, reducing their efficiency.

Figure 7. Modules per String

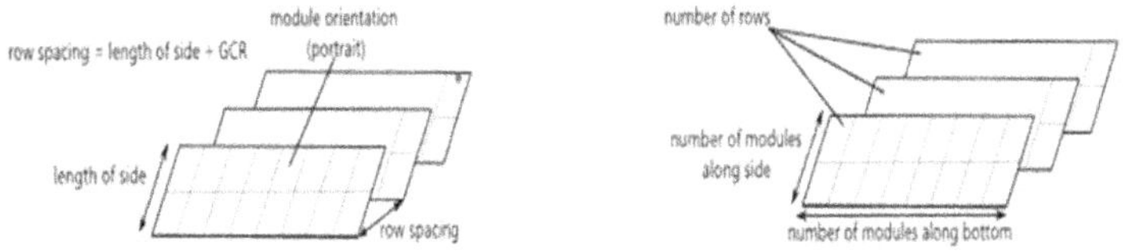

The module aspect ratio is 1.7, representing the ratio of its length to width. The module has a length of 1.665 meters and a width of 0.979 meters, resulting in a total area of 1.631 square meters.

3.6 Losses Considerations

Losses in a PV system refer to energy wastage due to various inefficiencies. These include:

- **Irradiance losses** – A reduction in incident solar irradiance, often caused by dust accumulation or seasonal soiling on the module surface.
- **DC losses** – Energy losses occurring on the DC side of the system due to electrical inefficiencies.
- **AC losses** – Electrical losses in wiring on the AC side, particularly between the inverter and the grid connection point.
- **Transformer losses** – Energy dissipated as heat in transformers, which occurs in large photovoltaic systems when power is transmitted from the inverter through a distribution or substation transformer.
- **System availability losses** – Reductions in energy output due to operational factors such as maintenance downtime or other constraints that prevent the system from functioning optimally.

3.7 Installation Cost and Operation Cost

Installation costs comprise both direct and indirect capital expenses associated with the system. These values can be adjusted based on financial requirements in the SAM software (Khan A.A. et. al., 2022) Direct capital costs include expenses for modules and inverters, while indirect costs cover land acquisition. Additionally, operational costs account for the maintenance and upkeep of the system (Akpolat et. al., 2019).

Table 6. Installation Cost

	Units	/Wdc	Cost ( )
Direct capital cost			
Module	**18**	**34.03**	**2,05,326.23**
Inverter	**1**	**20.75**	**1,25,198.86**
Balance system equipment		**25.73**	**1,55,246.52**

continued on following page

Table 6. Continued

	Units	/Wdc	Cost ( )
Installation Labour		**15.77**	**95,151.20**
Installer margin and overhead		**22.41**	**1,35,215.30**
Total direct cost			**7,16,138.94**
Indirect capital cost			
Permitting and environmental studies		**19.92**	**1,20,191.47**
Engineering and developer overhead		**81.34**	**4,90,780.66**
Total indirect cost			**6,10,971.30**
Total installed cost			**13,62,917.27**

The table outlines the expenses involved in setting up a solar power system. Direct costs include purchasing solar panels, inverters, and other essential equipment, along with labor charges for installation and overhead expenses, amounting to 7, 16, 138.94. Indirect costs cover non-hardware-related expenditures, such as permits, environmental assessments, and engineering services, totalling 6, 10, 971.30. The total installation cost, which is the sum of both direct and indirect costs, comes to 13, 62, 917.27, representing the overall investment required to make the system operational.

3.8 Electrical Loads Assessments

The table presents data on monthly energy consumption and peak power demand throughout the year. Energy usage fluctuates seasonally, with the highest consumption recorded in June, July, August, and September. Additionally, peak power demand signifies the maximum power required by the system at any given time. The total annual energy consumption is 10,829.34 kWh, while the peak power demand reaches 4.3 kW.

Table 7. Electrical load

Month	Energy (kWh)	Peak (kW)
JAN	752.19	1.85
FEB	642.38	1.76
MAR	647.75	1.82
APR	643.76	2.3
MAY	777.22	2.69
JUN	1,151.70	4.05
JUL	1,594.78	4.3
AUG	1,393.36	4.2
SEP	1,016.16	3.67
OCT	837.85	2.55
NOV	640.38	1.71

continued on following page

Table 7. Continued

Month	Energy (kWh)	Peak (kW)
DEC	731.81	1.87
Annual	10,829.34	4.3

4. RESULTS AND DISCUSSION

The table provides insights into energy production, efficiency, cost-effectiveness, and financial returns of the solar system. Annual AC energy production represents the total electricity generated in the first year, crucial for evaluating system performance. Among the selected locations, Delhi generates the highest annual AC energy, while Imphal produces the least. The DC capacity factor, which indicates the system's efficiency in converting sunlight into electricity, ranges from 17.40% in Imphal to 20.00% in Delhi. Similarly, energy yield, measuring electricity generation per unit of capacity, varies between 1,526 kWh/kW in Imphal and 1,750 kWh/kW in Delhi. The performance ratio, which reflects overall system efficiency, remains fairly consistent at approximately **0.8** across locations.

The economic assessment considers the levelized cost of energy (LCOE) in both nominal and real terms, a key metric that determines the cost per unit of electricity produced. A comparison of electricity bills with and without solar systems highlights substantial savings when solar energy is utilized. Other financial indicators include net savings, net present value (NPV), simple payback period, net capital cost, equity, and debt, which help evaluate the viability and return on investment of the solar projects.

Table 8. Comparison of Results

Metric	Gandhi Nagar	Imphal	Hyderabad	Bhopal	Delhi
Annual AC energy in Year 1	9,772 kWh	9,207 kWh	10,005 kWh	10,258 kWh	10,559 kWh
DC capacity factor in Year 1	18.50%	17.40%	18.90%	19.40%	20.00%
Energy yield in Year 1	1,620 kWh/kW	1,526 kWh/kW	1,658 kWh/kW	1,700 kWh/kW	1,750 kWh/kW
Performance ratio in Year 1	0.79	0.79	0.8	0.8	0.8
LCOE Levelized cost of energy nominal	6.74 /kWh	7.49 /kWh	6.88 /kWh	6.66 /kWh	6.56 /kWh
LCOE Levelized cost of energy real	5.38 /kWh	5.98 /kWh	5.49 /kWh	5.32 /kWh	5.23 /kWh
Electricity bill without system (year 1)	1,19,188	1,12,714	1,25,662	1,09,240	1,38,610
Electricity bill with system (year 1)	39,840	54,863	45,899	43,824	44,903
Net savings with system (year 1)	79,348	57,851	79,763	62,499	93,707
Net present value	2,92,907	51,543	3,20,297	1,19,188	4,93,850
Simple payback period	13.8 years	19.4years	13.4 years	17.4years	11.3 years
Net capital cost	13,62,943	13,62,943	13,62,943	13,62,943	13,62,943
Debt	13,62,943	13,62,943	13,62,943	13,62,943	13,62,943

Figure 8. Annual AC Energy Generation

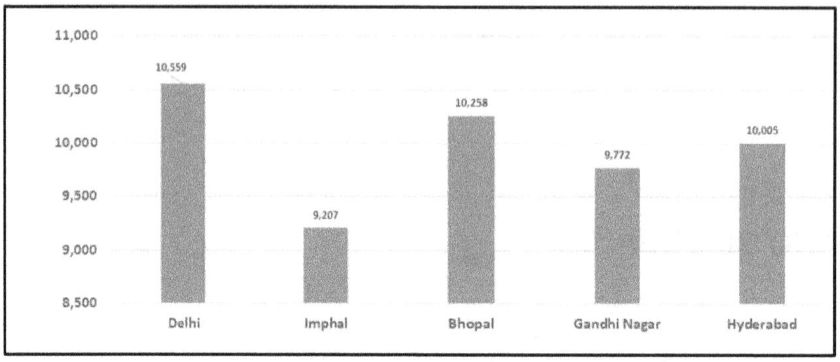

The graph illustrates the annual AC energy production across five different regions. Delhi records the highest annual AC energy production at 10,559 kWh, highlighting its strong potential for solar energy generation. In contrast, Imphal ranks the lowest with an annual AC energy production of 9,207 kWh, indicating relatively lower generation capacity. The average annual AC energy production across all five regions is 9,159 kWh.

Figure 9. Performance Ratio

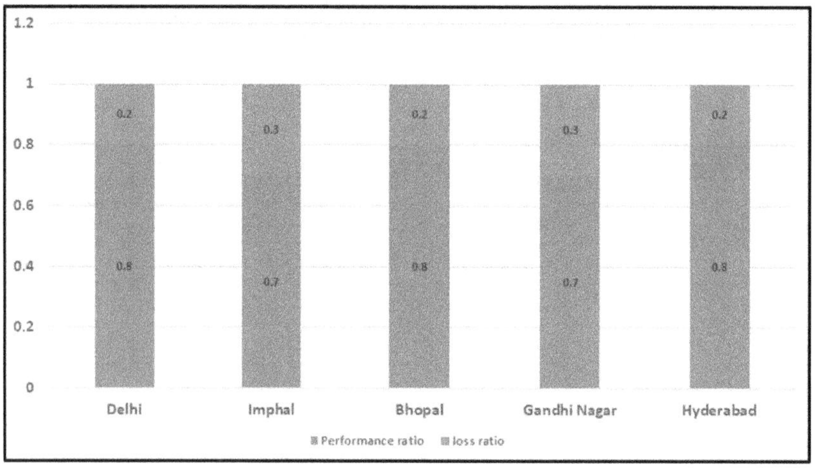

A higher performance ratio reflects a more efficient and productive solar system. This ratio serves as a key measure of a system's capability to convert sunlight into electricity, with higher values signifying better efficiency. Delhi, Bhopal, and Hyderabad achieve the highest performance ratio of 0.8, with a corresponding loss ratio of 0.2, indicating a well-functioning solar system. The average performance ratio across the five regions is 0.76.

Figure 10. Return on Investment Duration

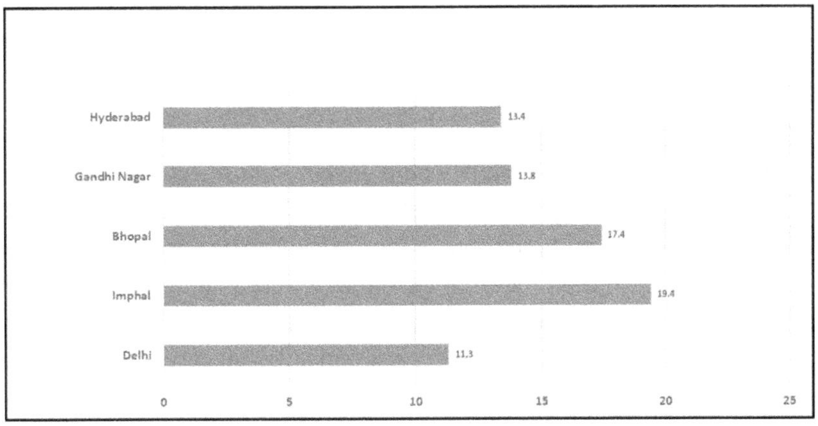

A shorter payback period typically signifies a more favourable financial investment. Among the five regions, Delhi has the shortest payback period of 11.3 years, making it a promising choice for solar energy investments. In contrast, Imphal has the longest payback period at 19.4 years. On average, the payback period across all five regions is approximately 15.06 years, indicating a relatively extended duration to recover the initial investment in solar energy systems.

Table 9. Ranking based on different performance parameters

SI.NO	Region	Annual AC Energy (kwh)	Performance ratio	Simple payback period	Ranking based on AC energy	Ranking based on Performance Ratio	Ranking based on Payback period	COMMENTS
1	Delhi	10,559	0.8	11.3 years	1	1	1	Highly recommended location
2	Hyderabad	10,005	0.8	13.4 years	3	1	2	
3	Bhopal	10,258	0.8	17.4 years	2	1	4	
4	Gandhi Nagar	9,772	0.7	13.8 years	4	2	3	
5	Imphal	9,207	0.7	19.4 years	5	2	5	Least preferred location

The overall analysis indicates that Delhi is the most favourable location for solar energy projects due to its high energy generation potential and short payback period, making it an attractive investment. Bhopal and Hyderabad also demonstrate strong performance in terms of energy production and system efficiency, positioning them as promising options for solar investments. Gandhi Nagar ranks moderately in both energy output and financial returns. Imphal, while contributing to solar energy generation, has a longer payback period and slightly lower efficiency. These insights provide valuable guidance for stakeholders and decision-makers in evaluating solar energy projects, considering factors such as energy production, efficiency, and financial viability.

5. CONCLUSION

The findings highlight the need to tailor renewable energy solutions to specific regional conditions. By acknowledging the varying solar energy potential across different parts of India, this study enables a more context-sensitive approach to implementing solar projects. The use of SAM software in the comparative analysis enhances the accuracy of solar potential assessments and underscores the necessity for continuous advancements in modeling tools and data sources. As India moves toward a sustainable energy future, these results emphasize the dynamic nature of solar resources and the importance of data-driven, adaptable decision-making in the renewable energy sector. The identified high-potential regions serve not only as prime investment opportunities but also as reference points for ongoing evaluation and enhancement of solar energy policies. To fully harness the potential of solar energy, collaboration among researchers, policymakers, and industry leaders is essential in driving India towards a cleaner, more sustainable, and resilient energy landscape.

REFERENCES

Abdin, Z., & Mérida, W. (2019). A techno-economic analysis of hybrid renewable energy-based off-grid power supply and hydrogen production. *Energy Conversion and Management, 196*, 1068–1079. DOI: 10.1016/j.enconman.2019.06.068

Akpolat, A. N., Dursun, E., Kuzucuoğlu, A. E., Yang, Y., Blaabjerg, F., & Baba, A. F. (2019). Performance Analysis of a Grid-Connected Rooftop Solar Photovoltaic System. *Electronics (Basel), 8*(8), 905. DOI: 10.3390/electronics8080905

Aniello, G., Shamon, H., & Kuckshinrichs, W. (2021). Micro-economic assessment of residential PV and battery systems: The underrated role of financial and fiscal aspects. *Applied Energy, 281*, 115667. DOI: 10.1016/j.apenergy.2020.115667

Carlo, J., Melendez, J., Rossi, M., & Renzi, M. (2020). Optimal sizing of a hybrid renewable energy system: Importance of data selection with highly variable renewable energy sources. *Energy Conversion and Management, 223*, 113303. DOI: 10.1016/j.enconman.2020.113303

Chennaif, M., Zahboune, H., & Elhafyani, M. (2020). *Techno-economic sizing of a stand-alone hybrid energy and storage system for a water pumping application. Advances in Smart Technologies and Applications*. Springer., DOI: 10.1007/978-3-030-53187-4_33

Ekren, O., & Ekren, B. Y. (2010). Size optimization of a PV/wind hybrid energy conversion system with battery storage using simulated annealing. *Applied Energy, 87*(2), 592–598. DOI: 10.1016/j.apenergy.2009.05.022

Firtina-Ertis, I., Acar, C., & Erturk, E. (2020). Optimal sizing design of an isolated stand-alone hybrid wind-hydrogen system for a zero-energy house. *Applied Energy, 274*, 115244. DOI: 10.1016/j.apenergy.2020.115244

Khan, A. A., Minai, A. F., Pachauri, R. K., & Malik, H. (2022). Optimal Sizing, Control, and Management Strategies for Hybrid Renewable Energy Systems: A Comprehensive Review. *Energies, 15*(17), 6249. DOI: 10.3390/en15176249

Krishnamoorthy, M., Raj, P. A. D. V., Suresh, S., & Natarajan, K. (2020). Techno Economic Analysis of Hybrid Renewable Electrification System in Different Climatic Zones. In Hitendra Sarma, T., Sankar, V., & Shaik, R. (Eds.), *Emerging Trends in Electrical, Communications, and Information Technologies. Lecture Notes in Electrical Engineering* (Vol. 569). Springer., DOI: 10.1007/978-981-13-8942-9_14

Lian, J., Zhang, Y., Ma, C., Yang, Y., & Chaima, E. (2019). A review on recent sizing methodologies of hybrid renewable energy systems. *Energy Conversion and Management, 199*, 112027. DOI: 10.1016/j.enconman.2019.112027

Liu, J., Wang, M., Peng, J., Chen, X., Cao, S., & Yang, H. (2020). Techno-economic design optimization of hybrid renewable energy applications for high-rise residential buildings. *Energy Conversion and Management, 213*, 112868. DOI: 10.1016/j.enconman.2020.112868

Liu, M., Zhang, B., Wang, J., Liu, H., Wang, J., Liu, C., Zhao, J., Sun, Y., Zhai, R., & Zhu, Y. (2023). Optimal Configuration of Wind-PV and Energy Storage in Large Clean Energy Bases. *Sustainability (Basel)*, *15*(17), 12895. DOI: 10.3390/su151712895

Loukatou, A., Johnson, P., & Howell, S. (2019). Evaluation of wind energy project worth when paired with battery storage. *Applied Energy*, *283*, 116247. DOI: 10.1016/j.apenergy.2020.116247

Madziga, M., Rahil, A., & Mansoor, R. (2018). Comparison between Three Off-Grid Hybrid Systems (Solar Photovoltaic, Diesel Generator and Battery Storage System) for Electrification for Gwakwani Village, South Africa. *Environments (Basel, Switzerland)*, *5*(5), 57. DOI: 10.3390/environments5050057

Naaim, S., Ouhammou, B., Aggour, M., Daouchi, B., El Mers, E. M., & Mihi, M. (2024). Multi-Utility Solar Thermal Systems: Harnessing Parabolic Trough Concentrator Using SAM Software for Diverse Industrial and Residential Applications. *Energies*, *17*(15), 3685. DOI: 10.3390/en17153685

Ogunjuyigbe, A. S. O., Ayodele, T. R., & Akinola, O. A. (2016). Optimal allocation and sizing of PV/wind/split-diesel/battery hybrid energy system for minimizing life cycle cost, carbon emission, and dump energy of remote residential buildings. *Applied Energy*, *171*, 153–171. DOI: 10.1016/j.apenergy.2016.03.051

Papadopoulos, V., Knockaert, J., Develder, C., & Desmet, J. (2019). Study of wind energy system measurements for industrial applications incorporating battery storage. *Applied Energy*, *247*, •••. DOI: 10.1016/j.apenergy.2019.04.051

Ridha, H. M., Gomes, C., Hizam, H., Ahmadipour, M., Muhsen, D. H., & Ethaib, S. (2020). Optimum Design of a Standalone Solar Photovoltaic System Based on Novel Integration of Iterative-PESA-II and AHP-VIKOR Methods. *Processes (Basel, Switzerland)*, *8*(3), 367. DOI: 10.3390/pr8030367

Rus-Casas, C., Gilabert-Torres, C., & Fernández-Carrasco, J. I. (2024). Optimizing Energy Management and Sizing of Photovoltaic Batteries for a Household in Granada, Spain: A Novel Approach Considering Time Resolution. *Batteries*, *10*(10), 358. DOI: 10.3390/batteries10100358

Shabani, M., Dahlquist, E., Wallin, F., & Yan, J. (2020). Techno-economic investigation of renewable-battery storage and renewable micro pumped hydro storage power supply systems in Sweden. *Applied Energy*, •••, 279.

Zahboune, H., Zouggar, S., Krajacic, G., Varbanov, P. S., Elhafyani, M., & Ziani, E. (2016). Optimal hybrid renewable energy design in autonomous system using Modified Electric System Cascade Analysis and Homer software. *Energy Conversion and Management*, *126*, 909–922. DOI: 10.1016/j.enconman.2016.08.061

Zhang, Y., Ma, C., Lian, J., Pang, X., Qiao, Y., & Chaima, E. (2019). Optimal photovoltaic capacity of large-scale hydro-photovoltaic complementary systems considering electricity delivery demand and reservoir characteristics. *Energy Conversion and Management*, *195*, 597–608. DOI: 10.1016/j.enconman.2019.05.036

Ziyati, D., Dollet, A., Flamant, G., Volut, Y., Guillot, E., & Vossier, A. (2021). A Multiphysics model of large-scale compact PV–CSP hybrid plants. *Applied Energy*, *288*, 116644. Advance online publication. DOI: 10.1016/j.apenergy.2021.116644

Chapter 3
MPPT Control for Optimized Energy Extraction in Wind Turbines

Soukayna Tantani
https://orcid.org/0009-0000-2011-9314
Faculty of Science and Techniques of Mohammedia, Morocco

Mohamed Bezza
Faculty of Sciences and Techniques of Mohammedia, Morocco

Sara Sghiouri
https://orcid.org/0009-0003-4466-7798
Faculty of Sciences and Techniques of Mohammedia, Morocco

ABSTRACT

This document presents an in-depth study on the modeling and control of wind turbines, aimed at optimizing their energy efficiency. The fundamental aerodynamic equations, including the power (Cp) and torque (Ct) coefficients, are analyzed to determine their impact on the extraction of wind energy. Control strategies, such as MPPT (Maximum Power Point Tracking) and indirect speed control, are implemented to keep the peak speed ratio (λ) close to its optimal value (λopt). Simulations in MATLAB/Simulink demonstrate the stability of the system, with a rotor rotation speed converging towards 171.5 rad/s and balanced sinusoidal stator currents. The results confirm the effectiveness of the methods proposed to maximize energy conversion while ensuring structural reliability. This study highlights the importance of dynamic models and advanced control algorithms in the optimization of modern wind farms.

1. INTRODUCTION

Wind energy is establishing itself as a cornerstone of the global energy transition, offering a sustainable alternative to fossil fuels, which still account for over 80% of global CO_2 emissions and significantly contribute to climate change (IPCC, 2023). The urgency of decarbonization has led to widespread investment in renewable energy sources, including solar, hydropower, and biomass. However, wind

DOI: 10.4018/979-8-3693-9924-8.ch003

energy stands out due to its scalability, technological advancements, and rapidly declining costs, with installed capacity surpassing 900 GW worldwide in 2023 (IRENA, 2023). Despite these advancements, fully harnessing the potential of wind energy remains a complex challenge, as it requires overcoming critical limitations inherent to wind technologies, including energy intermittency, mechanical stress on turbine components, and challenges in grid integration due to fluctuating wind speeds. Among the key challenges, aerodynamic optimization and dynamic control strategies are crucial to adapting turbines to variable wind conditions (Burton et al., 2021). Modern wind turbines employ advanced control techniques, such as Maximum Power Point Tracking (MPPT), to dynamically adjust rotor speed and blade pitch (β), ensuring optimal power extraction even in turbulent conditions (Li et al., 2021). The efficiency of MPPT algorithms relies on their ability to maximize power coefficient (Cp) while mitigating mechanical fatigue caused by rapid wind fluctuations. While variable-pitch turbines provide greater adaptability, fixed-pitch turbines remain widely used for their simplicity and cost-effectiveness, though they struggle to maintain efficiency across diverse wind profiles (Jonkman & Sclavounos, 2020). These challenges highlight the necessity of intelligent control frameworks capable of balancing energy capture and structural durability under dynamic wind conditions.

This study aims to address these challenges through a dual approach that combines aerodynamic modeling and intelligent control strategies. First, we analyze the fundamental equations governing wind-to-power conversion, including the power coefficient (Cp), torque coefficient (Ct), and their relationship with the tip-speed ratio (λ)and pitch angle (β). These parameters play a pivotal role in determining the efficiency of energy extraction (Manwell et al., 2010). Second, we propose a robust MPPT-based control framework, integrated with indirect speed regulation, designed to stabilize rotor dynamics while maximizing energy yield. To validate these approaches, MATLAB/Simulink simulations are employed, illustrating the turbine's response to wind fluctuations and its ability to converge to optimal operating points (Hansen et al., 2015). By bridging theoretical aerodynamic models with practical control strategies, this research aims to enhance the reliability and efficiency of wind turbines, particularly for onshore and offshore installations. The findings contribute not only to the academic discourse but also to industrial priorities, such as reducing the levelized cost of energy (LCOE) and extending turbine lifespans under harsh environmental conditions (Veers et al., 2019). This study ultimately supports the advancement of next-generation wind energy systems, enabling them to operate more efficiently while addressing sustainability, reliability, and economic feasibility in the evolving energy landscape.

2. MODELING AND CONTROL STRATEGIES

2.1 Wind Turbine Modeling

To further illustrate the impact of Cp, a numerical example can be included. For instance, considering an optimal tip-speed ratio λ=8λ=8 and pitch angle β=0, if the empirical equation gives Cp=0.45, and assuming a wind speed of Vw=10V, the power captured by a turbine with a rotor radius of 50 m and air density ρ=1.225 kg/m^3 can be calculated using the power equation:

The aerodynamic model of a wind turbine is based on the wind power equation:

$$P_{vent} = \frac{1}{2}\rho A V_v^3 C_p$$

P_{vent}: is the power contained in the wind (W)
ρ: is the air density (kg/m3)
A: is the surface swept by the rotor (m2)
V_v^3 : is the wind speed (m/s)

However, a turbine can only extract part of this power due to the aerodynamic limitations defined by the Betz limit, which imposes a maximum efficiency of 59.3%. Consequently, the mechanical power recovered by the turbine is given by:

$$P_{turbine} = C_p * P_{vent}$$

where C_p is the power coefficient, depending on the characteristics of the turbine and the wind conditions

2.1.1 Calculation of the Power Coefficient

The power coefficient (Cp) quantifies the efficiency of a wind turbine in converting the kinetic energy of the wind into mechanical energy at the rotor. It serves as a crucial parameter in wind turbine performance analysis and is typically expressed as a function of the tip-speed ratio (λ) and the blade pitch angle (β). The power coefficient varies depending on turbine design, aerodynamic properties, and control mechanisms. A commonly used empirical expression for Cp is given as:

$$C_p(\lambda, \beta) = K_1\left(\frac{K_2}{\lambda_i} - K_3\beta - K_4 - \beta^{K_5} - K_6\right)e^{\frac{K_7}{\lambda_i}}$$

With:

$$\lambda_i = \frac{1}{\frac{1}{\lambda + K_8}}$$

where:

- k1,k2,...,k8: are empirical constants specific to the turbine,
- λi: is a factor for correcting the top speed ratio.

where λi is an intermediate function of λ, is experimentally determined coefficients that vary with the turbine design. These coefficients are typically obtained through wind tunnel experiments and computational fluid dynamics (CFD) simulations, ensuring that the theoretical model aligns with actual turbine performance. The constants k1,k2,...,k8k1,k2,...,k8 in some formulations are derived from manufacturer data and reflect the specific aerodynamic properties of a given turbine model. Clarifying the origins of these equations is essential, as different empirical models exist in the literature. Some formulations are derived from Betz's Law, which states that the maximum theoretical power extraction from wind energy is limited to 59.3% of the available wind power (Manwell et al., 2010). However, real-world turbines achieve lower Cp values due to blade inefficiencies, wake effects, and mechanical losses.

The ratio of top speed (λ) is a key parameter influencing Cp and is defined by:

$$\lambda = \frac{R\Omega t}{V_v}$$

where R is the rotor radius, Ω is the angular velocity of the turbine, and Vw is the wind speed. The optimal value of λ ensures maximum energy extraction by keeping the rotor blades in synchronization with the incoming wind stream. If λ is too high, the blades move too quickly, creating excessive drag and reducing efficiency. Conversely, if λ is too low, the blades do not capture enough wind energy. The pitch angle (β) is another key aerodynamic parameter, referring to the angle at which the blades are oriented relative to the wind.

2.1.2 Relationship Between Cp, the Peak Speed Ratio λ and the Pitch Angle β

Optimization of the power coefficient Cp is based on the appropriate choice of the peak speed ratio λ and the pitch angle β. These two parameters directly influence the turbine's ability to capture and convert wind energy into mechanical power. Good management of λ and β is essential to maximize energy efficiency and improve the overall performance of the wind turbine. In the case of fixed pitch wind turbines, the pitch angle β remains constant and does not adapt to variations in wind speed. The power coefficient Cp reaches its maximum for a specific value of the tip speed ratio, called λopt, which corresponds to the best aerodynamic efficiency of the rotor. This type of wind turbine is generally simpler in terms of design and control.

2.2 Wind Turbine Modeling

2.2.1 The Coefficient Curve

The coefficient of torque Ct is an essential parameter that reflects the efficiency of converting wind into mechanical torque applied to the rotor. It is defined by the following equation:

$$T_a = \frac{1}{2}\rho\pi R^3 v_v^2 C_t(\lambda)$$

Figure 1. the evolution of the torque coefficient Ct as a function of the gear ratio and variation of the power generated as a function of the wind speed.

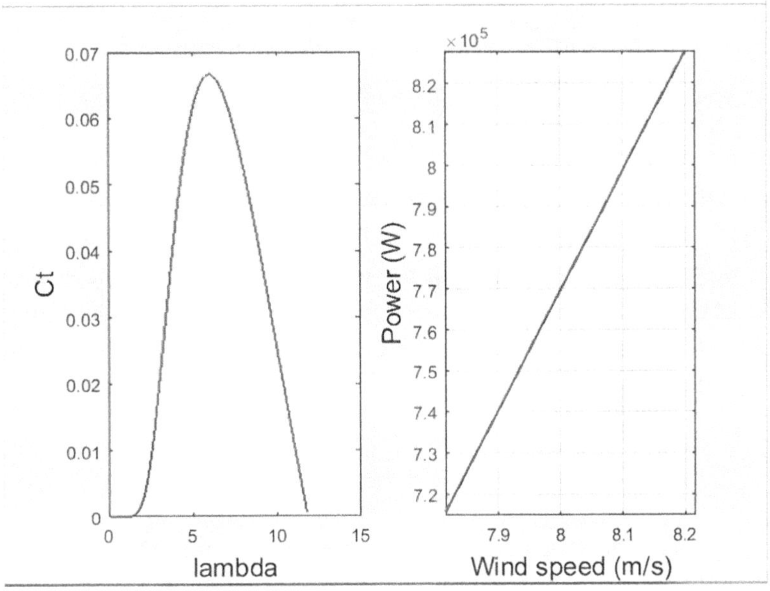

The energy conversion model of a wind turbine describes how the kinetic energy of the wind is transformed into mechanical power at the rotor and then converted into electrical power by the generator. This conversion efficiency is largely influenced by three key aerodynamic parameters: the power coefficient (Cp), the tip-speed ratio (λ), and the pitch angle (β). Optimizing these parameters is crucial to maximizing power extraction while ensuring the durability and stability of the wind turbine system. The power coefficient (Cp) represents the efficiency with which the wind turbine extracts energy from the wind. It is dependent on the tip-speed ratio (λ), which determines the relative speed of the blade tips compared to the incoming wind, and the pitch angle (β), which regulates the aerodynamic interaction between the blades and the wind. In the graph, we can see a linear relationship between the power and the speed of the wind in the operating range studied. This trend is expected for moderate wind regimes where the turbine operates at its optimal efficiency, close to Cpmax. However, in higher wind regimes, the power no longer follows this linear trend, as the turbine reaches its nominal power and regulation mechanisms (such as pitch control) limit excess power to avoid structural damage. The figure shown corresponds to a MATLAB/Simulink model for calculating the aerodynamic torque Tt transmitted to the rotor as a function of the wind speed Vw and the rotor rotation speed Ωm. This model is based on the fundamental equations of wind turbine aerodynamics and incorporates blocks of mathematical calculations and correspondence tables. This model begins by taking as input the wind speed Vw and the rotational speed of the rotor Ωm, two essential parameters for determining the tip speed ratio λ. The tip speed ratio is a key factor in assessing the aerodynamic performance of the turbine and is defined by the following relationship:

Figure 2. Wind turbine modeling

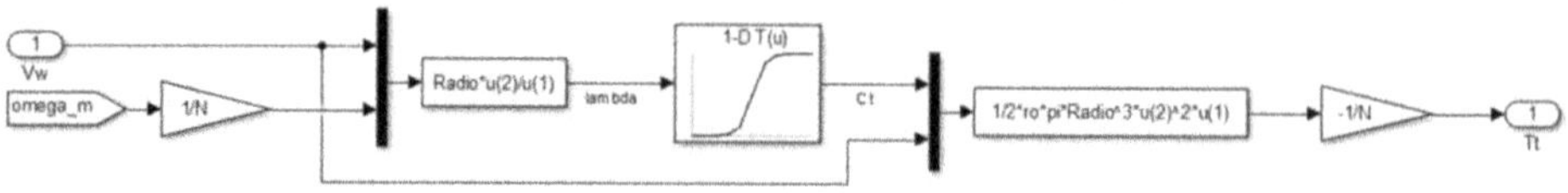

Source: generated by ourselves (Matlab output)

where R represents the radius of the turbine blades. In the Simulink model, this calculation is performed via a division block that divides the angular velocity of the rotor by the wind speed. A scale factor of 1/N is applied to ensure a correct correspondence with the dynamics of the system. Once λ has been calculated, it is sent to a non-linear correspondence table (1-D T(u)) which allows the value of the torque coefficient Ct to be extracted. This coefficient, which is a function of λ, is determined experimentally or by aerodynamic simulations.

$$T_a = \frac{1}{2}\rho\pi R^3 v_v^2 C_t(\lambda)$$

where ρ is the air density. This calculation is implemented in a multiplication block that applies the physical coefficients and geometric terms to the Ct, Vw and R variables. This equation reflects the dependence of the generated torque on the wind speed and the aerodynamic performance of the turbine. Finally, to adapt the value of the torque to the generator, a factor −1/N is applied.

2.3 The Wind Turbine Control Strategy

The functioning of the indirect speed control is based on several essential steps to ensure optimal regulation of the system. First, the detection of the rotational speed Ωm is achieved by measuring the rotor speed, which is then used as a main input to the control system. Once the speed has been measured, the calculation of the torque optimization term is performed. Speed optimization is based on an optimal tuning coefficient kopt, which is determined from the turbine's aerodynamic parameters and the maximum power point (MPPT). This coefficient makes it possible to dynamically adapt the generator torque in order to maintain optimal operation. The equation used to express kopt is as follows:

$$K_{opt} = \frac{0.5 . \rho . \pi . R^5 . C_{pmax}}{\lambda_{opt}^3}$$

where R is the radius of the rotor blades, ρ is the air density, Cpmax is the maximum power coefficient and λopt represents the optimal tip speed ratio for extracting the maximum energy from the wind.

Figure 3. Indirect speed control (D_{tm}=D_t+D_m)

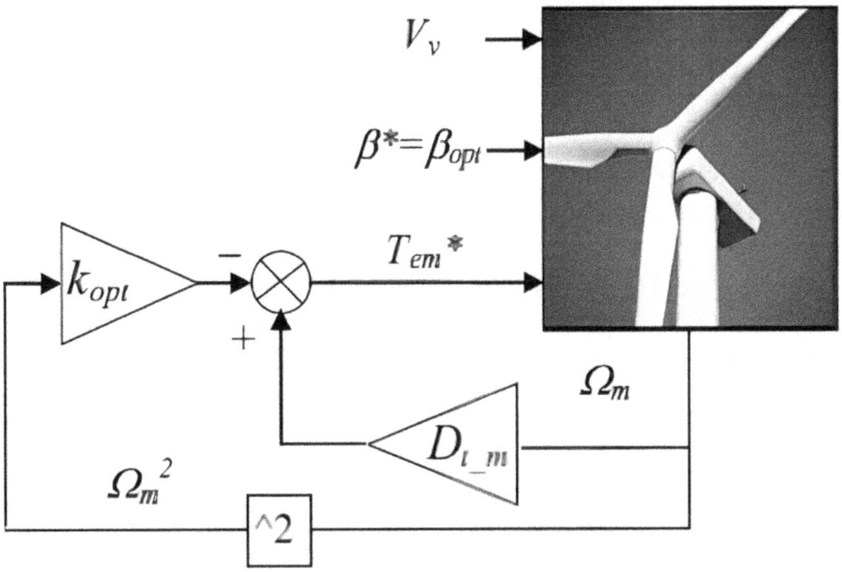

Indirect speed control is a crucial approach for optimizing wind turbine performance by dynamically adjusting the electromagnetic torque to maintain an optimal tip-speed ratio (λ). Unlike direct speed control, which directly regulates the rotational speed of the rotor through mechanical braking or pitch adjustments, indirect control relies on modifying the generator torque, making it more suitable for modern variable-speed wind turbines. A key element in indirect speed control is the optimization of the coefficient kopt, which defines the optimal relationship between generator torque and rotor speed. This coefficient is determined through experimental calibration and numerical simulations, considering aerodynamic characteristics, turbine geometry, and wind conditions. The optimal torque setpoint is expressed as:

$$T^*_{em} = kopt\,.\,\Omega^2_m$$

where Tem∗ is the reference electromagnetic torque, Ωm is the rotor speed, and kopt is a system-specific coefficient derived from wind tunnel experiments and computational fluid dynamics (CFD) simulations. Adjusting kopt ensures that λ remains close to its optimal value, preventing efficiency losses due to deviations from the maximum power point.

2.4 The Dynamic Resistance of the System

In the context of indirect speed control, the dynamic resistance of the system is represented by the Dtm parameter, which includes two essential components. The first, Dt, is related to aerodynamic losses and rotor drag forces, representing the effect of aerodynamic friction on blade rotation and their interaction with the wind. The second, Dm, concerns mechanical and electromagnetic losses at the generator, thus including losses due to bearings, gears in the case of a speed multiplier, as well as the electromagnetic effects of the generator.

The dynamics of the system are described by the following equation:

$$J\frac{d\Omega_m}{dt} = T_a - T_{em} - D_{tm}.\Omega_m$$

where J represents the rotor's moment of inertia, Ta is the aerodynamic torque produced by the turbine, Tem corresponds to the electromagnetic torque applied by the generator, and Dtm is defined as the sum of the dynamic losses, Dtm=Dt+Dm. This Dtm parameter plays a fundamental role in the stability and regulation of the turbine's speed. An increase in Dtm leads to a faster dissipation of mechanical energy, which can be advantageous in strong winds in order to limit the mechanical stresses on the system. Conversely, a decrease in Dtm promotes better energy recovery by reducing losses, thus optimizing the overall efficiency of the wind turbine.

2.5 MPPT (Maximum Power Point Tracking)

The process begins by measuring the rotational speed Ωm, which is then divided by a factor NN, generally used to normalize the speed according to the transmission ratio in the case of a speed multiplier system. This normalized value is then squared in the next step, where the output is expressed as u(1)·u(1), thus reflecting the quadratic relationship between rotor speed and optimal torque. The result is then multiplied by an optimal coefficient Kopt, which is a key parameter defined on the basis of the aerodynamic characteristics of the turbine and the principle of maximizing the power captured.

Figure 4. Turbine control modeling

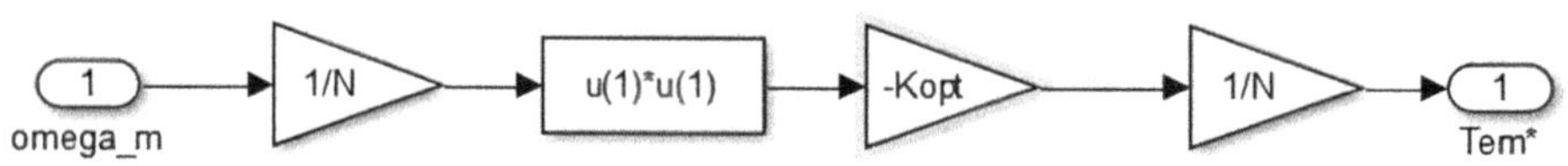

Source: generated by ourselves (Matlab output)

The value obtained is divided again by N, in order to adapt the scale of the calculated torque, thus producing the electromagnetic torque instruction Tem∗. This reference torque is then used to control the electric generator, adjusting its braking torque to keep the rotor at the optimum speed to maximize the power captured by the wind turbine.

3. THE SIMULATION AND RESULTS

The figure represents three sinusoidal curves out of phase by 120°, corresponding to the three phases of the system. The amplitude of the currents varies between a maximum value of 2000 A and a minimum value of -2000 A. This representation indicates that the generator is operating stably and that the electromechanical conversion is efficient. The sinusoidal nature of the currents testifies to the proper functioning of the generator and an adequate energy conversion. Balancing the three phases demonstrates good regulation of the electrical system, while the constant frequency of the signal reflects the stability of the generator speed. However, the possible presence of abnormal distortions or variations could indicate undesirable harmonics, resulting from electromagnetic disturbances or malfunctions of the energy converter.

3.1 Initialization of the Wind Turbine Parameters

The moment of inertia of the rotor is set to $J=0.21$ kg·m^2, which determines the turbine's resistance to variations in rotational speed. A lower inertia allows faster adaptation to wind changes, whereas higher inertia slows down adjustments. The friction coefficient f=0.0001 represents mechanical losses due to friction in the turbine's drivetrain, with a low value indicating minimal energy dissipation. Regarding geometric parameters, the rotor blade radius is set to R=3 meters, which directly influences the swept area and, consequently, the amount of wind energy captured by the turbine. The swept area is calculated as S=πR2, playing a crucial role in determining wind power extraction. In terms of transmission parameters, the gear ratio is defined as G=6, meaning that the generator rotates six times faster than the main rotor. This parameter is essential for adapting the turbine's mechanical power to the electrical generator's operating range.

Table 1. Wind turbine parameters

Paramertes	Values
f	0.21
J	0.0001
R	3
G	6
Ro	1.22
Cpmax	0.39
Lamdopt	10

The aerodynamic parameters include the air density $\rho=1.22$ kg/m^3, which is essential for computing wind power. This value corresponds to air density at sea level. The maximum power coefficient Cp-Max=0.39 represents the highest efficiency at which the turbine can convert wind energy into mechanical energy. Given that the theoretical Betz limit is 0.593, a value of 0.39 is realistic for a well-designed wind turbine. The optimal tip speed ratio (TSR) is set to λopt=10, indicating that the rotor blades move ten times faster than the wind speed, which is typical for high-performance wind turbines.

3.2 Entry Signal

The graph displayed in the Scope window represents a step signal, indicating a sudden transition at t≈15s. The x-axis corresponds to time in seconds, while the y-axis represents an input variable, most likely the wind speed or a control setpoint of the MPPT system. The background of the image highlights the presence of a Step block, confirming that the observed signal is a step input applied to the model. This type of signal is used to test the system's response to a sudden disturbance. In the context of MPPT for a wind turbine, it can represent a variation in wind speed. Before t=15s, the wind speed remains stable at 10 m/s, then undergoes a sudden increase to 12 m/s, simulating a change in environmental conditions. This variation helps evaluate how the system adjusts the rotational speed and optimizes the harvested power. It is also possible that this input corresponds to a modification of the MPPT controller setpoint, involving a change in the optimal tip speed ratio or blade angle, which serves to observe the system's ability to maintain optimal energy efficiency.

Figure 5. Entry signal

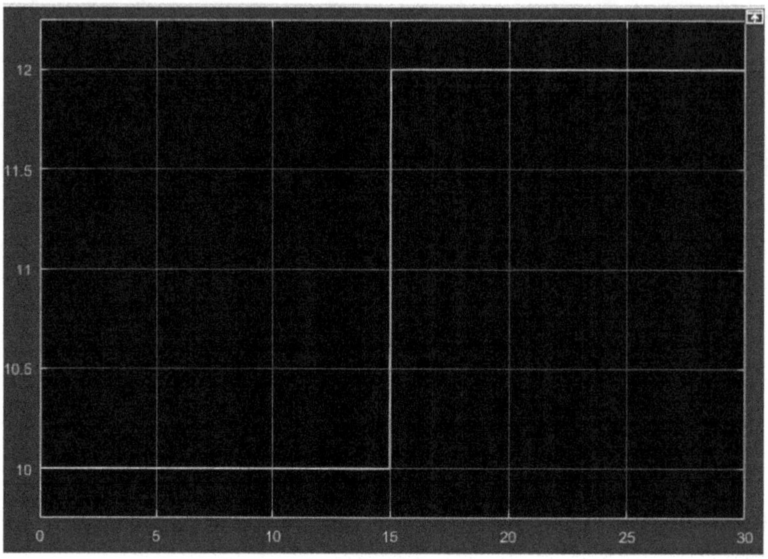

Source: generated by ourselves (Matlab output)

The response analysis shows that the system operates in a stable state before t=15s, then undergoes a sudden disturbance that requires adaptation. After this change, the new value reaches a stable plateau, indicating that the system successfully accommodates the variation. If this input corresponds to a change in wind speed, a well-tuned MPPT will quickly adjust the rotational speed to maximize the captured power, whereas a poorly configured MPPT may cause oscillations or an excessive stabilization time, leading to energy losses.

3.3 Dynamic Response of the Wind Turbine

The displayed curve follows a progressive evolution with an S-shaped rise before reaching a stable plateau. The x-axis represents time in seconds, while the y-axis illustrates the rotational speed of the wind turbine, expressed in rad/s. In the context of MPPT for a wind turbine, this response corresponds to the evolution of the rotor's rotational speed in reaction to an increase in wind speed. At the beginning of the simulation, the rotational speed is low, then it gradually increases before reaching a new dynamic equilibrium. This trend indicates that the MPPT algorithm progressively adjusts the turbine speed to maximize the captured energy by maintaining the rotor at an optimal speed.

Figure 6. Dynamic response of the wind turbine

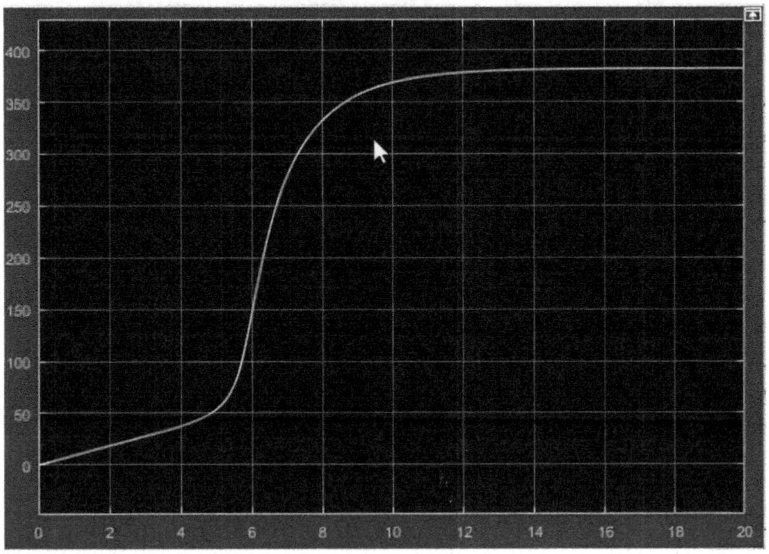

Source: generated by ourselves (Matlab output)

The S-shaped form of the curve indicates a well-controlled dynamic response, without excessive oscillations. Three phases can be distinguished: an initial phase where the rotational speed evolves slowly, corresponding to a transient acceleration phase of the rotor; a rising phase where the speed increases rapidly, indicating an active response of the system to a variation in wind conditions or load; and a stabilization phase where the rotational speed reaches a plateau, showing that the system has converged to a stable equilibrium state.

3.4 Analysis of the Oscillations and Stability of the MPPT System

The curve displays two sharp peaks followed by a gradual decline and stabilization. The x-axis represents time in seconds, while the y-axis likely corresponds to a dynamic variable of the system, such as the mechanical torque applied to the turbine shaft or a transient power response. Such a response is characteristic of transient phenomena in dynamic systems, particularly when a system undergoes a sudden disturbance or a load variation. In the context of MPPT applied to a wind turbine, this response could be attributed to a sudden change in operating conditions, such as a rapid variation in wind speed or an adjustment in the blade pitch angle (β). Before t=10s, the variable remains close to zero, indicating an initial stable state. A first sharp peak appears shortly after, suggesting a sudden increase in the measured parameter, likely due to a change in wind speed or a new MPPT control setpoint. A second peak follows shortly after, which may represent the system's reaction before settling into a new equilibrium state. After t=20s, the amplitude gradually decreases, indicating that the system is stabilizing over time.

Figure 7. Transient behavior

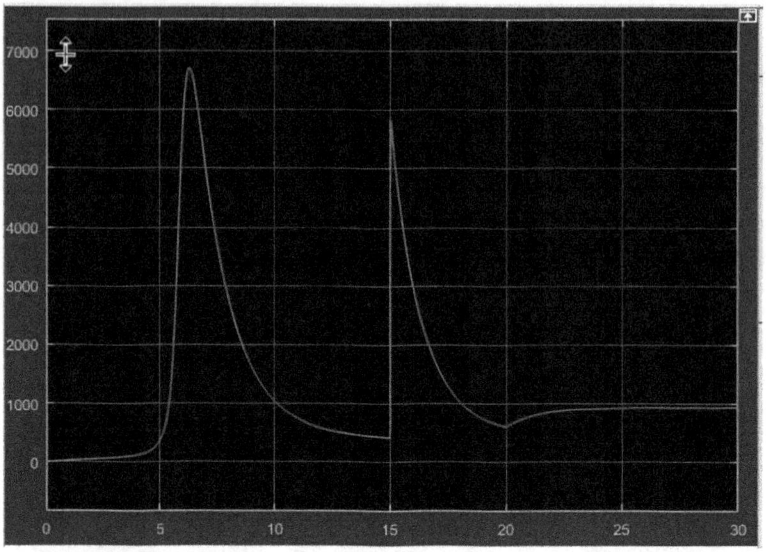

Source: generated by ourselves (Matlab output)

This response suggests that the system exhibits oscillatory transient behavior before reaching a steady state. Such a phenomenon can arise from suboptimal MPPT regulation, leading to temporary fluctuations around the Maximum Power Point (MPP). If these oscillations are excessive, they may cause energy losses or excessive mechanical stress on the turbine shaft.

4. CONCLUSION

This study emphasizes the critical importance of precise modeling and advanced control strategies in optimizing the performance of wind turbines. The detailed analysis of the power coefficient (**Cp**) and torque coefficient (**Ct**), as well as the dynamic adaptation of the tip-speed ratio (**λ**) and pitch angle (**β**), underscores their direct influence on energy efficiency. By fine-tuning these parameters, wind turbines can operate closer to their optimal power point, maximizing energy extraction while reducing mechanical stress on components. The simulations conducted using MATLAB/Simulink provide a comprehensive validation of the proposed control methodologies, demonstrating system stability, efficient rotor speed regulation, and optimal electromechanical conversion. The results highlight the effectiveness of indirect speed control and MPPT-based optimization, which ensure that the generator operates at maximum efficiency while mitigating fluctuations in wind speed. A key indicator of this efficiency is the balanced sinusoidal stator currents, which confirm the quality of energy conversion and grid integration.

The study also highlights the advantages of variable-pitch wind turbines when combined with MPPT algorithms, making them particularly suitable for environments with highly variable wind conditions. The ability to dynamically adjust the blade pitch angle enables these systems to efficiently regulate power output and reduce mechanical loads, significantly extending the lifespan of turbine components. However, despite these advancements, continuous monitoring and adaptive control remain essential to anticipate disturbances, optimize system response, and maintain high operational efficiency in real-world conditions. Beyond the current findings, this work paves the way for further advancements in wind energy optimization, particularly through the integration of artificial intelligence (AI) and machine learning. AI-driven control strategies have the potential to refine real-time performance prediction, adaptive tuning of operational parameters, and predictive maintenance techniques, further enhancing system reliability and energy yield. As wind energy continues to expand its share in the global energy mix, these innovations will play a crucial role in strengthening the competitiveness, efficiency, and sustainability of wind power generation. Ultimately, the integration of intelligent control systems will reinforce wind energy's role as a key pillar of the global transition toward a cleaner, more resilient, and economically viable energy future.

REFERENCES

Burton, T., Jenkins, N., Sharpe, D., & Bossanyi, E. (2021). *Wind Energy Handbook* (3rd ed.). Wiley.

Hansen, M. O. L., Sørensen, N. N., & Voutsinas, S. (2015). State of the Art in Wind Turbine Aerodynamics and Aeroelasticity. *Progress in Aerospace Sciences*, *74*, 54–88.

IPCC (2023) - International Panel on Climate Change, *Sixth Assessment Report.*

IRENA (2023) - International Renewable Energy Agency, *Global Wind Energy Outlook 2023.*

Jonkman, J., & Sclavounos, P. (2020). Effect of Control Strategies on the Dynamics of Floating Offshore Wind Turbines. *Journal of Wind Engineering and Industrial Aerodynamics*, *198*, 104093.

Li, H., Chen, Z., & Blaabjerg, F. (2021). Maximum Power Point Tracking Control Strategies for Wind Energy Systems: A Review. *IEEE Transactions on Sustainable Energy*, *12*(2), 872–890.

Manwell, J. F., McGowan, J. G., & Rogers, A. L. (2010). *Wind Energy Explained: Theory, Design and Application.* John Wiley & Sons.

Rhiat, M., Karrouchi, M., Atmane, I., Touhafi, A., Bossoufi, B., Almalki, M. M., & Hirech, K. (2024). Maximizing solar energy efficiency: Optimized DC power conversion for resistive loads. *Computers & Electrical Engineering*, *120*, 109867.

Rhiat, M., Karrouchi, M., Hassari, A., Melhaoui, M., Atmane, I., Ouariachi, M. E., & Hirech, K. (2024). Developing and simulating an inexpensive solar irradiance metre for photovoltaic utilisation. *International Journal of Power and Energy Conversion*, *15*(4), 376–388.

Rhiat, M., Karrouchi, M., Hassari, A., Melhaoui, M., Atmane, I., Ouariachi, M. E., Schmitz, P., Benhala, B., & Hirech, K. (2024). Developing and simulating an inexpensive solar irradiance metre for photovoltaic utilisation. *International Journal of Power and Energy Conversion*, *15*(4), 376–388. DOI: 10.1504/IJPEC.2024.141867

Veers, P., Dykes, K., Lantz, E., Barth, S., & Bottasso, C. L. (2019). Grand Challenges in the Design and Control of Large Wind Turbines. *Wind Energy Science*, *4*(1), 1–37.

Chapter 4
Optimized Transformer-Less PV Inverter Design for Grid Operation Using Super Junction MOSFETs and SiC Diodes

D. Lakshmi
https://orcid.org/0000-0002-8071-805X
AMET University, India

G. Jegadeeswari
https://orcid.org/0000-0002-7865-4026
Saveetha Engineering College, India

B. Kirubadurai
Vel Tech Dr. Rangarajan Dr. Sagunthala R&D Institute of Science and Technology, India

ABSTRACT

This paper presents a high-efficiency transformer-less single-phase sinusoidal pulse-width modulation (SP-SPWM) full-bridge photovoltaic (PV) inverter optimized for grid-connected applications. Traditional designs using super junction MOSFETs and SiC diodes face challenges due to reverse recovery in MOSFET body diodes, leading to switching losses and leakage currents. To overcome these issues, the proposed inverter reduces the free-rail voltage to half of the DC input during freewheeling, ensuring a stable common-mode voltage and significantly minimizing leakage current. Additionally, it eliminates switching time constraints, reducing waveform distortions at high frequencies and grid zero-crossing points. A 1 kW prototype was developed and tested, achieving an efficiency of 98.5%, surpassing the European benchmark of 98.32%. The experimental results demonstrate improved efficiency, lower leakage current, and enhanced reliability, making the proposed design a strong candidate for next-generation grid-connected PV systems.

DOI: 10.4018/979-8-3693-9924-8.ch004

I.INTRODUCTION

Photovoltaic (PV) micro-inverters are broadly categorized into three topologies: single-phase, two-phase, and multi-phase. Multi-stage micro-inverters, under maximum power point tracking (MPPT) control, typically comprise a step-forward DC–DC converter, a high-frequency intermediate stage to achieve a rectified sinusoidal waveform, and a low-frequency development stage for grid connection. However, the multi-stage architecture, along with its numerous components, leads to increased complexity and higher costs.

A two-phase topology can simplify the architecture by cascading an MPPT-controlled step-up DC–DC converter with a grid-tied high-frequency inverter. However, in single-phase systems, the architecture must integrate voltage stepping, MPPT, and DC–AC inversion within a single stage. For effective grid connection, the low voltage generated by PV panels must first be significantly boosted to meet the grid energy supply levels (Islam, Mekhilef, & Hasan, 2019).

This voltage boosting requirement presents a significant challenge for PV inverter designers, as conventional boost converters struggle to achieve the necessary efficiency and performance gains. Consequently, extensive research has been directed towards the development of advanced high-stage DC–DC converters (Patrao et al., 2021), (Li, Kai, Lanlan, Hongfei, & Yan, 2019). These converters are being optimized for integration with half-bridge and full-bridge inverter topologies, enabling more efficient solar power generation systems that address both performance and cost challenges.

The adoption of distributed photovoltaic (PV) systems for commercial and residential use has grown rapidly in recent years. Despite a significant reduction in panel prices, the overall cost of PV grid system investments and energy output remains high compared to other renewable energy sources. This cost disparity highlights the need for innovative designs that optimize performance and reduce expenses. Grid-connected inverters, especially in single-phase systems with outputs below 5 kW, play a critical role in achieving high efficiency, reduced costs, compact size, and lightweight construction (Islam & Mekhilef, 2019). Historically, line-frequency transformers have been widely employed in PV grid inverters to ensure galvanic isolation for safety. However, these transformers are large and heavy, resulting in cumbersome installations and increased overall system weight.

To address these challenges, inverters with high-frequency isolation transformers have been developed. These designs offer several advantages over their line-frequency counterparts, including lower costs, reduced size, and lighter weight, while maintaining the necessary safety and performance standards. This shift toward high-frequency transformer-based inverters is driving the evolution of more efficient and practical PV grid systems for distributed applications.

Figure 1. Leakage current paths for transformer-less PV inverters.

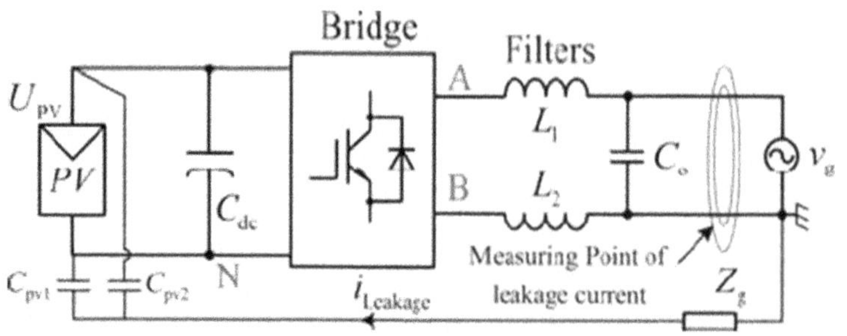

High-frequency transformer-based inverters, while offering advantages in size and weight reduction, often incorporate multiple power stages. This increases system complexity and reduces overall efficiency. Consequently, these inverters are less suitable for grid-mounted applications, particularly in low-power PV systems.

In transformer-less PV systems, the absence of a transformer introduces the potential for common-mode (CM) leakage currents. These currents flow through parasitic capacitances between PV panels and the ground, which can lead to significant safety and electromagnetic interference issues. Figure 1 illustrates the leakage current path, which involves parasitic capacitances (CPV1 and CPV2), the inverter bridge, filters (L1 and L2), the grid, and the ground impedance (Zg).

Leakage currents must be limited to acceptable levels to ensure system safety and compliance with electromagnetic interference standards. Effective design strategies, such as maintaining a stable common-mode voltage and minimizing parasitic capacitance effects, are critical to addressing these challenges and ensuring the reliable operation of transformer-less PV systems.

II.ENERGY CONVERSION FROM SOLAR ENERGY

Solar power may be transformed to thermal (or heat) energy for the usage in houses or buildings and in pools of heat water: Heating areas - inside greenhouses, residences, etc. Two techniques of converting solar energy to electricity:

- Photovoltaic or solar cells (PV devices)

This transforms the sun into electricity immediately. In rural places that have not been linked to the electric grid, PV systems are commonly employed. The watches, calculators and LED road signs are also utilized for power.

- Solar Power Plants

In these devices, sunlight is indirectly transformed into power. It is turned into mechanical energy and then into electric power. When solar thermal heat is utilized for heating a fluid that creates steam used to generate power. Out of the 15 known U.S. solar power generators, 10 of them are in California, and 5 in Arizona, at the end of 2006 (Xiao, Liu, & Lan, 2024). There is no statistics on solar power plants that produce less than 1 megawatt of energy, therefore in a number of different states there can be smaller solar power plants. Let us start with a detailed description of how solar energy is turned into electricity.

II.I Photovoltaic Cell

Photovoltaic are the direct transformation of light into atomic electricity. Some materials have a photoelectric effect, which causes photons of light to absorb and electrons to release. When these free electrons are collected, electricity may be used as electricity. Figure 2 shows photovoltaic cell.

Figure 2. Photovoltaic cell

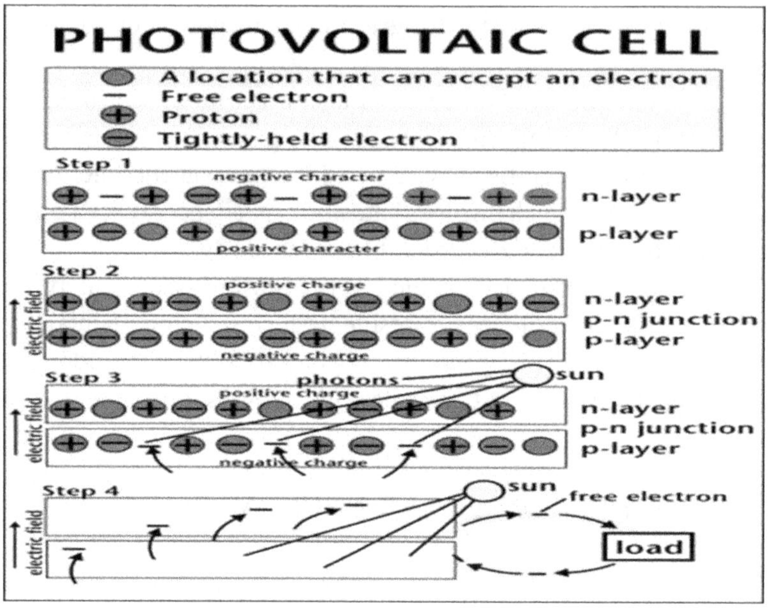

The photoelectric effect was originally seen as being the result of the exposure of certain materials to light by a French scientist, Edmund Bequerel, in 1839. In 1905, Albert Einstein outlined the nature of light and the photoelectric effects that photovoltaic technology was built on (Lakshmi, Baskaran, Ezhilarasi, & Valli, 2020). Bell Laboratories constructed in 1954 the first photovoltaic module. It was considered a solar battery and most of the time it was only a curiosity, as it was too costly to be used. In the sixties, the space industry started to employ the technology for power supply on spacecraft as first seriously. The space programmes enhanced technology, demonstrated its trustworthiness and started to reduce costs. Photovoltaic were recognized as a power source for non-space applications in the energy crisis of the 1970s (Lakshmi, Baskaran, Ezhilarasi, & Valli, 2020).

II.II Solar Power Inverter

In a solar energy system, the solar inverter is an essential component. The fluctuating DC output of the PV module(s) is converted to a clean sinusoidal AC current of 50 Hz or 60 Hz which is then supplied straight to the commercial mains or to a local grid. Communication capabilities are typically incorporated so that users may monitor the inverter and report on power and operational conditions, give firmware updates and regulate the grid connection to the inverter (Jegadeeswari, 2019). Wired or networked (Bluetooth, ZigBee/IEEE802.15.4, 6lo WPAN) network alternatives may be utilized dependent on the grid infrastructure (RS-485, CAN, power line communication, Ethernet).

A real-time microprocessor is at the core of the inverter. The controller performs the highly accurate algorithms necessary to reverse the solar module DC voltage. This controller is designed for the essential power management features including DC/DC and DC/AC to conduct the control loops. The controls additionally maximize the PV output by use of sophisticated algorithms known as maximum power points tracking (MPPT).

The PV maximum output capacity depends on the operating circumstances, and is varied momentarily, because of the temperature, the shade, the soil, cloud cover and time of day. The controller can regulate the charging and switch to battery power for systems with battery power storage when the sun has set or cloud cover limit the photovoltaic output power. The controller features sophisticated peripherals for controller loop implementation, such high-precision PWM outputs and ADCs. The ADC monitors factors such as the PV output voltage and currents and, by altering the PWM working cycle, changes the DC/DC or the DC/AC converter (Manikandan, Sivabalan, & Sundar, 2020). Figure 3 (a) illustrate solar power inverter.

Figure 3a. Solar power inverter

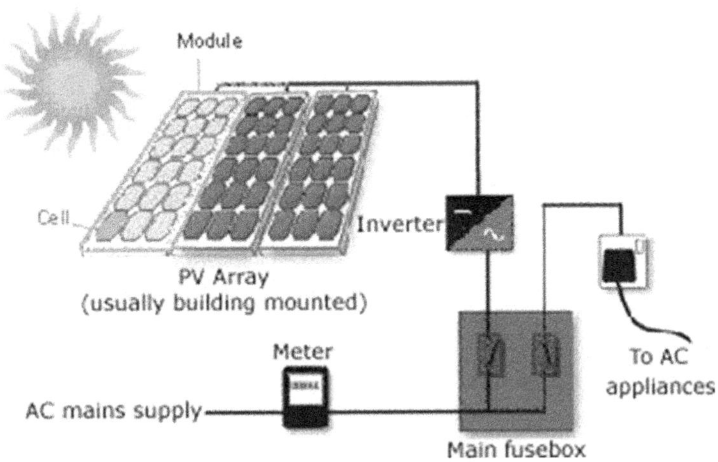

Figure 3b. Typical PV system

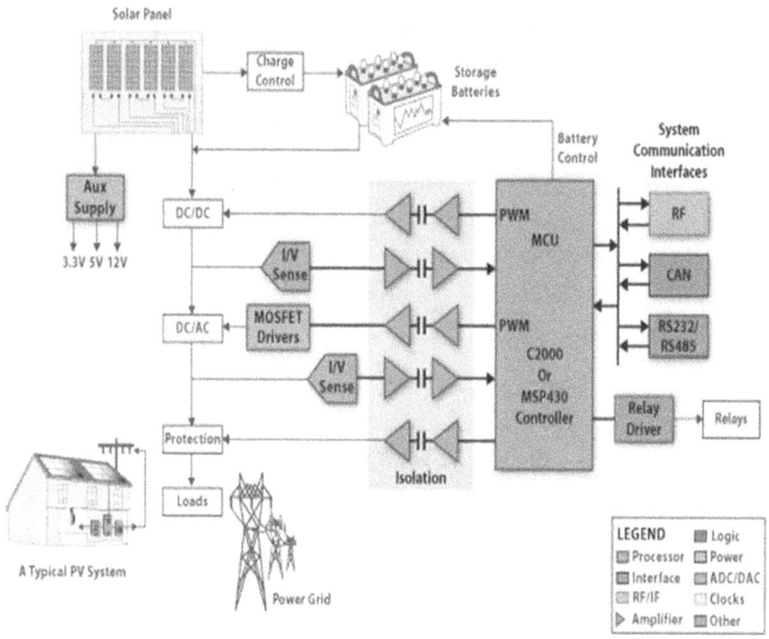

You can utilize the solar light system when you encounter the power or load scheduling challenge. It may be used for house lighting, apartment common lighting, garden lighting, emergency light, etc. And solar water heaters are perfect for all applications that demand hot water. Continuous water heating costs for conventional fuel such as power, kerosene, timber, diesel etc. are insured in the existing modes.

It is appropriate to utilize energy saving equipment that saves your recurrent expenditures and also reduces conventional energy, in accordance with the government policy (Veerakumar, Dheepak, & Saravanan, n.d.). Typical PV system is illustrated in figure 3(b)

The C2000 is intended to read the ADC and modify the PWM in one cycle, to make it feasible to monitor the ADC in real time. A single processor can manage communication on a simple system, more advanced systems with sophisticated displays, and consumption reporting and feed-in tariff payments may require a separate processor that might perhaps have Ethernet capabilities such as the Cortex M3 Stellaris components (Balaji & Sasilatha, 2019). It is also necessary to isolate the CPU from the current and the voltage and also the communication link to the outside world for safety reasons.

II.III Transformer-Less PV Systems

A transformer-less PV system is a photovoltaic power system that directly converts the DC output from solar panels into AC without using a transformer for isolation. This design improves efficiency by eliminating energy losses associated with transformers, typically achieving efficiencies above 97%. It is lightweight, compact, and cost-effective, making it ideal for residential and commercial applications. Transformer-less inverters use advanced circuitry, such as high-frequency switching and sophisticated filtering with inductors and capacitors, to ensure a smooth sinusoidal AC output. They also incorporate ground fault detection and residual current protection to address safety concerns, as they lack galvanic isolation. Despite challenges like managing leakage currents and meeting regulatory standards, transformer-less PV systems are widely adopted due to their enhanced efficiency, reduced costs, and suitability for modern grid-tied and standalone solar applications. Layout of simple transformer-less PV system is illustrated in figure 4.

- In case of failure, the efficiency of the entire PV system can be enhanced by extra 1%-2%.
- Lower transformer efficiencies, lower weights and dimensions compared to their galvanic separation counterparts.

Figure 4. Transformer-less PV systems

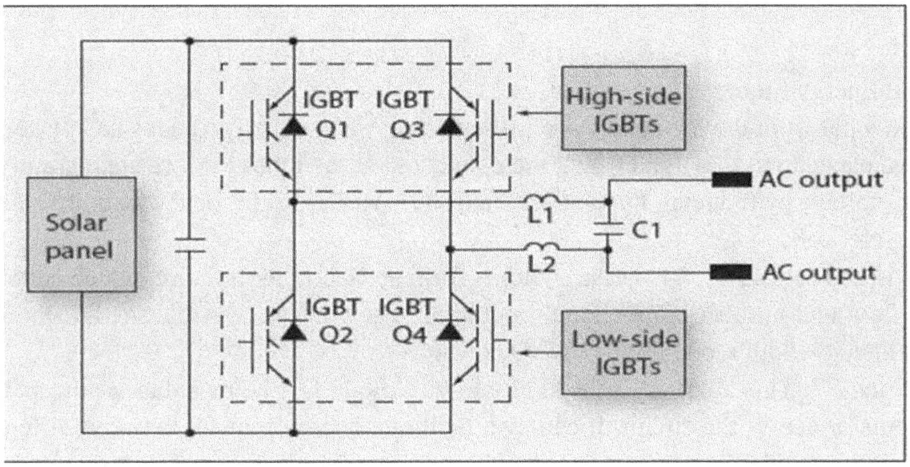

Issue in Transformer-less PV Inverters

- When the transformer is disconnected, the inverter topology's common mode behavior has an effect on the ground leakage current through the PV parasite capacity (Jegadeeswari, Kirubadurai, & Lakshmi, 2020), (Jegadeeswari, Dheepak, Saravanan, & Vanisree, 2018).
- Several ways have been used to decrease ground leakage current using the PV Array's Parasite capabilities.

III. PROPOSED SYSTEM

The family of recommended transformers with a low PV inverter topology is illustrated, derived from the preceding section's derivative method. This circuit diagram represents a dual-input photovoltaic (PV) inverter system, commonly used in renewable energy applications to convert DC power from PV panels into AC power suitable for grid connection or standalone use. Circuit diagram of the Proposed system is shown in figure 5.

Figure 5. Proposed system

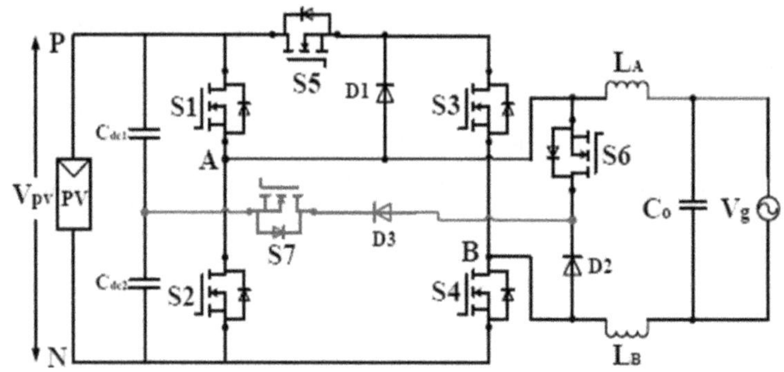

1. Inputs and Energy Storage
   - The input from the PV panels is represented as VPV, which provides DC power.
   - Two capacitors, Cdc1 and Cdc2, are connected at the input to store energy and stabilize the DC voltage by reducing fluctuations caused by variations in the PV output.
2. Switching Network
   - Switches S1, S2, S3, S4: These switches form an H-bridge structure, which controls the power flow and ensures bidirectional operation. They are responsible for DC-to-AC inversion and pulse-width modulation (PWM) control.
   - Switch S7: This auxiliary switch, along with diode D3, helps balance power flow between inputs or across the circuit. It can also facilitate energy transfer between different nodes or operating modes.
   - Switch S5: Used in conjunction with S6to control the direction and flow of energy, particularly in applications with dual inputs or complex power flow requirements.
   - D1, D2: Protect against backflow of current, ensuring unidirectional operation and preventing damage to the components.
   - D3: Works with S7 to manage the auxiliary power path, enabling advanced modes of operation, such as power balancing or additional protection.
3. Filter Network
   - Inductors LA and LB: These inductors serve as energy storage elements in the AC side, smoothing out current flow and reducing harmonics.
   - Output Capacitor Co: Works with the inductors to filter out high-frequency noise, ensuring the output voltage Vg is stable and of high quality.
4. Operation Modes
   - The circuit supports multiple modes of operation, such as:
   - Dual-input operation: If two different PV sources are used, the circuit can manage their integration.
   - Boost mode: By controlling the switching of S7, the voltage can be stepped up before inversion.
   - Inversion mode: The H-bridge formed by S1 to S4 converts the DC input into an AC waveform.

5. Output
    - The final output is Vg, a sinusoidal AC voltage suitable for powering loads or feeding into the grid.
    - The filter network ensures the output meets the quality standards for grid connection, reducing total harmonic distortion (THD).
6. Advantages of the Design
    - Efficiency: By utilizing advanced switching mechanisms, the system minimizes energy loss.
    - Flexibility: The inclusion of S7 and D3 enables the circuit to handle dual inputs or operate in different modes, making it adaptable to various renewable energy setups.
    - Power Quality: The filter network ensures a smooth and stable output, critical for grid-tied applications.

This proposed system is well-suited for renewable energy systems, particularly in scenarios requiring efficient energy transfer, grid integration, or hybrid energy sources.

IV. RESULTS AND DISCUSSION

The image illustrates the hardware setup of a photovoltaic (PV) inverter system that integrates a solar panel and electronic circuitry for energy conversion and control. Here is a detailed explanation of the components and their roles. Hardware setup is shown in figure 6.

Figure 6. Hardware setup

1. Solar Panel (Left Side)
    - Function: The solar panel converts sunlight into DC electricity. This is the primary energy source for the system.

- Integration: The generated DC power is fed into the inverter circuitry for further processing, such as DC-DC conversion or DC-AC inversion.

2. Power Conversion Circuits (Right Side)

The setup includes multiple printed circuit boards (PCBs), each serving a specific function:

a. Input Control Board

Components: Likely includes capacitors, diodes, and input regulators.
Function: Stabilizes the input from the solar panel, filtering out fluctuations in voltage or current.

b. Power Stage Board

Components: Contains MOSFETs, IGBTs, or other switching devices.
Function: This is the core inverter circuit responsible for: DC-AC conversion using pulse-width modulation (PWM) or similar techniques. Regulating the power flow to ensure the desired output voltage and frequency.

c. Driver Circuit

Components: May include microcontrollers, gate drivers, and signal conditioning components.
Function: Controls the switching of MOSFETs/IGBTs based on the input commands or algorithms. It ensures precise timing for the switches to generate a smooth AC output.

d. Filter and Protection Circuit

Components: Includes inductors, capacitors, and diodes.
Function:

- Reduces harmonic distortion in the AC output.
- Protects the system against surges, overvoltage, or short circuits.

3. Transformers (Top Right)

Function:

- Provide galvanic isolation between the input and output for safety.
- Step up or step down the voltage based on the requirements of the load or grid.

4. Interconnections and Wires

Purpose: These connect the solar panel, various PCBs, and output terminals, ensuring proper communication and power transfer between components.

5. Indicators and Connectors (Bottom or Side Panels)

Purpose: Likely includes LEDs, connectors, or switches for monitoring, testing, and interfacing with external devices.

Overall Purpose

This hardware setup is a practical implementation of a PV inverter system, designed to:

- Convert DC power from the solar panel into AC power suitable for household appliances or grid integration.
- Regulate and stabilize the power output to ensure reliability and efficiency.
- Serve as a prototype or experimental model to test the circuit's performance, efficiency, and safety features.

This setup is often used in laboratories or small-scale renewable energy projects.

Figure 7. Gate pulses zoomed view

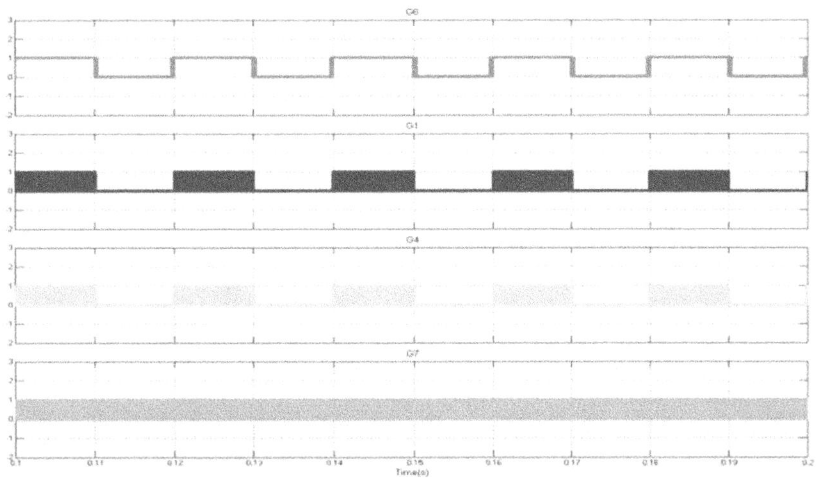

Figure 7 shows a set of waveform plots representing the switching states of various power switches (S2, S1, S4, and S7) in a photovoltaic inverter or power converter system. Let's explain these waveforms in detail:

The pink waveform shows the gate signal applied to switch S2. The signal is periodic, alternating between high (ON) and low (OFF) states. This pattern indicates that S2 is being controlled using Pulse Width Modulation (PWM) to regulate power flow.

The blue waveform represents the gate signal for S1. Similar to S2, S1 is controlled periodically, with an ON-OFF pattern. The phase difference between S1 and S2 likely indicates their complementary switching behavior, crucial in H-bridge or inverter circuits.

The green waveform corresponds to the gate signal for S4. The switching pattern of S4 is periodic and phase-shifted relative to S1, suggesting that S4is part of the complementary pair with S1 in an H-bridge topology.

The yellow waveform is a steady high signal, indicating that S7 remains ON throughout the operation. S7's role is likely auxiliary, such as enabling current flow or power balancing, without active modulation during this period.

Key Observations:

- Complementary Switching: The waveforms for S1, S2, and S4 suggest complementary switching, which is standard in inverter circuits to prevent shoot-through (simultaneous ON states causing short circuits).
- Fixed ON State: S7's continuous ON state shows that it supports auxiliary functions without interfering with the main power flow.

Figure 8. IG, IS5, IS3, IS2 Figure 9. IG, IS1, IS6, IS4

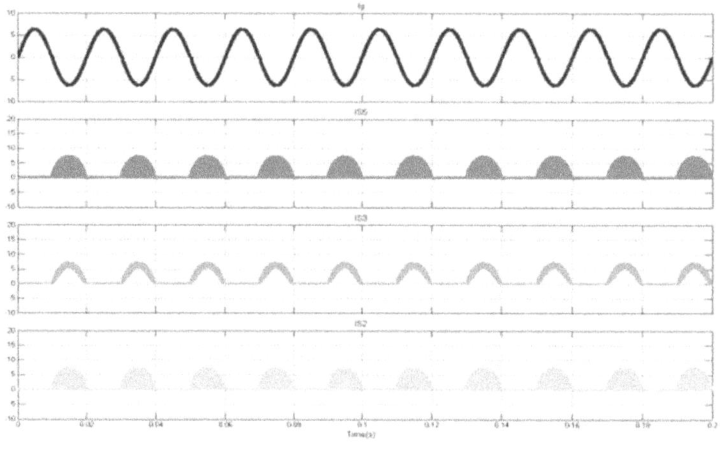

This set of waveforms illustrates in the figure 8 is the operation of an inverter system, showcasing the gate current Ig and output current (I) behavior of switches (S5, S3, S2). The top plot displays a smooth sinusoidal output current (Ig), indicating the inverter successfully converts DC power into a high-quality AC voltage suitable for grid or load applications. The second, third and fourth plots show the current through switches S5, S3 and S2, respectively, highlighting their periodic switching behavior in response to pulse-width modulation (PWM) control. These currents are complementary and correspond to the alternation of conduction phases in the H-bridge inverter.

This set of waveforms illustrates in the figure 9 is the operation of an inverter system, showcasing the gate current Ig and output current (I) behavior of switches (S1, S6, S4). The top plot displays a smooth sinusoidal output current (Ig), indicating the inverter successfully converts DC power into a high-quality AC voltage suitable for grid or load applications. The second, third and fourth plots show the current through switches S1, S6 and S4, respectively, highlighting their periodic switching behavior in response to pulse-width modulation (PWM) control.

Figure 10(a). IG, VDS1, VDS6, VDS4

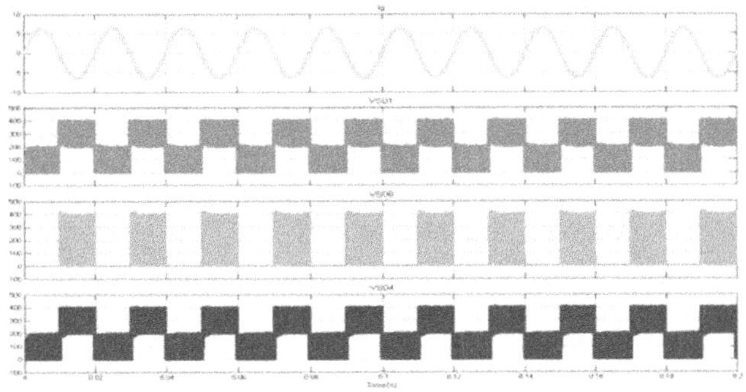

Figure 10(b) VG, IG, VAN, VBN, VAB

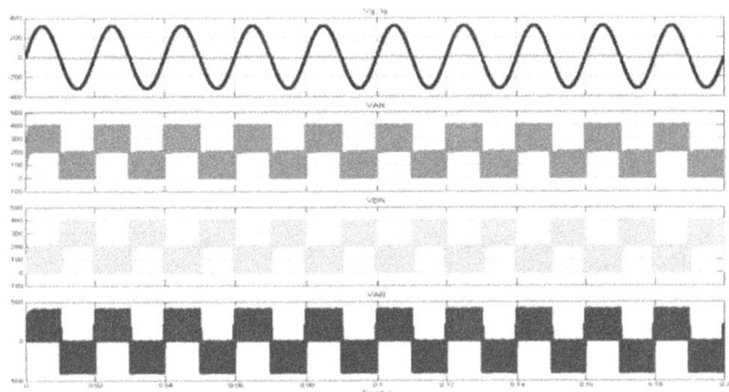

This set of waveforms represents in figure 10(a) is the operation of an inverter circuit showing the output voltage and switching behaviors. The top plot displays the gate current (Ig) which is a sinusoidal waveform, demonstrating successful DC-to-AC conversion with a stable and high-quality AC output. The subsequent plots (VDS1, VDS6, VDS4) are square waveforms, representing the **switching signals** for different switches in the inverter. These square waves alternate between ON and OFF states, following a pulse-width modulation (PWM) control strategy to generate the desired output.

This set of waveforms represents in figure 10(b) is the operation of an inverter circuit showing the output voltage and switching behaviors. The top plot displays the gate current and voltage (Vg and Ig) which is a sinusoidal waveform, demonstrating successful DC-to-AC conversion with a stable and high-quality AC output. The subsequent plots (VAN, VBN, VAB) are square waveforms, representing the **switching signals** for different switches in the inverter. These square waves alternate between ON and OFF states, following a pulse-width modulation (PWM) control strategy to generate the desired output.

Figure 11(a). Output voltage

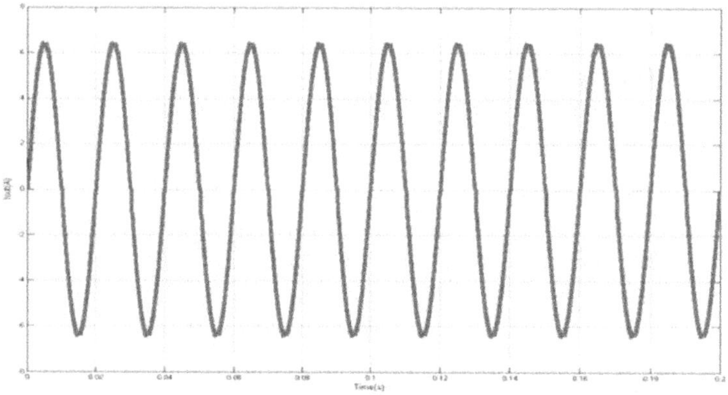

Figure 11(b). Output voltage

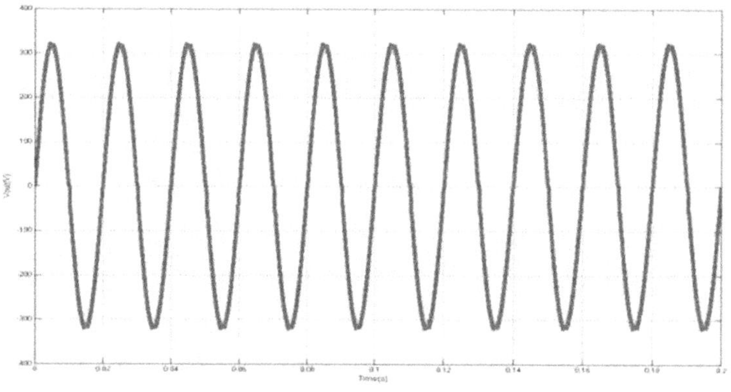

This waveform represents in figure 11(a) is an AC output voltage (Vg) from an inverter circuit. The sinusoidal shape indicates successful DC-to-AC conversion with minimal distortion, showcasing a clean and stable AC output. The periodicity and amplitude consistency demonstrate that the inverter is functioning efficiently, producing a high-quality waveform suitable for grid-tied or standalone applications.

The waveform shown in Figure 11(b) represents the AC output current (Ig) from an inverter circuit. Its sinusoidal shape indicates a successful DC-to-AC conversion, with minimal distortion, showcasing a clean and stable current output. The consistent periodicity and amplitude demonstrate the inverter's efficiency and proper synchronization with the load or grid requirements.

Figure 11(c). Output power

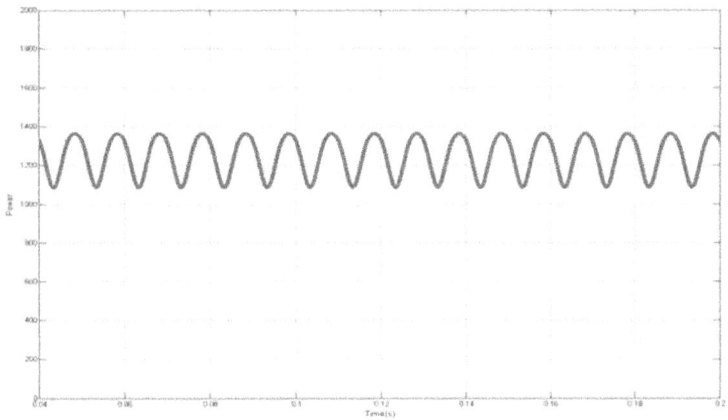

The output power waveform in an inverter system is shown in figure 11(c) is typically derived from the product of the output voltage (Vg) and output current (Ig). Since both Vg and Ig are sinusoidal, the output power waveform exhibits a periodic pattern, often sinusoidal in nature but with a frequency twice that of the voltage or current waveforms.

V. CONCLUSION

In conclusion, the proposed MOSFET Inverter Family Design for a grid-connected solar system demonstrates significant advantages in efficiency, performance, and practicality. Using MOSFET super junctions and SiC diodes, the design achieves high efficiency across a wide load-speed range, with a maximum efficiency of 98.5% and a European efficiency of 98.32%. The constant common-mode (CM) voltage minimizes leakage current, while excellent differential-mode (DM) characteristics are ensured through unipolar SPWM, similar to isolated full-bridge inverters. Additionally, the design eliminates the need for dead time PWM, further simplifying control and enhancing reliability. The performance was validated using a 240/50 Hz 1 kW prototype, confirming its suitability for grid-connected single-phase PV applications. This innovative topology is a highly efficient and effective solution for modern solar energy systems.

REFERENCES

Balaji, S., & Sasilatha, T. (2019, November). Detection of denial of service attacks by domination graph application in wireless sensor networks. *Cluster Computing*, *22*(6), 15121–15126. DOI: 10.1007/s10586-018-2504-5

Islam, M., & Mekhilef, S. (2019). H6-type transformerless single-phase inverter for grid-tied photovoltaic system. *IET Power Electronics*, *8*(4), 636–644. DOI: 10.1049/iet-pel.2014.0251

Islam, M., Mekhilef, S., & Hasan, M. (2019). Single phase transformer less inverter topologies for grid-tied photovoltaic system: A review. *Renewable & Sustainable Energy Reviews*, *45*, 69–86. DOI: 10.1016/j.rser.2015.01.009

Jegadeeswari, G. (2019). Performance analysis of power quality improvement using shunt active power filter. [IJRTE]. *International Journal of Recent Technology and Engineering*, *7*(5), 1–3.

Jegadeeswari, G., Dheepak, M., Saravanan, S. V., & Vanisree, V. (2018). The Power Factor Correction Improvement For A Single Phase Ac/Dc Converter Using An Enabling Window Control. [IJMPERD]. *International Journal of Mechanical and Production Engineering Research and Development*, *8*(5), 41–48.

Jegadeeswari, G., Kirubadurai, B., & Lakshmi, D. (2020). Multi carrier based new random pulse width modulation for three phase inverters. at International Journal of Scientific and Technology Research (IJSTR), 9(2), 5425-5432.

Lakshmi, D., Baskaran, S., Ezhilarasi, G., & Valli, C. (2020). Direct AC/DC power converter using auxiliary circuits. *International Journal of Advanced Science and Technology: Vol*, *29*, 785–793.

Lakshmi, D., Ezhilarasi, G., Baskaran, S., & Siva Prasad, S. (2020). Analysis of various pwm schemes for the design of asymmetric single phase 31 level cascaded mli. *International Journal of Advanced Science and Technology, Vol*, *29*, 1050–1066.

Li, Z., Kai, S., Lanlan, F., Hongfei, W., & Yan, X. (2019, February). A family of neutral point clamped full-bridge topologies for transformerless photovoltaic grid-tied inverters. *IEEE Transactions on Power Electronics*, *28*(2), 730–739.

Manikandan, K., Sivabalan, A., Sundar, R., & Surya, P. (2020, March). A study of landsman, sepic and zeta converter by particle swarm optimization technique. In 2020 6th international conference on advanced computing and communication systems (ICACCS) (pp. 1035-1038). IEEE.

Patrao, I., Figueres, E., Gonz'alez-Esp'ın, F., & Garcer'a, G. (2021). Transformerless topologies for grid connected single-phase photovoltaic inverters. *Renewable & Sustainable Energy Reviews*, *15*(7), 3423–3431. DOI: 10.1016/j.rser.2011.03.034

Veerakumar P., Dheepak M., Saravanan S.V., (n.d.). "PLC based automatic control for onboard ship gangway conveyor system" International Journal of Mechanical Engineering and Technology Volume 8, Issue 3.

Xiao, H., Liu, X. P., & Lan, K. (2024, October). Zero-voltage-transition full bridge topologies for transformerless photovoltaic grid-connected inverter. *IEEE Transactions on Industrial Electronics*, *61*(10), 5393–5401. DOI: 10.1109/TIE.2014.2300044

Chapter 5
Off-Grid Photovoltaic Solar Design Systems for Improved Grid Performance:
IoT Software-Defined Network (SDN) Approach

Odubola Abel Oluwatimilehin
https://orcid.org/0009-0002-1361-547X
Obafemi Awolowo University, Nigeria

Aderinola Musefiu
https://orcid.org/0009-0000-6610-8186
Hussaini Adamu Federal Polytechnic, Nigeria

Chukwuka Michael Oforgu
https://orcid.org/0009-0005-1216-1957
University of the West of England, UK

Jazuli Sanusi Kazaure
https://orcid.org/0000-0001-7681-2177
Hussaini Adamu Federal Polytechnic, Nigeria

Olasubomi Asuni
https://orcid.org/0009-0007-7891-6210
Northumbria University Newcastle, UK

Oluwaseun Jeremiah Adesina
https://orcid.org/0009-0007-4742-6323
De Montfort University, UK

Oyedemi Oluyemisi Adenike
https://orcid.org/0009-0005-8496-736X
Federal University of Lavras, Brazil

Ugochukwu Okwudili Matthew
https://orcid.org/0000-0003-0828-9710
Federal University of Lavras, Brazil

ABSTRACT

A solar powered system, often known as a photovoltaic (PV) system, is an electric power system that uses solar radiation to generate usable solar energy for household consumption. In this paper, the authors designed an off-grid energy distribution system, utilizing an artificial neural network (ANN) algorithm to build a model that predicts, on a monthly basis, the average daily global sun radiation on a horizontal region of ground base using meteorological data which include the earth's temperature, relative humidity, wind speed, rainfall and ambient air quality. The authors discussed the issues in implementing software defined network (SDN) in the energy distribution, energy demand response, energy optimization and real-time monitoring required for load balancing. The SDN integration offered solutions that renewable energy technology could bring in the future technology-based electricity network driven by automation,

DOI: 10.4018/979-8-3693-9924-8.ch005

communication, and internet of things (IoT) connectivity in the off grid implementation.

INTRODUCTION

The development of off-grid electrical systems for various locations and purposes, particularly in the rural areas are made possible by the adaptability and flexibility of solar photovoltaic (PV) technology(Agajie et al., 2024). It is now feasible to choose a clever design strategy for the off-grid solar PV hybrid system, although in the past, off-grid PV-based systems were put up in numerous locations for a variety of purposes using traditional straightforward designs. Depending on site-specific energy resources requirements, and electrical load balancing characteristics, a variety of off-grid system configurations are feasible. Due to the development of new large solar power plants and the rise in rooftop solar installations by individuals and companies, solar PV alone is anticipated to provide 80% of the global expansion in renewable capacity between now and 2030(Khalid, Awais, Jamil, & Khan, 2024). Currently, the majority of PV systems worldwide are running at maximum power and making several meaningful contribution to the network's energy management. Large-scale PV generator deployment in an off-grid system may result in issues including voltage swings, frequency deviations, and network power quality issues, configuration changes if it is not effectively managed and controlled(Khalid et al., 2024). The idea of an active generator is one potential remedy for these issues. The PV array-based active generator will be highly adaptable and capable of controlling the power supply. It will work in tandem with energy storage technologies and sufficient power conditioning devices to guarantee the resilience of the solar grid.

Similar to a conventional generator system, the active generator will have great adaptability and control over power supply. Given that the PV array output is weather dependent, the predictability of the PV power output is essential for off-grid system operating planning(Ohabuiro, Matthew, Umar, Tonga, & Onyebuchi, 2022). New types of inverters are being developed by several manufacturers of PV power condition devices, which could distribute power from PV systems without the need for energy storage batteries, as is the case with conventional power plants(Bullich-Massagué et al., 2020). The wind renewable technology industry is predicted to recover despite ongoing challenges, with the rate of growth doubling from 2024 to 2030 in comparison to 2017–2023(Swadi et al., 2024). In almost every nation, wind renewable energy and solar photovoltaics are now the most economical ways to increase the amount of electricity generated(Hassan et al., 2024). The off-gird system design should aim to offer the highest level of efficiency, dependability, and flexibility at a reasonable cost. Because of these advancements, almost 70 nations which produce 80% of the world's renewable energy, are expected to reach or exceed their current renewable targets by 2030. This is entirely in line with the goal of triple global renewable energy capacity project in the current century, which was established by nearly 200 governments at the COP28 climate change meeting in December 2023(Acen et al., 2024). But according to International Energy Agency (IEA) analysis, the doubling objective is completely achievable if countries seize short-term possibilities for action in green economy implementation towards improving energy security, energy efficiency, lower energy costs, and move us closer to achieving climate sustainability goals(Matthew, Asuni, & Fatai, 2024). Energy efficiency is one important area where governments meeting under the auspices of the UN Framework on Climate Change (UNFCCC) have been discussing ways to increase ambition in reducing greenhouse gas emissions before 2020, when a new universal climate change agreement and its accompanying technology resolution are set to come into effect(Ihemeson, 2024),(Torney & O'Gorman, 2020).

Climate technologies refer to the technology we employ to combat climate change. Climate technologies that lower greenhouse gas emissions include renewable energy sources like hydropower, solar electricity, biomass gasification and wind power(Matthew et al., 2022),(Kazaure, Matthew, & Nwamouh, 2022). To achieve this, it is necessary to present bold plans in the next round of the Paris Agreement's nationally determined contributions, and to encourage international cooperation to reduce the high financing costs in developing an emerging economies that will enable expansion of renewable energy in high-potential regions such as Southeast Asia and Africa. Renewable energy is advancing more quickly than national governments can set goals for. The primary reason for this is not only initiatives to reduce emissions or improve energy security, but also the fact that renewable energy sources are currently the most affordable way to build new power plants in practically every nation on the planet, according to IEA Executive Director Fatih Birol(Smith, 2024). This study demonstrates how the expansion of renewable energy sources, particularly off grid solar, will change electrical systems worldwide. The entire world is expected to install about 5500 gigawatts of renewable energy power capacity between now and 2030, which is about equal to the combined power capacity of the US, China, the EU, and India(Schönfisch, Dasgupta, & Wanner, 2024),(Sayigh, 2023). By 2030, we anticipate that half of the world's electricity will come from renewable sources. It is predicted that the proportion of wind and solar PV alone in the world's electricity generation would treble to 30% in a decade(Yalılı, Menlik, & Boran, 2024). Governments must, however, step up their efforts to safely incorporate these variable renewable sources into electricity infrastructure in line with the technology progression. In addition to increasing the deployment of renewable power, the adoption of sustainable biofuels, biogases, hydrogen, and e-fuels would need to be significantly accelerated in order to fulfill global climate targets. Global solar production capacity is expected to surpass 1100 megawatts by the end of 2024, more than tripling the projected demand(M. S. Hossain, Wadi Al-Fatlawi, Kumar, Fang, & Assad, 2024). In order to support global diversification, solar PV manufacturing capacity is predicted to triple in both India and the US by 2030 due to the increased emphasis on industrial competitiveness on a worldwide scale. However, the cost of manufacturing solar panels in the US is twice as high as in India and three times higher than in China(Abhyankar et al., 2023). In addition to taking important issues like employment growth and energy security into account, governments should explore how to balance the increased costs and advantages of local manufacturing for even distribution.

The community micro grids have played a critical role in extending access to off-grid rural electrification in many developing countries, particularly sub-Saharan Africa(Bhattacharyya & Palit, 2021). On the other hand, the potential of community micro grids in sub-Saharan Africa is virtually untapped. This research looks into the possibilities and limitations of community micro grids as a means of extending access to off-grid rural electricity in Sub-Saharan Africa. This research's key contribution is to demonstrate the circumstances in which community micro grids might help to rural development, as well as those circumstances under which consumers are likely to initiate and collaboratively contribute in the ongoing off grid renewable projects. There are two policy implications from this study. The potential for community micro grids in rural areas with adequate micro-PV power sites appearing significant, to the point that projects aiming at spreading rural electrification in Nigeria and other African nations with similar settings appear to have overlooked this possibility. Meanwhile, there is a real risk that community involvement, without further technical aid from outside sources, might push the rural populations into employing more expensive, low-quality installations. After that, a methodical approach to the design and sizing of these systems was described. There are recommendations for system sizing and PV component selection. Electrical properties, energy resource allocation, and installation site load requirements can all have an impact on an off-grid solar system's physical design. An off-grid solar system should be designed with the utmost flexibility, dependability, and efficiency in mind, as well as a reasonable installation cost. Forecasting PV power generation is crucial for operational planning of off-grid system operation since weather and climate conditions affect how well solar energy systems work.

BACKGROUND OF THE STUDY

The goal of the new renewable energy project is to establish a uniform framework for promoting energy derived from global renewable energy sources(Umeh, Nwankwo, Oluka, Umeh, & Ogbonnaya, 2024). The prior statute, which had undergone numerous significant amendments since its initial formulation was modified. Its primary goal is to set a legally enforceable goal for the UN member states as a whole, that by 2030, 32% of the UN member states gross final energy consumption must come from renewable energy sources(Torney & O'Gorman, 2020). Member states must set national contributions in accordance with Article 3(2) in order to jointly reach the legally mandated overall UN climate mitigation aim. In 2023–2024, Member States must update their national energy and climate plans (NECPs) in accordance with the governance regulation, which includes their national commitment to meeting the legally mandated renewable energy target. This provision provides a great deal of flexibility in the framework for accomplishing the worldwide renewable energy objective, especially when combined with the review procedures in provision 3(2) of the renewable energy directives(Trevisan, Ghiani, & Pilo, 2023). Furthermore, the Paris Agreement's worldwide stock-take process is intimately, if not totally, linked to this flexibility on renewable energy demand(Matthew et al., 2022). In order to lift people out of poverty on a planet, we must have the guts to provide them with affordable clean electricity. The Swiss Federal Institute of Technology (ETH) in Zurich established the 2000-Watt Society approximately ten years ago with the goal of creating an energy policy model that restricts use within global reserves while maintaining a tolerable environmental impact. Every year, people around the world use 17,500 kilowatt-hours, or 2000 watts continuously, of energy. The average person in Switzerland is thought to need 6000 watts per person, while in some Asian or African nations, the amount is much lower. This ratio, however, is predicated on an uneven distribution of energy with impact on economic and business sustainability. As a result, the objective of a 2000-Watt Society is to achieve an energy consumption balance between industrialized and developing countries, allowing everyone to live in dignity and with a high quality of life(Fuchs et al., 2021).

Consequently, the African Development Bank and the World Bank Group have partnered on the Mission 300 project to provide electricity to 300 million Africans over the course of the next six years(-Cust & Zeufack, 2023). In order to facilitate cross-sector development, off-grid solar will be essential for getting electricity to residences and for accelerating the availability of electricity for businesses, schools, and health facilities. Off-grid solar is the most efficient means of reaching the over 400 million rural Africans who lack access to electricity, offering them cutting-edge energy solutions to power their homes, farms, businesses, and public services, according to the Off-Grid Solar Market Trends Report(Trompette, Etienne, & Francius, 2022),(Trompette et al., 2022). The solar energy industry has shown remarkable resilience in the face of challenging macroeconomic conditions. The $21 billion needed to create a commercially sustainable off-grid solar sector that can grow, reach the most distant regions, and support ambitious projects like M300 as well as energy access and climate goals must be unlocked by companies, investors, governments, and development partners. The off-grid renewable energy project is an endeavor for the off-grid solar energy industry, which stands for economic and environmental activism and aims to change lives by providing high-quality, affordable, and clean solar products and services(Nurhidayah, Alam, Utomo, & Suntoro, 2024). Off-grid solar now powers the homes, businesses, agriculture, and public infrastructure of over 560 million rural climate-vulnerable people. The off-grid sector has the potential to grow quickly with the right support, aiming to improve the lives of one billion people by 2030. By serving as a focal point for the industry, supplying vital market data, advocating for advantageous regulations and boosting investment, and offering value-added services to rural communities, the Off-grid renewable energy initiative drives this development. There are six sections in this study, the first Section talked about the introduction, Section two focused on the background of the study, Section three presented the objectives of the study. Section four focused on the theoretical framework (literature review), Section five presented the in-depth discussion of SDN's incorporation into the off-grids renewable energy system. Section six delineates the methodology and data collection, Section seven look at the discussion of research findings, Section eight focused on the future research, while Section nine presented the conclusion.

OBJECTIVES OF THE STUDY

Estimating the energy demand and evaluating the realistic availability of solar radiation are the most crucial jobs that must be completed in the off grid standalone solar PV system. As a matter of fact, the choice to install an off-grid solar system is entirely based on the needs of the home and degree of power independence. Off-grid solar systems have the benefit of complete independence from conventional energy sources, but they come with a somewhat elevated installation cost. This study evaluates off-grid solar PV systems, concentrating on the impact on system performance based on the global sun radiation. To ensure optimal grid operation, an artificial neural network (ANN) model was developed that predicts global solar radiation and matches it with energy demand. In this study, the following goals were pursued:

i. To create an ANN model that forecasts the mean monthly values of global solar radiation based on temperature, relative humidity, wind speed, rainfall, ambient air quality, and sunshine hours, which is required for the study area's optimal use of solar renewable energy installation.
ii. To construct an ANN model for solar energy system installation, optimization, forecasting, prediction, and control that calculated mean monthly values of global solar radiation, the work established an ANN model that uses temperature and sunlight hours as input parameters.
iii. The study reaches major conclusions about the optimal solar renewable energy generation, contrasts and compares the results of the ANN model to empirical models.
iv. The results revealed that, in places without meteorological monitoring stations, the ANN model could accurately anticipate the monthly average daily global sun radiation.

THEORETICAL FRAMEWORK

Off-grid systems are ideal for households and business organizations in isolated locations or those seeking complete energy independence because they are not connected to the utility grid(Utoh, Ekpotu, & Obialor, 2024). These systems employ batteries to store extra solar energy produced during the day, which is then used to power electronics and appliances at night or during cloudy weather. The battery storage system is crucial to the performance and dependability of off-grid solar PV systems, delivering a steady and dependable source of electricity. Battery storage is required to power these systems when the sun is not shining and raising to its peak. Off-grid systems provide steady electricity and might be especially beneficial in areas with frequent power outages. Off-grid systems are those that are self-contained and do not connect to the grid. These systems employ solar panels to generate electricity and store the excess current in batteries for later use. Solar panels are used to generate electricity in off-grid systems, as are batteries for energy storage, charge controllers, and inverters to convert direct current(DC) to usable alternating current (AC) power(Dorel, Gmal Osman, Strejoiu, & Lazaroiu, 2023). Furthermore, these systems can dramatically save energy expenses over time. Off-grid solutions are frequently useful for households in distant areas where the electric grid is inaccessible. However, it is also ideal for individuals who want to carefully regulate their energy consumption and avoid relying on external power sources. Off-grid solar PV systems are becoming increasingly popular as the world shifts toward renewable energy. These devices provide a consistent and sustainable source of electricity to remote places without grid connectivity. Off-grid solar PV systems require an efficient energy storage solution, which is often in the form of a battery, to function successfully. Several elements influence the performance of the battery storage system, including the battery charging technique, which has a substantial impact on overall off-grid solar PV system performance. Off-grid solar PV systems use a variety of battery charging methodologies, each with benefits and drawbacks.

Effective battery charging techniques are necessary for off-grid solar PV systems to preserve the longevity and peak performance of their batteries(Dorel et al., 2023). There are several ways to charge batteries, such as float charging, pulse charging, constant voltage, and constant current. Each method has pros and cons that might impact system efficiency, battery performance, and total system cost. Off-grid solar PV systems are independent renewable energy sources that employ solar panels to produce electricity and store extra energy in batteries for later use. Unlike grid-tied systems, off-grid solar PV systems are not connected to the utility grid and are frequently used in remote locations where grid electricity is unreliable or unavailable(Dorel et al., 2023). The off grid network's ability to detect faults, perform load balancing, and perform predictive maintenance is made possible by its unprecedented granularity and agility(Ohabuiro et al., 2022). Since they are more resilient to outages, have lower energy losses, better power quality, and support the integration of renewable energy sources, smart grids have various advantages, although the switch to smart grids has brought with it additional difficulties for network management. Because of these systems' size, complexity and dynamic resource allocation, creative solutions are required to guarantee top performance. Given this, Software-Defined Networking (SDN) appears to be a promising technology capable of completely changing the management of smart grids(Cunha et al., 2024). Network resource programmability and centralized control are made possible by SDN's well-known ability to divide the control plane from the data plane that facilitated broadly smart grid communication support(Hakiri, Sellami, & Yahia, 2024). Numerous industries, including as data centers, telecommunications, and the energy industry have used this architectural paradigm to achieve efficiency(Pandey, Sircar, Bist, Solanki, & Yadav, 2023). SDN's guiding principles are well suited to smart grid needs, which demand dynamic, real-time network resource management. This paper combined SDN and smart grid technology in renewable energy operation, emphasizing the developments and problems faced in this developing collaboration. Providing a thorough summary of the research innovations, highlighting significant developments, and resolving roadblocks are among the key objectives of this paper.

Through an analysis of the mutually beneficial link between these two areas, it becomes evident how SDN redefines the management of smart grids and stimulates innovation within them(Abdi et al., 2024). The novel developments in renewable energy with SDN and smart grids offer a multitude of benefits that surpass those of conventional grid systems, presenting a revolutionary approaches to energy management and distribution. There are several maintenance plans in this context, including SDN enabled condition monitoring, using the network of sensors and smart devices for wireless sensors intercommunication (Darwish, Hassan, Abdel-Gawad, & Mansour, 2022), showing that smart grids are able to better diagnose faults with their constituent parts by continuously gathering real-time data about the componentry of the grid. In this paper, the literature offers optimum plans for predictive maintenance under the framework of minimizing grid interruptions and guaranteeing a more dependable power supply. SDN allows utilities to dynamically manage their assets in response to demands and conditions in real time, resulting in another maintenance paradigm known as smart maintenance(Velasquez, Moreira-Moreira, & Alvarez-Alvarado, 2024). Certain problems, including rerouting power flow or isolating malfunctioning components while preserving service to unaffected areas, can occasionally be handled automatically by SDN(Syrmakesis, Alcaraz, & Hatziargyriou, 2022).

The SDN is essential to enable the integration of renewable energy sources, which is a crucial area in the advancement of smart grids(Kaur et al., 2019). We look at how SDN makes it possible for renewable energy sources to be seamlessly integrated into the electrical grid in this section. By offering real-time management and dynamic adaptability, SDN helps create a more efficient and sustainable energy landscape. Effective resource management in SDN-based IoT ecosystems involves the dynamic allocation of bandwidth, processing capacity, and memory resources to meet the specific requirements of IoT devices and applications(Jeyaraj et al., 2023). Traditional resource allocation methods often fall short in these environments due to their static nature and inability to adapt to real-time changes in network conditions. This inadequacy can lead to resource underutilization or over-provisioning, negatively impacting network performance and energy efficiency (Zormati & Lakhlef, 2023). To address these challenges, we propose an SDN in IoT ecosystems for renewable energy resource management that leverages ANN machine learning algorithm for the solar radiation energy model training.

Among the emerging information technology paradigm that is enabling a wide range of applications, including smart grid network maintenance, system resilience and quality of service (QoS) optimization in the solar energy distribution is SDN(Mwangi, Sahay, Fumagalli, Gryning, & Gibescu, 2024). Through computational integration of SDN with IoT, several data-acquiring sensors are connected via the internet by dividing the control plane from the data plane(Siddiqui et al., 2022). Using dynamic software programs that use software-based controllers or application programming interfaces (APIs) to communicate with underlying hardware infrastructure and direct traffic on a network, SDN's separation of the control plane and data plane offers flexibility in managing, configuring, securing and optimizing grid network resources(Gonzalez, Charfadine, Flauzac, & Nolot, 2016). The smart grid infrastructure anticipate that there will be billions of connected devices in IoT networks in the future, which will present significant management and control issues(Matthew, Olofin, et al., 2024). This development requires flexibility approaches without changing the core architecture of existing implementations, SDN can make an

IoT network programmable, scalable and perhaps provide solutions to the indicated IoT management issues(Sarabia-Jácome et al., 2024).

In smart grids, SDN has also been proposed for network administration and QoS as opined by (Velasquez et al., 2024), who examined three use scenarios of SDN, with a focus on virtual networks for distributed energy resource, content-based data sharing, and smart grid management. Advances have been proposed for the architecture and execution of SDN-based IoT management frameworks and solutions for a range of smart grid applications utilizing renewable solar energy sources(Sellami, Hakiri, Yahia, & Berthou, 2022). By prioritizing traffic according to grid priorities, SDN optimizes traffic flows inside smart grid networks, ensuring dependable and effective communication between grid components(Sharma, Tokekar, & Varma, 2023). SDN also improves grid resilience and security by making it easier to deploy cutting-edge security measures and by giving utilities better tools for identifying and thwarting cyber threats. In order to optimize QoS and enable resilience support, the SDN controller reconfigures the communication network at the appropriate times by executing different SDN applications(Al-Rubaye, Kadhum, Ni, & Anpalagan, 2017). According to(Dong, Lin, Tan, Iyer, & Kalbarczyk, 2015),SDN technologies allow the SDN controller to handle the communication network shared by the supervisory control and data acquisition(SCADA) master, SCADA slaves, and control devices, sensors, and actuators which enable circuit breakers and tap changers to get commands from intelligent electronic devices (IEDs) and remote terminal units (RTUs), which also gather data from sensors like standard meters and phasor measurement units (PMUs).

When used with IoT for communication in a smart grid context, SDN completely transforms how utilities oversee and maximize grid operations(Velasquez et al., 2024). Whereas IoT devices offer real-time data collecting and transmission capabilities, SDN offers a centralized and programmable approach to network administration. With its ability to centralize control and enable dynamic resource allocation based on grid conditions, SDN helps smart grids operate communication networks more effectively(Kaur et al., 2019). Utility companies are able to prioritize important data traffic, manage network performance, and react quickly to grid events as a result(Kaur et al., 2019). A multitude of real-time data from different grid assets, including smart meters, substation sensors, and other grid-connected devices, are made available by IoT devices, which improve this communication infrastructure. Renewable energy utility companies can accomplish multiple important goals by integrating SDN with IoT in power systems communication(Wu, Wu, Guerrero, & Vasquez, 2021). The main management of the grid is carried out by the SCADA system, which is made up of computers, networks, control devices, and software. According to (Dong et al., 2015), the SCADA master computers and servers are operated by the control center to perform various grid control applications, including under-frequency load shedding, frequency and voltage modifications, and grid condition monitoring. In the smart grid SCADA systems, the SCADA master interacts with the grid's SCADA slaves through the grid communication network, transmitting control commands and gathering measurement data that helps build more adaptable, accurate, and effective preventative and countermeasures against threats to critical smart grid devices(Enescu, Bizon, & Moraru, 2019).

Figure 1. An architecture of SCADA master and the SDN controller enabled grid

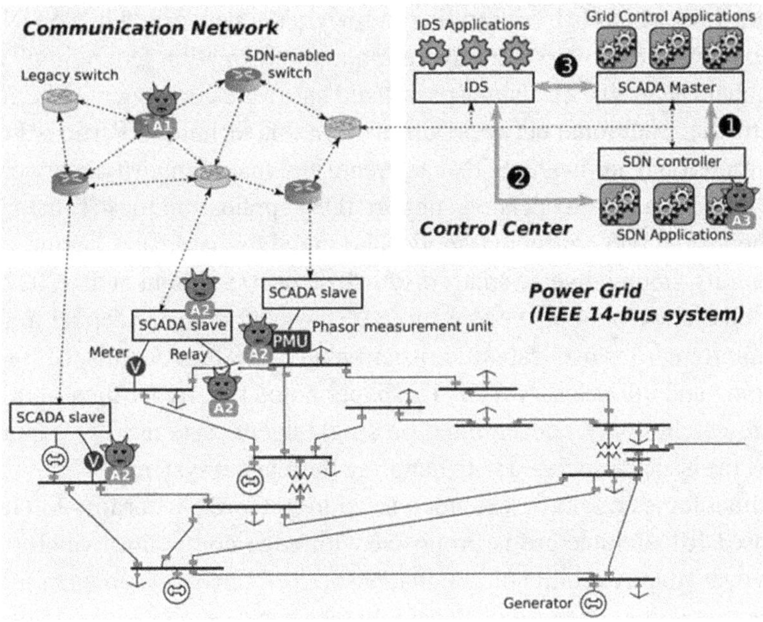

(Dong et al., 2015).

The figure 1 above depicted an SDN-enabled smart grid with three main components: a control center, a communication network, and a power grid, represented by the IEEE 14-bus coupled scheme(Dong et al., 2015). The SCADA master and the SDN controller, for instance, can work together to automatically create a route so that control commands are only transmitted when needed. The PMUs component depicted in Figure 1, is a device utilized to determine the phase angle and magnitude of an electrical phasor quantity such as voltage or current in the electrical grid by utilizing a shared time source for synchronization(Biswal, Sahu, Mishra, & Rout, 2023). In order to provide synchronized real-time measurements at numerous remote places on the grid, time synchronization is typically supplied by Global Positioning System (GPS) or IEEE 1588 Precision Time Protocol(Idrees et al., 2020). In the electrical grid, PMUs may also monitor frequency. They typically provide findings with a very high degree of precision, up to 120 readings per second necessitated by SDN enabled SCADA which performs better than standard SCADA. The standard SCADA produces one measurement every two or four seconds which is impractical for engineers to evaluate dynamic events in the smart grids(Abdelkader et al., 2024). PMUs are among the most significant measuring devices in power systems of the future because they enable grid utilities to monitor and control operations more effectively.

According to(Sharma et al., 2023), a three-tiered strategy comprising an application plane, a control plane, and a data forwarding plane is suggested for the SDN framework. An overview of SDN-based IoT architecture for smart grid that allows the network to monitor and prioritize network traffic using centrally located programmable APIs was developed by (Velasquez et al., 2024). By enabling dynamic adaptation to run-time changes in the network environment, SDN seeks to improve the agility, flexibility, and intelligence of network design. An IoT network must adapt to its own needs because it is very dynamic by nature, mostly because of its limited resources like battery, processing power, and storage capacity. SDN is the most advantageous networking architecture because it offers customizable network APIs that enable such agility. An IoT design based on SDN sensors data capturing as part of grid SCADA were shown in figure 3, in which IoT seek to link items via the network, while SDN decouples control and data planes to orchestrate network administration and information federation(Tache, Păscu oiu, & Borcoci, 2024),(Kabbara et al., 2022). Using specialized hardware components like switches and routers to control network traffic, traditional networks differ from this technique(Kirubasri et al., 2022). In addition to managing traditional hardware, SDN can create and maintain virtual networks via software. In an intelligent SDN-based intelligent decision support(IDS) application for IoT, the system is composed of two core components: the IDS conceptual model and the SDN-based architecture, designed to ensure seamless interoperability among diverse smart grid IoT devices(Siddiqui et al., 2022), refer to figure 1.

According to (Mohammed, Sultan, Lee, & Lim, 2023), who employed deep reinforcement learning to capture and clean important IoT sensor data since it can enhance the functionality of smart IoT applications by removing damaging and unnecessary data. The paper noted that rather than multiagent systems, the majority of recent researches have concentrated on single-agent systems. The paper suggested further studies to investigate the issues associated with handling multiagent systems and how deep reinforcement learning based methodologies might be extended to address them. According to (Javadpour & Wang, 2022), the SDN-based IoT architecture is proposed with edge computing which outperformed some traditional methods over adaptive computation, that ensured effective resource management and latency in industrial wireless networks. Because of these inherent benefits, SDN is well-suited to address this problem of energy conservation(R. Kumar & Agrawal, 2024). Nevertheless, some of its most potent and sophisticated routing features, such dynamic multi-path scheduling, have not yet been completely utilized. More significantly, previous researches have mostly concentrated on the forwarding plane of SDN and have not taken into account the energy-saving aspects of the control plane, especially in the context of multi-controller SCADA SDN environments(Kanellopoulos & Sharma, 2022).

GLOBAL SOLAR RADIATION MODELLING

There is practically no limit to the amount of solar energy that can be used on earth's surface(Karathanasis & Karathanasis, 2019). Creating new solar power technologies is thought to be one of the most important ways to meet the growing need for energy in the globe. Compared to other renewable energy sources, solar energy is superior in terms of availability, cost-effectiveness, accessibility, capacity, and efficiency, making it one of the greatest solutions to fulfill future energy demand(Ahmadizadeh et al., 2024). On the other hand, direct solar radiation is the only source of energy used by solar radiation concentrating systems. As such, in order to properly plan and construct them, we require appropriate computation and calculating tools that enable us to predict the amount of direct solar radiation incident on the collectors of the system. It would therefore be advantageous to investigate the characteristics of solar radiation and the geometrical factors influencing its quantity at the boundary of earth's atmosphere. Further modifying the solar radiation that reaches the earth's surface are the processes of radiation-matter interaction. In order to select or create the most suitable direct solar radiation model, several phenomena must be considered. The electromagnetic waves that the Sun emits are called solar radiation, and life on earth depends on it(E. Hossain, 2022), refer to fig 2. Among other things, it makes plant photosynthesis possible and regulates atmospheric and climatological dynamics and trends. There wouldn't be life on earth without solar radiation, and it also enables us to generate photovoltaic energy(Ohabuiro et al.,

2022), which is crucial in the fight against climate change. The energy released by the Sun in the form of electromagnetic waves that travel through space in all directions is known as solar radiation(Karathanasis & Karathanasis, 2019). This radiation, which the Sun's surface emits, affects meteorological and atmospheric processes of solar renewable energy generation(Yousef, Radwan, Olabi, & Abdelkareem, 2023).

Figure 2. Solar radiation energy emitted by the Sun, sent in all directions through space as electromagnetic waves

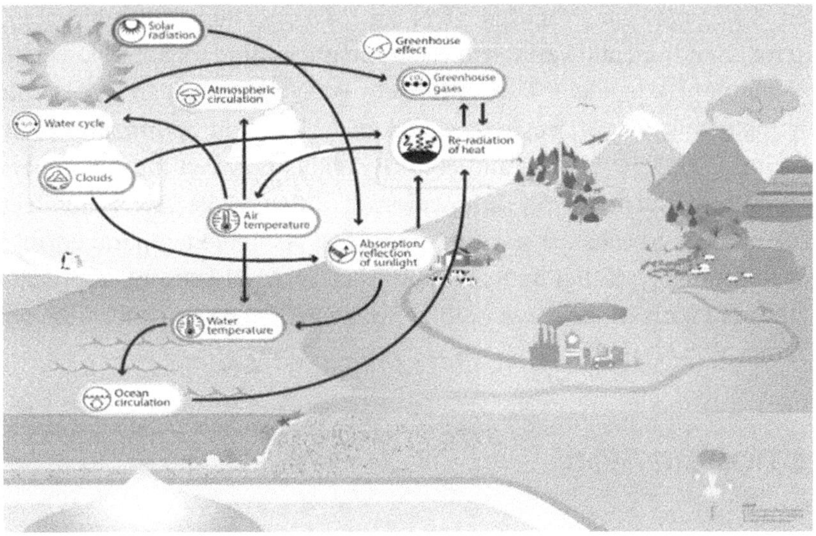

(Ohabuiro et al., 2022).

The Sun generates short-wave radiation, which is absorbed by gas molecules or suspended particles and diminished in the atmosphere due to cloud cover(Kosmopoulos, 2023). Solar radiation travels through the atmosphere before being reflected or absorbed at the surface of the oceans and continents. Ultimately, it is returned to space by the surface as long-wave radiation. In addition, it has a direct or indirect role in everyday occurrences like plant photosynthesis, maintaining the planet's temperature where life is possible, and wind formation which is necessary to produce wind power. An ANN based algorithm for calculating the monthly mean daily and hourly values of solar global radiation is presented in this section. The hourly global solar radiation data are essential for conducting an efficient investigation into solar energy consumption and estimating solar renewable energy resources. This research builds a model using ANN to tackle the solar energy distribution problem in order to forecast monthly average daily global sun irradiance on a horizontal area in North West Nigeria (Kazaure).

To help with the forecast of the solar energy distribution pattern, five empirical correlations are generated using the 42-month data. The program is built around the multilayer perceptron under tabbed categories, a feed forward ANN that converts sets of input data into a set of appropriate outputs in the neural network toolbox of MATLAB 9.7 version(Ohabuiro et al., 2022). It uses three or more layers of neurons (nodes) with nonlinear activation functions, which sets it apart from traditional linear perception. In addition, it performs better than perceptrons at recognizing input that cannot be linearly separated by a linear hyper-plane. The global sun irradiation levels that were computed and measured agree well, according to results that were obtained using the proposed framework. Because of its tiny noise margin, the ANN model is demonstrated to be better than empirical models.

METHODOLOGY

This section provides an overview of the methodology, including the architecture, software tools used, algorithms, and the detailed ANN training steps involved in the data collection and resource allocation process. The Figure 1 depicted the Proposed SDN for IoT ecosystems for renewable energy resource management setup for environmental variable measurement required for solar power utility (smart grid) and distribution. The Figure 3 depicts IoT senor devices generating data, an SDN controller (labeled as SDN) dynamically managing network resources, and an ANN agent optimizing data computation and sensor value allocation. The flow of data and control signals between these components is indicated by arrows. The figure also illustrates the integration of a dual-stage ANN architecture, comprising primary and target networks, to enhance learning stability and handle temporal correlations in network state observations. Additionally, it highlights the multi-agent ANN learning approach, where multiple agents operate in different parts of the network to optimize local resource allocation while coordinating for global optimization.

DATA COLLECTION METHOD

The most accurate way to gather information on global solar radiation is through remote measurements conducted at a specific place with appropriate equipment(Pazikadin et al., 2020). However, because it is costly to calibrate and maintain these devices, the availability of solar irradiation data is limited to a small number of meteorological stations globally(D. S. Kumar, Yagli, Kashyap, & Srinivasan, 2020). The Nigerian Meteorological Agency's statistical data on global sun radiation and climate is used in the current study. Numerous meteorological factors, such as minimum and maximum temperatures, relative humidity, sunshine hours, latitude, longitude, altitude, wind speed, humidity, rainfall quality and global solar radiation(Zhang, Zhao, Deng, Xu, & Zhang, 2017), can be used to forecast the monthly average global solar radiation for North-West Nigeria. Utilizing a few frequently recorded meteorological parameters, such as mean, maximum, and lowest air temperature, relative humidity, cloud cover, etc., several models and algorithms have been created to estimate global sun irradiation. In particular, ANNs have seen a lot of use recently in the application of artificial intelligence and computational methods to real-world problem solving. Thus, the current work involves creating a global solar radiation estimation model that attempts to estimate radiation from more readily seen climate variables collected concurrently with the desired projection. In this particular study, meteorological data from Nigerian Meteorological Agency (NiMet), being an agency of the government that exists to provide efficient and reliable meteorological information by collecting, processing, archiving and dissemination of meteorological information to end-users were utilized as a predictor.

Figure 3. IoT SDN-based architecture for data gathering and analytics

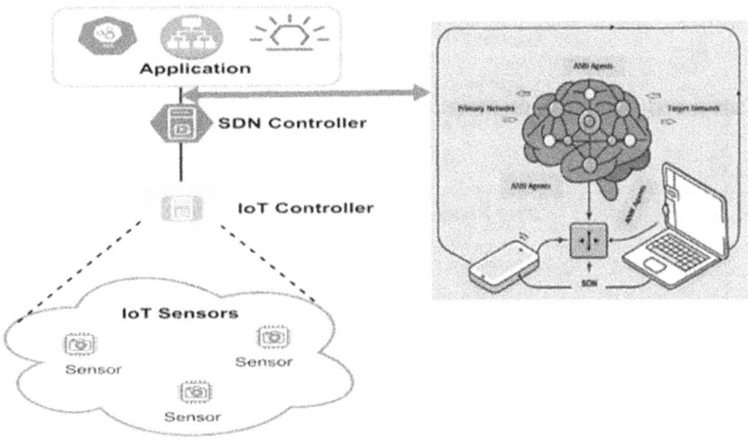

(Siddiqui et al., 2022).

Sensor data offers important information on how real-world data collection systems and processes operate, refer to figure 3. On the other hand, there are a lot of abnormalities in this data collection, like missing data imputation, outliers, noisy data, and irregular sample rates, which make it difficult to interpret most times. A significant volume of data from several data loggers, tests, systems, etc., may make these difficulties worse. When compiling sensor data for computation in MATLAB, the following procedures are needed to gather the necessary information: Using time series sensor data in MATLAB datatypes, managing substantial telemetry data sets, preprocessing functions to identify and manage anomalies, Spectral analysis using the signal analyzer application, as well as smoothing and filtering noisy data, using MATLAB Live Editor to effect the computation and data analyses.

Figure 4. IoT Architecture Layers for SDN: The application Layer, Data Processing Layer, Network Layer, and Sensing Layer

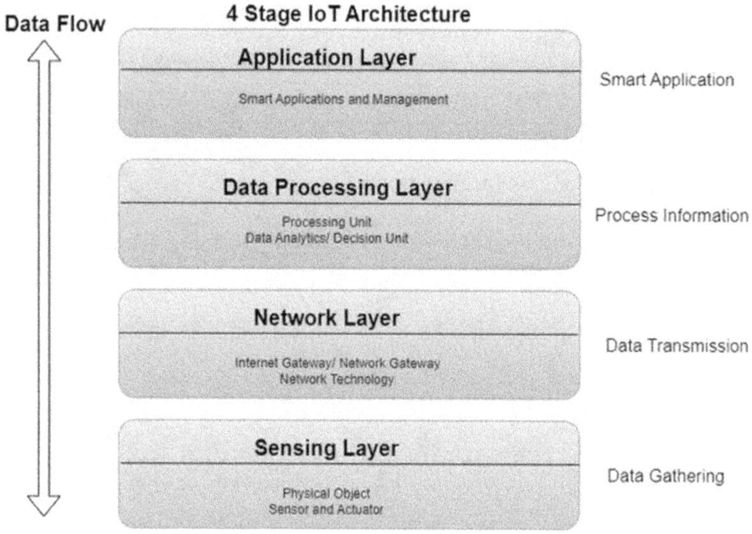

(Vijay et al., 2024).

From **Figure 4**, IoT network monitoring software was primarily used to retrieve basic performance indicators and verify device uptime with the introduction of SDN. While this is still necessary, sophisticated performance indicators, defect identification, sensor data collection, and root cause analysis are now the main drivers. As a result, selecting software for sensor network monitoring has become a crucial decision. In order to monitor network performance, the setup requires a program that can absorb vast amounts of data from various IT sensor infrastructure at rapid speeds. With reference to Figure 3 and Figure 4, there are essential layers of the architecture.

i. **Sensing Layer:** Data collection from various smart sensor sources is the responsibility of the sensing layer, which is the initial layer of the IoT architecture. This layer collects data about temperature, humidity, light, sound, and other physical characteristics using sensors and actuators positioned throughout the environment. The network layer is connected to these devices via wired or wireless communication methods(Okafor, Ingle, Matthew, Saunders, & Delaney, 2024).
ii. **Network Layer:** In an IoT design, the network layer is in charge of facilitating connectivity and communication amongst the system's devices. It consists of the technologies and protocols that let devices connect and talk to the internet as a whole as well as to one another. WiFi, Bluetooth, Zigbee, and cellular networks like 4G and 5G technology are a few examples of network technologies that are frequently utilized in the IoT (Kazaure et al., 2021). In order to prevent unwanted access, the network layer may also contain security features like encryption and authentication, as well as gateways and routers that serve as middlemen between devices and the larger internet.
iii. **Data processing Layer**: In IoT architecture, the hardware and software elements in charge of gathering, evaluating, and interpreting data from IoT devices are referred to as the data processing layer(Ugochukwu et al., 2024). This layer is in charge of taking in the devices' raw data, processing it, and releasing it for additional study or use. A range of tools and technologies, including ana-

lytics platforms, machine learning algorithms, and data management systems, are part of the data processing layer(Okafor, Mathew, & Delaney, 2022). These technologies are employed to glean valuable insights from the data and make informed judgments. A centralized repository for storing raw data from Internet of Things devices, a data lake is an example of a technology utilized in the data processing layer.

iv. **Application Layer:** In the IoT architecture, the application layer is the highest layer that communicates with users directly. It is in charge of offering functionalities and user-friendly interfaces that let consumers access and manage IoT devices. Multiple software and applications for mobile devices, and other user interfaces, are included in this layer for the purpose of interacting with the underlying IoT infrastructure. Middleware services are also included, which facilitate smooth data sharing and communication across various IoT systems and devices(Ebong et al., 2024). Additionally, the application layer has processing and analytics tools that enable data to be examined and turned into insightful knowledge. This can contain advanced analytics features like data visualization tools and machine learning algorithms.

ANN PRE-PROCESSING LOCATION MODEL

MATLAB 9.7 edition's Neural Network Toolbox was used to build multi-layer feed-forward back-propagation networks with a variety of architectural styles. The three layers that comprise the network are the input layer, output layer, and hidden layer. Models 1 and 2 have two input parameters, whereas model 3 has three. The output parameter for all three models is the average global sun radiation measured at North-West (Kazaure), Nigeria. Two separate algorithms are used with single and double hidden layer topologies to enhance the network's capacity to generalize: Levenberg-Marquardt (LM) and Gradient Descent (GD) methodologies(El Misilmani, Naous, & Al Khatib, 2020). While GD is the most straightforward technique, it requests information from the gradient vector, the LM is designed to work explicitly with loss functions that take the form of a sum of squared mistakes. A schematic of a feedforward ANN training inputs are presented in Figure 5 . An ANN is made up of fundamental processing units called neurons that are linked to one another via unidirectional linkages. It is specifically necessary to comprehend that every link has a weight of inputs in real numbers. All feedforward networks, however, will classify the neurons into one of the three categories (input layers, hidden layers and output layer).

Figure 5. An ANN framework for global solar radiation prediction

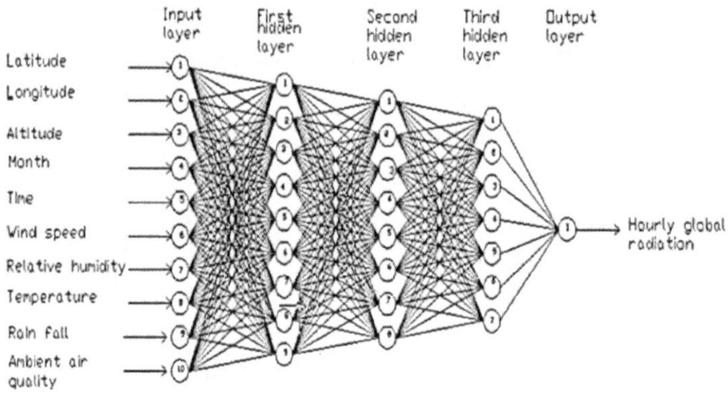

(Ohabuiro et al., 2022).

When the network is learning, a collection of training patterns is fed to it one pattern at a time. An input vector makes up the majority of the components that make up the training set pattern. The network receives the input vector after every training pattern, which causes all neuron activations to be updated and neuron outputs to be produced. Subsequently, the learning algorithm is applied to update the network weights based on the revised neuron outputs. The ANN is repeatedly retrained until satisfactory training is achieved, starting with a small number of training components and gradually increasing the number of these components. It is believed that the optimal quantity of hidden elements is the one needed for efficient training (Figure 5).

ANN APPLICATIONS IN FORECASTING SOLAR POWER GENERATION

On the account that ANNs can perform self-training by comparing expected and actual results, academic research and industry have concentrated on using ANNs for a variety of predicting applications. ANNs are widely used in forecasting because of their capacity for self-learning, which allows them to modify their weighting variables in order to reduce future inaccuracy between predicted and actual results. The capacity of ANNs to imitate learning in this way has made them highly adaptable to a wide range of applications, particularly in areas lacking well-defined models, such forecasting solar power generation. The performance of several ANN model evolutions was compared in the study using RMSE evaluation. Base on statistical and mathematical study, the ANN model yields the best results. The statistical error tests Mean Percentage Error (MPE), Root Mean Square Error (RMSE), and Mean Bias Error (MBE) are used to assess the performance of the models(Bayrakçı, Demircan, & Keçebaş, 2018). These are tests that are most frequently used to compare models used to estimate solar radiation. While a zero value for MBE is preferable, low values for RMSE and MPE are advised.

MBE, RMSE and MABE

The MBE is defined by;

```
<math xmlns="http://www.w3.org/1998/Math/MathML" xmlns:tps="http://www.typefi.com/
ContentXML" xmlns:xsi="http://www.w3.org/2001/XMLSchema-instance" id="m1" display="-
block">
 <mrow>
  <mrow>
   <mi>M</mi>
   <mi>B</mi>
   <mi>E</mi>
   <mo>=</mo>
   <mrow>
    <mo>(</mo>
    <mrow>
     <mrow>
      <munderover>
       <mo stretchy="false">∑</mo>
       <mrow>
        <mi>i</mi>
        <mo>=</mo>
        <mn>1</mn>
       </mrow>
       <mrow>
        <mi>N</mi>
       </mrow>
      </munderover>
      <mrow>
       <mfenced separators

                                        ="|">
        <mrow>
         <msub>
          <mrow>
           <mi>y</mi>
          </mrow>
          <mrow>
           <mi>i</mi>
          </mrow>
         </msub>
         <mo>-</mo>
         <msub>
          <mrow>
           <mi>x</mi>
          </mrow>
          <mrow>
           <mi>i</mi>
          </mrow>
         </msub>
```

</mrow>
</mfenced>
</mrow>
</mrow>
</mrow>
<mo>)</mo>
</mrow>
</mrow>
<mo>/</mo>
<mrow>
<mi>N</mi>
</mrow>
</mrow>
</math>

(1)

i is an index, y_i is the ith estimated value, x_i is the ith measured value and N is the number of examinations.

How well a certain correlation has held up over time can be determined by looking at the MBE test results. When the MBE is positive, it means that the computed or estimated value was exceeded; when it is negative, it means that the value was underestimated(Mahima, Karakoti, Nandan, & Pathak, 2024). While a low MBE is ideal, it should be understood that in an alternate observation, overestimating one item of data will cancel out underestimating another.

Establishing the value of RMSE:

$$RMSE = \sqrt{\left(\sum_{i=1}^{N}(y_i - x_i)^2\right)/N} \tag{2}$$

A thorough explanation of a correlation's performance over a brief period of time is provided by the RMSE test. It makes it possible to compare the actual differences between the numbers that were calculated and the ones that were actually measured term by term. In a perfect world, the RMSE number is zero, but it is always positive. Diminished RMSE improves the model's performance (the more accurate the estimate, the lower the RMSE). A few severe summation errors might considerably enhance the indicator, even though low RMSE values are recommended.

Furthermore, it is feasible to have a relatively small RMSE and a relatively large MBE, or a large RMSE number and a small MBE simultaneously.

Defining the MABE:

$$MABE = \left(\sum_{i=1}^{N}(\ y_i - x_i\ )\right)/N \tag{3}$$

A small value of MABE, which stands for the mean of the absolute bias error, is preferred. RMSE, correlation coefficient, regression coefficient, and ranking, however, will be used in this study to assess how well the empirical models perform.

DISCUSSION OF RESEARCH FINDINGS

Table 1 displayed the results, and Figure 6 displayed the best projected irradiation values for the ANN model in comparison to measured irradiation values. Table 1 presents the monthly average daily solar radiation for Kazaure derived from models 1 and model 3, in addition to the measured values. It is really satisfying to see a very excellent agreement between the measured values and the ANN. As illustrated in Figure 7, the ANN model performs the best when it comes to monthly estimates of Kazaure's average daily solar radiation. Table 2 shows the results of the statistical test. The Root Mean Square Error (RMSE) measurements quantify the accuracy of a given model or correlation. Table 2 displays the lowest ANN model value (4.74) for the current investigation. The models' MBE values show that there can be variation in the models' estimates of the global amount of solar radiation received, with some models having positive MBE values and others having negative MBE values.

The model has the least under estimate (-1.029), which is reasonable and predicted. The MPE is expected to be low; the MPE value of the ANN model was determined to be (- 2.345). One can precisely estimate the global solar radiation by using a number of suggested models and daily recorded meteorological variables, including cloud cover, rainfall, relative humidity, and temperature changes between the highest and lowest points. To produce some trustworthy estimates of solar radiation, all the climatological parameters must be thoroughly and mathematically modeled(Puga-Gil, Astray, Barreiro, Gálvez, & Mejuto, 2022). As illustrated in Figure 2, during the rainy season, a significant percentage of cloudy days with low solar energy and low temperatures occur across the latitudes, while during the dry season, there are comparatively few low-overcast days with high solar energy and high temperatures. Research has shown that ANN models perform well when statistical parameters such as RMSE, MBE, and MPE are considered(Kazem, 2021). It is demonstrated that the new model can adapt more readily to highly variable meteorological conditions and has the potential to provide more accurate daily estimates of solar radiation levels worldwide. As solar radiation estimates indicate, solar energy may compensate for insufficient energy sources quite well necessary for smart grid renewable installation and load balancing.

On the comparison test, in Figure 6, estimates from empirical models and ANNs are compared with measured values at the validation site, Kazaure in North West Nigeria. The graphical comparison shows that, in assessment to the measured data, the validation station's ANN estimates of the global sun irradiation during the second part of the year are significantly, but somewhat, underestimated. Conversely, the empirical model typically overestimates global solar radiation levels. In contrast to the availability of solar radiation and the natural world, the empirical model has a tendency toward linearity. Furthermore, the empirical model has a restricted number of independent parameters that it can employ. In search of a suitable model, the ANN model might accept more processed or unprocessed input parameters. Looking back, Figure 6 illustrates a few cases when the ANN model's predictions differ from the measured values.

This can be explained by the possibility that some of the input variables used to create the suggested ANN model did not fully capture the non-linearity of solar radiation around the globe. To a respectable degree of accuracy, the neural network model can investigate, learn from, and recreate the uncertain nature of solar radiation globally using GPS(Ghimire, Deo, Downs, & Raj, 2019). Since the results were favorable when taking into account a range of realistic climate circumstances, the obtained results showed that the ANN technique would be an appealing choice to predict worldwide daily solar radiation. These findings indicate that the suggested ANN model is an effective machine learning strategy for precise daily global solar radiation forecast; however, additional research into the updated version based on different learning machines algorithm developed is necessary. The model can still be improved, though, as there are numerous additional ways to combine satellite data with optimization techniques. Future research on this significant sub-tropical area may take this into consideration.

Table 1.

Months	RH (%)	RF(mm)	T_{max} (°C)	T_{min} (°C)	n/N	H_m
Jan.	33.4	0.00	35.1	19.7	0.79	19.9
Feb.	30.2	0.00	39.6	22.2	0.64	21.3
Mar.	27.7	0.00	42.8	26.3	0.61	22.5
Apri.	49.5	44.9	38.9	29.5	0.65	20.4
May	68.7	98.7	34.6	27.1	0.52	19.7
Jun.	88.3	150.4	32.9	25.9	0.53	18.1
Jul	89.5	160.9	31.6	24.7	0.58	17.6
Aug.	96.2	203.2	32.3	24.3	0.44	15.8
Sept.	94.8	167.8	34.5	24.7	0.54	17.6
Oct.	78.5	87.4	35.8	24.5	0.68	20.3
Nov.	65.2	34.98	38.6	23.9	0.44	22.9
Dec.	53.5	0.00	36.5	21.8	0.53	18.4

Figure 6. Correlation between the relative sunshine duration and the clearness index.

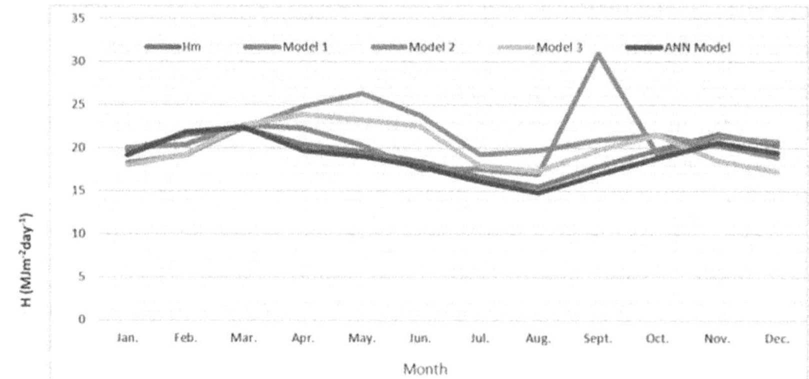

Figure 7. Monthly average daily global sun radiation, both estimated (Model 1) and observed values.

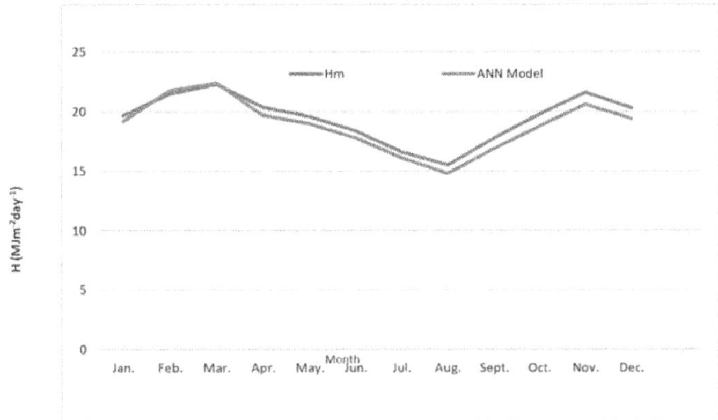

Table 2. Comparison of the Calculated RMSE, MBE, and MPE for the Monthly Average Daily Solar Radiation with the Measured Value.

Generated Models	Techniques for Statistical Analysis		
	RMSE	MBE	MPE
Model I	6.71	0.41	-1.195
Model II	10.38	2.38	0.377
Model III	9.82	1.77	2.255
ANN Model	4.73	-1.028	-2.346

In addition to investigating the application and analyzing the performance of one of the most accessible and straightforward meteorological-based solar models for assessing the monthly average values of daily GSR, this work attempts to produce quick and accurate global solar radiation models. Additionally, the study locations explore the generalization capacity of the built models. Based on the findings, the study's produced models both the local and general formulas can be used to evaluate the GSR more accurately at the chosen areas as well as throughout the entire Kazaure-Nigeria zone.

RESEARCH IMPLEMENTATION AND FUTURE FOCUS

Some of the major obstacles that could prevent the deployment of solar PV capacity over the next three decades include: lack of standards and quality measures; economies of scale; consumer awareness; and grid integration and adaptability. It is essential to mitigate the current obstacles as soon as possible through a variety of implementation strategies and supportive policies, such as creative financial instruments, business models and technology modernism, in order to increase the deployment of solar PV capacities in the future and facilitate the shift to a sustainable, low-carbon energy future. Flexibility must be utilized in every area of the energy system, including power generation, transmission and distribution networks, solar PV and electrical storage, and, more and more, flexible demand (demand-side management and sector coupling), in order to efficiently manage large-scale variable renewable energy sources

CONCLUSION

The work has improved the field of forecasting global sun irradiation by creating a far more user-friendly ANN model. Global solar radiation is an important factor in the design and operation of solar energy systems. The ANN-based model outperforms models based on regression techniques in terms of accuracy since, with the same variables, it not only generates suitable statistical averages for the three study regions taken together, but it also lowers monthly relative errors at the local level. In the end, the results showed that the artificial neural network model may be utilized to estimate the average monthly global sun radiation in Kazaure, Nigeria, whenever there are no monitoring facilities. The missing data resulting from malfunctioning measurement devices that have previously been placed can also be found using the ANN model. This ANN model allows for a precise estimation of Kazaure Nigeria's solar energy potential, which is necessary for designing solar energy installations, offering insightful analysis into the exciting prospects and the need for creative solutions, serving as a roadmap for scholars, practitioners, and policymakers looking to comprehend the changing environment of smart grids. Implementing solar energy projects requires in-depth understanding of solar radiation computation and evaluation of solar PV potential to energize the power grid. ANN models were used to estimate the solar radiation in Kazaure, North West, Nigeria which shows that while designing photovoltaic systems, global sun radiation is a crucial factor. Photovoltaic systems must be designed and used with efficiency, which requires a precise understanding of the location's global solar radiation. Based on meteorological data, the primary goal of the study is to forecast the daily global sun radiation at any place on a horizontal surface with clear atmospheres. The potential for enhancing resilience in smart grids through SDN implementation is explored in this research, illustrating how SDN can improve smart grid resilience even in the face of unfavorable climatic condition. The paper recommends additional research to increase the viability of SDN for managing resilient smart grids within the parameters of the current work.

CONFLICT OF INTEREST

There is no conflict of interest regarding this publication. The paper was supported by U&J Digital Consult Limited, a Global IT and Educational Consulting Firm base in Nigeria.

REFERENCES

Abdelkader, S., Amissah, J., Kinga, S., Mugerwa, G., Emmanuel, E., Mansour, D.-E. A., Bajaj, M., Blazek, V., & Prokop, L. (2024). Securing modern power systems: Implementing comprehensive strategies to enhance resilience and reliability against cyber-attacks. *Results in Engineering*, *23*, 102647. DOI: 10.1016/j.rineng.2024.102647

Abdi, A. H., Audah, L., Salh, A., Alhartomi, M. A., Rasheed, H., Ahmed, S., & Tahir, A. (2024). Security Control and Data Planes of SDN: A Comprehensive Review of Traditional, AI and MTD Approaches to Security Solutions. *IEEE Access : Practical Innovations, Open Solutions*.

Abhyankar, N., Mohanty, P., Deorah, S., Karali, N., Paliwal, U., Kersey, J., & Phadke, A. (2023). India's path towards energy independence and a clean future: Harnessing india's renewable edge for cost-effective energy independence by 2047. *The Electricity Journal*, *36*(5), 107273. DOI: 10.1016/j.tej.2023.107273

Acen, C., Bamisile, O., Cai, D., Ukwuoma, C. C., Obiora, S., Huang, Q., Uzun Ozsahin, D., & Adun, H. (2024). The complementary role of carbon dioxide removal: A catalyst for advancing the COP28 pledges towards the 1.5° C Paris Agreement target. *The Science of the Total Environment*, *947*, 174302. DOI: 10.1016/j.scitotenv.2024.174302 PMID: 38945244

Agajie, E. F., Agajie, T. F., Amoussou, I., Fopah-Lele, A., Nsanyuy, W. B., Khan, B., Bajaj, M., Zaitsev, I., & Tanyi, E. (2024). Optimization of off-grid hybrid renewable energy systems for cost-effective and reliable power supply in Gaita Selassie Ethiopia. *Scientific Reports*, *14*(1), 10929. DOI: 10.1038/s41598-024-61783-z PMID: 38740883

Ahmadizadeh, M., Heidari, M., Thangavel, S., Al Naamani, E., Khashehchi, M., Verma, V., & Kumar, A. (2024). *Technological advancements in sustainable and renewable solar energy systems Highly Efficient Thermal Renewable Energy Systems*. CRC Press.

Al-Rubaye, S., Kadhum, E., Ni, Q., & Anpalagan, A. (2017). Industrial internet of things driven by SDN platform for smart grid resiliency. *IEEE Internet of Things Journal*, *6*(1), 267–277. DOI: 10.1109/JIOT.2017.2734903

Bayrakçı, H. C., Demircan, C., & Keçebaş, A. (2018). The development of empirical models for estimating global solar radiation on horizontal surface: A case study. *Renewable & Sustainable Energy Reviews*, *81*, 2771–2782. DOI: 10.1016/j.rser.2017.06.082

Bhattacharyya, S., & Palit, D. (2021). A critical review of literature on the nexus between central grid and off-grid solutions for expanding access to electricity in Sub-Saharan Africa and South Asia. *Renewable & Sustainable Energy Reviews*, *141*, 110792. DOI: 10.1016/j.rser.2021.110792

Biswal, C., Sahu, B. K., Mishra, M., & Rout, P. K. (2023). Real-time grid monitoring and protection: A comprehensive survey on the advantages of phasor measurement units. *Energies*, *16*(10), 4054. DOI: 10.3390/en16104054

Bullich-Massagué, E., Cifuentes-García, F.-J., Glenny-Crende, I., Cheah-Mañé, M., Aragüés-Peñalba, M., Díaz-González, F., & Gomis-Bellmunt, O. (2020). A review of energy storage technologies for large scale photovoltaic power plants. *Applied Energy*, *274*, 115213. DOI: 10.1016/j.apenergy.2020.115213

Cunha, J., Ferreira, P., Castro, E. M., Oliveira, P. C., Nicolau, M. J., Núñez, I., Sousa, X. R., & Serôdio, C. (2024). Enhancing Network Slicing Security: Machine Learning, Software-Defined Networking, and Network Functions Virtualization-Driven Strategies. *Future Internet, 16*(7), 226. DOI: 10.3390/fi16070226

Cust, J., & Zeufack, A. (2023). *Africa's resource future: harnessing natural resources for economic transformation during the low-carbon transition*. World Bank Publications. DOI: 10.1596/978-1-4648-1743-4

Darwish, M. M., Hassan, M. H., Abdel-Gawad, N. M., & Mansour, D.-E. A. (2022). *Application of infrared spectroscopy for discrimination between electrical and thermal faults in transformer oil.* Paper presented at the 2022 9th International Conference on Condition Monitoring and Diagnosis (CMD). DOI: 10.23919/CMD54214.2022.9991616

Dong, X., Lin, H., Tan, R., Iyer, R. K., & Kalbarczyk, Z. (2015). *Software-defined networking for smart grid resilience: Opportunities and challenges.* Paper presented at the Proceedings of the 1st ACM workshop on cyber-physical system security. DOI: 10.1145/2732198.2732203

Dorel, S., Gmal Osman, M., Strejoiu, C.-V., & Lazaroiu, G. (2023). Exploring Optimal Charging Strategies for Off-Grid Solar Photovoltaic Systems: A Comparative Study on Battery Storage Techniques. *Batteries, 9*(9), 470. DOI: 10.3390/batteries9090470

Ebong, G. N., Matthew, U. O., Olofin, B., Andrew-Vitalis, N., Fatai, L. O., Waliu, A. O., & Oladipupo, M. A. (2024). *Multimedia Cloud Data Warehouse Design for Knowledge Sharing in the University Environment: A Proposed Digital Solution Implementing Interactive Learning Strategies in Higher Education.* IGI Global. DOI: 10.4018/979-8-3693-3559-8.ch010

El Misilmani, H. M., Naous, T., & Al Khatib, S. K. (2020). A review on the design and optimization of antennas using machine learning algorithms and techniques. *International Journal of RF and Microwave Computer-Aided Engineering, 30*(10), e22356. DOI: 10.1002/mmce.22356

Enescu, F. M., Bizon, N., & Moraru, C. M. (2019). Issues in Securing Critical Infrastructure Networks for Smart Grid Based on SCADA, Other Industrial Control and Communication Systems. *Power Systems Resilience: Modeling, Analysis and Practice*, 289-324.

Fuchs, D., Sahakian, M., Gumbert, T., Di Giulio, A., Maniates, M., Lorek, S., & Graf, A. (2021). *Consumption corridors: Living a good life within sustainable limits.*

Ghimire, S., Deo, R. C., Downs, N. J., & Raj, N. (2019). Global solar radiation prediction by ANN integrated with European Centre for medium range weather forecast fields in solar rich cities of Queensland Australia. *Journal of Cleaner Production, 216*, 288–310. DOI: 10.1016/j.jclepro.2019.01.158

Gonzalez, C., Charfadine, S. M., Flauzac, O., & Nolot, F. (2016). *SDN-based security framework for the IoT in distributed grid.* Paper presented at the 2016 international multidisciplinary conference on computer and energy science (SpliTech). DOI: 10.1109/SpliTech.2016.7555946

Hakiri, A., Sellami, B., & Yahia, S. B. (2024). Joint energy efficiency and network optimization for integrated blockchain-SDN-based internet of things networks. *Future Generation Computer Systems*, •••, 107519.

Hassan, Q., Viktor, P., Al-Musawi, T. J., Ali, B. M., Algburi, S., Alzoubi, H. M., & Jaszczur, M. (2024). The renewable energy role in the global energy Transformations. *Renewable Energy Focus, 48*, 100545. DOI: 10.1016/j.ref.2024.100545

Hossain, E. (2022). The Sun and the Earth *The Sun, Energy, and Climate Change* (pp. 1-67): Springer.

Hossain, M. S., Wadi Al-Fatlawi, A., Kumar, L., Fang, Y. R., & Assad, M. E. H. (2024). Solar PV high-penetration scenario: an overview of the global PV power status and future growth. *Energy Systems*, 1-57.

Idrees, Z., Granados, J., Sun, Y., Latif, S., Gong, L., Zou, Z., & Zheng, L. (2020). IEEE 1588 for clock synchronization in industrial IoT and related applications: A review on contributing technologies, protocols and enhancement methodologies. *IEEE Access : Practical Innovations, Open Solutions, 8*, 155660–155678. DOI: 10.1109/ACCESS.2020.3013669

Ihemeson, O. C. (2024). An Assessment of UN Policies on Fossil Emission and Climate Change: Implications on the National Security of US, 2010-2023. *Journal of Integrity in Ecosystems and Environment, 2*(5), 8–23.

Javadpour, A., & Wang, G. (2022). cTMvSDN: Improving resource management using combination of Markov-process and TDMA in software-defined networking. *The Journal of Supercomputing, 78*(3), 3477–3499. DOI: 10.1007/s11227-021-03871-9

Jeyaraj, R., & Balasubramaniam, A., MA, A. K., Guizani, N., & Paul, A. (2023). Resource management in cloud and cloud-influenced technologies for internet of things applications. *ACM Computing Surveys, 55*(12), 1–37.

Kabbara, N., Nait Belaid, M. O., Gibescu, M., Camargo, L. R., Cantenot, J., Coste, T., Audebert, V., & Morais, H. (2022). Towards software-defined protection, automation, and control in power systems: Concepts, state of the art, and future challenges. *Energies, 15*(24), 9362. DOI: 10.3390/en15249362

Kanellopoulos, D., & Sharma, V. K. (2022). Dynamic load balancing techniques in the IoT: A review. *Symmetry, 14*(12), 2554. DOI: 10.3390/sym14122554

Karathanasis, S., & Karathanasis, S. (2019). Solar Radiation. *Linear Fresnel Reflector Systems for Solar Radiation Concentration: Theoretical Analysis, Mathematical Formulation and Parameters' Computation using MATLAB*, 13-72.

Kaur, K., Garg, S., Kaddoum, G., Ahmed, S. H., Gagnon, F., & Atiquzzaman, M. (2019). Demand-response management using a fleet of electric vehicles: An opportunistic-SDN-based edge-cloud framework for smart grids. *IEEE Network, 33*(5), 46–53. DOI: 10.1109/MNET.001.1800496

Kazaure, J. S., Matthew, U. O., & Nwamouh, U. C. (2022). Performance of a Gasifier Coupled to Internal Combustion Engine and Fired Using Corn Cob Feedstock in Biomass Energy Production. *Energy, 11*(2), 35–46.

Kazaure, J. S., Matthew, U. O., Okafor, N. U., & Okey, O. D. (2021). Telecommunication network performances and evaluation of radio frequency electromagnetic radiation: Health effects of the RF-EMR GSM base stations. [IJICTHD]. *International Journal of Information Communication Technologies and Human Development, 13*(3), 16–37. DOI: 10.4018/IJICTHD.2021070102

Kazem, H. A. (2021). Evaluation of PV output in terms of environmental impact based on mathematical and artificial neural network models. *International Journal of Energy Research*, *45*(1), 396–412. DOI: 10.1002/er.5564

Khalid, W., Awais, Q., Jamil, M., & Khan, A. A. (2024). Dynamic Simulation and Optimization of Off-Grid Hybrid Power Systems for Sustainable Rural Development. *Electronics (Basel)*, *13*(13), 2487. DOI: 10.3390/electronics13132487

Kirubasri, G., Sankar, S., Pandey, D., Pandey, B. K., Nassa, V. K., & Dadheech, P. (2022). *Software-defined networking-based Ad hoc networks routing protocols Software defined networking for Ad Hoc networks*. Springer.

Kosmopoulos, P. (2023). *Planning and management of solar power from space*. Elsevier.

Kumar, D. S., Yagli, G. M., Kashyap, M., & Srinivasan, D. (2020). Solar irradiance resource and forecasting: A comprehensive review. *IET Renewable Power Generation*, *14*(10), 1641–1656. DOI: 10.1049/iet-rpg.2019.1227

Kumar, R., & Agrawal, N. (2024). Software defined networks (SDNs) for environmental surveillance: A Survey. *Multimedia Tools and Applications*, *83*(4), 11323–11365. DOI: 10.1007/s11042-023-15729-8

Mahima, K., Karakoti, I., Nandan, H., & Pathak, P. P. (2024). An empirical technique to predict monthly mean global solar radiation for PV applications in Indian context. *Environmental Progress & Sustainable Energy*, *43*(2), e14277. DOI: 10.1002/ep.14277

Matthew, U. O., Asuni, O., & Fatai, L. O. (2024). *Green Software Engineering Development Paradigm: An Approach to a Sustainable Renewable Energy Future Advancing Software Engineering Through AI, Federated Learning, and Large Language Models*. IGI Global. DOI: 10.4018/979-8-3693-3502-4.ch018

Matthew, U. O., Kazaure, J. S., Ohabuiro, J., Aderinola, M., Haladu, N. A., & Nwamouh, U. C. (2022). Biomass Renewable Energy Production from Corn Cobs Feedstock Gasifier as Energy Constituent in Internal Combustion Engines (ICEs). *American Journal of Modern Energy*, *8*(2), 25–35.

Matthew, U. O., Olofin, B., Kazaure, J. S., Fatai, L. O., Falebita, O. S., Nwanakwaugwu, A. C., & Okafor, N. U. (2024). *Smart IoT Drone Systems for News and Intelligence Gathering in Digital Journalism: Drone Evidences in Media Forensics Drone Applications for Industry 5.0*. IGI Global. DOI: 10.4018/979-8-3693-2093-8.ch023

Mohammed, A. F., Sultan, S. M., Lee, J., & Lim, S. (2023). Deep-reinforcement-learning-based IoT sensor data cleaning framework for enhanced data analytics. *Sensors (Basel)*, *23*(4), 1791. DOI: 10.3390/s23041791 PMID: 36850388

Mwangi, A., Sahay, R., Fumagalli, E., Gryning, M., & Gibescu, M. (2024). Towards a Software-Defined Industrial IoT-Edge Network for Next-Generation Offshore Wind Farms: State of the Art, Resilience, and Self-X Network and Service Management. *Energies*, *17*(12), 2897. DOI: 10.3390/en17122897

Nurhidayah, L., Alam, S., Utomo, N. A., & Suntoro, A. (2024). Indonesia's Just Energy Transition: The Societal Implications of Policy and Legislation on Renewable Energy. *Climate Law*, *14*(1), 36–66. DOI: 10.1163/18786561-bja10047

Ohabuiro, J., Matthew, U. O., Umar, S., Tonga, D. A., & Onyebuchi, A. (2022). Global Solar Radiation Modelling using an Artificial Neural Network for Kazaure, Jigawa State, Nigeria. *Journal of Electrical Engineering*, *4*(4), 316–331.

Okafor, N., Ingle, R., Matthew, U., Saunders, M., & Delaney, D. (2024). Assessing and Improving IoT Sensor Data Quality in Environmental Monitoring Networks: A Focus on Peatlands. *IEEE Internet of Things Journal*, *11*(24), 40727–40742. DOI: 10.1109/JIOT.2024.3454241

Okafor, N., Mathew, U., & Delaney, D. (2022). *ProxySense: An effective approach for gas concentration estimation using low-cost IoT sensors.* Paper presented at the 2022 IEEE/WIC/ACM International Joint Conference on Web Intelligence and Intelligent Agent Technology (WI-IAT). DOI: 10.1109/WI-IAT55865.2022.00058

Pandey, V., Sircar, A., Bist, N., Solanki, K., & Yadav, K. (2023). Accelerating the renewable energy sector through Industry 4.0: Optimization opportunities in the digital revolution. *International Journal of Innovation Studies*, *7*(2), 171–188. DOI: 10.1016/j.ijis.2023.03.003

Pazikadin, A. R., Rifai, D., Ali, K., Malik, M. Z., Abdalla, A. N., & Faraj, M. A. (2020). Solar irradiance measurement instrumentation and power solar generation forecasting based on Artificial Neural Networks (ANN): A review of five years research trend. *The Science of the Total Environment*, *715*, 136848. DOI: 10.1016/j.scitotenv.2020.136848 PMID: 32040994

Puga-Gil, D., Astray, G., Barreiro, E., Gálvez, J. F., & Mejuto, J. C. (2022). Global solar irradiation modelling and prediction using machine learning models for their potential use in renewable energy applications. *Mathematics*, *10*(24), 4746. DOI: 10.3390/math10244746

Sarabia-Jácome, D., Giménez-Antón, S., Liatifis, A., Grasa, E., Catalán, M., & Pliatsios, D. (2024). Progressive Adoption of RINA in IoT Networks: Enhancing Scalability and Network Management via SDN Integration. *Applied Sciences (Basel, Switzerland)*, *14*(6), 2300. DOI: 10.3390/app14062300

Sayigh, A. (2023). *Global Catastrophe: Climate Change Is Happening Now, Renewable Energy Can Reduce Its Impact.* Paper presented at the Mediterranean Architecture and the Green-Digital Transition: Selected Papers from the World Renewable Energy Congress Med Green Forum 2022. DOI: 10.1007/978-3-031-33148-0_24

Schönfisch, M., Dasgupta, A., & Wanner, B. (2024). Projected Global Demand for Energy Storage. *Emerging Battery Technologies to Boost the Clean Energy Transition*, 29.

Sellami, B., Hakiri, A., Yahia, S. B., & Berthou, P. (2022). Energy-aware task scheduling and offloading using deep reinforcement learning in SDN-enabled IoT network. *Computer Networks*, *210*, 108957. DOI: 10.1016/j.comnet.2022.108957

Sharma, A., Tokekar, S., & Varma, S. (2023). A Comprehensive Survey on Network Resource Management in SDN Enabled Data Centre Network. *6G Enabled Fog Computing in IoT: Applications and Opportunities*, 333-353.

Siddiqui, S., Hameed, S., Shah, S. A., Ahmad, I., Aneiba, A., Draheim, D., & Dustdar, S. (2022). Toward software-defined networking-based IoT frameworks: A systematic literature review, taxonomy, open challenges and prospects. *IEEE Access : Practical Innovations, Open Solutions, 10*, 70850–70901. DOI: 10.1109/ACCESS.2022.3188311

Smith, D. C. (2024). *'The'key 2024 election involving climate change and the energy transition: Biden vs Trump 2.0 and the matter of global trust in the US* (Vol. 42). Taylor & Francis.

Swadi, M., Kadhim, D. J., Salem, M., Tuaimah, F. M., Majeed, A. S., & Alrubaie, A. J. (2024). Investigating and predicting the role of photovoltaic, wind, and hydrogen energies in sustainable global energy evolution. *Global Energy Interconnection, 7*(4), 429–445. DOI: 10.1016/j.gloei.2024.08.009

Syrmakesis, A. D., Alcaraz, C., & Hatziargyriou, N. D. (2022). Classifying resilience approaches for protecting smart grids against cyber threats. *International Journal of Information Security, 21*(5), 1189–1210. DOI: 10.1007/s10207-022-00594-7

Tache, M. D., Păscu oiu, O., & Borcoci, E. (2024). Optimization Algorithms in SDN: Routing, Load Balancing, and Delay Optimization. *Applied Sciences (Basel, Switzerland), 14*(14), 5967. DOI: 10.3390/app14145967

Torney, D., & O'Gorman, R. (2020). Adaptability versus certainty in a carbon emissions reduction regime: An assessment of the EU's 2030 Climate and Energy Policy Framework. *Review of European, Comparative & International Environmental Law, 29*(2), 167–176. DOI: 10.1111/reel.12354

Trevisan, R., Ghiani, E., & Pilo, F. (2023). Renewable Energy Communities in Positive Energy Districts: A Governance and Realisation Framework in Compliance with the Italian Regulation. *Smart Cities, 6*(1), 563–585. DOI: 10.3390/smartcities6010026

Trompette, P., Etienne, E., & Francius, R. (2022). *At the margins of the grid: the politics of off-grid electrification in Senegal Off-Grid Solar Electrification in Africa: A Critical Perspective*. Springer.

Ugochukwu, O. M., Ayub, M. S., Adasme, P., Rosa, R. L., Rodriguez, D. Z., Saadi, M., & Guimaraes, F. G. (2024). *Adaptive resource management in software-defined networks for iot ecosystems.* Paper presented at the 2024 International Conference on Software, Telecommunications and Computer Networks (SoftCOM). DOI: 10.23919/SoftCOM62040.2024.10721707

Umeh, C. A., Nwankwo, A. U., Oluka, P. O., Umeh, C. L., & Ogbonnaya, A. L. (2024). The Role of Renewable Energies for Sustainable Energy Governance and Environmental Policies for the Mitigation of Climate Change in Nigeria. *European Journal of Applied Science. Engineering and Technology, 2*(2), 71–98.

Utoh, I.-O., Ekpotu, W., & Obialor, M. (2024). *Assessing the Viability and Impact of Off Grid Systems for Sustainable Electrification of Rural Communities in Sub-Saharan Africa.* Paper presented at the SPE Nigeria Annual International Conference and Exhibition. DOI: 10.2118/221709-MS

Velasquez, W., Moreira-Moreira, G. Z., & Alvarez-Alvarado, M. S. (2024). Smart Grids Empowered by Software-Defined Network: A Comprehensive Review of Advancements and Challenges. *IEEE Access : Practical Innovations, Open Solutions, 12*, 63400–63416. DOI: 10.1109/ACCESS.2024.3396402

Vijay, A., Kasiselvanathan, M., Sridhar, P., Sharan, R., Surya, G., & Surjith, M. V. (2024). *Recent developments in designing internet of things architectures.* Paper presented at the 2024 3rd international conference on applied artificial intelligence and computing (ICAAIC). DOI: 10.1109/ICAAIC60222.2024.10574910

Wu, Y., Wu, Y., Guerrero, J. M., & Vasquez, J. C. (2021). A comprehensive overview of framework for developing sustainable energy internet: From things-based energy network to services-based management system. *Renewable & Sustainable Energy Reviews*, *150*, 111409. DOI: 10.1016/j.rser.2021.111409

Yalılı, M., Menlik, T., & Boran, F. E. (2024). A novel multi-objective decision-making model to determine optimum resource and capacity configuration for hybrid electricity generation systems: A comparative case study in Türkiye. *Applied Energy*, *376*, 124338. DOI: 10.1016/j.apenergy.2024.124338

Yousef, B. A., Radwan, A., Olabi, A. G., & Abdelkareem, M. A. (2023). *Sun composition, solar angles, and estimation of solar radiation Renewable Energy-Volume 1: Solar, Wind, and Hydropower*. Elsevier.

Zhang, J., Zhao, L., Deng, S., Xu, W., & Zhang, Y. (2017). A critical review of the models used to estimate solar radiation. *Renewable & Sustainable Energy Reviews*, *70*, 314–329. DOI: 10.1016/j.rser.2016.11.124

Zormati, M. A., & Lakhlef, H. (2023). *An Overview of Machine Learning-Enabled Network Softwarization for the Internet of Things.* Paper presented at the 2023 International Conference on Software, Telecommunications and Computer Networks (SoftCOM). DOI: 10.23919/SoftCOM58365.2023.10271664

Chapter 6
Simulation and Analysis of PV Energy Harvesting Using Boost Converters: Performance Evaluation Under Dynamic Irradiance Conditions

Mohammed Rhiat
https://orcid.org/0000-0002-9046-2887
Higher School of Education and Training, Mohammed First University, Morocco

Anas Hassari
Higher School of Education and Training, Mohammed First University, Morocco

Mohammed Karrouchi
Higher School of Education and Training, Mohammed First University, Morocco

Ilias Atmane
Higher School of Education and Training, Mohammed First University, Morocco

Mostafa El Ouariachi
Higher School of Technology, Mohammed First University, Morocco

Elhadi Baghaz
Faculty of Sciences El Jadida, Chouaib Doukkali University, Morocco

Jaouad Benhamou
Higher School of Education and Training, Mohammed First University, Morocco

Assia Sadiqi
https://orcid.org/0000-0003-2033-2283
Higher School of Education and Training of Oujda, Mohammed First University, Oujda, Morocco

Kamal Hirech
https://orcid.org/0000-0002-3531-7791
Higher School of Education and Training, Mohammed First University, Morocco

ABSTRACT

The efficient harvesting of photovoltaic (PV) energy requires advanced power electronics and robust control strategies to maximize energy transfer to the load. In this chapter, we explore the simulation of PV energy harvesting using a boost converter in MATLAB/Simulink. A Maximum Power Point Tracking (MPPT) algorithm is implemented to ensure optimal energy extraction from the PV panel under varying irradiance conditions. The simulation evaluates the boost converter's input and output characteristics,

DOI: 10.4018/979-8-3693-9924-8.ch006

including voltage, current, and power, using two irradiance profiles: a smooth Gaussian profile representing a clear day and abrupt irradiance changes simulating a non-clear day. Results demonstrate the converter's efficiency and adaptability, providing insights for practical applications in renewable energy systems.

INTRODUCTION

The global transition towards sustainable energy has become a critical imperative in addressing climate change and ensuring energy security. Among the plethora of renewable energy sources, solar power stands out due to its vast availability and potential to significantly contribute to the global energy mix. Photovoltaic (PV) systems, in particular, have garnered widespread attention as an efficient and clean means of harnessing solar energy. These systems utilize the photovoltaic effect to directly convert sunlight into electricity, providing a renewable, environmentally friendly, and increasingly cost-effective energy solution, thanks to rapid advancements in technology and economies of scale (Kannan & Vakeesan, 2016; Khare et al., 2023; Victoria et al., 2021).

Despite their numerous advantages, the performance of PV systems is inherently influenced by external environmental factors. Solar irradiance and temperature, two critical variables, have a profound impact on the electrical output of PV panels (Carrasco et al., 2014; Chouay & Ouassaid, 2024; Rhiat et al., 2023; Rhiat et al., 2024). This variability poses a challenge in optimizing the energy yield of PV installations, particularly under fluctuating weather conditions. To address this challenge, power electronic converters play a vital role in the integration of PV panels with electrical loads or the grid. These converters are multifunctional devices designed to step up or step-down voltage levels, ensuring compatibility with load requirements. Furthermore, they incorporate sophisticated control strategies, such as Maximum Power Point Tracking (MPPT) algorithms, to dynamically optimize the extraction of energy from the PV panels (Gao et al., 2009; Rhiat et al., 2024).

However, the integration of PV panels with power electronic converters introduces new complexities. The efficiency of energy conversion is highly dependent on the effectiveness of the MPPT algorithm and the performance of the boost converter under varying environmental conditions. The dynamic nature of solar irradiance and temperature fluctuations can lead to transient effects that impact the stability and efficiency of the power transfer process. Therefore, a comprehensive analysis of the interaction between PV panels, boost converters, and MPPT algorithms is essential to enhance the overall efficiency and reliability of PV energy systems.

The significance of MPPT algorithms in enhancing the performance of PV systems cannot be overstated. These algorithms, which include Perturb and Observe (P&O), Incremental Conductance, and others, are instrumental in enabling PV panels to operate at their maximum power point (MPP) regardless of changes in irradiance or temperature. The P&O algorithm, in particular, has gained widespread acceptance due to its simplicity, ease of implementation, and reliable performance under most conditions. By continuously adjusting the operating voltage of the PV panel and observing changes in power output, this algorithm ensures that the system consistently extracts maximum energy (Gaikwad & Lokhande, 2016; Rezk & Eltamaly, 2015).

The role of power converters in this context is equally critical. Boost converters, a common type of DC-DC converter, are particularly suited for PV applications. They not only elevate the voltage output of the PV panels to meet the requirements of downstream systems but also serve as the platform for integrating MPPT algorithms. The dynamic behavior of boost converters, including their efficiency and response to varying input conditions, makes them a focal point of research in renewable energy systems. Their ability to provide stable and optimized energy output ensures that PV systems can reliably supply power to resistive loads or feed energy into the grid.

The importance of modeling and simulation in the development and optimization of PV systems cannot be overlooked. Simulation tools such as MATLAB/Simulink offer a robust platform for analyzing the performance of renewable energy systems. These tools allow researchers and engineers to replicate real-world conditions, evaluate system behavior, and identify potential areas for improvement without the need for extensive physical prototyping. By simulating various environmental scenarios, such as changes in irradiance and temperature, we can better understand how PV systems and their associated components interact and adapt. This knowledge is crucial for designing systems that are not only efficient but also resilient to the variability inherent in renewable energy sources.

In this chapter, we conduct an in-depth exploration of a PV system integrated with a boost converter, focusing on its ability to maximize power delivery to a resistive load. The analysis emphasizes the interplay between the PV panels, the power converter, and the control algorithms that govern their operation. Leveraging the capabilities of MATLAB/Simulink, we model the entire system and simulate its behavior under two distinct irradiance profiles:

- A smooth Gaussian profile that replicates the irradiance pattern of a clear, sunny day.
- Sudden steps in irradiance to represent the dynamic and unpredictable conditions of a non-clear day.

The simulation results provide valuable insights into the adaptability and efficiency of the PV system under varying environmental conditions. By examining the system's response to these profiles, we highlight the effectiveness of the boost converter in maintaining optimal energy transfer and the critical role of MPPT algorithms in ensuring consistent performance. This chapter aims to contribute to the understanding and advancement of PV energy systems, addressing key challenges and paving the way for enhanced renewable energy utilization.

BACKGROUND

Photovoltaic panels are sophisticated devices that convert sunlight into electricity using semiconductor materials, with silicon being the most widely used. These materials harness the photovoltaic effect, a phenomenon where absorbed sunlight excites electrons, generating a flow of electric current. When photons from sunlight strike the surface of the PV cells, they impart energy to electrons in the semiconductor material, creating electron-hole pairs. These pairs are then separated by the electric field within the cell, resulting in a directed flow of charge carriers and producing a usable electrical current. The efficiency and performance of PV panels are largely defined by their current-voltage (I-V) and power-voltage (P-V) characteristics, which are intrinsically linked to external environmental factors such as solar irradiance levels, temperature, and shading. This variability underscores the importance of implementing strategies to optimize their operation under real-world conditions.

The power output of a PV panel is not constant and depends heavily on the intensity of sunlight and ambient temperature. Higher irradiance levels increase the current generated by the PV cells, while temperature variations influence both the current and voltage outputs. For instance, an increase in temperature typically reduces the open-circuit voltage of a PV panel, which can lower its overall efficiency. Additionally, partial shading caused by obstacles such as trees, buildings, or clouds can significantly alter the panel's output, creating localized mismatches in current flow. These challenges necessitate the use of advanced power electronics and control strategies to ensure consistent and efficient energy harvesting.

Figure 1. Current-Voltage characteristics for a fixed temperature and variable solar irradiance

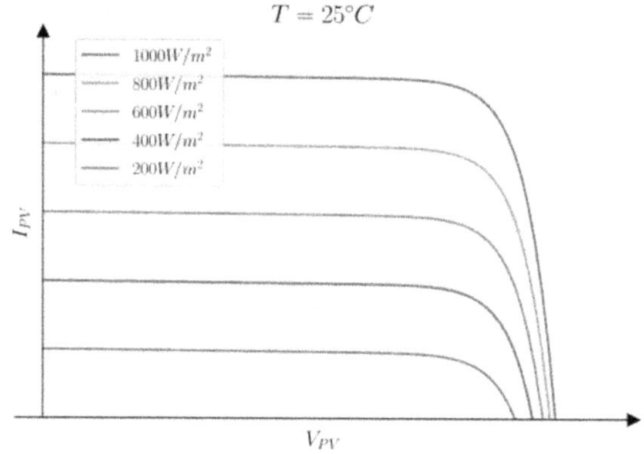

A boost converter is a type of DC-DC converter specifically designed to step up the input voltage to a higher output voltage, making it a critical component in PV systems. It achieves this voltage transformation through the coordinated operation of its core components: an inductor, a switch (commonly a transistor), a diode, and an output capacitor as shown in Figure 2. During operation, the inductor alternates between two states: storing energy when the switch is closed and releasing this energy to the load when the switch is open. This cyclical process occurs at high switching frequencies, enabling the boost converter to maintain a steady and efficient output voltage. The relationship between the input and output voltages in an ideal boost converter is determined by the duty cycle of the switching signal, where the duty cycle defines the proportion of time the switch remains closed during each cycle. Precise control of this duty cycle is essential to achieve optimal performance, particularly in applications where the input voltage varies dynamically.

In PV systems, the boost converter plays a dual role that is pivotal to their functionality and efficiency:

- **Voltage Elevation:** The boost converter increases the relatively low and variable voltage output from the PV panel to levels required by the connected load or energy storage system. This voltage elevation ensures compatibility with downstream components and enhances the overall utility of the PV system.
- **Maximum Power Point Tracking (MPPT):** The boost converter integrates advanced MPPT algorithms that enable the system to operate at the maximum power point (MPP) of the PV panel.

By continuously monitoring and adjusting the panel's operating point, these algorithms optimize the energy extraction process and compensate for fluctuations in environmental conditions.

The relationship between the input and output voltages/currents of an ideal boost converter is given by:

$$V_{out} = \frac{V_{in}}{1 - d} \tag{1}$$

$$I_{out} = (1 - d)I_{in} \tag{2}$$

Where d represents the duty cycle of the boost converter.

Figure 2. Classique boost converter

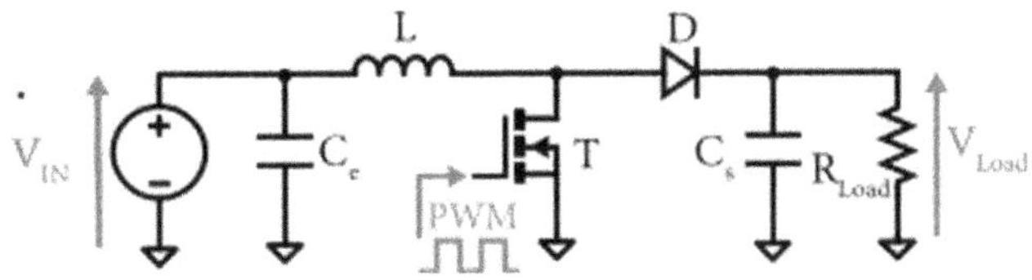

The integration of MPPT functionality within the boost converter is a cornerstone of modern PV system design. Algorithms such as Perturb and Observe (P&O) and Incremental Conductance are widely employed due to their ability to dynamically track the MPP under varying conditions. The P&O algorithm, for instance, adjusts the operating voltage of the PV panel incrementally and observes changes in power output to determine the optimal point of operation. This iterative process ensures that the system consistently harvests the maximum available energy, even during periods of rapid changes in irradiance or temperature.

Using this foundation, this chapter embarks on a detailed simulation study of a PV system integrated with a boost converter. The primary objective is to evaluate the system's performance in maximizing energy transfer to a resistive load under diverse weather conditions. By employing MATLAB/Simulink, we model the dynamic behavior of the system and analyze its efficiency and adaptability across various scenarios. Specifically, the focus is on assessing the converter's ability to:

- Maintain high efficiency by minimizing energy losses during voltage transformation.
- Effectively track the maximum power point under both stable and fluctuating environmental conditions.
- Ensure reliable and consistent energy harvesting despite challenges such as partial shading, rapid irradiance changes, and temperature variations.

The simulation explores the system's response to two distinct irradiance profiles: a smooth Gaussian curve representing a clear, sunny day and abrupt step changes simulating a cloudy, unpredictable day. These profiles provide valuable insights into the robustness and efficiency of the PV-boost converter system, highlighting its capacity to adapt to real-world conditions. By examining the interplay between the PV panel, boost converter, and MPPT algorithms, this study aims to contribute to the advancement of reliable and efficient renewable energy solutions.

Furthermore, the results of this simulation underscore the critical role of power electronics in the optimization of PV systems. As the global demand for renewable energy continues to rise, the development of intelligent, adaptive, and efficient power conversion systems will remain at the forefront of efforts to enhance the sustainability and reliability of solar energy technologies. This chapter provides a comprehensive analysis that not only addresses current challenges but also lays the groundwork for future innovations in PV energy harvesting and integration.

SIMULATION ENVIRONMENT UNDER SIMULINK

To evaluate the PV system in real-time under varying solar irradiance profiles, we developed a simulation model within the Simulink environment, as depicted in Figure 3. The model comprises the following components:

- A 280W PV Panel Simulated using a block that represents the electrical characteristics of a real PV module. The parameters of the PV block are configured based on manufacturer data, including maximum power, open-circuit voltage, and short-circuit current, as detailed in Table 1. The panel temperature is fixed at 25°C to focus exclusively on the effects of solar irradiance variations.
- A classical topology of Boost Converter consisting of an inductor, switch, diode, and capacitor, with component values provided in Table 2. The switching frequency is set to 10 kHz, and the duty cycle is dynamically adjusted by the MPPT algorithm.
- A Fixed Resistor of 20Ω is used to represent the energy consumption device.
- Voltage and Current Sensors that supply feedback to the MPPT controller, enabling precise tracking of the maximum power point.
- MPPT Algorithm Subsystem based on the Perturb and Observe (P&O) algorithm, implemented as a subsystem, is illustrated in Figure 4.

To ensure effective power management, our methodology involves continuously monitoring PV voltage and current to dynamically adjust the duty cycle of the power converter. This real-time adaptation allows the system to operate at the maximum power point, optimizing energy extraction from the PV panel. By simulating real-world conditions, this approach provides an accurate assessment of system performance under various environmental scenarios.

Figure 3. PV system model under Simulink

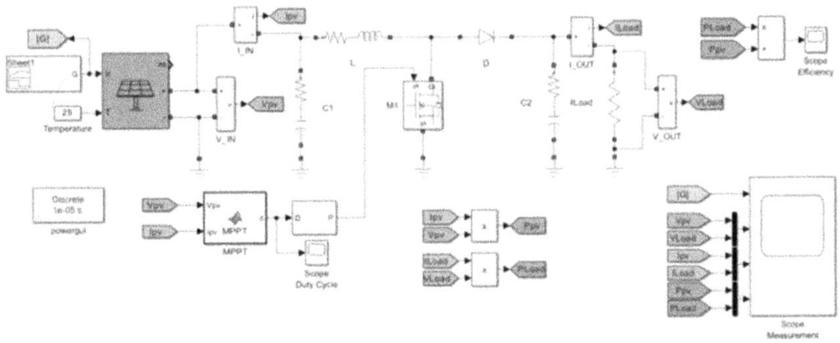

Figure 4. MPPT algorithm

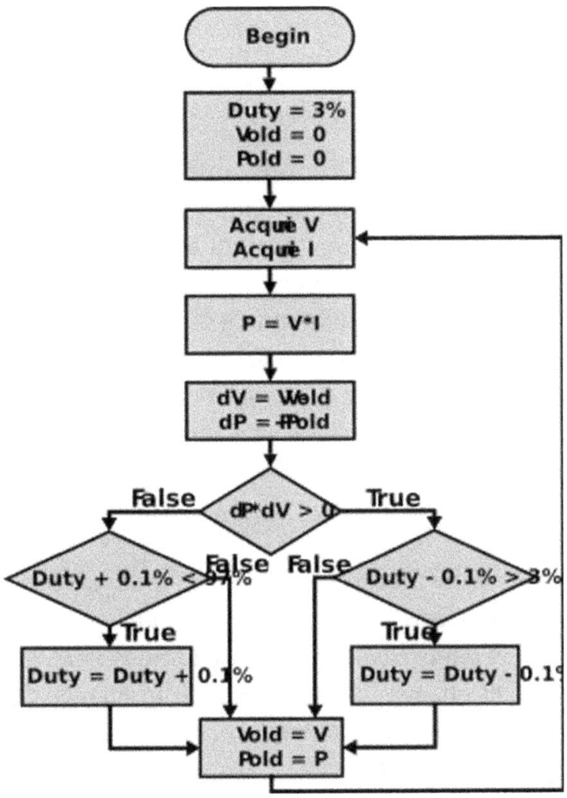

Table 1. Characteristics of the PV Panel at Standard Test Conditions

Model	HSL72P6-PA-0-280T
Optimal Power $\mathbf{P}_{MP}$	280W
Optimal Voltage V_{MP}	35.7 V
Optimal Current I_{MP}	7.84 A
Short Circuit Current $\mathbf{I}_{SC}$	8.43 A
Open Circuit Voltage $\mathbf{V}_{OC}$	44.6 V

Table 2. Values of components of the boost converter

Input and Output Capacitors	C = 1000 μF with ESR = 100mΩ
Inductor	L = 500 μH with ESR = 150mΩ
MOSFET	R_{dson} = 20mΩ
Schottky Diode	V_D = 0.4V

SIMULATION AND RESULTS

We conducted a comprehensive 24-hour simulation to evaluate the performance of the boost converter under conditions that mimic a clear, sunny day. For this purpose, the solar irradiance profile was modeled as a perfect Gaussian curve, which closely represents the behavior of solar irradiance during a day with no clouds. This profile, depicted in Figure 5, demonstrates a gradual increase in irradiance from sunrise, peaking at noon, followed by a symmetrical decrease until sunset.

The simulation results, illustrated in Figure 6, provide detailed insights into the variations in voltage, current, and power over the course of the day. These results are pivotal in assessing the behavior and efficiency of the boost converter in transferring energy from the photovoltaic (PV) panel to the load. Below are the key findings derived from the analysis of the simulation data:

- The boost converter successfully transfers energy from the solar panel to the load from sunrise to sunset. The maximum power output reaches 275W at noon, which matches the maximum power generated by the solar panel under Standard Test Conditions (STC) as outlined in Table 1.
- The boost converter operates effectively when the solar irradiance exceeds 250 W/m^2. Below this threshold, the input and output electrical properties (voltage, current, and power) remain identical, indicating minimal energy conversion.
- Above the irradiance threshold, the input voltage stabilizes at a nearly constant value, while the input current closely follows the trend of solar irradiance throughout the day.
- The output power is slightly lower than the input power, which can be attributed to modeled losses in the inductor and capacitors, represented by series resistive elements. Additionally, the efficiency of the boost converter is not constant and varies depending on the input voltage, current, and power levels.

Figure 5. Solar irradiance of a clear sunny day represented as perfect gaussian

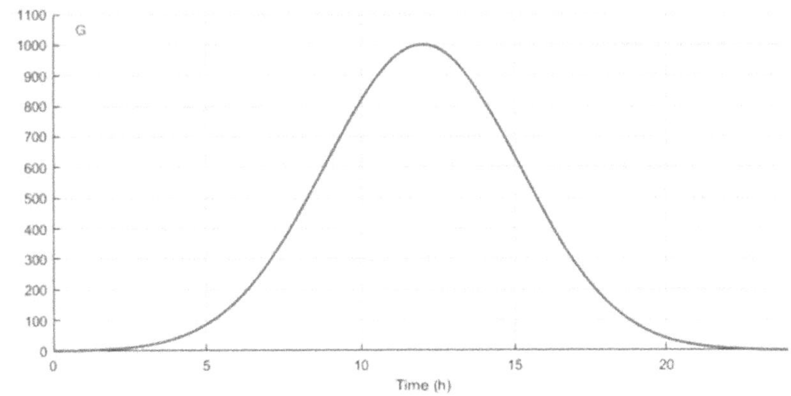

Figure 6. Electrical quantities at the input and output at a sunny day

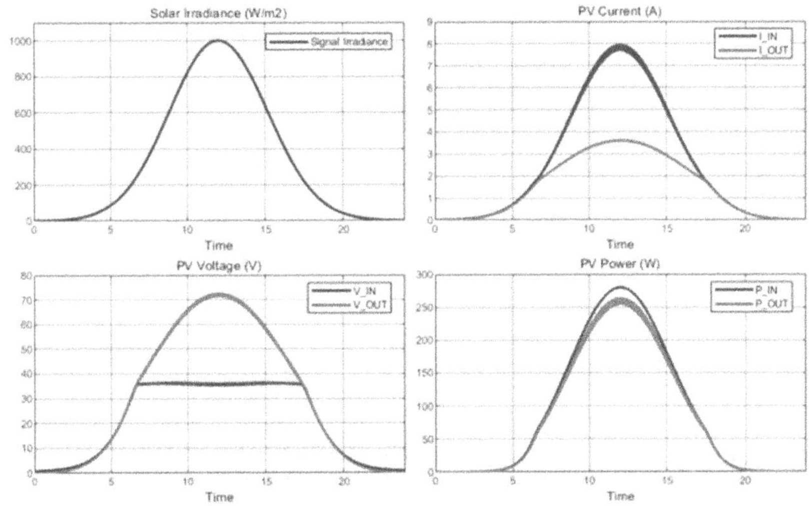

The second test aimed to evaluate the effectiveness of the boost converter and MPPT algorithm under sudden changes in solar irradiance. In this scenario, the solar irradiance varied in discontinuous steps, simulating cloud cover in adverse weather conditions, as shown in Figure 7. The interpretation of the results, illustrated in Figure 8, is as follows:

- The boost converter, assisted by the MPPT algorithm, quickly identifies and tracks the maximum power point almost instantaneously, even under abrupt changes in solar irradiance.
- When the solar irradiance reaches 200 W/m^2, the converter ceases to function as a boost converter. In this case, the input and output voltages, currents, and powers are equal, as previously observed in the first test. This behavior is consistent with the threshold effect explained earlier.

Figure 7. Solar irradiance of a day with bad weather represented as discontinuous steps

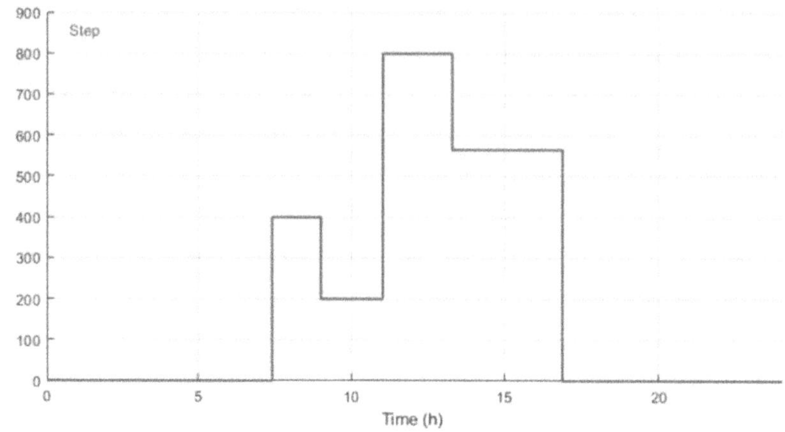

Figure 8. Electrical quantities at the input and output at a day with bad weather

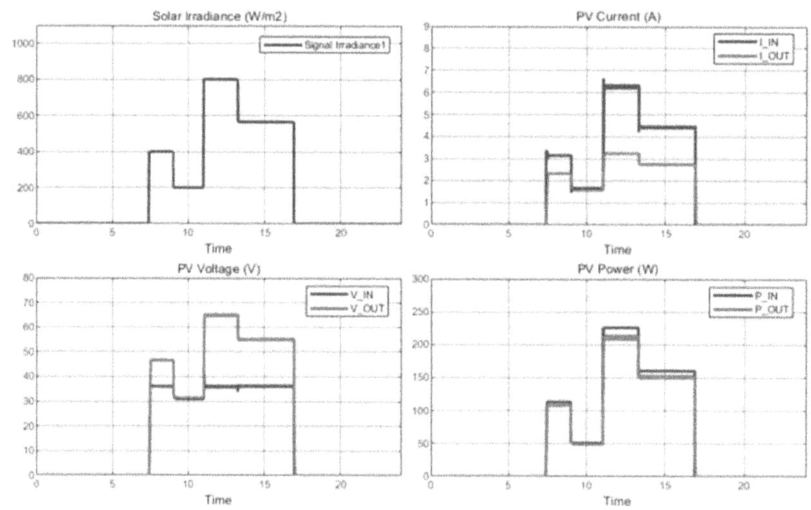

CONCLUSION

The simulation results demonstrate the effectiveness of the boost converter coupled with an MPPT algorithm in maximizing power extraction from photovoltaic panels under varying environmental conditions. In the clear-day scenario with a Gaussian irradiance profile, the boost converter successfully tracks the maximum power point, achieving peak power output of 275W at noon and maintaining efficient operation above the 250 W/m^2 irradiance threshold. The system's performance highlights the importance of precise control and energy conversion mechanisms in optimizing energy transfer to the load.

Under discontinuous irradiance conditions simulating adverse weather, the boost converter adapts rapidly to abrupt changes, ensuring minimal energy loss. The MPPT algorithm consistently identifies the maximum power point, enabling efficient energy harvesting despite fluctuating solar input. However, the system exhibits a threshold effect where the converter ceases boost operation below 200 W/m^2 irradiance, aligning with observations from the clear-day test.

In future work, additional factors influencing the performance of the boost converter can be explored. Specifically, analyzing the impact of temperature variations on the PV panel and converter efficiency would provide a more comprehensive understanding of system behavior under real-world conditions. Investigating different MPPT techniques, such as machine learning-based algorithms, could further enhance tracking accuracy and response speed under rapidly changing irradiance. Finally, experimental validation of the simulation results through hardware implementation and real-time testing would strengthen the findings and contribute to practical deployment strategies for PV energy harvesting systems.

REFERENCES

Carrasco, M., Mancilla-David, F., & Ortega, R. (2014). An Estimator of Solar Irradiance in Photovoltaic Arrays With Guaranteed Stability Properties. *IEEE Transactions on Industrial Electronics*, *61*(7), 3359–3366. DOI: 10.1109/TIE.2013.2281154

Chouay, Y., & Ouassaid, M. (2024). An accurate real time neural network based irradiance and temperature sensor for photovoltaic applications. *Results in Engineering*, *21*, 101766. DOI: 10.1016/j.rineng.2024.101766

Gaikwad, K., & Lokhande, S. (2016). Novel maximum power point tracking (MPPT) algorithm for solar tree application. *International Conference on Energy Systems and Applications, ICESA 2015*, 189–193. DOI: 10.1109/ICESA.2015.7503337

Gao, J., Yao, S., Xu, J., & Shi, Z. (2009). Design of a high efficiency boost DC-DC converter. *Transactions of Tianjin University*, *15*(5), 350–354. DOI: 10.1007/s12209-009-0061-y

Kannan, N., & Vakeesan, D. (2016). Solar energy for future world: - A review. *Renewable & Sustainable Energy Reviews*, *62*, 1092–1105. DOI: 10.1016/j.rser.2016.05.022

Khare, V., Chaturvedi, P., & Mishra, M. (2023). Solar energy system concept change from trending technology: A comprehensive review. *E-Prime - Advances in Electrical Engineering. Electronics and Energy*, *4*, 100183. DOI: 10.1016/j.prime.2023.100183

Rezk, H., & Eltamaly, A. M. (2015). A comprehensive comparison of different MPPT techniques for photovoltaic systems. *Solar Energy*, *112*, 1–11. DOI: 10.1016/j.solener.2014.11.010

Rhiat, M., Hassari, A., Karrouchi, M., Atmane, I., Abdellah, T., & Hirech, K. (2024). Educational High Power Photovoltaic Curve Tracer Using an IoT DC to DC Power Converter with Smartphone Integration. *2024 IEEE 11th International Conference on E-Learning in Industrial Electronics (ICELIE)*, 1–6. DOI: 10.1109/ICELIE62250.2024.10814775

Rhiat, M., Karrouchi, M., Atmane, I., Touhafi, A., Bossoufi, B., Almalki, M. M., Alghamdi, T. A. H., & Hirech, K. (2024). Maximizing solar energy efficiency: Optimized DC power conversion for resistive loads. *Computers & Electrical Engineering*, *120*, 109867. DOI: 10.1016/j.compeleceng.2024.109867

Rhiat, M., Karrouchi, M., Hassari, A., Melhaoui, M., Atmane, I., Azzaoui, H., Ouariachi, M. E., Bouchnaif, J., Schmitz, P., & Hirech, K. (2023). Design and simulation of a low-cost solar irradiance meter for PV applications. *E3S Web of Conferences, 469*, 00076. DOI: 10.1051/e3sconf/202346900076

Rhiat, M., Karrouchi, M., Hassari, A., Melhaoui, M., Atmane, I., Ouariachi, M. E., Schmitz, P., Benhala, B., & Hirech, K. (2024). Developing and simulating an inexpensive solar irradiance metre for photovoltaic utilisation. *International Journal of Power and Energy Conversion*, *15*(4), 376–388. DOI: 10.1504/IJPEC.2024.141867

Victoria, M., Haegel, N., Peters, I. M., Sinton, R., Jäger-Waldau, A., del Cañizo, C., Breyer, C., Stocks, M., Blakers, A., Kaizuka, I., Komoto, K., & Smets, A. (2021). Solar photovoltaics is ready to power a sustainable future. *Joule*, *5*(5), 1041–1056. DOI: 10.1016/j.joule.2021.03.005

Chapter 7
Comparison of Solar Irradiance Forecasting Performance With K-Means++ Clustering Combined With Hybrid Deep Learning Models

Madderla Chiranjeevi
https://orcid.org/0009-0002-7837-796X
National Institute of Technology, Karnataka, India

S Ramesh Torun
National Institute of Technology, Karnataka, India

Vaibhav Suresh Ghangale
National Institute of Technology, Karnataka, India

Abhinav Pundir
National Institute of Technology, Karnataka, India

Tukaram Moger
https://orcid.org/0000-0003-4176-5125
National Institute of Technology, Karnataka, India

Debashisha Jena
National Institute of Technology, Karnataka, India

ABSTRACT

Solar irradiance forecasting plays a crucial role in renewable energy, weather prediction, and climate modeling. Accurate forecasts are essential for optimizing solar power efficiency, grid integration, and energy planning. Traditional forecasting methods, based on physical and statistical models, struggle to capture the complex, nonlinear relationships inherent in solar irradiance. To address these challenges, this chapter presents a comparative analysis of advanced machine learning (ML) and deep learning (DL) models. Techniques like CNNs, RNNs, and hybrid models have demonstrated strong capabilities in extracting temporal and spatial patterns from Solar data. The integration of K-means++ clustering with DL frameworks further enhances model robustness, generalization, and interpretability. This chapter evaluates hybrid models, such as Temporal CNN-LSTM, CNN-GRU using metrics based on Solcast data. Results reveal that the TCNN-GRU model outperforms other state-of-the-art approaches, underscoring the value of clustering-enhanced DL frameworks for accurate solar irradiance forecasting.

DOI: 10.4018/979-8-3693-9924-8.ch007

1. INTRODUCTION

The increasing demand for cleaner energy alternatives to fossil fuels underscores the significance of exploring various Renewable Energy Resources (RES). Solar Energy, in particular, has seen significant advancements in recent years. However, its utilization is hindered by its unpredictable nature and seasonal variations. The dynamic upper limit of power generation associated with solar energy fluctuates constantly over time, adding complexity to its integration with the grid. Therefore, accurate prediction of solar radiation is indispensable for Independent System Operators (ISOs) to effectively manage the increased levels of power production while ensuring electric power quality and reliability. Solar energy represents an excellent alternatives of RES with the caliber to alleviate climatic variations and reduce resiliency on fossil fuels. Nevertheless, the intermittent and unpredictable nature of solar irradiance poses challenges for the efficient use of solar power systems. Precise prediction of solar radiation is critical for optimizing the performance of solar energy systems, facilitating integration with the power grid, and enabling effective energy management strategies.

Qu et al., (Qu et al., 2021) proposed a hybrid methodology provides a robust solution for accurate day-ahead hourly solar power forecasting, crucial for effective energy management and grid integration. The model integrates a novel multiple relevant and target variables prediction pattern (MRTPP) with hybrid deep learning networks called attention based long short term memory (ALSM). The approach performs better at apprehending complex temporal dependencies and multivariate relationships within the data. Notably, the method's ability to adapt to various prediction horizons is underscored by optimizing memory lengths, a critical factor influencing forecasting accuracy. The proposed method demonstrates its effectiveness compared to existing approaches through comprehensive validation exercises, affirming its potential as a reliable tool for solar power forecasting in real-world applications.

A novel deep learning approach called decomposition-clustering-ensemble (DCE) for solar radiation forecasting is introduced (Sun et al., 2018), emphasizing two distinct steps: data analysis and forecast modeling. Leveraging the ensemble empirical mode decomposition - least square support vector regression - K-means (EEMD-LSSVR-K-LSSVR) learning architecture, the method offers a sophisticated framework for predicting solar radiation levels. Empirical findings provide convincing evidence supporting the efficacy of this novel architecture, underscoring its potential to enhance forecasting accuracy and reliability.

Varanasi and Tripathi (Varanasi and Tripathi, 2019) introduces a novel approach for photovoltaic power forecasting, employing K-means clustering in conjunction with artificial neural networks, particle swarm optimization (PSO), and SVR. This comprehensive methodology integrates clustering techniques with recent developed machine learning algorithms to enhance the accuracy and efficiency of photovoltaic power predictions. Through the synergistic application of these methods, the proposed framework aims to provide more reliable forecasts, thereby contributing to the optimization of solar energy utilization and grid integration strategies.

Ali-Ou-Salah et al., (Ali-Ou-Salah et al., 2022) introduces a novel hybrid methodology for forecasting global solar radiation one hour ahead, combining seasonal clustering techniques with ANN. Initially, the fuzzy c-means method (FCM) is employed to cluster three years of monthly mean dataset into distinct seasons depending on solar and meteorological features specific to E´vora city. Subsequently, leveraging the results of seasonal clustering, the meteorological data is further partitioned into various training subsets, facilitating the development of accurate forecasting models tailored to different seasonal variations. A newly devised K-means-support vector regression (SVR) algorithm has been introduced (Ayodele et al., 2019) to forecast the GHI value. Comparative assessments against established models like ANN, ARMA, and Angstrom- Prescott models revealed the exceptional performance of the presented algorithm.

Zhang et al., (Zhang et al., 2021) introduces a novel K-means time series clustering (K- MTSC) model aimed at clustering solar irradiation intensity. This algorithm is then tested with both astronomy methods and traditional K-means clustering. Following the clustering process, various feature inputs are evaluated for distinct types of solar radiation intensity. Subsequently, the study compares various available kernel functions of Gaussian process regression (GPR) and identifies the best kernel function based on its performance in both deterministic and probabilistic forecasting across different categories.

Velasco et al., (Velasco et al., 2018) proposes a hybrid approach by integrating K-means with ANN is proposed for day-ahead and peak load forecasting. Initially, the collected data is clustered using K-means, and the resulting clusters are then inputted into an ANN model utilizing gaussian activation functions and the quick propagation training algorithm. The input data for clustering comprises date, time, and kilowatt consumption (load). The authors in (Wu et al., 2023) introduce five weather classification models: K-means, Self-Organizing Maps (SOM), diurnal segmentation based on Pearson correlation coefficients (morning/afternoon), classification using threshold-based criteria enhanced by optimization algorithms, and season-based classification. Furthermore, (PSO) technique is used to optimize the hyperparameters of Gated Recurrent Unit (GRU) and transformer models. Results indicate that the model utilizing the transformer architecture exhibits the lowest forecast errors.

Azimi et al., (Azimi et al., 2016) proposes an approach called Transformation Based K-Means (TB K-Means) to address the limitations observed in the original K-means algorithm. This method integrates an initialization strategy and a progressive data transformation technique to overcome the variability often encountered in traditional K-means implementations across multiple trials. Unlike standard K-means, TB K-means consistently produces stable outcomes across different algorithm runs. The study further employs an inverse transformation method to mitigate data variance towards incorrect cluster centroids. Similarly, Ghofrani et al., (Ghofrani et al., 2016) presents an innovative soft computing model aimed at improving the accuracy of solar radiation forecasting for better grid integration. This framework incorporates a modified clustering technique called time series clustering (T.S.C) K-Means, and the MLPNN model.

Feng et al., (Feng et al., 2018) introduced the Optimized Cross Validated Clustering (OC CUR) technique to evaluate the optimal number of clusters for improved outcomes. Day categorization is accomplished using a support vector pattern recognition technique applied to data from the initial four hours. Subsequently, various models are utilized to forecast Global Horizontal Irradiance (GHI) values within each cluster. The OCCUR method employs four unsupervised clustering (UC) techniques: Agglomerative Hierarchical Clustering (AHC), K-Means, K-Medoids, and Divisive Hierarchical Clustering (DHC). In contrast, (Ghayekhloo et al., 2015) proposes a novel approach for GHI forecasting by integrating the original self-organizing map (SOM) algorithm with game theory principles. This method aims to address limitations of the SOM algorithm in handling non-winning neurons and enhancing their competitiveness with winning neurons. The objective is to enhance the accuracy of solar radiation forecasts by incorporating a broader range of input patterns into the model. Chiranjeevi et al., (Chiranjeevi et al., 2024) proposes K-means clustering-based CNN-BiLSTM-Attention hybrid architecture with PSO optimization and achieves good performance on various datasets collected.

Building upon the insights gleaned from the literature, this chapter proposes an innovative methodology that amalgamates the K-means++ clustering algorithm with various hybrid deep learning models. The objective is to harness the synergies between clustering methods and deep learning frameworks to refine the precision and efficacy of solar irradiance forecasting. By integrating K-means++ with hybrid deep learning approaches, the proposed method aspires to grab the advantage on the unique strengths of each method, thereby improving the accuracy and efficiency of solar irradiance prediction. Through this novel integration, the chapter endeavors to introduce a pioneering framework for more precise and dependable solar irradiance forecasting, thereby give rise to the advancement of renewable energy prediction techniques.

2. METHODOLOGY INVOLVED IN THE HYBRID MODEL

2.1 K-Means

Within the realm of unsupervised clustering algorithms, one stands out as particularly prevalent. This algorithm, known as K-means, operates as a partitioning mechanism, categorizing data points based on their similarities. Given a dataset ϕ containing *n* d-dimensional vectors, the algorithm endeavors to create K disjoint subsets by optimizing the objective function outlined in Eq. 1:

$$\textbf{Objective function J} = \sum_{j=1}^{k}\sum_{i=1}^{n}\left|x_i^j - c_j\right|^2 \tag{1}$$

The K-means process unfolds as follows, as depicted in Figure 1:

Initialization Step: Initially, this technique considers K random centroids from the data.

Assignment Step: All elements in the data are allotted to the nearest centroid depending on a specified distance metric, typically Euclidean, Manhattan, or Chebyshev.

Updation Step: After the initial run, centroids are recalculated by computing the mean value of data elements allotted to every cluster.

Iteration: Steps 3 and 4 iterate until the convergence criteria are gratified.

Convergence: The algorithm halts when centroids stabilize, indicating minimal positional changes across successive runs.

Figure 1. K means clustering working flow diagram

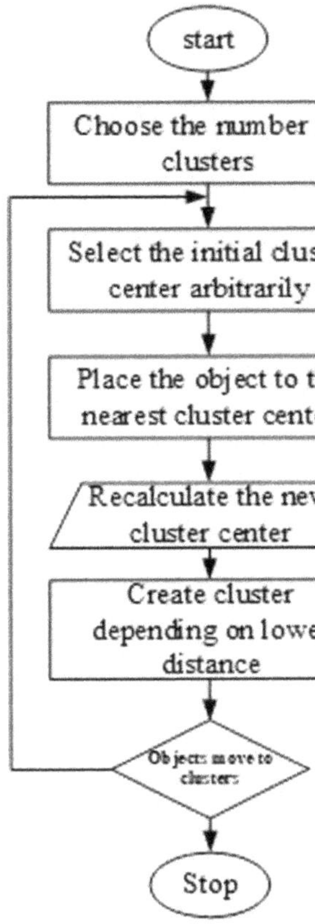

2.2 K-Means++

K-means++ is a technique used for initializing the centroids in the K-means clustering (KARPAGAM, 2022). Unlike the standard K-means model, which initializes centroids randomly, K-means++ aims to choose centroids that are well spread out across the dataset, leading to more stable and efficient convergence of the K-means. An overview of the K-means++ along with its mathematical equations is given as below:

Initialization: Select the first centroid randomly. For every subsequent centroid, compute the distance of each data element from the nearest centroid which is already selected. After that consider the next centroid from the data elements with a probability proportional to the squared distance from the nearest centroid.

Objective Function: Let $X = \{x_1, x_2, \ldots, x_n\}$ be the data samples with n data elements and $C = \{c_1, c_2, \ldots, c_k\}$ be the group of centroids. Then the objective function of the K-means clustering technique is to minimize the sum of squared distance between every data element and its nearby centroid as:

$$\mathbf{J} = \sum_{\mathrm{i}=1}^{\mathrm{n}} \mathbf{min} \, ||\mathbf{x}_\mathrm{i} - \mathbf{c}_\mathrm{j}||^2$$

Algorithm:

- **Initialize centroids using K-means++.**
- **Repeat until convergence:**
  - * Assign all the data elements to the near by centroid.
  - * Upgrade the centroid by computing the average of the data elements allocated to each centroid.
  - * The Euclidean distance between two points x_i and c_j is given by:

$$\|x_i - c_j\| = \sqrt{\sum_{d=1}^{D}\left(x_{id} - c_{jd}\right)^2} \tag{2}$$

where D denotes the dimensionality of the data.

* The probability $P(x_i)$ of choosing data point x_i as the next centroid is given by:

$$P\left(x_i\right) = \frac{D(x_i)^2}{\sum_{i=1}^{N} D(x_i)^2} \tag{3}$$

Here, $D(x_i)$ is the distance to the nearest existing centroid, and N is the total number of data elements. The steps are repeated until k centroids are selected. Once the centroids are initialized using K-means++, the standard K-means clustering technique is applied.

2.3 Hybrid Deep Learning Approaches

In recent years, machine learning (ML) techniques have became popular for solar irradiance forecasting due to their ability to capture complex relationships and patterns in data. Support vector regressors (SVR), artificial neural networks (ANN), random forests (RF), and ensemble methods are among the ML algorithms commonly employed for solar irradiance prediction. ANN based models, particularly deep learning architectures such as CNN and RNN, have proven excellency in adopting temporal dependencies and improving forecast accuracy. Additionally, hybrid models that combine physical models with ML techniques have evolved as a viable approach for enhancing the robustness and accuracy of solar irradiance forecasts.

In this comparative analysis, the chapter explores a variety set of ML and DL models for sequential data analysis. These algorithms encompass a range of architectures and techniques tailored to encapsulate temporal dependencies and patterns within sequential information. CNNs excel at gathering spatial features through convolutional operations, making them the best suitable for tasks such as image classification and time-series analysis. GRUs and LSTM models are part of recurrent neural networks (RNNs) equipped with gating mechanisms to mitigate the vanishing gradient problem and encapsulate long-range dependencies. Combining CNNs and RNNs, CNN- LSTM and CNN-GRU architectures supports the benefits of both convolutional and recurrent layers for enhanced performance. Temporal Convolutional Networks (TCNs) introduce dilated convolutions to capture long-range dependencies efficiently while maintaining parallelizability, offering an alternative to traditional RNNs. Additionally, bidirectional variants of LSTM and TCN models, such as BiLSTM and TCN-BiLSTM, enable the incorporation of future context during training, enhancing the models' ability to capture temporal dynamics. Through this comparative analysis, this chapter aims to evaluate the strengths and weaknesses of all algorithms in various sequential data analysis tasks, providing deeper understanding into their applicability and performance across different domains.

2.3.1 Convolutional Neural Network (CNN)

The CNN architecture is tailored to address structured grid like data, such as images or sequential data like time-series. Its core framework comprises several pivotal layers: input, convolutional, pooling, fully connected, and output layers (Liao and Coimbra, 2024).

Initially, the input layer takes the raw data, which, in this context, is interpreted as time-series data. The convolutional layers serve as the backbone of the CNN, where input data features are extricated using convolution kernels. These kernels traverse the input data, identifying patterns and capturing spatial relationships (Jonathan et al., 2024). It's essential to emphasize that each neuron in the convolutional layer exclusively connected to a localized region of the preceding layer. Through the convolutional operations, the convolutional layer produces feature maps, representing the learned features of the input data by the network. These feature maps encapsulate details about edges, textures, shapes, and other prominent features inherent in the input data. Determination of every element in the feature map is:

$$x_{i,j}^{out} = f_{cov}\left(\sum_{m=0}^{k}\sum_{n=0}^{k} w_{m,n} x_{i+m,j+n}^{in} + b\right) \quad (3)$$

where $x_{i,j}^{out}$ represents the output value in row i and column j of the feature map. $x_{i+m,j+n}^{in}$ represents the value in row i and column j of the input matrix, $f_{cov}(.)$ represents the activation function, $w_{m,n}$ represents the weight in row m and column n for the convolution kernel, and b represents the bias of the convolution kernel.

Figure 2. Feature extraction of GHI data

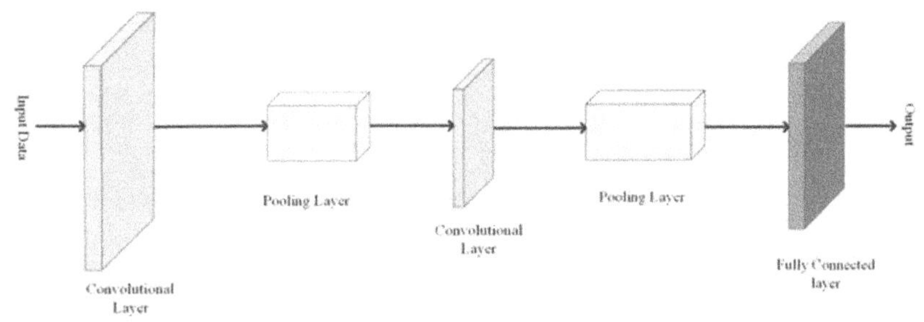

Subsequent to the convolutional layers, pooling layers plays a pivotal role in diminishing the spatial dimensions of the feature maps, as depicted in Fig. 2. By down-sampling the feature maps, pooling layers effectively diminish computational complexity and facilitate improved generalization. Common pooling techniques are max pooling and average pooling, which respectively preserve the maximum value or mean value within every pooling region. As the network traverses through multiple convolutional and pooling layers, it gradually acquires abstract information of the input, capturing higher-level features and patterns.

Following the convolutional and pooling layers, the feature maps are flattened into a vector and channeled into fully connected layers. In these layers, each neuron establishes connections with every neuron in the preceding layer, by enabling the network to execute high-level abstraction and regression tasks. Fully connected layers empower the network to discern intricate relationships between features extracted from the input data.

In scenarios involving regression tasks, the fully connected layers forecast continuous values predicated on the acquired features. The network's parameters undergo optimization throughout the training process using techniques like gradient descent, thereby regulating the weights and biases to diminish the disparity between the actual and predicted values. Ultimately, the output layer yields the required output, which encompasses continuous values for regression tasks.

2.3.2 Long Short-Term Memory (LSTM)

The LSTM stands out as a recurrent neural network (RNN) model substantially used in deep learning models for time series applications. Renowned for its prowess in capturing long-term dependencies, it emerges as an optimal choice for sequence prediction tasks. Diverging from CNNs, the LSTM model includes feedback connections, enabling it to analyze the whole data sequences rather than separate data elements exclusively. This distinctive capability renders it highly proficient in understanding and forecasting patterns inherent in sequential data types, including time series, textual data, and speech.

The LSTM cell units uphold their computations depending on the input and output from the hidden layers of the preceding level, while experiencing internal structural modifications while maintaining an unchanged external structure (Hendrikx et al., 2024). As depicted in Fig. 3 the internal composition of the LSTM cell unit encompasses the input gate *(i)*, forget gate *(f)*, output gate *(o)*, and memory unit *(c)*. The forget gate regulates the degree of forgetting for both the input *x(t)* and the output of the upper hidden layer *h(t-1)*. Concurrently, the input gate manages the efficacy of updating for the input *x(t)* and the output of the upper hidden layer *h(t-1)*. The memory unit *c* dictates which novel content is retained within the cell and refreshes the cell state accordingly. Lastly, the output gate oversees the quantity of data to be output from the cell. The mathematical background of the LSTM model can be understood from (Chiranjeevi et al., 2023a).

Figure 3. Architecture of LSTM

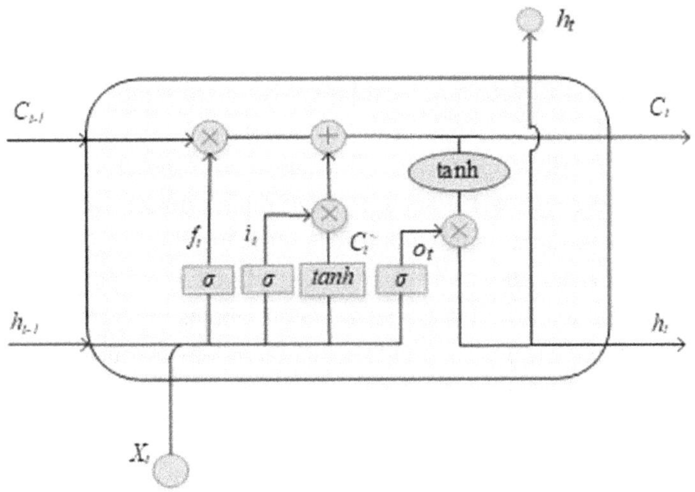

2.3.3. CNN-LSTM

The CNN-LSTM architecture blends CNN and LSTM layers, tailored for sequence modeling tasks, particularly adept at handling sequential data imbued with spatial characteristics, such as time series data or image sequences (Ibrahim et al., 2024).

In the CNN segment, the input, convolution, and pooling layers play pivotal roles in gathering spatial features from the input data. The input layer receives raw input data, like images or time series sequences. Subsequently, the convolutional layers consists of filters to discern spatial patterns within the data, while the pooling layers downsample the convolutional layer outputs, diminishing spatial dimensions while preserving crucial features. Following the CNN layers, the acquired features undergo flattening into 1-D arrays to ready them for input into the LSTM layer. This flattening procedure transforms spatial features into a format amenable for time series analysis.

In the LSTM segment, the flattened features are input as a time sequence into the LSTM layer, enabling the network to apprehend temporal dependencies and scrutinize the time sequence features of the input data. LSTM cells retain memory of past inputs, empowering the network to glean insights from sequential data and generate predictions depending on the context of prior inputs. Ultimately, the output of the LSTM layer traverses through fully connected layers and output layers to generate the desired predictions.

2.3.4. Gated Recurrent Unit (GRU)

The GRU is a specialized type of recurrent neural network (RNN) tailored for processing sequential data, especially time series analysis. It consists of interconnected layers, including input and GRU cells, equipped with specialized gates: update gate z and reset gate r to control information flow and the architecture is given as in Fig. 4.

The update gate balances retaining previous information with integrating new input, while the reset gate manages the extent of memory reset. Each GRU cell maintains an evolving internal memory state, capturing temporal dependencies and contextual information crucial for accurate predictions (Xu et al., 2024). During training, the model adjusts its parameters via back propagation, optimizing prediction accuracy. The GRU's potential to capture long-term dependencies and adapts changing patterns makes it widely used in tasks like forecasting and anomaly detection across various domains.

Figure 4. Architecture of GRU

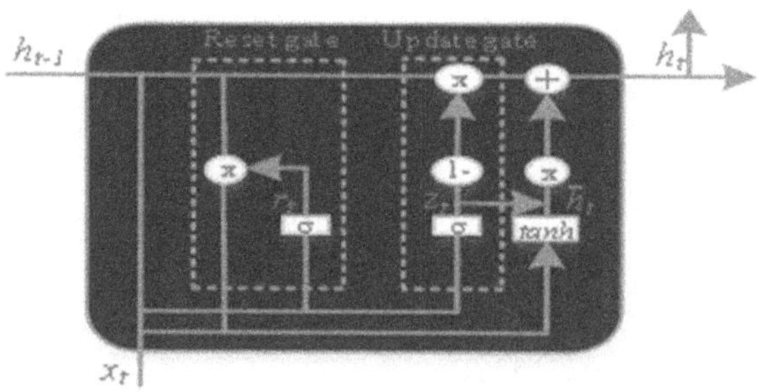

The transition functions within the hidden units of a GRU are expressed as follows:

$$z_t = \sigma\left(W^z x_t + V^z h_{t-1} + b^z\right)$$

$$r_t = \sigma\left(W^r x_t + V^r h_{t-1} + b^r\right)$$

$$\tilde{h}_t = tanh\left(W^c x_t + V^c\left(r_t \odot h_{t-1}\right)\right)$$

$$h_t = (1 - z_t) \odot h_{t-1} + z_t \odot \tilde{h}_t)$$

2.3.5 Convolutional Neural Network-Gated Recurrent Unit (CNN-GRU)

The CNN-GRU model stands as a potent fusion of CNN and GRU layers, adept at harnessing both spatial features and temporal dependencies in input data. The CNN segment employs convolution operations to extract pertinent features, pinpointing patterns and relationships within the data, particularly beneficial for tasks with spatial nuances. After convolution, pooling operations further condense the data's dimensionality while preserving key features, enhancing model efficiency. The same can be better understood from Fig. 5. To curb overfitting, a dropout layer is integrated post pooling, deactivating neurons at randomly during training to minimize reliance on specific features and foster better generalization. This regularization technique bolsters the model's resilience and performance. Subsequently, the GRU layer processes the data sequence, capturing temporal dependencies and sequential patterns. With GRU's proficiency in learning long-range dependencies, the model adeptly grasps the temporal structure of the input data, paving the way for accurate predictions. Finally, the output from GRU layer undergoes reshaping via a fully connected layer into a 1-D sequence.

Figure 5. Architecture of CNN-GRU

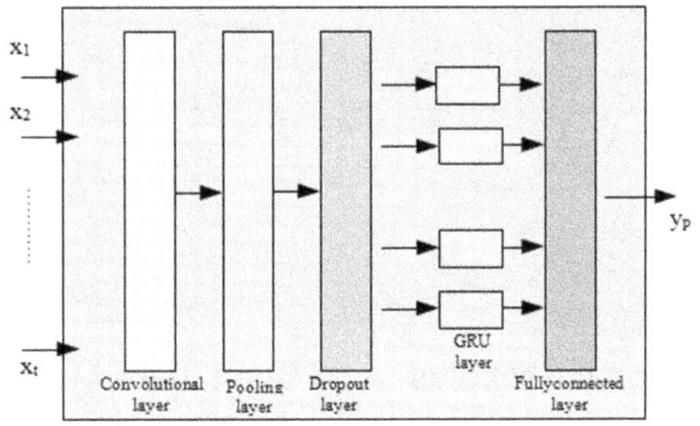

2.3.6 Temporal Convolutional Network (TCN)

Temporal Convolutional Networks (TCNs) represent a class of deep learning architectures specialized in analyzing sequential data, particularly time series. They rely on one-dimensional convolutional layers to effectively capture temporal patterns and dependencies within the data. TCNs are characterized by several key components, including dilated convolutions, which enable them to expand their receptive field without significantly rising the number of parameters (Elmousaid et al., 2024).

Additionally, TCNs often incorporate residual connections to make the training of deep networks and reduce the vanishing gradient problem and its architecture is given in Fig. 6. TCN's can also utilize techniques such as temporal pooling for downsampling and causal padding to ensure that predictions are based solely on past and current input data. Due to their ability to capture long-range dependencies and extract informative features from sequential data, TCNs have found applications in various domains, like time series prediction, and natural language processing.

Figure 6. Architecture of TCN

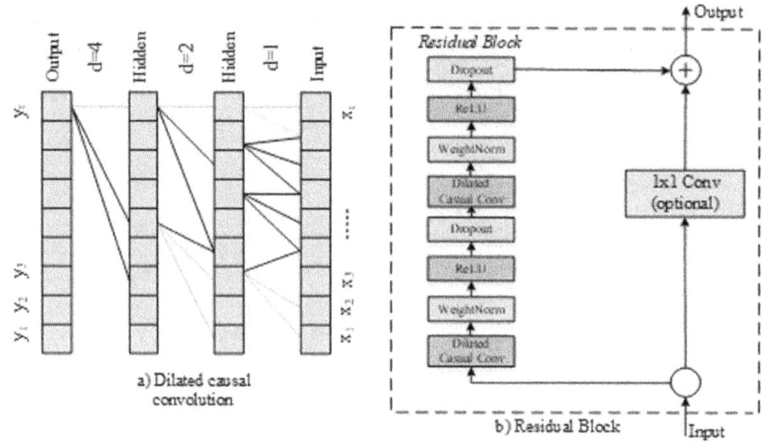

2.3.7 Temporal Convolutional Networks-Long Short-Term Memory (TCN-LSTM)

The TCN-LSTM architecturel represents an advanced neural network that merges the functionalities of Temporal Convolutional Networks (TCN) and Long Short-Term Memory (LSTM) layers. This hybrid framework is tailored for proficient feature extraction and prediction tasks, particularly in the realm of time series data.

Initially, the TCN component employs 1-D causal convolution to extract pertinent information from historical data while maintaining the temporal sequence. By ensuring that each output relies solely on past inputs, causal convolution effectively captures time dependencies with precision. Additionally, the integration of residual connections expedites convergence, aiding smoother training by facilitating gradient flow shortcuts. Furthermore, dilated convolutions play a crucial role in capturing temporal features efficiently, allowing the model to extract details across a broader range of time scales.

Following feature extraction via the TCN layer, the extracted features undergo normalization and Rectified Linear Unit (ReLU) activation functions, augmenting the model's capacity to identify meaningful patterns. Leveraging expansion convolution and causal convolution operations, the TCN layer produces sequence features, enabling the extraction of comprehensive temporal dependencies over time.

The output of the TCN layer feeds as an input to the following LSTM network layer. This integration enables further feature extraction while retaining the features captured by the TCN layer. By amalgamating the functionalities of both TCN and LSTM layers, the model can prevent feature degradation and effectively capture intricate temporal patterns. The fused features from the TCN and LSTM layers are then forwarded to a fully connected layer for additional processing. To mitigate overfitting, a dropout mechanism is introduced, randomly deactivating neurons during training to enhance generalization and expedite convergence.

2.3.8 Bidirectional Long Short Term Memory (BiLSTM)

BiLSTM, is an improved version of LSTM, enhances its capabilities by introducing bidirectional processing of input data. This architecture employs two LSTM layers: one processes data in the original forward direction, while the other handles data in the backward direction. These two layers' outputs are then merged to form the final output. This bidirectional approach overcomes the gradient vanishing problem encountered in standard RNNs by allowing information flow in both directions, thereby facilitating more efficient gradient flow during training. Moreover, bidirectional processing aids in feature extraction by capturing information from both past and future contexts.

By incorporating two LSTM layers with opposite input directions, BiLSTM effectively captures temporal features from input series, leveraging information from both past and future contexts. This capability often leads to superior prediction performance compared to unidirectional LSTM models. Consequently, BiLSTM finds wide application in various fields, particularly in tasks involving sequential data analysis, including time series forecasting, agricultural forecasting, speech recognition, and natural language processing.

2.3.9 CNN-BiLSTM

The CNN-BiLSTM architecture shown in Fig. 7 collects the strengths of CNNs and BiLSTM architecture to effectively process and analyze sequential information, such as time series data or sequences of images (Chiranjeevi et al., 2023b).

In this architecture, the CNN component is accountable for gathering spatial features from the input data. CNNs are adept at establishing patterns and spatial associations within the data, making them well-suited for tasks involving structured data like images or sequential data with spatial characteristics. The CNN layers apply convolutional operations to the input data, followed by pooling operations to reduce dimensionality and retain essential features.

However, the BiLSTM component focuses on apprehending temporal dependencies within the data. BiLSTM networks leverage the bidirectional processing of input sequences, allowing them to incorporate details from both past and future contexts. This bidirectional flow enables the architecture to better understand and predict patterns in sequential data by apprehending long-range dependencies effectively.

By aggregating CNNs for spatial feature collection and BiLSTM for temporal modeling, the CNN-BiLSTM architecture can successfully analyze sequential data with both spatial and temporal features. This hybrid approach enables the model to grasp complex patterns and dependencies within the data, enhancing its performance in tasks like time series analysis, image sequence analysis, and natural language processing.

Figure 7. Architecture of CNN-BiLSTM

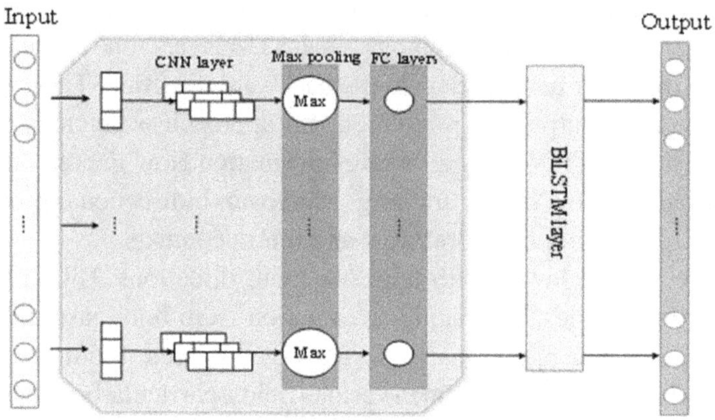

3. RESULTS AND DISCUSSIONS

The subsequent part details the specifics of data collection from the weather station, the method of data preprocessing, and the evaluation of performance for all models.

3.1 Data Collection

The experiments utilize publicly available dataset from the Solcast website (Solcast, 2024). The dataset encompass various features including Global Horizontal Irradiance (GHI), clear-sky GHI, air temperature, cloud opacity, relative humidity, dewpoint temperature, wind direction, and wind speed. The dataset corresponds to Maharashtra location, with coordinates at 19.751480 latitude and 75.713888 longitude.

The dataset cover a substantial time frame from January 1, 2010, to December 31, 2015, offering a comprehensive insight into solar irradiation patterns over six years. For model training and evaluation, data up to the end of 2014 were allocated for training purposes. The remaining data from 2015 onward were reserved for testing, ensuring a rigorous assessment of the models' predictive capabilities on unseen data. Table 1. provides a comprehensive summary of the data.

Table 1. Features variables and its Statistics summary

Attribute	Air Temp	Clearsky GHI	Cloud Opacity	Wind Speed	Dewpoint Temp	Relative Humidity
Max	47	1016	97	14	27.3	100
Min	9	0	0	0.1	-6	7.2
Mean	27.9238	250.2338	14.9119	3.98117	57.9104	57.9104
Std. Dev	6.0643	332.159	24.274	1.9732	22.8351	22.83617

3.2 Data Preprocessing

The system encounters messy data due to various operational challenges, including intermittent inverter performance, module damage necessitating replacement, temporary system disconnections during cable maintenance, and occasional data loss. Within the photovoltaic dataset, anomalies such as missing data and negative values emerge, demanding rectification before accurate solar irradiation forecasting.

To address missing data, seasonal values from previous periods are utilized as substitutes. Negative values in the Global Horizontal Irradiance (GHI) data during early morning hours are reset to zero to maintain data integrity. Outliers are observed using Z-score technique and updated with the corresponding seasonal values to ensure dataset consistency. Furthermore, data normalization is crucial to standardize values across different scales and mitigate undue influence on forecasting models. Normalization, expressed in Eq. 5 ensures that each feature's values are within a comparable range:

$$\acute{x} = \frac{x - x_{min}}{x_{max} - x_{min}} \quad (5)$$

Here, x_{min} and x_{max} represent the minimum and maximum values of the respective parameter, facilitating uniform data representation for accurate forecasting.

3.3 Performance Metrices

The proposed methodology undergoes training and analysis utilizing a dataset comprising solar irradiance measurements from real-world scenarios. To gauge the forecasting accuracy, standard performance metrics such as Root Mean Squared Error (RMSE), Mean Absolute Error (MAE), and R^2 score will be utilized. However, in the analysis of nonlinear and nonstationary series, relying solely on a single evaluation metric may not yield precise measurements. Therefore, it is advisable to compare the error using many metrics presented in the provided Table. 2.

Table 2. Performance evaluation metrices

Metric	Formula	Definition
MAE	$\frac{1}{N}\sum_{i=1}^{N}\lvert y_i - p_i\rvert$	Calculates MAE of *N* forecasting results
MSE	$\frac{\sum_{i=1}^{N}(y_i - p_i)^2}{N}$	Calculates average of error squares
RMSE	$\sqrt{\frac{\sum_{i=1}^{N}(y_i - p_i)^2}{N}}$	Calculates the square root of MSE
R^2 score	$1 - \frac{\sum_{i=1}^{N}(y_i - p_i)^2}{\sum_{i=1}^{N}(y_i - \bar{y}_i)^2}$	Denotes the percentage of variance in the target data

3.4 Performance Comparison of the Hybrid Architecture in Univariate Analysis

Univariate analysis pertains to the examination of a single variable, focusing solely on historical patterns and trends of solar irradiance data over time. This hybrid framework facilitates the detection of cyclic patterns and seasonal trends in solar radiation levels.

Threshold values for Global Horizontal Irradiance (GHI) were crucially employed throughout this study to classify distinct weather conditions across seasons. Fig. 8 shows the various categorization of the seasons based on the GHI value. For the Summer classification, a GHI threshold value was utilized to differentiate between clear and non-clear conditions, aiding in the precise categorization of instances. Similarly, in the Winter classification, the GHI threshold value played a pivotal role in distinguishing between moist and dry conditions, contributing to the refinement of seasonal categorization. Moreover, for the Monsoon classification, GHI threshold values were instrumental in discerning between cloudy and partly cloudy conditions, facilitating accurate classification within this seasonal category. This strategic use of GHI threshold values ensured the robustness and accuracy of our classification approach across different weather conditions and seasons.

It can be observed from Table 3 that in case of summer days, BiLSTM has the best performance than other basic models, having a R^2 score of 0.9701 for single step ahead (i.e., 1 hour) prediction. The R^2 score gradually decreases as the number of steps ahead values (from 1 to 24) increases. The single step ahead prediction plot for the individual models can be inferred from Fig. 9a where the predicted values almost matches with the actual values.

It can be observed from Table 4 that in case of summer days, TCN-LSTM has the best performance than other hybrid models considered, having a R^2 score of 0.9765 for single step ahead (i.e 1 hour) prediction. The R^2 score gradually decreases as the number of steps ahead (from 1 to 24 h) increases. The single step ahead prediction plot for the considered hybrid models can be inferred from Fig. 9b.

It can be concluded from Table 5 that in case of winter days BiLSTM has the best performance than other basic models, having a R^2 score of 0.9474 for single step ahead (i.e 1 hour) prediction. The R^2 score gradually decreases as the number of steps ahead (from 1 to 24 h) increases. The single step ahead prediction plot for the considered models can be inferred from the Fig.10a.

Table 3. Performance evaluation of basic models (Summer)

Forecasting horizon		Evaluation Index	Forecasting Models				
			CNN	GRU	BiLSTM	LSTM	TCN
		MAE	0.0035	0.0031	0.0027	0.0032	0.0039
1-Step	(1 hour)	RMSE	0.0562	0.0518	0.0492	0.0551	0.0625
		R^2	0.9685	0.9688	0.9701	0.9692	0.9659
		MAE	0.0041	0.0035	0.0026	0.0034	0.0045
3-Step	(3 hour)	RMSE	0.0567	0.0561	0.0452	0.0575	0.0586
		R^2	0.9602	0.9635	0.9613	0.9632	0.9621
		MAE	0.0035	0.0038	0.0032	0.0037	0.0043
6-Step	(6 hour)	RMSE	0.0621	0.0589	0.0535	0.0568	0.0605
		R^2	0.9620	0.9654	0.9703	0.9675	0.9635

continued on following page

Table 3. Continued

Forecasting horizon		Evaluation Index	Forecasting Models				
			CNN	GRU	BiLSTM	LSTM	TCN
		MAE	0.0049	0.0037	0.0033	0.0035	0.0042
12-Step	(12 hour)	RMSE	0.0698	0.0652	0.0624	0.0655	0.0637
		R^2	0.9614	0.9642	0.9715	0.9611	0.9614
		MAE	0.0049	0.0043	0.0038	0.0043	0.0050
24-Step (1 day)		RMSE	0.0692	0.0672	0.0610	0.0648	0.0698
		R^2	0.9455	0.9524	0.9651	0.9455	0.9482

Table 4. Performance evaluation of hybrid models (Summer)

Forecasting horizon		Evaluation Index	Forecasting Models				
			CNN-LSTM	CNN-GRU	CNN-BiLSTM	TCN-LSTM	TCN-GRU
		MAE	0.0033	0.0029	0.0026	0.0028	0.0036
1-Step	(1 hour)	RMSE	0.0559	0.0502	0.0485	0.0532	0.0602
		R^2	0.9711	0.9702	0.9735	0.9765	0.9751
		MAE	0.0038	0.0033	0.0021	0.0030	0.0039
3-Step	(3 hour)	RMSE	0.0552	0.0558	0.0512	0.0523	0.0541
		R^2	0.9635	0.9638	0.9685	0.9721	0.9659
		MAE	0.0032	0.0033	0.0030	0.0031	0.0039
6-Step	(6 hour)	RMSE	0.0611	0.0572	0.0531	0.0552	0.0583
		R^2	0.9612	0.9602	0.9624	0.9645	0.9612
		MAE	0.0051	0.0042	0.0038	0.0039	0.0042
12-Step	(12 hour)	RMSE	0.0698	0.0625	0.0614	0.0645	0.0626
		R^2	0.9594	0.9598	0.9601	0.9618	0.9627
		MAE	0.0054	0.0049	0.0045	0.0047	0.0051
24-Step (1 day)		RMSE	0.0725	0.0716	0.0684	0.0693	0.0709
		R^2	0.9468	0.9542	0.9653	0.9464	0.9487

Table 5. Performance evaluation of basic models (Winter)

Forecasting horizon		Evaluation Index	Forecasting Models				
			CNN	GRU	Bi-LSTM	LSTM	TCN
		MAE	0.0045	0.0042	0.0044	0.0040	0.0044
1-Step	(1 hour)	RMSE	0.0686	0.0679	0.0672	0.0675	0.0678
		R^2	0.9418	0.9462	0.9474	0.9408	0.9437
		MAE	0.0053	0.0052	0.0051	0.0052	0.0055

continued on following page

Table 5. Continued

Forecasting horizon		Evaluation Index	Forecasting Models				
			CNN	GRU	Bi-LSTM	LSTM	TCN
3-Step	(3 hour)	RMSE	0.0758	0.0751	0.0735	0.0747	0.0761
		R^2	0.9325	0.9354	0.9384	0.9326	0.9331
		MAE	0.0065	0.0062	0.0058	0.0062	0.0064
6-Step	(6 hour)	RMSE	0.0854	0.0834	0.0829	0.0839	0.0844
		R^2	0.9258	0.9259	0.9294	0.9256	0.9257
		MAE	0.0072	0.0076	0.0072	0.0073	0.0075
12-Step	(12 hour)	RMSE	0.0959	0.0958	0.0928	0.0946	0.0953
		R^2	0.9156	0.9154	0.9158	0.9152	0.9153
		MAE	0.0085	0.0084	0.0080	0.0083	0.0081
24-Step (1 day)		RMSE	0.0984	0.0986	0.0949	0.0963	0.0976
		R^2	0.9102	0.9132	0.9135	0.9095	0.9098

Table 6. Performance evaluation of hybrid models (Winter)

Forecasting horizon		Evaluation Index	Forecasting Models				
			CNN-LSTM	CNN-GRU	CNN-BiLSTM	TCN-LSTM	TCN-GRU
		MAE	0.0043	0.0041	0.0042	0.0038	0.0041
1-Step	(1 hour)	RMSE	0.0638	0.0642	0.0635	0.0659	0.0638
		R^2	0.9451	0.9468	0.9495	0.9502	0.9524
		MAE	0.0050	0.0049	0.0051	0.0050	0.0048
3-Step	(3 hour)	RMSE	0.0714	0.0719	0.0702	0.0708	0.0711
		R^2	0.9415	0.9421	0.9454	0.9452	0.9418
		MAE	0.0058	0.0052	0.0050	0.0055	0.0053
6-Step	(6 hour)	RMSE	0.0784	0.0762	0.0748	0.0737	0.0743
		R^2	0.9357	0.9361	0.9375	0.9358	0.9241
		MAE	0.0064	0.0062	0.0059	0.0060	0.0062
12-Step	(12 hour)	RMSE	0.0835	0.0837	0.0802	0.0821	0.0815
		R^2	0.9201	0.9285	0.9305	0.9295	0.9286
		MAE	0.0072	0.0070	0.0068	0.0070	0.0069
24-Step (1 day)		RMSE	0.0921	0.0915	0.0907	0.0921	0.0912
		R^2	0.9157	0.9150	0.9242	0.9156	0.9187

Table 7. Performance evaluation of basic models (Monsoon)

Forecasting horizon		Evaluation Index	Forecasting Models				
			CNN	GRU	Bi-LSTM	LSTM	TCN
		MAE	0.0048	0.0046	0.0042	0.0045	0.0046
1-Step	(1 hour)	RMSE	0.0725	0.0718	0.0711	0.0720	0.0714
		R^2	0.9422	0.9425	0.9434	0.9405	0.9418
		MAE	0.0059	0.0055	0.0052	0.0056	0.0056
3-Step	(3 hour)	RMSE	0.0758	0.0745	0.0738	0.0755	0.0750
		R^2	0.9356	0.9351	0.9359	0.9355	0.9352
		MAE	0.0065	0.0064	0.0058	0.0062	0.0064
6-Step	(6 hour)	RMSE	0.0821	0.0814	0.0805	0.0818	0.0809
		R^2	0.9238	0.9232	0.9355	0.9282	0.9314
		MAE	0.0072	0.0074	0.0069	0.0072	0.0070
12-Step	(12 hour)	RMSE	0.0872	0.0868	0.0855	0.0862	0.0869
		R^2	0.9206	0.9215	0.9312	0.9264	0.9244
		MAE	0.0084	0.0083	0.0078	0.0080	0.0082
24-Step (1 day)		RMSE	0.0915	0.0918	0.0902	0.0924	0.0922
		R^2	0.9154	0.9122	0.9221	0.9185	0.9162

Table 8. Performance evaluation of hybrid models (Monsoon)

Forecasting horizon		Evaluation Index	Forecasting Models				
			CNN-LSTM	CNN-GRU	CNN-BiLSTM	TCN-LSTM	TCN-GRU
		MAE	0.0036	0.0038	0.0034	0.0038	0.0035
1-Step	(1 hour)	RMSE	0.0683	0.0658	0.0624	0.0635	0.0642
		R^2	0.9625	0.9637	0.9649	0.9635	0.9676
		MAE	0.0047	0.0045	0.0041	0.0044	0.0042
3-Step	(3 hour)	RMSE	0.0758	0.0734	0.0712	0.0725	0.0728
		R^2	0.9564	0.9572	0.9601	0.9597	0.9578
		MAE	0.0058	0.0055	0.0049	0.0053	0.0056
6-Step	(6 hour)	RMSE	0.0866	0.0852	0.0817	0.0842	0.0846
		R^2	0.9428	0.9431	0.9452	0.9438	0.9429
		MAE	0.0065	0.0067	0.0062	0.0063	0.0066
12-Step	(12 hour)	RMSE	0.0871	0.0868	0.0854	0.0853	0.0867
		R^2	0.9325	0.9354	0.9386	0.9348	0.9375
		MAE	0.0076	0.0071	0.0068	0.0075	0.0077
24-Step (1 day)		RMSE	0.0902	0.0895	0.0887	0.0912	0.0908
		R^2	0.9248	0.9237	0.9335	0.9249	0.9257

Figure 8a. Season wise clusters under univariate case: Summer Cluster

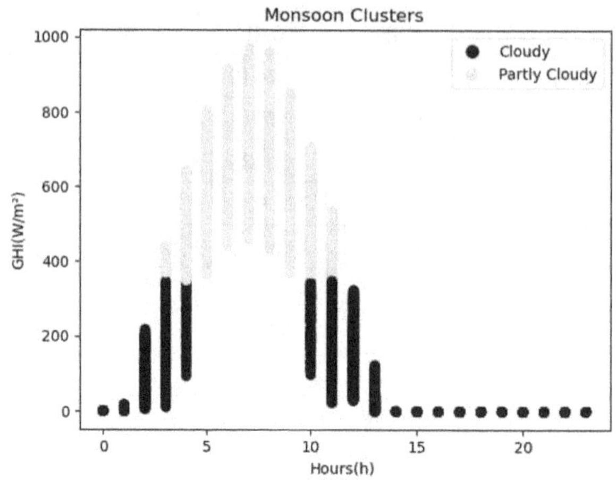

Figure 8b. Season wise clusters under univariate case: Winter Cluster

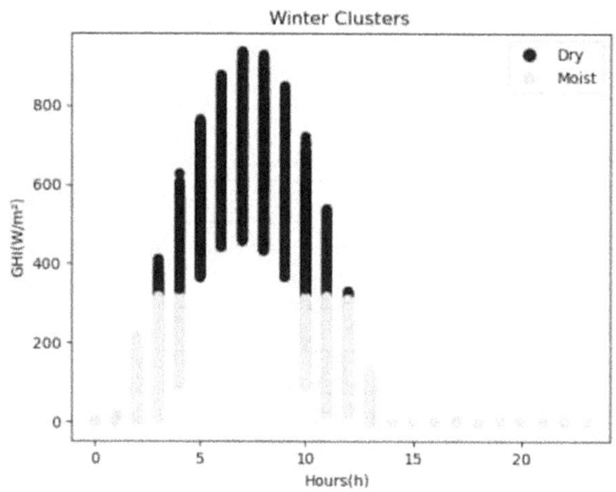

Figure 8c. Season wise clusters under univariate case: Monsoon Cluster

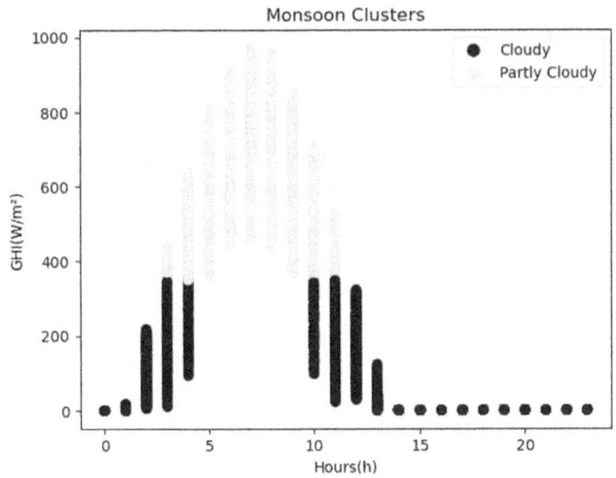

Figure 9a. GHI Predictions of Summer season for 1-h ahead under univariate case: Predictions with basic models for 1-h ahead

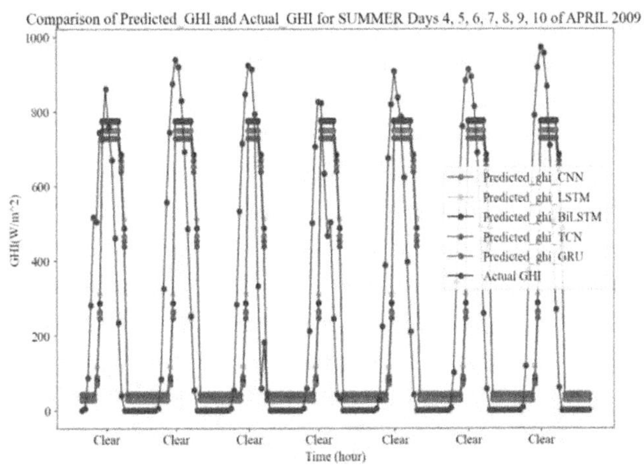

Figure 9b. GHI Predictions of Summer season for 1-h ahead under univariate case: Predictions with hybrid models for 1-h ahead

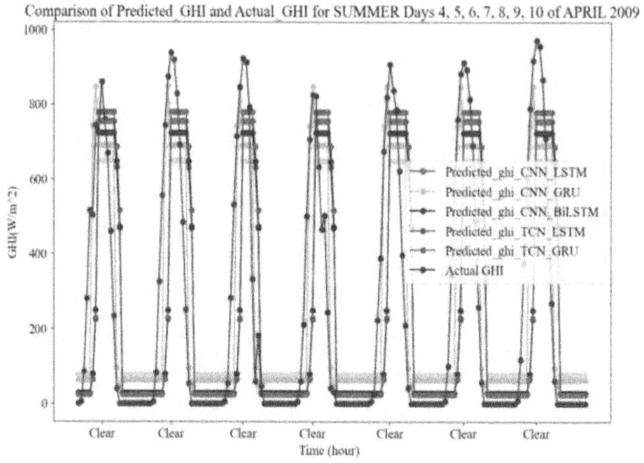

Figure 10a. GHI Predictions of Winter season for 1-h ahead under univariate case: Predictions with basic models for 1-h ahead

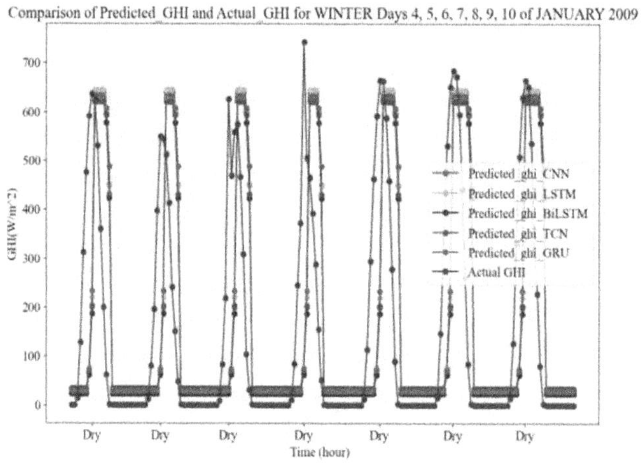

Figure 10b. GHI Predictions of Winter season for 1-h ahead under univariate case: Predictions with hybrid models for 1-h ahead

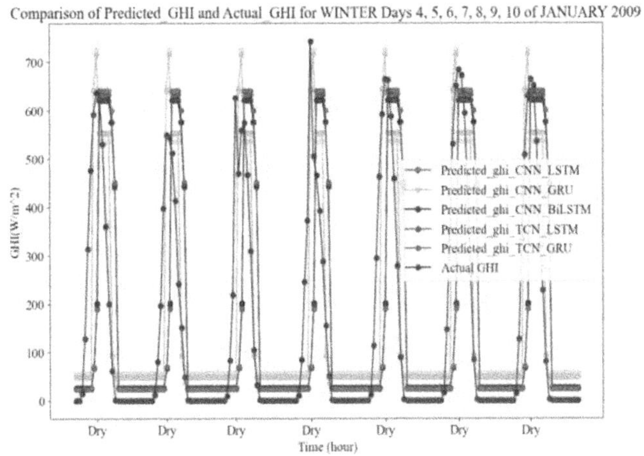

It can be concluded from Table 6 that in case of winter days TCN-GRU has the best performance than other hybrid models, having a R^2 score of 0.9524 for single step ahead (i.e 1 hour) prediction. The R^2 score gradually decreases as the number of steps ahead (from 1 to 24 h) increases. The single step ahead prediction plot for the considered hybrid models can be inferred from Fig. 10b.

It can be observed from Table 7 that in case of monsoon days BiLSTM has the best performance than other basic models, having a R^2 score of 0.9434 for single step ahead (i.e 1 hour) prediction. The R^2 score gradually decreases as the number of steps ahead (from 1 to 24 h) increases. The single step ahead prediction plot for the considered basic models can be inferred from Fig. 11a.

It can be observed from Table 8 that in case of monsoon days TCN-GRU has the best performance than other hybrid models, having a R^2 score of 0.9676 for single step ahead (i.e 1 hour) prediction. The R^2 score gradually decreases as the number of steps ahead (from 1 to 24 h) increases. The single step ahead prediction plot for the considered hybrid models can be inferred from Fig. 11b.

Figure 11a. GHI Predictions of monsoon season for 1-h ahead under univariate case: Predictions with basic models for 1-h ahead

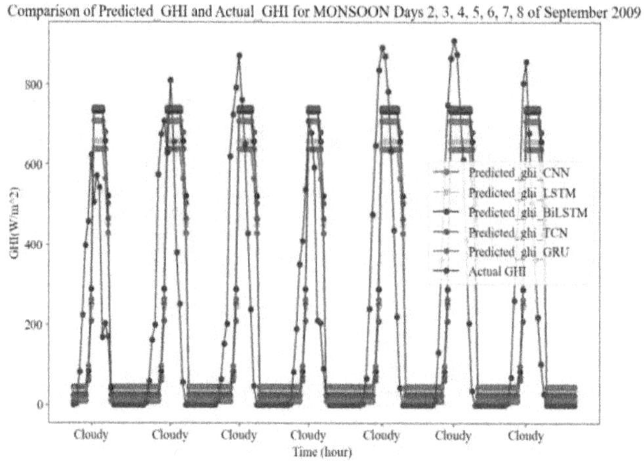

Figure 11b. GHI Predictions of monsoon season for 1-h ahead under univariate case: Predictions with hybrid models for 1-h ahead

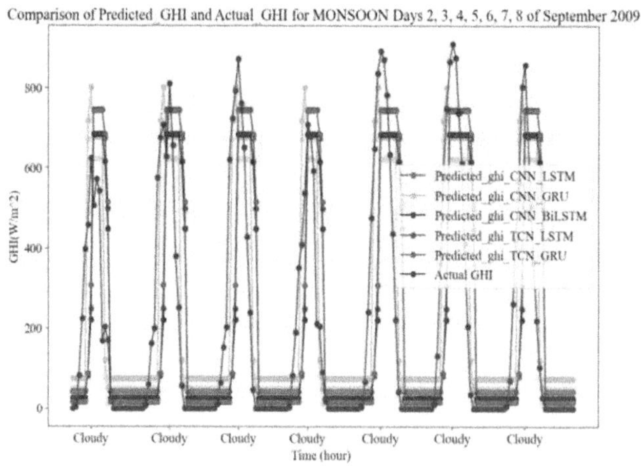

3.5 Multivariate Clustering

Multivariate analysis involves the consideration of multiple variables and their collective impact on solar irradiance. In addition to historical solar irradiance data, multivariate analysis integrates other relevant factors such as meteorological conditions (e.g., temperature, humidity, cloud cover), geographical parameters (e.g., latitude, longitude), and temporal factors (e.g., time of day, season). By incorporating diverse sources of information, multivariate approaches aim to capture the complex interactions influencing solar radiation levels, leading to more comprehensive and accurate forecasting models. In summary, while univariate analysis gives insights into the inherent variability of solar irradiance, multivariate analysis offers a more holistic understanding by accounting for the multifaceted factors influencing solar radiation, ultimately enhancing the accuracy of solar irradiance forecasts.

The study employed distinct weather parameters for seasonal classification: clearsky-ghi and air temperature differentiated clear and non-clear conditions, leading to Summer clusters. Subsequent use of relative humidity and dewpoint temperature refined Winter clusters, distin- guishing moist and dry conditions. Lastly, wind-speed and cloud-opacity discerned cloudy and partly cloudy, categorizing instances into monsoon clusters. The clustering is clearly observed in Fig. 12.

It can be observed from Table 9 that in case of summer days BiLSTM has the best performance than other basic models, having a R^2 score of 0.9785 for single step ahead (i.e 1 hour) prediction. The R^2 score gradually decreases as the number of steps ahead (from 1 to 24 h) increases. The single step ahead prediction plot for the considered models can be inferred from Fig. 13a.

It can be observed from Table 14 that in case of summer days TCN-GRU has the best performance than other hybrid models, having a R^2 score of 0.9789 for single step ahead (i.e 1 h) prediction. The R^2 score gradually decreases as the number of steps ahead (from 1 to 24-h) increases. The single step ahead prediction plot for the considered models can be inferred from Fig. 13b.

It can be observed from Table 11 that in case of winter days LSTM has the best performance than other basic models, having a R^2 score of 0.9630 for single step ahead (i.e 1 hour) prediction. The R^2 score gradually decreases as the number of steps ahead (from 1 to 24 h) increases. The single step ahead prediction plot for the considered models can be inferred from Fig. 14a.

It can be observed from Table 11 that in case of winter days TCN-GRU has the best performance than other hybrid models, having a R^2 score of 0.9623 for single step ahead (i.e 1 hour) prediction. The R^2 score gradually decreases as the number of steps ahead (from 1 to 24 h) increases. The single step ahead prediction plot for the considered models can be inferred from Fig. 14b.

It can be observed from Table 13 that in case of monsoon days TCN has the best performance than other basic models, having a R^2 score of 0.9768 for single step ahead (i.e 1 hour) prediction. The R^2 score gradually decreases as the number of steps ahead (from 1 to 24 h) increases. The single step ahead prediction plot for the considered models can be inferred from Fig. 15a.

It can be observed from Table 14 that in case of monsoon days TCN-GRU has the best performance than other hybrid models, having a R^2 score of 0.9776 for single step ahead (i.e 1 h) prediction. The R^2 score gradually decreases as the number of steps ahead (from 1 to 24 h) increases. The single step ahead prediction plot for the considered models can be inferred from Fig. 15b.

Figure 12a. Season wise Clusters under multivariate case: Summer Cluster

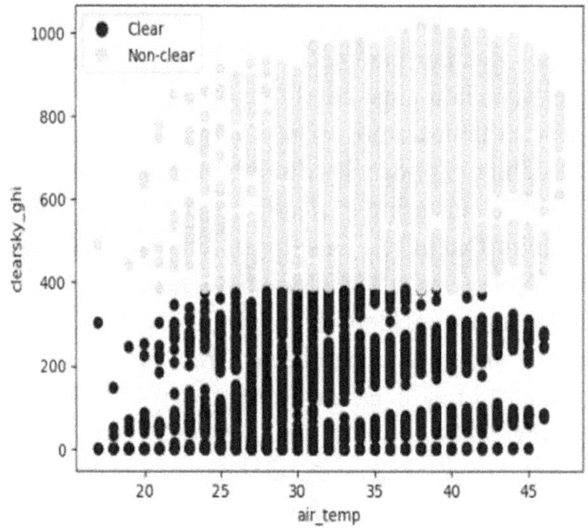

Figure 12b. Season wise Clusters under multivariate case: Winter Cluster

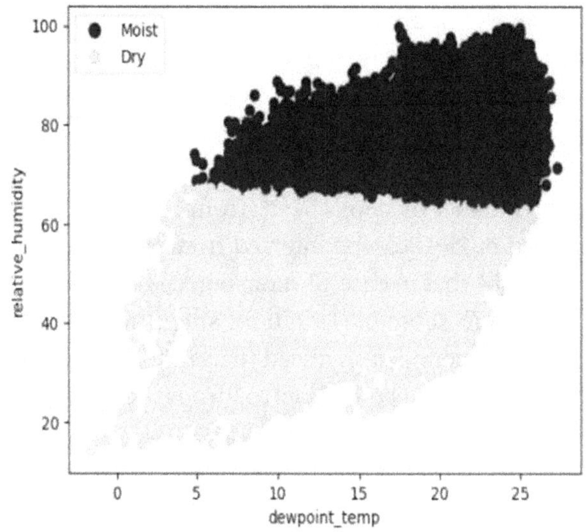

Figure 12c. Season wise Clusters under multivariate case: Monsoon Cluster

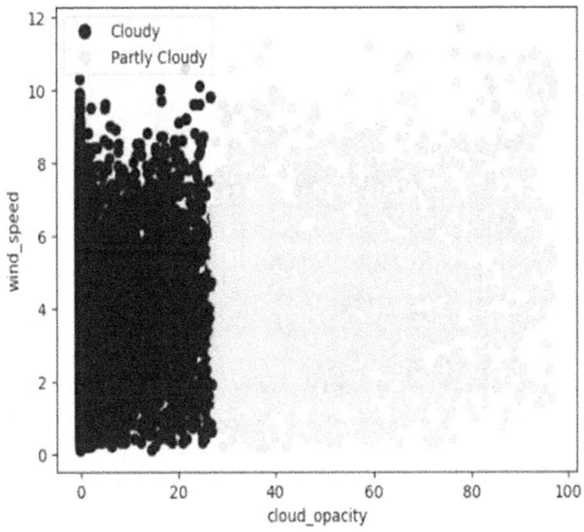

Figure 13a. GHI Predictions of Summer season for 1-h ahead under Multivariate case: Predictions with basic models for 1-h ahead

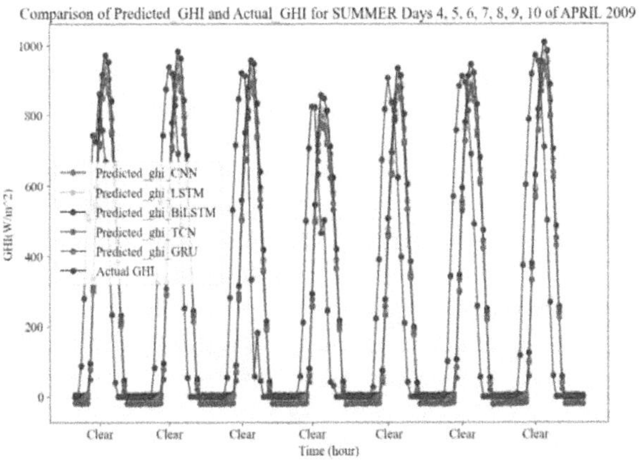

Figure 13b. GHI Predictions of Summer season for 1-h ahead under Multivariate case: Predictions with hybrid models for 1-h ahead

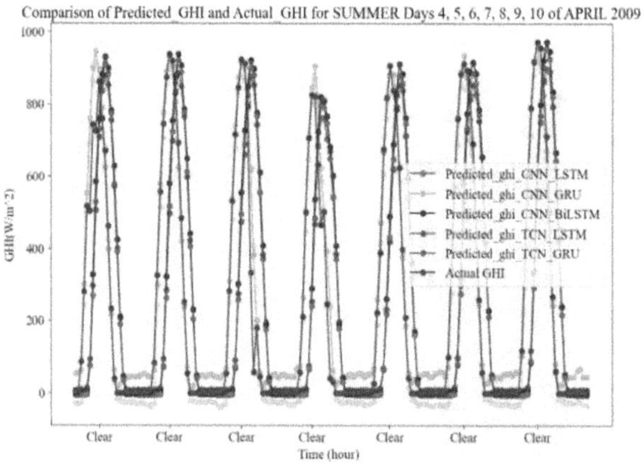

Table 9. Performance evaluation of basic models (Summer)

Forecasting horizon		Evaluation Index	Forecasting Models				
			CNN	GRU	Bi-LSTM	LSTM	TCN
		MAE	0.0030	0.0026	0.0023	0.0029	0.0037
1-Step	(1 hour)	RMSE	0.0552	0.0515	0.0489	0.0543	0.0613
		R^2	0.9727	0.9763	0.9785	0.9736	0.9664
		MAE	0.0030	0.0029	0.0020	0.0032	0.0042
3-Step	(3 hour)	RMSE	0.0554	0.0552	0.0448	0.0562	0.0648
		R^2	0.9725	0.9712	0.9817	0.9712	0.9631
		MAE	0.0035	0.0031	0.0025	0.0035	0.0042
6-Step	(6 hour)	RMSE	0.0598	0.0562	0.0509	0.0592	0.0651
		R^2	0.9680	0.9691	0.9773	0.9692	0.9622
		MAE	0.0042	0.0035	0.0030	0.0039	0.0045
12-Step	(12 hour)	RMSE	0.0651	0.0612	0.0548	0.0652	0.0692
		R^2	0.9652	0.9652	0.9731	0.9612	0.9610
		MAE	0.0045	0.0039	0.0033	0.0041	0.0051
24-Step (1 day)		RMSE	0.0672	0.0692	0.0579	0.0692	0.0712
		R^2	0.9493	0.9575	0.9700	0.9514	0.9513

Table 10. Performance evaluation of hybrid models (Summer)

Forecasting horizon		Evaluation Index	Forecasting Models				
			CNN-LSTM	CNN-GRU	CNN-BiLSTM	TCN-LSTM	TCN-GRU
		MAE	0.0028	0.0026	0.0029	0.0024	0.0026
1-Step	(1 hour)	RMSE	0.0532	0.0510	0.0544	0.0436	0.0515
		R^2	0.9746	0.9767	0.9735	0.9782	0.9789
		MAE	0.0029	0.0038	0.0031	0.0026	0.0029
3-Step	(3 hour)	RMSE	0.0581	0.0616	0.0551	0.0511	0.0535
		R^2	0.9736	0.9660	0.9710	0.9762	0.9710
		MAE	0.0031	0.0041	0.0035	0.0028	0.0031
6-Step	(6 hour)	RMSE	0.0592	0.0656	0.0591	0.0531	0.0565
		R^2	0.9710	0.9610	0.9652	0.9753	0.9692
		MAE	0.0033	0.0047	0.0039	00.0037	0.0038
12-Step	(12 hour)	RMSE	0.0606	0697	0638	0.0644	0.0592
		R^2	0.9682	0.9582	0.9603	0.9630	0.9650
		MAE	0.0039	0.0051	0.0045	0.0045	0.0041
24-Step (1 day)		RMSE	0.0617	0.0712	0.0692	0.0691	0.0615
		R^2	0.9654	0.9498	0.9543	0.9582	0.9602

Figure 14a. GHI Predictions of Winter season for 1-h ahead under Multivariate case: Predictions with basic models for 1-h ahead

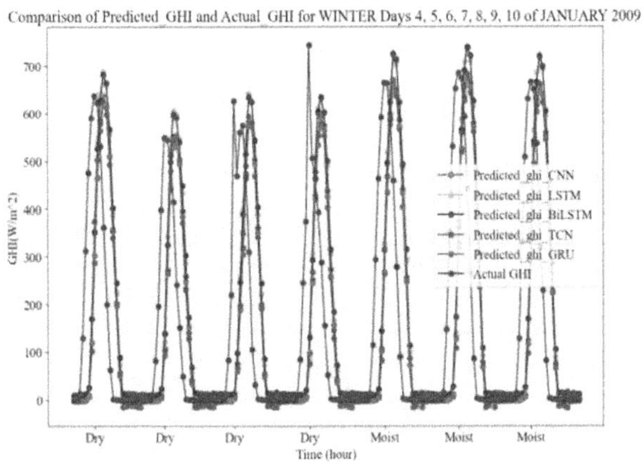

Figure 14b. GHI Predictions of Winter season for 1-h ahead under Multivariate case: Predictions with hybrid models for 1-h ahead

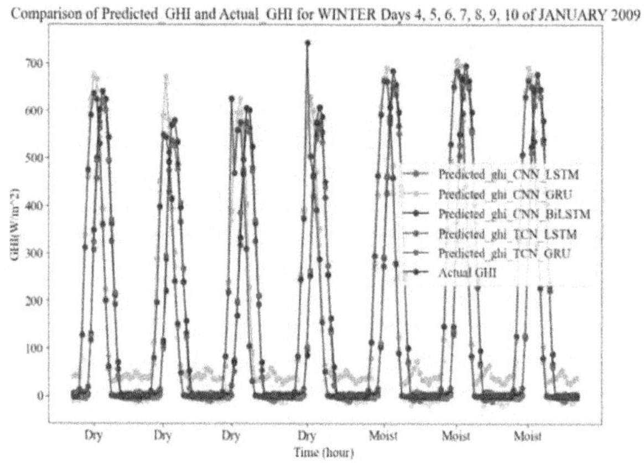

Table 11. Performance evaluation of basic models (Winter)

Forecasting horizon		Evaluation Index	Forecasting Models				
			CNN	GRU	Bi-LSTM	LSTM	TCN
		MAE	0.0040	0.0041	0.0044	0.0038	0.0042
1-Step	(1 hour)	RMSE	0.0635	0.0643	0.0666	0.0619	0.0655
		R^2	0.9531	0.9519	0.9486	0.9630	0.9502
		MAE	0.0040	0.0045	0.0049	0.0039	0.0044
3-Step	(3 hour)	RMSE	0.0637	0.0652	0.0700	0.0625	0.0664
		R^2	0.9529	0.9502	0.9421	0.9535	0.9489
		MAE	0.0041	0.0049	0.0053	0.0040	0.0051
6-Step	(6 hour)	RMSE	0.0646	0.0682	0.0712	0.0639	0.0714
		R^2	0.9508	0.9492	0.9406	0.9526	0.9398
		MAE	0.0048	0.0052	0.0055	0.0040	0.0056
12-Step	(12 hour)	RMSE	0.0699	0.0698	0.0715	0.0639	0.0752
		R^2	0.9423	0.9453	0.9392	0.9518	0.9298
		MAE	0.0052	0.0059	0.0059	0.0043	0.0061
24-Step (1 day)		RMSE	0.0712	0.0792	0.0719	0.0650	0.0792
		R^2	0.9392	0.9423	0.9305	0.9502	0.9301

Table 12. Performance evaluation of hybrid models (Winter)

Forecasting horizon		Evaluation Index	Forecasting Models				
			CNN-LSTM	CNN-GRU	CNN-BiLSTM	TCN-LSTM	TCN-GRU
		MAE	0.0038	0.0040	0.0039	0.0035	0.0039
1-Step	(1 hour)	RMSE	0.0619	0.0636	0.0632	0.0595	0.0630
		R^2	0.9555	0.9531	0.9537	0.9589	0.9623
		MAE	0.0041	0.0042	0.0044	0.0036	0.0034
3-Step	(3 hour)	RMSE	0.0640	0.0648	0.0695	0.0607	0.0587
		R^2	0.9524	0.9513	0.9502	0.9562	0.9589
		MAE	0.0045	0.0046	0.0051	0.0040	0.0036
6-Step	(6 hour)	RMSE	0.0710	0.0698	0.0703	0.0638	0.0600
		R^2	0.9482	0.9483	0.9563	0.9520	0.9575
		MAE	0.0049	0.0049	0.0058	0.0046	0.0036
12-Step	(12 hour)	RMSE	0.0752	0.0702	0.0753	0.0678	0.0600
		R^2	0.9401	0.9456	0.9603	0.9466	0.9575
		MAE	0.0061	0.0052	0.0061	0.0047	0.0036
24-Step (1 day)		RMSE	0.0810	0.0756	0.0715	0.0692	0.0607
		R^2	0.9375	0.9403	0.9475	0.9434	0.9572

Figure 15a. GHI Predictions of Monsoon season for 1-h ahead under Multivariate case: Predictions with basic models for 1-h ahead

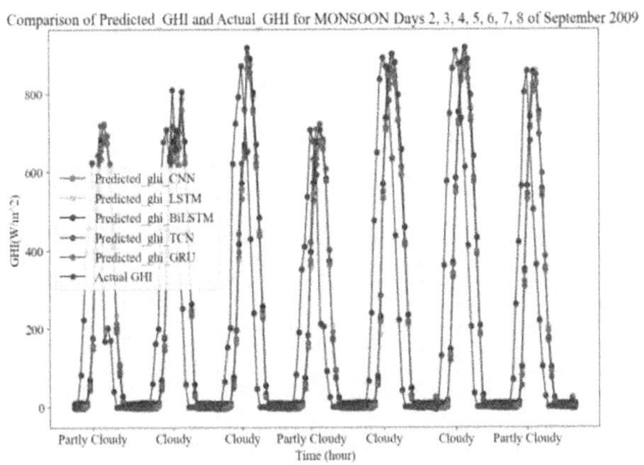

Figure 15b. GHI Predictions of Monsoon season for 1-h ahead under Multivariate case: Predictions with hybrid models for 1-h ahead

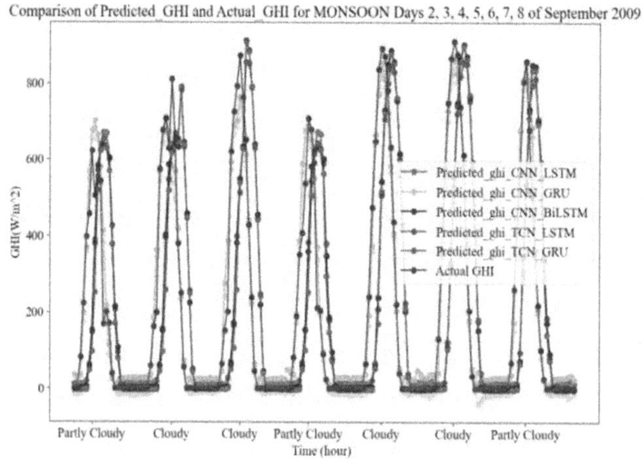

Table 13. Performance evaluation of basic models (Monsoon)

Forecasting horizon		Evaluation Index	Forecasting Models				
			CNN	**GRU**	**Bi-LSTM**	**LSTM**	**TCN**
		MAE	0.0025	0.0024	0.0026	0.0030	0.0022
1-Step	(1 hour)	RMSE	0.0501	0.0493	0.0512	0.0552	0.0479
		R^2	0.9746	0.9754	0.9734	0.9692	0.9768
		MAE	0.0027	0.0024	0.0028	0.0032	0.0024
3-Step	(3 hour)	RMSE	0.0507	0.0494	0.0531	0.0561	0.0499
		R^2	0.9702	0.9753	0.9709	0.9612	0.9748
		MAE	0.0045	0.0024	0.0031	0.0035	0.0049
6-Step	(6 hour)	RMSE	0.0677	0.0498	0.0551	0.0592	0.0703
		R^2	0.9536	0.9745	0.9692	0.9592	0.9499
		MAE	0.0051	0.0026	0.0035	0.0042	0.0049
12-Step	(12 hour)	RMSE	0.0681	0.0523	0.0565	0.0621	0.0723
		R^2	0.9493	0.9692	0.9652	0.9552	0.9480
		MAE	0.0062	0.0025	0.0039	0.0047	0.0052
24-Step (1 day)		RMSE	0.0692	0.0552	0.0692	0.0652	0.0726
		R^2	0.9392	0.9623	0.9614	0.9512	0.9312

Table 14. Performance evaluation of hybrid models (Monsoon)

Forecasting horizon		Evaluation Index	Forecasting Models				
			CNN-LSTM	**CNN-GRU**	**CNN-BiLSTM**	**TCN-LSTM**	**TCN-GRU**
		MAE	0.0023	0.0026	0.0031	0.0027	0.0022
1-Step	(1 hour)	RMSE	0.0487	0.0516	0.0559	0.0521	0.0470
		R^2	0.9760	0.9730	0.9684	0.9725	0.9776
		MAE	0.0024	0.0028	0.0035	0.0030	0.0023
3-Step	(3 hour)	RMSE	0.0494	0.0531	0.0602	0.0550	0.0489
		R^2	0.9752	0.9714	0.9605	0.9694	0.9758
		MAE	0.0025	0.0031	0.0042	0.0033	0.0024
6-Step	(6 hour)	RMSE	0.0510	0.0552	0.0615	0.0562	0.0496
		R^2	0.9712	0.9701	0.9582	0.9652	0.9750
		MAE	0.0025	0.0035	0.0049	0.0039	0.0026
12-Step	(12 hour)	RMSE	0.0552	0.0652	0.0612	0.0610	0.0535
		R^2	0.9692	0.9624	0.9502	0.9592	0.9692
		MAE	0.0026	0.0041	0.0053	0.0045	0.0029
24-Step (1 day)		RMSE	0.0592	0.0692	0.0695	0.0621	0.0592
		R^2	0.9501	0.9551	0.9432	0.9514	0.9612

4. CONCLUSION

The chapter presents a novel methodology for forecasting mean hourly solar radiation on a horizontal surface across multiple time steps. This approach integrates the K-means++ clustering technique with a suite of hybrid deep learning models, applied to both univariate and multi- variate cases. Throughout the analysis, the time series under investigation exhibits notable characteristics, including periodic cycles and intermittent discontinuities.

Our findings underscore the superior predictive capabilities of hybrid deep learning models when compared to their basic counterparts, particularly across diverse seasonal variations. These models demonstrate a remarkable proficiency in capturing the inherent cyclic patterns within the solar radiation data, while effectively accommodating the stochastic fluctuations inherent in the process.

Of particular significance is the comparative assessment of the various models, which consistently identifies the multivariate K-means++ TCN-GRU model as delivering the most accurate predictions across all observed seasons compared to its univariate counterpart models. For instance, in case of multivariate analysis of summer days, TCN-GRU has the best performance than other hybrid models, having a R^2 score of 0.9789, MAE of 0.0026 W/m^2, and RMSE of 0.0515 W/m^2 for single step ahead (i.e 1 hour) prediction. The R^2 score gradually decreases as the number of steps ahead (from 1 to 24 h) increases.This underscores the robustness and effectiveness of the proposed methodology in forecasting solar irradiation, offering meaningful insights for advancing research and practical applications in renewable energy forecasting.

REFERENCES

Ali-Ou-Salah, H., Oukarfi, B., & Mouhaydine, T. (2022). Short-term solar radiation forecasting using a new seasonal clustering technique and artificial neural network. *International Journal of Green Energy*, *19*(4), 424–434.

Ayodele, T., Ogunjuyigbe, A., Amedu, A., & Munda, J. (2019). Prediction of global solar irradiation using hybridized k-means and support vector regression algorithms. *Renewable Energy Focus*, *29*, 78–93. DOI: 10.1016/j.ref.2019.03.003

Azimi, R., Ghayekhloo, M., & Ghofrani, M. (2016). A hybrid method based on a new clustering technique and multilayer perceptron neural networks for hourly solar radiation forecasting. *Energy Conversion and Management*, *118*, 331–344. DOI: 10.1016/j.enconman.2016.04.009

Chiranjeevi, M., Karlamangal, S., Moger, T., & Jena, D. (2023a). Solar irradiation forecast enhancement using hybrid architecture. In *2023 5th International Conference on Energy, Power and Environment: Towards Flexible Green Energy Technologies (ICEPE)*, pages 1–6. IEEE.

Chiranjeevi, M., Karlamangal, S., Moger, T., & Jena, D. (2023b). Solar irradiation prediction hybrid framework using regularized convolutional bilstm-based autoencoder approach. *IEEE Access : Practical Innovations, Open Solutions*, *11*, 131362–131375.

Chiranjeevi, M., Madyastha, A., Maurya, A. K., Moger, T., & Jena, D. (2024). Solar irra- diation forecast enhancement using clustering based cnn-bilstm-attention hybrid architecture with pso. *International Journal of Ambient Energy*, *45*(1), 2414924.

Elmousaid, R., Drioui, N., Elgouri, R., Agueny, H., & Adnani, Y. (2024). Ultra-short-term global horizontal irradiance forecasting based on a novel and hybrid gru-tcn model. *Results in Engineering*, *23*, 102817.

Feng, C., Cui, M., Hodge, B.-M., Lu, S., Hamann, H. F., & Zhang, J. (2018). Unsuper- vised clustering-based short-term solar forecasting. *IEEE Transactions on Sustainable Energy*, *10*(4), 2174–2185. DOI: 10.1109/TSTE.2018.2881531

Ghayekhloo, M., Ghofrani, M., Menhaj, M., & Azimi, R. (2015). A novel clustering approach for short-term solar radiation forecasting. *Solar Energy*, *122*, 1371–1383. DOI: 10.1016/j.solener.2015.10.053

Ghofrani, M., Ghayekhloo, M., & Azimi, R. (2016). A novel soft computing framework for solar radiation forecasting. *Applied Soft Computing*, *48*, 207–216. DOI: 10.1016/j.asoc.2016.07.022

Hendrikx, N., Barhmi, K., Visser, L., de Bruin, T., P'o, M., Salah, A., & van Sark, W. (2024). All sky imaging-based short-term solar irradiance forecasting with long short-term memory networks. *Solar Energy*, *272*, 112463. DOI: 10.1016/j.solener.2024.112463

Ibrahim, M. S., Gharghory, S. M., & Kamal, H. A. (2024). A hybrid model of cnn and lstm autoencoder-based short-term pv power generation forecasting. *Electrical Engineering*, *106*(4), 1–17. DOI: 10.1007/s00202-023-02220-8

Jonathan, A. L., Cai, D., Ukwuoma, C. C., Nkou, N. J. J., Huang, Q., & Bamisile, O. (2024). A radiant shift: Attention-embedded cnns for accurate solar irradiance forecasting and prediction from sky images. *Renewable Energy*, *234*, 121133. DOI: 10.1016/j.renene.2024.121133

Karpagam, K. (2022). An improved k-means clustering algorithm for pattern discovery in data mining. *Journal of Positive School Psychology*, ●●●, 2387–2395.

Liao, Z., & Coimbra, C. F. (2024). Hybrid solar irradiance nowcasting and forecasting with the scope method and convolutional neural networks. *Renewable Energy*, *232*, 121055. DOI: 10.1016/j.renene.2024.121055

Qu, J., Qian, Z., & Pei, Y. (2021). Day-ahead hourly photovoltaic power forecasting using attention-based cnn-lstm neural network embedded with multiple relevant and target variables prediction pattern. *Energy*, *232*, 120996. DOI: 10.1016/j.energy.2021.120996

Solcast (2024). https://solcast.com/. Accessed: October 29, 2024.

Sun, S., Wang, S., Zhang, G., & Zheng, J. (2018). A decomposition-clustering-ensemble learning approach for solar radiation forecasting. *Solar Energy*, *163*, 189–199. DOI: 10.1016/j.solener.2018.02.006

Varanasi, J. and Tripathi, M. (2019). K-means clustering based photo voltaic power forecasting using artificial neural network, particle swarm optimization and support vector regression. *Journal of information and optimization sciences*, 40(2):309–328.

Velasco, L. C. P., Estoperez, N. R., Jayson, R. J. R., Sabijon, C. J. T., & Sayles, V. C. (2018). Day-ahead base, intermediate, and peak load forecasting using k-means and artificial neural networks. *International Journal of Advanced Computer Science and Applications*, *9*(2).

Wu, Y.-K., Phan, Q.-T., & Zhong, Y.-J. (2023). Overview of day-ahead solar power forecasts based on weather classifications and a case study in taiwan. *IEEE Transactions on Industry Applications*.

Xu, Y., Zheng, S., Zhu, Q., Wong, K., Wang, X., & Lin, Q. (2024). A complementary fused method using gru and xgboost models for long-term solar energy hourly forecasting. *Expert Systems with Applications*, *254*, 124286. DOI: 10.1016/j.eswa.2024.124286

Zhang, Z., Wang, C., Peng, X., Qin, H., Lv, H., Fu, J., & Wang, H. (2021). Solar radiation intensity probabilistic forecasting based on k-means time series clustering and gaussian process regression. *IEEE Access : Practical Innovations, Open Solutions*, *9*, 89079–89092. DOI: 10.1109/ACCESS.2021.3077475

Chapter 8
A Power Management Strategy for a Wind Energy Conversion System and Plug-In Electric Vehicles in Residential Applications

Siwar Khemakhem
Faculty of Sciences of Gafsa, Tunisia

Mouna Rekik
https://orcid.org/0000-0003-1869-2394
National Engineering School of Sfax, Tunisia

Lotfi Krichen
National Engineering School of Sfax, Tunisia

ABSTRACT

Green energy systems offer a promising solution to meet the increasing energy demands. In this paper, we introduce a supervisory approach for intelligent household power control. This approach integrates Plug-in Electric Vehicles (PEVs) with Lithium-ion battery packs and wind energy systems, linking them to the smart grid. The key objective is to investigate a power management strategy that utilizes a Wind turbine Generator system (WG) as a renewable energy source and PEVs not only as energy storage but also as an emergency power supply during peak demand periods. By smoothing the household power curve, this approach improves grid stability by filling valleys during off-peak times and shaving peaks during rush periods. The strategy adjusts power demand based on PEV storage levels and WG power generation, ensuring optimal utilization. Simulation results across nine different residential load scenarios demonstrate the effectiveness of this strategy in achieving a balance between demand and supply, fostering harmony within the system.

DOI: 10.4018/979-8-3693-9924-8.ch008

INTRODUCTION

In today's era of globalization and technological advancement, energy has become an indispensable necessity. The escalating demand for electricity underscores the urgent need to address the decline in conventional energy sources, particularly due to their adverse environmental effects. Concerns over the detrimental impacts of fossil fuels, petroleum products and natural gas have spurred significant endeavours to transition away from reliance on these resources for electricity generation. Instead, there is a growing emphasis on embracing new, innovative alternatives (Enderami et al., 2023; Cyril et al., 2024). Non-conventional options incorporating wind energy and Plug-in Electric Vehicles (PEVs) have emerged as promising solutions that are not only environmentally friendly but also viable for long-term energy production (Kazemi et al., 2022; Huo et al., 2024; AlHajri et al., 2021; Hashemi et al., 2023). The main benefits lie in providing a renewable energy source and reducing emissions of greenhouse gases. Furthermore, the widespread adoption of these modern production systems, which have seen rapid growth in recent years, focuses primarily on environmental friendliness and energy efficiency. They present a new avenue for mitigating the challenges posed by energy crises, leading to significant savings for households on fuel costs and garnering increased interest from researchers (Khemakhem et al., 2019; Lin et al., 2018; Salah et al., 2019). New strategies have been investigated to enhance the integration of these environmentally friendly sources into the smart grid infrastructure, ensuring a seamless household power load profile and enabling efficient smart grid operation. This becomes particularly critical with the deployment of bidirectional information and communication technologies, such as advanced metering infrastructure (AMI) and advanced sensing technologies (Mitra et al., 2024; Liu et al., 2020; Munoz et al., 2024).

Certainly, this evolving bidirectional flow of information and energy is designed to effectively manage daily power loads by leveraging data on individual power production and consumption from each participant in the smart grid. Its aim is to harmonize the balance between energy demand and supply, ensuring an optimal energy equilibrium (Mbungu et al., 2024; Mbungu et al., 2020). Expanding on these principles, specific researchers (Palma-Behnke et al., 2013) have devised a rolling horizon strategy for an energy management system focused on a renewable microgrid. This microgrid incorporates two wind turbines, photovoltaic panels (PV), a diesel power system, and a storage unit. Other researchers, as indicated in (Wang et al., 2016), have investigated how clients' response capabilities affect the economic incorporation of photovoltaic power in smart houses, along with the influence of PV and battery status on energy expenditures employing the non-cooperation game theory-based energy market complementarity approach. Studies in (Verma et al., 2015; Khemakhem et al., 2017) have focused on the rapid adoption of plug-in electric vehicles (PEVs) within residential settings, capitalizing on their storage capacities and power flexibility. Likewise, (Aghaei et al., 2016) have delved into the significant role of a large number of plug-in hybrid electric vehicles (PHEVs)/PEVs, which serve not only as transportation options but also as manageable source for the electric system. This research addresses main challenges and examines their capacity to store and/or provide energy from/into the power grid. Additionally, in (Zhenga et al., 2019), the concept of vehicle-to-grid (V2G) is examined as an economical and cost-effective option for PEVs into the electrical grid. The study delves into the current state of related fields, exploring the energy interactions between PEVs and the electrical system, as well as the incorporating of V2G scheduling. The authors specifically address technical limitations, grid management, and PEV battery degradation for various power transactions scenarios, considering both centralized and decentralized supervising methodologies. Moreover, within the same domain, vehicle-to-home (V2H) technology is discussed as a scaled-down version of V2G operation, enabling PEVs parked at residences to supply energy from their battery storage systems (Ou et al., 2018). As well, (Yang et al., 2017) have highlighted the development of a mixed integer-programming method to optimize residential energy management, considering the existence of smart domestic devices and PEVs. This model aligns with the demand-side management strategy to achieve the most favourable PEV integration in residential buildings. Similarly, research in (Zhang et al., 2015) has focused on developing a predictive control procedure aimed at minimizing operational costs by analyzing load demand uncertainties and storage influences in residential applications. In (Jian et al., 2013) others have developed an algorithm aimed at reducing household load power variance by leveraging the charging/discharging profiles of PEVs in a smart micro-grid. Furthermore, to minimize charging costs and achieve valley filling objectives, a novel electric vehicle (EV) charging method is investigated in (Zhao et al., 2019), integrating three charging modes within a single framework (quick charging, slow charging and battery swapping). This approach evaluates the efficacy of a distributed neurodynamic strategy to determine an optimal approach. To reach the same target, in (Sheikhi et al., 2013), an innovative practical Plug in Hybrid Electric Vehicles (PHEV) charging scheduling system is proposed to enhance clients' charging price, taking into account both generation capacity constraints and electricity prices throughout various periods of the day. This process incorporates a stochastic method for determining charging start times and durations within the optimization procedure, aiming to provide a practical approach for modelling vehicle owners' charging behaviour while also facilitating peak load demand shaving. The research described in (Wu et al., 2016) focuses on implementing stochastic power management in a smart house equipped with PEV storage system and photovoltaic panels (PV), aiming to reduce a client's energy costs based on time-of-use tariffs. This involves meeting both household demand and PEV power charging constraints, while also addressing the fluctuation of solar power generation. The authors of this study have developed random-variable models using a Markov Chain approach of PEV mobility and predictive models for household energy demand and photovoltaic energy delivery. On one hand, they establish a mathematical formulation for a stochastic control challenge to regulate the electricity flow between energy suppliers in the smart home. On the other hand, they evaluate the effectiveness of this control approach based on time-varying electricity prices. (Wu et al., 2017) have concentrated on enhancing energy management and sizing smart home elements, which include home batteries, PEVs, and PV systems, with the goal of improving household economy. A convex programming framework is conceived to efficiently deal with the control strategy and optimize various characteristics of the residential battery energy storage unit, according to a smart home nano grid system model. This investigation encompasses the optimization of diverse time horizons, battery energy storage costs, categories and switching modes of PEVs, house energy storage parameters, and electric costs.

All of the aforementioned studies have collectively contributed to a comprehensive approach for scheduling PEV charging, introducing a novel concept to refine the power profile of smart home energy consumption, enhancing power system quality, reducing greenhouse gas emissions, and minimizing energy loss in the residential sector. Briefly, managing energy stability in a residential smart grid requires the adoption of various approaches. However, while renewable resources are crucial for improving energy efficiency, their power intermittency and susceptibility to weather conditions pose significant challenges (Pourbehzadi et al., 2019; Wang et al., 2019).

To the best of our knowledge, it is noteworthy to mention that the recent advancements in the integration of two-way vehicle-to-home (V2H) and home-to-vehicle (H2V) capabilities of PEVs offer the potential to mitigate peak demand intensity and achieve a more balanced energy distribution.

In this context, our study aims to accomplish corresponding goals through the introduction of an innovative power management strategy, which integrates a Wind turbine Generator system (WG) with Plug-in Electric Vehicles (PEVs) within domestic environments. The principal contribution of this research lies in establishing a dynamic and adaptable relationship between WG and PEVs, facilitating bidirectional power exchange, smart home interaction, and seamless integration with the smart grid. This study focuses on a smart home endowed with a WG conversion system, accommodating a variable number of PEVs parked at home, capable of participating in both V2H and H2V operations, while effectively managing their charging and discharging power. A model for both WG and PEVs is implemented to facilitate bidirectional power exchange with the smart home, considering factors such as the household base energy consumption, PEV requirements and charging, state of charge (SOC), and battery model and capacity. The attained results highlight its usefulness in mitigating peak demands, minimizing fluctuations in smart home load demand, enhancing power system functionality, and ensuring stability to thwart electrical grid blackouts.

The remainder of this study is arranged as follows: Section. **2** outlines the structure of the studied system and details the modelling of the WG and PEVs. Section. **3** elucidates the suggested power supervision strategy aimed at achieving smooth power load profile. To validate the effectiveness of this control algorithm, simulation results are highlighted and discussed in Section. **4.** Finally, Section. **5** presents the conclusions.

STUTIED SYSTEM CONFIGURATION

In this study, we focus on an intelligent home characterized by a unique power demand profile. It is equipped with a hybrid green energy production system comprising a WG energy conversion system and PEVs. These PEVs serve dual roles, acting not only as storage units but also as power suppliers. Furthermore, the smart home incorporates an array of smart domestic devices, including AC-powered appliances such as TVs, lights, microwaves, refrigerators, washing machines, computers, and air conditioners. Additionally, it features DC-powered devices like fire alarms, exhaust fans, and LED lighting. To ensure seamless operation, the setup includes AC-DC power converters, a smart meter, and smart sockets, as depicted in Figure 1.

Figure 1. General studied system configuration.

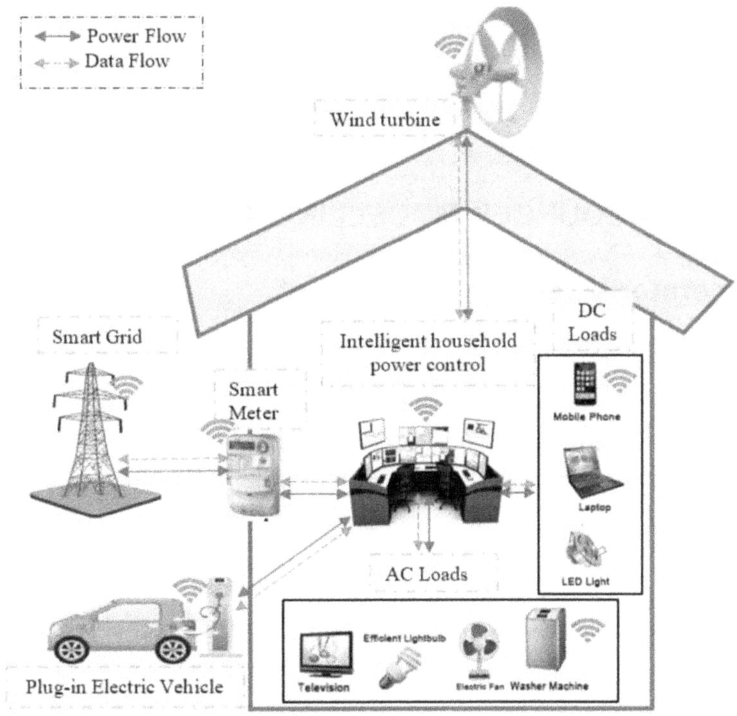

The cooperation between the WG and PEVs systems facilitates bi-directional energy and information flow, aiming to achieve optimal power transmission in the residential sector. This collaboration aims to smooth peak load demands, maintain a flattened load power profile to the best extent possible, achieve a balance between energy demand and supply, and enhance overall system stability within the V2H scheme and therefore Vehicle-to-Grid (V2G) concept.

In this context, the smart home is fortified with a smart meter to gather crucial information regarding daily household power consumption, WG power production, and PEV connectivity details such as PEV identity "PEV_i", type "PEV_{typ}", battery specifications (capacity "BAT_{cap}", battery SOC lowest restrict "SOC_{low}", battery SOC highest restrict "SOC_{high}"), connection "T_{con}"/disconnection "T_{disc}" times, and initial battery SOC "SOC_{int}". To ensure a flattened power consumption through the day, clients require data about energy demand, supply, smart socket, energy exchange, and smart meter usage. The installed smart meters facilitate the transmission of data between these units, receive real-time instructions, and relay commands to the control unit (Jin & Sanders, 2024; Cominola et al., 2015; Rafiq et al., 2023). Indeed, the proposed supervisory strategy enables the control of household power consumption, achieves power optimization, and demonstrates terminal intelligence, effectively participating in the V2H and H2V procedures within the smart city concept.

The WG utilizes a Permanent Magnetic Synchronous Generator (PMSG), with a static AC/DC converter established at the PMSG to control the dynamic speed operation of the WG and achieve Maximum Power Point Tracking (MPPT) for wind speed optimization. For managing PEV power during charging or discharging processes, a bidirectional buck-boost converter is used to handle power shortages or surpluses in demand for the studied home. Additionally, it adjusts the DC bus voltage when the PEV is parked. A two DC/AC converter setup is utilized to transmit power produced by both the WG and PEV unit through a line equipped with an RL filter. This later aims to mitigate harmonics emitted by the inverter, enhancing system efficiency. Such a configuration provides increased flexibility for both the power conversion system and the power charging/discharging process.

POWER GENERATION SOURCES MODELLING AND CONTROL

Wind turbine Generator Modelling and control

The WG is linked to the PMSG, converting wind into energy. The variable speed wind turbine operates via an AC/DC converter control. This process entails regulating the speed of the PMSG to match the electromagnetic torque with the reference rate, optimizing power extraction (Rekik et al., 2015; Abdelkafi &Krichen, 2014). The aerodynamic power applied to the turbine rotor is defined as follows:

$$P_{aer} = 0.5\rho\pi R^2 V^3 C_P \tag{1}$$

The Park reference model defines the expressions for the PMSG as follows (Rekik et al., 2015; Abdelkafi &Krichen, 2014):

$$V_{sd} = R_s I_{sd} + L_s \frac{dI_{sd}}{dt} - p\,\Omega\, L_s I_{sq} \tag{2}$$

$$V_{sq} = R_s I_{sq} + L_s \frac{dI_{sq}}{dt} + p\,\Omega\, L_s I_{sd} + p\,\Omega\,\phi_m \tag{3}$$

The expression for the electromagnetic torque is given as follows (Rekik et al., 2015; Abdelkafi &Krichen, 2014):

$$T_{em} = p\,\phi_m I_{sq} \tag{4}$$

This equation highlights the direct relationship between electromagnetic torque control and the current I_{sq}. The electromagnetic torque and current I_{sq} are directly proportional when the flux remains constant. Typically, the current I_{sd} is negligible because torque primarily depends on I_{sq} (Rekik et al., 2015; Abdelkafi &Krichen, 2014):

$$\begin{cases} I_{sd-ref} = 0 \\ I_{sq-ref} = \dfrac{T_{em}}{p\phi_m} \end{cases} \tag{5}$$

In this study, the PMSG is a smooth pole system, necessitating regulation of currents I_{sd} and I_{sq} along the d-q axes. To achieve this, reference voltage outputs V_{sd-reg} and V_{sq-reg} are computed. These voltages are then transformed using inverse Park transformation to obtain the three-phase voltages "V_a", "V_b" and "V_c".

Therefore, three duty cycles are established to produce three Pulse Width Modulated (PWM) signs for controlling the power converter.

The WG and PMSG characteristics are outlined as follows:

Nominal power $3kW$

Number of pole pairs 4

Resistance of the stator 0.82Ω

Inductance of the stator $15.1mH$

Inertia $99\ 10^{-4}Kgm^2$

Friction coefficient $10^{-3}Nms\ rad^{-1}$

Permanent magnetic flux $0.5Wb$

PEV Modelling and Control

PEVs play a crucial role in both smart homes and the broader smart grid context. Traditional PEV models primarily feature expanded battery packs equipped with intelligent charging systems. The modelling of PEVs has focused on their electric components, which not only serve as rechargeable loads to conclude stock energy but also act as power suppliers to introduce energy into the grid as needed. In recent years, Lithium-ion batteries have emerged as the most popular technology for PEV integration due to their ability to store energy effectively, high specific energy and voltage, high energy density, extended lifecycle, lightweight construction, and safety features (Salyer et al., 2021; Hossain et al., 2023; Singh et al., 2023; Abada et al., 2016). Consequently, PEVs in this study are equipped with associated bidirectional converters. The selected model for the simplest Lithium-ion battery packs is based on the Tremblay model, which consists of a voltage source in series with a resistance and is characterized as follows (Tremblay & Dessaint, 2009; Khemakhem et al., 2022):

$$V_{batt} = E_{batt} - R.i \tag{6}$$

The sign of the PEV current indicates whether the battery is undergoing discharging or charging processes. Specifically, a positive battery current denotes discharging, while a negative current signifies the PEV charging action. The equations for battery voltage during charging E_{batt}^{ch} and discharging E_{batt}^{disch} are as follows (Tremblay & Dessaint, 2009; Khemakhem et al., 2022):

$$E_{batt}^{ch} = U_0 - K\frac{Q}{i.t - 0.1Q}i^* - K\frac{Q}{Q - i.t}i.t + A\,e^{-B.i.t}$$

$$E_{batt}^{dich} = U_0 - K\frac{Q}{Q - i.t}i^* - K\frac{Q}{Q - i.t}i.t + A\,e^{-B.i.t} \tag{7}$$

A battery pack consisting of N_S cells in series and N_P cells in parallel is configured to determine the voltage and current requirements (Sadoun et al., 2011):

$$N_S = \frac{U_{DC}}{U_0}$$

$$N_P = \frac{E_{bat}}{\left(E_{cell} - \eta_{E-w}.w_{cell}\right).N_S}$$

With:

$$\eta_{E-w} = \alpha.1.4 \tag{8}$$

The State of Charge (SOC) of the PEV represents its existing state, which can involve discharging, charging, or outstanding in an idle state. It's crucial to maintain the SOC of the lithium-ion battery pack within a defined bounds to prevent deep discharging or full charging, thereby safeguarding battery performance and longevity. Consequently, the SOC is regulated to stay between 20% and 80%, and the initial battery SOC must be specified as indicated in the following equations (Khemakhem et al., 2022):

$$SOC(\%) = SOC_{int}(\%) - 100*\left(\frac{\int idt}{Q}\right)$$

$$SOC_{low} \leq SOC \leq SOC_{high}$$

With:

$$SOC_{low} = 20\%,\ SOC_{high} = 80\% \tag{9}$$

Additionally, the SOC varies taking into account the energy expected for each trip P_{trip}, which is determined by factors such as the driving distance "$AD_{driving}$" and the proficiency of the PEV "η_{PEV}". As well, the power for PEV charging/discharging must adhere to the prescribed minimum "P_{low}" and maximum "P_{hish}" power limits, as outlined below:

$$P_{trip} = \eta_{PEV}.AD_{drivng}$$

$$P_{low} \leq P_{PEV} \leq P_{high} \tag{10}$$

The characteristics of the considered PEV are outlined as follows:

U_0 3.366V
R 0.01Ω
K 0.0076Ω
A 0.26422V
B 26.5487$(Ah)^{-1}$
$\eta_{PEV}\ \frac{1}{6}(KW/Km)$
Battery capacity 14KWh
P_{low}- 2 KW
P_{hihg}- 2 KW

The considered PEV is equipped with a DC/DC static converter to oversee the bidirectional power flows of the PEVs, and a bidirectional DC/AC converter positioned between the DC bus and the RL line. This converter rectifies the AC power transferred to or from a smart grid during charging or discharging modes, respectively. A PI controller is employed to regulate the battery current and establish its reference value of the charge or discharge operation, fixed by the proposed supervisory strategy. The adjustment of the DC/DC converter involves regulating the duty ratio within the PWM control. This action modulates the charging phase as buck switches or the discharging phase as boost controls, thereby adapting the battery output to the suitable inverter input voltage.

Additionally, a PI regulator is employed to adjust the DC bus voltage and to control the voltage of the bidirectional AC/DC converter. This regulator commands the switching of both active and reactive powers and determines whether the converter functions as an inverter or a rectifier during the discharging or charging procedure, respectively. Accordingly, the PEV operates with its reference active power P_{PEV} derived from the developed supervisory strategy, while its reference reactive power Q_{PEV} is establish to zero, ensuring a power factor corresponding to one. As a result, these reference active and reactive powers generate the Park reference currents "i_{td-ref}, i_{tq-ref}", and the operation of the DC/AC converter is facilitated by calculating the reference d-q voltages "V_{md-reg}, i_{mq-reg}". The d-q voltages are then used to create the duty control signals for the DC/AC converter. Furthermore, an RL filter is integrated between the power grid and the DC/AC converter, with two PI regulators utilized to synchronize this line, thereby enhancing current quality. Figure 2 presents a fundamental modelling of the WG and PEV.

PROPOSED INTELLIGENT HOME POWER SUPERVISORY STRATEGY

Achieving an efficient V2H concept hinges greatly on implementing a well-developed control algorithm for managing power from WG system and PEVs located at residences. The primary objective is to optimize the load power profile by effectively performing peak shaving and valley filling. Aligned with this objective, an Intelligent Home Power Supervisory Strategy is elucidated in the flowchart depicted in Figure 3.

Figure 2. WG and PEV fundamental modelling

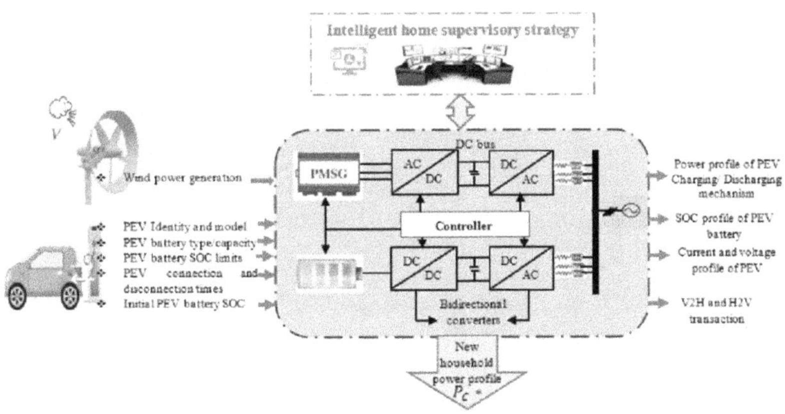

The management of the developed control strategy requires various inputs, including the smart home's power consumption "$P_C(t)$" and its average load demand "$P_A(t)$", the power production of the WG, the total number of PEVs stationed at the studied house "N_{PEV}", each PEV's connection

"$Tcon_{PEVj}$ " and disconnection time "$Tdisc_{PEVj}$ ", the current battery SOC value "SOC_{PEVj}", the higher "SOC_{high}" and the lower "SOC_{low}" battery SOC limits for the PEVs.

This algorithm is proposed to ensure a smooth power curve by utilizing the surplus power from the WG conversion system or by scheduling the PEVs' power charging/discharging to accumulate/supply energy from/to the electrical system as needed. To this end, this control algorithm targets to perform several key functions. Firstly, it detects whether there is an excess or a shortfall in demand. Secondly, it identifies the power generated by the WG. Then, it determines the presence and timing of PEV plug-in and plug-out events at the home. Subsequently, it assesses the availability of WG power to shave peak demand and later considers the accessibility of PEVs to engage in V2H or H2V processes. These processes aim to shave peak loads, principally during high-demand intervals (in cases of surplus in demand), or fill valleys, notably during off-peak demand (in cases of gap in consumption). Moreover, this strategy ensures the supervision of WG power generation and PEV battery SOC control within satisfactory charge and discharge thresholds. This is crucial for optimally managing PEV reference power, whether it's being drawn from the smart home or injected into the electrical grid, and for preventing PEVs from reaching deeply discharged or fully charged states.

Figure 3. Intelligent Home Power Supervisory Strategy Flowchart.

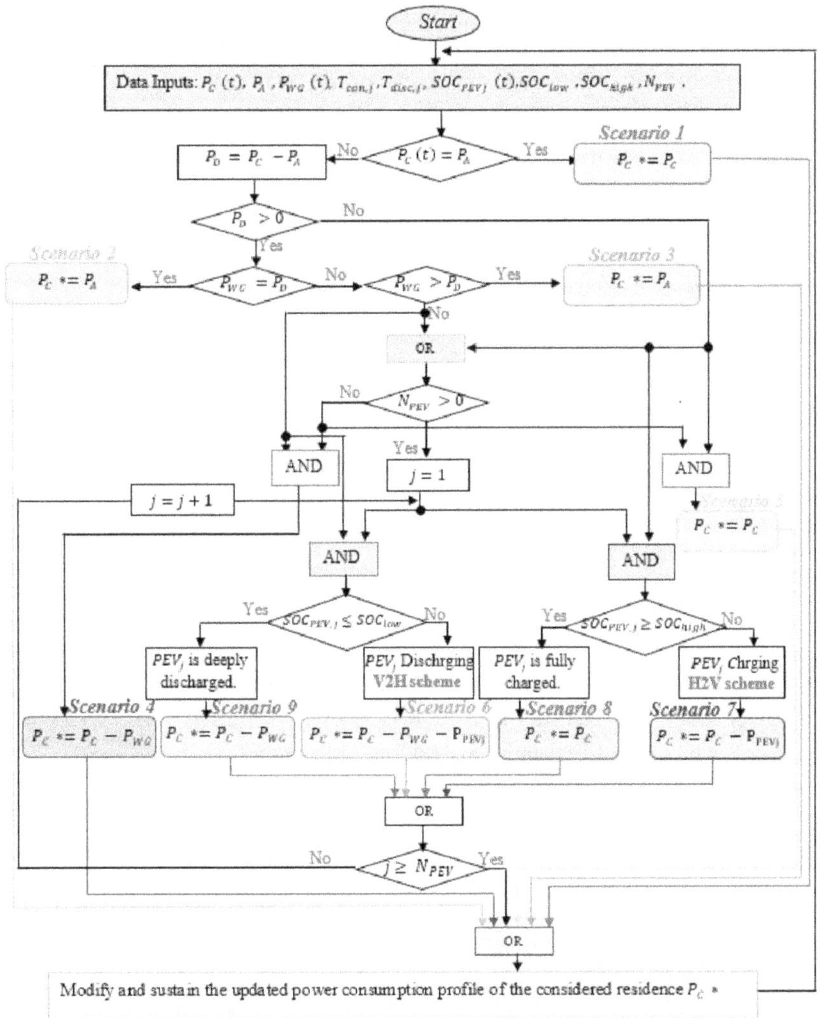

Concisely, the goal is to lessen fluctuations in base household demand and achieve a flattened power profile, thereby enhancing the efficiency and functioning of the main grid. The proposed power management strategy encompasses nine distinct scenarios, each of which is described in detail in the following sub-parts:

Scenario 1

This scenario occurs when the base load consumption "$P_C(t)$" matches its average demand "$P_A(t)$". Consequently, the controlled expression of the load demand "$P_C*(t)$" rests untouched:

$$P_C*(t) = P_C(t) \tag{11}$$

Scenario 2

We express "$P_D(t)$" as the difference in household power consumption between its base level "$P_C(t)$" and its average load "$P_A(t)$". This scenario is shown when the base consumption "$P_C(t)$" exceeds its average power demand "$P_A(t)$" and the WG power production "$P_{WG}(t)$" matches the difference in load power "$P_D(t)$". The daily household regulation will be ensured solely through WG power injection, preventing any excess WG production. Therefore, the updated demand "$P_C*(t)$" is formulated as follows:

$$P_C*(t) = P_A(t) \tag{12}$$

Scenario 3

This operational scenario will be achieved when "$P_C(t)$" surpasses "$P_A(t)$" and "$P_{WG}(t)$" exceeds the difference in load power "$P_D(t)$". In this scenario, there is an excess in consumption, and the WG produces more than necessary. Consequently, "$P_C*(t)$"is determined as follows:

$$P_C*(t) = P_A(t) \tag{13}$$

Scenario 4

This scenario is triggered when "$P_C(t)$" surpasses "$P_A(t)$", signifying a surplus in consumption. Additionally, "$P_{WG}(t)$" is lower than "$P_D(t)$", and there are no PEVs stationed at home "N_{PEV}=0". Consequently, the updated expression for household power "$P_C*(t)$" is as follows:

$$P_C*(t) = P_C(t) - P_{WG}(t) \tag{14}$$

Scenario 5

This operational scenario is activated when "$P_C(t)$" is less than the "$P_A(t)$", indicating a shortfall in demand, and there are no PEVs stationed at home "N_{PEV}=0" Accordingly, the modified expression for the demand power "$P_C*(t)$" remains untouched and is presented as follows:

$$P_C*(t) = P_C(t) \tag{15}$$

Scenario 6

This scenario appears when "$P_C(t)$" exceeds "$P_A(t)$", demonstrating an excess in consumption. Additionally,"$P_{WG}(t)$" is lower than "$P_D(t)$", there is one PEV parked at home "N_{PEV}=1", and the PEV battery SOC is above the low threshold "SOC_{low}". In this situation, the PEV functions to supply necessary power to the grid, a concept known as Vehicle-to-Home (V2H), aimed at reducing peak load in base demand and alleviating stress on the main grid. The updated expression for the demand power "$P_C*(t)$" is described as follows:

$$P_C*(t) = P_C(t) - P_{WG}(t) - P_{PEV}(t) \tag{16}$$

Scenario 7

This scenario is accomplished when there is a PEV present at home "N_{PEV}=1", "$P_C(t)$" is below "$P_A(t)$", signifying a deficit in demand, and the PEV battery SOC is inferior to the high threshold "SOC_{high}". In this context, the PEV aims to address this shortage in consumption by storing suitable power from the home, a concept known as Home-to-Vehicle (H2V). The updated expression for the base demand "$P_C{}^*(t)$" is outlined as follows:

$$P_C{}^*(t) = P_C(t) - P_{PEV}(t) \tag{17}$$

Scenario 8

This scenario is obtained if "$P_C(t)$" is below "$P_A(t)$" and there is one PEV present at home "N_{PEV}=1". This indicates a scarcity in consumption. However, the PEV cannot contribute to the smart home concept in this scenario as it is fully charged, exceeding its maximum battery "SOC_{high}". Consequently, the modified expression for the demand "$P_C{}^*(t)$" remains unchanged as follows:

$P_C{}^*(t) = P_C(t)$ (Eq.18)

Scenario 9

This scenario is operated when "$P_C(t)$" surpasses "$P_A(t)$", "$P_{WG}(t)$" is lower than "$P_D(t)$", and there is one PEV parked at home "N_{PEV}=1". This signifies an excess in consumption, where the wind generator injects all its generated power $P_{WT}(t)$ into the grid. However, in this scenario, the PEV cannot participate in the smart home concept as it is deeply discharged, reaching its minimum state of charge "SOC_{low}". This situation corresponds to the partial load power regulation using only the wind turbine generator. The modified expression for the daily demand power "$P_C{}^*(t)$" is provided as follows:

$$P_C{}^*(t) = P_C(t) - P_{WG}(t) \tag{19}$$

The developed algorithm is convenient for integrating a variable number of PEVs. If all the PEVs within the studied smart home were integrated and adhered to this energy management approach, the home would subsequently update and retain its updated daily load profile "$P_C{}^*(t)$" .

Eventually, this supervisory strategy is practical for application in numerous smart homes introducing a smart city context as an innovative notion in smart grid concept.

SIMULATION RESULTS AND DISCUSSION

To assess the efficacy of the suggested power management approach, we utilize the Matlab-Simulink software as our simulation platform. The simulation is conducted within a smart home area, which is characterized by a predefined daily load demand. This setup includes a WG power conversion system and an adaptive number of PEVs. The aim is to evaluate the effectiveness of the power management strategy in controlling the smart home's power consumption across different operating modes, ultimately aiming for a regulated power profile. In our investigation, we divide the day into 96 equally spaced periods. Specifically, we validate the day-ahead information and simulation outcomes at 15-minute periods. The WG used in our simulation has a rated power output of 3 kW. Additionally, we set the permissible higher and lower battery SOC values to 0.8 and 0.2, respectively.

Originally, both the WG and PEVs are not connected to the existing electrical grid. The household daily power consumption baseline is depicted in Figure 4. It is evident that this curve exhibits numerous spikes in demand, thereby adversely affecting grid stability and power quality. To address these issues, the proposed supervisory approach implemented in both WG and PEVs will be integrated into the power system. The WG and PEV serve as the primary sources in this investigated system.

Figure 4. Household daily power consumption.

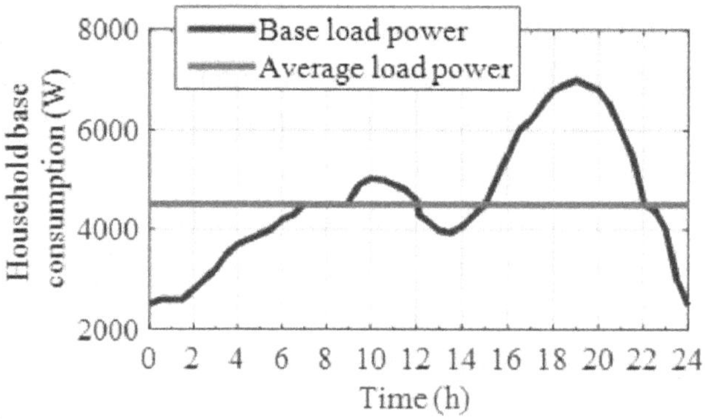

Figure 5 illustrates the wind speed and WG power production curves, confirming the effective operation of the WG. It is noted that the WG achieves its rated power output when it reaches its designated wind speed threshold set at 9 m/s.

Figure 6 illustrates the PEV's parking duration (time of plugging in/plugging out) and its readiness to adopt the smart home concept. It's noteworthy that the PEV is anticipated to commence its parking duration with its starting SOC value to guarantee bidirectional power flow exchange and to conclude its parking period with the necessary SOC. Indeed, the PEV remains disconnected at home during the time interval [7 am, 7 pm].

Figure 5. Operational characteristics of the WG.

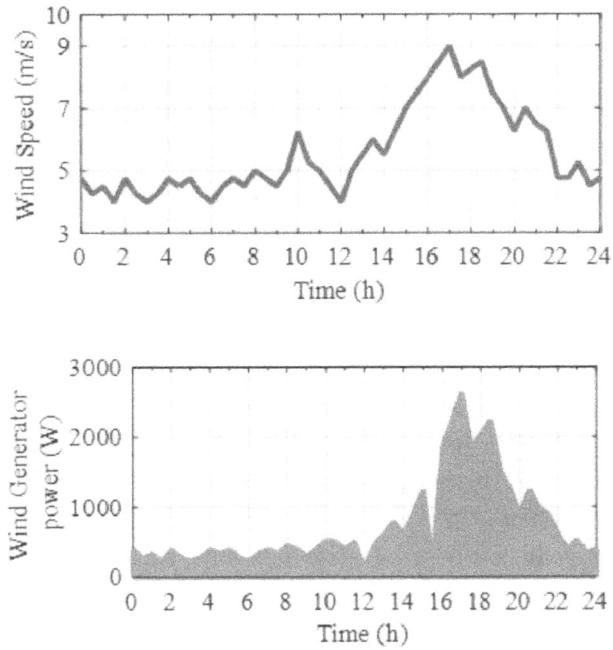

Figure 6. PEV connectivity status.

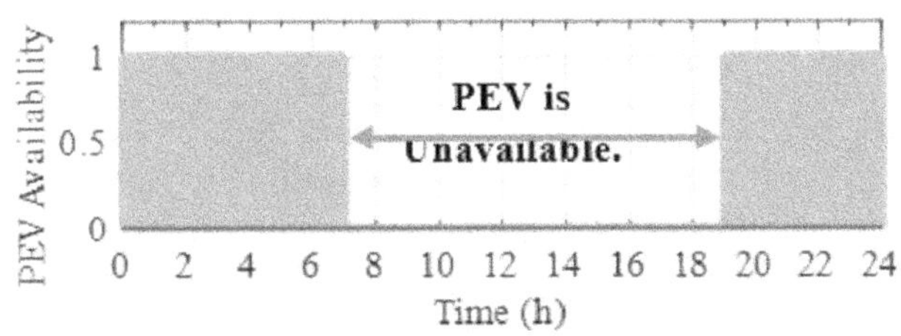

Figure 7 presents the designated PEV reference power profile in accordance with the power management algorithm and its transitions between special modes. It's noticed that negative PEV powers reveal the charging operation, during which the connected PEV has the capacity to retain energy from the grid during off-peak hours when consumption is low (H2V concept). On the other hand, positive PEV powers confirm the discharging procedure, where the PEV aims to supply electrical power to the home during peak hours when consumption is high (V2H concept).

Figure 7. Reference power profile of the studied PEV.

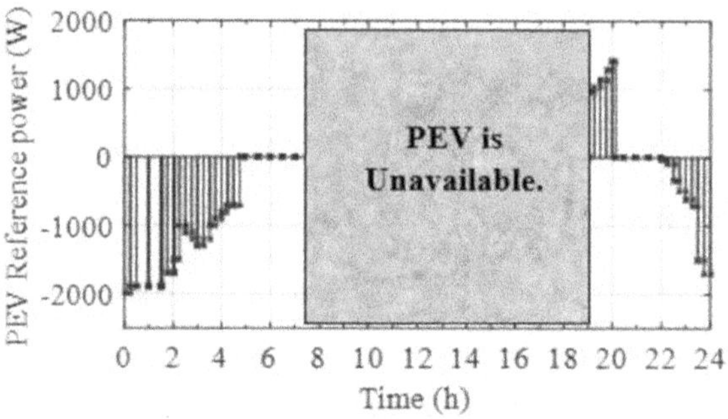

As depicted in Figure 8, the SOC profile for a typical day-ahead scenario of a PEV parked at home is simulated. This confirms that the PEV battery will undergo charging or discharging cycles while its SOC does not attain its highest or lowest thresholds, respectively. Once the PEVs are parked and reach SOC levels exceeding 80% or falling below 20%, there is no electricity transfer occurs between the PEV and the smart home. Consequently, this proposed strategy not only guarantees effective energy management but also boosts both the PEV battery's self-sufficiency and lifespan.

Figure 8. SOC profile of the studied PEV.

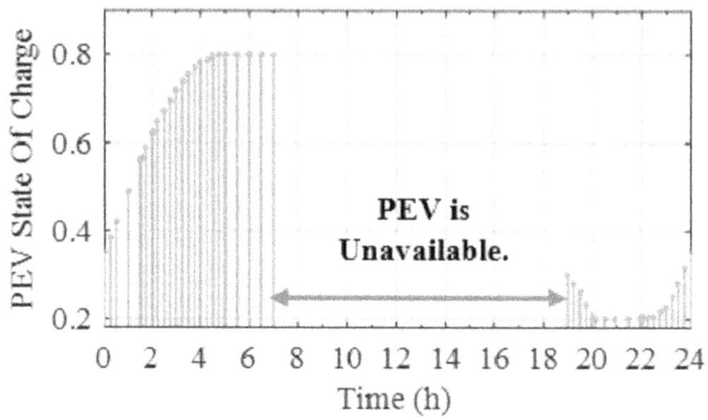

Figure 9 displays the voltage profiles of the PEV battery. It's evident from this figure that the dynamic PEV voltage increases during the H2V concept when the PEV stores power from the electrical grid, decreases during the V2H concept when the PEV injects power into the grid, and remains constant within the upper and lower SOC boundaries when it is not engaged in these concepts. This is demonstrated during the intervals [5 am, 7 pm] and [8 pm, 10 pm].

Figure 9. PEV battery voltage characteristic.

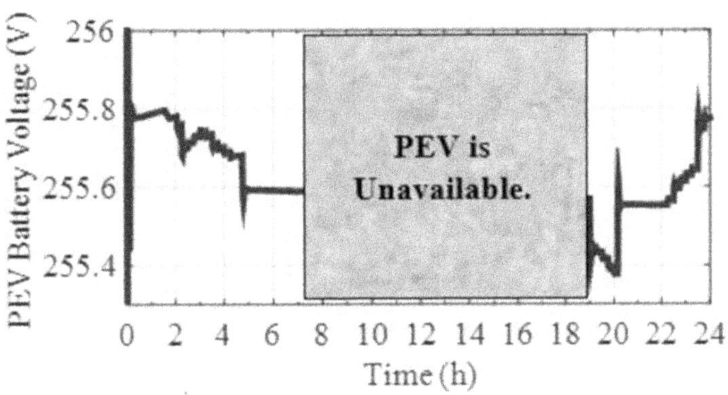

Figure 10 illustrates the curves of PEV battery currents. In Figure 10a, the PEV current profile throughout a single day cycle is depicted, affirming the convention where negative currents represent the charging technique, and positive currents represent the discharging scheme. As observed in Figure 10b, the modulated currents at the DC bus "I_{m-bat} "and "I_{m-inv}, exhibit close similarity, indicating that the current provided by the battery equals the current needed by the grid. Thus, this provides evidence of the effectiveness of DC bus management.

Figure 10. PEV battery currents curves: (a) Battery current, (b) Modulated currents.

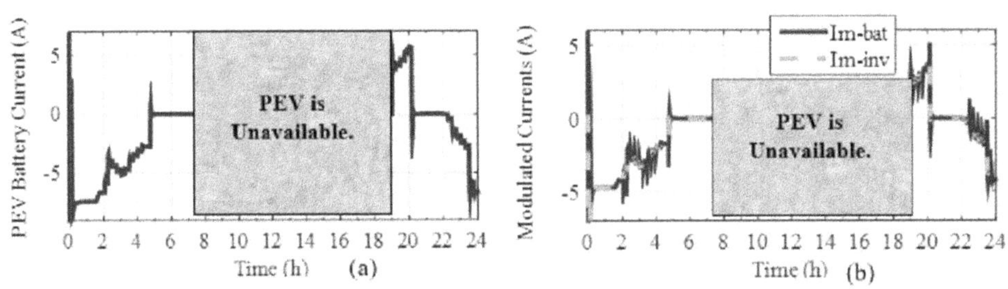

In line with these curves, maintaining control over the DC/DC and DC/AC converters necessitates regulating the DC bus to its reference voltage, "$U_{DC\_ref} = 400V$", with smallest variation following the transitions between various operational scenarios. This underscores the efficacy of the proposed supervisory algorithm, as illustrated in Figure 11.

Figure 11. DC bus voltage proving the control strategy effectiveness.

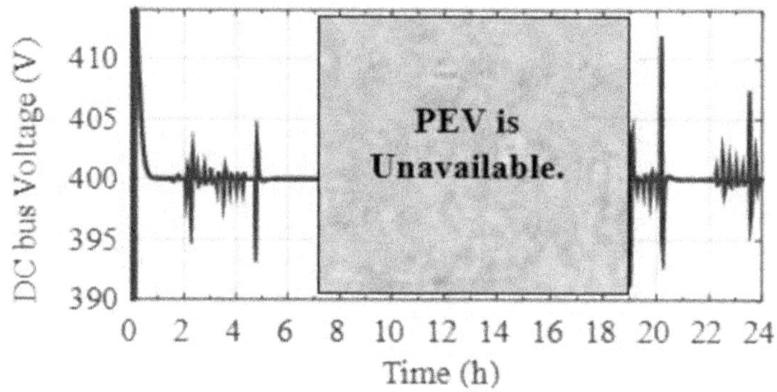

In this study, different scenarios of power consumption of the studied smart home were delineated. Specifically, Figure 12 delves into the transitions among the nine distinct scenarios derived from the "Intelligent Home Power Supervisory Strategy".

Figure 12. Specific transition scenarios.

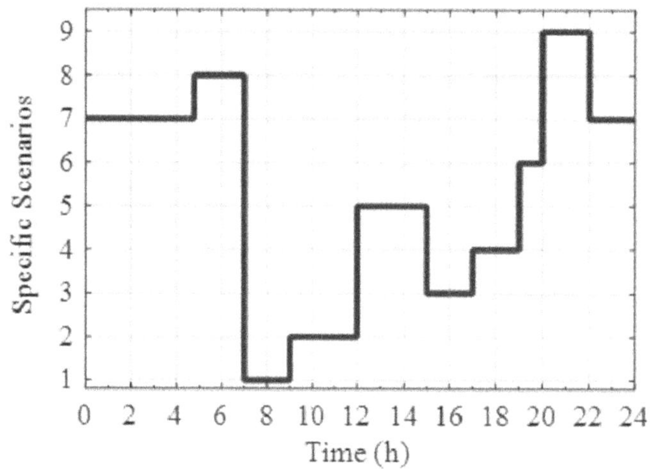

Figure 13 vividly illustrates the efficacy of the power management strategy. The load power demand within the analyzed home shows a significant flattening attributed to the WG power conversion system and the contributions from PEV charging/discharging. Consequently, the valley load of the home is reduced as the PEV absorbs power from the grid. Furthermore, WT injection power helps alleviate peak loads, further moderated by PEV power feedback to the grid. However, specific intervals maintain an unchanged smart home power load profile: [5 am, 7 am] when the PEV is fully charged, [12 pm, 3 pm] when no PEV is connected at home, and [8 pm, 10 pm] when there is no excess WG power production and the PEV parked at home is deeply discharged.

Figure 13. Comparison of the power consumption within the monitored household.

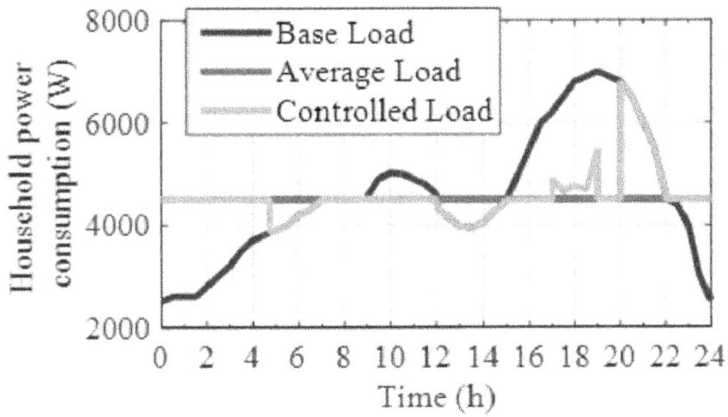

This control strategy is implemented across numerous households within a small smart city, as part of the smart grid technology, incorporating WG as a green renewable resource and large-scale coordinated PEVs. Figure 14 emphasizes the benefits of renewable energy resources coupled with V2H, aiming to swiftly smooth the base power consumption profile. This outcome is geared towards enhancing the smart city stability of while evenly minimizing stress on the power grid, thereby mitigating the risk of power grid blackouts.

Figure 14. Evolution of the power consumption in small smart city.

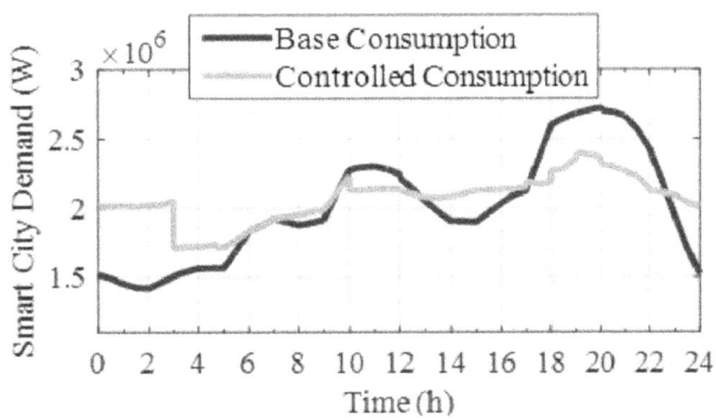

CONCLUSION

This study investigates a novel power supervisory strategy implemented within a smart home-connected framework, involving WG and PEVs to mitigate fluctuations in the overall load profile. The principal objective of this approach is to synchronize WG power generation and PEV charging/discharging with the actual daily consumption patterns of the considered home. This entails addressing various constraints, such as PEV availability, its connection and disconnection times, as well as its initial battery SOC. The main objective is to stock power from the grid during low-demand periods to fill in valleys in the load profile, while also employing WG power injection and/or PEV power feedback during periods of excessive demand to reduce load peaks. The effectiveness of this supervisory strategy was evaluated across nine transition modes for updating smart home load power. The outcomes underscore the efficacy of the proposed power management strategy in moderating severe load peaks, resulting in a flattened load profile. This not only enhances power grid quality and stability but also helps preventing electrical system blackouts. In the future, our attention will be directed towards assessing the investment cost associated with the practical application of the proposed Intelligent Home Power Supervisory Strategy, along with its payback period, as part of an overarching economic investigation. Additionally, we aim to provide a realistic evaluation of the cost of different components within the considered system for further enhancement of economic analysis in future studies.

REFERENCES

Abada, S., Marlair, G., Lecocq, A., Petit, M., Sauvant-Moynot, V., & Huet, F. (2016). Safety focused modeling of lithium-ion batteries: A review. *Journal of Power Sources*, *306*, 178–192. DOI: 10.1016/j.jpowsour.2015.11.100

Abdelkafi, A. & Krichen, L. (2014). Energy management optimization of a hybrid power production unit based renewable energies. International Journal of Electric Power & Energy, 1-9.

Aghaei, J., Nezhad, A. E., Rabiee, A., & Rahimi, E. (2016). Contribution of Plug-in Hybrid Electric Vehicles in power system uncertainty management. *Renewable & Sustainable Energy Reviews*, *59*, 450–458. DOI: 10.1016/j.rser.2015.12.207

AlHajri, I., Ahmadian, A., & Elkamel, A. (2021). Stochastic day-ahead unit commitment scheduling of integrated electricity and gas networks with hydrogen energy storage (HES), plug-in electric vehicles (PEVs) and renewable energies. *Sustainable Cities and Society*, *67*, 102736. DOI: 10.1016/j.scs.2021.102736

Cominola, A., Giuliani, M., Piga, D., Castelletti, A., & Rizzoli, A. E. (2015). Benefits and challenges of using smart meters for advancing residential water demand modeling and management: A review. *Environmental Modelling & Software*, *72*, 198–214. DOI: 10.1016/j.envsoft.2015.07.012

Cyril, A. O., Ujah, C. O., Ekwueme, B. N., & Asadu, C. O. (2024). Photovoltaic mini-grid incorporation: The panacea for electricity crisis in sub-Saharan Africa. *Unconventional Resources*, *4*, 100079. DOI: 10.1016/j.uncres.2024.100079

Enderami, S. A., Sutley, E. J., Mazumder, R. K., & Dumler, M. (2023). Virtual testbeds for community resilience analysis: Step-by-step development procedure and future orientation. *Resilient Cities and Structures*, *2*(2), 42–56. DOI: 10.1016/j.rcns.2023.07.002

Hashemi, B., Taheri, S., Cretu, A. M., & Pouresmaeil, E. (2023). Computational intelligence based PEVs aggregator scheduling with support for photovoltaic power penetrated distribution grid under snow conditions. Electric Power Systems Research Part A, 108922.

Hossain, M. H., Chowdhury, M. A., Hossain, N., Islam, M. A., & Mobark, M. H. (2023). Advances of lithium-ion batteries anode materials—A review. *Chemical Engineering Journal Advances*, *16*, 100569. DOI: 10.1016/j.ceja.2023.100569

Huo, L. F., Wang, Z., Reda, S. A., Tashkandi, M. A., Moria, H., Nag, K., Mouldi, A., Loukil, H., Lee, V. F., & Fang, H. (2024). Power generation enhancement and CO2 emission reduction of a biomass powered gas turbine via hydrogen injection and wind energy incorporation. *Process Safety and Environmental Protection*, *185*, 105–115. DOI: 10.1016/j.psep.2024.01.075

Jian, L., Xue, H., Xu, G., Zhu, X., Zhao, D., & Shao, Z. (2013). Regulated Charging of Plug-in Hybrid Electric Vehicles for Minimizing Load Variance in Household Smart Micro-Grid. *IEEE Transactions on Industrial Electronics*, *60*(8), 3218–3226. DOI: 10.1109/TIE.2012.2198037

Jin, A. S., & Sanders, K. T. (2024). Characterizing residential sector load curves from smart meter datasets. *Applied Energy*, *366*, 123316. DOI: 10.1016/j.apenergy.2024.123316

Kazemi, K., Ebrahimi, M., Lahonian, M., & Babamiri, A. (2022). Micro-scale heat and electricity generation by a hybrid solar collector-chimney, thermoelectric, and wind turbine. *Sustainable Energy Technologies and Assessments*, *53*, 102394. DOI: 10.1016/j.seta.2022.102394

Khemakhem, S., Rekik, M., & Krichen, L. (2017). A flexible control strategy of plug-in electric vehicles operating in seven modes for smoothing load power curves in smart grid. *Energy*, *118*, 197–208. DOI: 10.1016/j.energy.2016.12.039

Khemakhem, S., Rekik, M., & Krichen, L. (2019). Double layer home energy supervision strategies based on demand response and plug-in electric vehicle control for flattening power load curves in a smart grid. *Energy*, *167*, 312–324. DOI: 10.1016/j.energy.2018.10.187

Khemakhem, S., Rekik, M., & Krichen, L. (2022). A Smart Home Power Supervision Strategy Based on Optimal Plug-in Electric Vehicles Integration into Smart City Context for Peak Loads Profiles Flattening. *Arabian Journal for Science and Engineering*, *47*(3), 3383–3396. DOI: 10.1007/s13369-021-06315-x

Lin, H., Liu, Y., Sun, Q., Xiong, R., Li, H., & Wennersten, R. (2018). The Impact of Electric Vehicle Penetration and Charging Patterns on The Management of Energy Hub–A Multi-Agent System Simulation. *Applied Energy*, *230*, 189–206. DOI: 10.1016/j.apenergy.2018.08.083

Liu, H., Zhu, X., Xiao, Z., Wu, Y., Li, P., Li, D., Deng, K., & Liu, S. (2020). The Application of Smart Meter in the Management of Electric Vehicle Charging Facilities. *Procedia Computer Science*, *175*, 774–777. DOI: 10.1016/j.procs.2020.07.115

Mbungu, N. T., Bansal, R. C., Naidoo, R. M., Bettayeb, M., Siti, M. W., & Bipath, M. (2020). A dynamic energy management system using smart metering. *Applied Energy*, *280*, 115990. DOI: 10.1016/j.apenergy.2020.115990

Mbungu, N. T., Bansal, R. C., Naidoo, R. M., Siti, M. W., Ismail, A. A., Elnady, A., Abokhali, A. G., & Hamid, A. K. (2024). Performance analysis of different control models for smart demand–supply energy management system. *Journal of Energy Storage*, *90*, 111809. DOI: 10.1016/j.est.2024.111809

Mitra, S., Chakraborty, B., & Mitra, P. (2024). Smart meter data analytics applications for secure, reliable and robust grid system: Survey and future directions. *Energy*, *289*, 129920. DOI: 10.1016/j.energy.2023.129920

Munoz, O., Ruelas, A., Rosales-Escobedo, P. F., Acuña, A., Suastegui, A., Lara, F., Reyes-Zamora, R. A., & Rocha, A. (2024). Development of an IoT smart energy meter with power quality features for a smart grid architecture. *Sustainable Computing : Informatics and Systems*, *43*, 100990. DOI: 10.1016/j.suscom.2024.100990

Ou, S., Lin, Z., He, X., & Przesmitzki, S. (2018). Estimation of vehicle home parking availability in China and quantification of its potential impacts on plug-in electric vehicle ownership cost. *Transport Policy*, *68*, 107–117.

Palma-Behnke, R., Benavides, C., Lanas, F., Severino, B., Reyes, L., Llanos, J., & Saez, D. (2013). A microgrid energy management system based on the rolling horizon strategy. *IEEE Transactions on Smart Grid*, *4*, 996–1006.

Pourbehzadi, M., Niknam, T., Aghaei, J., Mokryani, G., Shafie-khah, M., & Catalão, J. P. S. (2019). Optimal operation of hybrid AC/DC microgrids under uncertainty of renewable energy resources: A comprehensive review. *International Journal of Electrical Power & Energy Systems*, *109*, 139–159. DOI: 10.1016/j.ijepes.2019.01.025

Rafiq, H., Manandhar, P., Rodriguez-Ubinas, E., Barbosa, J. D., & Qureshi, O. A. (2023). Analysis of residential electricity consumption patterns utilizing smart-meter data: Dubai as a case study. *Energy and Building*, *291*, 113103. DOI: 10.1016/j.enbuild.2023.113103

Rekik, M., Abdelkafi, A., & Krichen, L. (2015). Synchronization of Wind Farm Power System to Utility Grid under Voltage and Frequency Variations. *International Journal of Renewable Energy Research*, *5*, 70–81.

Sadoun, R., Rizoug, N., Bartholomeüs, P., Barbedette, B., & Le Moigne, P. (2011, September). Optimal Sizing of Hybrid Supply for Electric Vehicle Using Li-ion Battery and Supercapacitor. *2011 IEEE Vehicle Power and Propulsion Conference*, Chicago, IL, USA. DOI: 10.1109/VPPC.2011.6043183

Salah, W. A., Albreem, M. A., Alsayid, B., Zneid, B. A., Alkhasawneh, M., Anwar, A. M., & Sneineh, A. A, & Abu_AL_Aish, A. M. (2019). Electric vehicle technology impacts on energy. *International Journal of Power Electronics and Drive Systems*, *10*, 1–9.

Salyer, Z., D'Arpino, M., Canova, M., & Guezennec, Y. (2021). Optimal Health-Conscious Fast Charging of Lithium Ion Batteries. *IFAC-PapersOnLine*, *54*(20), 516–521. DOI: 10.1016/j.ifacol.2021.11.224

Sheikhi, A., Bahrami, S., Ranjbar, A. M., & Oraee, H. (2013). Strategic charging method for plugged in hybrid electric vehicles in smart grids; a game theoretic approach. *International Journal of Electrical Power & Energy Systems*, *53*, 499–506. DOI: 10.1016/j.ijepes.2013.04.025

Singh, A., Pal, K., & Vishwakarma, C. B. (2023). State of charge estimation techniques of Li-ion battery of electric vehicles. e-Prime - Advances in Electrical Engineering. *Electronics and Energy*, *6*, 100328.

Tremblay, O., & Dessaint, L. (2009). Experimental validation of a battery dynamic model for EV applications. *World Electric Vehicle Journal*, *3*(2), 289–298. DOI: 10.3390/wevj3020289

Verma, A., Asadi, A., Yang, K., & Tyagi, S. (2015). A data-driven approach to identify households with plug-in electrical vehicles (PEVs). *Applied Energy*, *160*, 71–79. DOI: 10.1016/j.apenergy.2015.09.013

Wang, F., Zhang, Z., Liu, C., Yu, Y., Pang, S., Duić, N., Shafie-khah, M., & Catalão, J. P. S. (2019). Generative adversarial networks and convolutional neural networks-based weather classification model for day ahead short-term photovoltaic power forecasting. *Energy Conversion and Management*, *181*, 443–462. DOI: 10.1016/j.enconman.2018.11.074

Wang, G., Zhang, Q., Li, H., McLellan, B. C., Chen, S., Li, Y., & Tian, Y. (2016). Study on the promotion impact of demand response on distributed PV penetration by using non-cooperative game theoretical analysis. *Applied Energy*, *185*, 1869–1878. DOI: 10.1016/j.apenergy.2016.01.016

Wu, X., Hu, X., Moura, S., Yin, X., & Pickert, V. (2016). Stochastic control of smart home energy management with plug-in electric vehicle battery energy storage and photovoltaic array. *Journal of Power Sources*, *333*, 203–212. DOI: 10.1016/j.jpowsour.2016.09.157

Wu, X., Hu, X., Teng, Y., Qian, S., & Cheng, R. (2017). Optimal integration of a hybrid solar-battery power source into smart home nanogrid with plug-in electric vehicle. *Journal of Power Sources*, *363*, 277–283. DOI: 10.1016/j.jpowsour.2017.07.086

Yang, X., Zhang, Y., Zhao, B., Huang, F., Chen, Y., & Ren, S. (2017). Optimal energy flow control strategy for a residential energy local network combined with demand-side management and real-time pricing. *Energy and Building*, *150*, 177–188. DOI: 10.1016/j.enbuild.2017.05.076

Zhang, Y., Zhang, T., Wang, R., Liu, Y., & Guo, B. (2015). Optimal operation of a smart residential microgrid based on model predictive control by considering uncertainties and storage impacts. *Solar Energy*, *122*, 1052–1065. DOI: 10.1016/j.solener.2015.10.027

Zhao, Y., He, X., Yao, Y., & Huang, J. (2019). Plug-in electric vehicle charging management via a distributed neurodynamic algorithm. *Applied Soft Computing*, *80*, 557–566. DOI: 10.1016/j.asoc.2019.01.053

Zhenga, Y., Niua, S., Shanga, Y., Shaoc, Z., & Jian, L. (2019). Integrating plug-in electric vehicles into power grids: A comprehensive review on power interaction mode, scheduling methodology and mathematical foundation. *Renewable & Sustainable Energy Reviews*, *112*, 424–439. DOI: 10.1016/j.rser.2019.05.059

ABBREVIATIONS

Ω: rotational speed.
η_{PEV}**:** Efficiency of the studied PEV (*KW/Km*)
: air density.
$_{m}$**:** permanent magnetic rotor flux.
*A***:** Exponential zone amplitude (V).
$AD_{driving}$**:** Distance running by the PEV(*Km*)
*B***:** Exponential zone inverse time (Ah^{-1}).
BAT_{cap}**:** PEV battery capacity (*KWh*)
C_p**:** WT power coefficient.
I^***:** Filtered current of the Battery (A).
*I.t***:** Actual battery charge (Ah).
*I:***:** Battery current (A).
I_{sd}, I_{sq}**:** *d* and *q* components of the stator currents.
*j***:** Counter of PEV number jϵ[1,N_{PEV}].
*K***:** Constant Polarization (Ω).
L_s**:** stator winding inductance.
N_{PEV}**:** Total number of PEVs connected to home.
*p***:** number of pole pairs.
P_A**:** Average home power demand.
P_C***:** Updated home power consumption.
P_C**:** Power consumption of a studied smart home.
P_{high}**:** Upper operating PEV power (*KW*)

P_{low}: Lower operating PEV power (KW).
P_{trip}: Necessary power for travelling(KW).
P_{PEVj}: Power of the PEV number "j".
P_{WG}: Power of the WG.
Q: Battery capacity (Ah).
R: Battery internal resistance (Ω).
R: blade radius.
R_s: stator winding resistance.
SOC_{high}: Tolerable upper battery SOC.
SOC_{low}: Tolerable lower battery SOC.
SOC_{PEVj}: Primary PEV SOC value.
$Tcon_{PEVj}$: PEV Connection time.
$Tdisc_{PEVj}$: PEV Disconnection time.
T_{em}: Electromagnetic torque.
U_0: Battery constant voltage (V).
V: wind speed.
V_{batt}: Battery voltage (V).
V_{sd}, V_{sq}: d and q components of the stator voltages.

Chapter 9
Bond Graph Modeling of an Off Grid Brackish Water Pumping/Desalination Unit Fed by Renewable Energy for Freshwater Production and Agricultural Irrigation

ines Ben ali
https://orcid.org/0000-0002-8152-1697
Université de Tunis El Manar, Tunisia

Jamel Belhadj
University of Tunis, Tunisia

ABSTRACT

This paper, is dedicated to a Brackish Water Reverse Osmosis (BWRO) pumping/desalination unit systemic modeling. The studied system is likely to operate in remote off-grid communities lacking access to freshwater. Without battery storage, the unit is powered by hybrid PV/Wind-turbine generator to provide freshwater for domestic and agricultural uses. The investigated system includes three concurrent processes: pumping, desalination and irrigation putting ahead different disciplinary notably hydraulic, electric, mechanical and thermal domains. Therefore, based on bond graph technique, a multidisciplinary approach is adopted for such a multi-physical system modeling. The established model is assessed using BWRO test bench through an experimental characterization of the studied processes. The experimental results demonstrated the significant accuracy of the bond graph models obtained from simulation drawn with a dedicated software tool.

DOI: 10.4018/979-8-3693-9924-8.ch009

INTRODUCTION

Regarding the big development in the industry, the lifestyle changes and Human population growth the energy demand is quickly increasing (Majeed et al., 2023). This critical situation was mitigated by the trends to meet the required energy using renewable sources. Recently, the worldwide registered a significant development of renewable energies. Their use is expanding in several common and emerging applications. The rush to renewable energies is imposed by conventional energies prices incessant surge owing to the international context high-strung and political crises on the one hand and technological developments in producing electricity by renewable energy sources such as solar and wind on the other hand.

In the future, renewable energy consumption in the power, heat and transport sectors is projected to increases near 60% over 2024-2030. In particular, electricity generation from said energy sources would account up more than 75% of the overall rise, due to continued governmental support in many countries, decreasing costs and the increasing use of electricity for road transport and heat pumps (IEA, 2024).

By 2028, renewable energy sources would account for over 42% of global electricity generation, mainly by wind and solar PV according to the IAE (IEA, 2024). As a result, Solar photovoltaic (PV) and wind energy systems are interesting energy alternatives either for small or large-scale electricity generation.

Closely related to the energy problem, global warming and drought are transforming life on earth in a significant way. Collectively, they favor freshwater scarcity which became a main concern for many countries.

Moreover, by 2025, water consumption is expected to increase by 30% in developing countries and by more than 10% in developed countries (Baggio et al., 2021). Consequently, the water and electrical energy demands in the countries concerned will follow demographic developments, combined with the increase in the standard of living of the populations. In addition to the dynamics of the populations, there were several communities in these countries which are isolated (remote villages and islands not served with either electrical energy or drinking water): More than 1 billion people are deprived of drinking water and about 2.3 billion people (28% of the world's population) live in regions that suffer from water scarcity (United Nations, 2019).

Thus, several million people lives is dramatically affected namely in underdeveloped countries where most of them are installed in rural areas working on farms and sheep breeding. Indeed, pasture and fresh water are becoming very scarce. Feeding and watering herds would be a daily challenge for breeders.

Besides, many water sources quality is more and more degraded because of salinity increase. The regions around these sources covered with plants, that used to be pasture, become arid and none useful for sowing varieties to feed sheep.

Considering this issue, comprehensively, there is no longer serval solutions. It's imperative to target a smart alternative for providing fresh water in small and medium scale in order to meet domestic freshwater demand of rural isolated communities and eventually for irrigating some parcel with the hope to moderate the lack of pasturage. The solution consists of sustainable brackish water pumping/desalination unit fed by renewable energy with low installation and operating cost. While taking in account the abundance of renewables energies sources Solar and Wind, it aims to produce enough electric energy and freshwater. The chosen architecture should be designed without conventional energy (electric or fossils) supply not only due to high provisioning cost in case of gasoil generator but also the lack of grid connection.

Based on literature review we can find several combinations of renewable energy technologies and desalination. However, we should put ahead the feasibility of each solution technically and economically and its adequacy to the target. Several factors should be taken in account such as renewable energy resources availability, electrochemical and/or hydraulic storage, and finally the nature (sea water/brackish water) and characteristics (salinity) of water feeding desalination unit (Mohamed & Papadakis, 2004; Garcia-Rodriguez, 2003; Gude et al., 2010; El-Ghonemy, 2012; Al-Karaghouli & Kazmerski, 2013, Park et al., 2009, Al-Karaghouli et al., 2009, Novosel et al., 2014, Greenlee et al., 2009) . One of the most spread methods for both brackish and sea water desalination is Reverse Osmosis (RO) desalination technology (Miranda & Infield, 2003, Schafer et al., 2007, Richards et al., 2008, Thomson et al., 2002). In particular, the above-mentioned technology can take advantage of abundant renewable energy resources in remote arid regions to feed the desalination unit (Ghaffour et al., 2013, Miranda & Infield, 2003, Kanematsu et al., 1988). Therefore, renewable energy is an attractive alternative for powering pumping/desalination systems instead of conventional energy sources. However, to cope with the intermittent nature of the available energy (solar PV and wind) many solutions can be implemented. One wide spread solution is combining hybrid renewable energy sources such as PV/Wind generator (Khan et al., 2018, Ghermandi & Messalem, 2009, Eltawil et al., 2009, Jurasz et al., 2020, Bett & Thornton, 2016).

Other designers consider electrochemical storage as battery bank to face the variability and the uncertainty of Res (Mokheimer et al., 2013, Tzen et al., 2008, Tzen & Morris, 2003, Fahmy et al., 2012). Nonetheless, the plants installing and operating (periodic maintenance) costs would impact the solution feasibility with the respect of economic et logistic constraints (Richards & Schafer, 2003, Thomson, 2003).

In this perspective, a brackish water pumping/desalination unit (BWRO) supplied by hybrid (photovoltaic wind) generator without batteries is considered as a viable application to provide fresh water for rural communities in remote areas suffering from fresh water scarcity and energy supplies.

The system design and the component sizing is expected to ensure, notably because of the use of hydraulic storage, the production of the required fresh water quantities and electric energy to water the community and their cattle in one hand and to irrigate Lucerne plants for pasture in the other hand.

Furthermore, in such RO pumping/desalination systems modelling we have to consider the following disciplinary fields: electrical, hydraulic, chemical, mechanical and thermal. Consequently, the modeling process of such multidisciplinary system became a challenging task when integrating these different kind of energy behavior due to most current software limits operating in general over a single domain. In this context, the Bond Graph formalism have the merit to be an adequate tool for such a multi-physical system dynamic modeling (Dauphin-Tanguy, 2000, Gawthrop & Bevan, 2007).

For this purpose, an experimental study of the energy behavior of said system through a small-scale experimental unit is carried to the energy characterization of the proposed system (Turki et al., 2009). This allowed us, in a second step, to proceed with the energy modeling of the different devices and power flows involved. At the end of the modeling phase, an energy/water management strategy would be conducted in future work.

The following part of this chapter would be organized in four sections: Section 2 presents global system description with focus on hydraulic architecture and system constraints. Section 3 is dedicated for the system Hydro-mechanical characterization to study its energetic behavior. In Section 4 an overview on bond graph modeling approach is given. Section 5 is dedicated for BWRO pumping/desalination processes modeling based on bond graph formalism. Besides, in order to evaluate the overall developed model, an experimental validation have been performed. In section 6, the energy management for the studied system is formulated. Finals, in section 7 some conclusions are given.

GLOBAL SYSTEM DESCRIPTION

Context

The scientific challenges that we contribute to addressing through this research study are:

- The integration of non-polluting renewable energy resources into dedicated electrical energy production systems for supplying water pumping and/or desalination units installed in isolated sites without electrical and hydraulic networks
- The development of energy control and management strategies adapted and optimized for these types of autonomous systems for producing fresh water using wind and sun with/without electrical storage

The optimized energy management is the main objective of the research work carried out. In this section, we will define the architecture of the system and describe the resulting problem in terms of energy-matter management of said system.

The investigated system consists of an autonomous small-scale (300 l/h), BWRO desalination unit supplied, via a DC link, with an intermittent generated power (PV-Wind) offered along wind speed and solar irradiation conditions without battery storage. (Figure.1) The system is dedicated for off-grid small communities without access to freshwater.

The desalination unit consists mainly of:

- A single-stage three-phase centrifugal motor pump: this is a Low Pressure Pump (PBP) dedicated to pumping brackish water which will be stored in a raised tank (C_1).
- A three-phase multi-stage centrifugal motor pump: this is a High Pressure Pump (PHP) dedicated to supplying the reverse osmosis system (hydraulic load) with raw water at high pressure
- A reverse osmosis system composed of two stages: two RO modules each containing a single spiral membrane

Two hydraulic storage tanks: the tank (C_1) for the storage of brackish water and tanks (C_2) for the storage of produced fresh water.

This unit consists of pumping and/or desalinating brackish water. The proposed architecture has the particularity of not using electrochemical storage, but rather of taking advantage of tanks hydraulic storage. This system is oriented towards autonomous applications of pumping and/or desalination of brackish water in isolated sites, without electrical and hydraulic networks, where wind and sun are abundant.

Furthermore, reformulating the energy management problem of such an autonomous system requires providing answers to the following questions:

- What are management objectives that must achieved?
- What constraints does this system undergo?
- What are the means of action to achieve the objectives?

Figure 1. Synoptic of the desalination system powered by a hybrid renewable energy source

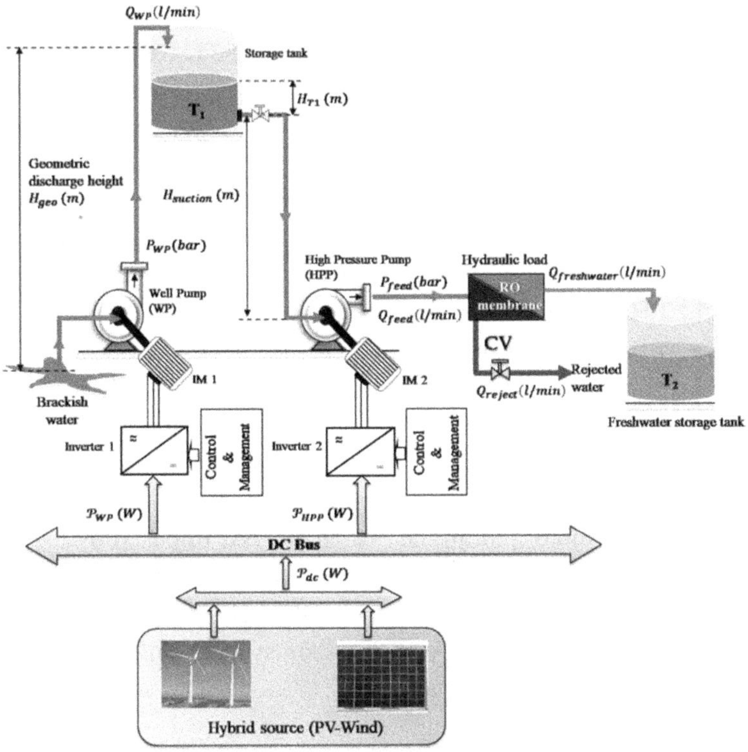

We must then identify the objectives, constraints and means of action for our proposed system:

Objectives

The objective sought through the development of an energy management strategy for the desalination system is to maximize the fresh water production based on the intermittent electrical power provided by the source.

Constraints

Any energy system is subject to different constraints during its operation. These constraints can be internal to the system and depend on the characteristics of its components, or external and depend on its environment. They can be of the technological, environmental, economic, normative type (Saudement, 2013).

The proposed management strategy for the said desalination system, depends mainly on the intermittent electrical power coming from the hybrid source, and on various studied system constraints, such as:

- The hydraulic stored reserves state in the tanks: the stored water levels (L) in the tanks must vary within the interval of the limit levels $[L_{min} - L_{max}]$ for each tank. In the particular case, these operating constraints (of the inequality type) are applied only for the storage tank C_1, such that: $L_{1_{min}} \leq L_1 \leq L_{1_{max}}$

- Technological constraints: these constraints are closely linked to the operating limits (power and pressure ranges) of pressure booster (motor pumps) and the reverse osmosis system. These constraints are defined as follows:
    - Respect the hydraulic operating range $(P_a - Q_a)$ for the desalination process hydraulic load (the osmosis unit): this operating range is highly variable depending on the characteristics of the osmosis unit, namely the conversion rate of the RO module, the temperature and salinity of the raw water to be desalinated, etc. According to the manufacturer, the maximum supply pressure of the studied RO membrane must not exceed 20 bars $\left(P_{a_{max}} \leq 20\, bars\right)$. On the other hand, the PHP pump must ensure a minimum pressure $\left(P_{a_{min}}\right)$ so that the membrane can produce a minimum of fresh water.
    - Respect the recommended power range for the correct operation of each motor pump: in fact, the hydraulic operating range (Pressure - Flow rate) of each motor pump is influenced by the associated hydraulic load, and is reflected in the electrical domain by the range of electrical power limits $[\mathscr{P}_{min} - \mathscr{P}_{max}]$ absorbed by the drive motor of each pump. Concerning the low pressure PBP motor pump, the power range is imposed by the discharge height and the hydraulic losses produced in the hydraulic network depending on the flow rate of this pump. The high pressure PHP motor pump is strongly influenced by the operating range of its complex hydraulic load (the reverse osmosis system) and also operates in an electrical power range imposed by the latter

Given the hydraulic load of each process (pumping and desalination process), the recommended electrical power range $[\mathscr{P}_{min} - \mathscr{P}_{max}]$ for each motor pump of the overall desalination system is determined based on experimental tests. In this way, the technological constraints of the system, in particular those of the reverse osmosis system, can be respected and taken into consideration, $[P_{min}\,[bar] - P_{max}\,[bar]]$ (Turki et al., 2011, Ben Ali et al., 2012, Ben Ali et al., 2014, Ben Ali et al., 2016).

Actions

The actions are defined by the references sent to specific elements within the studied system. Two means of action are then identified, such as: the references of the powers assigned to the two PBP and PHP motor pumps, which we will be denote *as* $\mathscr{P}_{PBP}$ and $\mathscr{P}_{PHP}$ respectively.

ARCHITECTURE

In order to study the energy behavior and modeling of the BWRO desalination process, we have designed an experimental test bench consisting in a reduced scale BWRO desalination plant.

Figure 2 presents an overview of the experimental model produced by (Turki, 2010) and improved by the author in the current study It was designed as a prototype and installed in our research laboratory (L.S.E at ENIT-UTM, Tunis, Tunisia). Its rate of freshwater production can be up to 300 liters/hour (for one RO membrane). Both motor-pumps are driven by variable-frequency three-phase inverters, for experimental characterizations under variable frequency (i.e. variable power supply) in order to emulate the unit operation under variable power supply in real sites. This prototype has been built in order to demonstrate operation under variable power supply and accordingly, operation of the RO module under variable feed pressure and flow rate;

Figure 2. Experimental desalination model overview

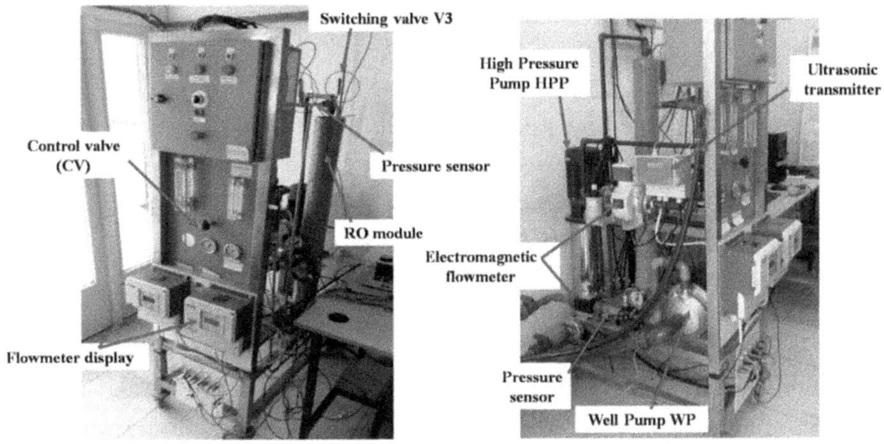

Figure 3 presents the hydraulic architecture of the experimental unit. It has been investigated and defined so that several advantageous can be drawn; It is equipped with the necessary measuring instruments and includes the following elements:

Figure 3. Synoptic of the experimental model of water desalination by RO

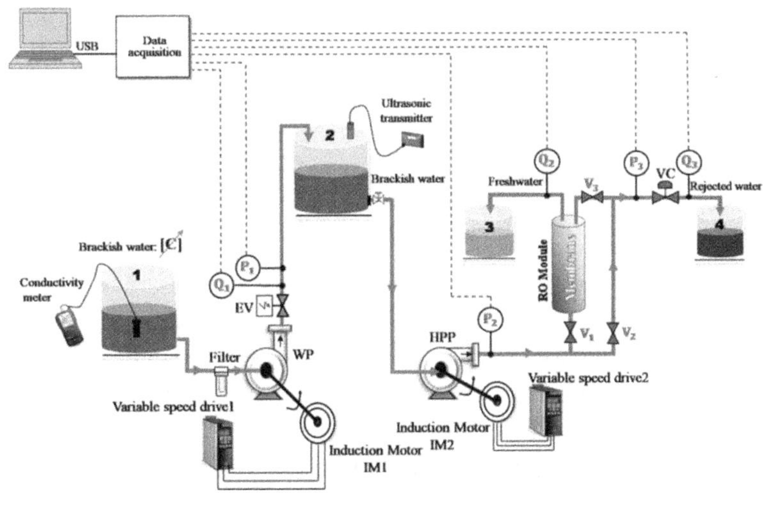

- **Hydraulic storage:** the storage device consists of two tanks dedicated for hydraulic storage, whether brackish water (2) with a capacity of 2000 liters or desalinated water (3) with a capacity of 1000 liters.
- **Booster pump system:** The Booster system of this unit consists of two centrifugal motor pumps decoupled through the tank 2
  - **pump PBP:** It is used to collect the water to be desalinated and to lift it from the tank (1) to the tank (2) for storage. It is characterized by
    - P_{utile}=0.75kW, 220–380V/50Hz and I_n=2.5A
    - Operating pressure: 2.5–3.4 bar
    - Maximum discharge flow rate: 5.4m^3/h=90l/min
  - **Motopompe PHP:** It is mainly dedicated to supply the RO module with brackish water with a high supply pressure. It is characterized by
    - P_{utile}=2.2kW, 220–380V/50Hz and I_n=2.5A
    - Operating pressure: 8.17–18.6 bar
    - Maximum discharge flow rate: 3.6m^3/h=60l/min
- **RO filtration system:** To desalinate raw water an RO membrane technique is used. The RO module has a cylindrical shape containing a TORAY-TM710 type composite polyamide spiral membrane. This membrane is dedicated to the desalination of low salinity water (brackish water).
- **Control /command system:** It includes the elements to regulate the unit operation unit and the freshwater production:
  - Two industrial frequency variators
  - Control Valve (CV): this is a control valve installed in the experimental unit discharge circuit. It allows to change the hydraulic load by throttling (vary the valve position) and consequently, the conversion rate of the desalination process t.
  - Switching valves (V_1, V_2, and V_3)
  - Pressure sensors (P_1, P_2, and P_3) and electromagnetic flow meters (Q_1, Q_2, and Q_3)
  - A conductivity meter and an ultrasonic level transmitter.
  - Acquisition unit: It consists of an analog/digital input/output module and a controller that ensures data communication with a computer where a graphical interface by "LabVIEW" is embedded. The signals measured by the flow meters and pressure sensors are then transmitted to the computer by the control unit as an analog voltage (0-10V)).
  - Solenoid valve (EV)

The raw water (brackish) treatment is accomplished in three stages:

- **Stage 1:** the raw water is discharged by the PBP motor pump with a flow rate Q_1, and stored in the intermediate tank (2). This phase constitutes the "pumping process" of the brackish water
- **Stage 2:** the PHP motor pump strives the brackish water stored in (2) to supply the RO module at the necessary pressure P_2 required for said module operation.
- **Stage 3:**it consists of adjusting the desired conversion rate $\left(\tau_c = \frac{Q_{fresh\ water}}{Q_{raw\ water}} = \frac{Q_2}{(Q_2+Q_3)}\right)$ of the RO module by acting on the position of the valve CV, and consequently on the production of fresh water Q_2 (the evolution of $Q_{freshwater}$ follows the evolution of the conversion rate τ_c).

Phases 2 and 3 constitute the "desalination process". Since the electrical power generated is sufficient to power both processes, they can operate simultaneously, if there is a sufficient quantity of stored water, otherwise, only one process is operated depending on the state of the storage tank (empty or overfull tank).

Given the energy challenges, managing energy in an intelligent and efficient way is essential. And, the formulation of the energy management problem is based on an approach that should allow optimal energy management.

Depending on the complex energy system nature, the design requirements are translated into needs, constraints and design criteria. Once the analysis of the energy system is done, the next step is the selection of the technology to be implemented and the choice of elements. Thus, depending on the tasks that this system should do, the optimization of the sizing is an essential step in order to achieve the desired objectives, satisfy the needs and respect the constraints. The latter are generally expressed in a standard way and can represent the capacity, availability or security of the system. The scientific challenge when implementing a tool for the optimal sizing of such a complex system is to develop an optimal design methodology in a systemic vision. Generally, systemic design is based on two approaches, namely "simultaneous design" and "sequential design". In the current study we are interested in "sequential systemic design". In this regard, several aspects should be addressed (Buzila Pankovits, 2015):

1- The first aspect concerns choosing energy architecture for the system
2- Depending on the architecture choosing, the modeling and simulation of the system constituent elements are then considered. Therefore, a study on optimal pre-sizing should be conducted in the second phase. This phase seeks to specify the nature and characteristics of the system elements, as well as the instantaneous energy transfers during an idealized operating cycle.
3- After defining the energy system elements characteristics, the next step is focused on optimizing different power flows involved in this system. The objective here is to find the optimal management for the power flows provided by the hybrid energy sources in order to satisfy the power required by the load on the one hand and, to ensure the optimal distribution of the provided power between the different load elements (case of multi-pump stations for pumping water or producing fresh water based on wind and sun, for example) on the other hand. This phase should make it possible to establish a framework of constraints and management rules necessary for the real time approach (online management strategy). These rules constitute operating recommendations for the real-time supervisor to ensure operations optimality.

In many cases, the optimization process is considered an iterative cycle. Indeed, the optimization of power flows in the energy system is directly linked to the overall energy management approach in the "design/control" system (offline management strategy) (Guemri, 2013, Levron et al., 2013, Tischer & Verbic, 2011).

The question that arises here is the following:

"How can we design a system without knowing its behavior?".

To do this, a crucial step is to focus on the in-depth system processes study, notably the energy behavior and the transfer of different flows (hydraulic and electric) between the interconnected components

The design of the model was carried out in such a way that it is possible to study and model the different water processes. Indeed, this flexibility allows to carry out the necessary tests for the characterization and modeling of the hydro-mechanical processes and to monitor the global desalination system energy behavior in order to control and manage energy flows.

HYDRO-MECHANICAL CHARACTERIZATION FOR OVERALL SYSTEM

The hydro-mechanical characterization of the studied system composed by centrifugal motor pump, hydraulic network and RO membrane is achieved based on two degrees of action allowed by the experimental model:

- The first degree of action is the "set point input frequency" of the speed variator. With this degree of action, we can obtain the hydro-mechanical characteristic of the centrifugal motor pump for different rotation frequencies (Figure.4 – a).
- The second degree of action is the variation of the valve position. We can identify the hydro-mechanical characteristic of the opposite load for the centrifugal motor pump (Figure.4 – b)

Figure 4. Possible characterization tests of a centrifugal motor pump

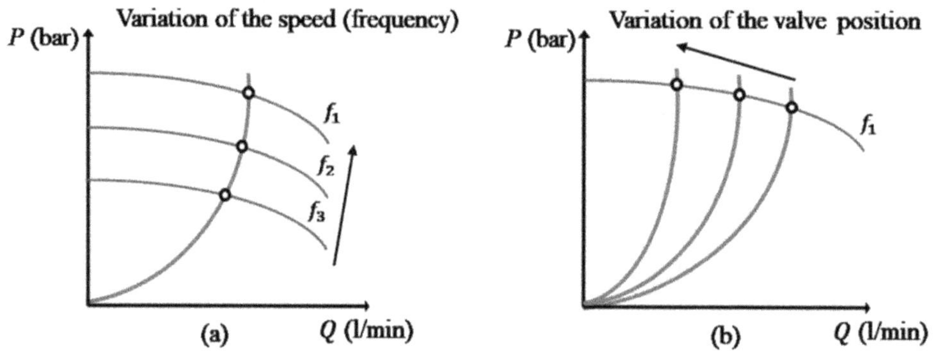

Figure 5. Electrical consumed power Behavior for a given hydraulic load

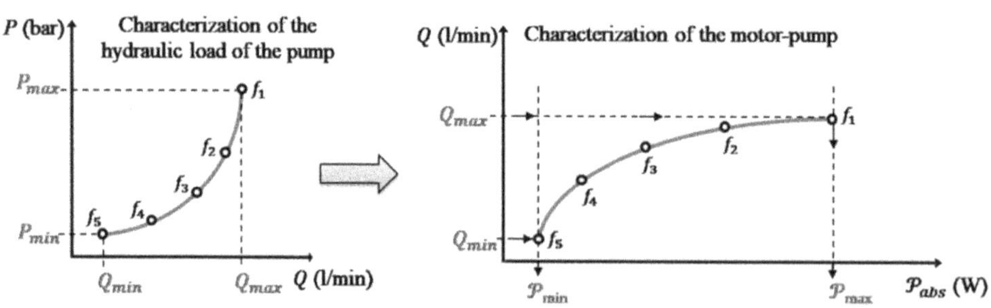

Through these experimental characterizations, at a given hydraulic load (Figure.4 – a), we can then deduce the adequate interval of the electrical power supply $[\mathscr{P}_{min} - \mathscr{P}_{max}]$ of each motor pump (Figure.5). This interval constitutes the power range (corresponding to a given hydraulic load) to be respected in the desalination system energy management in order to ensure the suitable functioning of these elements. Indeed, the recommended power interval for each of the motor pumps presents inequality constraints $\left(\mathcal{P}_{min} \leq \mathcal{P}_{pompe} \leq \mathcal{P}_{max}\right)$ defining the maximum and minimum limits on the electrical power that must be addressed to each motor pump. This interval is inherent not only to constraints of motor pumps correct functioning but also to energy efficiency considerations (efficiency range, cf. Figure.8).

CHARACTERIZATION OF THE BRACKISH WATER PUMPING PROCESS

In order to identify the hydraulic characteristic of the pump, we varied the valve position at a fixed rotation frequency. The experimental tests results are presented in Figure 6. From these illustrations, we note that the hydraulic curves (Pressure-Flow) verify the similarity laws. Through these hydraulic characterization curves, we can deduce the pump analytical model by identifying the different modeling equations parameters (see equation (28)).

Figure 6. Characteristic of variable hydraulic load centrifugal pump for different frequencies

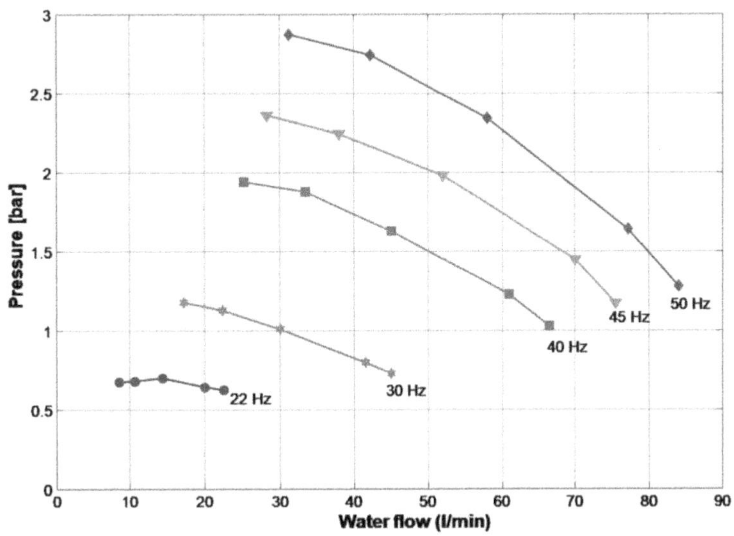

The hydraulic load characteristics is identified by varying the rotation frequency using the speed variator for a given valve position (fixed). The test results are shown in Figure.7.

It is noted that the centrifugal motor pump achieved the maximum water production when the valve is fully open (Figure 7). However, this valve position presents poor efficiencies throughout the operating range of the motor pump (Figure 8). Therefore, a compromise must be found between water production and the motor pump efficiency $\left(\frac{\mathcal{P}_{hyd}}{\mathcal{P}_{élec}}\right)$.

Figure 7. Variable speed centrifugal pump Characteristic for different hydraulic loads

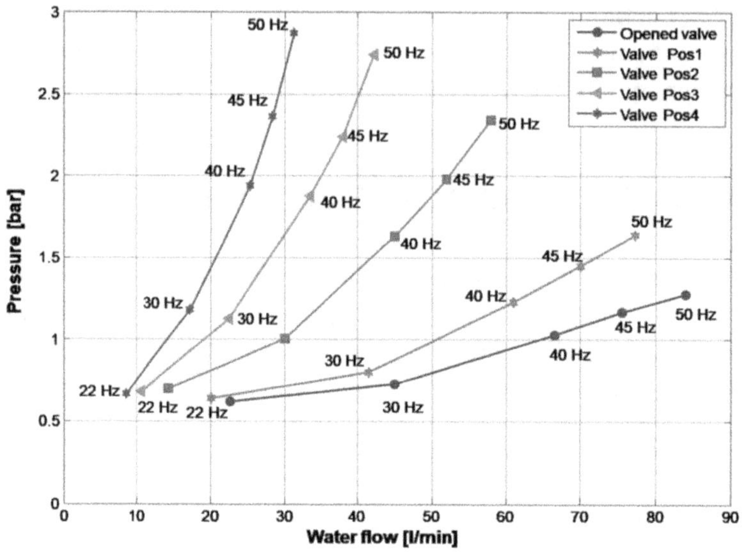

Figure 8. Total efficiency of the motor pump as a function of the absorbed electrical power

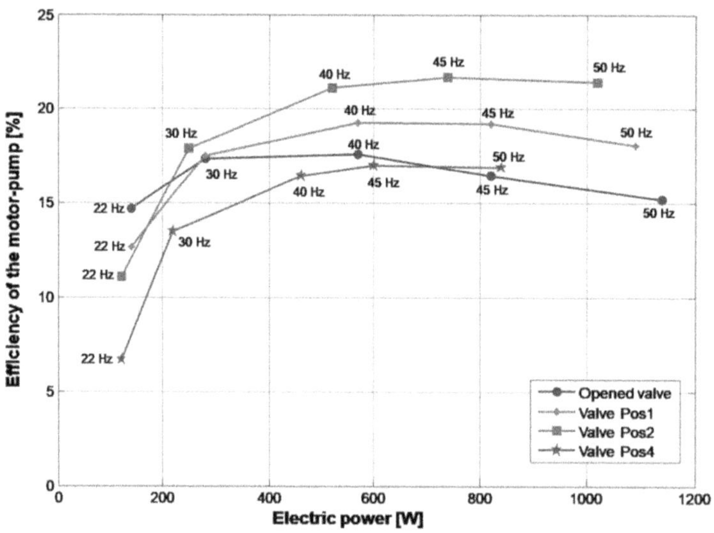

By observing the two mentioned above figures, we can conclude that the valve position Pos2 is considered adequate for an acceptable operation in terms of efficiency-production. By fixing the valve at this position, the motor pump can produce water with a flow rate ranging from 14.3-58 l/min for a variable efficiency of 13.9-22.18%. On the other hand, the experimental characteristics obtained present a quadratic form in accordance with the theoretical ones. Thus, we can deduce the analytical expression of a typical hydraulic load. The latter, during the liquid flow, causes pressure/load losses that are characterized by a pressure drop, such as:

$$\Delta P_{cond} = K \bullet Q_{PBP}{}^{2} + \rho \bullet g \bullet H_{géo} = P_{dynamique} + P_{statique} \quad (1)$$

with:
ΔP_{cond}: load losses in the hydraulic network in [m].
g: Gravitational constant ($9.8 m/s^2$).
ρ: water Density (1000 kg/m^3).
$H_{géo}$: Total geometric height [m] (cf. Figure 19).
Q_{PBP}: Motor pump discharge flow rate PBP in [m^3/s].
According to fluid mechanics, the characteristic of a typical hydraulic load depends on two factors, namely:

- The fluid flow total geometric height $H_{géo}$ describing the gravity pressure, this pressure can be considered as static pressure:

$$P_{statique} = \rho \bullet g \bullet H_{géo}.$$

- The so-called local or singular load losses in the installation which are caused by the obstacles and modifications of the pipeline, such as: narrowing, widening, flap, valve, elbow, flow meter, etc. These losses are proportional to the square of the average velocity of the fluid ($V = Q/S$). They are expressed as a "dynamic height" of fluid in the form:

$$h_{dynamique}\left(m\right) = k \bullet \frac{V^2}{2 \bullet g} \quad (2)$$

Or again:

$$h_{dynamique}(m) = k \bullet \frac{1}{S^2} \bullet \frac{Q^2}{2 \bullet g} \quad (3)$$

where:
Q : The flow rate of the fluid in [m^3/s].
S : The pipe section in [m^2].
k : Loss coefficient which depends on the pipe (without unit).
V: fluid Average velocity [m/s].

Figure 9. motor pump experimental characteristic curve for different valve positions

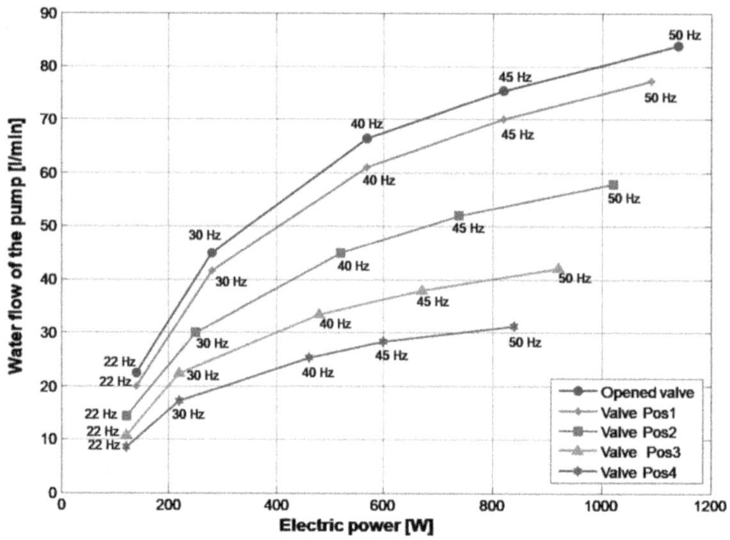

Figure 9 shows a series of curves describing the discharge flow rate variation (output variable) of the motor pump as a function of the asynchronous drive motor electrical power consumption (input variable) for different hydraulic loads (different valve positions). It can be noted that the motor pump electrical power consumption becomes clearly greater when the discharge flow rate increases. As a result, each time the pump production increases, the latter requires more electrical energy supplied.

This experimental relationship (faithful to reality in terms of load losses, mechanical friction, etc.) presents, in fact, the equivalent experimental quasi-static model, $Q_{PBP}[l/min] = f(\mathcal{P}_{PBP}[W])$, of a motor pump at a given hydraulic load,, $P_{PBP}[bar] = f(Q_{PBP}[l/min])$. This model can be used in a simple way for the modeling of the studied experimental desalination system in order to achieve the design the energy management strategy of the overall desalination system.

On the other hand, following these experimental results (Figure 9), we can in particular deduce the technological constraints of the motor pump imposed by the associated hydraulic load. These constraints are translated by the range of limits of electrical supply power suitable for each valve position.

Thus, the range of power limits recommended for the PBP motor pump (at valve position Pos2) is determined experimentally, such as: [120–1020W]. Characterization of the brackish water desalination process

BRACKISH WATER DESALINATION PROCESS CHARACTERIZATION

We recall that, to achieve the hydro-mechanical process characterization, we can act on the variation of the rotation frequency and the variation of the valve position. For the desalination process, we have a third degree of action, by varying the salinity of the raw water, which can be measured using the conductivity meter. Indeed, the osmosis unit is a complex and non-linear hydraulic load whose behavior depends on several factors (conversion rate "CV valve position", raw water concentration, raw water temperature, membrane permeability).

Thus, in order to carry out the desired experimental tests, we proceeded as follows:

- We fixed the position of the CV control valve such that the conversion rate $\left(\tau_c\left(\%\right) = \frac{Q_{eau\ douce}}{Q_{alimentation}} \times 100\right)$ is equal to 20%. This rate is imposed by the membrane manufacturer when using a single membrane without recovery.
- then, we carried out several tests for different water salinities ([2 g/l], [4 g/l] and [6 g/l]).
- For each salinity value, we varied the rotation the motor pump frequency in order to obtain the corresponding hydraulic load characteristic (i.e. each characteristic curve of the hydraulic load is imposed by the salinity of the raw water).

Indeed, these tests are carried out mainly in order to observe the salinity influence on the reverse osmosis system hydraulic operating range, and consequently that of the PHP motor pump translated by the limit interval $[\mathcal{P}_{min} - \mathcal{P}_{max}]$.

Below (Figure 10), a set of hydraulic load characteristic curves is presented for different feed water salinities.

Figure 10. Hydraulic load characteristics for different raw water salinities

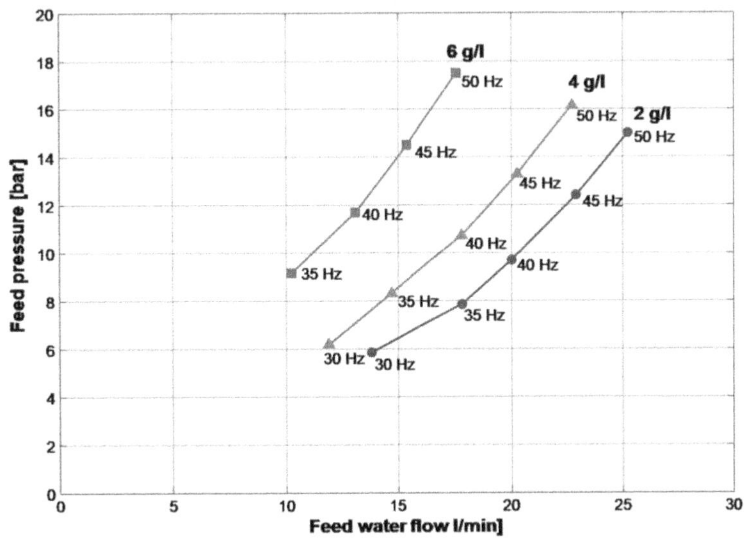

In fact, by fixing the valve position CV (second degree of action), the hydraulic load characteristic becomes proportional to the raw water salinity. It is quadratic in shape and varies according to the variation of the salinity.

We observe that, the higher the raw water salt concentration, the higher the membrane attack pressure is. Theoretically, this is explained by the fact that the feed pressure is closely linked to the osmotic pressure. The latter increases proportionally to the salinity of the water to be treated

Figure 11. Feed rate (Qa) as a function of the electrical power absorbed by the PHP motor pump

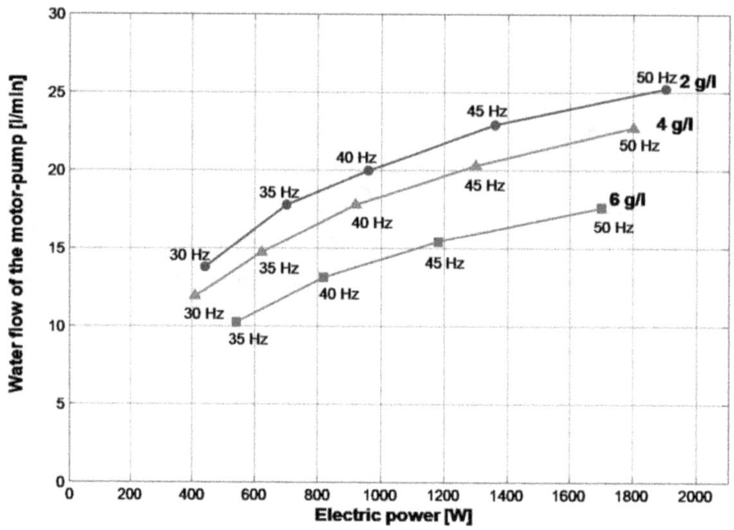

Figure 12. water flow rates evolution as a function of consumed electrical power (for raw water salinity 4 g/l)

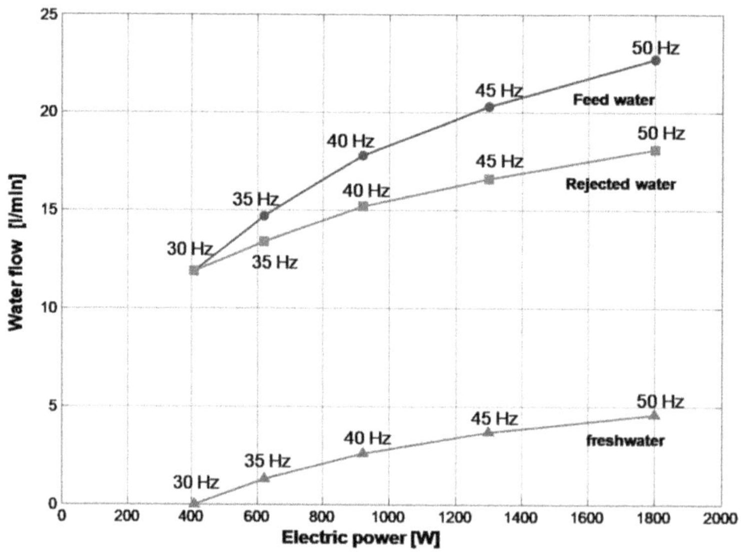

Figure 13. desalinated water production depending on the feed pressure (for raw water salinity 4 g/l)

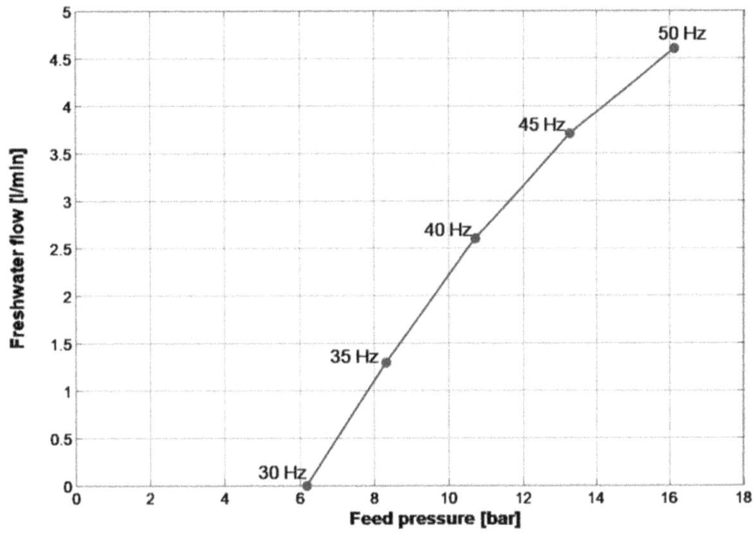

The curves presented in Figure 11 show the relationship between the reverse osmosis system feed flow rate Q_a e (provided by the PHP) and the electrical power absorbed $\mathscr{P}_{PHP}$ by the motor pump for different raw water salinities. Through these curves, we can deduce the suitable electrical operating range $[\mathscr{P}_{min} - \mathscr{P}_{max}]$ of the PHP motor pump for each salinity in order to ensure proper operation of the RO membrane. On the other hand, the electrical power range for the PHP motor pump and that of the reverse osmosis system are not always the same. For instance, let us focus on the case of a water salinity of 4 g/l. We present in Figure 12 the feed, discharge and fresh water flow rates of the reverse osmosis system as a function of the absorbed power P_PHP in order to show the difference between the electrical operating range of the PHP and the reverse osmosis system. Indeed, at 410 W the PHP can deliver water to feed the RO membrane. However, the latter cannot produce fresh water and the entire quantity of incoming water is sent as discharge. This is due to the feed pressure of the osmosis unit (Figure 13) which is insufficient (6.2 bar for 410 W) to produce desalinated water. At a power value equal to 620 W, the osmosis unit can produce fresh water because the feed pressure has increased. Therefore, the suitable electrical operating range of the desalination process is imposed by the operating range of the hydraulic load (the osmosis unit).

In the following, we have chosen 4 g/l as the salinity of the raw water to be desalinated. Thus, the technological constraints on the desalination process are translated across the power interval, namely [620-1800 W] (Figure 12).

Furthermore, the relation $(Q_a = f(\mathscr{P}_{PHP}))$ (Figure.11) describes the experimental quasi-static model of the desalination process {PHP + osmosis + VC} at a given hydraulic load (at fixed VC and any given salinity).

BOND GRAPH FORMALISM (Dauphin-Tanguy, 2000)

The Bond Graph methodology ("link graph" or "link graph") was introduced by H. Paynter and formalized by Karnopp and Rosenberg in 1975. It developed very quickly in Europe at the end of the 1970s, notably via the Netherlands (University of Twente) and France (Alsthom company). The Bond Graph tool is first of all an intermediate formalism between the physical system studied and the mathematical models associated with it. It is characterized by a unifying aspect whatever the physical field of application (electrical, mechanical, hydraulic, chemical,). This formalism is based on a power transfers analysis within the studied system and a hypotheses modeling definition on physical criteria.

The bond graph formalism has been selected to represent in a unified way the energetic transfer in multi-fields systems. A major benefit of using bond graph technique for such a system is that, besides being a compact representation they may be used to model energy transformation across many boundaries including mechanical, electrical, hydraulic, chemical thermodynamics fields (Dauphin-Tanguy, 2000, Gawthrop & Bevan, 2007). The so called "word" bond graph model of the overall structure is displayed in Figure 14.

Figure 14. "Word"bond graph model of t the RO desalination process

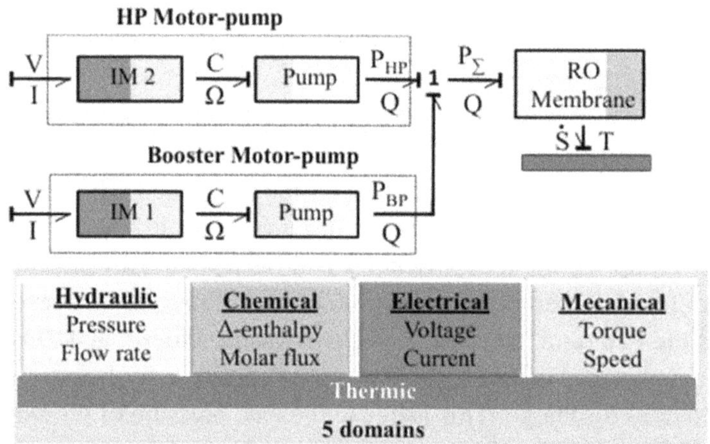

Notice that power variables (effort-flow) are supported bonds respectively corresponding to (Voltage-Current) electricity, (Torque-Speed) in on in mechanics, and (Pressure-flow) in hydraulic devices

It should be noted that modeling of the RO desalination process has been based on experimental parameters that are well identified following experimental characterizations (Turki et al., 2009, Turki et al., 2008). These parameters enabled us to effectively refine model the different developed models, and thereby validate the integral model of the whole RO process.

In the following we gives an overview on main concepts in Bond graph.

POWER TRANSFERS

Therefore, the Bond Graph is namely a power transfers representation within a global system composed of several subsystems. The said subsystems may be of different nature and characterized by energy conservation and the exchanged power continuity. The instantaneous power exchanged is the product of two variables which are the effort and the flow.

The effort and flow are represented on the Bond Graph by a half-arrow indicating the direction of the power exchange. Thus, Figure 15 represents the power transfer from subsystem A to subsystem B:

Figure 15. Bond Graph power transfer from A to B

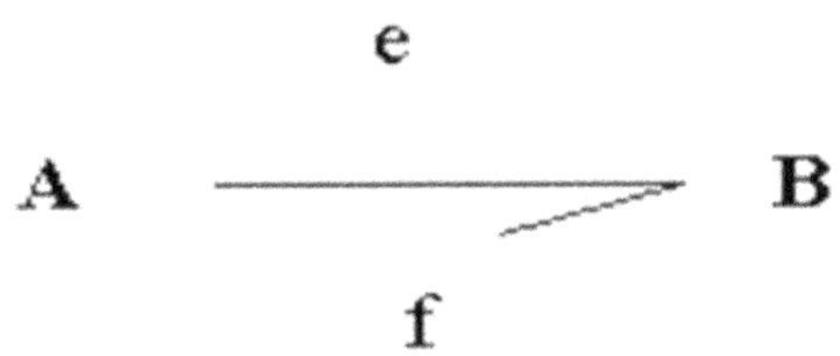

The following table presents the variables associated with generalized variables in several physical domains:

Table 1. Generalized variables equivalences in physical domains

Domain	Effort e	flow f	torque p	shift q
Electric	electric voltage u	Electric current i	Magnetic flow Φ	load q
Mechanic Translation	force F	velocity v	Pulse p	Shift x
Mechanic Rotation	torque T	Angular velocity ω	Angular Pulse h	Angle θ
Hydraulic	Pressure P	flow rate Volume Q	Pressure pulse Γ	Volume V
Chemical	Chemical potential	Molar fow		Moles number
Thermodynamics	Temperature	Entropy flow		Entropy
Magnetic	Magnetomotive force	Derivative of magnetic flow		Magnetic flow
Acoustic	Pressure	Volumetric velocity	Pulse	Volume

BOND GRAPH ELEMENTS

Basics Elements

Elements are used to represent phenomena that link generalized variables. They can be separated into three categories:

Table 2. Bond graph basics elements

Element	Symbol	Generic law	Examples
Active elements	$Se \rightharpoonup$	e independent of f	Ideal voltage generator
	$Sf \rightharpoonup$	f independent of e	Ideal current generator
Passive elements	$\rightharpoonup R$	$\Phi_R(e,f) = 0$	Electrical resistance, friction, shock absorber, hydraulic restriction
	$\rightharpoonup I$	$\Phi_I(p,f) = 0$	Coil, inertia, mass
	$\rightharpoonup C$	$\Phi_C(e,q) = 0$	Capacitor, spring, elasticity, tank
sensor	De		Voltmeter, pressure gauge
	Df		ammeter

JUNCTION ELEMENTS

These elements are used to couple the above-mentioned elements, namely the passive (R, C, I) and active (source) elements. They form four types of junction and are power conservative:

- Junctions 1 are iso-flow junctions.
- Junctions 0 are iso-effort junctions.
- TF junctions (transformer) transform the variables effort-effort, flow-flow.
- GY (gyrator) junctions transform the variables effort-flow, flow-effort

The following table shows the symbol and generic law of each type of junction:

Table 3. Bond graph Junction Elements

Elements	*Symbol*	*Generic law*	*Examples*
Junctions	1	$f_1 = f_2 = \dots = f_n$ $\sum_{i=0}^{n} a_i . e_i = 0$	Parallel coupling in mechanics, in series in electrical and hydraulics
	0	$e_1 = e_2 = \dots = e_n$ $\sum_{i=0}^{n} a_i . f_i = 0$	Series coupling in mechanics, in parallel in electrical and hydraulics
	$\xrightarrow{1}$ TF (m) $\xrightarrow{2}$	$e_1 = m\, e_2$ $f_2 = m\, f_1$	Electrical transformer, lever, mechanical reducer, coupling of physical systems
	$\xrightarrow{1}$ GY (r) $\xrightarrow{2}$	$e_1 = r\, f_2$ $e_2 = r\, f_1$	DC motor, coupling of physical systems

CAUSALITY CONCEPT

The causality concept makes it possible to determine for a Bond Graph element, which are the known and unknown power variable. Indeed, to be able to simulate the behavior of the system, it is first necessary to correctly determine its equations in mathematically solvable causal form. We will thus first determine the variables that are imposed on the elements and then those that these elements impose. The Bond Graph model thus makes it possible to define the calculation structure with highlighting of the **cause and effect** relationships within the system.

When two systems A and B are coupled and exchange power, two situations are possible:

- A applies a force e on B which reacts by sending a flow f

Figure 16. Bond Graph force applying from A to B

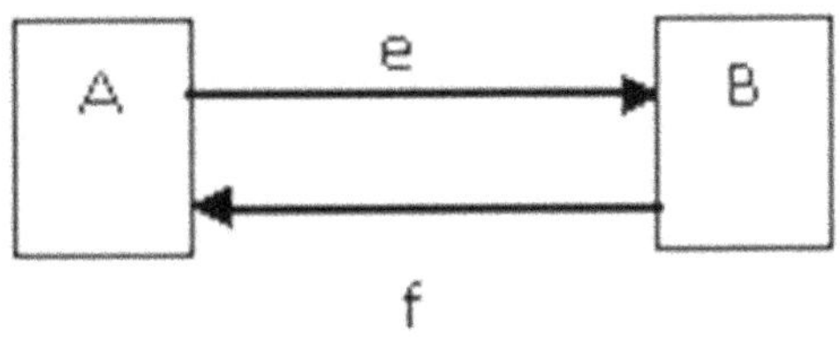

- A sends a flow f to B which responds with an effort e

Figure 17. Bond Graph effort response from B to A

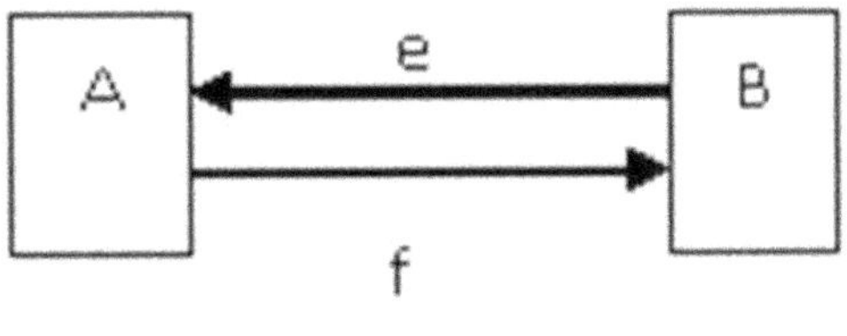

The following table presents the rules for assigning causality for each element and for each junction.

Table 4. Causality assignment rules

Element	Causality	Symbol	Generic law
Active elements	Mandatory	Se ——⇀|	*e* imposed by *Se*
		Sf |——⇀	*f* imposed by *Sf*
	Integral	——⇀| *I*	$f_I := \Psi_I(\int e_I(\tau)\, d\tau)$
		|——⇀ *C*	$e_C := \Psi_C(\int f_C(\tau)\, d\tau)$
Passive elements	derivative	|——⇀ *I*	$e_I := \Psi_I^{-1}\left(\frac{df_I}{dt}\right)$
		——⇀| *C*	$f_C := \Psi_C^{-1}\left(\frac{de_C}{dt}\right)$
	Arbitrary (linear case) or non-arbitrary (non-linear case)	|——⇀ *R*	$e_R := R.f_R$ $\quad e_R := \Psi_R(f_R)$
		——⇀| *R*	$f_R := \frac{1}{R} f_R$ $\quad f_R := \Psi_R^{-1}(e_R)$
Junctions	Causality restriction	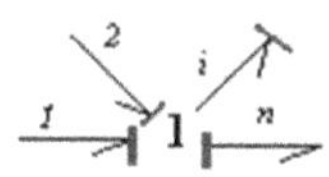 1 causal line far from J-1	$\begin{cases} f_1 := f_i \\ \vdots \\ f_n := f_i \end{cases}$ $e_i := -a_1 e_1 - \ldots - a_n e_n$
		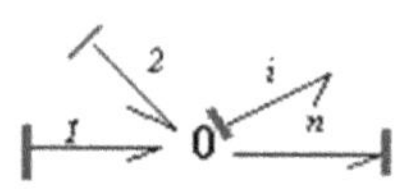 1 causal line near J-0	$\begin{cases} e_1 := e_i \\ \vdots \\ e_n := e_i \end{cases}$ $f_i := -a_1 f_1 - \ldots - a_n f_n$
		|—1—⇀ TF (m) |—2—⇀	$e_1 := m\, e_2$ $\quad f_2 := m\, f_1$
		—1—⇀| TF (m) —2—⇀|	$e_2 := m^{-1} e_1$ $\quad f_1 := m^{-1} f_2$
		|—1—⇀ GY (r) —2—⇀|	$e_1 := r\, f_2$ $\quad e_2 := r\, f_1$
		—1—⇀| GY (r) |—2—⇀	$f_1 := r^{-1} e_2$ $\quad f_2 := r^{-1} e_1$

BUILDING A MODEL PROCEDURE

For a simple system where the physical phenomena are uncoupled, for example in the case of a hydraulic system, the procedure consists of:

- choose a circulation direction for the fluid which will be taken as power transfer direction.
- Search for all the nodes of different pressures where a junction 0 will be placed.

- Place a junction 1 between two junctions 0 and attach the elements which are submitted to the difference in the corresponding pressures,
- Connect the junctions by links while respecting the direction of power transfer,
- Choose a reference pressure (atmospheric pressure for example)

In the electrical systems case, the procedure is very similar to this one, in particular because of the **flow-current** and **pressure-potential** analogy.

GLOBAL SYSTEM MODELING BASE ON BOND GRAPH

The two processes of pumping and desalination of brackish water each constitute a multidisciplinary energy conversion chain (electrical, mechanical, hydraulic, chemical, thermal domains). The modeling of such a system requires a good knowledge of the energy behavior of the different mechanisms (pump motorization, production of brackish water/fresh water, etc.) during the operation of said system. The different subsystems involved are inter-coupled through power flows (electrical, mechanical and hydraulic variables). In addition, the energy conversion in this system causes load losses, whether electrical, mechanical or hydraulic (caused by internal friction), which must be taken into account during modeling.

Figure 18 presents a synoptic diagram modeling the energy conversion for the different power flows throughout the chain of a hydro-mechanical process (from the source to the hydraulic load).

Figure 18. Block diagram of power flows in a hydro-mechanical process (pumping/desalination)

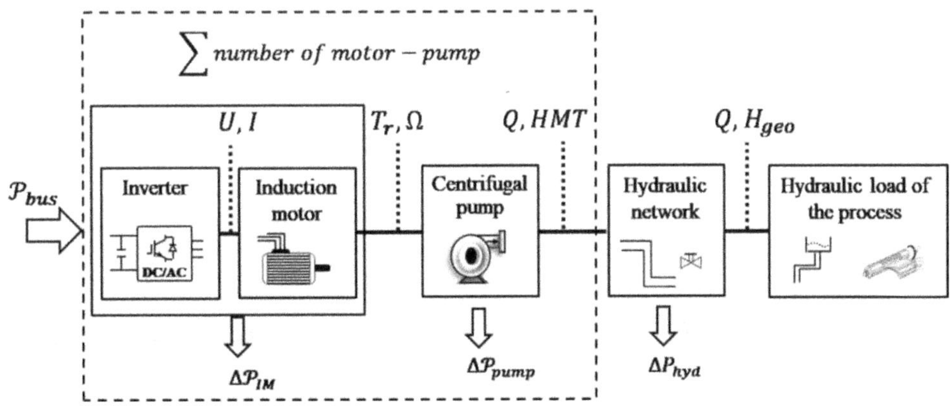

with:
Q: Discharge flow rate of the pump in [m^3/s].
H_{geo}: Geometric delivery height of the hydro-mechanical process in [m].
HMT: Total manometric head of the process in [m], such as: $HMT = H_{geo} + \sum \Delta P_{hyd}$.
ΔP_{hyd} : Hydraulic load losses due to water flowing through the pipes in [m].

Following the experimental characterization of the hydro-mechanical processes, we will approach the modeling of the experimental brackish water desalination unit.

BRACKISH WATER PUMPING PROCESS MODELING

The brackish water pumping process (Figure 19) involves a typical hydraulically loaded single-stage centrifugal motor pump (hydraulic line) supplying brackish water to a hydraulic storage tank (C_1). The centrifugal pump is driven by a three-phase asynchronous squirrel-cage motor with a useful power of 750W.

Figure 19. Schematic diagram of the brackish water pumping process hydraulic structure

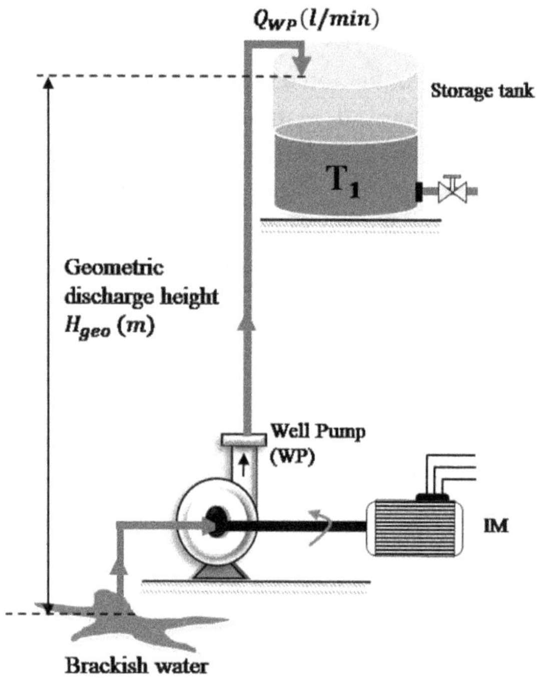

ASYNCHRONOUS MOTOR DRIVING THE CENTRIFUGAL PUMP MODELING

The two-phase asynchronous motor model consists of the following differential equations:

$$\begin{cases} V_{sd} = R_s \cdot I_{sd} + \dfrac{d\varphi_{sd}}{dt} - \dfrac{d\theta_a}{dt} \cdot \varphi_{sq} \\ V_{sq} = R_s \cdot I_{sq} + \dfrac{d\varphi_{sq}}{dt} + \dfrac{d\theta_a}{dt} \cdot \varphi_{sd} \\ V_{rd} = 0 = R_r \cdot I_{rd} + \dfrac{d\varphi_{rd}}{dt} - \dfrac{d(\theta_a - \theta)}{dt} \cdot \varphi_{rq} \\ V_{rq} = 0 = R_r \cdot I_{rq} + \dfrac{d\varphi_{rq}}{dt} + \dfrac{d(\theta_a - \theta)}{dt} \cdot \varphi_{rd} \end{cases} \tag{4}$$

Thus, the expressions for the stator and rotor flux with the Park transformation are given as follows:

$$\begin{cases} \varphi_{sd} = L_s \cdot I_{sd} + M_{sr} \cdot I_{rd} \\ \varphi_{sq} = L_s \cdot I_{sq} + M_{sr} \cdot I_{rq} \\ \varphi_{rd} = L_r \cdot I_{rd} + M_{sr} \cdot I_{sd} \\ \varphi_{rq} = L_r \cdot I_{rq} + M_{sr} \cdot I_{sq} \end{cases} \tag{5}$$

In the particular case where the asynchronous motor is modeled in the fixed frame (α,β), the Park transformation angle of this frame is zero,, $\theta_a = 0$ (the "d" axis is coincident with the "a" phase of the stator). The system of equations (4) becomes

$$\begin{cases} V_{s\alpha} = R_s \cdot I_{s\alpha} + \frac{d\varphi_{s\alpha}}{dt} \\ V_{s\beta} = R_s \cdot I_{sq} + \frac{d\varphi_{s\beta}}{dt} \\ 0 = R_r \cdot I_{r\alpha} + \frac{d\varphi_{rd}}{dt} - \omega_m \cdot \varphi_{r\beta} \\ 0 = R_r \cdot I_{r\beta} + \frac{d\varphi_{r\beta}}{dt} + \omega_m \cdot \varphi_{r\alpha} \end{cases} \tag{6}$$

The asynchronous motor was modeled using the bond graph formalism as part of Grace GANDANEGARA's PhD thesis (June Gandanegara, 2003). The Bond graph technique is a powerful tool used for systemic modeling. The bond graph provides a graphical description of the multi-physical systems dynamic behavior. It describes the power flow and models the transformation between domains (electrical, mechanical, and hydraulic…etc.). In addition, it provides a uniform representation that provides a unique and understandable way to deal with the physical model of such a system (Saisset, 2004, Ben Ali, 2010).

Thus, the bond graph model of the induction motor (IM) modeled in the (α,β) frame, with the totalized stator leakage inductances, is given as follows:

Figure 20. Induction motor Bond graph model in the reference frame (α, β)

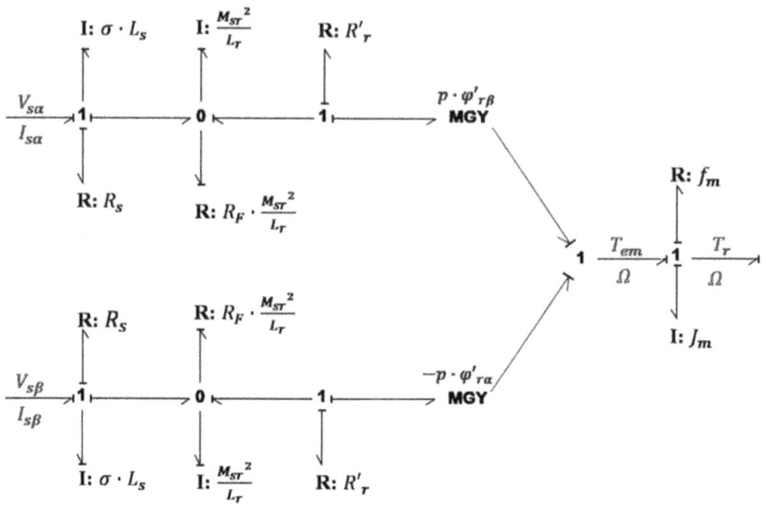

The resulting electromagnetic torque is the sum of the two torques obtained by the axes α and β:

$$T_{em} = T_{em_\alpha} + T_{em_\beta} = p \bullet \left(\varphi_{r\beta} \bullet I_{r\alpha} - \varphi_{r\alpha} \bullet I_{r\beta}\right) \tag{7}$$

Dynamic modeling of such a device is necessary for controlling energy flows in the overall desalination system. In order to obtain a quasi-static model for optimizing the energy management of this system we have to pass from the bond graph model (dynamic) of the hydro-mechanical process (motors, pumps, hydraulic loads, etc.) to simplified analytical equations. For this modeling, we have two approaches to follow:

- Those resulting from experimentation: $Q[l/min] = f(\mathscr{P}_{IM}[W])$, at a given hydraulic load $P[bar] = f(Q[l/min])$, (see experimental characterization section).
- Those $(\mathscr{P}_{IM}[W] = f(Q),\ d'où : Q[l/min] = f^{-1}(\mathscr{P}_{IM}))$ resulting from the analytical equations modeling the overall hydro-mechanical process in dynamic mode. In this case, we can neglect the electrical and mechanical dynamics (fast dynamics) and keep the hydraulic dynamics (slow dynamics). In this case, to calculate the flow rate as a function of the electrical power supply $\mathscr{P}_{IM} = f(Q)$, we used a dedicated function for solving the multi-variable nonlinear equations, $fsolve=f^{-1}$.

It should be noted that $\mathscr{P}_{IM}$ denotes the total electrical power of the set {voltage inverter + IM}.

Let us recall that our objective is to model the complete hydro-mechanical process. Therefore, in this work, we focus on establishing the quasi-static model of the motor pump following the second approach, and to taking advantage of the experimental model for static validation.

Let us proceed step by step and start with the induction motor (IM). Based on the original electrical equations (4) and mechanical equations of the stator and rotor quantities in the axis frame $(d-q)$ modeling the induction motor in dynamic mode, we can deduce its quasi-static model in steady state (IM with field-oriented control). This model is characterized by the analytical relationship between the electrical input power of the induction motor, the electromagnetic torque and the mechanical speed $\left(\mathcal{P}_{IM} = \mathcal{P}_{IM}\left(T_{em}, \Omega\right)\right)$.

The second step will be the expression of the mechanical variables of the induction motor $\left(T_{em}\ et\ \Omega\right)$ as a function of the pump flow rate $\left(T_{em}(Q)\ and\ \Omega(Q)\right)$ in order to define the complete motor pump quasi-static model, and consequently that of the hydro-mechanical process $\left(\mathcal{P}_{IM} = \mathcal{P}_{IM}(Q)\right)$.

i. IM Stator Voltage Vector Reformulation

From the system of electrical equations (4), we can express the stator voltages as follows:

$$\begin{cases} V_{sd} = R_{sr} \cdot I_{sd} + \sigma . L_s \cdot \frac{dI_{sd}}{dt} - \omega_{mr} \cdot \sigma \cdot L_s \cdot I_{sq} - R_r \cdot \frac{M_{sr}}{L_r^2} \cdot \varphi_{rd} \\ V_{sq} = R_{sr} \cdot I_{sq} + \sigma \cdot L_s \cdot \frac{dI_{sq}}{dt} + \omega_{mr} \cdot \sigma \cdot L_s \cdot I_{sd} + \omega_m \cdot \frac{M_{sr}}{L_r} \cdot \varphi_{rd} \end{cases} \tag{8}$$

where: $R_{sr} = R_s + R_r \bullet \left(\frac{M_{sr}}{L_r}\right)^2$

In steady state, system (8) is reduced to:

$$\begin{cases} V_{sd} = R_{sr} \cdot I_{sd} - \omega_{mr} \cdot \sigma \cdot L_s \cdot I_{sq} - R_r \cdot \frac{M_{sr}}{L_r^2} \cdot \varphi_{rd} \\ V_{sq} = R_{sr} \cdot I_{sq} + \omega_{mr} \cdot \sigma \cdot L_s \cdot I_{sd} + \omega_m \cdot \frac{M_{sr}}{L_r} \cdot \varphi_{rd} \end{cases} \tag{9}$$

and:

$$\omega_{mr} = \frac{d\theta_a}{dt} = \omega_m + \omega_r \tag{10}$$

where:

ω_r: Rotor pulsation.
ω_m: The electrical angular speed of the rotor.
ω_{mr} : The stator pulsation (or the electrical rotation speed of the rotor flux vector).

Thus, from equation (10) the stator voltages (9) can be expressed as a function of the stator currents, electrical speed of the rotor and the rotor pulsation:

$$\begin{cases} V_{sd} = V_{sd}\left(I_{sd}, I_{sq}, \omega_m, \omega_r\right) \\ V_{sq} = V_{sq}\left(I_{sd}, I_{sq}, \omega_m, \omega_r\right) \end{cases} \tag{11}$$

ii. Stator Currents and Rotor Pulsation Reformulation

From the expressions (5) of the fluxes, we deduce the relation between the stator and rotor currents in quadrature:

$$I_{sq} = -\frac{L_r}{M_{sr}} \bullet I_{rq} \tag{12}$$

From equation (12) and the general expression of the electromagnetic torque, the expression of the latter becomes:

$$T_{em} = \frac{p \bullet M_{sr}}{L_r} \bullet \varphi_{rd} \bullet I_{sq} \tag{13}$$

Thus, from (13) we can deduce that the expression of the stator current in quadrature is a function of the electromagnetic torque:

$$I_{sq} = \frac{L_r}{p \bullet M_{sr} \bullet \varphi_{rd}} \bullet T_{em} = I_{sq}(T_{em}) \tag{14}$$

The rotor pulsation is expressed as follows:

$$\omega_r = \frac{R_r \bullet M_{sr}}{L_r \bullet \varphi_{rd}} \bullet I_{sq}(T_{em}) = \omega_r(T_{em}) \tag{15}$$

In steady state, the direct rotor current is zero (I_{rd}=0). Furthermore, during transient state of the rotor flux, the variation law of the direct rotor current is as follows:

$$I_{rd} = \frac{\varphi_{rd}}{L_r} - \frac{M_{sr}}{L_r} \bullet I_{sd} \tag{16}$$

Thus, from (16) we obtain the relation between I_{sd} and φ_{rd}:

$$R_r \bullet \varphi_{rd} + L_r \bullet \frac{d\varphi_{rd}}{dt} = M_{sr} \bullet R_r \bullet I_{sd} \tag{17}$$

In steady state, this relationship is reduced to:

$$\varphi_{rd} = M_{sr} \bullet I_{sd} \tag{18}$$

so:

$$I_{sd} = \frac{\varphi_{rd}}{M_{sr}} \tag{19}$$

iii. Reformulation of the asynchronous motor input electrical power as a function of rotor electromagnetic torque and mechanical speed

From expressions (14), (15) and (19) we obtain the following system:

$$\begin{cases} I_{sd} = \frac{\varphi_{rd}}{M_{sr}} = constant \\ I_{sq} = \frac{L_r}{p \cdot M_{sr} \cdot \varphi_{rd}} \cdot C_{em} = I_{sq}(T_{em}) \\ \omega_r = \frac{R_r \cdot M_{sr}}{L_r \cdot \varphi_{rd}} \cdot I_{sq}(C_{em}) = \omega_r(T_{em}) \\ \omega_m = p \cdot \Omega = \omega_m(\Omega) \end{cases} \tag{20}$$

From this system, the stator voltages are expressed as a function of the electromagnetic torque and the mechanical speed:

$$\begin{cases} V_{sd} = V_{sd}\left(I_{sq}(T_{em}),\ \omega_r(T_{em}), \omega_m(\Omega)\right) = V_{sd}(T_{em}, \Omega) \\ V_{sq} = V_{sq}\left(I_{sq}(T_{em}),\ \omega_r(T_{em}), \omega_m(\Omega)\right) = V_{sd}(T_{em}, \Omega) \end{cases} \tag{21}$$

In our case, where the motor pump is powered by an electrical source via a DC bus, the input electrical power of the asynchronous motor is expressed as follows:

$$\mathscr{P}_{IM} = V_{bus} \bullet I_{pump} = V_{sd} \bullet I_{sd} + V_{sq} \bullet I_{sq} = \mathscr{P}_{IM}(T_{em}, \Omega) \tag{22}$$

Finally, the following analytical equation describes the quasi-static model of the asynchronous motor driving the centrifugal pump:

$$\mathscr{P}_{IM}(T_{em}, \Omega) = R_{sr} \bullet \left(\frac{\varphi_{rd}}{M_{sr}}\right)^2 - R_r \bullet \left(\frac{\varphi_{rd}}{L_r}\right)^2 + R_{sr} \bullet \left(\frac{L_r}{M_{sr} \bullet \varphi_{rd}}\right)^2 \bullet T_{em}{}^2 + \Omega \bullet T_{em} \tag{23}$$

It now remains to express each of the variables $(\Omega\ et\ T_{em})$ as a function of the motor pump flow rate to formulate the electrical power supplying the induction motor as a function of the discharge flow rate of the associated pump. This expression is valid for both hydro-mechanical processes of the desalination system

WP CENTRIFUGAL PUMP MODELING

The fundamental equations of the centrifugal pump, modeling the conversion from the mechanical domain to the hydraulic domain, under the assumption of power conservation are given as follows

$$P_{eff}(bar) = (a \bullet \Omega + b \bullet Q) \bullet \Omega \tag{24}$$

$$T_r(N \bullet m) = (a \bullet \Omega + b \bullet Q) \bullet Q \tag{25}$$

with:
P_{eff}: Effective pressure in [bar].
Q: Discharge flow of the pump in [l/min].

T_r: Load torque on the shaft in [N.m].

Ω: Angular speed of the pump impeller and the shaft in [rad/s].

Indeed, these two equations are established for a wheel equipped with an infinity of blades and a flow perfectly stuck and tangent to the blades (without pressure losses).

Furthermore, on practice, the centrifugal pump drive by an electric motor, and consequently the flow of the fluid inside this pump, causes dynamic phenomena and loads losses, whether mechanical (friction and inertia at the level of the pump shaft and its coupling with the drive motor (J_m+J_p, f_m+f_p) or hydraulic. The latter are caused by friction inside the wheel and shocks to the blades according to the variation of the flow rate. These are translated in the form of pressure loss as a function of the square of the fluid flow rate ($\Delta P = -c \bullet Q^2$).

From the power balance, the mechanical equation modeling the power transfer from the motor to the centrifugal pump via the shaft is given as follows:

$$J \bullet \frac{d\Omega}{dt} = T_{em} - T_r - \left(f_m + f_p\right) \bullet \Omega \tag{26}$$

With:

C_{em} : Electromechanical torque of the electric drive motor in [N.m].

J: Total inertia of the motor pump (J_m+J_p) en [Kg.m²].

f_p: Mechanical losses on the pump side in[N.m/(rad/s)].

f_m: Mechanical losses on the motor side in [N.m/(rad/s)].

In steady state, the mechanical equation becomes:

$$T_{em} - T_r - \left(f_m + f_p\right) \bullet \Omega = 0 \tag{27}$$

Thus, the equations for the pump discharge pressure (P) and the load torque (T_r), used to model the static part of the pump, are expressed as follows:

$$P\,(bar) = (a \bullet \Omega + b \bullet Q) \bullet \Omega + c \bullet Q^2 \tag{28}$$

$$T_r(N \bullet m) = (a \bullet \Omega + b \bullet Q) \bullet Q + f_p \bullet \Omega \tag{29}$$

The constants (a, b, c, and f_p) are determined through experimentation.

Figure 21 presents the dynamic model, in bond graph, of the single-stage centrifugal pump (a single impeller).

The centrifugal pump in this process is characterized by a typical hydraulic load, such as the discharge of water at a certain manometric height is imposed by the installation. This typical load constitutes an essential element in the modeling of such a process given the notable influence of this load, whatever its type, on the operation of this process. This has been proven through experimental characterization.

Figure 21. Bond graph model of the WP single-stage pump

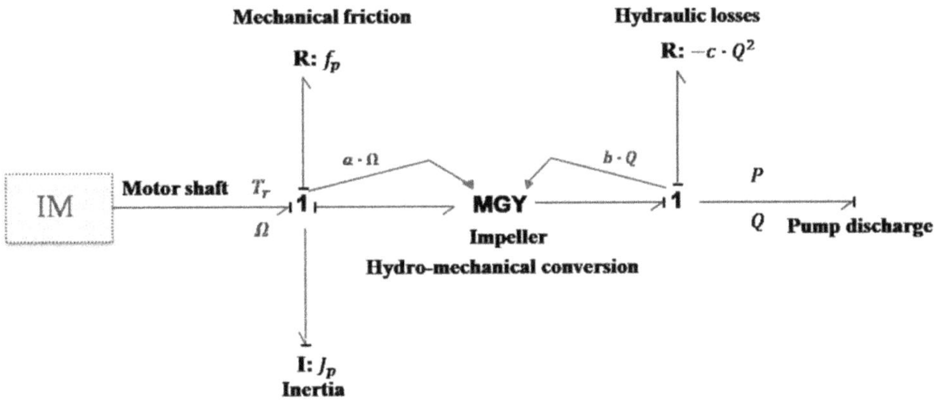

HYDRAULIC NETWORK AND THE STORAGE SYSTEM MODELING

Indeed, the hydraulic network is characterized by load losses as described in previous section of this chapter (Eq.1). This is modeled in bond graph by the combination of:

- A force source $(\boldsymbol{Se} : \rho . g . H_{geo})$ modeling the gravity pressure resulting from the total geometric height of the process.
- A resistive element ($\boldsymbol{R} : K \bullet Q^2$) modeling the hydraulic dynamic losses caused by the presence of accidents on the discharge pipe (brackish water pipe).

The bond graph model of the water pumping process is presented in the following figure:

Figure 22. water pumping process Bond graph model with storage tank

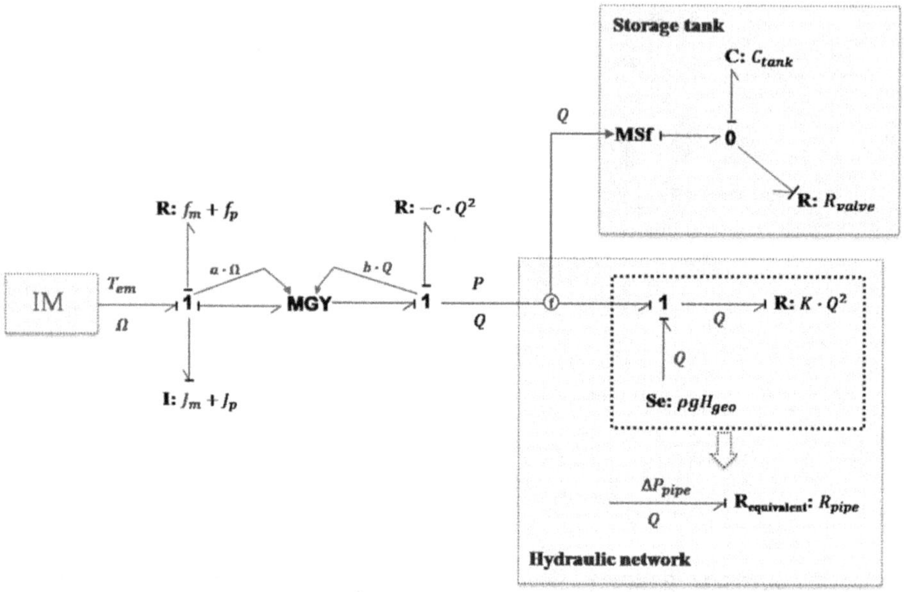

For the hydraulic storage tank, it was modeled by a capacitive storage element$(\boldsymbol{C} : C_{tank})$ and a resistive element for the valve $(\boldsymbol{R} : R_{valve})$. The water flow rate of the tank feed is modeled by a variable flow source($\boldsymbol{MSf} : Q$).

Indeed, the tank model and the hydraulic network model are not continuous but are coupled by a measurement (*f*) of the discharge flow rate of the motor pump.

Thus, through equation (29) we can deduce the following:

$$T_{em} = (a \bullet \Omega + b \bullet Q) \bullet Q + \left(f_m + f_p\right) \bullet \Omega = T_{em}(Q,\Omega) \; (30)$$

In the case of the pumping process (WP with typical hydraulic load), we have:

$$\begin{cases} \Delta P_{pipe} = K \cdot Q^2 + \rho \cdot g \cdot H_{geo}: Equationdechargehydrauliquetypique \\ P = (a \cdot \Omega + b \cdot Q) \cdot \Omega + c \cdot Q^2: Dischargepressure \\ Aupointdefonctionnementona: P = \Delta P_{pipe} \end{cases} \quad (31)$$

Therefore, the expression for the mechanical velocity, by equating the two pressure expressions from the system of equations (31), is given as follows:

$$\Omega = \frac{-b \bullet Q + \sqrt{(b \bullet Q)^2 - 4 \cdot a \cdot \left[(c - K) \bullet Q^2 - \rho \bullet g \bullet H_{geo}\right]}}{2 \cdot a} = \Omega(Q) \quad (32)$$

We can then express the electromagnetic torque as a function of the pump flow rate:

$$T_{em} = (a \bullet \Omega(Q) + b \bullet Q) \bullet Q + (f_m + f_p) \bullet \Omega(Q) = T_{em}(Q) \quad (33)$$

By replacing the expressions for the electromagnetic torque (33) and the rotation speed (32) of the induction motor in the expression for the electrical power (23), we obtain the quasi-static model describing the pumping process:

$$\mathscr{P}_{IM} = \mathscr{P}_{IM}(T_{em}(Q), \Omega(Q)) = \mathscr{P}_{IM}(Q) \quad (34)$$

Avec: $\mathscr{P}_{IM}$ is the total power of the set {inverter + IM}.

- *developed quasi-static model validation*

To validate the developed quasi-static model, we plotted the simulation results of this model in an electrical power range, with that from the experimental tests at different operating points (variable frequency) for a given hydraulic load (valve fixed in position Pos.2). Regarding the hydraulic load equation presented by equation Eq.31 integrated in the quasi-static model, we replaced it with an expression extrapolated from the experimental hydraulic load curve (for a valve position Pos.2 (see Figure.6)) to refine the model used. This expression is given as follows:

$$P_{WP} = 2.167 \bullet 10^{11} \times (Q_{WP})^2 - 3.477 \bullet 10^7 \times Q_{PBP} + 6.556 \bullet 10^4 \equiv K.(Q_{WP})^2 + \rho.g.H_{geo} \quad (35)$$

With:
P_{WP} : The discharge pressure of the WP motor pump in Pascal [Pa].
Q_{WP} : The discharge flow rate of the WP motor pump in [m^3/s].
Figure 23 shows the validation of the developed model.

Figure 23. Comparison of the developed and experimental WP motor pump quasi-static model at a valve position Pos 2

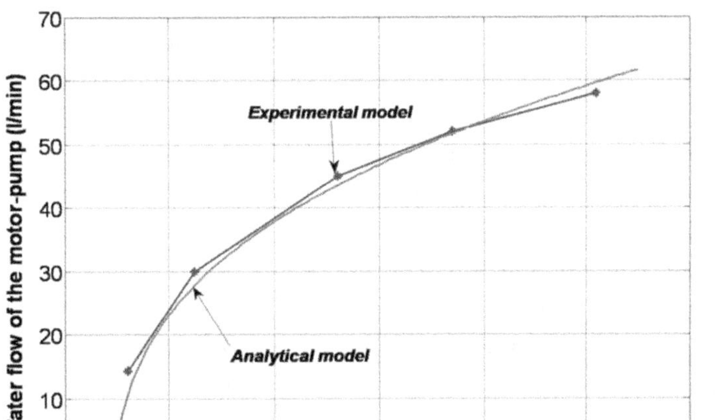

BRACKISH WATER DESALINATION PROCESS MODELING

The RO desalination process is illustrated in Figure 24. It is composed of a multi-stage high-pressure centrifugal motor pump (HPP), two hydraulic storage tanks $((T_1, T_2)$ and an RO module with a single membrane.

Figure 24. Schematic diagram of the water RO desalination process hydraulic structure

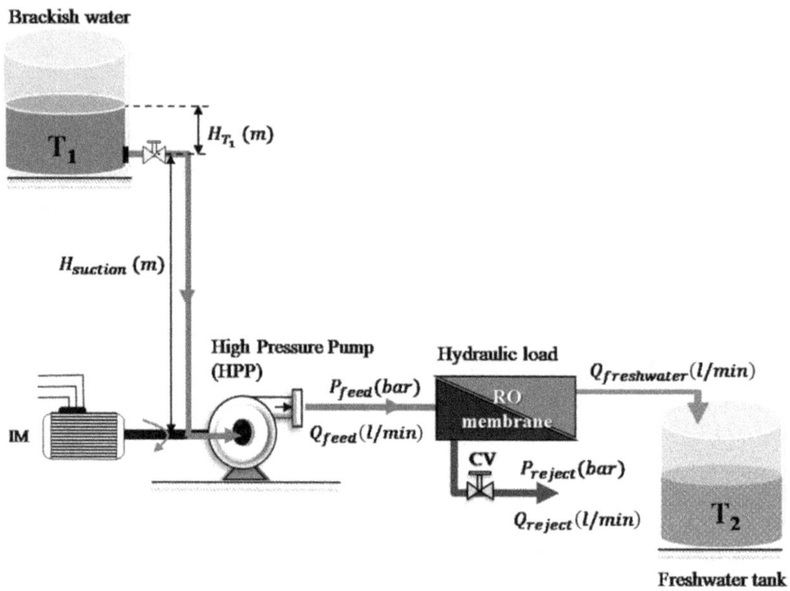

REVERSE OSMOSIS SYSTEM MODELING

The RO membrane is the main and most complex element in the desalination process. Indeed, the osmosis unit is a multidisciplinary system connecting various energy flows (hydraulic, chemical and thermal). Given its complexity, several chemistry and energy research laboratories, are investing in the characterization and modeling of this type of membrane phenomena at the microscopic scale. Thus, we present the bond graph model of the RO module in Figure 25. The different parameters of the osmosis unit model were determined and validated through the experimental characterization of the desalination process (Turki et al., 2008, Ben Ali et al., 2014).

In order to ensure the control of the motor pumps and the energy management of the overall system, we restricted ourselves to a purely "hydraulic" model of the reverse osmosis system. In fact, the reverse osmosis system involves energy losses in the pipes $(\boldsymbol{R} : R_{pipe})$, the Control Valve VC $(\boldsymbol{MR}{:}R_{valve})$ at the discharge (the variation in the position of the VC generates a variation in hydraulic load at the reverse osmosis system) and in the membrane. The latter is represented by a hydraulic storage element $(\boldsymbol{C} : C_{memb})$

(often neglected) and a variable resistive element of load loss $(\boldsymbol{MR}:R_{memb})$ which depends on the salinity, the temperature of the brackish water and the permeability of the membrane.

The equivalent dynamic hydraulic circuit, presented in Figure 26, allows to simplify the model and facilitates the task of identifying the different parameters. To do this, we made approximations of the elements that do not have a great influence on the RO process. Indeed, we neglected the pressure of the fresh water, the pressure of the load losses (which are generally low), and the osmotic pressure since the brackish water salinity is not very high in the current study (2 to 6 g/l). In static mode, the effect of the capacity C_{memb} becomes negligible. Therefore, the quasi-static model of the reverse osmosis system can be reestablished through the following equations:

$$P_{feed} - \Delta P_{pipe} = \left(R_{module} + R_{valve}\right) \bullet Q^2_{reject} \cong P_{feed} \tag{36}$$

$$Q_{freshwater} = \frac{P_{feed} - \Delta P_{pipe}}{R_{memb}} = \frac{P_{feed}}{R_{memb}} \tag{37}$$

$$Q_{feed} = Q_{freshwater} + Q_{reject} \tag{38}$$

With:

$Q_{feed}, Q_{freshwater}, Q_{reject}$: denote, respectively, the feed water flow rate of the membrane, the flow rate of the produced freshwater and the rejected water flow rate.

P_{feed}: The feed pressure of the membrane.

ΔP_{pipe}: Pressure drop at the hydraulic pipe. Generally, this loss is negligible compared to the feed pressure P_{feed}.

Figure 25. RO module Bond graph model

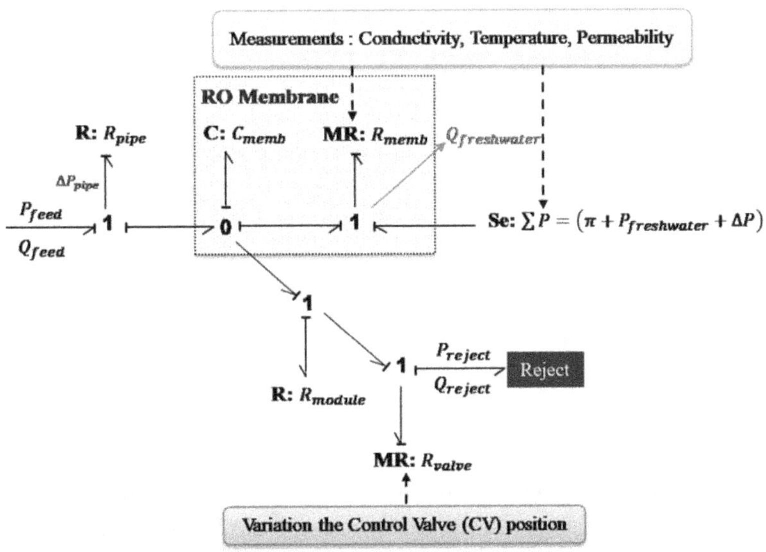

Figure 26. RO module Equivalent hydraulic circuit

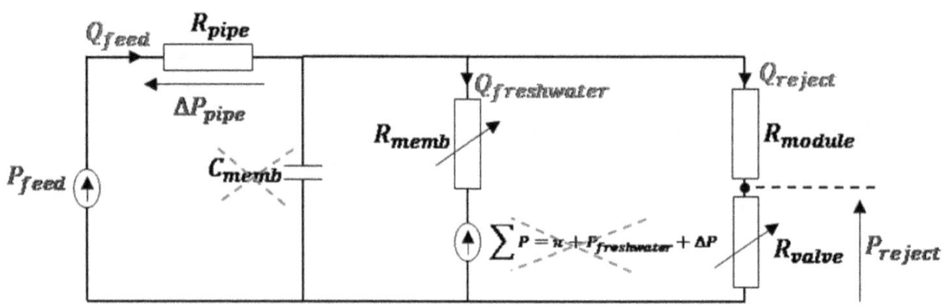

From (36) and (37), expression (38) becomes:

$$Q_{feed} = \frac{P_{feed}}{R_{memb}} + \sqrt{\frac{P_{feed}}{R_{module} + R_{valve}}} \tag{39}$$

From (39), we obtain a second degree equation. The solution is as follows:

$$\sqrt{P_{feed}} = \sqrt{\left(\frac{R_{memb}^{2}}{4(R_{module} + R_{valve})} + R_{memb} \bullet Q_{feed}\right)} - \frac{R_{memb}}{2\sqrt{R_{module} + R_{valve}}} \tag{40}$$

then:

$$P_{feed} = \left(\sqrt{\frac{R_{memb}^{2}}{4(R_{module} + R_{valve})} + R_{memb} \bullet Q_{feed}} - \frac{R_{memb}}{2\sqrt{R_{module} + R_{valve}}}\right)^{2} = P_{feed}\left(Q_{feed}\right) \tag{41}$$

This analytical expression (41) describes the *hydraulic load* analytical model (the osmosis unit) for the desalination process, at any given raw water salinity and at a *fixed* Control Valve (CV).

HPP PUMP MODELING

The multi-stage centrifugal pump modeling (Figure 27) is deduced from its real behavior. Since the pump impellers are grouped on the same shaft, the torque is divided according to the number of impellers (n).

The impeller speeds are very close (*Jonction* 1) given the mechanical stiffness between two successive impellers causing minimal speed differences. These stiffnesses are materialized by high values $\left(\textit{Jonction}\ 0 : \textit{élément}\ \boldsymbol{C} : \frac{1}{k}\right)$. Similarly, thanks to *Jonction* 1 at the output of the model, the different pressures provided by each impeller are added together:

$$P\left(bar\right) = P_1 + P_2 + \cdots + P_n = n \bullet P_{impeller} \tag{42}$$

with:

$$Q(l/min) = Q_{impeller} = Q_1 = Q_2 = \cdots = Q_n \quad (43)$$

Figure 27. HPP multistage pump bond graph model

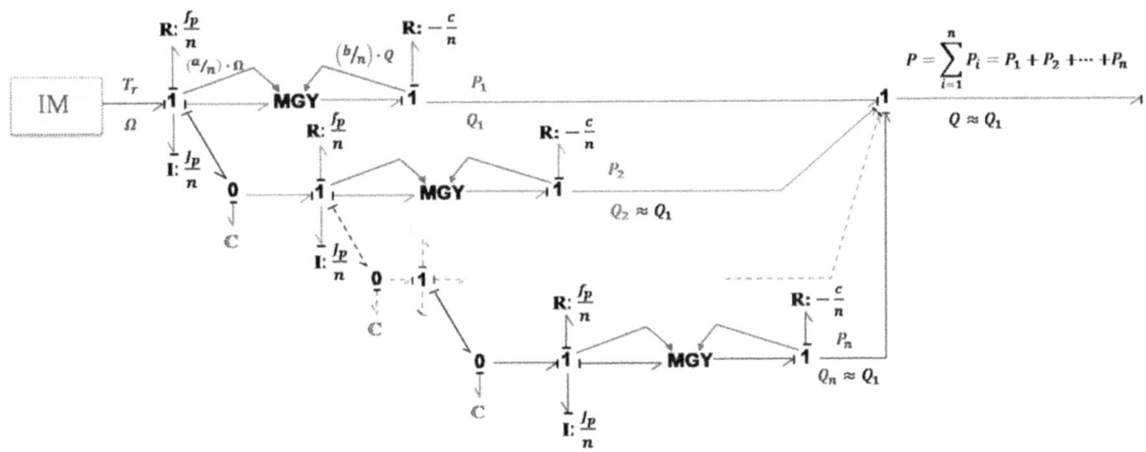

Therefore, the expression for the resulting pressure of the PHP pump is given as:

$$P_{HPP} = (a \cdot \Omega + b \cdot Q_{HPP}) \cdot \Omega + c \cdot Q_{HPP}{}^2 \quad (44)$$

This relationship describes the HPP motor pump hydraulic characteristic.

According to the desalination process hydraulic structure (see Figure 24) the reverse osmosis unit feed pressure is related to the different pressures as follows:

$$P_{feed} = P_{T_1} + P_{suction} + P_{HPP} \quad (45)$$

$$P_{T_1} = \rho \cdot g \cdot H_{T_1} = \rho \cdot g \cdot (L_1 - L_{1_{min}}) \quad (46)$$

$$P_{suction} = \rho \cdot g \cdot H_{suction} \quad (47)$$

with:

$P_{suction}$, $H_{suction}$: are respectively the pressure (gravity) in [bar] and the suction height of the brackish water by the HPP pump in [m].

P_{T_1}: The resulting pressure (gravity) of the water stored in the tank T_1 in [bar].

L_1: The total water stored level in T_1 in [m].

$L_{1_{min}}$: The minimum storage limit in T_1 in [m] (cf. Figure 30).

It should be noted that this hydraulic storage tank (T_1) installed at height allows to increase the feed pressure of the osmosis unit (P_{feed}) thanks to the free pressure provided by the gravitational effect, consequently improving the desalination process energy efficiency.

Finally, the bond graph model of the overall hydro-mechanical structure for the desalination unit is given by Figure 28.

Figure 28. Bond graph model for the overall hydro-mechanical desalination system process

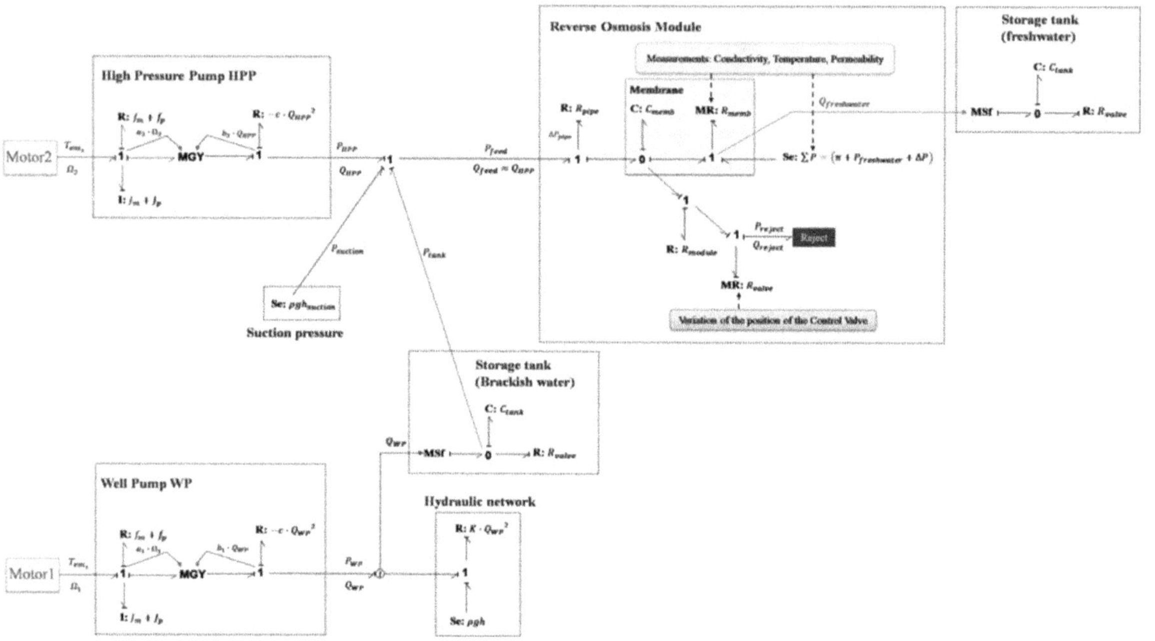

From expressions (41) and (44), (45) is expressed as follows:

$$P_{feed}\left(Q_{feed}\right) = a \bullet \Omega^2 + b \bullet Q_{feed} \bullet \Omega + c \bullet Q_{feed}{}^2 + \rho \bullet g \bullet \left(L_1 - L_{1\,min} + h_{suction}\right) \tag{48}$$

where: $Q_{feed} = Q_{HPP}$

The solution to the second degree equation (48) is:

$$\Omega\left(Q_{feed}\right) = \frac{-b \bullet Q_{feed} + \sqrt{\left(b \bullet Q_{feed}\right)^2 - 4a \bullet \left[c \bullet Q_{feed}{}^2 + \rho \bullet g \bullet \left(L_1 - L_{1\,min} + h_{suction}\right) - P_{feed}\left(Q_{feed}\right)\right]}}{2a} \tag{49}$$

Similarly, for the desalination process, the expression of the electrical power given by equation (23) is expressed as a function of the HP motor pump flow rate:

$$\mathscr{P}_{IM} = \mathscr{P}_{IM}\left(T_{em}\left(Q_{HPP}\right), \Omega\left(Q_{HPP}\right)\right) = \mathscr{P}_{IM}\left(Q_{HPP}\right) \tag{50}$$

This relationship, $\mathcal{P}_{IM}(Q_{HPP})$, describes the desalination process quasi-static model (the set {HP motor pump and the reverse osmosis unit}).

Where,, $\mathcal{P}_{IM}$ denotes the total electrical power of the set {inverter + IM} associated with the PHP pump.

- *Validation of the developed quasi-static model*

Figure 29 shows the validation of the hydro-mechanical desalination process quasi-static model through the model resulting from the experimental tests.

Indeed, in the quasi-static model we implemented an extrapolated expression for the hydraulic load curve $P_{feed}(Q_{feed})$ resulting from the experiments in order to refine the adopted model. This expression, equivalent to the developed equation Eq.41, is given as follows:

$$P_{feed} = 2.309 \bullet 10^5 \times exp\left(5155 \bullet Q_{feed}\right) \tag{51}$$

With, P_{feed} is expressed in [Pascal] and Q_{feed} is expressed in [m^3/s].

Figure 29. Comparison of the analytical and experimental HPP motor pump quasi-static model

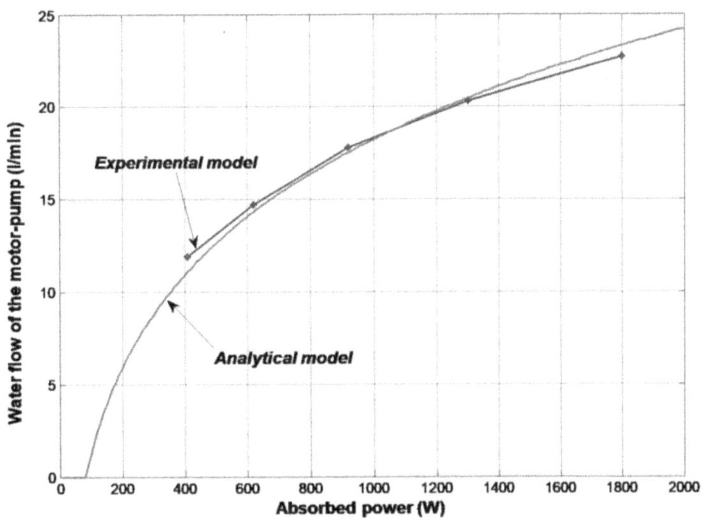

SETTING THE ENERGY MANAGEMENT PROBLEM

Hydraulic Storage Tank Modeling

The storage tank T_1 (Figure 30) is characterized by the stored brackish water level L_1 (Eq.46). This level is variable depending on the inlet flow rates Q_{WP} and outlet flow rates Q_{HPP} of this tank. It is calculated, at time $(t_1 + \Delta t)$, as follows:

$$L_1(t_1 + \Delta t) = L_1(t_1) + \frac{1}{S_1} \bullet \int_{t_1}^{t_1+\Delta t} \left(Q_{WP}(t) - Q_{HPP}(t)\right) dt \tag{52}$$

with, S_1 : being the surface area of the tank T_1 in [m^2].

Furthermore, the stored water level in this tank is limited by two levels (minimum and maximum). Indeed, the stored water volume is divided into two types of volume:

- Dead volume, where, $L_1 < L_{1_{min}}$. This volume of water will never be exploited because it is reserved for the sedimentation of brackish water raised from the well. Generally, 20 cm of height must be reserved for the dead volume (Figure 30).
- Useful volume, where $L_1 \in \left[L_{1_{min}} - L_{1_{max}}\right]$. This volume is limited by $L_{1_{max}}$ in order to avoid the reservoir overflowing. The useful volume is reserved to be exploited (by the desalination process in our case).

The inequality $L_{1_{min}} \leq L_1 \leq L_{1_{max}}$, presents the constraint on the element of the hydraulic storage T_1. It is necessary to restrict the evolution of the filling state of this tank in such a way that it remains in the recommended interval.

The tank T_2 is characterized by the stored freshwater level L_2. It is calculated, at time $((t_1 + \Delta t)$, as follows

$$L_2(t_1 + \Delta t) = L_2(t_1) + \frac{1}{S_2} \bullet \int_{t_1}^{t_1+\Delta t} Q_{freshwater}(t)\, dt \tag{53}$$

with, S_2 : being the surface area of the tank T_2 (m^2) (Figure 31).

Figure 30. water volume distribution in a tank

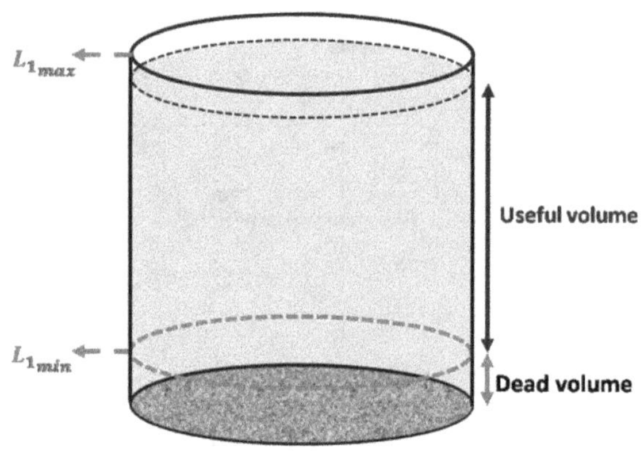

Figure 31. Hydraulic configuration for our case study

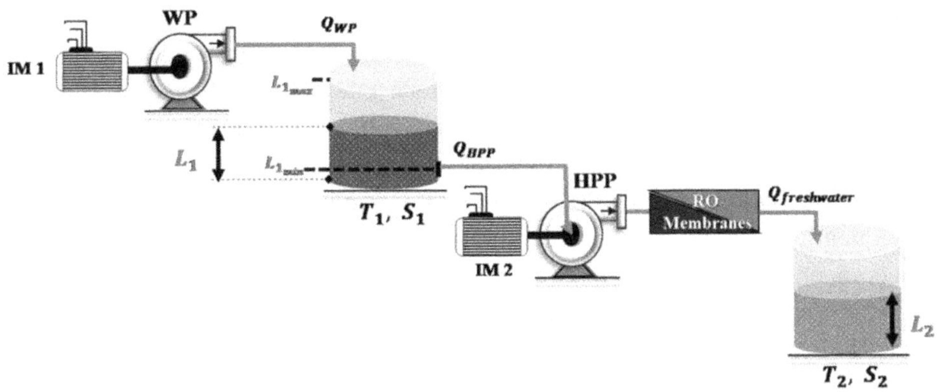

At this level, the C_2 tank is not taken into consideration in the energy manager since our configuration only includes two motor pumps. In the case of a third motor pump for the produced water distribution stored in C_2, the latter must be considered in the management strategy (through L_2 as the third input variable of the manager).

Energy Management Problem

Finally, the developed quasi-static model' of the BWRO desalination unit is the main component for developing the energy-water management strategy. Thus it will allow us to maintain the stability of the continuous bus and to control the flow of electrical power transferred from the hybrid source to the different hydro-mechanical processes (pumping and desalination) of the desalination system by ensuring the hydraulic level stored in the tanks , and respecting the technological constraints of the system.. To answer this essential question, we have defined an energy management strategy allowing, given the generated electrical power " according to the climatic conditions, and given the characteristic of the hydraulic load , to make the adequate distribution of the given electrical power between the different motor pumps according to a "power sharing factor, "α" defined as follows:

$$\begin{cases} P_{HPP}^{*} = \alpha \cdot P_{dc} \\ P_{WP}^{*} = (1-\alpha) \cdot P_{dc} \end{cases}$$
$$\text{subject to:}$$
$$0 \leq \alpha \leq 1 \tag{54}$$
$$P_{pump_i}^{min} \leq P_{pump_i} \leq P_{pump_i}^{max};\ i = \{1,2\}$$
$$L_1^{min} \leq L_1 \leq L_1^{max}$$

This factor should be determined according to both the instantaneous generated power and the current filling state of the brackish water storage and fresh water tanks given by (52,53)

CONCLUSION

In this study, a bond graph modeling approach is presented for a BWRO desalination unit powered by a hybrid renewable energy source (PV-wind system). The choice of the bond graph formalism is justified as it is a powerful dynamic modeling technique that enables the unified representation of various energy transfers, particularly in complex multidisciplinary systems involving different domains (electrical, mechanical, hydraulic, chemical, thermal, etc.). In this context, for each element (BP, HPP, and RO membrane), the hydraulic equivalent sub-model was derived based on its prevailing hydraulic behaviors. Subsequently, the bond graph model of the entire desalination system, including the electrical and mechanical components, was constructed by integrating the aforementioned sub-models to perform simulations. To validate the developed model, the experimental results were compared with simulations conducted using 20-Sim. The accuracy of the bond graph-based modeling was confirmed by demonstrating adherence to the "Pump Affinity Laws" (Energy Design Resources, 2010). Based on these results, the prerequisites for developing control strategies and energy management for the autonomous BWRO desalination system can be established, taking into account system constraints and objectives."

ACKNOWLEDGMENT

This work was supported by the Tunisian Ministry of High Education and Research under Grant LSE-ENIT-LR 11ES15, and CMCU-12G1103 project for financial support.

REFERENCES

Al-Karaghouli, A., & Kazmerski, L. (2013). Energy consumption and water production cost of conventional and renewable-energy-powered desalination processes. *Renewable & Sustainable Energy Reviews, 24*, 343–356. DOI: 10.1016/j.rser.2012.12.064

Al-Karaghouli, A., Renne, D., & Kazmerski, L. (2009). Solar and wind opportunities for water desalination in the Arab regions. *Renewable & Sustainable Energy Reviews, 13*(9), 2397–2407. DOI: 10.1016/j.rser.2009.04.003

Baggio, G., Qadir, M., & Smakhtin, V. (2021). Freshwater availability status across countries for human and ecosystem needs. *The Science of the Total Environment, 792*, 148230. DOI: 10.1016/j.scitotenv.2021.148230 PMID: 34147805

Ben Ali, I. (2010). Caractérisation expérimentale, commande et mise en place d'un gestionnaire d'énergie d'un procédé de dessalement de l'eau alimenté par une source hybride photovoltaïque-éolien. Master's thesis, LSE-ENIT Tunisie

Ben Ali, I., Turki, M., Belhadj, J., & Roboam, X. (2014). Systemic design of a reverse osmosis desalination process powered by hybrid energy system: Bond graph modeling approach & experimental validation. In *Proceedings of the 1st International Conference on Electrical Sciences and Technologies in Maghreb (CISTEM)*, Tunis, Tunisia, 3–6 November. DOI: 10.1109/CISTEM.2014.7076941

Ben Ali, I., Turki, M., Belhadj, J., & Roboam, X. (2012, March 25-28). Energy management of reverse osmosis desalination process powered by a hybrid renewable energy source. *16th IEEE Mediterranean Electrotechnical Conference (MELECON)*, Yasmine Hammamet, Tunisia, 800–805

Ben Ali, I., Turki, M., Belhadj, J., & Roboam, X. (2014, May 19-22). Fuzzy logic-based power management for reverse osmosis desalination process fed by intermittent hybrid renewable source. *11th International Conference on Modeling and Simulation of Electric Machines, Converters, and Systems (ElectrIMACS)*, Valencia, Spain

Ben Ali, I., Turki, M., Belhadj, J., & Roboam, X. (2016, March 22-24). Using quasi-static model for water/power management of a stand-alone wind/photovoltaic/BWRO desalination system without batteries. *7th International Renewable Energy Congress (IREC)*, Hammamet, Tunisia.

Bett, P. E., & Thornton, H. E. (2016). The climatological relationships between wind and solar energy supply in Britain. *Renewable Energy, 87*, 96–110. DOI: 10.1016/j.renene.2015.10.006

Buzila Pankovits, P. V. (2015). Gestion énergétique optimale des installations fixes de traction électrique ferroviaire hybrides. (PhD thesis, Université Lille 1, Sciences et Technologies, France, September 2015).

El-Ghonemy, A. M. K. (2012). Water desalination systems powered by renewable energy sources [Review]. *Renewable & Sustainable Energy Reviews, 16*(3), 1537–1556. DOI: 10.1016/j.rser.2011.12.014

Eltawil, M. A., Zhengming, Z., & Yuan, L. (2009). A review of renewable energy technologies integrated with desalination systems. *Renewable & Sustainable Energy Reviews, 13*(9), 2245–2262. DOI: 10.1016/j.rser.2009.06.011

Energy Design Resources. (2010). Design briefs: Centrifugal pump application and optimization. Available at www.energydesignresources.com

Fahmy, F. H., Ahmed, N. M., & Farghally, H. M. (2012). Optimization of renewable energy power system for a small-scale brackish reverse osmosis desalination unit and a tourism motel in Egypt. In *Proceedings of the 8th International Conference on Electrical Engineering ICEENG 2012*. Military Technical College. DOI: 10.4236/sgre.2012.31006

Garcia-Rodriguez, L. (2003). Renewable energy applications in desalination: State of the art. *Solar Energy, 75*(5), 381–393. DOI: 10.1016/j.solener.2003.07.005

Ghaffour, N., Missimer, T. M., & Amy, G. L. (2013). Technical review and evaluation of the economics of water desalination: Current and future challenges for better water supply sustainability. *Desalination, 309*, 197–207. DOI: 10.1016/j.desal.2012.10.015

Ghermandi, A., & Messalem, R. (2009). Solar-driven desalination with reverse osmosis: The state of the art. *Desalination and Water Treatment, 7*(1–3), 285–296. DOI: 10.5004/dwt.2009.723

Greenlee, L. F., Lawler, D. F., Freeman, B. D., Marrot, B., & Moulin, P. (2009). Reverse osmosis desalination: Water sources, technology, and today's challenges. *Water Research, 43*(9), 2317–2348. DOI: 10.1016/j.watres.2009.03.010 PMID: 19371922

Gude, V. G., Nirmalakhandan, N., & Deng, S. (2010). Renewable and sustainable approaches for desalination. *Renewable & Sustainable Energy Reviews, 14*(9), 2641–2654. DOI: 10.1016/j.rser.2010.06.008

Guemri, M. (2013). Heuristiques optimisées et robustes de résolution du problème de gestion d'énergie pour les véhicules électriques et hybrides. (PhD thesis, Institut National Polytechnique de Toulouse-INPT, France, 2013).

International Energy Agency (IEA). (2024). Renewables 2024: Global overview. Retrieved from https://www.iea.org/reports/renewables-2024/global-overview

June Gandanegara, G. (2003). Méthodologie de conception systémique en génie électrique à l'aide de l'outil Bond. Mémoire de thèse, Université des Sciences et Technologies de Lille.

Jurasz, J., Canales, F. A., Kies, A., Guezgouz, M., & Beluco, A. (2020). A review on the complementarity of renewable energy sources: Concept, metrics, application, and future research directions. *Solar Energy, 195*, 703–724. DOI: 10.1016/j.solener.2019.11.087

Kanematsu, K., Abe, T., Tanaka, A., & Hamakawa, Y. (1988). Simulation design method of stand-alone PV power system and its applications for desalination plant. In *Proceedings of the 20th IEEE Photovoltaic Specialists Conference*, Volume 2, 1081–1086. https://doi.org/DOI: 10.1109/PVSC.1988.1059527

Khan, M. A. M., Rehman, S., & Al-Sulaiman, F. A. (2018). A hybrid renewable energy system as a potential energy source for water desalination using reverse osmosis: A review. *Renewable & Sustainable Energy Reviews*, *97*, 456–477. DOI: 10.1016/j.rser.2018.08.049

Levron, Y., Guerrero, J. M., & Beck, Y. (2013). Optimal power flow in microgrids with energy storage. *IEEE Transactions on Power Systems*, *28*(3), 2546–2554. DOI: 10.1109/TPWRS.2012.2236106

Majeed, Y., Khan, M. U., Waseem, M., Zahid, U., Mahmood, F., Majeed, F., Sultan, M., & Raza, A. (2023). Renewable energy as an alternative source for energy management in agriculture. *Energy Reports*, *10*, 344–359. DOI: 10.1016/j.egyr.2023.06.032

Miranda, M. S., & Infield, D. (2003). A wind-powered seawater reverse-osmosis system without batteries. *Desalination*, *153*(1–3), 9–16. DOI: 10.1016/S0011-9164(02)01129-8

Mohamed, E. S., & Papadakis, G. (2004). Design, simulation, and economic analysis of a stand-alone reverse osmosis desalination unit powered by wind turbines and photovoltaics. *Desalination*, *164*(1), 87–97. DOI: 10.1016/S0011-9164(04)00049-6

Mokheimer, E. M. A., Sahin, A. Z., Al-Sharafi, A., & Ali, A. I. (2013). Modeling and optimization of a hybrid wind–solar-powered reverse osmosis water desalination system in Saudi Arabia. *Energy Conversion and Management*, *75*, 86–97. DOI: 10.1016/j.enconman.2013.06.002

Murray, T., & Infield, D. (2003). A photovoltaic-powered seawater reverse osmosis system without batteries. *Desalination*, •••, 1–8. DOI: 10.1016/S0011-9164(02)01129-8

Novosel, T., Cosic, B., Krajacic, G., Duic, N., Puksec, T., Mohsen, M. S., & Ashhab, M. S. (2014). (in press). The influence of reverse osmosis desalination in combination with pump storage on the penetration of wind and PV energy: A case study for Jordan. *Energy*. Advance online publication. DOI: 10.1016/j.energy.2014.01.009

Park, G. L., Schafer, A. I., & Richards, B. S. (2009). Potential of wind-powered renewable energy membrane systems for Ghana. *Desalination*, *248*(1–3), 169–176. DOI: 10.1016/j.desal.2008.05.062

Richards, B. S., Capao, D. P. S., & Schafer, A. I. (2008). Renewable energy powered membrane technology. 2. The effect of energy fluctuations on performance of a photovoltaic hybrid membrane system. *Environmental Science & Technology*, *42*(12), 4563–4569. DOI: 10.1021/es703157n PMID: 18605587

Richards, B. S., & Schafer, A. I. (2003). Photovoltaic-powered desalination system for remote Australian communities. *Renewable Energy*, *28*(13), 2013–2022. DOI: 10.1016/S0960-1481(03)00081-8

Saisset, R. (2004). Contribution à l'étude systémique de dispositifs énergétiques à composants électrochimiques. Formalisme Bond Graph appliqué aux piles à combustible, accumulateurs Lithium-Ion, Véhicule Solaire. (PhD thesis, Toulouse, France, April 2004).

Saudement, C. (2013). Méthodologie de développement de stratégies de supervision énergétiques en temps réel dans les systèmes électriques de puissance. Habilitation, Université de Lille 1

Schafer, A. I., Broeckmann, A., & Richards, B. S. (2007). Renewable energy powered membrane technology. 1. Development and characterization of a photovoltaic hybrid membrane system. *Environmental Science & Technology*, *41*(3), 998–1003. DOI: 10.1021/es0602277 PMID: 17328215

Thomson, A. M. (2003). Reverse-osmosis desalination of seawater powered by photovoltaics without batteries. Dissertation, Loughborough University

Dauphin-Tanguy, G. (2000). *Les Bond graphs*. Édition Hermes.

Gawthrop, P. J., & Bevan, G. P. (2007). Bond-graph modeling. *IEEE Control Systems Magazine*, (April), 24–45.

Thomson, M., Miranda, M. S., & Infield, D. (2002). A small-scale seawater reverse-osmosis system with excellent energy efficiency over a wide operating range. *Desalination*, *153*(1–3), 229–236. DOI: 10.1016/S0011-9164(02)01138-9

Tischer, H., & Verbic, G. (2011). Towards a smart home energy management system—A dynamic programming approach. In Innovative Smart Grid Technologies Asia (ISGT), IEEE PES (pp. 1–7). https://doi.org/DOI: 10.1109/ISGT-Asia.2011.6167004

Turki, M. (2010). Conception, analyse et commande d'un processus du dessalement par osmose inverse alimenté par un système hybride Photovoltaïque-Éolien (PhD thesis). LSE-ENIT, Tunisie.

Turki, M., Belhadj, J., & Roboam, X. (2008). Bond graph modeling and analysis of an autonomous reverse osmosis desalination process fed by a hybrid system (photovoltaic-wind). In Proceedings of Electrimacs, International Conference on Theory and Application of Modeling and Simulation in Electrical Power Engineering, Québec, Canada, 8–11

Turki, M., Ben Rhouma, A., & Belhadj, J. (2009). Experimental characterization of a reverse osmosis desalination process fed by a hybrid power source. In Proceedings of the International Multi-Conference on Systems, Signals & Devices (SSD), Jerba, Tunisia, 23–26 March.

Turki, M., Ben Rhouma, A., Ben Ali, I., Belhadj, J., & Roboam, X. (2011). Gestion énergétique sans stockage électrique d'un processus de pompage et/ou de dessalement alimenté 'au fil du vent et du soleil'. Electrotechnique du Futur, Belford-France, December

Tzen, E., & Morris, R. (2003). Renewable energy sources for desalination. *Solar Energy*, *75*(5), 375–379. DOI: 10.1016/j.solener.2003.07.010

Tzen, E., Theofilloyianakos, D., & Kologios, Z. (2008). Autonomous reverse osmosis units driven by RE sources: Experiences and lessons learned. *Desalination*, *221*(1–3), 29–36. DOI: 10.1016/j.desal.2007.02.048

UNICEF, & WHO. (2021, October 19). 1 in 3 people globally do not have access to safe drinking water. Retrieved from https://www.unwater.org/

United Nations. (2019). The United Nations World Water Development Report 2019. Retrieved May 8, 2024, from https://www.unwater.org/

Chapter 10
Material and Electrical Defects in Photovoltaic Systems:
Environmental Impacts and Challenges in the Oriental Region of Morocco

Abdel Hamid Adaliou
https://orcid.org/0000-0002-9504-2769
Mohammed First University, Morocco

Mostafa El Ouariachi
Mohammed First University, Morocco

Kamal Hirech
Mohammed First University, Morocco

ABSTRACT

Photovoltaic systems, crucial for sustainable energy generation, often suffer from various material and electrical defects that can significantly hinder their performance and longevity. Material defects such as microcracks, delamination, and degradation of encapsulants affect the mechanical integrity of PV modules, leading to efficiency losses and increased failure rates. Electrical defects, including hotspots, mismatched cells, and connection issues, further exacerbate these problems, reducing the overall power output and system reliability. These defects not only compromise the efficiency of energy conversion but also have broader environmental implications. Increased failure rates necessitate more frequent replacements and disposal of PV modules, contributing to electronic waste and resource consumption. Furthermore, the energy payback time of PV systems is extended, delaying the environmental benefits of transitioning to renewable energy sources. Understanding these defects' nature and impact is vital for improving PV durability and maximizing environmental contributions.

DOI: 10.4018/979-8-3693-9924-8.ch010

1. INTRODUCTION

In the quest for sustainable energy solutions, photovoltaic (PV) systems have emerged as one of the most promising technologies (Hussain et al., 2017)(Adaliou et al., 2023). By harnessing the abundant energy of the sun, PV systems offer a cleaner, renewable alternative to traditional fossil fuel-based energy production, playing a pivotal role in reducing greenhouse gas emissions and combating climate change (Shahsavari & Akbari, 2018)(Adaliou et al., 2023). As global energy demands continue to rise, the deployment of PV systems has surged, with installations ranging from small-scale residential setups to vast solar farms (Ukoba et al., 2024). However, despite their potential to revolutionize energy production, PV systems face significant technical challenges that can compromise their efficiency, longevity, and overall environmental impact (Al-Shahri et al., 2021)(Rhiat, Karrouchi, Atmane, et al., 2024).

One of the critical issues in PV technology is the occurrence of material and electrical defects that can significantly affect system performance (Aghaei et al., 2022a). Material defects, such as microcracks in the solar cells, delamination of protective layers, and degradation of encapsulants, undermine the mechanical integrity of the PV modules (Dimitriou et al., 2024). These issues often arise due to environmental stressors like temperature fluctuations, moisture, and UV exposure, which gradually degrade the materials over time (Schultz et al., 2010). These material flaws can lead to decreased energy conversion efficiency and, in severe cases, module failure, reducing the expected lifespan of the PV systems (Libra et al., 2023).

On the electrical side, defects like hotspots, cell mismatches, and connection problems further complicate the reliable operation of PV systems (Hijjawi et al., 2023). Electrical faults often stem from issues in the installation process, suboptimal design, or environmental factors such as partial shading, leading to uneven power generation across the PV array (Alkahtani, 2021). These electrical defects not only result in diminished energy output but can also increase the risk of permanent damage to the modules, exacerbating the need for early replacements (Aghaei et al., 2022b). Beyond their technical implications, these defects carry broader environmental consequences. As PV modules fail or degrade prematurely, the frequency of their replacement increases, contributing to the growing problem of electronic waste (Tsanakas et al., n.d.). This is particularly concerning given the complex nature of PV module recycling, which is still an emerging field with limited infrastructure globally (Farrell et al., 2020a). Additionally, the energy payback time the period required for a PV system to generate the amount of energy consumed in its production becomes extended (Schlaich et al., 2004). This delay in achieving energy neutrality undermines the environmental benefits that PV systems are supposed to provide, slowing down the transition to renewable energy sources and prolonging our reliance on carbon-intensive technologies (Sovacool et al., 2022).

The environmental and performance challenges posed by material and electrical defects in PV systems underscore the need for a deeper understanding of their causes, effects, and mitigation strategies (Abdulla et al., 2024). This review aims to explore the nature of these defects, the underlying factors contributing to their occurrence, and the associated environmental impacts. By analyzing current research, this paper seeks to identify opportunities for improving the durability and reliability of PV systems, with a particular focus on minimizing their environmental footprint. In doing so, it contributes to the ongoing efforts to optimize photovoltaic technology as a truly sustainable solution for future energy needs.

2. MATERIALS AND COMPONENTS IN PHOTOVOLTAIC SYSTEMS

2.1 Material of PV Cells

Photovoltaic (PV) cell technologies have undergone significant advancements, leading to the development of various types, each with unique characteristics and applications (Marques Lameirinhas et al., 2022). The most widely used technology is crystalline silicon (c-Si), which dominates the global market due to its high efficiency and long lifespan (Ballif et al., 2022). Crystalline silicon cells are further divided into monocrystalline (mono-Si) and polycrystalline (poly-Si) cells (Baghel & Chander, 2022). Monocrystalline cells, known for their uniform structure, offer the highest efficiency rates but are more expensive to produce (Pastuszak & Węgierek, 2022). In contrast, polycrystalline cells are less efficient but cost-effective, making them suitable for large-scale installations (R. Singh et al., 2013). Another important category is thin-film technologies, which include materials like cadmium telluride (CdTe) and copper indium gallium selenide (CIGS) (Hussain et al., 2017). Thin-film cells are lightweight, flexible, and can be integrated into unconventional surfaces such as building facades or wearable devices. However, their efficiency is typically lower than crystalline silicon cells, and some materials, like cadmium, raise environmental concerns (Hussain et al., 2017). Emerging technologies, such as perovskite solar cells, are gaining attention for their potential to revolutionize the industry (Li et al., 2022). These cells offer high efficiency, tunable bandgaps, and low production costs, but challenges related to stability and scalability remain (Li et al., 2022). Additionally, organic photovoltaics (OPVs) leverage organic materials to create lightweight and semi-transparent solar cells, ideal for niche applications (Bulavko, 2024). Lastly, multi-junction solar cells, often used in space applications, combine multiple layers of materials to achieve extremely high efficiencies by capturing a broader spectrum of sunlight (Yamaguchi et al., 2021). Each technology represents a step toward meeting diverse energy demands while addressing sustainability challenges, making the future of solar power increasingly promising.

Table 1. Summary table of different photovoltaic cell technologies

Solar cell type	Important milestones	Efficiency Range	Environmental Impact
Silicon solar cells	First commercial cell in 1954, efficiency over 25%	~15%–25%	High carbon footprint in production
Organic Solar Cells	Conductive polymers in 1977, efficiency around 10%	~3–12%	Lower environmental impact, concerns about long-term waste
Halide Perovskite Solar Cells	Introduced in 2009, efficiencies above 25%	Up to 25.5%	Toxicity, particularly from lead
CdTe Solar Cells	Developed in the 1950s, efficiencies over 22%	~10–22%	Cadmium toxicity, recycling challenges
GaAs Solar Cells	Invented in 1957, efficiencies over 29%	>29%	Costly production, arsenic toxicity concerns

Source: (Machín & Márquez, 2024)

2.2 Structure of PV Modules

A photovoltaic (PV) module is a complex assembly designed to convert sunlight into electrical energy (G. K. Singh, 2013). It consists of solar cells, typically made from crystalline silicon (monocrystalline or polycrystalline) or thin-film materials, which perform the photovoltaic effect (Green, 2003). These cells are encapsulated in layers of Ethylene Vinyl Acetate (EVA) to protect them from mechanical stress and environmental factors (Figure 1) (Jiang et al., 2015). A tempered glass layer serves as the front cover, providing mechanical protection and allowing maximum light transmission, while the backsheet, made from durable polymers like Tedlar, shields the module from moisture and UV radiation (Karahalli, n.d.). An anodized aluminum frame offers structural support and facilitates mounting, while the junction box houses electrical connections and diodes for safe energy transfer (Ahmed et al., 2023). Interconnections, usually composed of silver-plated copper, link the cells to form efficient circuits (R. Yang et al., 2017). To enhance performance, an anti-reflective coating minimizes sunlight reflection, ensuring optimal light absorption (Shanmugam et al., 2020)(Rhiat, Karrouchi, Atmane, et al., 2024). Together, these components ensure the module's efficiency, durability, and long-term reliability (Arriaga Arruti et al., 2023).

Figure 1. Components of PV panel

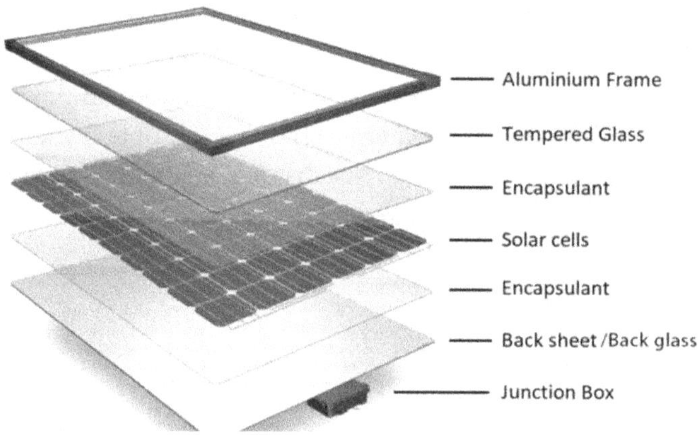

Source: (Papargyri et al., 2020a)

3. MATERIAL DEFECTS IN PHOTOVOLTAIC SYSTEMS

3.1 Microcracks

Microcracks are tiny fractures that develop in the crystalline silicon cells of photovoltaic (PV) panels, often too small to be visible to the naked eye (Papargyri et al., 2020a). These microcracks can disrupt the flow of electrons, reducing the panel's efficiency and power output over time (Ennemri et al., 2019). They also act as stress concentrators, potentially leading to larger cracks under mechanical or thermal stress, further compromising the panel's performance(Sun et al., 2025)(An et al., 2024). The causes of microcracks are diverse and include improper handling during transportation, manufacturing defects, and mechanical stress from environmental factors such as hail, wind, or heavy snow (Papargyri et al., 2020b). Thermal cycling, where the panel undergoes repeated expansion and contraction due to temperature changes, is another significant contributor (Jafari et al., 2019). Additionally, poor mounting practices or uneven surface pressure during installation can introduce stress that leads to microcracking. Proactive quality control and proper handling are essential to minimize these defects and maintain the long-term reliability of PV systems.

3.2 Snail Trails

Snail trails are faint, discolored streaks that appear on the surface of photovoltaic (PV) panels, often in a greyish or brownish hue (Daher et al., 2024). These trails form as a result of chemical reactions between the silver paste used in cell contacts, the encapsulant material (commonly EVA), and moisture that infiltrates the panel (Makhija & Bohra, 2023). While snail trails are primarily a cosmetic defect, they are often linked to underlying issues such as microcracks, which disrupt the electrical pathways within the cells and can reduce panel efficiency over time. Common causes of snail trails include poor manufacturing quality, such as defective encapsulant materials, insufficient lamination, or substandard cell metallization (Artificial Intelligence Applications in Sustainable Infrastructure: Solar Photovoltaic Inspection and Estimation of Indoor Environmental Variables in a Smart Building, n.d.). Environmental factors like prolonged exposure to high humidity, UV radiation, or mechanical stress during transportation and installation can also exacerbate the problem. Preventing snail trails requires stringent quality control during manufacturing, as well as proper handling and installation practices.

Figure 2. Snail trails default

Source: (Tanesab et al., 2020)

3.3 Delamination and corrosion

Delamination and corrosion are critical material defects that adversely affect the performance and longevity of photovoltaic (PV) panels (Aghaei et al., 2022a). Delamination in photovoltaic (PV) panels refers to the separation or detachment of layers within the panel, typically between the encapsulant, glass, and backsheet (Sinha, n.d.). This material default occurs when the layers within the PV module, such as the encapsulant, backsheet, or anti-reflective coating, separate due to thermal cycling, moisture ingress, or poor adhesion during manufacturing (Hasan et al., 2021). This separation can lead to the ingress of water and air, causing electrical insulation failure, reduced light transmission, and accelerated degradation of the solar cells (Lu et al., 2021). Corrosion, on the other hand, involves the electrochemical degradation of metallic components, such as the frame, interconnects, or busbars, often triggered by exposure to moisture, humidity, or corrosive pollutants (Alimi et al., 2022). This process can increase electrical resistance, cause contact failures, and compromise the structural integrity of the module. Both delamination and corrosion are particularly prevalent in harsh environments, such as coastal or industrial areas, where high humidity, temperature fluctuations, and salt exposure are common. Mitigating these defects requires advanced encapsulation materials, robust sealing techniques, and the use of corrosion-resistant alloys to ensure the durability and efficiency of PV systems under challenging operating conditions. Addressing these issues is essential for enhancing the reliability and economic viability of solar energy technologies (Elkelawy et al., 2024).

Figure 3. Delamination and corrosion

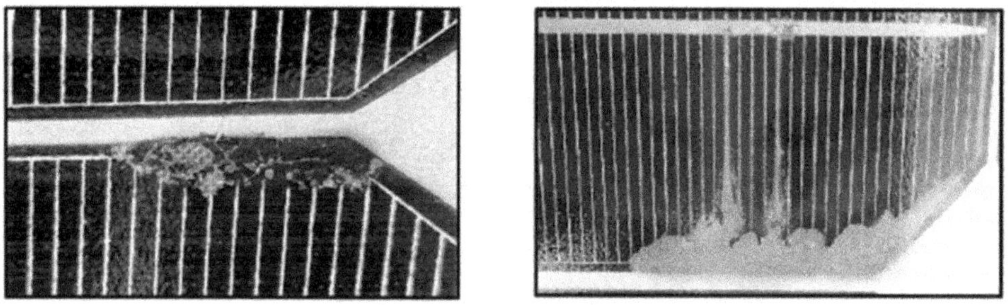

Source: (Nieto-Morone et al., 2024)

4. ELECTRICAL DEFECTS IN PHOTOVOLTAIC SYSTEMS

4.1 Hotspots

Hotspots are a significant electrical defect in photovoltaic (PV) panels, often arising from localized resistance imbalances or shading within the solar cell array (C. Yang et al., 2024). These imbalances can be caused by manufacturing defects, such as microcracks, poor solder connections, or impurities in the silicon material, as well as external factors like partial shading, soiling, or debris accumulation on the panel surface. When a cell or group of cells operates at a higher resistance than the surrounding cells, it dissipates power as heat instead of generating electricity, leading to elevated temperatures in the affected area (Sargunanathan et al., 2016). This localized overheating, known as a hotspot, can cause severe consequences, including permanent damage to the cell structure, degradation of the encapsulant material, and even delamination of the module layers. Over time, hotspots can significantly reduce the efficiency and power output of the PV panel, shorten its operational lifespan, and increase the risk of fire hazards in extreme cases. To mitigate hotspot formation, advanced bypass diodes, improved manufacturing quality control, and regular maintenance to remove shading or soiling are essential (Alami et al., 2022). Understanding and addressing the causes and consequences of hotspots is crucial for enhancing the reliability and performance of PV systems.

Figure 4. Thermography image

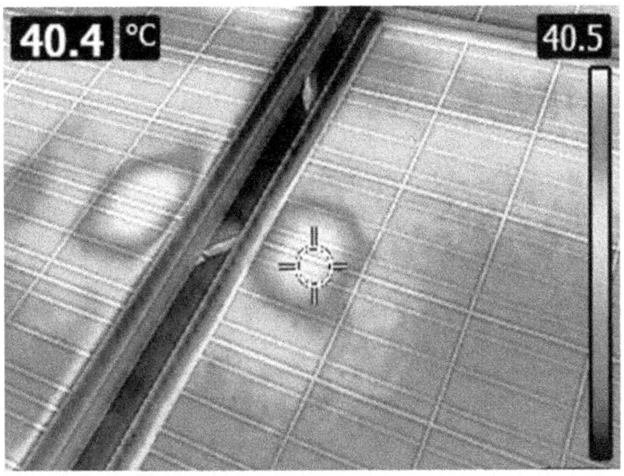

4.2 Cell Mismatch

Cell mismatch occurs in photovoltaic (PV) panels when individual solar cells within a module operate at different electrical parameters, such as current or voltage (Kim et al., 2015). This defect leads to a reduction in the overall performance and efficiency of the panel, as the mismatched cells can limit the output of the entire string of cells. Causes of cell mismatch include variations in cell manufacturing processes, leading to differences in cell quality or performance, as well as degradation over time due to environmental exposure, such as UV radiation or thermal cycling (Fouad et al., 2017). External factors, like partial shading, dirt accumulation, or uneven soiling, can also exacerbate cell mismatch (Alazbe et al., 2023). Additionally, poor interconnection design or improper handling during installation can introduce mechanical stress that damages certain cells, further contributing to the problem. Mitigating cell mismatch requires stringent quality control during manufacturing, regular panel maintenance, and careful installation practices to minimize damage and shading effects.

4.3 Electrical Connection Issues

Electrical connection issues in photovoltaic (PV) systems, such as arc faults, short circuits, and open circuits, are critical defects that can significantly compromise the performance, safety, and longevity of solar panels. Arc faults occur when high-voltage discharges form between conductive materials due to poor connections, damaged wiring, or insulation degradation. These discharges generate intense heat, posing a fire hazard and potentially causing irreversible damage to the PV module and surrounding components. Short circuits arise when an unintended low-resistance path forms between two points in the electrical circuit, often due to insulation failure, moisture ingress, or physical damage. This results in excessive current flow, overheating, and potential burnout of the affected cells or wiring. Open circuits, on the other hand, occur when a break in the electrical pathway interrupts current flow, often caused by broken interconnects, solder joint failures, or mechanical stress. This leads to a complete loss of power output from the affected module or string. These electrical connection issues are exacerbated by environmental factors such as temperature fluctuations, humidity, and UV exposure, which accelerate material degradation. To mitigate these problems, advanced monitoring systems, robust design practices, and the use of high-quality materials are essential. Addressing these electrical defects is crucial for ensuring the reliability, efficiency, and safety of PV systems, particularly in large-scale installations where the impact of such failures can be substantial.

Figure 5. Juction box after Arc fault

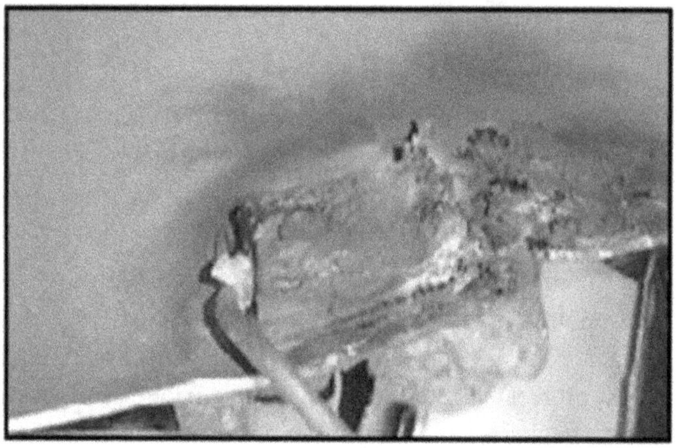

Source: (Johnson et al., n.d.)

Figure 6. Burned module after Arc fault

Source: (Johnson et al., n.d.)

5. IMPACT OF TEMPERATURE RISE AND IRRADIATION ON PHOTOVOLTAIC PERFORMANCE IN THE ORIENTAL REGION

The analysis of temperature and irradiation data from the PVGIS platform for the Oriental region of Morocco between 2018 and 2023 (Figure 7 and 8) reveals critical insights into the performance of photovoltaic (PV) installations and their environmental impact. The first figure shows a notable rise in monthly average temperature, with July 2019 recording 26°C and July 2023 reaching 28.9°C, indicating a clear warming trend likely influenced by global climate change and local environmental factors. The second figure illustrates that solar irradiation at the optimal angle has remained relatively stable, with values of 220.8 kWh/m^2 in June 2019 and 221.74 kWh/m^2 in July 2023, suggesting that the region maintains strong solar energy potential (Rhiat, Karrouchi, Hassari, et al., 2024). However, the increase in temperature presents challenges for PV systems, as elevated temperatures reduce module efficiency due to thermal losses and accelerate the degradation of materials, leading to hotspot formation, potential-induced degradation (PID), microcracks, and delamination. These defects not only lower energy yield but also contribute to electronic waste due to premature module replacements, raising environmental concerns. Moreover, increased panel surface temperatures necessitate more frequent cleaning in this arid region, leading to higher water consumption, which poses sustainability challenges. Despite these issues, the stable irradiation levels confirm that solar energy remains a viable and strategic solution for the region's energy transition, provided that adequate mitigation strategies are implemented. These include the adoption of heat-resistant PV materials, passive and active cooling techniques, predictive maintenance systems, and recycling initiatives to extend the lifespan of PV modules and minimize environmental impact. By addressing these challenges proactively, PV installations in the Oriental region can continue to provide sustainable energy while mitigating the risks associated with rising temperatures.

Figure 7. Monthly solar irradiation estimates in oriental region

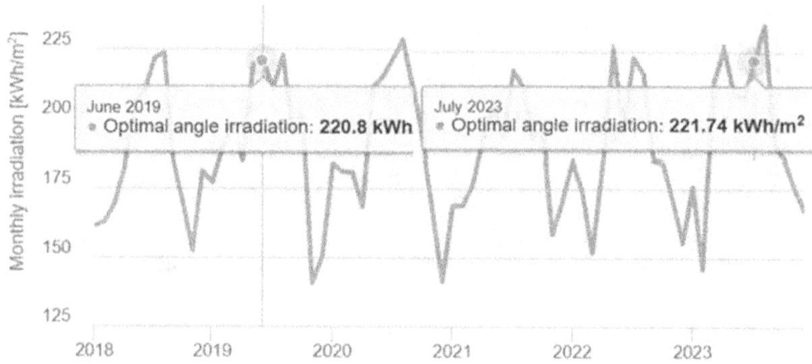

Source: https://rb.gy/nlk4sc

Figure 8. Monthly average temperature in oriental region

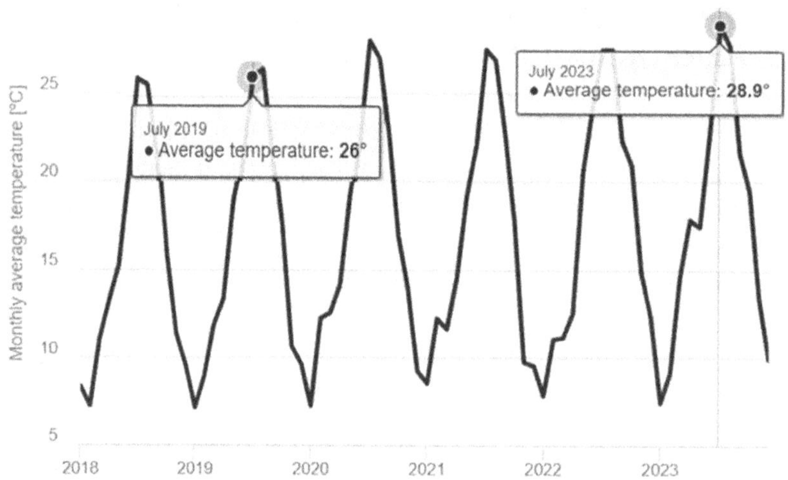

Source: https://rb.gy/nlk4sc

6. ENVIRONMENTAL IMPACT OF DEFECTS IN PHOTOVOLTAIC SYSTEMS

The environmental impact of defects in photovoltaic (PV) systems is a crucial issue, particularly in the Oriental region of Morocco, where solar energy plays a significant role in the region's transition towards sustainable energy sources. However, various defects such as hotspots, delamination, bypass diode failures, soiling, and electrical faults can severely affect the performance of PV installations, leading to both direct and indirect environmental consequences (Santhakumari & Sagar, 2019).

One of the most significant environmental concerns arises from the reduction in energy efficiency due to these defects (Tawalbeh et al., 2021). In the Oriental region, where solar irradiation is relatively high, defective PV modules can lead to a significant drop in energy production, forcing reliance on backup power sources, which are often fossil fuel-based (El Hadri et al., 2023). This increased dependency on conventional energy sources results in higher carbon dioxide (CO_2) emissions, undermining the prfimary goal of using solar energy to reduce greenhouse gas emissions and combat climate change (Rahman et al., 2017).

Furthermore, soiling and dust accumulation, particularly severe in the semi-arid climate of the Oriental region, can exacerbate performance losses (Zarei et al., 2022). The fine desert dust, carried by winds from the Sahara, settles on the PV panels, reducing their efficiency if left uncleaned (Simiyu, 2020). To counteract this, frequent panel cleaning is necessary, which leads to an increased demand for water resources. Given that the Oriental region faces water scarcity, the excessive use of water for PV cleaning contributes to the depletion of an already limited resource, creating a conflict between energy sustainability and water conservation. In some cases, chemical-based cleaning solutions are used, which can introduce chemical pollutants into the soil and nearby water bodies, further impacting local ecosystems (Sharma et al., 2019).

Another critical issue is PV module degradation and electronic waste (Farrell et al., 2020b). When defects such as microcracks or encapsulant browning occur, PV modules may reach the end of their operational life earlier than expected. In the absence of proper recycling facilities in the region, defective and decommissioned PV panels often end up in landfills, contributing to the accumulation of electronic waste (e-waste) (Farrell et al., 2020b). These panels contain materials like cadmium, lead, and other heavy metals, which, if improperly disposed of, can leach into the soil and groundwater, causing long-term contamination and environmental damage.

Moreover, electrical failures in PV systems such as arc faults, short circuits, and inverter malfunctions can pose fire hazards, particularly in rural and agricultural areas of the Oriental region (Fakoyede et al., 2024). Such incidents not only damage PV infrastructure but can also lead to wildfires, affecting local vegetation, biodiversity, and air quality.

To mitigate these environmental impacts, it is essential to implement advanced fault detection and predictive maintenance strategies. The use of AI-based monitoring systems, IoT sensors, and drone inspections can help identify defects early, reducing efficiency losses and extending the lifespan of PV modules. Additionally, promoting sustainable cleaning techniques, such as waterless robotic cleaning, can significantly reduce water consumption in panel maintenance.

In conclusion, while PV systems offer a promising clean energy solution for the Oriental region of Morocco, their environmental benefits can be compromised by various defects. Addressing these challenges through proactive maintenance, efficient resource management, and improved recycling policies is essential to ensure the long-term sustainability of solar energy in the region.

Table 2. Summary Table of the Impacts of Defects in Photovoltaic Systems

Type of Defect	Main Causes	Environmental Impact
Material Degradation	UV exposure, humidity, thermal cycling	Increased electronic waste, risk of toxic material leakage.
Hotspots	Partial shading, cell mismatching, manufacturing defects	Higher energy losses, increased maintenance needs.
Microcracks	Mechanical stress, temperature variations	More frequent panel replacements, material waste accumulation.
Delamination	Poor encapsulation, extreme weather conditions	Shortened panel lifespan, increased e-waste generation.
Connector and Wiring Faults	Poor installation, corrosion, loose connections	Increased risk of electrical fires, hazardous waste.
Arc Faults & Short Circuits	Insulation failure, cable damage	Potential wildfires, toxic emissions from burned components.

7. CONCLUSION

Material and electrical defects in photovoltaic (PV) systems, such as delamination, corrosion, hotspots, arc faults, short circuits, and open circuits, pose significant challenges to the efficiency, reliability, and longevity of solar energy systems. These defects are often exacerbated by environmental factors, including temperature fluctuations, humidity, UV exposure, and saline or polluted atmospheres, which accelerate the degradation of materials and electrical components. Delamination, for instance, compromises the structural integrity of PV modules by allowing moisture and air to penetrate, leading to cell degradation and power loss. Corrosion, on the other hand, weakens metallic components, increasing electrical resistance and the risk of system failure. Electrical defects like hotspots, arc faults, and short circuits not only reduce energy output but also pose serious safety hazards, including fire risks.

The environmental impacts of these defects are multifaceted. On one hand, they reduce the energy yield and operational lifespan of PV systems, undermining their economic viability and sustainability. On the other hand, the premature failure of PV modules contributes to electronic waste, raising concerns about the environmental footprint of solar energy technologies. Addressing these challenges requires a multi-faceted approach, including the development of advanced materials with enhanced durability, improved manufacturing processes to ensure quality control, and the implementation of robust monitoring and maintenance practices. Innovations such as anti-corrosive coatings, better encapsulation techniques, and smart monitoring systems can significantly mitigate these defects.

Furthermore, understanding the interplay between environmental conditions and material/electrical defects is crucial for designing PV systems that can withstand harsh operating environments. This is particularly important for large-scale installations in regions with extreme climates, where the economic and environmental stakes are highest. By addressing these material and electrical defects, the solar energy industry can enhance the reliability, efficiency, and sustainability of PV systems, contributing to the global transition towards renewable energy and a reduced carbon footprint. Future research should focus on developing cost-effective solutions and predictive models to further mitigate these challenges and optimize the performance of photovoltaic systems worldwide.

REFERENCES

Abdulla, H., Sleptchenko, A., & Nayfeh, A. (2024). Photovoltaic systems operation and maintenance: A review and future directions. In *Renewable and Sustainable Energy Reviews* (Vol. 195). Elsevier Ltd., DOI: 10.1016/j.rser.2024.114342

Adaliou, A. H., Lamreoua, A., Isknan, I., Doubabi, S., Melhaoui, M., El Ouariachi, M., & Hirech, K. (2023). High-Performance MPPT Based on Developed Fast Convergence for PV System-Experimental Validation. *International Conference on Digital Technologies and Applications*, 694–703. DOI: 10.1007/978-3-031-29857-8_69

Aghaei, M., Fairbrother, A., Gok, A., Ahmad, S., Kazim, S., Lobato, K., Oreski, G., Reinders, A., Schmitz, J., Theelen, M., Yilmaz, P., & Kettle, J. (2022a). Review of degradation and failure phenomena in photovoltaic modules. In *Renewable and Sustainable Energy Reviews* (Vol. 159). Elsevier Ltd., DOI: 10.1016/j.rser.2022.112160

Aghaei, M., Fairbrother, A., Gok, A., Ahmad, S., Kazim, S., Lobato, K., Oreski, G., Reinders, A., Schmitz, J., Theelen, M., Yilmaz, P., & Kettle, J. (2022b). Review of degradation and failure phenomena in photovoltaic modules. In *Renewable and Sustainable Energy Reviews* (Vol. 159). Elsevier Ltd., DOI: 10.1016/j.rser.2022.112160

Ahmed, A., Elsakka, M., Elminshawy, N., Mohamed, A., & Sundaram, S. (2023). Recent Advances in Floating Photovoltaic Systems. In *Chemical Record* (Vol. 23, Issue 12). John Wiley and Sons Inc. DOI: 10.1002/tcr.202300229

Al-Shahri, O. A., Ismail, F. B., Hannan, M. A., Lipu, M. S. H., Al-Shetwi, A. Q., Begum, R. A., Al-Muhsen, N. F. O., & Soujeri, E. (2021). Solar photovoltaic energy optimization methods, challenges and issues: A comprehensive review. In *Journal of Cleaner Production* (Vol. 284). Elsevier Ltd., DOI: 10.1016/j.jclepro.2020.125465

Alami, A. H., Rabaia, M. K. H., Sayed, E. T., Ramadan, M., Abdelkareem, M. A., Alasad, S., & Olabi, A.-G. (2022). Management of potential challenges of PV technology proliferation. *Sustainable Energy Technologies and Assessments*, *51*, 101942. DOI: 10.1016/j.seta.2021.101942

Alazbe, K., Roy, H. A., Alremeithi, A., AlKindi, S., AlHashmi, M., Daccache, L., & Alnuaimi, A. (2023). Insight into dust soiling characteristics and cleaning solutions to investigate impact on PV systems performance across the globe. *Progress in Energy*, *6*(1), 012004.

Alimi, O. A., Meyer, E. L., & Olayiwola, O. I. (2022). Solar Photovoltaic Modules' Performance Reliability and Degradation Analysis—A Review. In *Energies* (Vol. 15, Issue 16). MDPI. DOI: 10.3390/en15165964

Alkahtani, M. (2021). *Study for Non-uniform Aging Photovoltaic Array Performance Enhancements.*

An, D., Gong, G., Xu, D., Zhou, Z., Li, R., & Su, Y. (2024). Embracing Plasticity: Unlocking the Full Potential of Flexible and Stretchable Electronics Through the Elastoplastic Behavior of Metallic Materials. *Advanced Functional Materials*, •••, 2412796. DOI: 10.1002/adfm.202412796

Arriaga Arruti, O., Virtuani, A., & Ballif, C. (2023). Long-term performance and reliability of silicon heterojunction solar modules. In *Progress in Photovoltaics: Research and Applications* (Vol. 31, Issue 7, pp. 664–677). John Wiley and Sons Ltd. DOI: 10.1002/pip.3688

Abraham, K. (2024). Artificial Intelligence Applications in Sustainable Infrastructure: Solar Photovoltaic Inspection and Estimation of Indoor Environmental Variables in a Smart Building.

Baghel, N. S., & Chander, N. (2022). Performance comparison of mono and polycrystalline silicon solar photovoltaic modules under tropical wet and dry climatic conditions in east-central India. *Clean Energy*, *6*(1), 929–941. DOI: 10.1093/ce/zkac001

Ballif, C., Haug, F. J., Boccard, M., Verlinden, P. J., & Hahn, G. (2022). Status and perspectives of crystalline silicon photovoltaics in research and industry. In *Nature Reviews Materials* (Vol. 7, Issue 8, pp. 597–616). Nature Research. DOI: 10.1038/s41578-022-00423-2

Bulavko, G. V. (2024). Organic photovoltaics: A journey through time, advancements, and future opportunities. *History of Science and Technology*, *14*(1), 10–32. DOI: 10.32703/2415-7422-2024-14-1-10-32

Daher, D. H., Mathieu, A., Abdallah, A., Mouhoumed, D., Logerais, P. O., Gaillard, L., & Ménézo, C. (2024). Photovoltaic failure diagnosis using imaging techniques and electrical characterization. *EPJ Photovoltaics*, *15*, 25. Advance online publication. DOI: 10.1051/epjpv/2024022

Dimitriou, T., Skandalos, N., & Karamanis, D. (2024). Progress in Improving Photovoltaics Longevity. *Applied Sciences (Basel, Switzerland)*, *14*(22), 10373. Advance online publication. DOI: 10.3390/app142210373

El Hadri, Y., Khokhlov, V., & Slizhe, M. (2023). *Wind and Solar Energy Resources in Morocco: Current Status and Assessment up to 2050*. Springer.

Elkelawy, M., Saeed, A. M., & Seleem, H. (2024). Egypt's Solar Revolution: A Dual Approach to Clean Energy with CSP and PV Technologies. *Pharos Engineering Science Journal*, *1*(1), 39–50. DOI: 10.21608/pesj.2025.344888.1004

Ennemri, A., Logerais, P.-O., Balistrou, M., Belaidi, I., Ennemri, A., Logerais, P. O., Balistrou, M., Durastanti, J. F., & Belaidi, I. (2019). Cracks in silicon photovoltaic modules: A review. In *Article in Journal of Optoelectronics and Advanced Materials* (Vol. 21, Issue 2). https://www.researchgate.net/publication/331501514

Fakoyede, P. D., Innocent, A. O., Ajiboye, Q. O., Enabulele, E. C., Olaleye, T. G., & Ekunseitan, K. D. (2024). Assessing the Detrimental Effects of Solar Panel Installation on Residential Roofing Systems. *Traektoriâ Nauki*, *10*(7), 3023–3033. DOI: 10.22178/pos.106-15

Farrell, C. C., Osman, A. I., Doherty, R., Saad, M., Zhang, X., Murphy, A., Harrison, J., Vennard, A. S. M., Kumaravel, V., Al-Muhtaseb, A. H., & Rooney, D. W. (2020a). Technical challenges and opportunities in realising a circular economy for waste photovoltaic modules. In *Renewable and Sustainable Energy Reviews* (Vol. 128). Elsevier Ltd., DOI: 10.1016/j.rser.2020.109911

Farrell, C. C., Osman, A. I., Doherty, R., Saad, M., Zhang, X., Murphy, A., Harrison, J., Vennard, A. S. M., Kumaravel, V., Al-Muhtaseb, A. H., & Rooney, D. W. (2020b). Technical challenges and opportunities in realising a circular economy for waste photovoltaic modules. In *Renewable and Sustainable Energy Reviews* (Vol. 128). Elsevier Ltd., DOI: 10.1016/j.rser.2020.109911

Fouad, M. M., Shihata, L. A., & Morgan, E. I. (2017). An integrated review of factors influencing the perfomance of photovoltaic panels. *Renewable & Sustainable Energy Reviews*, *80*, 1499–1511. DOI: 10.1016/j.rser.2017.05.141

Green, M. A. (2003). Crystalline and thin-film silicon solar cells: State of the art and future potential. *Solar Energy*, *74*(3), 181–192. DOI: 10.1016/S0038-092X(03)00187-7

Hasan, A. A. Q., Alkahtani, A. A., Shahahmadi, S. A., Alam, M. N. E., Islam, M. A., & Amin, N. (2021). Delamination-and electromigration-related failures in solar panels—A review. *Sustainability (Switzerland)*, *13*(12). Advance online publication. DOI: 10.3390/su13126882

Hijjawi, U., Lakshminarayana, S., Xu, T., Piero Malfense Fierro, G., & Rahman, M. (2023). A review of automated solar photovoltaic defect detection systems: Approaches, challenges, and future orientations. In *Solar Energy* (Vol. 266). Elsevier Ltd., DOI: 10.1016/j.solener.2023.112186

Hussain, A., Arif, S. M., & Aslam, M. (2017). Emerging renewable and sustainable energy technologies: State of the art. In *Renewable and Sustainable Energy Reviews* (Vol. 71, pp. 12–28). Elsevier Ltd., DOI: 10.1016/j.rser.2016.12.033

Jafari, A., Ashrafi, H., Bazli, M., & Ozbakkaloglu, T. (2019). Effect of thermal cycles on mechanical response of pultruded glass fiber reinforced polymer profiles of different geometries. *Composite Structures*, *223*, 110959. DOI: 10.1016/j.compstruct.2019.110959

Jiang, S., Wang, K., Zhang, H., Ding, Y., & Yu, Q. (2015). Encapsulation of PV Modules Using Ethylene Vinyl Acetate Copolymer as the Encapsulant. *Macromolecular Reaction Engineering*, *9*(5), 522–529. DOI: 10.1002/mren.201400065

Johnson, J., Bower, W., & Quintana, M. (n.d.). *SANDIA REPORT Electrical and Thermal Finite Element Modeling of Arc Faults in Photovoltaic Bypass Diodes*. http://www.ntis.gov/help/ordermethods.asp?loc=7-4-0#online

Karahalli, M. E. (n.d.). *Performance, design, and interconnection optimization of shingled solar cell for a photovoltaic module application a thesissubmitted tothe graduate school of natural and applied sciences of middle east technical university*.

Kim, K. A., Seo, G.-S., Cho, B.-H., & Krein, P. T. (2015). Photovoltaic hot-spot detection for solar panel substrings using AC parameter characterization. *IEEE Transactions on Power Electronics*, *31*(2), 1121–1130. DOI: 10.1109/TPEL.2015.2417548

Li, X., Xie, Q., & Huang, L. (2022). Identifying the Development Trends of Emerging Technologies Using Patent Analysis and Web News Data Mining: The Case of Perovskite Solar Cell Technology. *IEEE Transactions on Engineering Management*, *69*(6), 2603–2618. DOI: 10.1109/TEM.2019.2949124

Libra, M., Mrázek, D., Tyukhov, I., Severová, L., Poulek, V., Mach, J., Šubrt, T., Beránek, V., Svoboda, R., & Sedláček, J. (2023). Reduced real lifetime of PV panels – Economic consequences. *Solar Energy*, *259*, 229–234. DOI: 10.1016/j.solener.2023.04.063

Lu, Q., Yang, Z., Meng, X., Yue, Y., Ahmad, M. A., Zhang, W., Zhang, S., Zhang, Y., Liu, Z., & Chen, W. (2021). A review on encapsulation technology from organic light emitting diodes to organic and perovskite solar cells. *Advanced Functional Materials*, *31*(23), 2100151. DOI: 10.1002/adfm.202100151

Machín, A., & Márquez, F. (2024). Advancements in Photovoltaic Cell Materials: Silicon, Organic, and Perovskite Solar Cells. In *Materials* (Vol. 17, Issue 5). Multidisciplinary Digital Publishing Institute (MDPI). DOI: 10.3390/ma17051165

Makhija, A. S., & Bohra, S. S. (2023). Performance and degradation analysis for different solar photovoltaic technologies under hot and humid environment: A review. *Progress in Energy*, *5*(4), 042002. DOI: 10.1088/2516-1083/acfc92

Marques Lameirinhas, R. A., Torres, J. P. N., & de Melo Cunha, J. P. (2022). A Photovoltaic Technology Review: History, Fundamentals and Applications. In *Energies* (Vol. 15, Issue 5). MDPI. DOI: 10.3390/en15051823

Nieto-Morone, M. B., Rosillo, F. G., Muñoz-García, M. A., & Alonso-García, M. C. (2024). Enhancing photovoltaic module sustainability: Defect analysis on partially repaired modules from Spanish PV plants. *Journal of Cleaner Production*, *461*, 142575. DOI: 10.1016/j.jclepro.2024.142575

Papargyri, L., Theristis, M., Kubicek, B., Krametz, T., Mayr, C., Papanastasiou, P., & Georghiou, G. E. (2020a). Modelling and experimental investigations of microcracks in crystalline silicon photovoltaics: A review. In *Renewable Energy* (Vol. 145, pp. 2387–2408). Elsevier Ltd., DOI: 10.1016/j.renene.2019.07.138

Papargyri, L., Theristis, M., Kubicek, B., Krametz, T., Mayr, C., Papanastasiou, P., & Georghiou, G. E. (2020b). Modelling and experimental investigations of microcracks in crystalline silicon photovoltaics: A review. *Renewable Energy*, *145*, 2387–2408. DOI: 10.1016/j.renene.2019.07.138

Pastuszak, J., & Węgierek, P. (2022). Photovoltaic Cell Generations and Current Research Directions for Their Development. In *Materials* (Vol. 15, Issue 16). MDPI. DOI: 10.3390/ma15165542

Rahman, F. A., Aziz, M. M. A., Saidur, R., Bakar, W. A. W. A., Hainin, M. R., Putrajaya, R., & Hassan, N. A. (2017). Pollution to solution: Capture and sequestration of carbon dioxide (CO2) and its utilization as a renewable energy source for a sustainable future. In *Renewable and Sustainable Energy Reviews* (Vol. 71, pp. 112–126). Elsevier Ltd., DOI: 10.1016/j.rser.2017.01.011

Rhiat, M., Karrouchi, M., Atmane, I., Touhafi, A., Bossoufi, B., Almalki, M. M., Alghamdi, T. A. H., & Hirech, K. (2024). Maximizing solar energy efficiency: Optimized DC power conversion for resistive loads. *Computers & Electrical Engineering*, *120*, 109867. DOI: 10.1016/j.compeleceng.2024.109867

Rhiat, M., Karrouchi, M., Hassari, A., Melhaoui, M., Atmane, I., El Ouariachi, M., Schmitz, P., Benhala, B., & Hirech, K. (2024). Developing and simulating an inexpensive solar irradiance metre for photovoltaic utilisation. *International Journal of Power and Energy Conversion*, *15*(4), 376–388. DOI: 10.1504/IJPEC.2024.141867

Santhakumari, M., & Sagar, N. (2019). A review of the environmental factors degrading the performance of silicon wafer-based photovoltaic modules: Failure detection methods and essential mitigation techniques. In *Renewable and Sustainable Energy Reviews* (Vol. 110, pp. 83–100). Elsevier Ltd., DOI: 10.1016/j.rser.2019.04.024

Sargunanathan, S., Elango, A., & Mohideen, S. T. (2016). Performance enhancement of solar photovoltaic cells using effective cooling methods: A review. *Renewable & Sustainable Energy Reviews*, *64*, 382–393. DOI: 10.1016/j.rser.2016.06.024

Schlaich, J., Bergermann, R., Schiel, W., & Weinrebe, G. (2004). Sustainable electricity generation with solar updraft towers. *Structural Engineering International*, *14*(3), 225–229. DOI: 10.2749/101686604777963883

Schultz, H. R., Stoll, M., & Schultz, H. R. (2010). Some critical issues in environmental physiology of grapevines: Future challenges and current limitations _74 4.24. *Critical Issues in Grapevine Environmental Physiology Australian Journal of Grape and Wine Research*, *16*, 4–24. DOI: 10.1111/j.1755-0238.2009.00074.x

Shahsavari, A., & Akbari, M. (2018). Potential of solar energy in developing countries for reducing energy-related emissions. In *Renewable and Sustainable Energy Reviews* (Vol. 90, pp. 275–291). Elsevier Ltd., DOI: 10.1016/j.rser.2018.03.065

Shanmugam, N., Pugazhendhi, R., Elavarasan, R. M., Kasiviswanathan, P., & Das, N. (2020). Anti-reflective coating materials: A holistic review from PV perspective. *Energies*, *13*(10), 2631. Advance online publication. DOI: 10.3390/en13102631

Sharma, A., Kumar, V., Shahzad, B., Tanveer, M., Sidhu, G. P. S., Handa, N., Kohli, S. K., Yadav, P., Bali, A. S., Parihar, R. D., Dar, O. I., Singh, K., Jasrotia, S., Bakshi, P., Ramakrishnan, M., Kumar, S., Bhardwaj, R., & Thukral, A. K. (2019). Worldwide pesticide usage and its impacts on ecosystem. In *SN Applied Sciences* (Vol. 1, Issue 11). Springer Nature. DOI: 10.1007/s42452-019-1485-1

Simiyu, D. S. N. (2020). *Optimal Cleaning Strategy of Large-Scale Solar Photovoltaic Arrays Considering Non-Uniform Dust Deposition*. University of Pretoria.

Singh, G. K. (2013). Solar power generation by PV (photovoltaic) technology: A review. In *Energy* (Vol. 53, pp. 1–13). Elsevier Ltd., DOI: 10.1016/j.energy.2013.02.057

Singh, R., Alapatt, G. F., & Lakhtakia, A. (2013). Making solar cells a reality in every home: Opportunities and challenges for photovoltaic device design. *IEEE Journal of the Electron Devices Society*, *1*(6), 129–144. DOI: 10.1109/JEDS.2013.2280887

Sinha, A. (n.d.). To cite this article: Archana Sinha et al 2021. In *J. Phys. D: Appl. Phys* (Vol. 54).

Sovacool, B. K., Hess, D. J., Cantoni, R., Lee, D., Claire Brisbois, M., Jakob Walnum, H., Freng Dale, R., Johnsen Rygg, B., Korsnes, M., Goswami, A., Kedia, S., & Goel, S. (2022). Conflicted transitions: Exploring the actors, tactics, and outcomes of social opposition against energy infrastructure. *Global Environmental Change*, *73*, 102473. Advance online publication. DOI: 10.1016/j.gloenvcha.2022.102473

Sun, L., Fukuda, K., Guo, R., Castriotta, L. A., Forberich, K., Zhou, Y., Someya, T., Brabec, C. J., & Almora, O. (2025). A Flexible Photovoltaic Fatigue Factor for Quantification of Mechanical Device Performance. *Advanced Functional Materials*, •••, 2422706. DOI: 10.1002/adfm.202422706

Tanesab, J. A., Amheka, A., Sinaga, R., Mauta, J. J., & Hattu, E. (2020). A preliminary study of common defects of photovoltaic modules in West Timor, Indonesia. *IOP Conference Series. Earth and Environmental Science*, *542*(1), 012041. Advance online publication. DOI: 10.1088/1755-1315/542/1/012041

Tawalbeh, M., Al-Othman, A., Kafiah, F., Abdelsalam, E., Almomani, F., & Alkasrawi, M. (2021). Environmental impacts of solar photovoltaic systems: A critical review of recent progress and future outlook. *The Science of the Total Environment*, *759*, 143528. Advance online publication. DOI: 10.1016/j.scitotenv.2020.143528 PMID: 33234276

Tsanakas, J. A., Van Der Heide, A., Radavičius, T., Denafas, J., Lemaire, E., Wang, K., Poortmans, J., & Voroshazi, E. (n.d.). *Towards a circular supply chain for PV modules: Review of today's challenges in PV recycling, refurbishment and re-certification.*

Ukoba, K., Yoro, K. O., Eterigho-Ikelegbe, O., Ibegbulam, C., & Jen, T. C. (2024). Adaptation of solar energy in the Global South: Prospects, challenges and opportunities. In *Heliyon* (Vol. 10, Issue 7). Elsevier Ltd. DOI: 10.1016/j.heliyon.2024.e28009

Yamaguchi, M., Dimroth, F., Geisz, J. F., & Ekins-Daukes, N. J. (2021). Multi-junction solar cells paving the way for super high-efficiency. In *Journal of Applied Physics* (Vol. 129, Issue 24). American Institute of Physics Inc. DOI: 10.1063/5.0048653

Yang, C., Sun, F., Zou, Y., Lv, Z., Xue, L., Jiang, C., Liu, S., Zhao, B., & Cui, H. (2024). A Survey of Photovoltaic Panel Overlay and Fault Detection Methods †. In *Energies* (Vol. 17, Issue 4). Multidisciplinary Digital Publishing Institute (MDPI). DOI: 10.3390/en17040837

Yang, R., Wang, Y., Wu, D., Deng, Y., Luo, Y., Cui, X., Wang, X., Shu, Z., & Yang, C. (2017). Low-Temperature Fusible Silver Micro/Nanodendrites-Based Electrically Conductive Composites for Next-Generation Printed Fuse-Links. *ACS Nano*, *11*(8), 7710–7718. DOI: 10.1021/acsnano.7b00935 PMID: 28719748

Zarei, T., Abdolzadeh, M., & Yaghoubi, M. (2022). Comparing the impact of climate on dust accumulation and power generation of PV modules: A comprehensive review. *Energy for Sustainable Development*, *66*, 238–270. DOI: 10.1016/j.esd.2021.12.005

Chapter 11
Optimal Control Strategies for Wind Energy Systems Based on DFIG:
LQR–GA and Robust LQMinMax Approaches

Ines Zgarni
https://orcid.org/0000-0001-8259-6191
National Engineering School of Carthage, Tunisia

Lilia El Amraoui
College of Engineering, Princess Nourah bint Abdulrahman University, Saudi Arabia

ABSTRACT

This chapter focuses on the optimal control of a wind energy system based on DFIG by highlighting two optimal control approaches: Linear Quadratic Regulator based on a genetic algorithm, and robust LQMinMax regulator. The study aims to enhance the system's dynamic response, stability, and robustness under disturbances in stator voltages and wind speed. The LQR regulator minimizes errors in active/reactive powers and DC-link voltage, with the genetic algorithm ensuring optimal tuning of the Q and R weighting matrices. The LQMinMax regulator minimizes errors and rejects disturbances in the worst-case scenarios. A linearized state-space model of the system, including the generator, wind turbine, grid-side filter, and DC-link, is developed in the d-q rotating frame. The dynamic performances of the proposed controllers are compared to PI controller and standard LQR regulator. Simulation results demonstrate that proposed controllers significantly improve time-domain specifications, stability, and robustness, with eigenvalue and participation factor analyses confirming these enhancements.

DOI: 10.4018/979-8-3693-9924-8.ch011

INTRODUCTION

Among the objectives of this chapter's work is to propose optimal control strategies for wind energy systems based on the Doubly-Fed Induction Generator (DFIG).

In the first part of this chapter, the studied wind energy system configuration will be presented, including the main components such as the generator, grid-side filter, wind turbine, and DC link. The overall structure and function of the system will be described, setting the stage for further analysis.

In the second part, detailed modeling of each subsystem will be provided. This includes the mathematical representation of the generator, grid-side filter, wind turbine, and DC link, which will form the basis for the system's state-space model. This modeling is essential for understanding the dynamic behavior of the system under various operating conditions.

A third part will describe the linearization of the system model and the implementation of different control strategies. Various controllers, including the Proportional-Integral (PI) controller, Linear Quadratic Regulator (LQR), LQR with Genetic Algorithm (LQR-GA), and the robust LQMinMax regulator, will be introduced. The objectives of these control strategies will be to enhance the dynamic response, stability, and robustness of the wind energy system. To test the different control strategies, a comparative study will be conducted.

In the fourth part, the performance of each controller will be evaluated through simulation, considering various disturbances at the stator voltage and wind speed levels. The results will demonstrate the effectiveness of each strategy in improving the system's dynamic performance and robustness.

Finally, the stability and robustness of the proposed control strategies will be analyzed. Eigenvalue analysis and participation factors will be used to assess the system's stability, while robustness will be evaluated through the system's response to different levels of disturbances. This comprehensive approach aims to ensure that the proposed controllers can maintain optimal performance under a wide range of operating conditions.

DESCRIPTION OF STUDIED SYSTEM CONFIGURATION

The electrical generator in wind turbines plays a crucial role in energy conversion and overall production efficiency, reliability, and costs (Guida, 2020). The DFIG configuration represents a fast-growing segment in the wind turbine market due to its cost-effectiveness and size advantages (Djoudi, 2016).

The typical DFIG configuration involves direct stator grid connection and rotor connection via back-to-back power electronic converters (Azzouz, 2015). The converters, comprising the Rotor-Side Converter (RSC) and Grid-Side Converter (GSC), play crucial roles in system stability and control.

The RSC provides magnetizing current to the rotor windings and controls both active and reactive power at the stator terminals, while the GSC stabilizes the DC-link voltage and controls rotor reactive power (Puchalapalli & Singh, 2020). Notably, the inclusion of GSC filter dynamics is essential for a comprehensive understanding of the system's dynamic behavior.

Additionally, the Park transformation simplifies system equations and facilitates analysis (Derbel & Zhu, 2018). Overall, this section outlines the key components and configurations of the studied wind energy conversion system, setting the stage for further analysis and optimization, figure 1.

Figure 1. Configuration of a WECS-DFIG connected to the grid

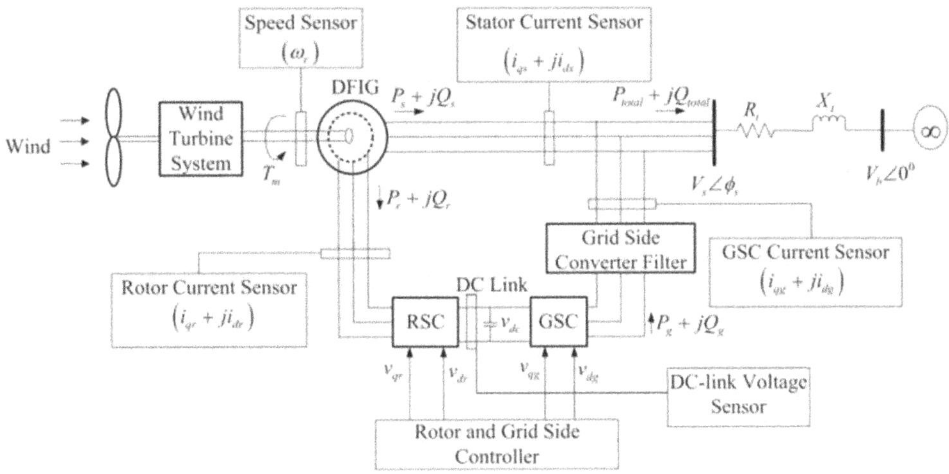

(Yaramasu et al., 2016)

STUDIED SYSTEM MODELING

The study system encompasses the generator, the grid-side filter, the wind turbine, and the DC link, with each subsystem being separately modeled.

Generator Modeling

To study the electrical equations of the DFIG, the relationships between the different machine frequencies are considered. Thus, the equation relating ω_s (stator voltage and current frequency), ω_r (rotor voltage and current frequency), and ω_m (rotor electrical speed) is described as follows (Abdeen et al., 2021):

$$\omega_s = \omega_r + \omega_m \tag{1}$$

The relationship between the mechanical speed of the shaft Ω_m and the electrical speed (expressed in rad/s), defined by equation (2), depends on the number of pole pairs of the machine p:

$$\omega_m = p\Omega_m \tag{2}$$

The slip g of the machine is given by equation (3):

$$g = \frac{\omega_s - \omega_m}{\omega_s} = \frac{\omega_r}{\omega_s} \tag{3}$$

In most applications, the stator windings are directly connected to the grid, and thus, ω_s is constant. This frequency is also referred to as synchronous frequency. However, ω_r obviously depends on the electrical speed of the shaft ω_m leading to three machine operating modes depending on speed (Masaud, 2013):

- $\omega_m < \omega_s \Rightarrow \omega_r > 0 \Rightarrow g > 0 \Rightarrow$ Sub-synchronous operation,
- $\omega_m > \omega_s \Rightarrow \omega_r < 0 \Rightarrow g < 0 \Rightarrow$ Super-synchronous operation,
- $\omega_m = \omega_s \Rightarrow \omega_r = 0 \Rightarrow g = 0 \Rightarrow$ Synchronous operation.

In addition, the DFIG space vector model is represented in the dq synchronous frame. For this purpose, the voltage equations in the dq frame are described by equations (4) and (5) (Zgarni & El Amraoui, 2022).

$$\begin{cases} U_{ds} = -R_s i_{ds} - \omega_s \psi_{qs} - \frac{1}{\omega_B}\frac{d\psi_{ds}}{dt} \\ U_{qs} = -R_s i_{qs} + \omega_s \psi_{ds} - \frac{1}{\omega_B}\frac{d\psi_{qs}}{dt} \end{cases} \tag{4}$$

$$\begin{cases} U_{dr} = -R_r i_{dr} - (\omega_s - \omega_r)\psi_{qr} - \frac{1}{\omega_B}\frac{d\psi_{dr}}{dt} \\ U_{qr} = -R_r i_{qr} - (\omega_s - \omega_r)\psi_{dr} - \frac{1}{\omega_B}\frac{d\psi_{qr}}{dt} \end{cases} \tag{5}$$

With:

- (U_{ds}, U_{qs}) and (U_{dr}, U_{qr}) denote, respectively, the stator and rotor voltages in frame dq,
- (i_{ds}, i_{qs}) and (i_{dr}, i_{qr}) represent, respectively, the stator and rotor currents in frame dq,
- R_r and R_s describe rotor and stator resistances, respectively.

The stator and rotor fluxes, in the d-q reference frame, are given by equations (6) and (7) (Hemeyine et al., 2020).

$$\begin{cases} \psi_{ds} = L_{ss} i_{ds} + L_m i_{dr} \\ \psi_{qs} = L_{ss} i_{qs} + L_m i_{qr} \end{cases} \tag{6}$$

$$\begin{cases} \psi_{dr} = L_{rr} i_{dr} + L_m i_{ds} \\ \psi_{qs} = L_{rr} i_{qr} + L_m i_{qs} \end{cases} \tag{7}$$

With:

$$\begin{cases} L_{ss} = L_s + L_m \\ L_{rr} = L_r + L_m \end{cases} \tag{8}$$

- L_{ss} and L_{rr} are the stator and rotor self-inductances,
- L_s and L_r represent stator and rotor leakage inductances,
- L_m is the mutual inductance.

For small-signal stability analysis and precise controller design in the next section, a linearized model of the DFIG system is developed from the set of derived equations. Figure 2 (A and B) shows the equivalent circuit of the DFIG system in transient operation, for the d and q axes, respectively, in the dq synchronous reference frame (Liu & Kong, 2013).

Figure 2. DFIG equivalent circuit: (A) for d axis and (B) for q axis

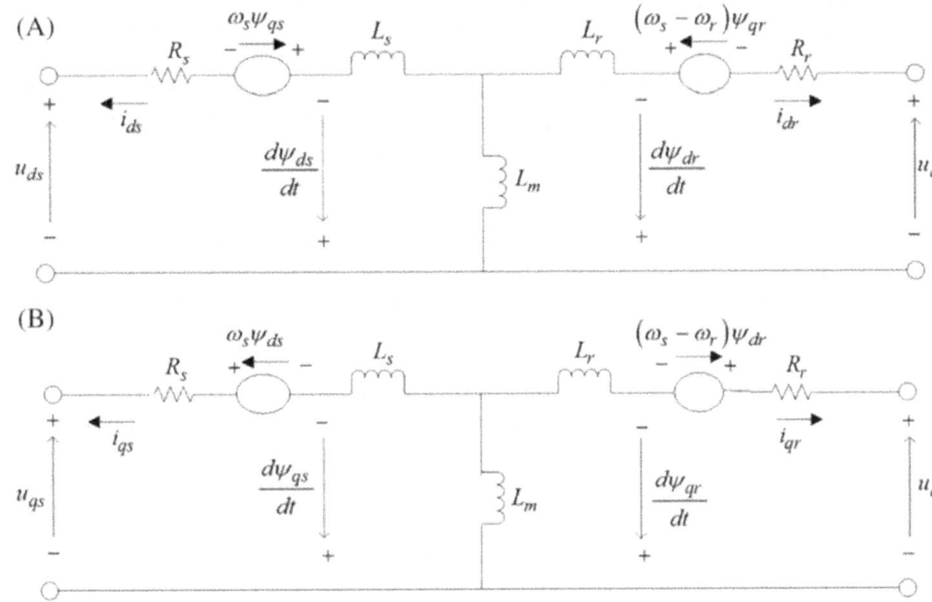

The following common assumptions are considered in the DFIG modeling process:

- The d-axis is 90° ahead of the q-axis in the same direction as rotation,
- The sign convention for the self-flux link and the current is considered positive for the outgoing flux,
- Hysteresis, eddy currents and magnetic saturation are neglected,
- Flux wave distribution is assumed to be purely sinusoidal.

After solving and rearranging equations (4) to (8), the derivatives of the stator and rotor currents in the d-q frame are written as follows:

$$\begin{cases} \frac{di_{qs}}{dt} = \sigma R_s L_{rr} i_{qs} + \sigma\left[\omega_{sr} L_m^2 - \omega_s L_{ss} L_{rr}\right] i_{ds} - \sigma R_r L_m i_{qr} + \sigma\left[\omega_{sr} L_m L_{rr} - \omega_s L_m L_{rr}\right] i_{dr} - \sigma L_m U_{qr} + \sigma L_{rr} U_{qs} \\ \frac{di_{ds}}{dt} = \sigma\left[\omega_s L_{ss} L_{rr} - \omega_{sr} L_m^2\right] i_{qs} + \sigma R_s L_{rr} i_{ds} - \sigma R_r L_m i_{dr} + \sigma\left[\omega_s L_m L_{rr} - \omega_{sr} L_m L_{rr}\right] i_{qr} - \sigma L_m U_{dr} + \sigma L_{rr} U_{ds} \end{cases} \tag{9}$$

$$\begin{cases} \frac{di_{qr}}{dt} = -\sigma R_s L_m i_{qs} + \sigma\left[\omega_s L_m L_{ss} - \omega_{sr} L_m L_{ss}\right] i_{ds} + \sigma R_r L_{ss} i_{qr} + \sigma\left[\omega_s L_m^2 - \omega_{sr} L_{ss} L_{rr}\right] i_{dr} + \sigma L_{ss} U_{qr} - \sigma L_m U_{qs} \\ \frac{di_{dr}}{dt} = \sigma\left[\omega_{sr} L_m L_{ss} - \omega_s L_m L_{ss}\right] i_{qs} - \sigma R_s L_m i_{ds} + \sigma R_r L_{ss} i_{dr} + \sigma\left[\omega_{sr} L_{ss} L_{rr} - \omega_s L_m^2\right] i_{qr} + \sigma L_{ss} U_{dr} - \sigma L_m U_{ds} \end{cases} \tag{10}$$

With:

- $\sigma = \left(\frac{\omega_B}{L_m^2 - L_{ss}L_{rr}}\right)$ where $\omega_B = 2\pi f$ (ω_B denotes the angular velocity and $f = 50Hz$ represents the nominal frequency),
- $\omega_{sr} = \omega_s - \omega_r$

Grid-Side Filter Modeling

The grid-side filter is inserted at the output of the GSC converter. It is used to modify the amplitude and phase of the voltage imposed by the GSC, and to attenuate harmonics present in the system. To ensure accurate results, the dynamics of the grid-side filter, i.e. the currents flowing through the inductance present in it, are also considered in the modeling process (Salhi et al., 2019). Figure 3 illustrates the equivalent circuit of the GSC filter connected between the GSC and the point of common coupling (PCC).

Figure 3. Equivalent circuit of grid-side-filter

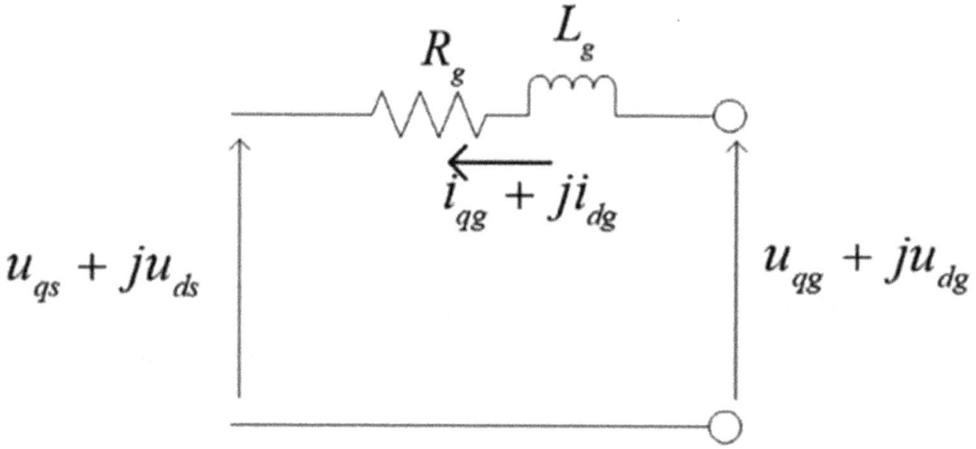

Where R_g and L_g identify the resistance and inductance of the mains-side filter, respectively. (i_{dg}, i_{qg}) and (V_{dg}, V_{qg}) represent the dq components of the GSC current and voltage, respectively.

The derivatives of the GSC converter currents, in the dq reference frame, are given by equations (11).

$$\begin{cases} \frac{di_{qg}}{dt} = -\frac{\omega_B R_g}{L_g} i_{qg} + \omega_B \omega_s i_{dg} + \frac{\omega_B}{L_g} U_{qg} - \frac{\omega_B}{L_g} U_{qs} \\ \frac{di_{dg}}{dt} = -\omega_B \omega_s i_{qg} - \frac{\omega_B R_g}{L_g} i_{dg} + \frac{\omega_B}{L_g} U_{dg} - \frac{\omega_B}{L_g} U_{ds} \end{cases} \tag{11}$$

Wind Turbine Modeling

Since the static characteristics of a wind turbine rotor are derived from the relationships between the total wind power and the mechanical power of the wind turbine. Some parameters of the variable-speed wind turbine are taken into consideration to model the DFIG (Ouhssain et al., 2024). In this study, the proposed system is simplified to a single-mass model since the effects of the gearbox and shaft are not considered. In other words, the angular velocity of the turbine and that of the generator are considered equal. They will also be available for feedback to ensure control design. The dynamics of the pitch angle controller (i.e. the mechanical control dynamics) are neglected due to its slow response compared to the electrical control dynamics. Next, the differential equation revealing the rotation equation is written as follows:

$$\frac{d\omega_r}{dt} = \frac{1}{2H}T_m - \frac{1}{2H}T_e \tag{12}$$

With:

- ω_r identifies the angular speed of the generator,
- H denotes the total equivalent inertia (the equivalent inertia reduced to the shaft),
- T_m and T_e represent, respectively, the mechanical torque and the electromagnetic torque.

Wind turbine modeling corresponds to the extraction of rotor power by determining the mechanical torque as a function of the air flow ρ_{air} over the blades as follows:

$$T_t = \frac{1}{2}\rho_{air}\pi R^3 V_w^{\ 2} C_p(\lambda,\beta) \tag{13}$$

Where T_t is the torque generated by the rotor, R identifies the blade radius, V_w designates the wind speed and $C_p(\lambda,\beta)$ represents the power coefficient which depends on the blade pitch angle β and the peak speed ratio λ.

In order to achieve maximum power extraction, a control approach is used by applying the MPPT strategy based on the Indirect Speed Control technique (Shutari et al., 2022), illustrated in figure 4.

Figure 4. Indirect speed control strategy

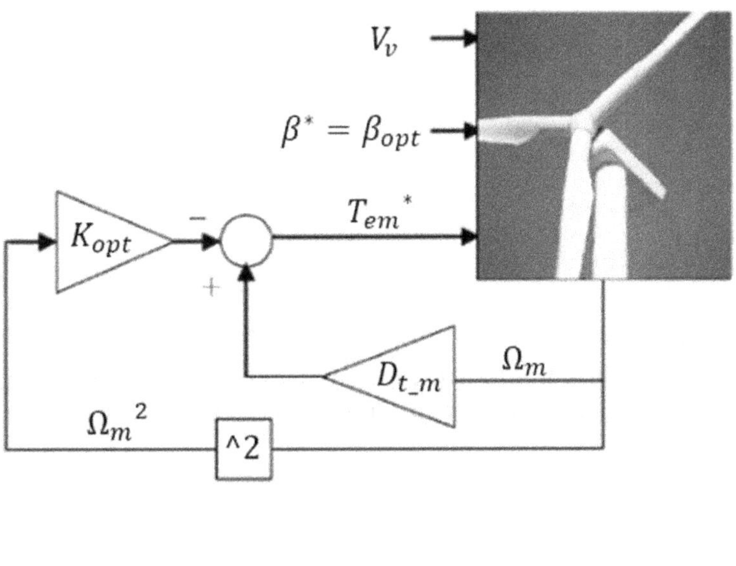

With the aim of optimizing energy efficiency, indirect speed control, which is applied in most variable-speed wind turbines, indirectly controls the rotational speed and operates with a specific speed λ_{opt} to take the maximum electromagnetic torque for each value of turbine rotational speed Ω_t (Sitharthan et al., 2020) (Dbaghi et al., 2021). λ_{opt} and C_p will be described as follows:

$$\begin{cases} \lambda_{opt} = \dfrac{R\omega_r}{V_w} \\ C_p = C_{p\_max} \end{cases} \tag{14}$$

The optimum power control range for WT is obtained when: $[\omega_{rmin}, \omega_{r\_nominal}]$ and $V_w \in [V_{wmin}, V_{w\_nominal}]$. Then, at any turbine speed, the extracted turbine power P_m (or mechanical torque T_m) is a function of the measured rotor angular velocity and always maintained within the maximum power curve as follows:

$$\begin{cases} P_m = K_{opt}\omega_r^{\,3} \\ T_m = K_{opt}\omega_r^{\,2} \end{cases} \tag{15}$$

Knowing that the constant K_{opt}, expressed by equation (16), depends on turbine characteristics and environmental conditions (Hammami et al., 2018).

$$K_{opt} = \frac{1}{2}\rho_{air}\,\pi\frac{R^5}{\lambda_{opt}^{\,3}}C_{p\_max} \tag{16}$$

Rotor speed disturbances affect the wind turbine's mechanical power and regulate its speed around its steady-state value by adjusting the electromagnetic torque. The electromagnetic torque T_e is then given by equation (17).

$$T_e = L_m i_{qs} i_{dr} - L_m i_{ds} i_{qr} \tag{17}$$

The derivative of the angular velocity of the generator, expressed by equation (18), is thus obtained from equations (12), (15) and (17).

$$\frac{\mathrm{d}\omega_r}{\mathrm{dt}} = \frac{K_{opt}}{2H}\omega_r^{\,2} - \frac{L_m}{2H}i_{qs}i_{dr} + \frac{L_m}{2H}i_{ds}i_{qr} \tag{18}$$

DC Link Modeling

The two converters of the wind energy conversion system (RSC and GSC) are interconnected via a DC link capacitor C_{dC}, which is used as an energy storage device (Prajapat et al., 2019). Acting as a reactive power source, the rotor-side magnetizing current is supplied by the DC link. Consequently, stator-side power factor control is established. For this chapter work, converter losses and switching dynamics are assumed to be negligible. Consequently, the active power of the DC link (P_{dC}) is formulated as follows:

$$P_{dc} = P_r - P_g \tag{19}$$

Where P_r and P_g identify, respectively, the active powers at the AC terminals of the RSC and GSC converters. Also, the active power of the DC link can be written as:

$$P_{dc} = U_{dc} i_{dc} = U_{dc} C_{dc} \frac{dU_{dc}}{dt} \tag{20}$$

Where U_{dC} and i_{dC} represent, respectively, the voltage across the DC link capacitor and the current flowing through it. The derivative of U_{dC} is determined as a function of P_g and P_r as follows:

$$\frac{dU_{DC}}{dt} = \frac{1}{C_{dc} U_{dc}} \left(P_r - P_g\right) \tag{21}$$

To study the power flow of the studied system, the apparent power supplied to the DFIG via the stator and rotor circuit is determined by equation 22.

$$P_s + jQ_s = U_{dqs} i_{dqs}^{\ *} = \left(U_{qs} + jU_{ds}\right)\left(i_{qs} + ji_{ds}\right)^* \tag{22}$$

The stator active power P_s represents the real part of equation (22) and the stator reactive power Q_s designates the imaginary part (Gurung et al., 2017), equation 23.

$$\begin{bmatrix} P_s \\ Q_s \end{bmatrix} = \begin{bmatrix} U_{qs} & U_{ds} \\ U_{ds} & -U_{qs} \end{bmatrix} \begin{bmatrix} i_{qs} \\ i_{ds} \end{bmatrix} \tag{23}$$

Based on the same principle, the rotor active and reactive powers and those of the GSC converter are obtained from equations (24) and (25) respectively.

$$\begin{bmatrix} P_r \\ Q_r \end{bmatrix} = \begin{bmatrix} U_{qr} & U_{dr} \\ U_{dr} & -U_{qr} \end{bmatrix} \begin{bmatrix} i_{qr} \\ i_{dr} \end{bmatrix} \tag{24}$$

$$\begin{bmatrix} P_g \\ Q_g \end{bmatrix} = \begin{bmatrix} U_{qg} & U_{dg} \\ U_{dg} & -U_{qg} \end{bmatrix} \begin{bmatrix} i_{qg} \\ i_{dg} \end{bmatrix} \tag{25}$$

As a result, equation (21) is rewritten using equations (24) to (25) as follows:

$$\frac{dU_{DC}}{dt} = \frac{1}{C_{dc} U_{dc}} \left(U_{qr} i_{qr} + U_{dr} i_{dr} - U_{qg} i_{qg} - U_{dg} i_{dg}\right) \tag{26}$$

LINEARIZED MODEL OF STUDIED SYSTEM

To analyze the inherent dynamic characteristics of the electrical system, a robust approach utilizing small-signal analysis is employed, incorporating linearization and eigenvalue analysis techniques (Amin et al., 2017). The presence of non-linearities in the DFIG-based WECS system significantly affects its dynamic performance. Therefore, linearizing the previously described equations around a nominal operating point becomes essential for designing an optimal linear controller. In this context, the mathematical models of the individual subsystems are combined to formulate a comprehensive set of state-space equations. Consequently, these state-space equations can be represented as a generalized set of nonlinear differential equations, expressed as follows:

$$\begin{cases} \dot{x} = f(x,y) \\ y = g(x,y) \end{cases} \tag{27}$$

With:

- $f= [f_1;f_2;\ldots;f_n]$ and $g= [g_1;g_2;\ldots;g_m]$ represent non-linear functions,
- $x= [x_1;x_2;\ldots;x_n]$ denotes the state vector, where n is the number of state variables,
- $u= [u_1;u_2;\ldots;u_r]$ denotes the control vector where r is the number of inputs,
- $y= [y_1;y_2;\ldots;y_m]$ is the output vector, where m is the number of outputs.

In addition, the small-signal model is developed by considering a small disturbance in the input variables u. This generates changes in state variables x and output variables y as follows:

$$\begin{cases} x = x_0 + \Delta x \\ u = u_0 + \Delta u \\ y = y_0 + \Delta y \end{cases} \tag{28}$$

With Δ indicates the variation of the respective variable.

Moreover, model initialization is critical for achieving a rapid return to equilibrium following a disturbance, as all state derivatives are zero at equilibrium. The nominal operating point is determined through the study and analysis of power or load flows, utilizing simplified concepts such as the "per-unit system" and the "one-line diagram." These methods provide insight into the amplitude and phase angle of active and reactive voltages and powers. The complete system model, which encompasses the 4th-order DFIG, the 2nd-order GSC filter, the 1st-order wind turbine, and the 1st-order DC link, is derived by integrating the models of the above-mentioned subsystems. The complete set of non-linear equations (9 to 11, 18, and 23 to 26) are linearized around an operating point (i_{qs0}, i_{ds0}, i_{qr0}, i_{dr0}, i_{qg0}, i_{dg0}, ω_{r0}, U_{dc0}) and described by the following state-space equations:

$$\begin{cases} \Delta\dot{x} = A\,\Delta x + B\,\Delta u \\ \Delta y = C\,\Delta x + D\,\Delta u \end{cases} \tag{29}$$

Where Δx, Δu, and Δy represent, respectively, variations in states, inputs and outputs. They are expressed by equations (30 to 32).

$$\Delta x = \begin{bmatrix} \Delta i_{qs} & \Delta i_{ds} & \Delta i_{qr} & \Delta i_{dr} & \Delta i_{qg} & \Delta i_{dg} & \Delta\omega_r & \Delta U_{dc} \end{bmatrix}^T \tag{30}$$

$$\Delta u = \begin{bmatrix} \Delta U_{qr} & \Delta U_{dr} & \Delta U_{qg} & \Delta U_{dg} & \Delta U_{qs} & \Delta U_{ds} \end{bmatrix}^T \tag{31}$$

$$\Delta y = \begin{bmatrix} \Delta P_s & \Delta Q_s & \Delta P_r & \Delta Q_r & \Delta P_g & \Delta Q_g & \Delta U_{dc} \end{bmatrix}^T \tag{32}$$

Coefficients A, B, C and D denote, respectively, the state matrix, the control or input matrix, the output matrix and the feedforward matrix. The analytical values of these matrices are expressed by equations (33 to 36).

$$A = \begin{bmatrix} A_{11} & A_{12} & A_{13} & A_{14} & A_{15} & A_{16} & A_{17} & A_{18} \\ A_{21} & A_{22} & A_{23} & A_{24} & A_{25} & A_{26} & A_{27} & A_{28} \\ A_{31} & A_{32} & A_{33} & A_{34} & A_{35} & A_{36} & A_{37} & A_{38} \\ A_{41} & A_{42} & A_{43} & A_{44} & A_{45} & A_{46} & A_{47} & A_{48} \\ A_{51} & A_{52} & A_{53} & A_{54} & A_{55} & A_{56} & A_{57} & A_{58} \\ A_{61} & A_{62} & A_{63} & A_{64} & A_{65} & A_{66} & A_{67} & A_{68} \\ A_{71} & A_{72} & A_{73} & A_{74} & A_{75} & A_{76} & A_{77} & A_{78} \\ A_{81} & A_{82} & A_{83} & A_{84} & A_{85} & A_{86} & A_{87} & A_{88} \end{bmatrix} \tag{33}$$

$$B = \begin{bmatrix} -\sigma L_m & 0 & 0 & 0 & \sigma L_{rr} & 0 \\ 0 & -\sigma L_m & 0 & 0 & 0 & \sigma L_{rr} \\ \sigma L_{ss} & 0 & 0 & 0 & -\sigma L_m & 0 \\ 0 & \sigma L_{ss} & 0 & 0 & 0 & -\sigma L_m \\ 0 & 0 & \frac{\omega_B}{L_g} & 0 & \frac{-\omega_B}{L_g} & 0 \\ 0 & 0 & 0 & \frac{\omega_B}{L_g} & 0 & \frac{-\omega_B}{L_g} \\ 0 & 0 & 0 & 0 & 0 & 0 \\ \frac{i_{qr0}}{C_{dc}U_{dc0}} & \frac{i_{dr0}}{C_{dc}U_{dc0}} & \frac{-i_{qg0}}{C_{dc}U_{dc0}} & \frac{-i_{dg0}}{C_{dc}U_{dc0}} & 0 & 0 \end{bmatrix} \tag{34}$$

$$C = \begin{bmatrix} U_{qs0} & U_{ds0} & 0 & 0 & 0 & 0 & 0 & 0 \\ U_{ds0} & -U_{qs0} & 0 & 0 & 0 & 0 & 0 & 0 \\ 0 & 0 & U_{qr0} & U_{dr0} & 0 & 0 & 0 & 0 \\ 0 & 0 & U_{dr0} & -U_{qr0} & 0 & 0 & 0 & 0 \\ 0 & 0 & 0 & 0 & U_{qg0} & U_{dg0} & 0 & 0 \\ 0 & 0 & 0 & 0 & U_{dg0} & -U_{qg0} & 0 & 0 \\ 0 & 0 & 0 & 0 & 0 & 0 & 0 & 1 \end{bmatrix} \tag{35}$$

$$D = \begin{bmatrix} 0 & 0 & 0 & 0 & i_{qs0} & i_{ds0} \\ 0 & 0 & 0 & 0 & -i_{ds0} & i_{qs0} \\ i_{qr0} & i_{dr0} & 0 & 0 & 0 & 0 \\ -i_{dr0} & i_{qr0} & 0 & 0 & 0 & 0 \\ 0 & 0 & i_{qg0} & i_{dg0} & 0 & 0 \\ 0 & 0 & -i_{dg0} & i_{qg0} & 0 & 0 \\ 0 & 0 & 0 & 0 & 0 & 0 \end{bmatrix} \quad (36)$$

As a result, the system analysis considers eight states, six inputs and seven output variables. The linearized equilibrium model is employed to evaluate the system's small-signal stability by determining the eigenvalues using the state matrix. The dimensions of A, B, C and D are (8 × 8), (8 × 6), (7 × 8) and (7 × 6) respectively. Eigenvalues are computed to assess system stability, while participation factors are analyzed to identify the dominant states contributing to oscillations associated with specific eigenvalues.

CONTROL STRATEGIES OF THE STUDIED SYSTEM

In the proposed system, linearized around an equilibrium point, stability is influenced by wind speed variations (mechanical parameter) and load disturbances (stator voltage variations, electrical parameter). Stability analysis focuses on stator voltage disturbances, while robustness and performance are evaluated under mechanical perturbations, requiring suitable MIMO control models to mitigate these effects.

Objectives of Control Strategy

The primary goal of the controllers is to restore the steady-state conditions of both the state and output variables after disturbances in stator terminal voltages. This requires nullifying deviations in DFIG currents and generator angular speed, as well as ensuring zero variations in active/reactive power and DC-link voltage. Peak values and stabilization times must remain within acceptable ranges. This is achieved by mitigating system disturbances through RSC and GSC voltage component control. The input variables of Equation (31) are divided into two groups: the control inputs ($\Delta u_n = [\Delta U_{qr} \;\; \Delta U_{dr} \;\; \Delta U_{qg} \;\; \Delta U_{dg}]^T$) for the RSC and GSC, and the disturbance inputs ($\Delta w = [\Delta U_{qs} \;\; \Delta U_{ds}]^T$) for the system.

The active and reactive powers of the DFIG, flowing through the stator and rotor, as well as the generator's angular speed, are controlled via the RSC by adjusting the d-q components of the rotor voltage (ΔU_{qr}, ΔU_{dr}). Meanwhile, the exchange of active and reactive powers between the GSC and PCC, along with the DC-link voltage, is managed via the GSC by controlling the d-q components of the GSC terminal voltages (ΔU_{qg}, ΔU_{dg}). To maintain a constant DC-link voltage, the power delivered to the RSC must equal the power delivered by the GSC under steady-state conditions, assuming negligible converter losses. Accordingly, the corresponding changes in the matrices $[B]$ and $[D]$ from Equations (29) are adjusted as $[B_1 F_1]$ and $[D_1 F_2]$, respectively. The state-space equations (29) are then updated as follows:

$$\begin{cases} \Delta\dot{x} = A\,\Delta x + \begin{bmatrix} B_1 & F_1 \end{bmatrix} \begin{bmatrix} \Delta u_n \\ \Delta w \end{bmatrix} \\ \Delta y = C\,\Delta x + \begin{bmatrix} D_1 & F_2 \end{bmatrix} \begin{bmatrix} \Delta u_n \\ \Delta w \end{bmatrix} \end{cases} \tag{37}$$

The matrices A and C remain unchanged. However, the updated matrices B_1, F_1, D_1 and F_2 are defined by Equations (38) to (41).

$$B_1 = \begin{bmatrix} -\sigma L_m & 0 & 0 & 0 \\ 0 & -\sigma L_m & 0 & 0 \\ \sigma L_{ss} & 0 & 0 & 0 \\ 0 & \sigma L_{ss} & 0 & 0 \\ 0 & 0 & \frac{\omega_B}{L_g} & 0 \\ 0 & 0 & 0 & \frac{\omega_B}{L_g} \\ 0 & 0 & 0 & 0 \\ \frac{i_{qr0}}{C_{dc}U_{dc0}} & \frac{i_{dr0}}{C_{dc}U_{dc0}} & \frac{-i_{qg0}}{C_{dc}U_{dc0}} & \frac{-i_{dg0}}{C_{dc}U_{dc0}} \end{bmatrix} \tag{38}$$

$$F_1 = \begin{bmatrix} \sigma L_{rr} & 0 \\ 0 & \sigma L_{rr} \\ -\sigma L_m & 0 \\ 0 & -\sigma L_m \\ \frac{-\omega_B}{L_g} & 0 \\ 0 & \frac{-\omega_B}{L_g} \\ 0 & 0 \\ 0 & 0 \end{bmatrix} \tag{39}$$

$$D_1 = \begin{bmatrix} 0 & 0 & 0 & 0 \\ 0 & 0 & 0 & 0 \\ i_{qr0} & i_{dr0} & 0 & 0 \\ -i_{dr0} & i_{qr0} & 0 & 0 \\ 0 & 0 & i_{qg0} & i_{dg0} \\ 0 & 0 & -i_{dg0} & i_{qg0} \\ 0 & 0 & 0 & 0 \end{bmatrix} \tag{40}$$

$$F_2 = \begin{bmatrix} i_{qs0} & i_{ds0} \\ -i_{ds0} & i_{qs0} \\ 0 & 0 \\ 0 & 0 \\ 0 & 0 \\ 0 & 0 \\ 0 & 0 \end{bmatrix} \quad (41)$$

The functional diagram of the linearized system, shown in figure 5, represents a small-signal model of the linearized system.

Figure 5. Block diagram of linearized studied system

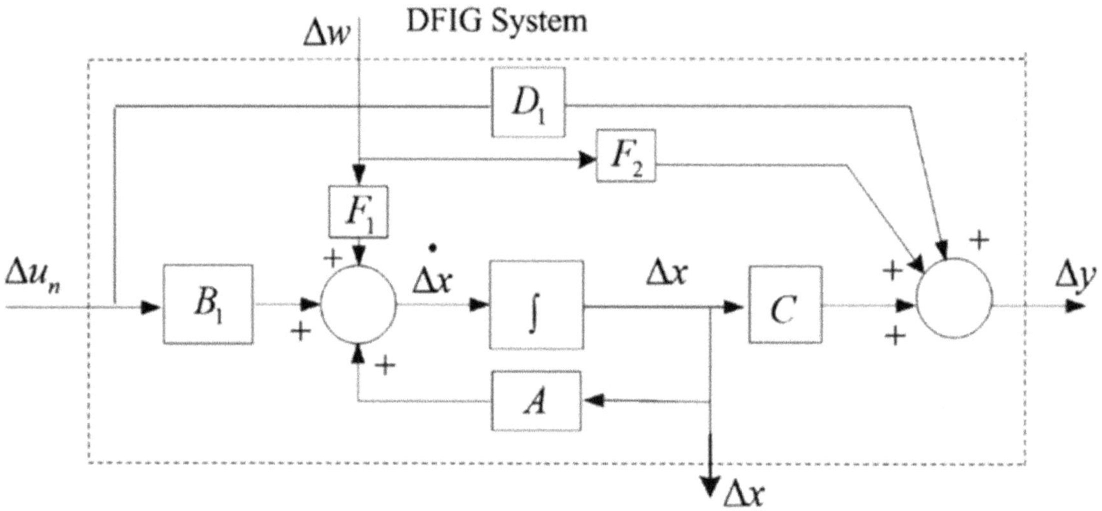

Moreover, Δw and Δu_n represent the disturbance and control inputs, respectively, which are used to introduce a disturbance in the system and control the perturbing parameters (Δx and Δy). Subsequently, studies on the effects of the operating point and the interactions of the adopted control design on the overall system stability will be developed. In this study, all states and output variables are assumed to be measurable and available for feedback. The vector control strategy is then employed for DFIG-based wind systems.

The alignment is chosen so that the Q-axis of the DFIG coincides with the voltage of the "slack bus" Vb ∠ 0°, and the q-axis of the DFIG aligns with the stator terminal voltage Us ∠ ϕs. As a result, the quadrature component of the stator voltage equals the total voltage (U_{qs}=U and U_{ds}=0).

The next section will focus on the study of various control designs for the DFIG-based WECS, governed by the state-space model. The control techniques used include the conventional PI controller, the LQR regulator with and without AG, and the robust LQMinMax regulator.

PI Controller Implementation

Given that all measured output variables (Δy) are available for feedback, the state-space-based PI controller for the MIMO system under study is integrated to minimize output deviations. In this case, four state variables ($PI_{U_{qr}}$, $PI_{U_{dr}}$, $PI_{U_{qg}}$ and $PI_{U_{dg}}$) are added due to the integral component of the PI controllers within the system. The control structure consists of four PI loops, where the error signals are processed through a decoupler, providing the necessary control signals based on the number of required control variables. Figure 6 shows the block diagram of the studied system with the conventional PI controller.

Figure 6. Block diagram of the studied system with PI controller

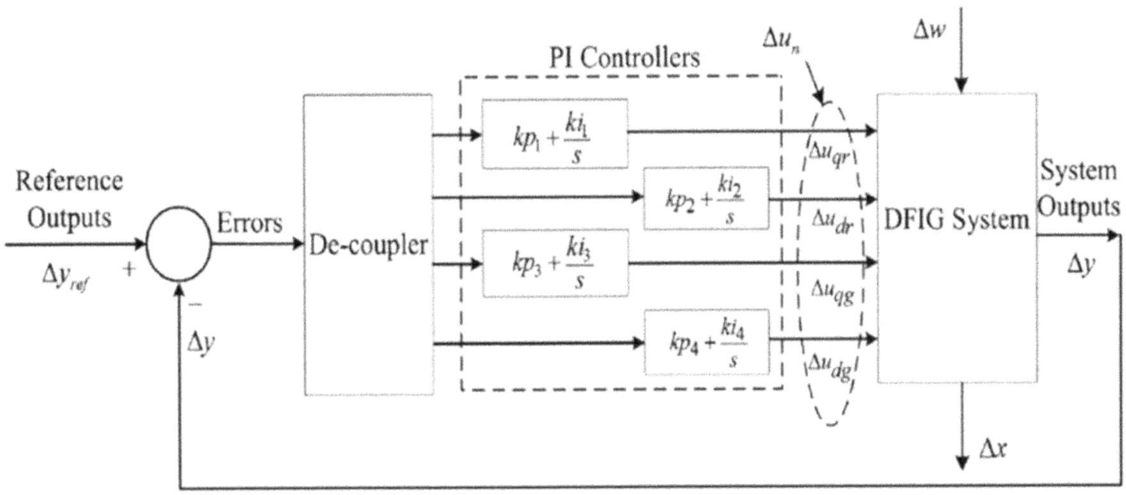

The controller's gains are adjusted according to the reference signals or the measured output signals (P_s, Q_s, P_r, Q_r, P_g, Q_g, U_{dC}) using the loop-tuning method to ensure satisfactory disturbance response and performance. To tune the PI controller gains using Matlab's "looptune" command (Bhushan et al., 2021), the state-space model of the system is first created, followed by the development of adjustable controller component models. A target bandwidth range of [0.1, 1] is defined, setting the crossover frequency of each loop within [0.1, 1 rad/min]. The "looptune" command is then used to adjust the controller's gains, and the system's performance is evaluated using a success criterion to ensure the desired results are achieved.

Implementation of LQR Regulator

In this study, all states and outputs of the system must be controllable and observable in the state-space model. The LQR controller is applied with full-state feedback using the system's eight available states, figure 7.

Figure 7. Block diagram of studied system with LQR controller

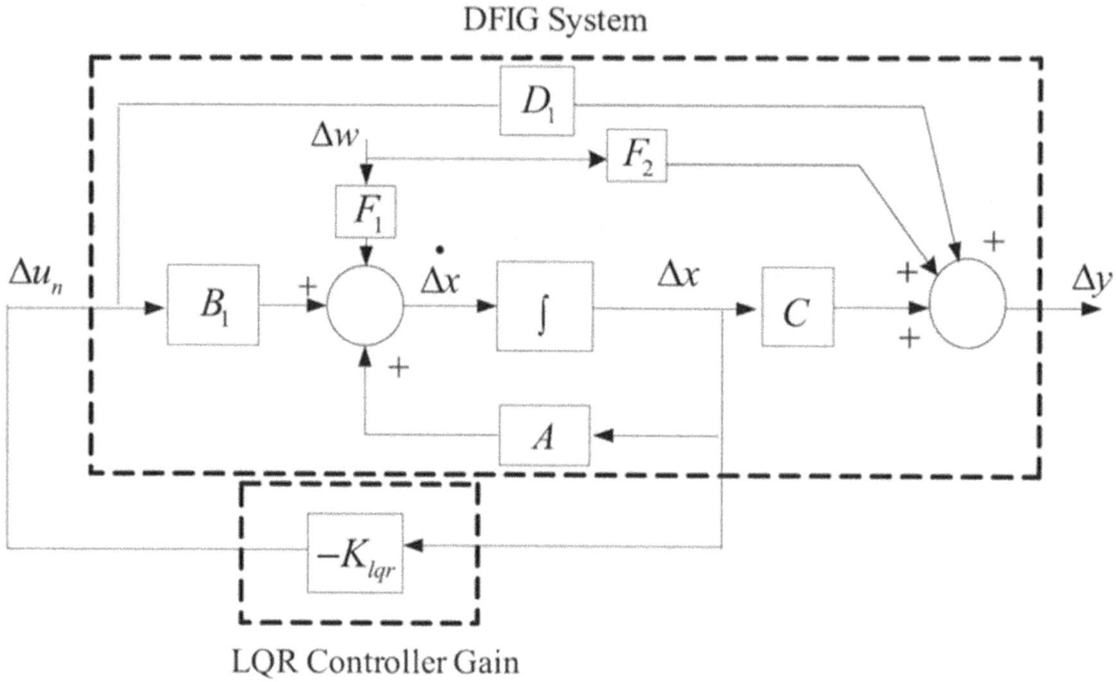

The LQR approach provides an optimal method to solve constrained optimization problems with linear dynamics and quadratic cost functions (Mehta et al., 2015).

The cost function J is defined by equation 42 to penalize deviations in states and control inputs, with the state-weighting matrix Q_{LQ} and control-weighting matrix R_{LQ}.

$$J = \int_0^\infty \left([\Delta x]^T Q_{LQ}[\Delta x] + [\Delta u_n]^T R_{LQ}[\Delta u_n]\right)dt \tag{42}$$

With T denotes the transpose of the associated matrix.

The LQR controller computes the optimal state feedback gain matrix K_{LQ}, which minimizes the cost function, resulting in the control law revealed by the equation (43).

$$\Delta u_n = -K_{LQ}\Delta x \tag{43}$$

The matrices Q_{LQ} and R_{LQ} are chosen to penalize state deviations and control effort, with typical values as follows:

$$\begin{cases} Q_{LQ} = W \times eye(8) \\ R_{LQ} = eye(4) \end{cases} \tag{44}$$

With W=0.1 and "*eye*()" denotes an identity matrix.

The optimal gain matrix K_{LQ} is computed by solving the Riccati equation. The stability of the closed-loop system is ensured by checking the controllability and observability of the system.

After calculating K_{LQ}, step responses to various perturbations are used to validate the controller's performance in optimizing the system's behavior.

Implementation of LQR-GA Regulator

To overcome the limitations of the traditional "trial-and-error" method for selecting the weighting matrices Q_{LQ} and R_{LQ}, a Genetic Algorithm (GA)-based optimization approach is implemented. This technique adjusts the matrices Q_{LQGA} and R_{LQGA} to minimize the cost function J, while adhering to the principles of the standard LQR.

The weighting matrices are defined as follows:

$$\begin{cases} Q_{LQAG} = diag\left(Q_{LQAG1},\ldots,Q_{LQAG8}\right) \\ R_{LQAG} = diag\left(R_{LQAG1},\ldots,R_{LQAG4}\right) \end{cases} \tag{45}$$

The cost function J comprises two key components: the deviation of the system states (Δx) from zero, and the energy of the control inputs (Δu). The optimization process seeks to achieve a rapid reduction in state deviations while simultaneously minimizing the energy required for control actions, thereby ensuring efficient and balanced system performance.

The Genetic Algorithm (GA) process begins with the initialization of a random population of chromosomes, each containing 12 variables (Q_{LQAG1} to R_{LQAG4}), encoded with 10 bits per variable, resulting in a total chromosome length of 120 bits. Key parameters include a population size of n_p=20, a crossover probability of P_c=0.8, a mutation probability of P_{mu}=0.05, and a maximum of N_{max}=100° generations. Fitness evaluation is performed by computing the cost function J for each chromosome. Genetic operations involve selecting the best-performing chromosomes, applying crossover with a probability of 0.8 to generate offspring, and introducing random mutations with a probability of 0.05. This iterative process continues until a stopping criterion, such as $N=N_{max}$ or convergence of the cost function, is achieved.

Figure 8 illustrates the block diagram of the studied system incorporating the LQR-GA regulator. Its structure closely resembles that of figure 7, with the only distinction being the inclusion of the Genetic Algorithm. As a result, the state variables remain unchanged.

Figure 8. Block diagram of the studied system incorporating the LQR-AG regulator

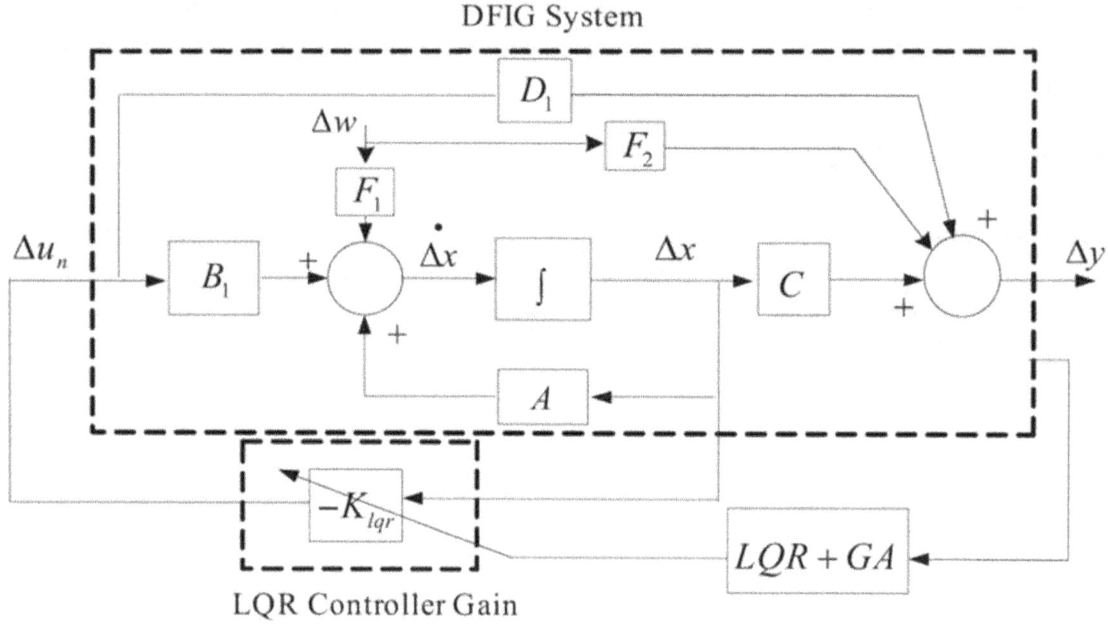

The optimized feedback gain matrix K_{LQGA} (4×8) is calculated to achieve the desired system performance. The GA-based approach ensures robustness by requiring only the cost function evaluation, without detailed knowledge of the overall system dynamics.

Implementation of LQMinMax Regulator

Figure 9 illustrates the closed-loop control structure incorporating the LQMinMax regulator. The system inputs remain unchanged (i.e., $\Delta u = \begin{bmatrix} \Delta u_n \\ \Delta w \end{bmatrix}$).

Figure 9. Block diagram of studied system with LQMinMax regulator

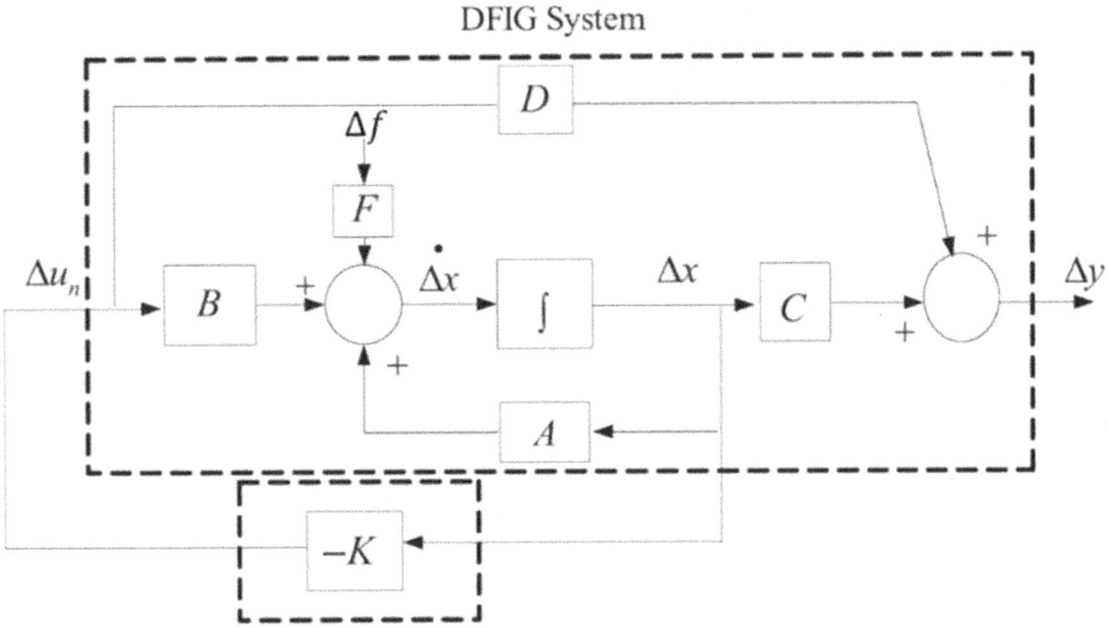

This controller builds upon the standard LQR principles but includes the system's perturbations in its model. The new state-space model of the perturbed system is expressed as:

$$\begin{cases} \Delta\dot{x}(t) = A\,\Delta x(t) + B\,\Delta u(t) + F\,\Delta f(t) \\ \Delta y(t) = C\,\Delta x(t) + D\,\Delta u(t) \end{cases} \tag{46}$$

where F is the disturbance matrix and $\Delta f(t)$ represents unknown disturbances or modeling uncertainties. The state variables Δx are compared with their reference values Δx_{ref}, resulting in an error $\Delta\varepsilon$ given by equation (47).

$$\Delta\varepsilon = \Delta x - \Delta x_{ref} \tag{47}$$

The new state-space model is then updated to include $\Delta\varepsilon$, with the error dynamics and outputs expressed as:

$$\begin{cases} \Delta\dot{\varepsilon}(t) = A\,\Delta\varepsilon(t) + B\,\Delta u(t) + F\,\Delta f(t) + d \\ \Delta y(t) = C\,\Delta\varepsilon(t) + D\,\Delta u(t) \end{cases} \tag{48}$$

Where:

$$d = A\,\Delta x_{ref}(t) - \Delta\dot{x}_{ref}(t) \tag{49}$$

$\Delta\varepsilon$ represents the system's state variables under the robust LQMinMax controller and includes deviations between measured and reference state variables. The control problem focuses on determining the optimal control input u^* and worst-case disturbance, f^* such that (Mao et al., 2020):

$$(u^*,f^*) = \min_u \max_f J(u,f) \tag{50}$$

The quadratic cost function is defined as:

$$J(u) = \int_{t_0}^{T} \left[\varepsilon(t)^T Q\, \varepsilon(t) + u(t)^T R_1\, u(t) - \gamma^2 f^* R_2 f\right] dt \tag{51}$$

where $Q°$, R_1, and R_2 are weighting matrices, and γ is the robustness parameter.

The problem, in this case, is to determine the two state feedback gains K and K_d expressed by the following equations:

$$\begin{cases} K = R_1^{-1} B^T P \\ K_d = \gamma^{-2} R_2^{-1} F^T P \end{cases} \tag{52}$$

with P being the solution to the Riccati MinMax differential equation:

$$-\dot{P} = A^T P + PA - PBR_1^{-1} B^T P + \gamma^{-2} PFR_2^{-1} F^T P + Q \tag{53}$$

For simulations, the weighting matrices Q, R_1, and R_2 follow those of the LQR-AG controller, with R_2 set higher than R_1 to account for the minmax problem. The minimum γ value (γ_{min}=0.0007) is determined via stability analysis or MATLAB's μ-Toolbox. γ must satisfy the following constraints:

- If ($\gamma \ll \gamma_{min}$), the closed-loop system risks instability.
- If ($\gamma \gg \gamma_{min}$), the system loses robustness against worst-case disturbances and behaves like a standard LQR controller.
- Higher γ values improve system dynamics but reduce robustness.

Simulations use γ values of -0.1, -0.2, and -0.3*pu*. Additionally, an increase in R_2 reduces K_d in steady state, limiting disturbance rejection, while a decrease in γ increases K_d and improves disturbance attenuation. The dynamic performance, including closed-loop gains and poles, is evaluated using the following updated closed-loop model:

$$\dot{x}(t) = (A - BK + FK_d)x(t) + (BK - FK_d)\, x_{ref}(t) \tag{54}$$

STUDIED SYSTEM IMPLEMENTATION

In this section, the functional diagram of the proposed model for the studied system, implemented using MATLAB Simulink and SimPowerSystems, is illustrated in figure 10.

Figure 10. Block diagram of the studied system simulation model

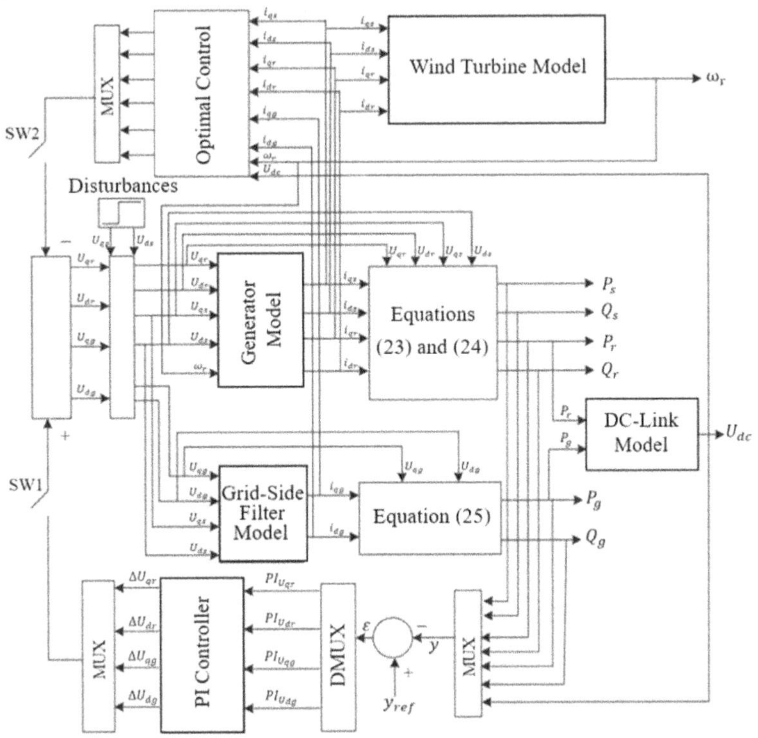

This model includes the generator, wind turbine, grid-side filter, and the DC link, as well as the configurations of the controllers under study. To assess the effectiveness of the controllers, both the stabilization and tracking performances will be analyzed in the next section, first without and then with the aforementioned regulators: the PI regulator, the standard LQR regulator, the LQR-GA regulator, and finally, the robust LQMinMax regulator. As shown in figure 10, the system model, including the PI controller, is implemented by closing switch SW1 and opening switch SW2 simultaneously. In contrast, for the optimal controllers (LQR, LQR-AG, and LQMinMax), switch SW1 is open while switch SW2 is closed.

For the LQR-GA regulator, the appropriate feedback gain, K_{LQGA}, is required. Similarly, for the LQMinMax regulator, the feedback gain, K_{LQMM}, is used. Furthermore, the principle of the proposed simulation model is based on two key aspects:

- The simulation provides the angular speed of the generator, the stator, rotor, and GSC currents, the DC link voltage, and the active and reactive powers of the system.
- The active power, reactive power, and DC link voltage are fed back through the PI controller when switch SW1 is active. Meanwhile, the acquired states (i.e., generator speed, stator and rotor currents, GSC currents, and DC link voltage) are fed back through the feedback gain of the LQR controller when switch SW2 is active.

Additionally, the performance and robustness of the linearized models of the proposed WECS system are studied with 10% (corresponding to 0.1 pu) disturbances in the stator voltages, to validate the proposed method for regulating both the power and the DC link voltage. The stator voltage disturbances are applied at time t=0s. The proposed model has a distinct advantage in its ability to simultaneously control the entire system when subjected to disturbances. Therefore, the proposed system facilitates the design of a controller capable of managing all variables concurrently. The stability of the proposed system is analyzed through state-space representation, examining the eigenvalues of the system (using the pole placement method). These eigenvalues are calculated from the system's state matrix A using the characteristic equation:

$$|A - \lambda I| = 0 \tag{55}$$

Where: λ denotes the eigenvalues of the system and I represents the identity matrix of the same dimension as the state matrix A.

The number of eigenvalues corresponds to the number of first-order differential equations considered for linearization. Each eigenvalue characterizes a specific dynamic behavior of the system. The eigenvalues can be real or complex. A positive real eigenvalue indicates an unstable system with exponential growth, while a negative real eigenvalue indicates stability with a non-oscillatory mode. Complex eigenvalues always occur in conjugate pairs, which are oscillatory in nature and can be expressed as:

$$\lambda = \rho \pm j\omega \tag{56}$$

Where:

- ρ is the real part of the eigenvalue, which must be negative for system stability,
- ω is the imaginary part of the eigenvalue.

If ρ is positive, even a small perturbation would result in exponentially increasing oscillations, rendering the system unstable. The oscillation frequency is determined as follows:

$$Damped\,frequency = \frac{\omega}{2\pi}\,[Hz] \tag{57}$$

Subsequently, the decay rate of the system's oscillation amplitude is quantified using the damping ratio, expressed as:

$$Damping\ ratio = \frac{-\rho}{\sqrt{\rho^2 + \omega^2}} \tag{58}$$

In the next section, the eigenvalues of the system, oscillation frequencies, damping ratios, and participation factors (i.e., the dominant states associated with specific eigenvalues) are provided for the five control scenarios mentioned earlier. Stability verification is performed for each case of the studied system.

STABILITY ANALYSIS OF STUDIED SYSTEM

This section presents a comparison of the dynamic responses of various controllers applied to the study system. This work focuses on the implementation of two optimal controllers: the LQR-AG controller and the robust LQMinMax controller. The LQR-AG controller aims to minimize the errors in active and reactive powers, as well as the DC-link voltage, caused by stator voltage disturbances. On the other hand, the LQMinMax robust controller not only eliminates tracking errors but also ensures disturbance rejection under the worst-case scenario. Additionally, the dynamic responses of the studied controllers are compared with those of conventional PI and LQR controllers. This comparative analysis evaluates the controllers in terms of eigenvalue analysis, participation factors, and the system's time-domain response. Stability and robustness are assessed for each controller to determine their effectiveness in maintaining system performance under varying conditions. Simulations were conducted in MATLAB to observe the efficiency of the designed controllers, as illustrated in figure 10. The simulations involved a $0.1pu$ voltage drop (i.e., 10%) at the stator terminals of the DFIG, applied at t = 0 s. The stabilization and tracking performances were analyzed for five different cases:

(i): Without controller,
(ii): With PI controller,
(iii): With a standard LQR controller,
(iv): With LQR-GA controller,
(v): With LQMinMax controller.

Overview of "Without Controller" Case

In this scenario, eight state variables are present in the system. The eigenvalues, damped frequency, damping ratio, and dominant states of the system's outputs are summarized in Table 1.

Table 1. Eigenvalues, Damped Frequency, Damping, and Dominant States for the "Without Controller" Case

Eigenvalues	Damped Frequency [Hz]	Damping [%]	Dominant States
-16.2 ± 313.5i	49.895	5.15	$\Delta i_{qr}, \Delta i_{dr}, \Delta i_{ds}, \Delta i_{qs}$
-14.4 ± 120i	19.098	11.91	$\Delta i_{dr}, \Delta i_{ds}$
-4.8	0	100	$\Delta i_{qr}, \Delta i_{qs}$
-167.5 ± 314i	49.974	47.06	$\Delta i_{qg}, \Delta i_{dg}$
+0.00000071	--	--	ΔU_{dC}

As shown in Table 1, the eight eigenvalues consist of six complex conjugates, one negative real value, and one positive real value. The complex conjugate eigenvalues indicate oscillatory modes, which oscillate at frequencies of 49.895 Hz, 19.098 Hz, and 49.974 Hz. The negative real eigenvalue corresponds to a non-oscillatory mode, while the positive real eigenvalue indicates an unstable mode. Additionally, the system's response to a -0.1 pu disturbance is illustrated through the steady-state errors in the output variables in the "No Controller" case. The steady-state errors of the active and reactive powers of the rotor and stator, as well as the grid-side filter, and the steady-state error of the DC link voltage in the "No Controller" case are illustrated in the following figures.

Figure 11. Steady-state errors of the output variables in the "No Controller" case

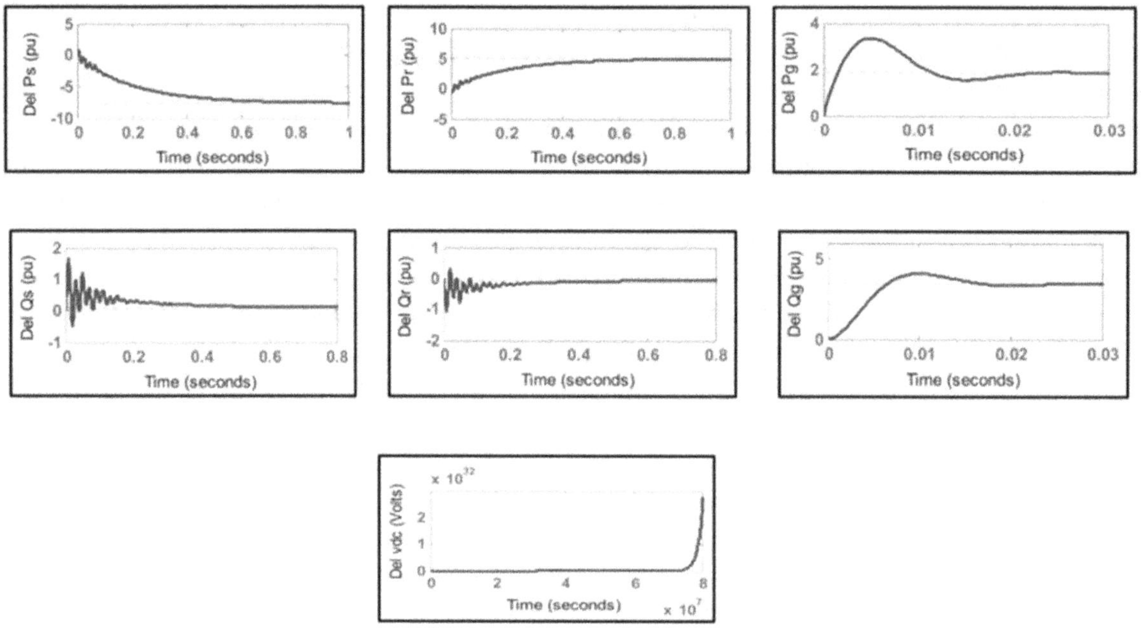

Therefore, based on figure 11, the time-domain-specifications of the output deviations due to a -0.1*pu* disturbance in the "Without Controller" case, as illustrated in Table 2, are determined.

Table 2. Time-domain-specifications of the output deviations due to a -0.1pu disturbance in the "Without Controller" case

	Peak Value (PV) [pu]	Rise Time (RT) [s]	Settling Time (ST) [s]	Steady-State Error (SSE) [pu]
ΔP_s	-7.659	0.468	0.788	-7.66
ΔQ_s	1.641	0.002	0.592	0.0877
ΔP_r	5.093	0.465	0.791	5.06
ΔQ_r	-1.076	0.001	0.596	-0.0391
ΔP_g	3.35	0.0012	0.029	1.89
ΔQ_g	4.12	0.0045	0.022	3.45
ΔU_{dC}	∞	Undefined	Undefined	∞

Thus, according to figures 11 and Table 2, the instability of the system in this case primarily arises from the DC link voltage. Furthermore, the system's damping is very low. As a result, the system outputs exhibit less "Settling Time," but there are significant peaks and steady-state errors.

Overview of "With PI Controller" Case

In this case, four new state variables ($PI_{U_{qr}}$, $PI_{U_{dr}}$, $PI_{U_{qg}}$ et $PI_{U_{dg}}$) are added to the studied system, bringing the total number of state variables to twelve. The eigenvalues, damped frequency, damping, and dominant states of the system's outputs are presented in Table 3.

Table 3. Eigenvalues, Damped Frequency, Damping, and Dominant States in the "With PI Controller" Case

Eigenvalues	Damped Frequency [Hz]	Damping [%]	Dominant States
-16.21 ± 313.5i	49.895	5.163	$\Delta i_{qr}, \Delta i_{dr}, \Delta i_{ds}, \Delta i_{qs}$
-167.5 ± 314.01i	49.976	53.33	$\Delta i_{qg}, \Delta i_{dg}$
-14.31 ± 120.02i	19.101	11.83	$\Delta i_{dr}, \Delta i_{ds}$
-4.18	0	100	$\Delta i_{qr}, \Delta i_{qs}$
-0.62	0	100	$\Delta i_{qr}, \Delta i_{qs}, PI_{U_{qr}}$
-0.03	0	100	ΔU_{dc}
-0.01± 0.003i	4,774e−04	95.78	$PI_{U_{qg}}, PI_{U_{dg}}$
-0.008	0	100	$PI_{U_{dr}}$

In Table 3, four oscillatory modes and four non-oscillatory modes are observed. The first three oscillatory modes oscillate at frequencies of 49.895, 49.976, and 19.101*Hz*, which are almost identical to those in the "Without Controller" case. The fourth remaining oscillatory mode oscillates at a frequency of 4.774*e*–04*Hz*, which is very low. However, in this case, the system's damping has been increased.

Moreover, the steady-state of the active and reactive powers of the rotor and stator, as well as the grid-side filter, and the steady-state error of the DC link voltage due to a -0.1 pu disturbance in the "With PI Controller" case are determined as follows:

Figure 12. Steady-state errors of the output variables in the "With PI Controller" case

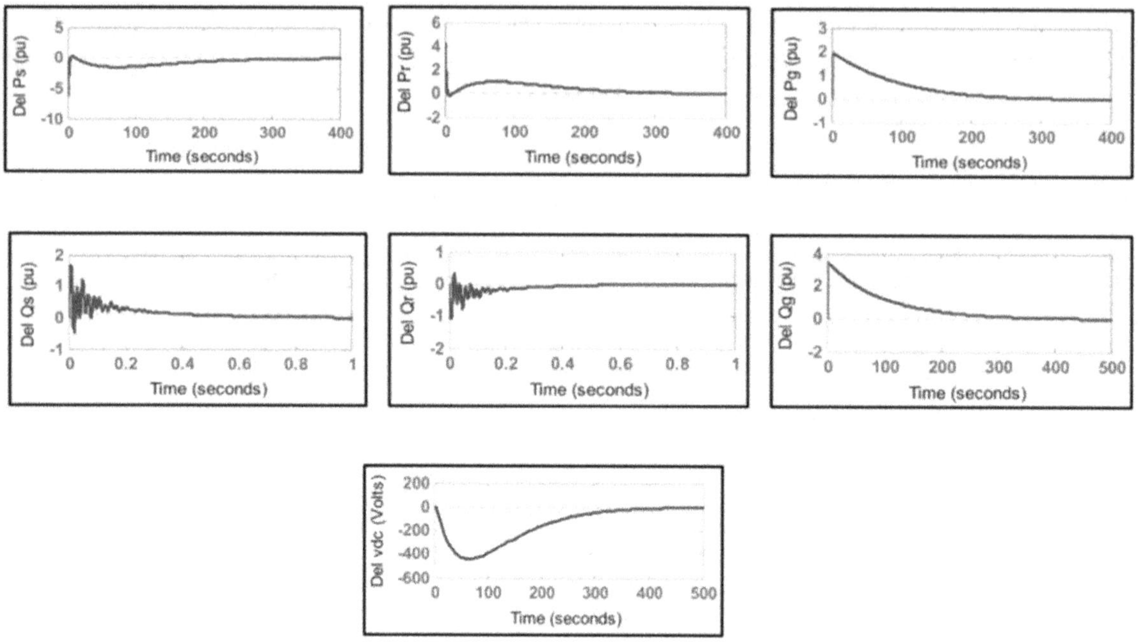

Therefore, the time-domain specifications of the output deviations due to a -0.1 pu disturbance in the "With PI Controller" case, as shown in Table 4, are determined.

Table 4. Time-domain specifications of the output deviations due to a -0.1 pu disturbance in the "With PI Controller" case

	Peak Value (PV)	Rise Time (RT) [s]	Settling Time (ST) [s]	Steady-State Error (SSE)
ΔP_s	-6.21 pu	0.271	325	-0.005 pu
ΔQ_s	1.642 pu	0.013	Indéfini	0.109 pu
ΔP_r	4.16 pu	9,49e–04	321	0.00113 pu
ΔQ_r	-1.09 pu	0.0009	Indéfini	-0.0608 pu
ΔP_g	1.92 pu	1.02e–05	296.708	0.00113 pu
ΔQ_g	3.51 pu	6.08e–06	364.12	0.00122 pu
ΔU_{dc}	-442 V	0.895	416.567	-0.574 V

In summary, based on figure 12 and Table 4, the system outputs with the PI controller show fewer static errors. However, the rise time and settling time are very long. In other words, only the static error values are acceptable, but the other parameters are poor. Therefore, the PI controller is not sufficient to achieve the desired results.

Synthesis of LQR, LQR-GA, and LQMinMax Controllers

Case of "With Standard LQR Controller"

For the "With LQR controller" case, the number of state variables is the same as for the "Without controller" case, i.e. eight. The eigenvalues, damped frequency, damping and dominant states of the outputs of the system under study are shown in Table 5.

Table 5. Eigenvalues, Damped Frequency, Damping, and Dominant States in the "With standard LQR controller" Case

Eigenvalues	Damped Frequency [Hz]	Damping [%]	Dominant States
-13 963.6 ± 314i	49.97	99.974	Δi_{qg}, Δi_{dg}
-4 429 ± 94.25i	15.0003	99.977	Δi_{qr}, Δi_{dr}, Δi_{ds}, Δi_{qs}
-1.4 ± 314i	49.97	0.445	Δi_{qr}, Δi_{dr}, Δi_{ds}, Δi_{qs}
-9.2	0	100	$\Delta \omega_r$
-1.33	0	100	ΔU_{dc}

According to Table 5, three oscillatory modes and two non-oscillatory modes are present. The oscillatory modes oscillate at frequencies of 49.97 and 15.0003 Hz. In this case, there is no significant effect on the damped frequency, but the system damping is much higher compared to the two previous cases ("no controller" and "with PI controller").

Case of "With LQR-GA Controller"

In this case, the AG approach is included to select the appropriate weighting matrices Q and R for the studied LQR controller. The number of state variables remains eight, as no additional states have been added. Table 6 represents the parameters of the weighting matrices Q_{LQGA} and R_{LQGA} for the LQR-GA controller.

Table 6. Parameters of the weighting matrices Q_{LQGA} et R_{LQGA} of the LQR-GA controller

Q_{LQGA_i}	Value	R_{LQAG_j}	Value
Q_{LQAG_1}	1	R_{LQAG_1}	0.512
Q_{LQAG_2}	0.1	R_{LQAG_2}	0.14
Q_{LQAG_3}	0.001	R_{LQAG_3}	0.67
Q_{LQAG_4}	0.01	R_{LQAG_4}	0.021
Q_{LQAG_5}	3.34		

continued on following page

Table 6. Continued

Q_{LQGA_i}	Value	R_{LQAG_j}	Value
Q_{LQAG_6}	66.7		
Q_{LQAG_7}	6.67		
Q_{LQAG_8}	0.1		

Indeed, the eigenvalues, damped frequency, damping, and dominant states of the study system's outputs in this case are presented in Table 7.

Table 7. Eigenvalues, Damped Frequency, Damping, and Dominant States in the "With LQR-GA controller" Case

Eigenvalues	Damped Frequency [Hz]	Damping [%]	Dominant States
-12 903.26	0	100	Δi_{qg}
-10 076.31	0	100	Δi_{dg}
-950.15	0	100	Δi_{qr}, Δi_{qs}
-547.91	0	100	Δi_{dr}, Δi_{ds}
-8.08± 314i	49.974	2.572	Δi_{qr}, Δi_{dr}, Δi_{ds}, Δi_{qs}
-145.15	0	100	$\Delta \omega_r$
-10.76	0	100	ΔU_{dc}

According to Table 7, two oscillatory modes and six non-oscillatory modes are present. The oscillatory mode oscillates at a frequency of 49.974 Hz. Consequently, the system's damping is greatly improved (almost 100%) in this case, compared to the previously mentioned cases.

Case of "With LQMinMax Controller"

The LQMinMax controller technique focuses on minimizing errors in the worst-case disturbance scenarios. To achieve this, the dynamic responses of the active and reactive powers, as well as the DC link voltage, with this type of controller, are determined under various disturbance values, as illustrated in the following sections.

In this case, the number of state variables remains the same as with the standard LQR controller (eight state variables). Indeed, the eigenvalues, damped frequency, damping, and dominant states of the system's outputs are shown in Table 8.

Table 8. Eigenvalues, Damped Frequency, Damping, and Dominant States in the "With LQMinMax controller" Case

Eigenvalues	Damped Frequency [Hz]	Damping [%]	Dominant States
-786 901.83	0	100	Δi_{dg}
-31 175.36	0	100	Δi_{qg}
-4348.09	0	100	Δi_{qr}
-2775.56	0	100	Δi_{ds}, Δi_{dr}
-6.17± 5.98i	0.95	71.8	ΔU_{dc}
-4.65-	0	100	$\Delta \omega_r$
-1.38	0	100	Δi_{qg}, Δi_{dg}

According to Table 8, two oscillatory modes and six non-oscillatory modes are present. The oscillatory mode oscillates at a low frequency of approximately 0.95 Hz. As a result, the system's damping is significantly improved (nearly 100%) in this case compared to the other previously mentioned cases.

Comparative Study of LQR, LQR-AG and LQMinMax Controllers

First, the comparative analysis of eigenvalues in the four cases studied in this research is summarized in Table 9.

Table 9. Comparative Analysis of Eigenvalues for Different Controllers in the Study System

PI Controller	LQR Regulator	LQR-GA Regulator	LQMinMax Regulator
-16.21 ± 313.5i	-13 963.6 ± 314i	-12 903.26	-786 901.83
-167.5 ± 314.01i	-4 429 ± 94.25i	-10 076.31	-31 175.36
-14.31 ± 120.02i	-1.4 ± 314i	-950.15	-4348.09
-4.18	-9.2	-547.91	-2775.56
-0.62	-1.33	-8.08± 314i	-6.17± 5.98i
-0.03		-145.15	-4.65-
-0.01± 0.003i		-10.76	-1.38
-0.008			

Given that the system is unstable due to the eigenvalue having a positive real part (Table 1) in the "no controller" case, Table 9 shows that the use of any controller ensures the stability of the study system, as all eigenvalues for the controllers considered have a negative real part.

Furthermore, Table 9 reveals that in the cases of the proposed regulators (LQR-AG and LQMinMax), the real parts of all eigenvalues are more negative compared to the LQR and PI controllers, and only one eigenvalue remains oscillatory.

Consequently, the magnitude of the negative real parts of all eigenvalues is significantly improved, indicating that the system's stability has increased, and the amplitude of oscillations has slightly decreased.

The simulation results in terms of the dynamic responses of active and reactive powers, as well as the DC link voltage, with the LQR, LQR-GA, and LQMinMax controllers are illustrated in figure 13.

Figure 13. Comparative dynamic response of the output variables

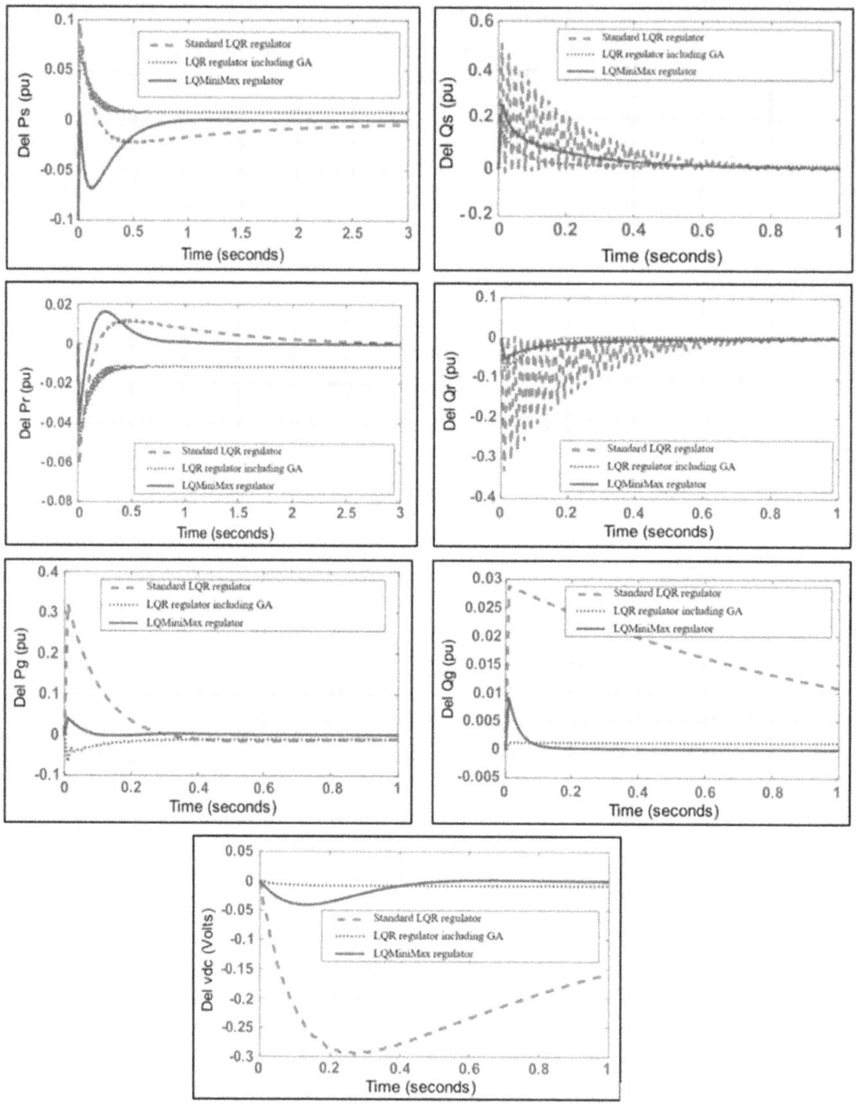

In order to carry out a comparative analysis of the different simulation results for the LQR, LQR-GA, and LQMinMax controllers, given by figure 13, Table 10 illustrates the time-domain specifications of the output deviations at a -0.1 pu disturbance in the stator voltages for each controller.

Table 10. Comparative analysis of the time-domain specifications of the system outputs at a -0.1 pu disturbance for the different studied controllers.

		LQR Regulator	**LQR-GA Regulator**	**LQMinMax Regulator**
ΔP_s	*VC [pu]*	-0.0999	-0.0999	-0.099998
	TM [s]	9.88e-04	3.02e-06	1.27e-07
	TS [s]	2.958	0.466	0.546
	ES [pu]	0	0.00818	0
ΔQ_s	*VC [pu]*	0.50987	0.153	0.2566
	TM [s]	3.84e-03	4.98e-04	2.37e-04
	TS [s]	0.612	0.213	0.388
	ES [pu]	0.0001	0.00905	0
ΔP_r	*VC [pu]*	-0.06516	-0.0524	-0.04033
	TM [s]	2.32e-04	3.47e-05	5.27e-09
	TS [s]	1.426	0.455	0.475
	ES [pu]	0.000002	-0.0115	0
ΔQ_r	*VC [pu]*	-0.333	-0.0805	-0.0498
	TM [s]	3.57e-04	7.16e-05	3.21e-06
	TS [s]	0.654	0.202	0.155
	ES [pu]	0.00005	0.0026	0
ΔP_g	*VC [pu]*	0.3191	-0.0668	0.0409
	TM [s]	5.33e-03	4.06e-03	1.19e-03
	TS [s]	0.365	0.349	0.133
	ES [pu]	0.0006	-0.0115	0
ΔQ_g	*VC [pu]*	0.0289	0.0013	0.009079
	TM [s]	0.0917	2.39e-03	0.0129
	TS [s]	1.639	4.25e-04	0.082
	ES [pu]	0.0001	0.0011	0
ΔU_{dC}	*VC [V]*	-0.2948	-0.0087	-0.04058
	TM [s]	0.0319	5.78e-03	5.01e-03
	TS [s]	2.939	0.333	0.388
	ES [V]	0.0001	-0.0087	0

Therefore, Table 10 shows that the LQMinMax controller achieved the desired zero steady-state error for the system outputs. Meanwhile, the settling time, rise time, and peak value of the dynamic response outputs are increased compared to the LQR-AG controller. However, the results in Table 10 reveal that the difficulties encountered with the standard LQR controller are resolved for the proposed controllers, meaning that they have achieved zero static errors and, simultaneously, obtained lower peak values. The settling time and rise time are, notably, within acceptable limits. Moreover, the total power deviation in steady state for the different controllers studied at a -0.1 *pu* disturbance at the stator voltages is also presented in Table 11.

Table 11. Total Power Deviation in Steady State for the Different Controllers Studied

	Without controller	With PI controller	With LQR controller	With LQR-GA controller	With LQMinMax controller
ΔP_s	-7.66	-0.005	0.00818	0	0
ΔP_r	5.06	0.00113	-0.0115	0.000002	0
ΔP_{totale}	-2.6	-0.00387	-0.00332	0.000002	0

with:

$$\Delta P_{totale} = \Delta P_s + \Delta P_r \tag{59}$$

Consequently, table 11 shows that the proposed controllers exhibit the smallest total power deviation and display the best performance among the four controllers in terms of power extraction in steady state.

ROBUSTNESS ANALYSIS OF STUDIED SYSTEM

In this section, the robustness of the study system is analyzed for the different proposed controllers. Two cases are considered:

- Study of the robustness of the different proposed controllers (PI, LQR, LQR-AG, and LQMinMax) under various perturbations at the stator voltage levels,
- Study of the robustness of the different proposed controllers (PI, LQR, LQR-AG, and LQMinMax) in the presence of perturbations at the wind speed level.

Analysis of Robustness to Various Stator Voltage Disturbances

In order to assess the efficiency and robustness of the proposed controllers, the dynamics of the DC-link voltage and the generator's angular speed are also simulated for higher voltage perturbations (i.e., −0.1, −0.2, and −0.3 pu) at the stator terminals.

The comparative dynamic responses of the DC-link voltage and the generator's angular speed are then analyzed for the different controllers under various perturbations at the stator voltage levels, figures 14 and 15.

Figure 14. Dynamic responses of ΔU_{dC} and $\Delta\omega_r$ for PI and LQR controllers under various perturbations at the stator voltage levels

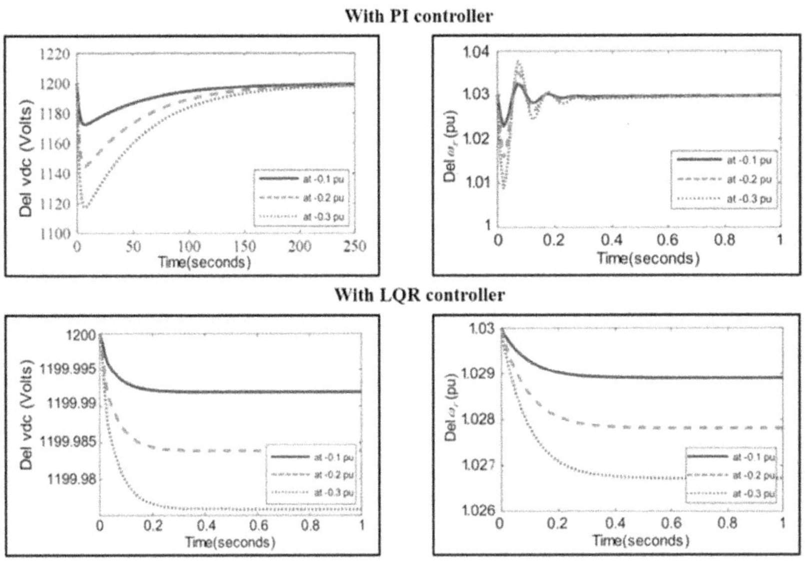

Figure 15. Dynamic responses of ΔU_{dC} and $\Delta\omega_r$ for LQR-GA and LQMinMax controllers under various perturbations at the stator voltage levels

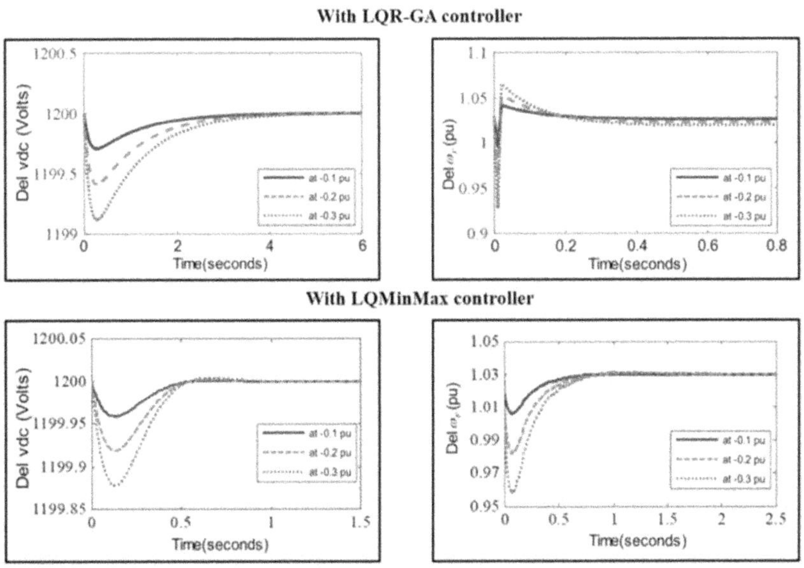

The robustness analysis of the different controllers under stator voltage disturbances reveals that the PI controller ensures proper tracking of the DC-link voltage and generator angular speed. However, the response of the DC-link voltage is unacceptable due to high peaks and a prolonged settling time. The standard LQR controller, while showing increased steady-state errors, fails to achieve the desired levels and requires optimization to enhance its performance. In contrast, the LQR-AG controller significantly improves the peak values and settling time, while the LQMinMax controller delivers the best performance with reduced deviations and acceptable settling times. Therefore, the LQR-AG and LQMinMax controllers are the most suitable for handling stator voltage disturbances.

Analysis of Robustness to Various Wind Speed Disturbances

The robustness study under wind speed variations evaluates the performance of the proposed system when subject to mechanical disturbances in the wind turbine (i.e., variable wind speeds) while maintaining a constant stator voltage. Since the turbine operates using the MPPT technique, any increase or decrease in wind speed leads to corresponding changes in the generator's angular speed to maintain a constant tip speed ratio. Consequently, the mechanical torque T_m (or turbine power P_m) increases or decreases accordingly. The effectiveness and robustness of the controllers are assessed under varying mechanical torque disturbances. For this purpose, specific system equations are modified as follows:

$$\left(\frac{d\omega_r}{dt}\right)_{modified} = \frac{1}{2H}T_m - \frac{L_m}{2H}i_{qs}i_{dr} + \frac{L_m}{2H}i_{ds}i_{qr} \tag{60}$$

$$(\Delta u)_{modified} = \left[\Delta U_{qr} \quad \Delta U_{dr} \quad \Delta U_{qg} \quad \Delta U_{dg} \quad \Delta T_m\right]^T \tag{61}$$

$$(\Delta w)_{modified} = \left[\Delta T_m\right] \tag{62}$$

Which resulting in changes to the dimensions and elements of the system matrices B and D, equations (63) and (64). These modified matrices accommodate the introduced mechanical disturbances. Furthermore, the matrix element a_{77} in A is adjusted to zero.

$$B_{modified} = \begin{bmatrix} -\sigma L_m & 0 & 0 & 0 & 0 \\ 0 & -\sigma L_m & 0 & 0 & 0 \\ \sigma L_{ss} & 0 & 0 & 0 & 0 \\ 0 & \sigma L_{ss} & 0 & 0 & 0 \\ 0 & 0 & \frac{\omega_B}{L_g} & 0 & 0 \\ 0 & 0 & 0 & \frac{\omega_B}{L_g} & 0 \\ 0 & 0 & 0 & 0 & \frac{1}{2H} \\ \frac{i_{qr0}}{C_{dc}U_{dc0}} & \frac{i_{dr0}}{C_{dc}U_{dc0}} & \frac{-i_{qg0}}{C_{dc}U_{dc0}} & \frac{-i_{dg0}}{C_{dc}U_{dc0}} & 0 \end{bmatrix} \tag{63}$$

$$D_{modified} = \begin{bmatrix} 0 & 0 & 0 & 0 & 0 \\ 0 & 0 & 0 & 0 & 0 \\ i_{qr0} & i_{dr0} & 0 & 0 & 0 \\ -i_{dr0} & i_{qr0} & 0 & 0 & 0 \\ 0 & 0 & i_{qg0} & i_{dg0} & 0 \\ 0 & 0 & -i_{dg0} & i_{qg0} & 0 \\ 0 & 0 & 0 & 0 & 0 \end{bmatrix} \tag{64}$$

Finally, comparative dynamic responses of the DC-link voltage and the generator's angular speed are analyzed under mechanical torque disturbances (-0.1, -0.2, and -0.3 pu). The system's responses under these conditions are illustrated in Figures 16 and 17 for the PI controller, LQR controller, LQR-AG controller, and LQMinMax controller, respectively.

Figure 16. Dynamic response of ΔU_{dC} and $\Delta\omega_r$ for PI and LQR controllers under various mechanical torque disturbances

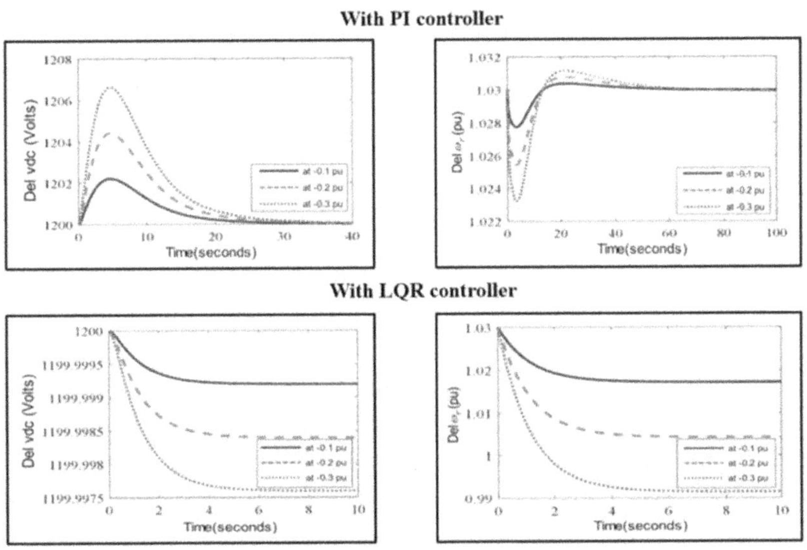

Figure 17. Dynamic response of ΔU_{dC} and $\Delta\omega_r$ for LQR-GA and LQMinMax controllers under various mechanical torque disturbances

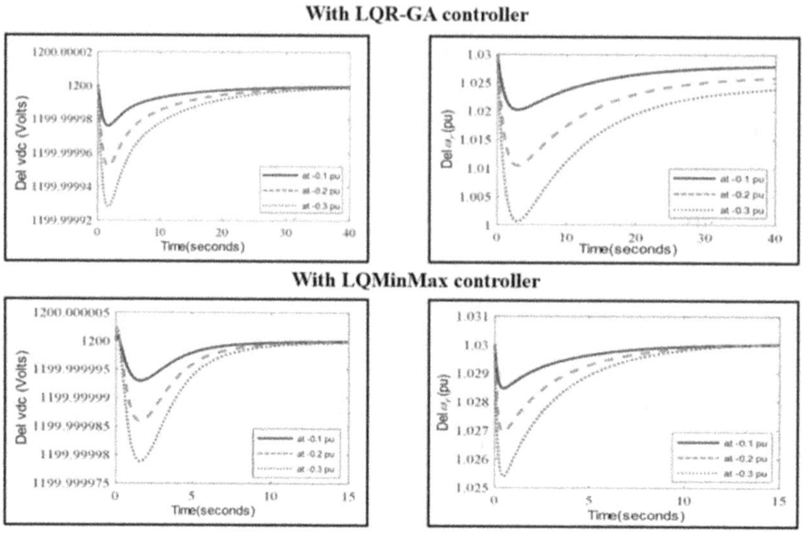

Based on figures 16 and17, the dynamic responses are significantly improved, with peak values and oscillations substantially reduced when using the proposed controllers. A comparative simulation study between the LQR-AG, LQMinMax, and PI controllers demonstrates that the LQR-AG and LQMinMax controllers outperform the PI controller in minmizing errors and rejecting disturbances. Although the PI controller is simpler and easier to design compared to the LQR-AG and LQMinMax controllers, the major advantage of the proposed controllers lies in their ability to provide state-specific control, resulting in superior performance. Generally, state-feedback control offers a more comprehensive and flexible solution, serving as a generalization of PID control. Furthermore, incorporating the genetic algorithm into the LQR controller enhances the system's performance compared to standard LQR and PI controllers. The two simulations (under stator voltage disturbances and wind speed variations) confirm that the LQR controller is less effective than the LQMinMax controller in minimizing uncertainties and rejecting disturbances. This is because the LQR structure only considers the system model for control synthesis, whereas the LQMinMax structure accounts for worst-case disturbances. Consequently, the LQMinMax controller provides a robust and high-performing control structure, effectively handling model uncertainties, noise rejection, and disturbance attenuation. In summary, the proposed controllers greatly enhance the system's dynamics and robustness when subjected to both stator voltage disturbances and mechanical torque variations.

CONCLUSION

In conclusion, this chapter has presented a comprehensive analysis of optimal control strategies for wind energy systems based on DFIG. The detailed modeling and linearization of the system provided a solid foundation for implementing various control strategies. Through simulation and comparative analysis, the effectiveness of the PI controller, LQR, LQR-GA, and LQMinMax regulators was demonstrated. The results highlighted the advantage of the LQR-GA and LQMinMax controllers in enhancing the system's stability and robustness against disturbances. This study underscores the importance of advanced control techniques in improving the performance of wind energy systems, paving the way for more reliable and efficient renewable energy solutions.

Future research could examine system sensitivity to parameter variations, explore alternative optimization methods such as Artificial Bee Colony (ABC) and Particle Swarm Optimization (PSO) for tuning weighting matrices, integrate an integral action into LQR-based controllers for improved disturbance rejection and reference tracking, and incorporate fuzzy logic for enhanced adaptability.

REFERENCES

Abdeen, M., Li, H., Kamel, S., Khaled, A., El-Dabah, M., Kharrich, M., & Sindi, H. F. (2021). A recent analytical approach for analysis of sub-synchronous resonance in doubly-fed induction generator-based wind farm. *IEEE Access : Practical Innovations, Open Solutions*, *9*, 68888–68897. DOI: 10.1109/ACCESS.2021.3075965

Amin, M., Rygg, A., & Molinas, M. (2017). Self-synchronization of wind farm in an MMC-based HVDC system: A stability investigation. *IEEE Transactions on Energy Conversion*, *32*(2), 458–470. DOI: 10.1109/TEC.2017.2661540

Azzouz, T. (2015). Modélisation et commande d'un système de conversion d'énergie éolienne à base d'une MADA (Doctoral dissertation, Université Mohamed Khider-Biskra).

Bhushan, R., Kumar, D., & Chatterjee, K. (2021). Disturbance rejection of the powers and DC-link voltage of a doubly-fed induction generator using state-space based linear quadratic integral optimal control approach. *International Transactions on Electrical Energy Systems*, *31*(5), e12865. DOI: 10.1002/2050-7038.12865

Dbaghi, Y., Farhat, S., Mediouni, M., Essakhi, H., & Elmoudden, A. (2021). Indirect power control of DFIG based on wind turbine operating in MPPT using backstepping approach. *International Journal of Electrical and Computer Engineering*, *11*(3), 1951. DOI: 10.11591/ijece.v11i3.pp1951-1961

Derbel, N., & Zhu, Q. (Eds.). (2018). *Modeling, identification and control methods in renewable energy systems*. Springer.

Djoudi, A. (2016). Contribution à la Conduite Robuste d'une Éolienne Basée sur une Machine Asynchrone à Double Alimentation, Connectée au Réseau Électrique (Doctoral dissertation, Ecole Nationale Polytechnique (ENP); Laboratoire de Génie Electrique de Grenoble (G2Elab)).

Guida, V. (2020). Conception et réalisation d'un convertisseur DC/DC à haut rapport de conversion pour électrolyseurs (Doctoral dissertation, Université de Lorraine).

Gurung, N., Bhattarai, R., & Kamalasadan, S. (2017, July). Optimal linear-quadratic-integral controller design for doubly-fed induction generator. In 2017 IEEE Power & Energy Society General Meeting (pp. 1-5). IEEE.

Hammami, A., Saidi, I., & Soudani, D. (2018). Comparative study of PMSG controllers for variable wind turbine power optimization. *International Journal of Advanced Computer Science and Applications*, *9*(8). Advance online publication. DOI: 10.14569/IJACSA.2018.090831

Hemeyine, A. V., Abbou, A., Tidjani, N., Mokhlis, M., & Bakouri, A. (2020). Robust takagi sugeno fuzzy models control for a variable speed wind turbine based a DFI-generator. *International Journal of Intelligent Engineering and Systems*, *13*(3), 90–100. DOI: 10.22266/ijies2020.0630.09

Khajah, A., & Ghazi, R. (2013). GA-based optimal LQR controller to improve LVRT capability of DFIG wind turbine. *Iranian Journal of Electrical and Electronic Engineering*, *9*(3), 167–176.

Liu, X., & Kong, X. (2013). Nonlinear model predictive control for DFIG-based wind power generation. *IEEE Transactions on Automation Science and Engineering*, *11*(4), 1046–1055. DOI: 10.1109/TASE.2013.2284066

Mao, J., Yi, H., Gao, C., Zhang, R., & Jiang, F. (2020). A New Minimax Optimization Design of H∞ LQ Fault-Tolerant Tracking Control for Multi-Phase Batch Processes. *IEEE Access : Practical Innovations, Open Solutions*, *8*, 148345–148356. DOI: 10.1109/ACCESS.2020.3015524

Masaud, T. (2013). Modeling, analysis, control and design application guidelines of doubly fed induction generator (DFIG) for wind power applications. 2013-Mines Theses & Dissertations.

Mehta, B., Bhatt, P., & Pandya, V. (2015). Small signal stability enhancement of DFIG based wind power system using optimized controllers parameters. *International Journal of Electrical Power & Energy Systems*, *70*, 70–82. DOI: 10.1016/j.ijepes.2015.01.039

Ouhssain, S., Chojaa, H., Aljarhizi, Y., Al Ibrahmi, E., Hadoune, A., Maarif, A., & Mossa, M. A. (2024). Performance Optimization of a DFIG-based Variable Speed Wind Turbines by IVC-ANFIS Controller. [JRC]. *Journal of Robotics and Control*, *5*(5), 1492–1501.

Prajapat, G. P., Bhui, P., Kumar, P., & Varma, S. (2019, March). Estimation based maximum power point control of DFIG based wind turbine systems. In 2019 IEEE PES GTD Grand International Conference and Exposition Asia (GTD Asia) (pp. 673-678). IEEE. DOI: 10.1109/GTDAsia.2019.8715955

Puchalapalli, S., & Singh, B. (2020, January). Adaptive RSC control through rotor currents dynamics linearization for WT coupled DFIG. In *2020 IEEE International Conference on Power Electronics, Smart Grid and Renewable Energy (PESGRE2020)* (pp. 1-6). IEEE. DOI: 10.1109/PESGRE45664.2020.9070743

Salhi, S., & Salhi, S. (2019, March). LQR control of a grid side converter of a DFIG based WECS: LMI approach based on Lyapunov condition. In 2019 16th International Multi-Conference on Systems, Signals & Devices (SSD) (pp. 595-600). IEEE.

Shutari, H., Ibrahim, T., Nor, N. B. M., Saad, N., Tajuddin, M. F. N., & Abdulrab, H. Q. (2022). Development of a novel efficient maximum power extraction technique for grid-tied VSWT system. *IEEE Access : Practical Innovations, Open Solutions*, *10*, 101922–101935. DOI: 10.1109/ACCESS.2022.3208583

Sitharthan, R., Karthikeyan, M., Sundar, D. S., & Rajasekaran, S. (2020). Adaptive hybrid intelligent MPPT controller to approximate effectual wind speed and optimal rotor speed of variable speed wind turbine. *ISA Transactions*, *96*, 479–489. DOI: 10.1016/j.isatra.2019.05.029 PMID: 31202532

Yaramasu, V., & Wu, B. (2016). *Model predictive control of wind energy conversion systems*. John Wiley & Sons.

Zgarni, I., & El Amraoui, L. (2022, March). Dynamic Modeling and Vector control of DFIG-based Wind Turbine. In 2022 5th International Conference on Advanced Systems and Emergent Technologies (IC_ASET) (pp. 262-267). IEEE.

Chapter 12
Opto-Thermal Characterization of GaAsPN Materials Using the Photothermal Deflection Technique for Tandem Solar Cells on Silicon

Ilahi Soufiene
https://orcid.org/0000-0001-9479-7476
Monastir University, Tunisia

ABSTRACT

This study investigates the optical absorption and thermal conductivity properties of GaAsPN/GaP samples, fabricated under various growth conditions and post-growth annealing treatments. The results show that optical absorption is strongly influenced by arsenic content, with GaAsPN alloys containing 10% arsenic and 2% nitrogen offering optimal characteristics for tandem solar cells on silicon, due to a perfect lattice match and absorption in the target energy range of 1.7-1.8 eV. Thermal annealing significantly improves both optical absorption and thermal conductivity, reaching 38,000 cm^{-1} and 4 W/mK, respectively. These properties were characterized using Photothermal Deflection Spectroscopy (PDS), a crucial technique for assessing thermal conductivity. These findings highlight the importance of GaAsPN materials for developing high-performance multijunction solar cells, where efficient thermal management and a broad absorption range are key to maximizing photovoltaic conversion efficiency.

INTRODUCTION:

The photovoltaic industry faces a range of challenges and opportunities, with significant goals centered around increasing the efficiency of solar energy conversion and reducing production costs. Currently, first-generation solar cells made from crystalline silicon are the dominant technology in the market, accounting for nearly 93% of global solar cell production (Adaine et al. 2016). Silicon-based solar cells are highly regarded for their efficiency and durability, making them the most widely used in commercial applications. Their high performance, especially in terms of energy conversion, has made

DOI: 10.4018/979-8-3693-9924-8.ch012

them a staple in both residential and commercial solar power systems. However, the production of these cells is complex, energy-intensive, and costly. This is primarily due to the extensive energy and labor involved in extracting and purifying silicon from sand, followed by the energy-intensive process of crystal growth. Despite these high production costs, first-generation cells are still the most effective in terms of efficiency. In laboratory settings, the highest recorded efficiency for these cells has reached 25.6% . In contrast, second-generation solar cells, commonly known as thin-film cells, offer a much more cost-effective production process. These cells are made by depositing very thin layers of photovoltaic materials onto a substrate, which requires significantly less raw material compared to traditional crystalline silicon cells. While their efficiency tends to be lower, around 9% on average (Sahoo 2016)), they have made strides in providing an affordable alternative for large-scale solar power generation. A notable achievement in this area came from "Solar Frontier," the largest solar energy provider globally, which recently set a new efficiency record of 22.3% for thin-film cells. This breakthrough was achieved through a collaborative effort with the Japanese Organization for the Development of New Energy and Industrial Technology (NEDO) (Sahoo 2016). Since 2008, second-generation thin-film solar technologies have gradually increased their market share, primarily because of their lower manufacturing costs and reduced material requirements. In addition to cost savings, one of the significant advantages of thin-film cells is their flexibility, which allows them to be used in a variety of applications, including on curved surfaces or integrated into building materials.

The third generation of solar cells, known as multi-junction solar cells (MJSC), represents a significant leap forward in efficiency and performance (Vladimir V. Fedorov et al 2023) Soline Boyer-Richard et al 2023). These cells are constructed using multiple layers of semiconductor materials, each optimized to absorb light from a specific part of the solar spectrum. By carefully selecting materials with different absorption properties, multi-junction cells can capture a broader range of wavelengths from sunlight, thus increasing their overall energy conversion efficiency. Most of the technologies in this generation are still in the research phase and are not yet commercially available. However, they hold immense potential for future solar power applications. Currently, the efficiency of multi-junction cells can reach up to 46%, allowing for the capture and conversion of nearly 80% of the solar energy incident upon them (F. Dimroth et al. 2016). This impressive efficiency is achieved by stacking several layers of semiconductors, each tuned to absorb different portions of the solar spectrum, thus minimizing the energy loss that typically occurs in traditional single-junction solar cell. The development of third-generation solar cells represents a major breakthrough in photovoltaic technology, significantly enhancing the potential for higher energy yields from solar panels. However, the widespread commercial adoption of these cells is still in progress, with ongoing research aimed at improving their cost-effectiveness and scalability. The figure 1 provided showcases the highest efficiencies achieved by various types of solar cells, as reported by the National Renewable Energy Laboratory (NREL), which serves as a reference for the most advanced photovoltaic technologies currently available. Figure 1 presents a vast amount of data that may seem overwhelming at first but is valuable for understanding the evolution of photovoltaic technologies. It illustrates efficiency trends over time, revealing that many solar technologies have plateaued or shown minimal improvement over the past decade. Silicon-based solar cells, for instance, remain stagnant at approximately 20.28% efficiency.

Figure 1. Best efficiencies achieved for solar cells between 1976 and 2017"

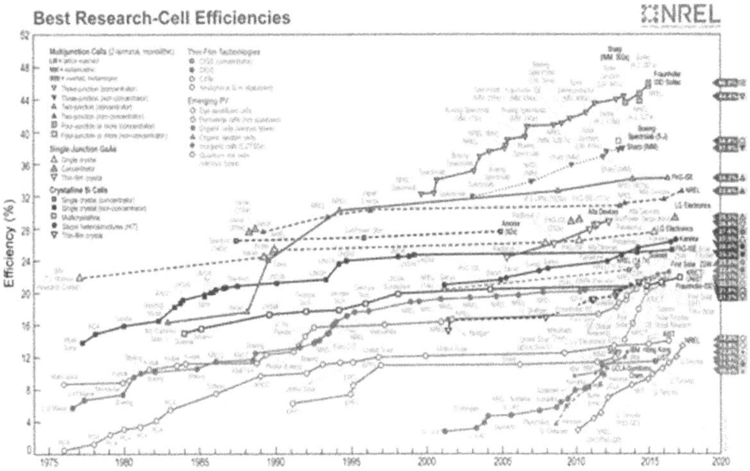

However, two technologies continue to advance: high-performance multijunction concentrator cells and organic photovoltaics. The former has reached nearly 43% efficiency, while the latter has improved to around 8%. Nanotechnology is expected to play a crucial role in transforming the production of solar cells (Pramita Nath et al. 2023). Traditional silicon-based solar cells rely on costly manufacturing processes that involve depositing material layers under high vacuum conditions. These inorganic devices are typically made from silicon, either in single-crystal or amorphous form. A major challenge in their production is the significant energy required to purify and process silicon into the high-quality material needed for efficient solar cell performance, particularly when using single-crystal silicon. However, the development of amorphous silicon has enabled the use of much thinner silicon layers in photovoltaic (PV) devices, reducing material consumption and potentially lowering costs. Amorphous silicon is more effective at absorbing light than single-crystal silicon. If manufacturing processes for amorphous silicon-based solar cells became more streamlined and continuous, production costs could drop significantly. This is where self-assembly techniques could play a transformative role in solar cell fabrication. It involves interactions that drive molecules into specific orientations, structures, or levels of order with minimal external intervention. If nanoscale structures capable of efficiently converting sunlight into electricity could be created through self-assembly, solar energy harvesting costs could be greatly reduced. Another cost-effective alternative to traditional silicon-based technology is the use of copper indium gallium diselenide (CIGS). This material can now be deposited using a simplified ink-based process that eliminates the need for ultra-high vacuum conditions. The company Nanosolar successfully commercialized this approach, producing over 115 MW of solar panels with 11% efficiency in 2011.Additionally, plastic or polymer-based solar films offer another low-cost solution. These can be manufactured using roll-to-roll printing techniques, significantly reducing fabrication time and energy consumption. Unlike conventional inorganic solar cells, which require multiple layers to be deposited under vacuum, polymer-based solar cells can be produced through a continuous printing process similar to newspaper production. These films can be lightweight, flexible, and adaptable to various surfaces, opening up new possibilities for integrating solar power into everyday applications, such as wearable technology, portable energy solutions, and building-integrated photovoltaics. Despite these advancements, challenges remain in improving efficiency and stability. Organic photovoltaics, for instance, have struggled with long-term performance due to material degradation when exposed to environmental factors like moisture and oxygen. Research is ongoing to develop new materials with higher stability and enhanced light absorption properties. Emerging perovskite solar cells have also gained attention due to their rapid increase in efficiency, with some laboratory prototypes exceeding 25%. However, issues related to lead toxicity and stability under prolonged exposure to heat and humidity must be resolved before commercialization. Looking toward the future, hybrid solar cells combining multiple technologies could offer a pathway to higher efficiencies. Tandem solar cells, which stack multiple layers with different bandgaps to capture a broader spectrum of sunlight, have demonstrated potential to exceed 50% efficiency. By integrating materials such as perovskites with traditional silicon or CIGS layers, researchers aim to develop next-generation solar panels that maximize energy conversion. Ultimately, advancements in solar cell technology will depend on a combination of scientific innovation, scalable manufacturing processes, and economic feasibility. As global demand for renewable energy sources grows, continued research in nanotechnology, material science, and production techniques will be essential for making solar energy more affordable and accessible. By leveraging breakthroughs in self-assembly, thin-film fabrication, and hybrid structures, the photovoltaic industry can push the boundaries of efficiency and cost-effectiveness, paving the way for a more sustainable energy future. In the past, the use of silicon (Si) substrates for high-efficiency III-V multijunction solar cells (MJSCs) was significantly limited due to substantial mismatches in crystal lattice constants and thermal expansion coefficients between the Si substrate and conventional III-V cell alloys. These discrepancies led to the formation of a high density of dislocations, which degraded performance and reduced the operational lifespan of solar cells. The challenge of integrating III-V materials on Si persisted for decades, restricting the development of cost-effective, high-efficiency solar devices based on this approach. However, recent advances in epitaxial growth techniques have renewed interest in this field, particularly with the successful development of GaP/Si materials and the incorporation of diluted nitrides (W. Shan et al., 2000). These developments have made it feasible to grow III-N-V materials directly on silicon, thus paving the way for novel solar cell architectures. One of the most promising materials emerging from this research is the GaAsPN alloy, which has been engineered to achieve a direct bandgap of approximately 1.7 eV by incorporating controlled amounts of arsenic (As) and nitrogen (N) into GaP. This tunability in composition enables the material to closely match the lattice parameters of GaP, AlP, and, most crucially, silicon, as illustrated in Figure 2. This compatibility is critical, as it reduces strain-induced defects, allowing for the fabrication of high-quality heterostructures with superior optical and electronic properties. Given the inherent bandgap of silicon at 1.12 eV, integrating GaAsPN as a top subcell in a tandem solar configuration could potentially lead to the realization of highly efficient silicon-based MJSCs (J. F. Geisz and D. J. Friedman, 2002). Beyond its potential for tandem solar applications, GaAsPN has recently gained attention as a candidate material for intermediate band solar cells (IBSCs), which belong to the third generation of photovoltaic technologies. These devices aim to surpass the Shockley-Queisser efficiency limit by incorporating an intermediate energy band within the bandgap, thereby enabling sub-bandgap photon absorption and improved carrier generation. Generally, intermediate band materials and structures fall into three main categories: (i) nanostructures, such as quantum dots (QDs) (Hatch et al., 2014), (ii) bulk semiconductor materials containing a high density of deep-level impurities, and (iii) highly mismatched alloys that introduce localized states within the bandgap (Ramiro et al., 2014). Among these, quantum dot-based IBSCs and highly mismatched alloys have shown particular promise due to their ability to engineer the density and distribution of intermediate states effectively. The ability to develop GaAsPN directly on silicon substrates represents a significant breakthrough for not only IBSCs but also MJSCs and optoelectronic applications such as laser emitters. The high arsenic content in GaAsPN, when combined with QD structures incorporating small amounts of nitrogen, has demonstrated notable improvements in optoelectronic performance. As a result, GaAsPN alloys and quantum dot-based approaches have been extensively studied, in recent years to optimize their structural and electronic properties (B. Kunert et al., 2006). These advancements hold considerable promise for the next generation of high-efficiency solar cells, particularly in silicon-based tandem architectures, where bandgap engineering is essential for achieving optimal power conversion efficiency. The overall objective is to develop a third-generation multi-junction solar cell using GaAsPN deposited on a silicon substrate. The use of silicon as the substrate offers several advantages, including well-established silicon technology and the relatively low cost of silicon compared to III-V materials and germanium. Additionally, as shown in figure 2, the lattice parameter of GaAsPN aligns well with that of silicon, which helps minimize the defect density at the interface.

Figure 2. Bandgap energy as a function of the lattice parameter for certain IV and III-V group semiconductors.

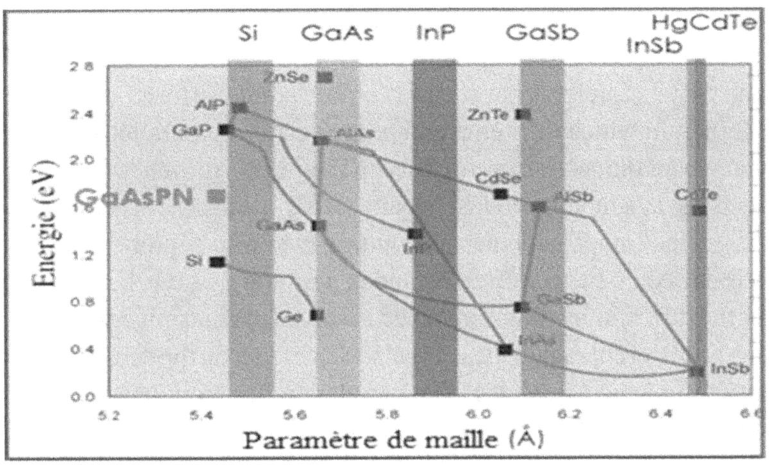

These factors have motivated researchers to explore the material's properties further and improve its physical characteristics to create a competitive silicon-based tandem solar cell. While the inclusion of nitrogen in GaAsP has been shown to contribute to its excellent optical emission properties (I. A. Buyanova et al. 2002), the optical absorption properties of this alloy have not been thoroughly studied, especially for compositions that are relevant to the development of tandem solar cells (R. Kudrawiec et al. 2014). Furthermore, the thermal properties of GaAsPN remain unmeasured, even though these are crucial for thermal dissipation in photovoltaic applications. Notably, the thermal conductivity of GaAsP and GaAsN alloys has been measured (Adachi, 2007) (Zhao et al. 2018). Increasing nitrogen content is known to degrade the crystalline quality and alter the physical properties of the alloy. This could be due to the uneven distribution of nitrogen and the formation of various defects such as vacancies, interstitial nitrogen atoms, and nitrogen clusters, which promote non-radiative recombination and carrier localization. According to several studies in the literature (A. Utsumi et al. 2003) and (M. Izadifard et al. 2004), rapid thermal annealing (RTA) could be an effective method to mitigate these effects and enhance the optical and thermal properties of GaPN-based alloys. Photoluminescence measurements, both temperature-dependent and time-resolved, have shown a reduction in intensity loss, indicating a significant improvement in carrier lifetime at room temperature following thermal treatment (M. Izadifard et al. 2004). This study focuses on optimizing the optical and thermal properties of the GaAsPN alloy through thermal annealing, utilizing both classical optical absorption spectroscopy and photothermal deflection spectroscopy (PDS). The effects of annealing on the optical and thermal properties have been clearly demonstrated (S. Ilahi et al. 2015).

THEORY OF PHOTOTHERMAL DEFLECTION TECHNIQUE PDS:

The photothermal deflection, also known as "Mirage Detection," uses two beams: one to create a disturbance at the surface of the sample (called the pump beam) and a second (called the probe beam) that passes through the refractive index gradient near the heated surface of the sample. Measuring the deflection of the probe beam allows one to determine the thermal properties of the sample (S. Ilahi et al. 2013a), (S. Ilahi et al. 2013b) and (S. Ilahi al. 2020). This technique is used in many fields, such as spectroscopy (Khalfaoui et al., 2017), and the thermal and optical characterization of solids and liquids (Loubiri et al.2015a), (D. Loubiri et al. 2015b), (S Ilahi et al. 2018) and (S. Ilahi et al. 2021). This method is highly sensitive and easy to interpret but requires a flat sample surface. Two types of configurations are distinguished: the first, in which the pump and probe beams are parallel, is called the collinear configuration; the second, where the two beams are perpendicular, corresponds to the perpendicular configuration. Using this technique, a modulated laser beam evenly illuminates the sample, creating a thermal wave that travels through the sample and the surrounding medium. A probe laser beam that passes near the sample's surface is deflected by the refractive index gradient that this heat wave creates in the fluid. The heat equation for the different mediums involved is solved to determine the probe beam's deflection. Along with the sample's optical characteristics, it is also dependent on the substrate, the surrounding fluid, and the sample's temperature characteristics. The correlation between the sample's optical and thermal characteristics and the deflection Ψ has been covered in earlier (D Loubiri et al.2015a), (D. Loubiri et al. 2015b), (Khalfaoui et al., 2017), (S Ilahi et al. 2017), (S Ilahi et al. 2018) and (S. Ilahi et al. 2021)

.

$$\Psi = \frac{L}{n}\frac{dn}{dT}\sigma_f T_0 \exp(-\sigma_f x_0) \tag{1}$$

L, n, σ_f, T_0 and x_0 are respectively, sample length, refractive index and distance between sample and probe laser beam. The thermal diffusion length in the fluid, denoted σ_f, is given by $\sigma_f = (1+j)/\mu_f$. The temperature distribution is determined by solving the one-dimensional heat equation for the sample, the backing material, and the surrounding fluid. The corresponding heat equations for the different media are illustrated in *Figure 3*

Figure 3. Schematic of different media

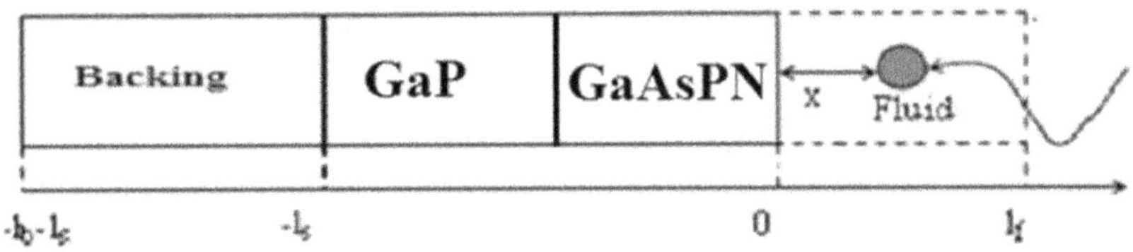

$$\frac{\partial^2 T_f}{\partial x^2} - \sigma_f^2 T_f = 0$$

;

$$h \le x \le h + h_f \qquad (2)$$

$$\frac{\partial^2 T_c}{\partial x^2} - \sigma_c^2 T_c = A_c$$

;

$$0 \le x \le h \qquad (3)$$

$$\frac{\partial^2 T_s}{\partial x^2} - \sigma_s^2 T_s = A_s$$

;

$$-l \le x \le 0 \qquad (4)$$

a ;

$$-l_b - l \le x \le -l \qquad (5)$$

The thermal diffusion length in the fluid, denoted σ_f, is given by $\sigma_f = (1 + j)/\mu_f$, where μ_f is the thermal diffusion length in the fluid and is expressed as and $\mu_f = (D/\pi F)^{1/2}$. Here, D represents the thermal diffusivity and F is the modulation frequency of the heating beam $\sigma_i = (1 + j)/\mu_i$, $\sigma_i = (1 + j)/\mu_i$ and $\mu_i = (D_i/\pi F)^{1/2}$ corresponds to the i medium number,

$A_c = \frac{-dI_c}{2K\,dx}$, $A_s = \frac{-dI_s}{2K\,dx}$ the quantities Ai describe the thermal source due to the absorption of light by the sample, with $I_c(x) = I_0 \exp\left[-\alpha_c(h - x)\right]$ in the layer and $I_s(x) = I_0 \exp(\alpha_s x)\exp(-\alpha_c x)$ in the substrate. The periodic solutions of the heat equation can be written as:

$$T_f = T_0 \exp(-\sigma_f(x - h))$$

;

$h \leq x \leq h + h_f$ (8)

$T_c = U_1 \exp(-\sigma_c(h - x) + V_1 \exp(\sigma_c(h - x)) - Z \exp(-\alpha_c(h - x))$

;

$0 \leq x \leq h$ (9)

$T_s = U \exp(\sigma_s x) + V \exp(-\sigma_s x) - E \exp(\alpha_s x)$

;

$-l \leq x \leq 0$ (10)

$T_b = W \exp(\sigma_b(x + l))$

;

$-l_b - l \leq x \leq -l$ (11)

T_0 represents the temperature at the sample surface (x=h), while E and Z are the particular solutions to equations (3) and (4), respectively. The constants U1, V1, and W have been determined using the boundary conditions for temperature and heat flux at the various interfaces.

$T_f(h) = T_s(h)$

(12),

$T_c(0) = T_s(0)$

(13)

$T_s(-l) = T_b(-l)$

(14)

$K_f \frac{\partial T_f}{\partial x}(h) = K_c \frac{\partial T_c}{\partial x}(h)$

(15)

$K_c \frac{\partial T_c}{\partial x}(0) = K_s \frac{\partial T_s}{\partial x}(0)$

,(16)

$$K_s \frac{\partial T_s}{\partial x}(-l) = K_b \frac{\partial T_b}{\partial x}(-l) \tag{17}$$

In reality, the layer thickness h is significantly smaller than the thermal diffusion length $h << \mu_c$. As a result, we can write $\exp(\sigma_c h) \approx 1$, which yields a simplified analytical expression for T_0 as detailed in [18].

$$T_0 = \frac{\left\{ \begin{array}{c} E[(1-r_s)(b+1)\exp(\sigma_s l) - (1-b)(1+r_s)\exp(-\sigma_s l) - 2(b-r_s)\exp(-\alpha_s l)] \\ +Z[(1-r_c)(b+1)\exp(\sigma_s l) - (1-b)(1+r_s)\exp(-\sigma_s l)][1-\exp(-\alpha_c h)] \end{array} \right\}}{[(1-g)(1-b)\exp(-\sigma_s l) + (1+g)(1+b)\exp(\sigma_s l)]}$$

$$E = \frac{\alpha_s I_0 \exp(-\alpha_c h)}{2K_s(\alpha_s^2 - \sigma_s^2)}$$

,

$$Z = \frac{\alpha_c I_0}{2K_c(\alpha_c^2 - \sigma_c^2)}$$

,

$$g = \frac{K_f \sigma_f}{K_s \sigma_s}$$

,

$$b = \frac{K_b \sigma_b}{K_s \sigma_s}$$

,

$$r_s = \frac{\alpha_s}{\sigma_s}$$

and

$$r_c = \frac{\alpha_c K_c}{\sigma_s K_s}$$

The deflection is given by the following expression:

$$\Psi = \frac{L}{n}\frac{dn}{dT}\sigma_f T_0 \exp(-\sigma_f x_0) \tag{19}$$

Let n be the refractive index of the fluid, x_0 the distance between the probe beam and the sample surface, and L the length of the sample.

Figure 4 depicts the experimental setup for the photothermal deflection technique For the PDS measurements (S. Ilahi et al. (2015)), the sensitivity of the photothermal signal was enhanced by placing the sample in a cell filled with paraffin oil. As depicted in Figure 4, the sample is heated using mechanically chopped monochromatic light from a halogen lamp, which is first passed through a monochromator. The deflection of the probe laser beam (He-Ne) is then detected by a four-quadrant photodetector linked to a lock-in amplifier, which records the amplitude and phase of the photothermal signal. The normalized amplitude is determined by comparing the measured amplitude from the semiconductor sample with the reference amplitude obtained from a carbon black sample.

Figure 4. Schematic experimental setup of the PDS technique.

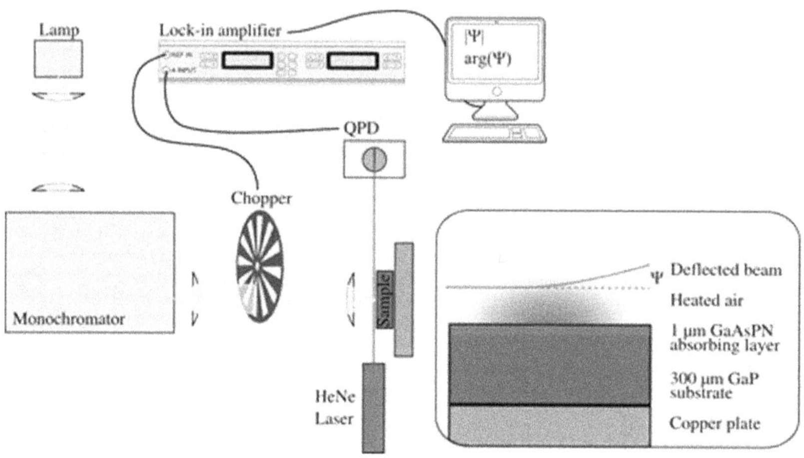

FABRICATION OF GAASPN/GAP ABSORBER

The GaAsPN samples were fabricated by our collaborators at the FOTON laboratory (Optical Function for Information Technology) at INSA (National Institute of Applied Sciences) in Rennes. The GaAsPN absorbers were grown on GaP (001) substrates using molecular beam epitaxy (MBE) (S. Almosni et al. 2013). The first series, referred to as S1, consists of four samples with different nominal arsenic compositions of 5%, 10%, 30%, and 40%, with a nitrogen content of 2%. The thickness of the GaAsPN layers is approximately 100 nm. All samples were annealed under an N2H2 flux at 800 °C for 1 minute in an oven. The second series, labeled S2, includes two samples that are 1 µm thick and have a nominal arsenic concentration of 10%, with about 2% nitrogen content (i.e., a GaP0.88As0.10N0.02 alloy). The nitrogen concentration is not exactly the same for both samples, as the incorporation of nitrogen (in terms of amount and type of incorporation, such as interstitial or substitutional) depends on the growth temperature. For both samples, the growth parameters were kept consistent, except for the growth temperatures, which were set to 480 °C and 450 °C, respectively (measured using an optical pyrometer). Each sample was divided into three pieces: (i) annealed under an N2H2 flux at 800 °C for 1 minute, (ii) annealed under an N2H2 flux at 730 °C for 30 minutes, and (iii) left unannealed for comparison.

IMPACT OF ANNEALING TEMPERATURE ON THE OPTICAL AND THERMAL PROPERTIES OF GAASPN/GAP.

Effect of Annealing Temperature on Optical Properties of GaAsPN

Measurements were conducted using advanced Reflection/Transmission techniques, which are commonly employed to analyze the optical properties of semiconductor materials. These measurements are crucial for understanding how the material's absorption characteristics change as a function of energy, particularly in relation to the arsenic composition.As anticipated, the absorption edge energy exhibits a clear trend of increasing as the arsenic content decreases. This observation is consistent with previous studies on similar materials, where the reduction in arsenic concentration leads to a shift in the absorption edge. For instance, the absorption edge energy increases from 1.4 eV for a composition with 40% arsenic to 1.55 eV for a composition with 30% As. Similarly, a shift is observed from 1.8 eV for a composition with 10% arsenic to 1.9 eV for a composition with 5% arsenic. These shifts reflect the material's electronic structure and provide insight into how the arsenic content affects the bandgap and optical properties of the material. One particularly noteworthy observation is the absorption coefficient, which reaches an impressive value of 38,000 cm^{-1} for higher arsenic concentrations. This increase in the absorption coefficient below the bandgap indicates a significant enhancement in the material's ability to absorb light, which is crucial for applications in optoelectronics and photovoltaics. The increase in absorption below the bandgap also suggests that the fractional character of the conduction band is increasing, as predicted by theoretical calculations by C. Robert et al. (2012). This increase in the absorption coefficient below the bandgap is a critical factor in determining the efficiency of semiconductor materials for use in devices such as solar cells and photodetectors. The observed trends in the absorption spectra are also in agreement with the predicted reduction in the bandgap, which can be attributed to various factors such as compositional changes and strain effects in the material. The measured absorption spectra further corroborate previous studies on the reduction in the bandgap and absorption measurements at various nitrogen concentrations, which are also known to influence the material's optical properties.In the context of tandem solar cell applications, particularly those based on silicon substrates, the sample with 10% arsenic emerges as the most promising candidate. This sample provides a perfect lattice match with silicon, which is critical for minimizing defects and ensuring efficient charge carrier transport across the junction in the tandem solar cell structure. Additionally, the absorption edge of the 10% arsenic sample falls within the targeted energy range of 1.7–1.8 eV, making it an ideal candidate for use in tandem solar cells. This range is particularly important for optimizing the performance of solar cells, as it ensures that the absorber material can efficiently absorb sunlight while also matching the energy levels of the silicon substrate. The theoretical efficiency of a GaAsPN/Si tandem solar cell, utilizing this particular arsenic composition, could reach as high as 37%, as predicted by S. Almosni et al. (2014). This high potential efficiency highlights the importance of optimizing the material properties, such as the arsenic content, in order to achieve the desired performance in tandem solar cells. The combination of a perfect lattice match with silicon and an appropriate absorption edge is crucial for achieving high efficiency in multi-junction solar cells, where each layer absorbs a different portion of the solar spectrum. For the subsequent analysis, the arsenic composition is fixed at 10%, and the S2 sample series is studied in greater detail. This choice allows for a more focused investigation into the effect of annealing conditions on the material's optical properties. Figures 4 (a) and 5 (a) illustrate the variations in the normalized experimental amplitudes of the photothermal signal as a function of wavelength, near the bandgap energy, for the two GaAsPN samples that were annealed at different temperatures and under various annealing conditions. These measurements provide valuable insight into how annealing affects the material's optical absorption properties and its performance in device applications. The measured amplitudes exhibit two distinct saturation regions: one for wavelengths shorter than 640 nm and another for wavelengths longer than 690 nm. Between these two regions, there is a noticeable decrease in the amplitude. This behavior is indicative of the material's response to light in different wavelength ranges and provides useful information about the material's absorption characteristics near the bandgap. The amplitude variations can be attributed to changes in the material's electronic structure as a result of the annealing process, which can influence factors such as defect density, crystal quality, and the distribution of alloy compositions within the material. *Figures 5 (a) and 6 (a)* respectively show the variations in the normalized experimental photothermal deflection spectroscopy (PDS) signal amplitudes as a function of wavelength near the bandgap energy of the GaAsPN absorbers.

Figure 5. Determination of the absorption spectrum by comparing the theoretical and experimental amplitudes of GaAsPN absorbers annealed at 450°C (a), along with the corresponding absorption spectra (b)."

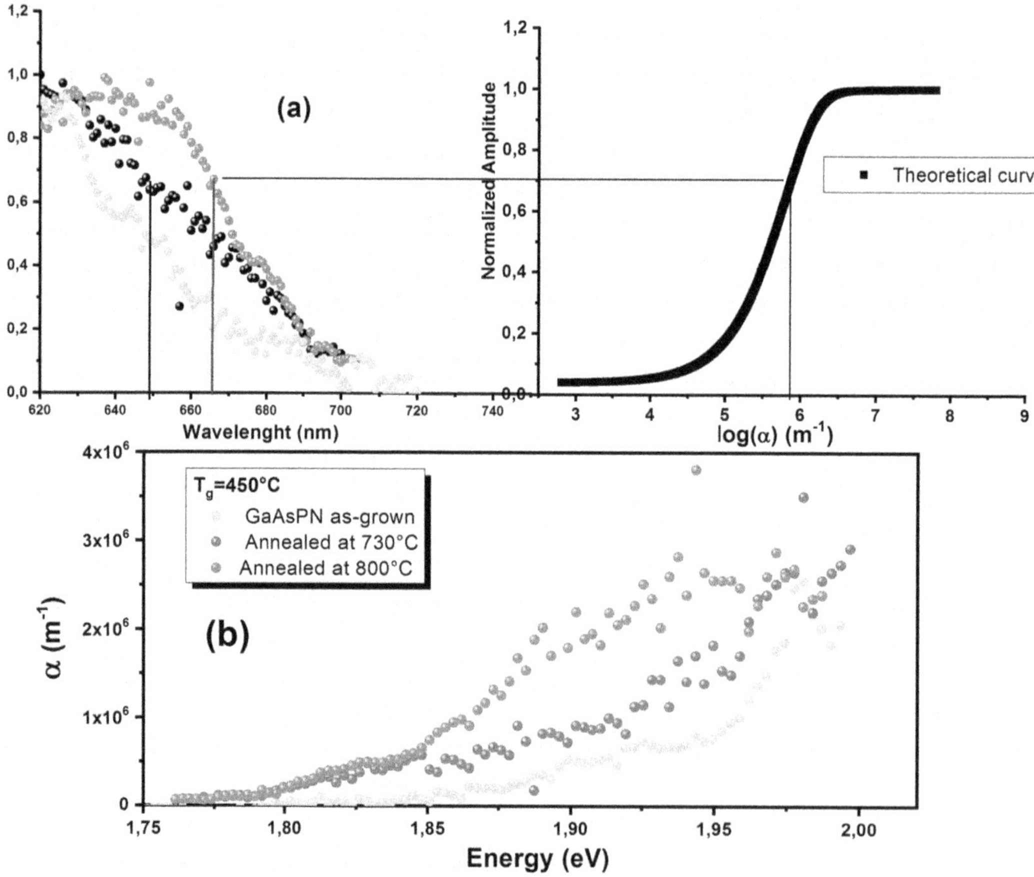

These experimental results are compared with the theoretical variations in the normalized amplitude, which are based on the optical absorption coefficient. The comparison between the experimental measurements and the theoretical model allows for a more accurate determination of the optical absorption coefficient at each wavelength. This enables the calculation of the absorption spectra, which provides critical information for understanding the material's performance in optoelectronic devices. The theoretical model, based on the optical absorption coefficient, provides a framework for interpreting the experimental data and allows for the extraction of key parameters such as the absorption edge and the material's response to light at different wavelengths. A comparison between the experimental and theoretical results also helps validate the model and ensures that the material properties are accurately captured. By assigning the corresponding optical absorption coefficient to each wavelength, it becomes possible to construct a detailed absorption spectrum, which is an essential tool for optimizing the design of semiconductor devices. This approach to analyzing the optical properties of GaAsPN materials using both experimental measurements and theoretical models is crucial for advancing the development of high-efficiency solar cells and other optoelectronic devices. The ability to tune the absorption characteristics of the material through variations in arsenic composition, annealing conditions, and other processing parameters offers significant opportunities for optimizing material performance and achieving the desired device efficiency. The combination of experimental techniques such as Reflection/Transmission and Photothermal Deflection Spectroscopy (PDS), along with theoretical modeling, provides a comprehensive understanding of the material's properties. This approach is vital for the design and development of next-generation optoelectronic devices that require precise control over material properties to achieve high performance and efficiency. Figures 5 (b) and Figures 6 (b) illustrate the absorption spectra of two GaAsPN samples grown under varying conditions of growth temperature and annealing processes.

Figure 6. Determination of the absorption spectrum by comparing the theoretical and experimental amplitudes of GaAsPN absorbers annealed at 480°C (a), along with the corresponding absorption spectra (b)."

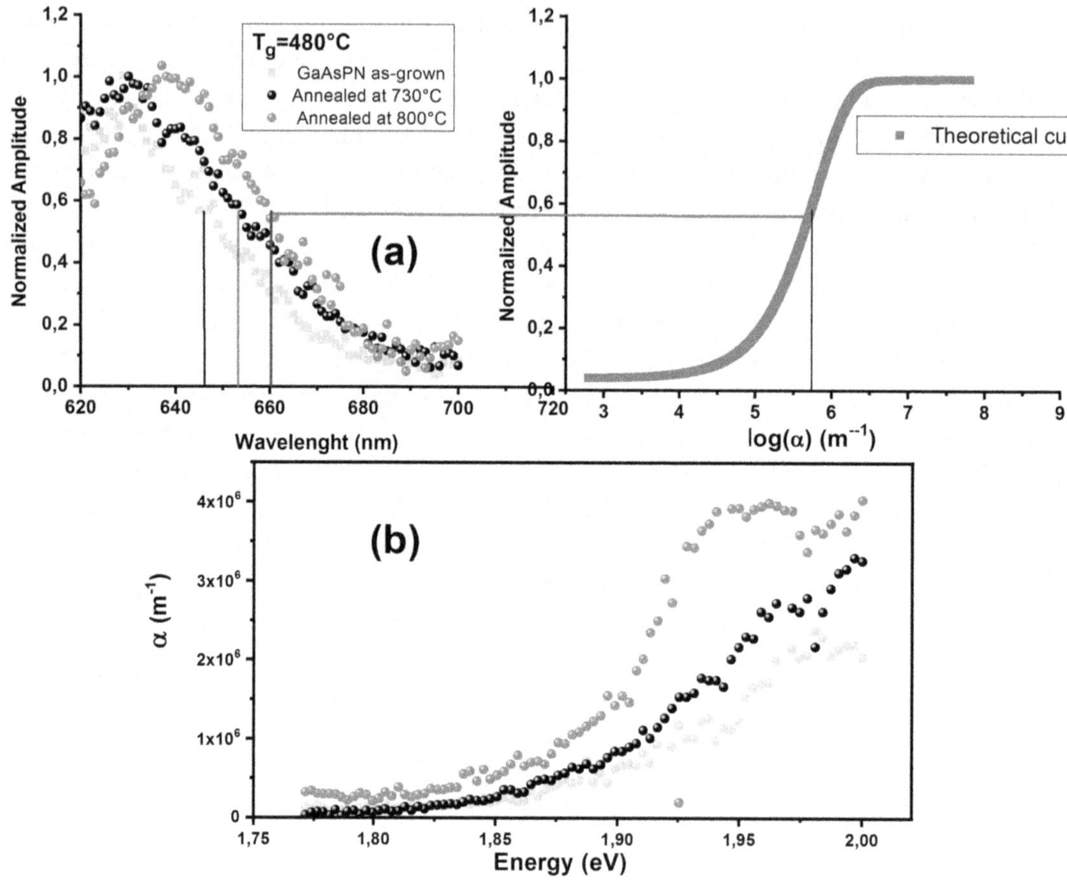

These figures provide essential insights into the relationship between the material's optical properties and its processing conditions. It is evident from some figures that the absorbance of the GaAsPN samples increases with the annealing temperature, even though a shorter annealing time of just 1 minute at 800 °C was used. This result underscores the significant impact that annealing has on the material's optical absorption, despite the relatively brief exposure to high temperature. Notably, samples that were annealed at 730 °C for 1 minute showed no significant difference in their optical properties compared to the unannealed samples. This is in stark contrast to the expected results and to findings from previous studies, such as the one presented in reference (R. Kudrawiec et al. 2014), where annealing did not seem to enhance the absorption coefficient. This discrepancy may be attributed to differences in experimental conditions, particularly in the nitrogen content of the samples. In this study, the nitrogen concentration was relatively higher, around 2%, and the annealing temperature applied was lower compared to the conditions in the aforementioned reference. It is possible that the lower annealing temperature of 730 °C did not provide sufficient thermal energy to induce significant changes in the absorption characteristics of the material, especially when compared to the higher temperature annealing at 800 °C, which led to more pronounced effects. Indeed, annealing at temperatures exceeding 800 °C was found to have detrimental effects on the samples. At these elevated temperatures, the material experienced damage that prevented any further improvement in photoluminescence properties. The high thermal energy likely caused defects or disorder within the crystal structure, which in turn degraded the optical properties of the material. This observation highlights the importance of carefully controlling the annealing temperature to avoid unwanted damage to the semiconductor material, particularly when targeting improvements in optical characteristics such as the absorption coefficient. Furthermore, Figures 5 (b) and Figures 6 (b) demonstrate that samples grown at 480 °C and annealed at 800 °C exhibit a higher optical absorption coefficient (approximately 40,000 cm^{-1} at 1.95 eV) compared to those grown at 450 °C (around 25,000 cm^{-1} at 1.95 eV). This difference in absorption properties suggests that the growth temperature plays a crucial role in determining the material's optical characteristics, particularly in terms of the efficiency of nitrogen incorporation into the crystal structure. Growth at 480 °C appears to facilitate the incorporation of nitrogen atoms into crystallographic sites or antisites that are more responsive to annealing processes. However, this interpretation still requires further experimental confirmation. The increased optical absorption coefficient observed in the 480 °C-grown and 800 °C-annealed samples is significant, as it suggests that these conditions lead to a material with superior optical properties, making it more suitable for optoelectronic applications. The higher absorption coefficient indicates that the material is more efficient at absorbing light, which is an important factor in the performance of devices such as solar cells, light-emitting diodes, and photodetectors. This improvement in absorption properties can be attributed to several factors, including enhanced nitrogen incorporation and the resulting changes in the material's electronic structure and defect landscape. However, it is important to note that these interpretations still need confirmation through further studies. The precise mechanisms by which growth temperature and annealing conditions affect nitrogen incorporation, crystallinity, and optical properties remain complex and are the subject of ongoing research in the field of semiconductor materials science. More detailed analyses, such as those involving deep-level transient spectroscopy (DLTS) or atomic-level imaging techniques, could provide further insights into the nature of the defects and the role of nitrogen atoms in determining the material's optical and electronic properties. Figure 6 further demonstrates the relationship between the photoluminescence and optical absorption spectra of GaAsPN samples grown at 480 °C from the S2 series and annealed at 800 °C. The spectra are shown as a function of excitation energy, allowing for a comprehensive comparison of the material's optical characteristics under different excitation conditions. The values for the optical absorption coefficients obtained from these measurements align well with those measured directly for the samples in the S1 series, indicating consistency in the material's behavior across different experimental setups. While the overall trends observed in the absorption spectra are similar, slight differences between the S1 and S2 samples could be attributed to variations in growth conditions or other factors such as light scattering effects associated with the roughness of the sample surface. These surface effects can influence the intensity of the measured absorption signals, especially at certain wavelengths, and might account for the small discrepancies observed between the two sample series. The growth conditions for the samples in the S2 series may have led to slightly different surface morphologies, which in turn affected the scattering and absorption properties of the material. Surface roughness can impact optical measurements, particularly for thin films or layered materials, by introducing additional scattering or interference effects that need to be taken into account when analyzing the absorption spectra. The direct optical absorption measurements of the 480 °C-grown and 800 °C-annealed samples (S2 series), as shown in Figure 6, provide further validation of the observed trends in the optical properties. These results offer a more detailed understanding of how the material behaves under different annealing conditions and provide additional evidence for the role of nitrogen incorporation and thermal treatment in enhancing the material's optical absorption. One notable feature of the photo lumniscence spectrum is the low-energy tail, which is generally associated with localized nitrogen states or alloy fluctuations, as described by C. Karcher et al. (2010). This low-energy tail is a typical signature of materials that contain a significant amount of nitrogen or other alloying elements, as they often lead to localized states in the bandgap that can result in extended absorption tails in the photoluminescence spectra. In addition to the low-energy tail, the spectrum also includes a distinct peak at higher energy, around 1.9 eV. This peak is of particular interest, as it corresponds to the absorption edge of the material and is indicative of the material's fundamental electronic transitions. The intensity of this higher energy peak is enhanced during the annealing process, which suggests that the annealing treatment plays a role in improving the material's electronic properties, particularly in terms of increasing the emission efficiency or enhancing the radiative recombination processes. The position of this peak corresponds well with the position of the absorption edge, which is a crucial factor for optoelectronic devices. Moreover, a Stokes shift of less than 100 meV is observed in this system, indicating that the energy difference between the absorption and emission processes is relatively small. This small Stokes shift is desirable for certain applications, as it suggests efficient energy conversion and minimal loss of excitation energy. The absorption and photoluminescence spectra of GaAsPN samples grown at different temperatures and annealed under varying conditions provide valuable insights into the relationship between growth and annealing conditions and the material's optical properties. These findings highlight the importance of optimizing both growth and post-growth treatment processes to achieve desired material characteristics for use in optoelectronic devices. Further studies and more precise characterization techniques are needed to fully understand the mechanisms at play and to confirm the role of nitrogen incorporation, crystallinity, and annealing in enhancing the material's performance. Additionally, it can be seen from this figure that the absorption range of interest for tandem solar cells using this GaAsPN absorber spans between 1.8 and 2.3 eV, which is directly associated with the delocalized states of the GaAsPN alloy in its bulk form. This particular absorption range is crucial for tandem solar cell applications, as it enables the material to absorb light efficiently within the ideal energy range for a complementary tandem configuration. In these devices, the top cell typically absorbs the higher-energy photons, while the bottom cell absorbs the lower-energy photons, leading to an overall enhancement in efficiency. This range of interest (1.8–2.3 eV) matches well with the bandgap energy of GaAsPN, which is tunable by adjusting the nitrogen content and growth conditions. Such tunability in the bandgap is one of the factors that make GaAsPN alloys highly appealing for use in next-generation solar cell technologies. The effects of annealing on the optical absorption near the band edge are not entirely unexpected. Annealing plays a significant role in altering the material's crystal structure and improving its optical properties. Specifically, the incorporation of nitrogen into GaAsP leads to the formation of randomly distributed nitrogen clusters and isolated nitrogen impurity states within the crystals, as indicated by L. Grenouillet et al. (2000). These nitrogen-related defects and clusters are often responsible for scattering and absorption losses in the material, which can adversely affect the efficiency of devices in which the material is used. Nitrogen incorporation, while essential for tuning the material's bandgap, also introduces localized states that can act as recombination centers, leading to non-radiative losses. However, the annealing process provides thermal energy that facilitates the reconfiguration of these group V atoms, including nitrogen, thus reducing the concentration of defects and improving the overall quality of the crystal structure. This reduction in defects leads to a significant improvement in the optical absorption properties of the GaAsPN layers, making them more suitable for use in high-performance optoelectronic applications such as solar cells, light-emitting diodes, and photodetectors. The annealing process works by providing the material with additional energy that allows atoms to move or realign, helping to reduce inhomogeneities that exist within the material's crystal structure. This process can effectively alter the nearest-neighbor environment of gallium atoms within the GaAsPN lattice, leading to a reduction in defect density and a subsequent enhancement in the material's absorption and emission properties. The result is a more homogeneous material that exhibits improved optoelectronic characteristics, which are crucial for the performance of devices fabricated from these materials. It is worth noting that annealing also helps to reduce localized nitrogen states and alloy inhomogeneity, which in turn decreases the number of non-radiative recombination centers that can negatively impact the efficiency of devices. As a result, the absorption properties of the GaAsPN layers improve, particularly near the bandgap energy, as delocalized electronic states become more prominent in the material's electronic structure. These findings are consistent with previous studies that have explored the effects of annealing on similar materials. For instance, in the case of the quaternary alloy GaInAsN, O. Rubel et al. (2005) observed that annealing led to improved material quality and enhanced optical absorption properties. Additionally, studies involving GaAsPN/GaP quantum wells (B. Kunert et al., 2006) have shown that annealing enhances the photoluminescence efficiency of the wells, further supporting the idea that annealing can be used to improve the optical properties of GaAsPN alloys. This improvement in photoluminescence efficiency can be attributed to the reduction in defect densities and the reconfiguration of the material's atomic structure during annealing. From the experimental data presented here, the following key interpretations can be made: (i) Annealing tends to reduce sample inhomogeneity, particularly by addressing localized nitrogen states, alloy inhomogeneity, and non-radiative centers, all of which contribute to the reduction in material performance. By promoting the emission and absorption of delocalized states in GaAsPN, annealing enhances the material's overall optical properties, particularly the transitions around 1.85 eV observed in this work. These delocalized states are important for achieving the desired electronic and optical behaviors, making the material more efficient for practical applications. (ii) The effect of annealing on the optical properties is highly dependent on both the annealing temperature and time, with higher temperatures and longer times leading to more significant improvements in the material's homogeneity and absorption efficiency. However, it is also crucial to balance these parameters, as excessive annealing can lead to material degradation, particularly at temperatures that exceed the stability limits of the material. (iii) The sample grown at a higher temperature (480 °C) is more homogeneous than both the unannealed sample and the sample grown at a lower temperature (450 °C), highlighting the importance of the growth temperature in determining the material's overall quality. Higher growth temperatures promote better nitrogen incorporation and reduce the number of defects, leading to a more uniform crystal structure and improved optical performance. These results suggest that careful control of both the growth and annealing processes is essential for optimizing the optical properties of GaAsPN alloys. The findings from this study contribute to a deeper understanding of how annealing influences the material's structure and performance, which will ultimately inform the development of more efficient optoelectronic devices. Further studies are needed to explore the underlying mechanisms that govern the interaction between nitrogen incorporation, crystal structure, and optical properties, as well as the optimal conditions for achieving the best performance in GaAsPN-based devices. In conclusion, the ability to fine-tune the optical absorption range of GaAsPN alloys through annealing is a critical aspect of their potential use in tandem solar cells and other optoelectronic applications. By understanding the impact of growth conditions, annealing temperature, and time on the material's properties, researchers can develop more efficient materials that are tailored for specific applications. This study paves the way for further exploration of GaAsPN as a promising material for next-generation solar cells and other high-performance devices, with the ultimate goal of enhancing energy conversion efficiencies and optimizing material properties for various optoelectronic applications.

Effect of Annealing Temperature on Thermal Conductivity

Thermal conductivity plays a crucial role in enhancing the quality and performance of solar cells and other optoelectronic devices. It is one of the key parameters that directly impacts the efficiency and reliability of these devices. High thermal conductivity enables efficient heat dissipation during device operation, which is essential to prevent overheating, maintain stable performance, and extend the lifespan of devices such as solar cells. In optoelectronic and photonic devices, the ability to manage heat effectively is critical, especially as the devices operate at high power levels. As the demand for high-performance devices continues to grow, understanding and improving thermal conductivity has become a focal point of material research for solar energy applications.

In this study, we evaluated the thermal conductivity of two GaAsPN absorbers from the S2 sample series, which shows great promise for use in tandem solar cells. Tandem solar cells, which consist of multiple layers designed to absorb different portions of the solar spectrum, can achieve significantly higher efficiency than conventional single-junction solar cells. Table 1, contains thermal conductivity of GaAsPN at different annealing temperature

Table 1. Thermal conductivity of GaAsPN grown at 450°C and 480°C as a function of annealing temperature.

	K (W/mK) GaAsPN	K (W/mK) GaAsPN (730°C)	K (W/mK) of GaAsPN (800°C)
Tg=480°C	2.40 (1.25%)	3.80 (0.9%)	4.30 (0.86%)
Tg=450°C	1.75 (1.1%)	2.10 (1%)	2.30 (1.3%)

The GaAsPN material system is particularly attractive for tandem solar cells due to its tunable bandgap, which can be adjusted to optimize light absorption. The ability to achieve high thermal conductivity in these materials is essential for their integration into tandem solar cells, as it ensures efficient heat dissipation during operation, thereby maintaining the performance and stability of the device.

We focused on the thermal conductivity of the GaAsPN samples by analyzing the phase of the photothermal signal near the bandgap energy. The phase of the photothermal deflection signal provides valuable information about the thermal properties of the material. For both low and high absorption coefficients, we observed that the phase of the photothermal signal saturates near the bandgap energy. In thin semiconductor layers, as demonstrated in previous studies (N. Yacoubi et al. 1991; L. Grenouillet et al. 2000; Terry M. Tritt 2004; O. Rubel et al. 2005; F. M. Gaitho et al. 2009; C. Karcher et al. 2010; S. Ilahi et al. 2013).

The phase difference near the bandgap between two saturation zone, is highly sensitive to the thermal conductivity of the material. This relationship between phase difference and thermal conductivity is key to determining the thermal properties of the samples. To measure the thermal conductivity, we varied the theoretical phase difference and adjusted it until it matched the experimental phase difference observed in the photothermal signal. Figures 7 (a) and (b) present the theoretical and experimental phases of the photothermal deflection spectroscopy (PDS) signal for samples grown at 480 °C and 450 °C, respectively. In fact, the phase of the photothermal signal in the PDS approach indicates the time required for a thermal wave to travel from the heat source to the detection point. Therefore, it is directly influenced by the properties of the studied material. When the heat source moves closer to the detection point, the phase increases; conversely, as the distance increases, the phase decreases. A maximum phase value is observed when the heat source is at its closest position to the detection site. These figures clearly illustrate the relationship between annealing conditions, phase difference, and thermal conductivity, allowing us to extract the thermal conductivity values for both samples. From this analysis, we concluded that the thermal conductivity of the GaAsPN samples increases with the annealing temperature. Both samples reached thermal conductivity values above 4 W/mK under the optimal annealing conditions (800 °C) when the growth temperature was 480 °C.

Figure 7. Comparison between experimental and theoretical phase difference of GaAsPN absorver developed at 450 ° C (a) and 480 °C (b)

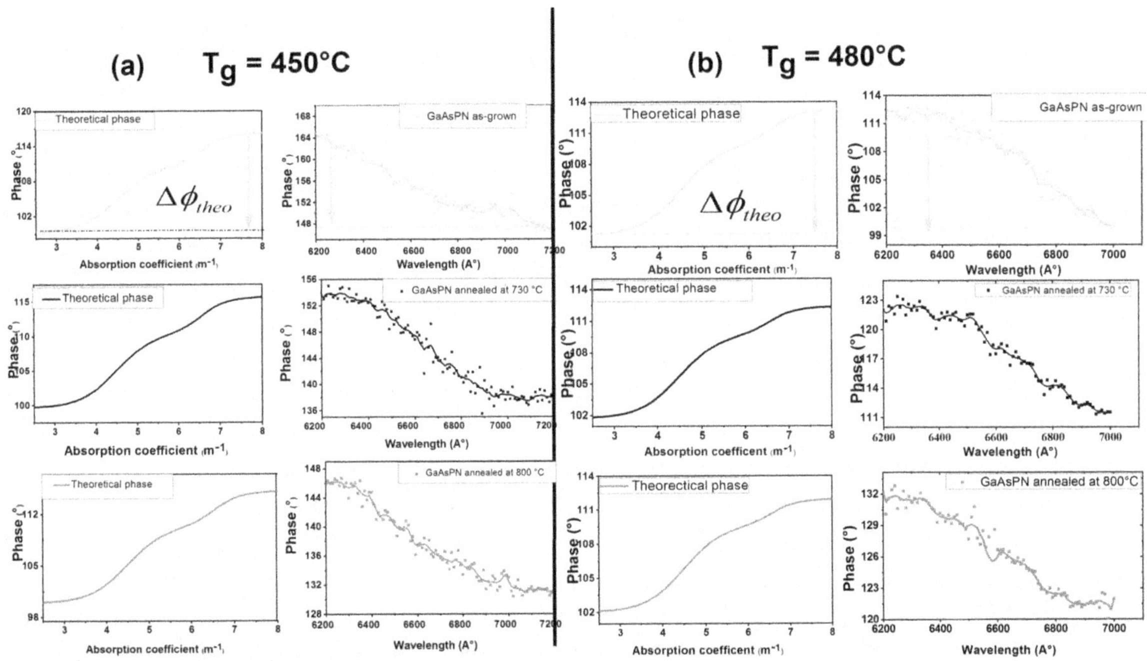

This finding is significant because it indicates that thermal conductivity can be effectively enhanced by controlling the annealing process. The improvement in thermal conductivity with increasing annealing temperature is linked to changes in the material's crystalline structure and the reduction in defects, which can impede phonon transport. The enhancement of thermal conductivity is a direct result of better crystalline quality, which is facilitated by the annealing process. Thermal conductivity is generally influenced by several factors, including phonon scattering at grain boundaries and crystal defects. These factors create additional resistance to heat flow, reducing the material's overall conductivity. The presence of crystal defects, such as vacancies or dislocations, leads to scattering events that hinder the movement of phonons, the primary carriers of heat in semiconductors. This scattering reduces the efficiency of heat dissipation and can contribute to the heating of the device during operation. In addition to phonon scattering, crystal ordering during the alloying process, as well as the interfaces in superlattices, can also impact thermal conductivity. As reported by X. Y. Yu et al. (1995), crystal ordering during alloying and the interface structures in multilayered materials can significantly influence the thermal properties of the material. In our study, the improvement in thermal conductivity observed with annealing is consistent with the findings from absorption measurements, which suggest that the materials grown at 480 °C and annealed at 800 °C exhibit better crystalline quality. The reduction in alloy inhomogeneity, non-radiative centers, and localized states likely plays a crucial role in improving thermal conductivity. As the material becomes more homogeneous, phonon scattering is reduced, allowing for more efficient heat transport through the crystal. The annealing process promotes this homogenization by allowing atoms to move and rearrange, ultimately improving the overall crystal structure and reducing defects that hinder phonon transport.Furthermore, several studies have reported values for the thermal conductivity of GaAs-based alloys that are relevant to our findings. For example, S. Adachi (2007) reported a thermal conductivity of 30 W/mK for the GaAs0.1P0.9 alloy, which is a quaternary alloy of GaAsP. This value is significantly higher than the thermal conductivity observed in our GaAsPN samples, which is expected due to the presence of nitrogen in the latter material. Nitrogen incorporation into GaAs-based alloys generally leads to a reduction in thermal conductivity because nitrogen creates additional point defects and disorder within the crystal structure, which impedes phonon transport. Y. Zhao et al. (2004) also measured the thermal conductivity of GaAsN with 2% nitrogen, finding a value of approximately 27 W/mK, which is about 60% of the thermal conductivity of GaAs (45 W/mK) in its binary form. This reduction in thermal conductivity is consistent with the alloying disorder introduced by the crystal structure of GaAsPN, which contains both nitrogen and phosphorus. In our study, the observed thermal conductivity values of approximately 4 W/mK are in line with those reported in the literature for GaAs-based alloys containing nitrogen and phosphorus. This confirms that the thermal conductivity of GaAsPN alloys is lower than that of pure GaAs due to the presence of nitrogen-related defects and disorder in the crystal structure. However, the values we measured are still significant for optoelectronic applications, especially when compared to other materials with lower thermal conductivity. The ability to tune the thermal conductivity of GaAsPN alloys through growth and annealing conditions makes these materials highly versatile for a range of applications, including high-performance tandem solar cells. This study provides valuable insights into the thermal conductivity of GaAsPN alloys and their potential for use in tandem solar cells. By controlling the annealing temperature and growth conditions, we can enhance the thermal conductivity of these materials, improving their performance in optoelectronic devices. The findings highlight the importance of optimizing the crystalline quality of the material, as well as the role of phonon scattering and defect density in determining thermal properties. As the demand for high-efficiency solar cells and optoelectronic devices increases, further research into the thermal conductivity of these materials will be essential for developing advanced materials with superior performance and reliability

CONCLUSION

This study presents a comprehensive analysis of the optical absorption and thermal conductivity properties of various GaAsPN/GaP samples, considering different growth conditions and post-growth annealing treatments. By employing optical absorption spectroscopy and photothermal deflection spectroscopy (PDS), we demonstrated that the optical absorption characteristics are highly sensitive to the arsenic content in the alloy. Specifically, GaAsPN absorbers with 10% arsenic and 2% nitrogen exhibit the most promising properties, as they offer a perfect lattice match with silicon, a crucial factor for the development of tandem solar cells. Moreover, these absorbers possess an absorption range within the targeted energy interval of [1.7–1.8 eV], making them ideally suited for efficient energy conversion in tandem solar cells. Our findings highlight the significant impact of annealing on both the optical absorption and thermal conductivity of the GaAsPN material. The annealing process enhances the material's optical properties, notably increasing the absorption coefficient, which reaches up to 38,000 cm^{-1} just below the GaP bandgap energy. The observed improvement in thermal conductivity, which reaches 4 W/mK, further underscores the benefits of annealing, suggesting better crystal quality and reduced defects in the material. These properties are essential for the effective performance of solar cells, as high thermal conductivity helps dissipate heat during operation, thus improving the overall efficiency and longevity of the device. The role of annealing in improving the structural and optoelectronic properties of semiconductor materials is well documented in the literature. During the annealing process, atoms gain sufficient thermal energy to rearrange themselves into a more ordered and energetically favorable configuration. This results in a reduction in defect density, including vacancies, dislocations, and impurity clusters, which are known to degrade the optical and thermal properties of the material. In GaAsPN alloys, annealing not only enhances crystal quality but also reduces the non-radiative recombination centers, thereby increasing the material's ability to efficiently absorb and convert solar energy. The presence of localized nitrogen states in the alloy before annealing can act as recombination centers, leading to lower absorption efficiency. However, post-growth annealing allows for the redistribution of nitrogen atoms, promoting more delocalized states and enhancing optical absorption. From a thermal perspective, annealing improves the material's ability to conduct heat by reducing phonon scattering at defect sites. In semiconductor materials, thermal conductivity is largely governed by phonon transport, which is hindered by defects and alloy disorder. By reducing inhomogeneity and optimizing the atomic arrangement within the GaAsPN lattice, annealing facilitates more efficient heat dissipation. This is particularly important for optoelectronic applications, where excessive heat buildup can degrade performance and shorten the lifespan of devices. The observed increase in thermal conductivity after annealing aligns with prior studies on similar III-V semiconductor alloys, further confirming the effectiveness of post-growth thermal treatments. In particular, these results are highly relevant for the development of GaAsPN absorbers in tandem solar cells, where materials with optimal optical and thermal properties are crucial for achieving high efficiency. Tandem solar cells, which combine multiple layers with different bandgaps to capture a wider spectrum of sunlight, stand to benefit from GaAsPN materials with tunable bandgaps and enhanced performance characteristics. The potential of GaAsPN alloys to serve as efficient middle-cell absorbers in silicon-based tandem solar cells is particularly noteworthy. With a bandgap that can be engineered to fall within the optimal range for tandem integration, GaAsPN absorbers can significantly improve the power conversion efficiency of these solar cells compared to traditional single-junction devices. Moreover, the compatibility of GaAsPN with silicon substrates provides a significant advantage for large-scale photovoltaic applications. Unlike other III-V materials that require costly substrates, GaAsPN can be directly grown on silicon, making it a cost-effective option for high-efficiency solar cells. The ability to achieve lattice matching with silicon while maintaining excellent optical and thermal properties makes GaAsPN an attractive candidate for future photovoltaic technologies. By further optimizing growth and annealing conditions, it may be possible to enhance the material's performance even further, leading to next-generation solar cells with superior efficiency and durability. The improved absorption and thermal conductivity in the GaAsPN alloys studied here could significantly contribute to the development of more efficient, high-performance tandem solar cells, paving the way for next-generation photovoltaic technologies. As solar energy continues to be a major focus of renewable energy research, advancements in semiconductor materials like GaAsPN will be crucial for achieving higher efficiency and reducing the cost of solar power generation. By leveraging the unique properties of GaAsPN, researchers and engineers can develop new solar cell architectures that push the boundaries of energy conversion efficiency. Future research directions could explore additional factors influencing the performance of GaAsPN absorbers, such as doping strategies, strain engineering, and interface optimization with other semiconductor materials. Additionally, further investigations into the long-term stability of GaAsPN-based solar cells under real-world operating conditions will be essential for commercial viability. Understanding how environmental factors such as temperature fluctuations, humidity, and prolonged light exposure affect the material's properties will provide valuable insights into the practical implementation of GaAsPN in solar energy applications. This study underscores the importance of optimizing both optical and thermal properties in semiconductor materials for photovoltaic applications. Through careful control of composition, growth conditions, and annealing treatments, GaAsPN alloys can be tailored to achieve superior performance in tandem solar cells. The demonstrated improvements in absorption efficiency and thermal conductivity highlight the potential of GaAsPN as a key material for next-generation photovoltaic technologies. By continuing to refine and develop these materials, the solar energy industry can move closer to achieving more efficient, cost-effective, and sustainable solar power solutions for the future. Furthermore, future research should explore refinements in annealing techniques to further enhance optical and structural properties. Investigating alternative substrate materials could also provide additional avenues for improving device efficiency, thermal stability, and mechanical integrity. In addition, expanding studies to include long-term stability assessments under real-world operating conditions will be crucial for evaluating the practical viability of GaAsPN-based solar cells. Exploring doping strategies, defect passivation methods, and interface engineering could further optimize performance and unlock new possibilities for tandem solar cell architectures. Overall, this study underscores the potential of GaAsPN materials in next-generation photovoltaic applications. By addressing key challenges in material processing, scalability, and integration, these findings lay the groundwork for further advancements in high-efficiency solar energy conversion. Future efforts in optimizing these materials could significantly contribute to the transition toward more efficient, durable, and commercially viable tandem solar cells, ultimately accelerating progress in the field of renewable energy.

REFERENCES

Adachi, S. (2007). Lattice thermal conductivity of group-IV and III–V semiconductor alloys. *Journal of Applied Physics, 102*(6), 063502. DOI: 10.1063/1.2779259

Adachi, S. (2007). Lattice thermal conductivity of group-IV and III–V semiconductor alloys. *Journal of Applied Physics, 06350*(6), 102. DOI: 10.1063/1.2779259

Adaine, A., Sidi, O. S. H., & Fressengeas, N. (2016). Simulation study of a new InGaN p-layer free Schottky based solar cell. *Superlattices and Microstructures, 96*, 121–133. DOI: 10.1016/j.spmi.2016.05.020

Almosni, S., Robert, C., Nguyen Thanh, T., Cornet, C., Létoublon, A., Quinci, T., Levallois, C., Perrin, M., Kuyyalil, J., Pedesseau, L., Balocchi, A., Barate, P., Even, J., Jancu, J. M., Bertru, N., Marie, X., Durand, O., & Le Corre, A. (2013). Evaluation of InGaPN and GaAsPN materials lattice-matched to Si for multi-junction solar cells. *Journal of Applied Physics, 113*(12), 123509. DOI: 10.1063/1.4798363

Bimberg, D., Heinrichsdorff, F., Bauer, R. K., Gerthsen, D., Stenkamp, D., Mars, D. E., & Miller, J. N. (1992). Binary AlAs/GaAs versus ternary GaAlAs/GaAs interfaces: A dramatic difference of perfection. *Journal of Vacuum Science & Technology B Microelectronics and Nanometer Structures, 10*, 1793.

Buyanova, I. A., Pozina, G., Bergman, J. P., Chen, W. M., Xin, H. P., & Tu, C. W. (2002). Time-resolved studies of photoluminescence in GaNxP1−x alloys: Evidence for indirect-direct band gap crossover. *Applied Physics Letters, 81*(1), 52–54. DOI: 10.1063/1.1491286

Dimroth, T. N. D., Tibbits, T. N. D., Niemeyer, M., Predan, F., Beutel, P., Karcher, C., Oliva, E., Siefer, G., Lackner, D., Fus-Kailuweit, P., Bett, A. W., Krause, R., Drazek, C., Guiot, E., Wasselin, J., Tauzin, A., & Signamarcheix, T. (2016). Four-Junction Wafer-Bonded Concentrator Solar Cells. *IEEE Journal of Photovoltaics, 6*(1), 343–349. DOI: 10.1109/JPHOTOV.2015.2501729

Fedorov, V. V., Dvoretckaia, L. N., Mozharov, A. M., Fedina, S. V., Kirilenko, D. A., Berezovskaya, T. N., Faleev, N. N., Yunin, P. A., Drozdov, M. N., & Mukhin, I. S. (2023). Dual-functional light-emitting and photo-detecting GaAsPN heterostructures on silicon. *Materials Science in Semiconductor Processing, 168*, 107867. DOI: 10.1016/j.mssp.2023.107867

Gaitho, F. M., Ndiritu, F. G., Muriithi, P. M., Ngumbu, R. G., & Ngareh, J. K. (2009). Effect of thermal conductivity on the efficiency of single crystal silicon solar cell coated with an anti-reflective thin film. *Solar Energy, 83*(8), 1290–1293. DOI: 10.1016/j.solener.2009.03.003

Geisz, J. F., & Friedman, D. J. (2002). III–N–V semiconductors for solar photovoltaic applications. *Semiconductor Science and Technology, 17*(8), 769–777. DOI: 10.1088/0268-1242/17/8/305

Grenouillet, L., Bru-Chevallier, C., Guillot, G., Gilet, P., Duvaut, P., Vannuffel, C., Million, A., & Chenevas-Paule, A. (2000). Evidence of strong carrier localization below 100 K in a GaInNAs/GaAs single quantum well. *Applied Physics Letters, 76*(16), 2241–2243. DOI: 10.1063/1.126308

Hatch, S., Wu, J., Sablon, K., Lam, P., Tang, M., Jiang, Q., & Liu, H. (2014). InAs/GaAsSb quantum dot solar cells. *Optics Express, 22*(S3), A679. DOI: 10.1364/OE.22.00A679 PMID: 24922376

Ilahi, S., Almosni, S., Chouchane, F., Perrin, M., Zelazna, K., Yacoubi, N., Kudrawiec, R., Râle, P., Lombez, L., Guillemoles, J.-F., Cornet, C., & Durand, O. (2015). Optical absorption and thermal conductivity of GaAsPN absorbers grown on GaP in view of their use in multijunction solar cells. *Solar Energy Materials and Solar Cells*, *141*, 291–298. DOI: 10.1016/j.solmat.2015.06.003

Ilahi, S., Almosni, S., Chouchane, F., Perrin, M., Zelazna, K., Yacoubi, N., Kudrawiec, R., Râle, P., Lombez, L., Guillemoles, J. F., Durand, O., & Cornet, C. (2015). Optical absorption and thermal conductivity of GaAsPN absorbers grown on GaP in view of their use in multijunction solar cells. *Solar Energy Materials and Solar Cells*, *141*, 291–298. DOI: 10.1016/j.solmat.2015.06.003

Ilahi, S., Baira, M., Saidi, F., Yacoubi, N., Auvray, L., & Maaref, H. (2013b). Non-radiative recombination process in BGaAs/GaAs alloys: Two layer photothermal deflection model. *Journal of Alloys and Compounds*, *581*, 358–362. DOI: 10.1016/j.jallcom.2013.07.101

Ilahi, S., Cornet, C., Yacoubi, N., & Durand, O. (2020). Improvement of carriers diffusion length and mobility in annealed GaAsPN materials for intermediate band solar cells. *Solar Energy Materials and Solar Cells*, *215*, 110622. DOI: 10.1016/j.solmat.2020.110622

Ilahi, S., Khalfaoui, A., Genty, F., & Yacoubi, N. (2021). Significant improvement of thermal conductivity and bandgap energy blue shift in annealed GaInAsSb/GaSb. *Optics & Laser Technology*, *137*, 106819. DOI: 10.1016/j.optlastec.2020.106819

Ilahi, S., Saadalah, F., & Yacoubi, N. (2013a). Investigation of the thermal annealing effect on the optical, thermal and electrical properties of Sn 2 Sb 2 S 5 evaporated thin films. *Applied Physics. A, Materials Science & Processing*, *110*, 459–464. DOI: 10.1007/s00339-012-7242-6

Ilahi, S., Saidi, F., Hamila, R., Yacoubi, N., Auvray, L., & Maaref, H. (2013). Photothermal deflection spectroscopy PDS investigation of optical and thermal properties of BGaAs/GaAs alloys. *Current Applied Physics*, *13*(3), 610–613. DOI: 10.1016/j.cap.2012.11.002

Ilahi, S., Saidi, F., Hamila, R., Yacoubi, N., Maaref, H., & Auvray, L. (2013). L Auvray. Shift of the gap energy and thermal conductivity in BGaAs/GaAs alloys. *Physica B, Condensed Matter*, *421*, 105–109. DOI: 10.1016/j.physb.2013.03.045

Ilahi, S., & Yacoubi, N. (2014). Interface recombination velocity measurement in SiO 2/Si. *Semiconductors*, *48*(3), 302–306. DOI: 10.1134/S1063782614030130

Ilahi, S., Yacoubi, N., & Genty, F. (2017). Thermal annealing effects on AlGaAsSb/GaSb laser structure: Bandgap energy blueshift and thermal conductivity enhancement. *Optical Materials*, *69*, 226–229. DOI: 10.1016/j.optmat.2017.04.050

Ilahi, S., Yacoubi, N., & Genty, F. (2018). Photothermal deflection technique investigation of thermal annealing effects of AlGaAsSb/GaSb laser structure: Non-radiative recombination parameters enhancement. *Materials Research Bulletin*, *106*, 332–336. DOI: 10.1016/j.materresbull.2018.06.007

Izadifard, M., Buyanova, I. A., Bergman, J. P., Chen, W. M., Utsumi, A., Furukawa, Y., Wakahara, A., & Yonezu, H. (2004). Effects of rapid thermal annealing on optical quality of GaNP alloys. *IEE Proceedings. Optoelectronics*, *151*(5), 335–337. DOI: 10.1049/ip-opt:20040872

Khalfaoui, A., Ilahi, S., Abdel-Rahman, M., Zia, M. F., Alduraibi, M., & Ilahi, B. (2017). Photothermal deflection technique investigation of annealing temperature and time effects on optical and thermal conductivity of V/V2O5 alternating layers structure. *Physica B, Condensed Matter*, *522*(26–30).

Kudrawiec, R., Luce, A. V., Gladysiewicz, M., Ting, M., Kuang, Y. J., Tu, C. W., Dubon, O. D., Yu, K. M., & Walukiewicz, W. (2014). Electronic Band Structure of Highly Mismatched Alloys: Suitability for Intermediate-Band Solar Cells. *Physical Review Applied*, *1*(3), 034007. DOI: 10.1103/PhysRevApplied.1.034007

Kunert, B., Volz, K., Koch, J., & Stolz, W. (2006). Direct-band-gap Ga(NAsP)-material system pseudomorphically grown on GaP substrate. *Applied Physics Letters*, *88*(18), 182108. DOI: 10.1063/1.2200758

Loubiri, D., Ben Hamed, Z., Ilahi, S., Sanhoury, M. A., Kouki, F., & Yacoubi, N. (2015b). Photothermal investigation of poly (3-hexylthiophene): ZnSe nanocomposites. *Applied Physics. A, Materials Science & Processing*, *119*(2), 581–588. DOI: 10.1007/s00339-015-8995-5

Loubiri, D., Hamed, Z. B., Ilahi, S., Sanhoury, M. A., Kouki, F., & Yacoubi, N. (2015a). Effect of TBPO-capped CdSe nanoparticles concentration on sub-bandgap absorption in poly (3-hexylthiophene) thin films studied by photothermal deflection spectroscopy. *Synthetic Metals*, *206*, 1–7. DOI: 10.1016/j.synthmet.2015.04.018

Nath, P., & Biswas, A. (2023). Probing performance of p-GaAsP/i-GaAs/n-GaAsP solar cells through compositional and geometrical variations. *Microsystem Technologies*, *29*(5), 773–778. DOI: 10.1007/s00542-023-05456-3

Ramiro, Í., Martí, A., Antolín, E., & Luque, A. (2014). Review of Experimental Results Related to the Operation of Intermediate Band Solar Cells. *IEEE Journal of Photovoltaics*, *4*(2), 736–748. DOI: 10.1109/JPHOTOV.2014.2299402

Robert, C., Perrin, M., Cornet, C., Even, J., & Jancu, J. M. (2012). Atomistic calculations of Ga (NAsP)/GaP (N) quantum wells on silicon substrate: Band structure and optical gain. *Applied Physics Letters*, *100*(11), 111901. DOI: 10.1063/1.3694028

Rubel, O., Galluppi, M., Baranovskii, S. D., Volz, K., Geelhaar, L., Riechert, H., Thomas, P., & Stolz, W. (2005). Quantitative description of disorder parameters in (GaIn)(NAs) quantum wells from the temperature-dependent photoluminescence spectroscopy. *Journal of Applied Physics*, *98*(6), 063518. DOI: 10.1063/1.2058192

Sahoo, S. K. (2016). Renewable and sustainable energy reviews solar photovoltaic energy progress in India: A review. *Renewable & Sustainable Energy Reviews*, *59*, 927–939. DOI: 10.1016/j.rser.2016.01.049

Shan, W., Walukiewicz, W., Yu, K. M., Wu, J., Ager, J. W.III, Haller, E. E., Xin, H. P., & Tu, C. W. (2000). Nature of the fundamental band gap in GaNxP1−x alloys. *Applied Physics Letters*, *76*(22), 3251–3253. DOI: 10.1063/1.126597

Sidi, O. S. H., Adaine, A., & Fressengeas, N. (2016). Numerical simulation of InGaN Schottky solar cell. *Materials Science in Semiconductor Processing*, *41*, 219–225. DOI: 10.1016/j.mssp.2015.09.001

Boyer-Richard, S., Fan, F., Beck, A., Levallois, C., Tavernier, K., Rohel, T., & Durand, O. (2023). Investigation of III-V GaP solar cell on silicon substrate. *EPJ Photovoltaics*, *14*, 31.

Utsumi, A., Yonezu, H., Furukawa, Y., Momose, K., & Kuroki, K. (2003). Increase in luminescence efficiency of GaPN layers by thermal annealing. [c]. *Physica Status Solidi. C, Conferences and Critical Reviews*, *0*(7), 2741–2744. DOI: 10.1002/pssc.200303346

Yacoubi, N., & Alibert, C. (1991). Determination of very thin semiconductor layer thickness by a photothermal method. *Journal of Applied Physics*, *69*(12), 8310–8312. DOI: 10.1063/1.347441

Yu, X. Y., Chen, G., Verma, A., & Smith, J. S. (1995). Temperature dependence of thermophysical properties of GaAs/AlAs periodic structure. *Applied Physics Letters*, *67*(24), 3554–3556. DOI: 10.1063/1.114919

Yu, X. Y., Chen, G., Verma, A., & Smith, J. S. (1995). Temperature dependence of thermophysical properties of GaAs/AlAs periodic structure. *Applied Physics Letters*, *67*(24), 3554–3556. DOI: 10.1063/1.114919

Zelazna, K., Gladysiewicz, M., Polak, M. P., Almosni, S., Letoublon, A., Cornet, C., Durand, O., Walukiewicz, W., & Kudrawiec, R. (2017). Nitrogen-related intermediate band in P-rich GaNxPyAs1−x−y alloys. *Scientific Reports*, *7*(1), 15703. DOI: 10.1038/s41598-017-15933-1 PMID: 29147023

Zhao, C.-Z., Zhu, M.-M., Sun, X.-D., Wang, S.-S., & Wang, J. (2018). The band gap energy of the dilute nitride alloy GaNxAsyP_{1-x-y} depending on content. *Applied Physics. A, Materials Science & Processing*, *124*(2), 216. DOI: 10.1007/s00339-018-1654-x

Zhao, Y. M., Chen, G., Wang, S. Z., & Yoon, S. F. (2004). Thermal characterization of gallium arsenic nitride epilayer on gallium arsenide substrate using pulsed photothermal reflectance technique. *Thin Solid Films*, *450*(2), 352–356. DOI: 10.1016/j.tsf.2003.11.289

Chapter 13

Renewable Energy, Economic Growth, and Environmental Sustainability Links in South Asian Countries:

Evidence Using Panel ARDL

Punica Mahajan
Shri Mata Vaishno Devi University, Katra, India

Meenakshi Gupta
https://orcid.org/0000-0002-7866-4961
Shri Mata Vaishno Devi University, Katra, India

ABSTRACT

The present study focuses on how the variables like energy consumption, domestic material consumption, access to clean fuels for cooking and trade openness affect the economic growth and other explanatory variables like renewable energy use and access to clean fuels for cooking affect the Carbon dioxide emissions in the short run and long run of South Asian Economies. So, the study made an attempt to evaluate the impact of above-mentioned variables on economic growth and carbon dioxide emission using the Panel Autoregressive Distributed Lag (ARDL) method. All the preliminary tests including unit root assessments, cross-sectional dependence evaluation, Kao-co-integration etc. were used before applying the ARDL panel model. There is a positive impact of energy consumption, access to clean fuels along with the trade openness (measured by trade¬ as % of GDP) on economic growth. However, the increased domestic material consumption has a negative impact on GDP per capita. Addition to this, the renewable energy consumption is significant and negatively impacts the emissions.

DOI: 10.4018/979-8-3693-9924-8.ch013

INTRODUCTION

Since the starting of the industrialization, the growth of economy is somehow, depend upon the fossil fuels as they are useful in providing the fuels for planes, ships and automobiles. Analyzing the dependency between economic growth and carbon dioxide is considered in this study. But with time their relationship has evolved. The energy consumption of economic development is slowly decreasing which implies a less amount of energy is required to generate one more unit of GDP. Recently, there has been an increase in the consumption of clean energies leading to a difference between the GDP growth and carbon dioxide emissions in the majority of the economies in the world. In addition, the GDP growth and the carbon dioxide emissions reached their heights in 2007 in developed economies and then began to go downwards. Now, a days there has been many significant changes in the consumption of renewable energy. In 2007, the carbon dioxide emissions and GDP per capita achieved their highest and then started decreasing within time. In recent times, it has been seen that South Asia countries have a correlation between carbon dioxide emissions, and economic growth. The present dissertation chooses related variables for South Asian countries whereas, Bangladesh, India, Pakistan, Nepal, Sri Lanka, and Bhutan are to be developed countries and also experience some drastic climate changes. These countries aim to enhance their environmental quality. The economies of country including Bangladesh, Bhutan and India analyzed the positive forecasts of GDP per capita while India ranked to be the top most growing economy and third most pollutant country in the world.

BACKGROUND

For the growth of the economy, non-renewable energy sources are determined. The people in South Asian countries resides below the poverty line and due to this problem, the majority of the South Asian countries are putting efforts to increase their economies using trade openness, industrialization, and globalization. So that the countries can recover from poverty. Furthermore, global warming and climate change are the main problems affecting the world's environment. The temperature is going upward due to global warming. In the Polar Regions, snow and ice are melting. But the only goal in these countries is to achieve a long-run development for their economies. The environment is in poor condition because of the degradation of land, the development of more industries, and many more. Carbon dioxide emissions are the main reason for these problems. $C0_2$ gasses stays in the biosphere and cannot be dismissed before thousands of years. The study shows that approximately, 90% of the CO_2 are caused from burning fossil fuels. The percentage of Carbon dioxide emissions are generated from burning fossil fuels. The percentage of Carbon dioxide emissions rises in developing countries due to this, ocean water forms into carbonic acid that affects marine life. Furthermore, the ecosystem can be enhanced by consuming more green energies and support sustainable related policies. Countries in South Asia faces variability in the environment, economic growth and technological upgradation. Moreover, technological digitalization is helping the economy to grow with sustainability. However, the Nexus between Carbon dioxide emissions and economic growth plays an important role in making effective policies for sustainability.

Table 1. Programs launched by South Asian countries in relation to environment sustainability

AFGHANISTAN	• Strengthening environmental governance and building institutions • Providing technical assistance in fulfilling the administrative obligations of each ratified convention • Preserving the country's diverse landscape
BANGLADESH	• Sustainable Forests & Livelihoods Project (SUFAL) • Sustainable Coastal and Marine Fisheries Project (SCMFP) • Sustainable Enterprises Project (SEP)
BHUTAN	• Advancing Climate Resilience of the Water Sector in Bhutan (ACREWAS)
INDIA	• National River Conservation Plan • Eco mark Scheme of India (ECOMARK) – Eco mark Labelling • National Clean Air Programme (NCAP) • National Adaptation Fund for Climate Fund (NAFCC)
MALDIVES	• National Adaptation Plan (NAP)
NEPAL	• Forestry Sector Policies and Strategies • Environment-Friendly Vehicle and Transport Policy • Environment Friendly Local Governance Framework
PAKISTAN	• Clean and Green Pakistan • Ten Billion Tree Tsunami • Recharge Pakistan
SRI-LANKA	• USAID's Clean Cities, Blue Ocean (CCBO) Program

Metrics of renewable energy use, GDP and CO_2 emissions:South Asian Nations

Figure 1. Renewable Energy Use

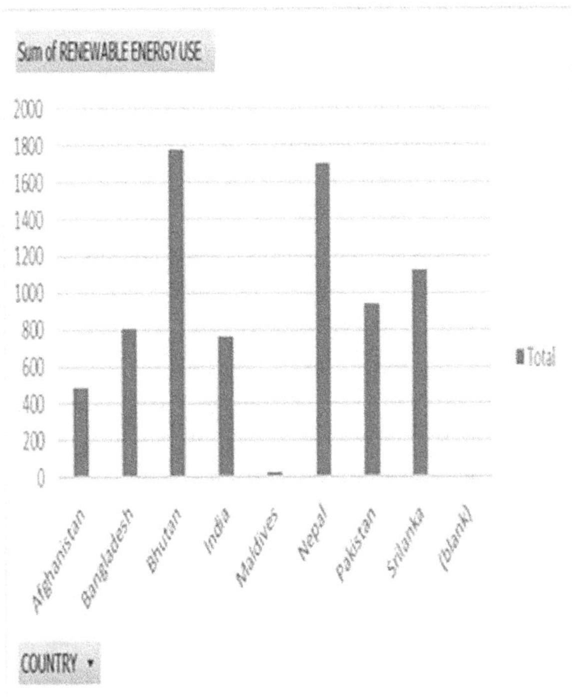

Figure 2. GDP

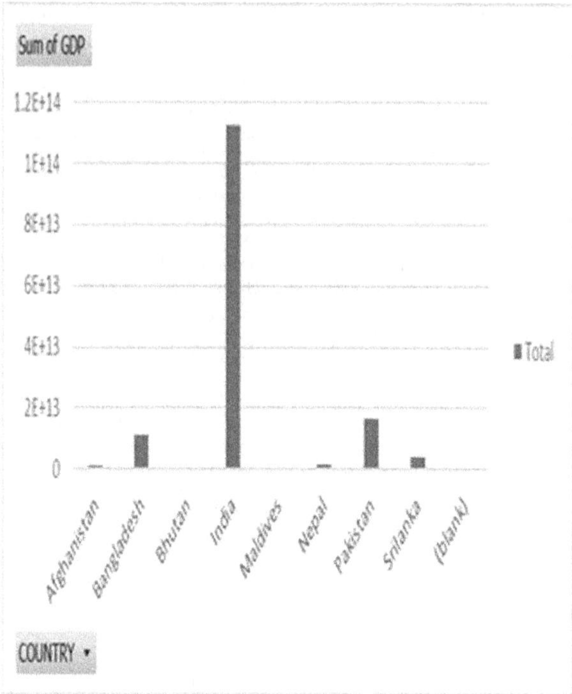

Figure 3. CO_2 Emission

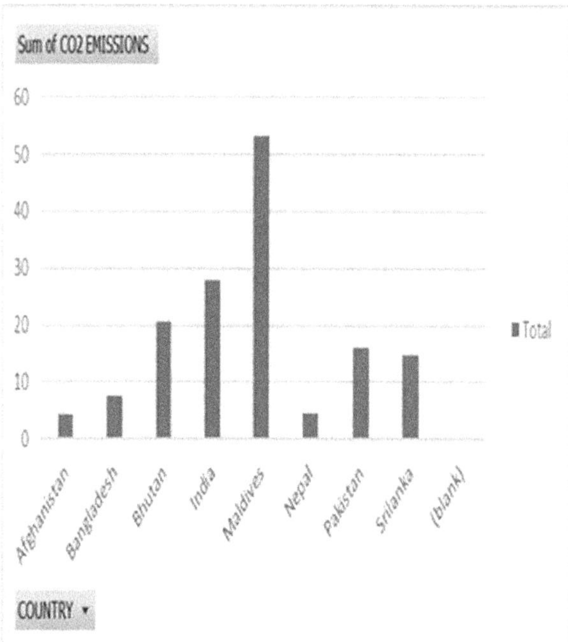

Source: https://ourworldindata.org/

The data for eight South Asian Countries shows that both Bhutan and Nepal are the one who adopt more green energies while Afghanistan followed by Maldives are the countries, which shows least concern towards the utilization of renewable energy. The given data for South Asian Countries shows that India has the highest GDP per capita whereas, Afghanistan and Maldives almost follow the same pattern for attaining GDP per capita. However, Bhutan is at the lowest among all the South Asian Countries. The given data for South Asian Countries shows Maldives emits the highest Carbon dioxide emissions followed by India and Bhutan whereas, Pakistan and Sri Lanka are on the same scale. Furthermore, Nepal and Afghanistan emit the lowest Carbon dioxide emissions among all the 8 South Asian Countries.

Metrics of socioeconomic, sustainable energy and environmental indicators indicators: South Asian nations

The economy of South Asia is rapidly expanding but this expansion leads to environmental degradation. The reason is that the fossil fuels are the major driver of economic growth in this region and these poses long term environmental challenges. However the use of renewable energy presents an opportunity to decouple environmental damage from economic growth. So, the metrics of indicators including energy consumption, trade openness, domestic material consumption, population having aces to clean fuels, Renewable energy use and CO_2 emissions etc. are important in understanding the concept of environmental sustainability. Here are some of data for South Asian nations for the period 2019 collected from Our World in Data

Figure 4. Socioeconomic and Environmental indicators

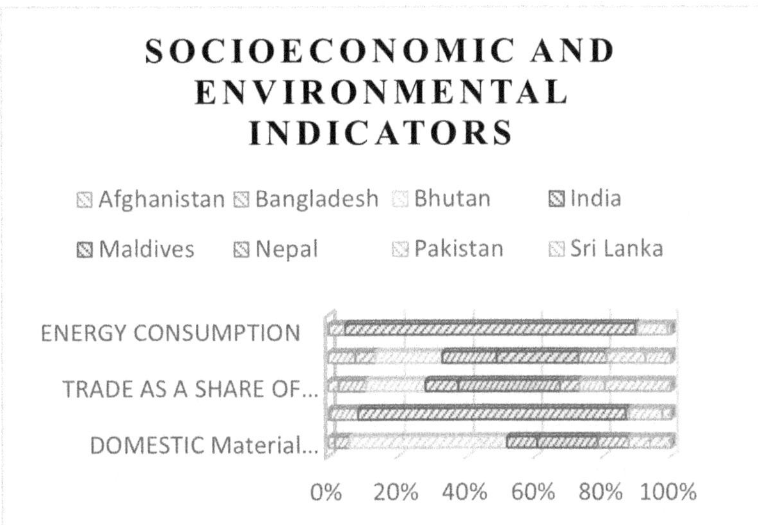

Figure 5. Metrics for Sustainable energy

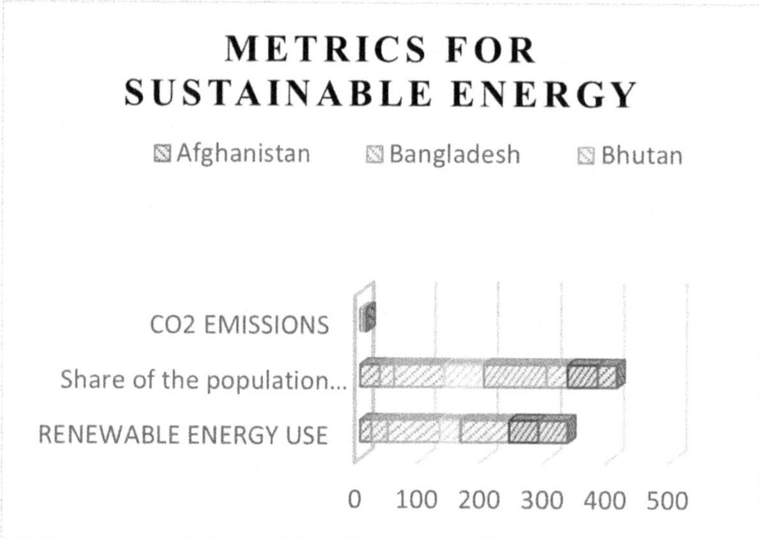

Source: https://ourworldindata.org/

PROBLEM STATEMENT

As an economy grows there is a surge in energy demand and to meet this increasing demand, fossil fuels are increasingly used which results in carbon emissions and thus negatively impacts the environment. The negative impacts of the use of fossil fuels has already resulted in climate change throughout South Asia. Therefore, in order to mitigate the negative impacts of there is a need to shift towards renewable energy. Access to clean fuels impacts the environment through reduced carbon emissions. The present study is an attempt to find out how the both consumption of renewable energy along with the clean energy affects carbon emissions.

REVIEW OF LITERATURE

Saqib et al. (2024) on the topic of "Harnessing the synergistic impacts of environmental innovations, financial development, green growth, and ecological footprint through the lens of SDGs policies for countries exhibiting high ecological footprints." discussed about the top 10 Asian countries from the period of 1990 to 2017 in order to determine the environmental technologies, energy use, and economic growth and how it impacts the ecological footprint. Keeping an eye on these two specific goals i.e., SDG-3and SDG-7, the given study provides a complete detailed relationship between economic growths, improvement in the environment along with the use of non-renewable energy. The researchers who worked on environmental innovation, ecological footprint, and the use of renewable and non-renewable energies determines that the using renewable energy is significant for the environment and also reveals out that the degradation in environment increases when the economy grows and non-renewable energy is used.

Mngumi et al. (2022) in their study "Does green finance mitigate the effects of climate variability: role of renewable energy investment and infrastructure?" highlighted the green finance and its impact on carbon dioxide emission in BRICS countries. The research considers green finance development index by determining green insurance, green securities, green investment, and green credit. The study is done from 2005 to 2019 which states increase the utilization of renewable energy and enhancement in the green finance decreases the carbon dioxide emissions in these countries. The study focuses to significantly use the sustainable technologies. On the other hand, the carbon dioxide emissions slowed down the growth of green investment and renewable energy which emphasized the importance to have policies that encourages green finance and clean energy sources. As per study, the equator principles and green innovation is helpful to avoid environmental issues. Therefore, these countries need a good green finance system to enhance.

Kazemzadeh et al. (2022) in their research on "Factors driving CO2 emissions: the role of energy transition and brain drain" discussed the analysis of energy transition and brain drain on carbon dioxide emissions using a two-fold study. First study indicates brain drain impacts degradation to the environment, when CO2 emission per capita is low and second includes the analysis of rain drain and energy transition on carbon dioxide emissions for the countries that follow same convergence patters. The study indicates of 75-nation from 2006 to 2020. The study emphasis that brain drain, economic growth and trade openers increases CO2 emissions. The study shows all these variables function in complex manner and demanding for more research and policies.

Alharbi et al. (2023) in their research on "Green Finance and Renewable Energy: A Worldwide Evidence" investigated 44 countries in the period of 2007-2020 depicts that green finance significantly impacts the renewable energy implies making investment in sustainable initiatives positive enhance the utilization of clean energy. Current technology is positively impacting the green finance on the production of renewable energy in the long run. The research examined study with possible challenges like structural break and cross-sectional dependence in the data.

Kazemzadeh et al. (2022) in their research on "Do energy efficiency and export quality affect the ecological footprint in emerging countries? A two-step approach using the SBM–DEA model and panel quantile regression" investigated 16 emerging countries from the period of 1990 to 2014 by considering a two-step model including SBM-DEA model and panel quantile regression. The SBM-DEA model reveals the energy efficiency and resources of a country depicting Hungary and Turkey were at the good rank whereas, India and China were at the last rank under these criteria on the other hand, panel quantile regression reveals the impact of resources and export quality and other variables on the ecological footprint. The study depicts the trade openers, resources and energy efficiency decrease the ecological footprint whereas, GDP, utilization of fossil fuels and urbanization also plays the same role in term of environmental footprint.

Rasoulinezhad & Taghizadeh-Hesary (2022) in their study on "Role of green finance in improving energy efficiency and renewable energy development" review that under sustainable Development goal (SGDs), green energy is related to environment and energy. The study analyzed the correlation between green finance, CO2 emissions, green efficiency, and green energy Index. The study uses STIRPAT model to examine the correlation considering top 10 economies pillar green finance. The research depicts green bond is the considerable method to enhance green energy projects and reduce the impact of CO2 emissions. For sustainable achievement, the governments should execute some policies to increase the indulgent of private organization for the investment in such projects. The policy may be relevant in the post and during era of covid-19.

Sun et al. (2023) in their research on "Green finance, renewable energy development, and climate change: evidence from regions of China" investigated from 2010 to 2020 to determine the relationship of green financing and consumption of renewable energy on carbon dioxide. The study emphasized on China and its provinces. The study is done by the analysis of STIRPAT model and GMM econometric model and analyzed green financing reduces the pollution and with increase of 1% in renewable energy, CO_2 emissions decrease by 0.103%. For accelerating the green growth, Chinese govt. should introduce more measures that supports environmental sustainability too.

Ahmed & Shimada (2019) in their research on the topic of "The Effect of Renewable Energy Consumption on Sustainable Economic Development: Evidence from Emerging and Developing Economies" Investigated the relationship between renewable energy consumption and sustainable economic development in the period between 2000 and 2020. The study is based on 30 emerging developing countries which is selected randomly and acc to some indicators like world development indicators and renewable energy counting attractiveness Index. Countries like Ghana, Tunisia, Zimbabwe, South Africa, and Cameroon have a positive relationship between renewable energy consumption and sustainable economic growth, but for Latin American and Caribbean countries, sustainable economic development depends on non-renewable energy consumption. The study emphasized by many econometrics model like Pesaran cross-section dependence test, panel unit root test, panel cointegration test and many more. Under the same study, the researchers investigated another relationship between renewable energy consumption and carbon dioxide emission for countries like Asia, South Asia, Latin America, and African countries depicts the Gross Domestic product and utilization of non-renewable energy consumption increases the carbon dioxide emissions and using more renewable energy cause carbon dioxide emission to decrease.

Wan et al. (2022) in their empirical study on "Green finance and carbon reduction: Implications for green recovery" investigated on 100 developed and emerging countries in 2016 to study the correlation between green property financing and the construction sector emission of carbon dioxide. The construction sector holds 40% of the total greenhouse gas emission globally. With the increase in green property financing, the emission of CO_2 in the construction sector would decrease. As the countries taken had huge dependence on oil and COVID-19 delayed the progress and availability of green funding.

Soundarrajan & Vivek (2016) on the topic of "Green finance for sustainable green economic growth in India" review that green finance plays a vital role to lowering carbon green growth. Green finance links the improvement in the environment, financial industry and growth in the economy. The researcher focuses on the ecological damage because of the release of Carbon-dioxide emissions. Simultaneously, analyzed the impact of green finance in Indian industries. The researchers also work on to analyze the trends and challenges in green finance in India.

Dey et al. (2022) in their study on "Renewable energy present status and future potentials in India: An Overview" describes the relationship between renewable energy and sustainable development and their future. The renewable energy is very friendly to the environment. The renewable save energy, enhance the environment. Population is increasing leads to increase in the energy consumption. The research considers India, that has the fourth most level of renewable energy market in the world. The government of India is supporting the clean energy by introducing many sustainable development projects. The study shows India has increased the consumption of renewable energy by analyzing on problems like energy efficiency, economic growth, and carbon dioxide emission challenges in green finance in India.

Ye et al. (2022) in their study on the topic "The nexus among green financial development and renewable energy: investment in the wake of the Covid-19 pandemic" emphasized to determine how green financial development include green credit, green investment, green securities, and corporate social responsibility impacts the investment in renewable energy and environmental related problems. The study considers the time-period from 1976 to 2020. The findings states that green credit, green investment, corporate social responsibility, and green securities have strong and significant impact on renewable energy investment. The econometric model used were Error Correlation model, autoregressive distributed lag and many more.

Peimani (2018) in their empirical study on "Financial barriers to development of renewable and green energy projects in Asia" highlighted the issue in Asia's bank that dominates the financial system due to its immature capital market. Asia's bank is responsible for the funding related to green energy projects depicting the growth of green energy has not developed yet. The bank considers the related projects are very hazardous with less rate of return. It is the basic obstacle for the expansion of renewable projects in Asia. It is necessary to tackle the problem of financing green projects in Asian Countries.

Raghutla & Kolati (2023) in their research on "Public-private partnerships investment in energy as new determinant of renewable energy: The role of political cooperation in China and India" review that, the degradation of land and drastic change in climate are the global problems and the only way to deal these issues is to shift towards renewable energy projects. On the other hand, government has insufficient funds for green projects leads to the declaration of Private-Public Partnership. The study emphasis to determine the impact of investment in renewable by Private-Public Partnership in India and China from the period of 1990 to 2021. The study determines the impact of investment in renewable energy and make special consideration on political cooperation, technological innovation and R&D expenditures. The research also determines the significant effect of R&D investment, Private-Public Partnership and political cooperation on renewable energy's investment.

Cedrick & Long (2017) on the topic of "Investment Motivation in Renewable Energy: A PPP Approach" discussed technology related to renewable energy requires huge investment, innovation and the support of Private-Public Partnership to help the public sector. The government has already access to the clean energies with increasing rate. Clean energies are the priority for almost every country as it is cheap, reliable, clean and enduring more power. The poor effect of fossil fuels leads to the adaptation of renewable energy. The research also considers the factors that influence to increase in renewable energy projects.

Chen & Zhang (2023) in their study on "Does financial globalization promote renewable energy investment? Empirical insights from China" discussed the correlation between financial globalization and investment in renewable energy in China from 1995 to 2001. There is an increase in the renewable energy investment as the financial markets are connected with each other. In Linear ARDL, study emphasis a significant impact of financial globalization on clean energies investment both in the long-run and the short-run. The non- linear ARDL model demonstrates the positive impact on the same variables. Moreover, Private-Public Partnership enhance the chance for more investment especially in clean energy projects.

Appiah-Otoo et al. (2023) in their study on "Modelling the impact of renewable energy investment on global carbon dioxide emissions" investigated the relationship of renewable energy investment and carbon dioxide emissions. The Ordinary Least Squares Model is used to analyze the correlation between the investment in renewable energy and carbon dioxide emissions. The results depicts that the investment in wind and solar energy have inverse relationship with carbon dioxide emissions. On the other hand, energies like biofuel and hydropower have a positive relationship with carbon dioxide emissions. So, investment in solar and wind energies is recommended to decrease the effect of carbon emissions from the environment.

Ren et al. (2023) on the topic of "Country risk and its impact on renewable energy investment in highly polluted economies" discussed the correlation between the country risk and investment in renewable energy. The researchers worked on the high polluted countries where different econometrics models are used including GMM, 2SLS, OLS and panel quantile regression. Taking into consideration, the effect of OLS, 2SLS and GMM shows insignificant impact of country's risk on renewable energy investment whereas, the effect of panel quantile regression model is also significant between these two variables. Technological innovation, Carbon dioxide emissions and GDP encourages the renewable energy investment.

He (2023) in his study on "Effects of the Green Policy Environment on Renewable Energy Investment and Effect Evaluation of Green Policies" discussed the relationship between the green policy for environment and the investment in the renewable energy. It is important to invest in the clean energy as it will help in the economic growth. Strategy like Energy Planning and Investment was introduced against green bonds and low carbon framework. Furthermore, a multistage optimization model was constructed for renewable energy investment. The model worked properly and the impact of green policies is also analyzed.

Ajide & Mesagan (2022) on the topic "Heterogeneous analysis of pollution abatement via renewable and non-renewable energy: lessons from investment in G20 nations" reviewed that non-renewable energies degrade the environment in the long term. The study emphasizes on how capital investment impact the relationship between energy consumption and pollution in G-20 countries. Factors affecting pollution includes foreign direct investment, trade openness and energy consumption. The pooled mean group estimator is used to estimate the heterogeneity and cross-sectional dependence among the countries. The result shows that the capital investment in long run is not good for the environment because it creates more pollution whereas, the short run capital investment reduces the pollution.

Azhgaliyeva et al. (2023) in their study on "What matters for private investment in renewable energy?" reviewed that researchers worked on 13 economies from 2008 to 2018 to know the impact different source of financing in private renewable investment shows that the expenditure of government on R&D is significantly impact the private investment. The feed-in-tariffs also have a significant impact on renewable investment that are financed through public markets. Moreover, the tax has a different impact on various kinds of funding. Additionally, the cost of technology and the price of energy have a significant effect on the investment which is used to finance renewable projects.

Abolhosseini & Heshmati (2014) in their study "The Main Support Mechanisms to Finance Renewable Energy Development" reviewed that different policies are used to decrease the effect of carbon emissions i.e. by increasing the creation of green jobs and promote technologies. There are three support mechanisms for renewable energy which includes tax incentives, green certificates and feed-in-tariffs. For the enhancement of technology, heavy investment is needed. These mechanisms are useful to decrease the impact of carbon emissions.

Research Gap

As far as the literature cited, it was found that, very few studies clarify the concern between economic growth, environmental innovation and ecological footprint which further leads to the deficiency in broad perspective of collision over environmental sustainability beside to the energy transition policies.

Moreover, in order to achieve the sustainable development goals (SDG-3 and 7) along with the initiatives taken by various organizations, it becomes necessary to understand the linkage between renewable energy, economic growth and environmental sustainability. So the present study focusses on analyzing the impact of renewable energy consumption amid environmental on economic growth and how the outcomes of clean energy and adoption of renewable energy affects the sustainability of environment.

METHODOLOGY

Our empirical seeks to address the effectiveness of (economic growth and emissions over the selected determinants across the South Asian economies) can be evaluated by examining the relationship between the Economic growth and Carbon emission along with the factors affecting these two variables. Here is the possible methodology to evaluate this relationship:

Sample period: The sample period was taken from 2001 to 2020 for South Asian economies, which includes Afghanistan, Bangladesh, Bhutan, India, the Maldives, Nepal, Pakistan, and Sri Lanka.

Table 2a. CHOICE of VARIABLES: a) Dependent Variables: The study analyzes 2 dependent variables in South Asian Countries from 2001 to 2020

DEPENDENT VARIABLE	EXPLANATION
GDP PER CAPITA	Gross Domestic Product (GDP) per capita is an indicator of economic performance and is used as a measure of average living standards or economic well-being. In South Asian Countries, some countries have high GDP growth as these countries already technological upgraded while other countries have less GDP growth because of excessive population growth and many other factors.
CARBON-DIOXIDE EMISSIONS	CO2 emissions are emissions release from the burning of fossil fuels and the manufacture of cement; they include carbon dioxide produced during consumption of solid, liquid, and gas fuels. In, South Asian Countries, almost every country is facing carbon emissions due to the growth in the industries and massive population.

Table 2b. Independent Variable: The study analyzes the impact of all nine explanatory variables on energy consumption and carbon dioxide emissions in South Asian countries from 2001–2020.

INDEPENDENT VARIABLE	EXPLANATION
RENEWABLE ENERGY USE	Renewable energy is energy derived from natural sources that are replenished at a higher rate than they are consumed. Sunlight and wind, for example, are such sources that are constantly being replenished. In every South Asian country, the consumption of renewable energy is increasing on daily basis due to the awareness and policies introduced by the Government.
ENERGY CONSUMPTION	Energy consumption is the total amount of energy required for a given process and is measured in kilowatt-hours (kWh). This includes the use of electricity, gas, diesel, oil, and biomass. Energy consumption in South Asian countries is not static and consumption depends on the growth of the economy and the population. Investment in renewables and providing subsidies is the only solution to increase energy consumption. The energy consumption in South Asian
DOMESTIC MATERIAL CONSUMPTION	Domestic material consumption, abbreviated as DMC, measures the total amount of materials directly used by an economy and is defined as the annual quantity of raw materials extracted from the domestic territory, plus all physical imports minus all physical exports. In South Asian countries, the domestic material consumption shows the problem of increase in population and increase in industrialization resulting in increase in pollution.

INDEPENDENT VARIABLE	EXPLANATION
TRADE OPENNESS	Trade openness is one measure of the extent to which a country is engaged in the global trading system. Trade openness is usually measured by the ratio between the sum of exports and imports and Gross Domestic Product. Some countries in South Asia supports trade and encourages investment from foreign while some countries do not encourage trade due to trade barriers.
SHARE OF POPULATION WITH ACCESS TO CLEAN FUELS	The proportion of the population with primary reliance on clean fuels and technology is calculated as the number of people using clean fuels and technologies for cooking, heating, and lighting divided by the total population reporting that any cooking, heating, or lighting, is expressed as a percentage. In some South Asian Countries, some countries have successfully accessed the clean fuels while other lacks behind due to the economic efficiency still there is a positive improvement.

SOURCE OF DATA

The data in this chapter has been collected from a secondary source. The secondary data for this chapter is collected from the main database: Our World in Data (https://ourworldindata.org/)

Table 3. Variables, symbols, measurement units, and sources of data

S.NO	VARIABLE	SYMBOL	MEASURING UNIT	SOURCE
1	RENEWABLE ENERGY USE	RE	kilowatt-hours (kWh)	Our World in Data
2	DOMESTIC MATERIAL CONSUMPTION	DMC	kilograms, tons, or cubic meters,	Our World in Data
3	GROSS DOMESTIC PRODUCT	GDP	US dollars (USD), euros (EUR), or Japanese yen (JPY) or any Monetary	Our World in Data
4	TRADE OPENNESS	TO	Ratio or Percentage	Our World in Data
5	SHARE OF THE POPULATION WITH ACCESS TO CLEAN FUELS	CFA	percentage	Our World in Data
6	ENERGY CONSUMPTION	EC	kilowatt-hours (kWh)	Our World in Data
7	CARBON-DIOXIDE EMISSIONS	$C0_2$ EMISSIONS	kilograms (kg)	Our World in Data

MODEL SPECIFICATION:

The Panel ARDL approach will be used to capture the objective of this study. The model of this study will follow the work of "(Ogunsola, 2016), (Mehrara, 2011)". Also, this method of estimation would be used to estimate the long & short-run analysis of the study. Therefore, the model for this study is specified below:

$$GDP = f(EC + DMC + CFA + TO). \quad \text{(i)}$$

$$CO2 = f(RE + CFA) \quad \text{(ii)}$$

Now, we convert equations I & ii) into the natural logarithmic form to estimate the effect of various components on EC &CO2. The equation can be written as follows:

$$GDP_{it} = \beta_0 + \beta_1 EC_{it} + \beta_2 DMC_{it} + \beta_3 CFA_{it} + \beta_4 TO_{it} \quad \text{(iii)}$$

$$CO_{2_{it}} = \beta_0 + \beta_1 RE_{it} + \beta_2 CFA_{it} \quad \text{(iv)}$$

GDP represents the Gross domestic product for entity i at time t

EC, DMC, CFA and T0 are the explanatory variables representing Energy consumption, domestic material consumption, the share of the population with access to clean fuels and trade openness respectively, for entity i at time t and, are coefficients and u is the error term.

where, C02 presents carbon dioxide emissions for entity i at time t, RE and CFAs are the explanatory variables representing renewable energy and the share of the population with access to clean fuels for entity i at time t and are coefficients and u is the error term.

According to Pesaran & Shin, in ARDL methodology, the basic ARDL form of the above equation is:

$$\Delta \text{GDP}_{\text{it}} = \alpha_0 + \sum_{j=0}^{p} \varphi_j \Delta EC_{it-j} + \sum_{j=0}^{p} \varnothing_j \Delta DMC_{it-j} + \sum_{j=0}^{p} \gamma_j \Delta CFA_{it-j} \quad \text{(v)}$$

$$+ \sum_{j=0}^{p} \vartheta_j \Delta TO_{it-j} + \beta_1 EC_{it} + \beta_2 DMC_{it} + \beta_3 CFA_{it} + \beta_4 TO_{it}$$

$$\Delta \text{CO}_{2_{\text{it}}} = \alpha_0 + \sum_{j=0}^{p} \varphi_j \Delta RE_{it-j} + \sum_{j=0}^{p} \varnothing_j \Delta CFA_{it-j} + \beta_1 RE_{it} + \beta_2 CFA_{it} \quad \text{(vi)}$$

Where, $\varnothing$, φ, γ,ϑ refer to short-run & $\beta_1 to \beta_4$ to long-run parameters. The null hypothesis thesis of no cointegration is; $H_0 = \beta_1 = \beta_2 = \beta_3 = \beta_4 = 0$ against alternative hypothesis: $H_1 = \beta_1 \neq \beta_2 \neq \beta_3 \neq \beta_4 \neq 0$.

Where, $\varnothing$, φ refer to short-run & $\beta_1 to \beta_2$ to long-run parameters. The null hypothesis thesis of no cointegration is; $H_0 = \beta_1 = \beta_2 = 0$ against alternative hypothesis: $H_1 = \beta_1 \neq \beta_2 \neq 0$.

ARDL method estimate $(p+1)^k$, the number of regressions to obtain the optimal lags for each variable, where p+1, is the maximum number of lags to be used & k is the number of variables in the equation.

The ARDL specification of short-run dynamics is investigated using the ECM version of the ARDL model of the following:

$$\Delta \text{GDP}_{\text{it}} = \alpha_0 + \sum_{j=0}^{p} \varphi_j \Delta EC_{it-j} + \sum_{j=0}^{p} \varnothing_j \Delta DMC_{it-j} + \sum_{j=0}^{p} \gamma_j \Delta CFA_{it-j} + \sum_{j=0}^{p} \vartheta_j \Delta TO_{it-j} + \rho ECT + \varepsilon_{it} \quad \text{(vii)}$$

$$\Delta \text{CO}_{2_{\text{it}}} = \alpha_0 + \sum_{j=0}^{p} \varphi_j \Delta RE_{it-j} + \sum_{j=0}^{p} \varnothing_j \Delta CFA_{it-j} + \rho ECT + \varepsilon_{it} \quad \text{(viii)}$$

ECT is an Error Component Term.

Results and Discussion

Figure 6. Geographical Analysis of energy consumption, Carbon dioxide emissions, renewable energy use and domestic material consumption

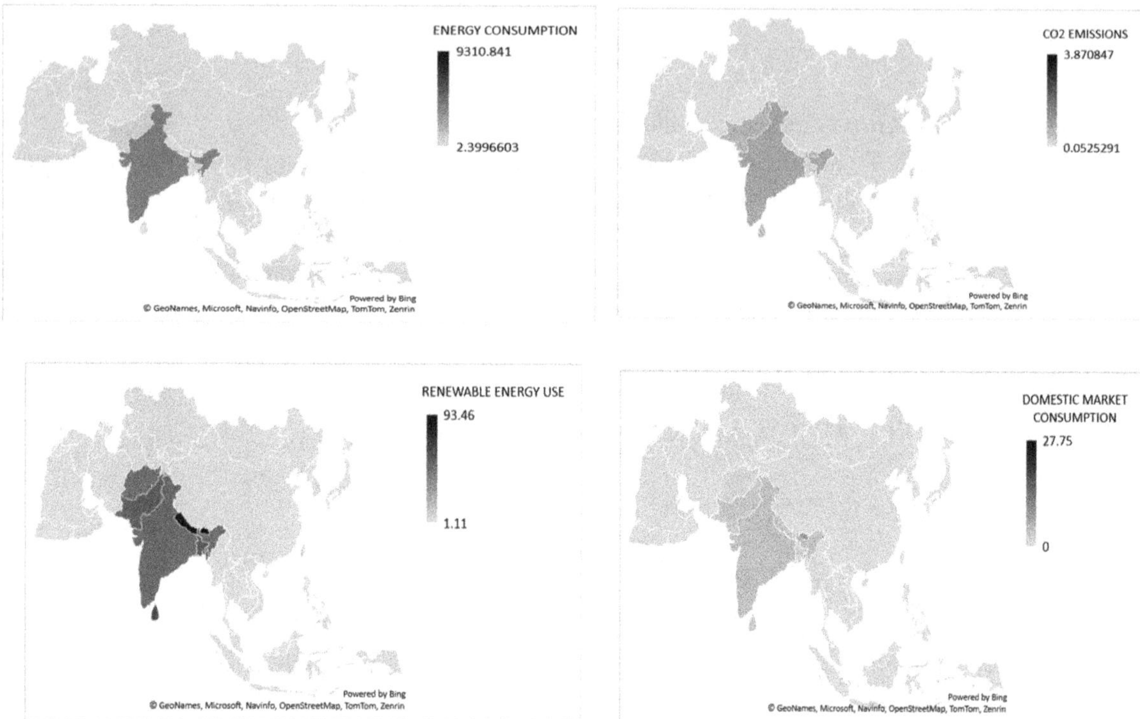

The study of Energy Consumption shows the lower consumption for Bhutan and Nepal at 5.49 and 15.52 and India and Pakistan have high consumption level at 3743.29 and 495.15. The energy consumption for Sri Lanka is balanced at 52.03. The representation for Carbon dioxide emissions shows the India has the highest emissions at 0.91843957 due to its highest population and industrialization whereas, Pakistan also shows the same effect. Bangladesh and Nepal have less carbon emissions at 0.23570 and 0.12967 respectively. Sri Lanka has the lowest emissions because of the less population.The study analyzes the adoption of Renewable Energy shows Bhutan and Nepal have high adoption rate at 91.75% and 87.76% whereas, India has the highest adoption of clean energy at 47.1%, followed by Sri Lanka at 62.9%. Afghanistan, Bangladesh and Pakistan have low adoption rate due to the ineffective policies and less investment. The graphical study of Domestic Material Consumption shows the Bhutan has the highest rate of the consumption at 8.32, followed by India at 3.69 whereas, Pakistan and Sri Lanka have a consumption rate at 2.78 and 3.09 respectively. Afghanistan has the lowest domestic consumption rate at 1.08 depicts Afghanistan as an undeveloped economy.

Figure 7. Geographical Analysis of Trade openness, GDP and Share of population with access to clean fuels

The geographical study for Trade as a share of GDP (Trade openness) in South Asian countries shows Sri Lanka has the highest ratio at 6016.4 followed by Bhutan, Pakistan and Bangladesh because of the dependency on international trade. Nepal and India have a balanced dependence at 2084.2 and 2646.87 respectively. Afghanistan lacks behind with a trade share of 0 due to lack of efficiency in production. The graphical study of the Gross Domestic Product in South Asian countries shows that y Bangladesh's GDP have significant growth in economy at 3.0776 followed by India's GDP at 2.8559. On the other hand, Sri Lanka's GDP is at 1.9378 depicting insignificant growth in the economy.The representation for Share of the Population with Access to Clean Fuels for Cooking in South Asian countries represent Bhutan is leading at 30.9%, followed by India and Pakistan at 23.4% and 24.4% respectively whereas, Nepal has low percentage at 6.9% outlines the need to improve the access of clean fuels while Sri Lanka plays a moderate role. Access to Clean Fuels for Cooking can be attained by fulfilling Sustainable Development Goals.

EMPIRICAL ANALYSIS

We first carry out the Unit Root Test to examine the stationarity of variables. For that, we applied both ADF & Philips Perron tests and their results are summarized in the following table

Table 4. Unit Root Analysis

Variable	Level				First Difference			
	ADF-Fisher Chi square		PP-Fisher Chi-Square		ADF-Fisher Chi square		PP-Fisher Chi-Square	
	Statistics	P-value	Statistics	P-value	Statistics	P-value	Statistics	P-value
Co_2 emission	4.94549	0.9960	1.99943	1.0000	42.7132	0.0003	93.1587	0.0000
Domestic Material Consumption	18.2312	0.3105	18.5759	0.2913	39.6008	0.0009	74.0190	0.0000
Energy Consumption	2.63169	0.9999	0.92641	1.0000	44.6494	0.0002	78.2872	0.0000
GDP	11.0315	0.8075	0.59438	1.0000	29.6125	0.0201	49.4470	0.0000
Renewable Energy Use	32.7962	0.0079	54.2693	0.0000	51.3282	0.0000	89.1333	0.0000
Share of the population with access to clean fuels	8.32668	0.9385	0.03571	1.0000	25.9254	0.0551	22.2668	0.1348
Trade as share of GDP	8.03049	0.9480	1.40117	1.0000	45.7017	0.0001	49.42235	0.0000

Author's Estimation Software: E-views12

The unit-root test results for above mentioned variables manifests that CO_2 emissions, domestic material consumption, energy consumption, GDP, renewable energy use and trade as a share of GDP, for the both the Augmented Dickey-Fuller (ADF) and Phillips-Perron (PP) tests demonstrates the presence of unit roots at the level, suggesting test statistics are significant. Moreover, their first differences reveal that results are not statistically significant, indicating no evidence of unit root. Addition, to this the share of the population with access to clean fuels, shows the presence of unit root at first difference but only for ADF-Fisher Chi square. Overall, these results highlights that most of the variables require differencing to achieve stationarity i.e., I(1) for further analysis.

CROSS-SECTION DEPENDENCE:

The follow up of the cross-section dependence tests indicates whether presence of CSD in the data or not. Cross-sectional dependence refers to the situation where the error terms of the observations are not independent. In such a case, traditional panel data analysis techniques may not be appropriate because the assumptions underlying these techniques are violated.

$$H_0 = No\ Cross - Section\ Dependence\ among\ variables.$$
$$H_1 = Cross - Section\ Dependence\ among\ variables.$$

Table 5. **Cross-Section Dependence Test**

	Breusch-Pagan LM		Pesaran Scaled LM		Bias-Correlated Scaled LM		Pesaran CD	
	Statistics	**P-value**	**Statistics**	**P-value**	**Statistics**	**P-value**	**Statistics**	**P-value**
Co_2 emission	381.7174	0.0000	47.26747	0.0000	47.05694	0.0000	19.31966	0.0000
Domestic Material Consumption	301.2987	0.0000	36.52108	0.0000	36.31055	0.0000	16.99007	0.0000
Energy Consumption	428.7636	0.0000	53.55429	0.0000	53.34376	0.0000	20.56394	0.0000
GDP	519.4434	0.0000	65.67188	0.0000	65.46136	0.0000	22.78024	0.0000
Renewable Energy Use	337.1507	0.0000	41.31200	0.0000	41.10148	0.0000	18.10911	0.0000
Share of the population with access to clean fuels	502.3745	0.0000	63.39097	0.0000	63.18044	0.0000	22.36971	0.0000
Trade as share of GDP	427.5105	0.0000	53.38683	0.0000	53.17630	0.0000	20.45747	0.0000

Author's Estimation Software: E-views12

The Cross-sectional dependence tests were used to examine the interdependence among variables by using GDP per capita and carbon dioxide emissions as a dependent variable. The statistical values for Breusch-Pagan LM, Pesaran Scaled LM, Bias-Correlated Scaled LM, and Pesaran CD all show significant results with a probability value of 0.0000 depicted greatest evidence of cross-sectional dependence.

KAO-COINTEGRATION:

Cointegration among a range of variables means that there is a long-run relationship between them and shows a statistical concept which implies that two or more time series are linked by a long-term relationship, even if they may not be directly correlated in the short run.

$$H0 = There\ is\ no\ Existence\ of\ Long - Run\ Cointegration.$$
$$H1 = There\ is\ Existence\ of\ Long - Run\ Cointegration$$

Table 6. Kao-cointegration test for GDP as dependent variable

ADF	**t-statistics** **-3.292373**	**Prob** **0.0005**
Residual variance	2.04E+21	
HAC variance	2.99E+21	

Author's Estimation Software: E-views12

The Cointegration test for GDP per capita indicates the t-statistics with a p-value of 0.0005 depicts a positive correlation between the variables. Furthermore, the ADF test having value of -3.292373, implies stationarity between variables. Overall, the given model suggests that there will be long term relation between variables i.e., GDP per capita and its determinants.

Table 7. Kao-cointegration test for carbon dioxide emissions as dependent variable

ADF	t-statistics -1.665632	Prob 0.0479
Residual variance	0.016489	
HAC variance	0.011208	

Author's Estimation Software: E-views12

The Cointegration test for Carbon-dioxide emissions states the ADF's t-statistics value is -1.665632 with a probability value of 0.0479 indicating a significant and long-term relationship between the variables.

PANEL ARDL:

Panel ARDL, a statistical technique which is generally used in econometrics in order to determine long term connectivity between variables of panel data. In conclusion allows to examine an interlinkage between the both short- and long-term analysis.

$H0 = No\ long -$ term linkage $between\ Dependent\ \&\ Independent\ variables.$
$H1 = Long -$ term linkage $between\ Dependent\ \&\ Independent\ variable$

Table 8. Panel ardl test for gdp as a dependent variable

Variables	Coefficient	Std. Error	t-Statistics	Prob. value
Long Run Equation				
ENERGY_CONSUMPTION	2.02E+08	6926593	2.916998	0.0043
DOMESTIC_MATERIAL_CONSUMPTION	-1.11E+10	1.83E+09	-6.063551	0.0000
SHARE_OF_THE_POPULATION_WITH...	1.29E+09	1.16E+08	11.10782	0.0000
TRADE_AS_A_SHARE_OF_GDP...	20863594	2206103	9.457216	0.0000

Short Run Equation				
COINTEQ01 D(ENERGY_CONSUMPTION) D(DOMESTIC_MATERIAL_CONSUMPT... D(SHARE_OF_THE_POPULATION_WI... D(TRADE_AS_A_SHARE_OF_GDP... C	-0.016159 -10003553 2.69E+09 8.79E+08 2.11E+08 -1.41E+09	0..047346 16675354 1.91E+09 1.47E+09 1.57E+08 1.62E+09	-0.341291 -0.599900 1.411104 0.596891 1.347148 -0.872042	0.7335 0.5498 0.1611 0.5518 0.1808 0.3851
Root MSE	2.44E+09	Mean dependent var		4.75E+10
S.D dependent var	1.31E+11	S.E. of regression		2.97E+09
Akaike info criterion	41.05677	Sum of squared resid		9.52E+20
Schwarz criteria	42.05620	Log-likelihood		-3232.542
Hannan- Quinn criter.	41.46261			

*Note: p-values and any subsequent tests do not account for model selection.

Author's Estimation Software: E-views12

For both long run and short run, the results of ARDL shows that GDP per capita is sharply associated with its explanatory variables. The results from the long run estimation manifest significant and positive relation coefficients, implying positive influence over the GDP per capita except the domestic material consumption having negative t-statistics (-1.11E+10) demonstrating unfavorable effect over GDP per capita. Moreover, in short run all of the explanatory variables don't have significant impact over the GDP per capita. Additionally, the coefficients also implies that these variables are not statistically significant which shows that the effect of these variables over the GDP per capita is almost near to the negligible. Overall, the analysis from the concerned specification signifies that in long run variables like energy consumption, access to the clean fuels along with trade openness plays an important role during the determination of GDP per capita, while in short run these factors don't possess any kind of significant impact.

Table 9. Panel ARDL test for carbon dioxide emissions as dependent variable

Variables	Coefficient	Std. Error	t-Statistics	Prob. value
Long Run Equation				
RENEWABLE_ENERGY_USE SHARE_OF_THE_POPULATION_WITH...	-0.004040 0.010954	0.000876 0.001154	-4.611304 9.490496	0.0000 0.0000
Short Run Equation				
COINTEQ01 D (CO_2_EMISSION (-1)) D(RENEWABLE_ENERGY_USE) D(RENEWABLE_ENERGY_USE (-1)) D(SHARE_OF_THE_POPULATION_W... D(SHARE_OF_THE_POPULATION_W... C	-0.717957 0.235833 -0.208310 -0.082988 0.059920 -0.036794 0.375064	0.435966 0.248204 0.177246 0.077364 0.041311 0.063743 0.188909	-1.646820 0.950160 -1.175258 -1.072686 1.450488 -0.577230 1.985418	0.1027 0.3443 0.2426 0.2859 0.1500 0.5651 0.0498
Root MSE	0.052114	Mean dependent var		0.027996
S.D dependent var	0.136534	S.E. of regression		0.065269

Short Run Equation			
Akaike info criterion	-3.134310	Sum of squared resid	0.434531
Schwarz criteria	-2.019560	Log-likelihood	308.7448
Hannan- Quinn criter.	-2.6681649		

*Note: p-values and any subsequent tests do not account for model selection.

Author's Estimation Software: E-views12

From the results of panel ARDL model, the long run equation shows that the use of renewable energy is of negative coefficient albeit statistically significant suggesting negative but significant relationship with the carbon dioxide emission. Opposite to this, access to the clean fuel is positive and statistically significant to the emissions. However, in short run the changes in emission influenced by the lagged of CO_2 emissions, change in use of renewable energy along with the access to clean fuel. Moreover, only the coefficient of lagged CO_2 emissions is significant. Overall, the analysis from the model suggests that the use of energy plays a significant role in the reduction of CO_2 emissions during the long run.

CONCLUSION

With GDP per capita as the dependent, the results show a substantial and positive correlation between GDP per capita and its determinant listed under the specification, which is further supported by its p-value (0.0005). In addition, the ADF test indicates that the variables listed in the concept of a long-term relationship are stationary. It was also discovered during analysis that ARDL analysis shows a strong correlation between specifications in the short and long term. In the long term, for example, factors like energy consumption, access to clean fuels, and trade openness show significant and positive coefficients, suggesting a consistent influence on GDP per capita. Furthermore, negative t-statistics for domestic material consumption point to a negative impact on GDP per capita.

Analysis concludes that the factors that influence carbon dioxide emissions are significantly linked to each other. Co-integration emphasizes the existence of a substantial and enduring relationship between variables in this regard. This association is further supported by Panel ARDL analysis, but the utilization of renewable energy shows a negative relationship with emissions over the long term that is statistically significant. Additionally, Furthermore, by offering important coefficients, the lagged value of CO_2 emission also affects the short-term variations in CO2 emission emission. The overall model illustrates the long-term emission reduction benefits of using renewable energy.

SIGNIFICANCE AND NOVELTY OF THE CHAPTER

This chapter on relationship between renewable energy, economic growth and environmental sustainability in South Asian nations is significant in terms of the Sustainable Development Goals (SDGs). In order to reach SDG targets, South Asian nations must understand this relationship. This study has a sense of novelty. First of all it focus on the important issues in the South Asian region, especially development, climate change and sustainable growth. Secondly, as we know that SDG 7 is closely related to renewable energy and it emphasises that everyone has access to clean and affordable energy, so this study is important in South Asian context as South Asia has issues with fossil fuel dependence and energy availability. Making the transition to renewable energy is essential for its sustainability in the future due to its rapidly expanding population and rising energy consumption. Thirdly, in South Asia, where millions of people live in poverty, economic growth is crucial to reducing poverty. Putting money into renewable energy sources (including solar, wind, and hydro) can help create jobs in the green energy industry, promoting economic growth

POLICY SUGGESTION

The policy suggestions for **GDP per capita** as a dependent variable in this study:

- **Green Energy Investment:** Investment in clean energy can accelerate the economy and the shift from the non-renewable energy to renewable energy remarks a significant change for the economy and the ecological system.
- **Tax on Carbon**: Levied taxes on the carbon emissions would be a significant change in the society and it is also motivating the private organizations to invest in the clean energies which leads to achieve the positive growth in the environment. The revenue yield from the excise would help the government to fund the investment related to clean energy and also increases economic efficiency.
- **Environmental Jobs**: These jobs enhance the sustainability in the environment and decreases the issue related to unemployment. Green jobs provide the solution for decreasing the poor effect of emissions from the environment.
- **Private-Public Partnership:** The partnership between government and private organizations signifies a positive impact on society. It increases the job opportunities where government works on idea and give directions and private organizations utilize their funds and resource to work on a certain project.

The policy suggestions for **Carbon Emissions** as a dependent variable:

- **Public Awareness**: Government should arrange campaigns on daily basis to enhance the functioning of clean energies. These campaigns would educate the general public by making them understand the negative effect of using non-renewable energy and positive effect of renewable energy.
- **Promote Clean Energy**: South Asian Countries should always need to welcome those companies that supports renewable source of energy. As various initiatives are taken by these companies is helpful to reduce the impact of carbon emissions from the environment.

REFERENCES

Abolhosseini, S., & Heshmati, A. (2014). The main support mechanisms to finance renewable energy development. *Renewable & Sustainable Energy Reviews*, *40*, 876–885. DOI: 10.1016/j.rser.2014.08.013

Ahmed, M., & Shimada, K. (2019). The effect of renewable energy consumption on sustainable economic development: Evidence from emerging and developing economies. *Energies*, *12*(15), 2954. DOI: 10.3390/en12152954

Ajide, K. B., & Mesagan, E. P. (2022). Heterogeneous analysis of pollution abatement via renewable and non-renewable energy: Lessons from investment in G20 nations. *Environmental Science and Pollution Research International*, *29*(24), 36533–36546. DOI: 10.1007/s11356-022-18771-5 PMID: 35066851

Alharbi, S. S., Al Mamun, M., Boubaker, S., & Rizvi, S. K. A. (2023). Green finance and renewable energy: A worldwide evidence. *Energy Economics*, *118*, 106499. DOI: 10.1016/j.eneco.2022.106499

Appiah-Otoo, I., Chen, X., & Kursah, M. B. (2023). Modelling the impact of renewable energy investment on global carbon dioxide emissions. *Energy Reports*, *10*, 3787–3799. DOI: 10.1016/j.egyr.2023.10.043

Azhgaliyeva, D., Beirne, J., & Mishra, R. (2023). What matters for private investment in renewable energy? *Climate Policy*, *23*(1), 71–87. DOI: 10.1080/14693062.2022.2069664

Cedrick, B. Z. E., & Long, P. W. (2017). Investment motivation in renewable energy: A PPP approach. *Energy Procedia*, *115*, 229–238.

Chen, Y., & Zhang, X. (2023). Does financial globalization promote renewable energy investment? Empirical insights from China. *Environmental Science and Pollution Research International*, *30*(45), 101366–101378. DOI: 10.1007/s11356-023-29293-z PMID: 37651014

Dey, S., Sreenivasulu, A., Veerendra, G. T. N., Rao, K. V., & Babu, P. S. S. A. (2022). Renewable energy present status and future potentials in India: An overview. *Innovation and Green Development*, *1*(1), 100006. DOI: 10.1016/j.igd.2022.100006

He, X. (2023). Effects of the green policy environment on renewable energy investment and effect evaluation of green policies. *Discrete Dynamics in Nature and Society*, *2023*, 1–16. DOI: 10.1155/2023/8698548

Iheanacho, E. (2016). The impact of financial development on economic growth in nigeria: An ardl analysis. *Economies*, *4*(4), 26. DOI: 10.3390/economies4040026

Kazemzadeh, E., Fuinhas, J. A., Koengkan, M., Osmani, F., & Silva, N. (2022). Do energy efficiency and export quality affect the ecological footprint in emerging countries? A two-step approach using the SBM–DEA model and panel quantile regression. *Environment Systems & Decisions*, *42*(4), 608–625. DOI: 10.1007/s10669-022-09846-2

Kazemzadeh, E., Fuinhas, J. A., Salehnia, N., Koengkan, M., Shirazi, M., & Osmani, F. (2022). Factors driving CO2 emissions: The role of energy transition and brain drain. *Environment, Development and Sustainability*, *26*(1), 1673–1700. DOI: 10.1007/s10668-022-02780-y

Khan, M. B., Saleem, H., Shabbir, M. S., & Huobao, X. (2022). The effects of globalization, energy consumption and economic growth on carbon dioxide emissions in South Asian countries. *Energy & Environment*, *33*(1), 107–134. DOI: 10.1177/0958305X20986896

Mehmood, U., Tariq, S., Ul Haq, Z., Aslam, M. U., & Imran, A. (2023). How economic growth contributes to co2 emissions in the presence of globalization and eco-innovations in south asian countries? *WORLD (Oakland, Calif.)*, *4*(1), 202–213. DOI: 10.3390/world4010014

Mehrara, M. (2011a). Health expenditure and economic growth: An ardl approach for the case of iran. *Journal of Economics and Behavioral Studies*, *3*(4), 249–256. DOI: 10.22610/jebs.v3i4.277

Mehrara, M. (2011b). Health expenditure and economic growth: An ardl approach for the case of iran. *Journal of Economics and Behavioral Studies*, *3*(4), 249–256. DOI: 10.22610/jebs.v3i4.277

. Meyer, C. S. (2024, April 7). *The impact of technological innovations and economic growth on carbon dioxide emissions • free network*. FREE NETWORK.

Mngumi, F., Shaorong, S., Shair, F., & Waqas, M. (2022). Does green finance mitigate the effects of climate variability: Role of renewable energy investment and infrastructure? *Environmental Science and Pollution Research International*, *29*(39), 59287–59299. DOI: 10.1007/s11356-022-19839-y PMID: 35386082

. Ogunsola, A. (2016). *The impact of human capital development on economic growth in nigeria: Ardl approach.*

Panel®, E. (n.d.). *Council post: 15 'green' tech innovations and practices that can help protect the environment.* Forbes.

Peimani, H. (2018). *Financial barriers to development of renewable and green energy projects in Asia* (Working Paper 862). ADBI Working Paper.

Raghutla, C., & Kolati, Y. (2023). Public-private partnerships investment in energy as new determinant of renewable energy: The role of political cooperation in China and India. *Energy Reports*, *10*, 3092–3101. DOI: 10.1016/j.egyr.2023.09.139

Rasoulinezhad, E., & Taghizadeh-Hesary, F. (2022). Role of green finance in improving energy efficiency and renewable energy development. *Energy Efficiency*, *15*(2), 14. DOI: 10.1007/s12053-022-10021-4 PMID: 35529528

Ren, H., Qifeng, X., Yongjun, G., Yunbao, X., Gaole, H., & Sohail, M. T. (2023). Country risk and its impact on renewable energy investment in highly polluted economies. *Environmental Science and Pollution Research International*, *30*(25), 67699–67707. DOI: 10.1007/s11356-023-27062-6 PMID: 37115445

Rodrigues, J., Domingos, T., Conceição, P., & Belbute, J. (2005). Constraints on dematerialisation and allocation of natural capital along a sustainable growth path. *Ecological Economics*, *54*(4), 382–396. DOI: 10.1016/j.ecolecon.2004.07.019

Saqib, N., Usman, M., Ozturk, I., & Sharif, A. (2024). Harnessing the synergistic impacts of environmental innovations, financial development, green growth, and ecological footprint through the lens of SDGs policies for countries exhibiting high ecological footprints. *Energy Policy*, *184*, 113863. DOI: 10.1016/j.enpol.2023.113863

Soundarrajan, P., & Vivek, N. (2016). Green finance for sustainable green economic growth in India. [Zemědělská Ekonomika]. *Agricultural Economics*, *62*(1), 35–44.

Sun, Y., Bao, Q., & Taghizadeh-Hesary, F. (2023). Green finance, renewable energy development, and climate change: Evidence from regions of China. *Humanities & Social Sciences Communications*, *10*(1), 107. DOI: 10.1057/s41599-023-01595-0 PMID: 36938579

United Nations Development Programme (UNDP) (Ed.). (1993). *Human development report 1993 / united nations development programme(Undp)*. Oxford University Press.

Wan, Q., Qian, J., Baghirli, A., & Aghayev, A. (2022). Green finance and carbon reduction: Implications for green recovery. *Economic Analysis and Policy*, *76*, 901–913. DOI: 10.1016/j.eap.2022.09.022

Ye, J., Al-Fadly, A., Huy, P. Q., Ngo, T. Q., Hung, D. D. P., & Tien, N. H. (2022). The nexus among green financial development and renewable energy: Investment in the wake of the Covid-19 pandemic. *Ekonomska Istrazivanja*, *35*(1), 5650–5675. DOI: 10.1080/1331677X.2022.2035241

Zhu, Z., Liu, B., Yu, Z., & Cao, J. (2022). Effects of the digital economy on carbon emissions: Evidence from china. *International Journal of Environmental Research and Public Health*, *19*(15), 9450. DOI: 10.3390/ijerph19159450 PMID: 35954806

Chapter 14

Machine Learning Systems to Optimize Teaching of Sustainable Development and Renewable Energy

Sara Ouald Chaib
https://orcid.org/0009-0002-0528-5036
University Abdelmalek Essaadi, Morocco

Fouad Muheya
https://orcid.org/0009-0003-5761-8485
Sidi Mohamed Ben Abdellah University, Morocco

Samira Khoulji
University Abdelmalek Essaadi, Morocco

ABSTRACT

The integration of renewable energy into sustainable development strategies is a global challenge. Teaching these concepts requires innovative pedagogical approaches that combine educational effectiveness with technical complexity. This paper explores the use of machine learning (ML) systems to enhance the teaching of renewable energy and sustainable development. We propose a pedagogical architecture using adaptive learning algorithms to analyze student performance and interactions, personalizing learning paths. The system adjusts teaching modules in real-time, providing individualized support. We also implement predictive models simulating the impact of renewable energy on ecosystems and infrastructure through interactive simulations, helping students understand sustainability challenges. Preliminary results show improved knowledge retention and better understanding of energy transition concepts. We discuss the pedagogical implications of ML systems in higher education and suggest directions for their large-scale implementation

DOI: 10.4018/979-8-3693-9924-8.ch014

INTRODUCTION

Sustainable development and renewable energies are fundamental pillars for addressing the environmental, social, and economic challenges of the 21st century. These fields require a deep understanding of systemic issues and the ability to develop innovative solutions to ensure the energy transition while preserving ecosystem balance (Bin Abu Sofian et al., 2024). However, teaching these topics remains complex, as it necessitates covering multidisciplinary concepts ranging from the physical and technological principles of energy systems to their societal and ecological impacts (García-González et al., 2020).

Traditional education often struggles to adapt to the diversity of student profiles and the rapid evolution of scientific and technological knowledge (Gligorea et al., 2023). In this context, machine learning (ML) systems emerge as promising tools to transform pedagogical approaches by offering adaptive and interactive solutions (Hu, 2024). ML, which relies on the collection and analysis of large volumes of data, enhances the personalization of educational pathways based on the individual needs of learners while enriching learning experiences through immersive simulations and precise predictions (Chaib et al., 2023).

The global energy transition demands a shift toward renewable energies such as solar, wind, and hydropower. While these energy sources offer ecological benefits, they also require technical expertise and systemic understanding to optimize their use (Kiasari et al., 2024). However, the teaching of renewable energies, coupled with sustainable development, often remains fragmented and fails to fully engage students or develop their ability to solve complex problems. Additionally, current educational tools often lack the flexibility to adapt to the specific needs of each student, which can result in a superficial understanding of the concepts taught (Park & Jung, 2023).

The integration of machine learning technologies presents a unique opportunity to address these challenges. Adaptive algorithms can analyze student evaluation data and interactions with educational content to adjust teaching modules in real-time (Hu, 2024). Furthermore, predictive models enable the simulation of scenarios related to renewable energies, making learning experiences more concrete and engaging (Bin Abu Sofian et al., 2024). These approaches could revolutionize how sustainable development concepts are taught, making learning more personalized, immersive, and effective.

The primary goal of this article is to explore the application of machine learning systems to improve the teaching of renewable energies and sustainable development. Specifically, we aim to:

- Develop a pedagogical architecture based on adaptive learning algorithms to personalize educational pathways.
- Implement interactive simulations using predictive models to enhance the understanding of renewable energy impacts.
- Evaluate the preliminary results of these approaches in terms of improved knowledge retention and student engagement.

In the following sections, we first present a state-of-the-art review on the use of machine learning systems in education, followed by a detailed description of the proposed pedagogical architecture. We then discuss the methodologies employed for the collection and analysis of educational data as well as for the design of interactive simulations. Finally, we present and analyze the preliminary results before concluding with recommendations and perspectives for large-scale deployment.

STATE-OF-THE-ART REVIEW

The integration of machine learning (ML) into educational systems has seen significant advancements in recent years, driven by the exponential growth of data and computing power. ML, a subset of artificial intelligence (AI), involves the design and use of algorithms that learn patterns from data to make predictions or decisions. In the context of education, these algorithms are increasingly leveraged to personalize learning experiences, enhance engagement, and improve educational outcomes (Kabudi et al., 2021).

One of the most transformative uses of ML in education is adaptive learning systems. These systems dynamically adjust instructional content based on the individual progress and needs of learners. Recent research highlights the potential of AI-enhanced adaptive systems in fostering personalized educational pathways and improving student performance (Gyonyoru & Katona, 2024). Such approaches have been shown to significantly enhance learning efficiency by addressing specific areas where students face challenges.

Predictive analytics is another notable application. Predictive models are employed to forecast student performance, dropout risks, and learning trajectories. By analyzing historical and real-time data, these models enable educators to intervene proactively and provide targeted support, enhancing retention and success rates (Almalawi et al., 2024; Namoun & Alshanqiti, 2020).

Interactive simulations powered by ML have become key tools for teaching complex concepts in science, technology, engineering, and mathematics (STEM). These simulations help students visualize and manipulate variables in renewable energy systems, such as solar panel efficiency or wind turbine placement, making abstract concepts tangible while fostering critical thinking and problem-solving skills (Zhai & Krajcik, 2024; Fan et al., 2023).

Gamification, often integrated with ML, leverages reinforcement learning algorithms to adapt challenges and rewards to maintain optimal engagement levels. Adaptive gamification in e-learning has been effective in improving motivation and knowledge retention, particularly in technical fields like renewable energy (Bennani et al., 2022; Chaib et al., 2024).

The intersection of ML and sustainable development education is a rapidly growing area of research. AI models developed to simulate environmental and economic impacts of renewable energy scenarios allow students to experiment with sustainable solutions in risk-free environments (Fan et al., 2023). These tools provide a holistic view of the challenges and trade-offs involved in transitioning to renewable energy systems (Ahmad et al., 2021).

Natural language processing (NLP) techniques are also employed to analyze textual data, such as student feedback and research papers, identifying trends and gaps in sustainable development education. Sentiment analysis, for instance, helps educators gauge student attitudes toward sustainability topics and tailor curricula accordingly (Ouald Chaib et al., 2022).

Despite its potential, the integration of ML in education faces several challenges. Data privacy and ethical considerations remain paramount, particularly in ensuring that student data is collected, stored, and used responsibly (Farooqi et al., 2024). Addressing these concerns is critical to maintaining trust and compliance with regulations such as GDPR.

Accessibility is another limitation, particularly in resource-constrained settings. While these technologies hold promise for personalized education, their implementation often requires significant financial and technical resources, which may not be readily available in all institutions (Nookala, 2024).

Future research in this field is likely to focus on enhancing the scalability and affordability of ML-driven educational tools. Emerging technologies, such as federated learning, offer solutions to data privacy concerns by enabling decentralized data processing. Additionally, advancements in explainable AI (XAI) aim to make ML algorithms more transparent and interpretable for educators and students (Rachha & Seyam, 2023).

The integration of ML with cutting-edge technologies such as augmented reality (AR) and virtual reality (VR) also presents exciting possibilities for immersive and interactive learning experiences. AR-based simulations, for example, could allow students to explore renewable energy installations in virtual settings, bridging the gap between theory and practice (Alqallaf & Ghannam, 2024).

As the potential of machine learning (ML) in education continues to expand, it is essential to bridge the gap between theoretical insights and practical implementations. While existing studies have demonstrated the efficacy of adaptive algorithms, predictive analytics, and interactive simulations, there remains a pressing need to design comprehensive frameworks that integrate these innovations into cohesive pedagogical architectures. This requires a systematic approach to data collection, analysis, and the deployment of ML-driven educational tools tailored to the complexities of renewable energy and sustainable development education.

In the next section, we present the methodology for developing and implementing such a pedagogical architecture. This framework leverages the insights from the state-of-the-art to create a dynamic, adaptive, and interactive learning environment that addresses the diverse needs of students while fostering a deeper understanding of the multidisciplinary challenges associated with sustainability and renewable energy.

METHODOLOGY

The proposed pedagogical architecture is designed to leverage the power of machine learning (ML) in transforming the teaching of renewable energy and sustainable development. It focuses on creating a dynamic, adaptive, and interactive educational framework that addresses the diverse learning needs of students while enhancing engagement and retention. The architecture is structured to integrate real-time data analysis, personalized learning paths, and immersive simulations, fostering an effective and holistic learning experience (Sajja et al., 2024; Vassoyan et al., 2023).

At the foundation of this architecture lies a robust data collection and analysis process. Data is gathered from various sources, including student interactions, performance metrics, and qualitative feedback. These inputs, which capture a comprehensive picture of each learner's engagement and challenges, are preprocessed using advanced techniques such as normalization and feature extraction. This step ensures that the data is clean, relevant, and ready for analysis (Sari et al., 2024). Machine learning models are then employed to identify patterns in learning behaviors, enabling the system to recognize trends, cluster students with similar challenges, and predict potential obstacles (Strielkowski et al., 2024).

Building on this data, the architecture employs adaptive algorithms to personalize the learning experience for each student. These algorithms dynamically adjust content delivery based on individual progress, ensuring that learners focus on areas where they encounter difficulties. For example, a student struggling with solar energy principles might receive additional resources or targeted exercises to address those gaps. The system also uses real-time Bayesian optimization to fine-tune the complexity of quizzes and simulations, offering an optimal balance between challenge and accessibility (Vassoyan et al., 2023). Furthermore, a dynamic knowledge graph is maintained for each student, mapping out the relationships between concepts and identifying areas that need reinforcement before advancing to more complex topics (Sari et al., 2024).

To enrich the learning process, the architecture integrates interactive simulations and feedback mechanisms. Predictive models simulate real-world scenarios, allowing students to manipulate variables like wind turbine placement or solar panel angles and observe the immediate and long-term impacts on energy efficiency and sustainability. These simulations are designed to make abstract concepts tangible, fostering a deeper understanding of the material. Moreover, the incorporation of gamification elements, such as achieving milestones or earning rewards, enhances motivation and engagement, particularly in technically demanding subjects like renewable energy systems (Oladapo et al., 2024).

Feedback plays a critical role in the system, operating on both immediate and long-term timescales. Students receive real-time feedback during simulations, helping them understand the outcomes of their decisions and guiding them toward more effective strategies. Meanwhile, the system generates detailed summaries of performance trends and learning progress, enabling educators to refine instructional materials and strategies (Strielkowski et al., 2024). These feedback loops ensure that the system continually evolves to meet the changing needs of students.

The architecture operates seamlessly through a series of interconnected layers. Students interact with the platform, completing activities and providing data that feed into the system's preprocessing and analysis engine. Adaptive learning algorithms then use this data to adjust educational pathways, while the simulation layer delivers interactive, gamified learning experiences. The outputs—personalized content, adjusted simulations, and comprehensive feedback—are delivered back to students, creating a continuous cycle of improvement and engagement.

This framework, by combining data-driven insights with adaptive and interactive technologies, represents a significant advancement in the teaching of renewable energy and sustainable development. It not only personalizes the educational journey but also equips students with the tools and knowledge to address complex, real-world challenges (Sari et al., 2024; Oladapo et al., 2024).

Figure 1. proposed pedagogical architecture for teaching renewable energy and sustainability.

DATA COLLECTION AND ANALYSIS METHODS

The success of the proposed pedagogical architecture depends on a structured process of data collection and analysis. This ensures that the system has sufficient and reliable data to adapt to students' needs effectively and model real-world renewable energy scenarios accurately.

1. Collection of Student-Centric Data

Three main types of student data are collected within the system:

- **Interaction Data:** Includes navigation logs, time spent on resources, simulation usage, and quiz attempts. This data captures how students engage with the platform and identifies patterns in their learning behavior.
- **Performance Data:** Derived from student assessments, including quizzes, problem-solving activities, and scenario-based exercises. It provides a quantitative evaluation of their understanding and knowledge progression.
- **Feedback Data:** Collected through surveys and open-ended comments, enabling students to express their views on the content and overall learning experience.

This data is recorded in real time and stored securely in the system's centralized database for further processing.

2. Integration of Domain-Specific Datasets

To ensure alignment with the context of renewable energy and sustainability, domain-specific data is incorporated:

- **Renewable Energy Data:** Includes parameters such as solar irradiance, wind speed patterns, and energy production statistics sourced from credible databases like the International Renewable Energy Agency (IRENA).
- **Environmental Metrics:** Includes datasets on carbon emissions, ecological footprints, and energy lifecycle impacts.
- **Educational Resources:** Text-based and multimedia content are curated to create a robust knowledge graph that aids in delivering personalized recommendations.

3. Data Preprocessing and Cleaning

Data preprocessing steps include removing duplicates, imputing missing values, and normalizing features to ensure consistency. For textual feedback, natural language processing techniques are applied to analyze sentiment and identify key topics of interest.

4. Analytical Techniques

Advanced machine learning methods are used to analyze the data:

- **Clustering Algorithms:** To identify learning styles and group similar student profiles.
- **Predictive Models:** For forecasting performance trends and identifying at-risk students.
- **Knowledge Graphs:** Used to represent relationships between key concepts in renewable energy and sustainability, aiding in intelligent content recommendation.

5. Privacy and Ethical Considerations

To ensure compliance with data protection regulations, such as the General Data Protection Regulation (GDPR), all collected data is anonymized, and only aggregated results are used for analysis. The GDPR, a legal framework established by the European Union, ensures the protection of personal data and privacy for individuals. Adhering to these principles, the system employs rigorous anonymization techniques to prevent the identification of individual users and safeguards all data storage and analysis processes. Ethical guidelines are strictly followed to maintain trust and data security.

These methods form the backbone of the system's ability to personalize learning, simulate realistic energy scenarios, and provide targeted interventions, thereby fostering a deep understanding of sustainable energy systems among learners.

IMPLEMENTATION AND RESULTS

The implementation of machine learning (ML)-based solutions to optimize the teaching of renewable energies and sustainable development followed a systematic, iterative process. This methodology ensured robust integration of ML models into the pedagogical architecture, delivering accurate predictions, personalized learning pathways, and interactive simulations.

DEVELOPMENT ENVIRONMENT

The foundation of the system was built using Python, leveraging libraries such as TensorFlow and Scikit-learn for their robust ML capabilities. PostgreSQL and Neo4j were selected as databases to manage structured and graph-based data, respectively, while Amazon Web Services (AWS) provided the scalability required to handle large student cohorts.

Figure 2. High-Level System Architecture of the Learning Analytics Platform.

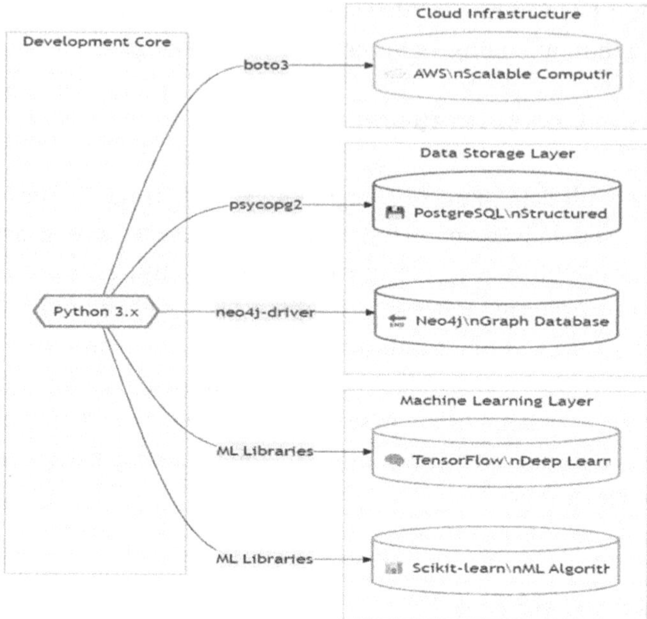

DATA PREPROCESSING

Data preprocessing was a critical step due to the volume and diversity of collected data, including:

- **Student Interaction Logs**: Time spent on modules, frequency of interactions, and quiz completion rates.
- **Simulation Results**: Metrics from experiments with renewable energy scenarios.
- **Textual Feedback**: Qualitative input from students about their learning experience.

To ensure compatibility with the ML models, data preprocessing included:

- **Data Cleaning**: Removing noise and handling missing values.
- **Normalization**: Standardizing numerical features such as quiz scores.
- **Feature Selection**: Identifying key predictors, e.g., time spent on specific topics.
- **Natural Language Processing (NLP)**: Tokenization, sentiment analysis, and keyword extraction were employed to process textual data.

Table 1. A summary of data preprocessing techniques and their objectives.

Step	Technique Used	Objective
Data Cleaning	Removal of outliers	Improve data quality and model accuracy.
Normalization	Min-Max Scaling	Standardize numerical values.
Feature Selection	Recursive Feature Elimination	Focus on the most relevant data points.
NLP	Sentiment Analysis	Analyze student feedback sentiment trends.

MODEL DEVELOPMENT AND TRAINING

Multiple machine learning models were developed to achieve system objectives:

- **Neural Networks**: Predicted the optimal sequence of learning modules for students based on their performance and interaction history. These networks were designed to learn from interaction logs and simulation results to generate personalized pathways.

The neural network model utilized in this system consists of multiple layers with ReLU activation functions for hidden layers and a sigmoid activation function for output. The model was trained using TensorFlow on a dataset of student interactions. Below is an excerpt from the implementation showing how the neural network model was built and trained.

Figure 3. Neural Network Model Implementation for Personalized Learning Path Prediction.

```
# Import necessary libraries
import tensorflow as tf
from tensorflow.keras.models import Sequential
from tensorflow.keras.layers import Dense
from tensorflow.keras.optimizers import Adam

# Define the model architecture
model = Sequential()

# Add the first dense layer (input layer)
model.add(Dense(units=64, activation='relu', input_dim=X_train.shape[1]))

# Add hidden layers
model.add(Dense(units=32, activation='relu'))
model.add(Dense(units=16, activation='relu'))

# Add the output layer (for predicting student performance)
model.add(Dense(units=1, activation='linear'))  # Regression to predict score

# Compile the model with an optimizer and loss function
model.compile(optimizer=Adam(), loss='mean_squared_error', metrics=['accuracy'])

# Train the model with training data
history = model.fit(X_train, y_train, epochs=50, batch_size=32, validation_data=(X_val, y_val))

# Make predictions on new data
predictions = model.predict(X_test)
```

Figure 4. Learning Curve of the Neural Network Model During Training.

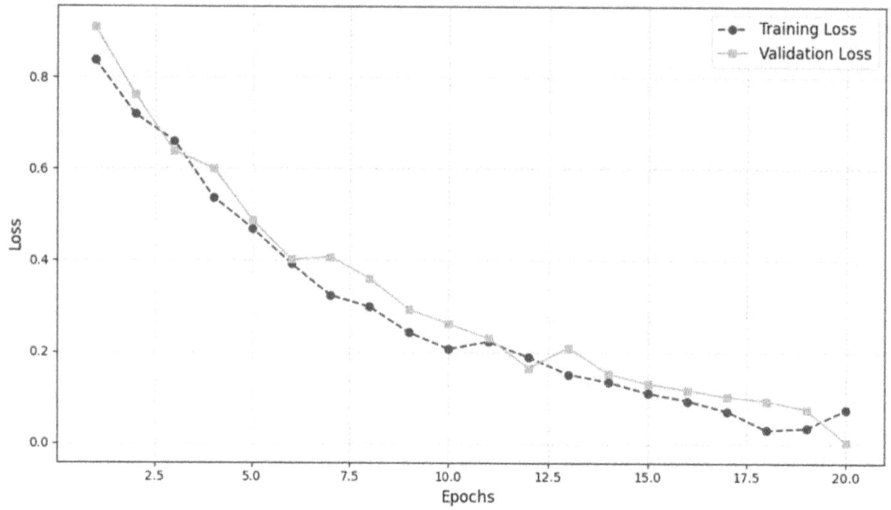

The graph illustrates the learning curve of the neural network model over 20 epochs. The training loss demonstrates a consistent decrease, indicating the model's ability to learn from the data effectively. The validation loss follows a similar trend, suggesting that the model generalizes well to unseen data. Minor fluctuations in the validation loss are attributed to the inherent variability in the validation set. This visualization underscores the convergence of the model and its robustness during training.

- **Random Forest Regressors**: Forecasted student success rates, allowing educators to identify at-risk students for proactive interventions.

Here is an excerpt from the implementation of the Random Forest Regressor used to forecast student success rates.

Figure 5. Random Forest Regressor Model Implementation for Student Success Prediction.

```
# Import necessary libraries
from sklearn.ensemble import RandomForestRegressor
from sklearn.model_selection import train_test_split
from sklearn.metrics import mean_squared_error

# Load dataset (X as features, y as target variable)
X = data.drop(columns=['target'])
y = data['target']

# Split data into training and test sets
X_train, X_test, y_train, y_test = train_test_split(X, y, test_size=0.2, random_state=42)

# Initialize the random forest regressor model
rf_model = RandomForestRegressor(n_estimators=100, random_state=42)

# Train the model on the training data
rf_model.fit(X_train, y_train)

# Make predictions on the test data
y_pred = rf_model.predict(X_test)

# Evaluate the model's performance using Mean Squared Error
mse = mean_squared_error(y_test, y_pred)
print(f"Mean Squared Error: {mse}")
```

- **Reinforcement Learning**: Optimized interactive simulations where students experimented with renewable energy scenarios (e.g., solar panel placement, wind turbine configurations).

PERFORMANCE METRICS

To evaluate the performance of the machine learning models, the following metrics were employed:

- **Accuracy**: Measures the overall correctness of predictions made by the models.
- **Precision**: Indicates the proportion of correctly predicted positive instances.
- **Recall (Sensitivity)**: Reflects the ability of the models to capture all positive instances.
- **F1-Score**: A balanced metric combining Precision and Recall to evaluate the overall prediction quality.
- **Mean Squared Error (MSE)**: Used to assess the prediction accuracy for regression models.
- **AUC-ROC**: Measures the performance of classification models in distinguishing between classes.

The results for each model are summarized in the table below:

Table 2. Performance Metrics of the Developed Models.

Metric	Neural Network Model	Random Forest Regressor	Reinforcement Learning Model
Accuracy	87%	83%	92%
Precision	85%	79%	90%
Recall	82%	75%	88%
F1-Score	83%	77%	89%
Mean Squared Error	0.04	0.06	0.03
AUC-ROC	0.92	0.89	0.95

These metrics indicate that the models performed well in predicting student behavior and optimizing their learning paths. Notably, the Reinforcement Learning Model achieved the highest scores, particularly in accuracy and AUC-ROC, showing its robustness in personalizing student interactions within the system. The Neural Network and Random Forest Regressor also demonstrated strong performance, particularly in predicting student success and engagement.

SYSTEM INTEGRATION

RESTful APIs were implemented to integrate the trained models with the educational content database and the user interface. This design ensured real-time updates to adaptive learning paths and simulations directly accessible to students.

Figure 6. User Interface Design for the Educational System.

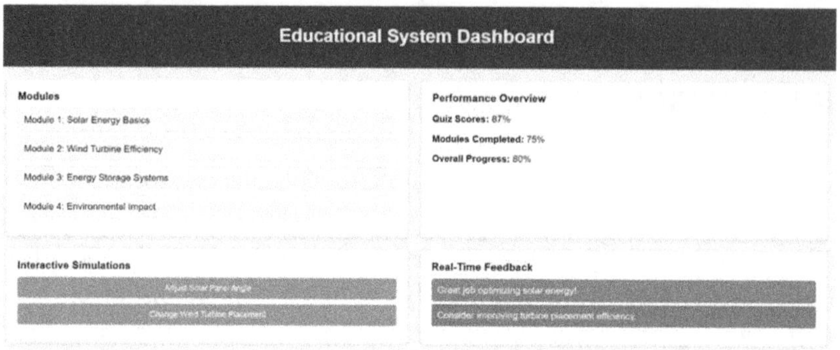

The figure illustrates the user interface of the developed educational system, which integrates RESTful APIs for real-time interaction with machine learning models and the content database. The dashboard provides access to personalized learning modules, displays performance metrics, allows manipulation of interactive renewable energy simulations, and delivers real-time feedback to enhance the learning experience. This design ensures seamless and adaptive engagement for students, promoting deeper understanding and retention of complex concepts.

VALIDATION AND TESTING

The implementation was validated through a series of tests:

1. **Functional Tests**: Verified the operation of individual components such as prediction engines and simulation tools.
2. **Performance Tests**: Measured system efficiency, including response times and scalability under high user loads.
3. **Pilot Study**: Conducted with 50 students, yielding critical insights into system usability and model accuracy, leading to iterative refinements.

Preliminary Results and Analysis

The system's effectiveness was evaluated through two key indicators: **knowledge retention improvements** and **enhanced student engagement and understanding**.

Knowledge Retention Improvements

The implementation of personalized learning paths significantly improved students' retention of complex concepts. Pre- and post-study assessments demonstrated an average increase of 23% in scores, particularly in understanding the systemic impacts of renewable energy decisions.

Table 3. Comparison of pre- and post-study scores.

Metric	Pre-study Average (%)	Post-study Average (%)	Improvement (%)
Conceptual Knowledge	56%	79%	+23%
Simulation Interaction	62%	84%	+22%

ENHANCED ENGAGEMENT AND UNDERSTANDING

The interactive simulations, optimized through reinforcement learning, fostered deeper engagement. Qualitative feedback revealed that 87% of participants found simulations intuitive and useful for understanding renewable energy trade-offs. Sentiment analysis of feedback further supported these findings, with a positive trend in comments related to system usability and learning value.

Figure 7. Comparison of Student Engagement Metrics Before and After ML-Based Implementation in Renewable Energy Education.

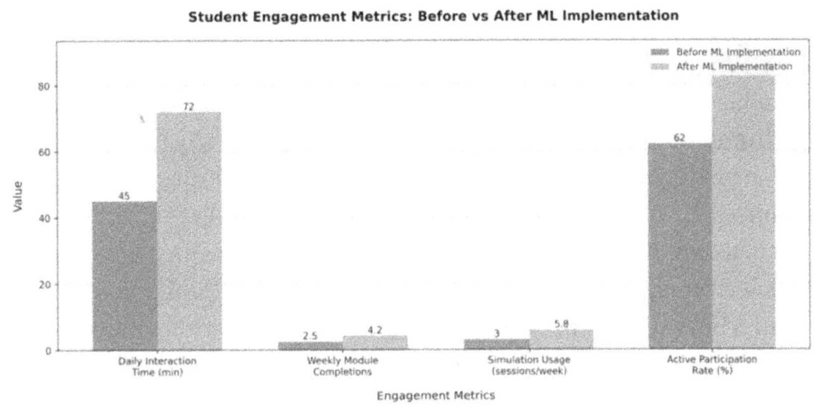

The comparative analysis illustrated in this graph demonstrates significant improvements across all student engagement metrics following the implementation of ML-based solutions. Most notably, daily interaction time increased by 60% (from 45 to 72 minutes), while the active participation rate showed a remarkable improvement from 62% to 89%. Weekly module completions rose from 2.5 to 4.2, and simulation usage nearly doubled from 3 to 5.8 sessions per week. These quantitative improvements align with the qualitative feedback indicating high levels of student engagement and validate the effectiveness of our ML-driven adaptive learning approach in teaching renewable energy concepts.

Observations from the Pilot Study

- Students appreciated the real-time feedback and adjustments provided by the adaptive learning system.

- The system's ability to simulate realistic renewable energy scenarios helped bridge the gap between theoretical knowledge and practical application.

DISCUSSION

The application of machine learning (ML) systems to optimize the teaching of renewable energy and sustainable development has profound implications for both the effectiveness and reach of education in these crucial fields. The preliminary results from the implementation of adaptive learning algorithms, interactive simulations, and predictive models suggest that this approach can significantly enhance how students engage with complex, interdisciplinary topics like sustainable development.

1. **Personalized Learning Paths**: One of the most notable impacts is the ability to offer highly personalized learning experiences. By using ML to adapt content based on individual student progress, the system ensures that learners can focus on areas where they need the most support, thereby improving overall learning outcomes. This dynamic adjustment allows for a more tailored educational experience that accommodates the varying learning paces and styles of students, which is particularly important in multidisciplinary fields such as sustainable development, where students may have different backgrounds and areas of expertise.
2. **Improved Knowledge Retention**: The results indicate a significant improvement in knowledge retention, with students showing up to a 23% increase in conceptual understanding after engaging with the ML-based learning system. This enhanced retention is particularly important for the sustainability field, where understanding complex systems and the long-term impacts of decisions is critical. By providing real-time feedback, targeted resources, and opportunities for repeated interaction with the material, students are better equipped to internalize key concepts, which is vital for the future leaders who will need to make informed decisions in the face of global environmental challenges.
3. **Engagement and Motivation**: The incorporation of interactive simulations, which allow students to experiment with renewable energy scenarios, significantly increases student engagement. These simulations not only make theoretical knowledge more tangible but also foster critical thinking and problem-solving skills. The fact that 87% of participants found these simulations intuitive and useful further highlights the potential of interactive learning tools to deepen students' understanding of sustainability. Engaged and motivated students are more likely to take the knowledge gained into real-world contexts, where they can apply it to address environmental challenges.
4. **Integration with Traditional Teaching Methods**: While the machine learning system provides valuable personalization and interactivity, it is essential to integrate these tools within a traditional teaching framework to maximize their effectiveness. Blended learning—combining ML with face-to-face instruction—creates a more holistic and comprehensive learning experience. In this approach, educators utilize data from the ML system to identify students who require additional support, allowing for targeted interventions. For example, insights from predictive models enable teachers to group students with similar challenges, facilitating collaborative learning and peer-to-peer exchanges. In addition, human instructors lead discussions on more complex sustainability issues, such as environmental justice and policy-making, which require nuanced understanding that current ML algorithms are not yet equipped to address. This integration ensures that both technological tools

and traditional pedagogical methods complement each other, providing students with a balanced, well-rounded educational experience.

5. **Real-World Application of Knowledge**: The use of predictive models to simulate real-world scenarios where students can explore the impacts of renewable energy on ecosystems and infrastructure gives them a clearer understanding of the complex trade-offs involved in the transition to sustainable energy systems. This approach encourages students to think about long-term impacts and the systemic nature of sustainability, moving beyond isolated technical solutions and fostering a broader, more integrated view of the challenges and opportunities in sustainable development.
6. **Scaling Education for Broader Impact**: The ability to scale this approach, as indicated by the positive results from the pilot study, presents an exciting opportunity to bring high-quality, interactive, and personalized sustainable development education to a wider audience. As the world faces increasingly urgent sustainability challenges, educating the next generation of policymakers, engineers, and citizens on these issues is more important than ever. By leveraging ML-driven educational systems, institutions can ensure that more students, regardless of geographical location or institutional resources, have access to high-quality learning experiences that prepare them to tackle these global challenges.
7. **Expanding the Pilot Study for Enhanced External Validity**: In order to further enhance the external validity of the findings, future work will focus on expanding the pilot study to include a more diverse student sample. This will involve participants from a range of educational backgrounds, geographic locations, and academic levels. Additionally, we plan to conduct longitudinal assessments to measure the long-term impact of the ML system on knowledge retention, problem-solving abilities, and real-world application of sustainable energy concepts. By extending the duration of the study to 12-24 months, we aim to capture a comprehensive understanding of how these ML-based learning systems influence students over time.

The integration of machine learning (ML) into educational practices, particularly in renewable energy and sustainable development, brings several distinct advantages. One of the most significant strengths is its ability to personalize learning experiences. Unlike traditional methods, an ML-based system adapts the content based on individual student performance and behavior, ensuring that each learner receives tailored support. This dynamic personalization caters to different learning styles and paces, which can lead to a deeper understanding of complex interdisciplinary concepts. In the context of sustainable development, where students may come with varying levels of prior knowledge, this individualized approach is essential. It allows students to progress at their own pace and grasp complex topics more effectively (Yao et al., 2023).

Another key strength is the enhanced engagement achieved through interactive simulations. These simulations allow students to experiment with real-world renewable energy scenarios, offering a hands-on learning experience. By exploring variables such as the placement of solar panels or optimizing wind turbine efficiency, students can see the immediate impact of their decisions. This immersive, interactive approach not only makes learning more engaging but also helps students understand the practical applications and long-term consequences of energy choices. The simulations create a tangible and memorable learning experience that reinforces theoretical knowledge, making the educational process both meaningful and enjoyable (Bukar et al., 2024).

The scalability and flexibility of ML-powered systems are also notable strengths. Once the system is set up and refined, it can be deployed across a large number of students with minimal effort. This makes it an invaluable tool for large-scale educational settings, where reaching a broad and diverse student population is a priority. Moreover, the adaptability of the ML approach means it can be used across different educational levels, from high schools to universities, providing a high-quality sustainable development curriculum to a global audience. This broad applicability helps ensure that educational resources can reach more students, regardless of location or institution size (Parra et al., 2024).

Furthermore, ML enables data-driven insights that allow for real-time monitoring of student progress. By analyzing student behavior and performance data, predictive models can anticipate potential challenges and identify students who may need additional support. This proactive approach helps educators intervene early, improving student retention and success rates. Over time, as more data is collected, the system's predictive capabilities improve, further enhancing its effectiveness. In this way, the use of data analytics continuously optimizes the learning experience for both students and educators (Eden et al., 2024).

While the ML-based approach offers numerous advantages, it also presents several challenges that must be addressed to ensure its effective implementation and to maximize its educational potential.

Dependence on Data Quality: One of the key limitations of an ML-based approach is its dependence on high-quality data. The effectiveness of adaptive learning algorithms and predictive models relies on the accuracy and completeness of the data collected from students. Incomplete or biased data can lead to inaccurate predictions and ineffective personalization, which could hinder the learning process rather than enhancing it. Data preprocessing and cleaning are critical to address this challenge, but it remains a concern that must be continuously monitored (Luan et al., 2020).

Data Privacy and Ethical Concerns: The collection and analysis of student data raise important privacy and ethical concerns. Ensuring that sensitive information is protected and that the data collection complies with regulations such as the General Data Protection Regulation (GDPR) is essential. Additionally, the transparency and accountability of ML algorithms must be maintained to ensure that the use of personal data does not inadvertently harm students or reinforce biases. Ethical guidelines and comprehensive frameworks are essential for maintaining trust and safeguarding student information (Taherdoost, 2023).

Resource and Infrastructure Constraints: While the system's scalability is an advantage, its implementation can be resource-intensive, particularly in terms of technical infrastructure and ongoing maintenance. Educational institutions with limited financial and technical resources may struggle to adopt such ML-based systems on a large scale. The costs associated with setting up and maintaining these platforms, as well as the need for staff training, may pose significant barriers, especially in lower-resource settings (Rashid & Kausik, 2024).

Potential for Reduced Human Interaction: Although ML systems can provide valuable personalized learning experiences, there is a risk that they may reduce the level of human interaction in education. Personalized content delivery and real-time feedback can replace some traditional forms of interaction with educators, which are essential for developing soft skills and critical thinking. Ensuring that these ML-based tools are integrated into a broader educational framework that retains a human-centered approach is crucial to avoid over-reliance on technology (Eden et al., 2024).

Challenges in Addressing Complex Ethical and Societal Issues: While ML can effectively teach technical aspects of sustainable development, it may struggle to address the more nuanced ethical, social, and policy-related dimensions of sustainability. For example, topics such as environmental justice, equity in energy access, and the socio-economic impacts of renewable energy transitions require deep contextual understanding that may not be fully captured through algorithms or simulations. These topics often involve complex trade-offs that are influenced by local, political, and cultural factors, which are difficult for algorithms to interpret accurately. Complementary approaches, such as discussions, debates, and critical engagement activities, will allow students to actively engage with diverse perspectives on these issues, fostering a more comprehensive understanding of the socio-political aspects of sustainability. Additionally, future curriculum iterations should ensure that real-world case studies are incorporated, providing students with opportunities to analyze the ethical implications of energy decisions on marginalized communities. These approaches are necessary to address these broader challenges effectively (Parra et al., 2024).

In summary, the ML-based approach offers significant strengths, particularly in personalizing learning and enhancing engagement through interactive tools. However, challenges related to data quality, ethical concerns, resource limitations, and the need for human interaction must be carefully considered to ensure the approach is implemented effectively and equitably in the context of sustainable development education.

CONCLUSION

In conclusion, the integration of machine learning (ML) into the teaching of renewable energy and sustainable development offers substantial potential to transform education. This approach enables the personalization of learning, enhancing engagement through interactive simulations, and provides scalability and flexibility in educational settings. Additionally, data-driven insights allow for proactive interventions that can improve student success. However, challenges remain, particularly in terms of data quality, privacy concerns, resource constraints, and the potential reduction in human interaction. Addressing these limitations is crucial to maximizing the effectiveness of ML-based learning systems.

To ensure successful large-scale deployment, it is recommended that educational institutions invest in robust infrastructure and prioritize data quality through consistent preprocessing. A strong emphasis should be placed on maintaining ethical standards and safeguarding student privacy to ensure trust in the system. Moreover, collaboration with experts in various fields of sustainable development is necessary to ensure the content remains relevant and comprehensive.

Future research should focus on refining ML algorithms to address complex societal issues within the context of sustainability, such as equity and social justice. Additionally, studies on the long-term impact of ML-based learning on student outcomes and its integration with traditional teaching methods will provide valuable insights. Further exploration into improving human-computer interaction and reducing the dependency on technology in education will also be key areas for development.

By addressing these recommendations and advancing research, ML-enhanced education has the potential to reshape the future of learning, particularly in the critical fields of renewable energy and sustainable development.

REFERENCES

Ahmad, T., Zhang, D., Huang, C., Zhang, H., Dai, N., Song, Y., & Chen, H. (2021). Artificial intelligence in sustainable energy industry: Status Quo, challenges and opportunities. *Journal of Cleaner Production, 289*, 125834. DOI: 10.1016/j.jclepro.2021.125834

Almalawi, A., Soh, B., Li, A., & Samra, H. (2024). Predictive Models for Educational Purposes: A Systematic Review. *Big Data and Cognitive Computing*, *8*(12), 187. DOI: 10.3390/bdcc8120187

Alqallaf, N., & Ghannam, R. (2024, March). Immersive Learning in Photovoltaic Energy Education: A Comprehensive Review of Virtual Reality Applications. In *Solar* (Vol. 4, No. 1, pp. 136-161). MDPI. DOI: 10.3390/solar4010006

Bennani, S., Maalel, A., & Ben Ghezala, H. (2022). Adaptive gamification in E-learning: A literature review and future challenges. *Computer Applications in Engineering Education, 30*(2), 628–642. DOI: 10.1002/cae.22477

Bin Abu Sofian, A. D. A., Lim, H. R., Siti Halimatul Munawaroh, H., Ma, Z., Chew, K. W., & Show, P. L. (2024). Machine learning and the renewable energy revolution: Exploring solar and wind energy solutions for a sustainable future including innovations in energy storage. *Sustainable Development (Bradford), 32*(4), 3953–3978. DOI: 10.1002/sd.2885

Bukar, U. A., Sayeed, M. S., Razak, S. F. A., Yogarayan, S., & Sneesl, R. (2024). Decision-making framework for the utilization of generative artificial intelligence in education: A case study of ChatGPT. *IEEE Access : Practical Innovations, Open Solutions, 12*, 95368–95389. DOI: 10.1109/ACCESS.2024.3425172

Chaib, S. O., Joti, I., & Khoulji, S. (2023). Design of a specialized search engine for university students dedicated to education and environment. In *E3S Web of Conferences* (Vol. 412, p. 01060). EDP Sciences.

Chaib, S. O., Joti, I., & Khoulji, S. (2023). Evaluation of a Computer-Assisted Language Learning System Based on Adaptive Learning Designed for Self-training in Scientific French Language. *International Journal of Information and Education Technology (IJIET), 13*(8), 1284–1296. DOI: 10.18178/ijiet.2023.13.8.1930

Chaib, S. O., Joti, I., & Khoulji, S. (2024, February). Gamification in mobile language learning applications: A case study examining a mobile learning tool in a Moroccan university context. In *E-Learning and Smart Engineering Systems (ELSES 2023)* (pp. 177-185). Atlantis Press.

Eden, C. A., Chisom, O. N., & Adeniyi, I. S. (2024). Integrating AI in education: Opportunities, challenges, and ethical considerations. *Magna Scientia Advanced Research and Reviews, 10*(2), 006-013.

Fan, Z., Yan, Z., & Wen, S. (2023). Deep learning and artificial intelligence in sustainability: A review of SDGs, renewable energy, and environmental health. *Sustainability (Basel), 15*(18), 13493. DOI: 10.3390/su151813493

Farooqi, M. T. K., Amanat, I., & Awan, S. M. (2024). Ethical considerations and challenges in the integration of artificial intelligence in education: A systematic review. *Journal of Excellence in Management Sciences*, *3*(4), 35–50. DOI: 10.69565/jems.v3i4.314

García-González, E., Jiménez-Fontana, R., & Azcárate, P. (2020). Education for sustainability and the sustainable development goals: Pre-service teachers' perceptions and knowledge. *Sustainability (Basel), 12*(18), 7741. DOI: 10.3390/su12187741

Gligorea, I., Cioca, M., Oancea, R., Gorski, A. T., Gorski, H., & Tudorache, P. (2023). Adaptive learning using artificial intelligence in e-learning: A literature review. *Education Sciences, 13*(12), 1216. DOI: 10.3390/educsci13121216

Gyonyoru, K. I. K., & Katona, J. (2024, October). Student Perceptions of AI-Enhanced Adaptive Learning Systems: A Pilot Survey. In *2024 IEEE 7th International Conference and Workshop Óbuda on Electrical and Power Engineering (CANDO-EPE)* (pp. 93-98). IEEE.

Hu, S. (2024). The effect of artificial intelligence-assisted personalized learning on student learning outcomes: A meta-analysis based on 31 empirical research papers. *Science Insights Education Frontiers, 24*(1), 3873–3894. DOI: 10.15354/sief.24.re395

Kabudi, T., Pappas, I., & Olsen, D. H. (2021). AI-enabled adaptive learning systems: A systematic mapping of the literature. *Computers and Education: Artificial Intelligence, 2*, 100017. DOI: 10.1016/j.caeai.2021.100017

Kiasari, M., Ghaffari, M., & Aly, H. H. (2024). A Comprehensive Review of the Current Status of Smart Grid Technologies for Renewable Energies Integration and Future Trends: The Role of Machine Learning and Energy Storage Systems. *Energies, 17*(16), 4128. DOI: 10.3390/en17164128

Luan, H., Geczy, P., Lai, H., Gobert, J., Yang, S. J., Ogata, H., Baltes, J., Guerra, R., Li, P., & Tsai, C. C. (2020). Challenges and future directions of big data and artificial intelligence in education. *Frontiers in Psychology, 11*, 580820. DOI: 10.3389/fpsyg.2020.580820 PMID: 33192896

Namoun, A., & Alshanqiti, A. (2020). Predicting student performance using data mining and learning analytics techniques: A systematic literature review. *Applied Sciences (Basel, Switzerland), 11*(1), 237. DOI: 10.3390/app11010237

Nookala, G. (2024). Adaptive Data Governance Frameworks for Data-Driven Digital Transformations. *Journal of Computational Innovation, 4*(1).

Oladapo, B. I., Olawumi, M. A., & Omigbodun, F. T. (2024). Machine Learning for Optimising Renewable Energy and Grid Efficiency. *Atmosphere (Basel), 15*(10), 1250. DOI: 10.3390/atmos15101250

Ouald Chaib, S., Joti, I., & Khoulji, S. (2022, November). Learning Analytics in the Teaching of French as a Foreign Language (FFL) and Big Data: What Resources? For What Skills? In *The International Conference on Artificial Intelligence and Smart Environment* (pp. 572-580). Cham: Springer International Publishing.

Park, S., & Jung, H. (2023). Bridging the Gap: Integrating Theory and Practice for Early Childhood Physical Education Teacher Education in Korea. *Sustainability (Basel), 15*(19), 14397. DOI: 10.3390/su151914397

Parra, J. A. B., Escobar, Á. E. S., & Martínez, A. P. P. (2024). Artificial Intelligence in Education: Benefits, Challenges, and Ethical Considerations According to Scientific Evidence. *Journal of International Crisis and Risk Communication Research*, 48-65.

Rachha, A., & Seyam, M. (2023). Explainable AI in education: Current trends, challenges, and opportunities. *SoutheastCon, 2023*, 232–239. DOI: 10.1109/SoutheastCon51012.2023.10115140

Rashid, A. B., & Kausik, A. K. (2024). AI revolutionizing industries worldwide: A comprehensive overview of its diverse applications. *Hybrid Advances*, 100277.

Sajja, R., Sermet, Y., Cikmaz, M., Cwiertny, D., & Demir, I. (2024). Artificial intelligence-enabled intelligent assistant for personalized and adaptive learning in higher education. *Information (Basel), 15*(10), 596. DOI: 10.3390/info15100596

Sari, H. E., Tumanggor, B., & Efron, D. (2024). Improving Educational Outcomes Through Adaptive Learning Systems using AI. *International Transactions on Artificial Intelligence*, *3*(1), 21–31. DOI: 10.33050/italic.v3i1.647

Strielkowski, W., Grebennikova, V., Lisovskiy, A., Rakhimova, G., & Vasileva, T. (2024). AI-driven adaptive learning for sustainable educational transformation. *Sustainable Development (Bradford)*, sd.3221. DOI: 10.1002/sd.3221

Taherdoost, H. (2023). Navigating the ethical and privacy concerns of big data and machine learning in decision making. *Intelligent and Converged Networks*, *4*(4), 280–295. DOI: 10.23919/ICN.2023.0023

Vassoyan, J., Vie, J. J., & Lemberger, P. (2023). Towards Scalable Adaptive Learning with Graph Neural Networks and Reinforcement Learning. *arXiv preprint arXiv:2305.06398.*

Yao, Z., Lum, Y., Johnston, A., Mejia-Mendoza, L. M., Zhou, X., Wen, Y., Aspuru-Guzik, A., Sargent, E. H., & Seh, Z. W. (2023). Machine learning for a sustainable energy future. *Nature Reviews. Materials*, *8*(3), 202–215. DOI: 10.1038/s41578-022-00490-5 PMID: 36277083

Zhai, X., & Krajcik, J. (2024). Artificial Intelligence-Based STEM. *Uses of Artificial Intelligence in STEM Education*, 1.

Chapter 15

Numerical Study of Natural Convection Energy Transfer in a Grooved Cavity Using the Hybrid Lattice Boltzmann Method

Jaouad Benhamou
https://orcid.org/0000-0002-1958-2843
Higher School of Education and Training, Mohammed First University, Oujda, Morocco

Salaheddine Channouf
Laboratory of Mechanics and Energetics, Faculty of Sciences, Mohammed First University, Oujda, Morocco

Youssef Admi
https://orcid.org/0000-0003-0920-0618
Laboratory of Mechanics and Energetics, Faculty of Sciences, Mohammed First University, Oujda, Morocco

El Bachir Lahmer
Laboratory of Mechanics and Energetics, Faculty of Sciences, Mohammed First University, Oujda, Morocco

Kamal Hirech
https://orcid.org/0000-0002-3531-7791
Higher School of Education and Training, Mohammed First University, Oujda, Morocco

Ahmed Kadhim Hussein
College of Engineering, University of Babylon, Iraq

Mohammed Jami
Laboratory of Mechanics and Energetics, Faculty of Sciences, Mohammed First University, Oujda, Morocco

Ahmed Mezrhab
Laboratory of Mechanics and Energetics, Faculty of Sciences, Mohammed First University, Oujda, Morocco

ABSTRACT

In this work, natural convection in a grooved rectangular cavity with aspect ratio R =2 is analyzed using a hybrid numerical approach combining the lattice Boltzmann method and the finite-difference technique. The introduction of two grooves in the cavity geometry aims to improve heat transfer passively, using air as the fluid with a Prandtl number of 0.71. The results show a significant improvement in heat transfer

DOI: 10.4018/979-8-3693-9924-8.ch015

efficiency, with a 76% gain in Nusselt number at a Rayleigh number (Ra) set at 10^3 and 43% at Ra=10^5, compared with the conventional Rayleigh-Bénard configuration. These improvements are attributed to modified flow structures and deformed isotherms near the grooves, favoring complex fluid dynamics and improved heat exchange. These results suggest that grooved geometries can be used effectively to enhance heat transfer in natural convection, making them suitable for a variety of industrial applications.

INTRODUCTION

Natural convection is an important factor in a variety of physical and industrial processes, such as cooling systems, natural ventilation, and thermal applications in buildings (Baïri et al., 2014; Buchberg et al., 1976; Rahimi et al., 2019). Among the geometric configurations frequently studied for natural convection are closed cavities, which occupy a central position due to their simplicity and suitability for simulating thermal flows in real-life contexts (Abdulkadhim et al., 2021; Ostrach, 1988). However, research in the literature has shown that internal modifications, such as adding partitions (Bilgen, 2002; Costa, 2012), or external modifications, such as deforming the shape of the geometry under study (Das et al., 2017; Rostami et al., 2020), can passively improve heat exchange efficiency by altering the flow structure. These complex geometries enable convection mechanisms to be intensified, generating more dynamic fluid movements and steeper temperature gradients, resulting in enhanced thermal performance.

The introduction of structured shapes, such as grooves, into the various geometries, can offer significant potential for modifying the characteristics of thermal flows. Indeed, these geometric perturbations lead to significant changes in the hydrodynamic and thermal behavior of the fluid. This can lead to a more efficient redistribution of heat, intensifying heat exchange across the entire domain. Grooves affect not only streamlines but also the distribution of isotherms, with densification of thermal gradients around grooved areas. This results in a significant improvement in heat transfer efficiency. Research into the addition of grooves focuses mainly on open geometries, where they are used to optimize heat transfer in systems such as heat exchangers and cooling devices (Miansari et al., 2023; Zontul et al., 2021). However, in confined geometries, such as closed cavities, studies are much more limited (Kwak & Song, 2000), although understanding them is essential for applications where heat management in confined spaces is particularly important.

In this work, the natural convection in a grooved rectangular cavity, characterized by an aspect ratio of $R = 2$, with grooves on the cold upper surface, is studied. This configuration simulates a situation where internal geometry directly influences thermal dynamics, making the analysis of flows and heat transfers more complex. The fluid studied is air, with a Prandtl number fixed at 0.71, while the Rayleigh number varies between 10^3 and 10^5, covering low and moderate natural convection regimes. The main objective is to quantify the impact of grooves on heat transfer by comparing results with an ungrooved Rayleigh-Bénard cavity. This study aims to understand the thermal and dynamic phenomena associated with this complex geometry and to explore the mechanisms by which grooves improve the thermal performance of the system.

METHODOLOGY: MRT-LATTICE BOLTZMANN METHOD FOR FLUID FLOW

The lattice Boltzmann method (LBM) based on the multiple relaxation time (MRT) model is used in this work to simulate fluid flow by solving the following Boltzmann equation (Benhamou et al., 2020; Mohamad, 2019):

$$f_i(x_i + c_i \Delta t, t + \Delta t) = f_i(x_i, t) + M^{-1} S \left[m_i^{eq} - m_i\right] + \Delta t F_i \quad (1)$$

where $f_i(x_i, t)$ is distribution function in the direction i at position x_i and time t.

Equation (1) is discretized according to the D2Q9-LBM lattice (Fig.1), which is characterized by the nine following velocities c_i:

$$c_i = c \begin{cases} (0,0) i = 0 \\ (1,0),(0,1),(-1,0),(0,-1) i = 1,..,4 \\ (1,1),(-1,1),(-1,-1),(1,-1) i = 5,..,8 \end{cases} \quad (2)$$

where $c = \Delta x / \Delta t$ is the unit velocity ($c = 1$). Δx and Δt are the space and time steps, respectively.

Figure 1. D2Q9 LBM-lattice

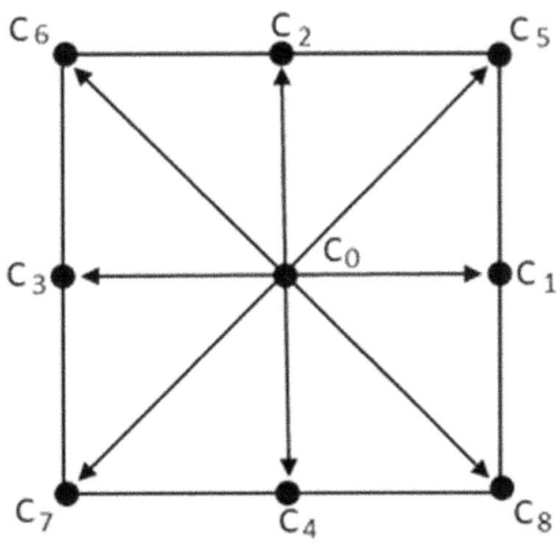

M^{-1} is the inverse matrix of the transformation matrix M:

$$M = \begin{pmatrix} 1 & 1 & 1 & 1 & 1 & 1 & 1 & 1 & 1 \\ -4 & -1 & -1 & -1 & -1 & 2 & 2 & 2 & 2 \\ 4 & -2 & -2 & -2 & -2 & 1 & 1 & 1 & 1 \\ 0 & 1 & 0 & -1 & 0 & 1 & -1 & -1 & 1 \\ 0 & -2 & 0 & 2 & 0 & 1 & -1 & -1 & 1 \\ 0 & 0 & 1 & 0 & -1 & 1 & 1 & -1 & -1 \\ 0 & 0 & -2 & 0 & 2 & 1 & 1 & -1 & -1 \\ 0 & 1 & -1 & 1 & -1 & 0 & 0 & 0 & 0 \\ 0 & 0 & 0 & 0 & 0 & 1 & -1 & 1 & -1 \end{pmatrix} \quad (3)$$

S is the relaxation matrix:

$$S = diag(s_0, s_1, s_2, s_3, s_4, s_5, s_6, s_7, s_8) \quad (4)$$

For all the simulations carried out in this paper, the relaxation times used are those mentioned in reference (Benhamou et al., 2020). We choose $s_0 = s_3 = s_5 = 1$, $s_1 = s_2 = 1.4$, $s_4 = s_6 = 1.2$ et $s_7 = s_8 = 1/(3\nu + 0.5)$.

For the D2Q9 scheme, the arrangement of the nine components of the m vector can be given in this order:

$$m = (\rho, e, \varepsilon, j_x, q_x, j_y, q_y, p_{xx}, p_{xy})^T \quad (5)$$

where ρ is the fluid density, e and ε are the moments of energy and the square of energy, q_x and q_y are related to the components of the energy flux and $j = (j_x, j_y) = \rho(u, v)$ is the momentum. p_{xx} and p_{xy} are related to the diagonal and off-diagonal components of the stress tensor, respectively.

The corresponding equilibrium moments m_i^{eq} are:

$$\begin{aligned} m_0^{eq} &= \rho \\ m_1^{eq} &= -2\rho + \frac{3}{\rho}(j_x^2 + j_y^2) \\ m_2^{eq} &= \rho - \frac{3}{\rho}(j_x^2 + j_y^2) \\ m_3^{eq} &= j_x \\ m_4^{eq} &= -j_x \\ m_5^{eq} &= j_y \\ m_6^{eq} &= -j_y \\ m_7^{eq} &= \frac{1}{\rho}(j_x^2 - j_y^2) \\ m_8^{eq} &= \frac{1}{\rho} j_x j_y \end{aligned} \quad (6)$$

F_i is the discretized buoyancy force (Mohamad, 2019):

$$F_i = 3\ w_i \rho g\ (T - T_m)\ c_{iy} \tag{7}$$

where w_i are the discretization weights of the D2Q9-LBM model, g is the gravitational vector, β is the coefficient of thermal expansion, T is the temperature and T_m is the mean temperature.

After solving the LBM equation (Eq. (1)), the macroscopic quantities such as density and velocities are calculated from the discrete distribution functions as:

$$\rho = \sum_{i=0}^{8} f_i,\ \rho \binom{u}{v} = \sum_{i=0}^{8} f_i \binom{c_{ix}}{c_{iy}} \tag{8}$$

FINITE DIFFERENCE METHOD FOR TEMPERATURE CALCULATION

The study of natural convection requires the calculation of temperature to determine the thermal behavior of the fluid in the system under study. Similar to the calculation of velocities, the temperature can be calculated using various numerical approaches, such as the lattice Boltzmann method, CFD techniques, and others. Our previous work has shown that using the CFD finite-difference method to calculate temperature, in combination with the LBM method, offers many advantages in terms of calculation time and accuracy (Benhamou et al., 2022). Thus, this CFD technique is also used in this work to determine the temperature distribution in the thermal problem considered. This is achieved by numerically solving the following energy equation:

$$\frac{\partial T}{\partial t} + u\frac{\partial T}{\partial x} + v\frac{\partial T}{\partial y} = \alpha\left(\frac{\partial^2 T}{\partial x^2} + \frac{\partial^2 T}{\partial y^2}\right) \tag{9}$$

Thermal diffusivity is linked to kinematic viscosity and Prandtl number as $\alpha = /Pr$.

As with other numerical methods, finite difference aims to replace a continuous field problem with infinite degrees of freedom by a discretized field with finite regular nodes. In this work, the partial differential heat equation (equation (9)) is transformed into a set of linear algebraic equations by discretizing the partial derivatives using the following second-order central difference approximations (Benhamou et al., 2022):

$$\frac{\partial T}{\partial x} = \frac{T(i+1,j) - T(i-1,j)}{2\Delta x},\ \frac{\partial T}{\partial y} = \frac{T(i,j+1) - T(i,j-1)}{2\Delta y} \tag{10}$$

$$\frac{\partial^2 T}{\partial x^2} = \frac{T(i+1,j) - 2T(i,j) + T(i-1,j)}{\Delta x^2},\ \frac{\partial^2 T}{\partial y^2} = \frac{T(i,j+1) - 2T(i,j) + T(i,j-1)}{\Delta y^2} \tag{11}$$

The energy equation (equation (9)) is solved directly in LBM units. To switch from the LBM system to dimensionless units, the following conversion has been adopted:

$$x^* = \frac{x}{H},\ y^* = \frac{y}{H},\ u^* = \frac{u}{\alpha}H,\ v^* = \frac{v}{\alpha}H, \theta = \frac{T - T_m}{T_h - T_c} \tag{12}$$

In physical units, H represents the height of the geometry studied. In LBM units, it corresponds to the number of points used along the $y-$axis.

BOUNDARY CONDITIONS

The boundary conditions used in this paper for the fluid flow are bounce-back boundary conditions to define the solid walls of the cavity simulated (Mohamad, 2019):

$$f_i(x_B, t) = f_{\bar{i}}(x_B, t) \tag{13}$$

where $f_i(x_B, t)$ is an unknown distribution function at the wall node (x_B) and $f_{\bar{i}}(x_B, t)$ is the known function in the opposite direction ($\bar{i} = -i$) to $f_i(x_B, t)$.

The thermal boundary conditions are used to define isothermal and adiabatic cavity walls. Adiabatic walls are characterized by a zero spatial deviation of the temperature $\partial\theta/\partial x = 0$ (Von Neumann boundary condition) (Benhamou et al., 2022; Mohamad, 2019):

$$\theta(x_B) = \theta(x_B \pm 1) \tag{14}$$

For isothermal walls, the cold side is defined directly by the dimensionless temperature $\theta_c = -0.5$, while the hot face is defined by the temperature $\theta_h = 0.5$ (Mezrhab et al., 2010).

VALIDATION

The physical problem considered to validate our numerical model for the simulation of natural convection is that studied by Corvaro and Paroncini (Corvaro & Paroncini, 2008). It consists of an air-filled square cavity whose vertical walls are cold, while 1/5 of the central part of the lower surface is hot (Fig. 2). The cold walls of the cavity have a dimensionless temperature $\theta_c = -0.5$ and the hot face is fixed at $\theta_h = 0.5$. The adiabatic walls are defined by Von Neumann boundary condition $\partial\theta/\partial y = 0$. All cavity walls are considered solid. The bounce-back boundary conditions are used in this validation and also in the rest of this work to define the fluid return.

Figure 2. Thermal problem studied for the validation.

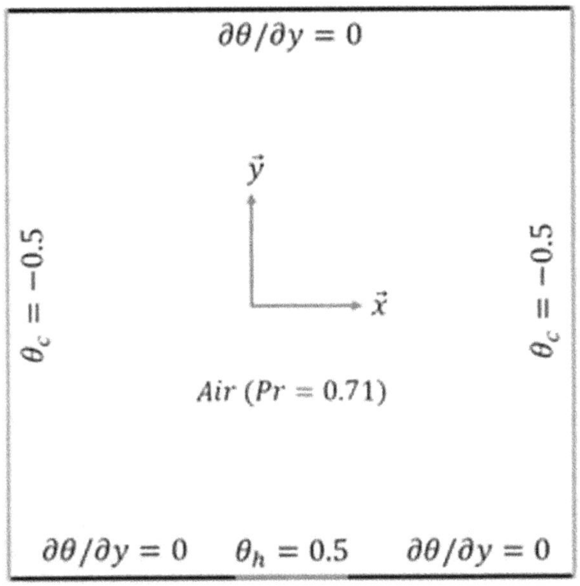

The first validation consists of comparing the average Nusselt number calculated at the hot section with the experimental data of Corvaro and Paroncini (Corvaro & Paroncini, 2008) for different values of the Rayleigh number (*Ra*). The obtained results are presented in Table 1. This table shows that our findings are close to those of the literature, with small deviations justified by a maximum relative error of less than 5%. A second comparison of isotherms and streamlines is presented in Figures 3 and 4, respectively. In Figure 3, the isotherms show horizontal and vertical temperature gradients. The streamlines show that the fluid forms two recirculation zones in the cavity. Overall, the isotherms and streamlines found by our numerical code are similar to those of Corvaro and Paroncini (Corvaro & Paroncini, 2008), showing good validation of our numerical method.

The Rayleigh number is defined as follows (Mohamad, 2019):

$$Ra = g\frac{(\theta_h - \theta_c)}{} H^3 \qquad (15)$$

where H is the vertical length of the cavity, α is the thermal diffusivity, and ν is the kinematic viscosity of the air.

Table 1. Comparison between the average Nusselt number calculated by our numerical method and measured experimentally by Corvaro and Paroncini (Corvaro & Paroncini, 2008) for different values of the Rayleigh (Ra) number

Ra	**Corvaro and Paroncini (Corvaro & Paroncini, 2008)**	**Present study**	**Error (%)**
7.5610^4	4.8	5.038	4.95
1.7110^5	6.3	6.5	3.17
2.510^5	6.81	6.98	2.49

Figure 3. Comparison between the isotherms obtained by Corvaro and Paroncini (Corvaro & Paroncini, 2008) (left) and our numerical code (right) for Ra = 2.02 10^5.

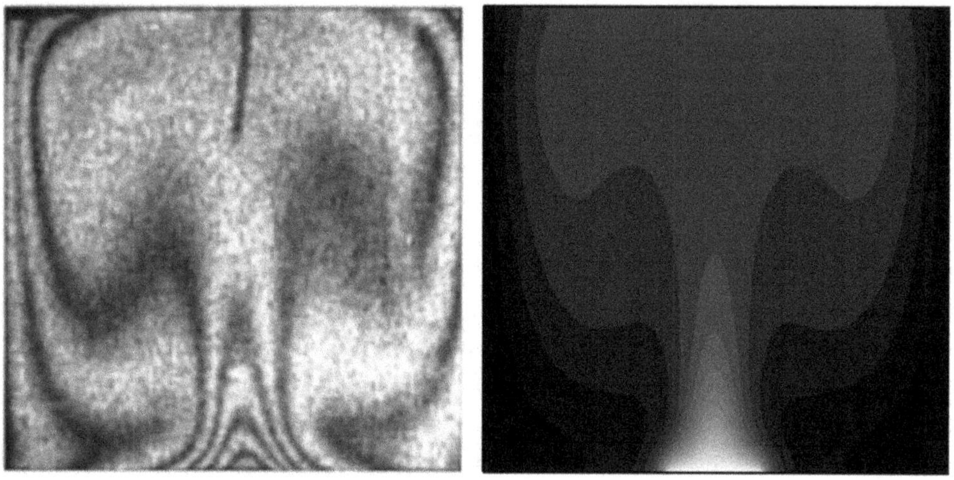

Figure 4. Comparison between the streamlines obtained by Corvaro and Paroncini (Corvaro & Paroncini, 2008) (left) and our numerical code (right) for Ra = 2.02 10^5.

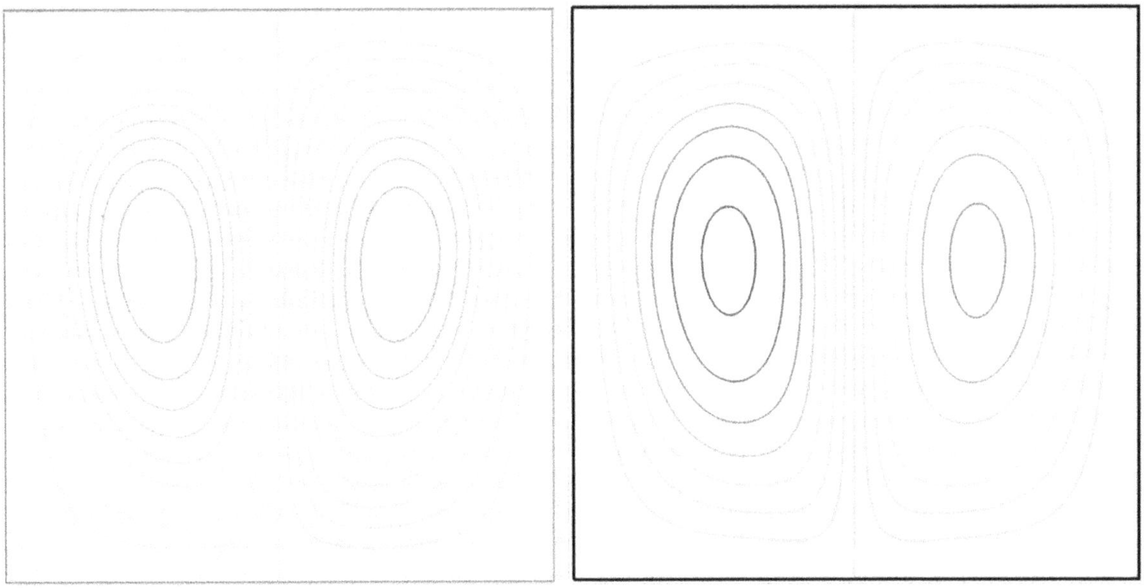

RESULTS AND DISCUSSION

In this work, natural convection in a grooved rectangular cavity, characterized by length L, height H, and aspect ratio $R = L/H = 2$, is studied (Fig.5). The lower surface of the cavity is hot and maintained at a dimensionless temperature (θ_h) of 0.5, while the upper surfaces are cold and defined by a temperature of -0.5. Vertical walls are considered adiabatic. The cavity is equipped with two grooves of length $l = L/5$ and height $h = H/2$, introducing geometric complexity conducive to the study of thermal flows. The fluid studied is air with a Prandtl (Pr) number set at 0.71. The primary aim of this configuration is to understand the thermal and dynamic phenomena associated with natural convection in grooved cavities, a model that is particularly relevant to many industrial and scientific applications. Secondly, this complex geometry is used to passively optimize heat exchange, offering a direct comparison with Rayleigh-Bénard convection in ungrooved cavities, enabling the impact of grooves on heat transfer efficiency to be assessed.

Figure 5. Geometry studied.

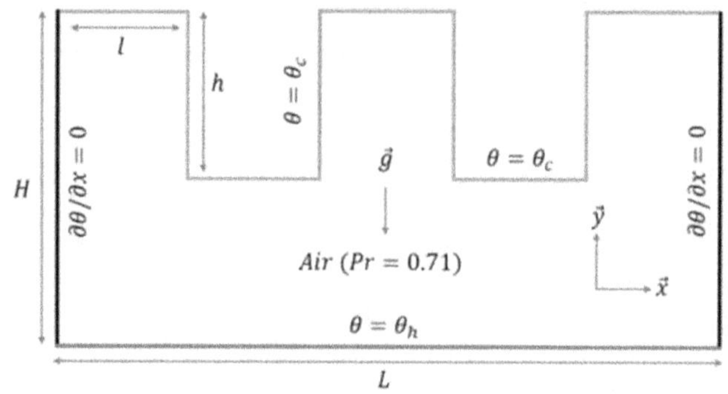

Table 2 shows the average Nusselt number values calculated along the hot bottom wall, comparing the thermal performance of the Rayleigh-Bénard and grooved cavity configurations for Rayleigh numbers of 10^3 and 10^5. The results show a significant increase in heat transfer efficiency, with an improvement of around 76% for $Ra = 10^3$ and 43% for $Ra = 10^5$, when moving from the conventional Rayleigh-Bénard cavity to a grooved cavity. This significant increase in the Nusselt number is directly linked to the alteration of flow structures and temperature profiles in the grooved cavity. Indeed, the presence of grooves disrupts the dynamics of natural convection flows, creating more complex fluid movements and promoting better heat redistribution. These disturbances also modify the isotherms, which deform and densify near the grooved zones, reinforcing local temperature gradients and improving overall heat exchange. Moreover, the effect of these grooves becomes more pronounced at lower Rayleigh numbers, which explains the greater difference for $Ra = 10^3$ compared to $Ra = 10^5$. At higher Rayleigh values, convection becomes more vigorous, which partially mitigates the influence of the grooves. However, even at $Ra = 10^5$, the addition of grooves continues to significantly improve heat transfer. These results confirm that grooved geometry can be effectively used to passively improve thermal efficiency in a variety of applications, taking advantage of flow and isotherm modifications induced by the internal cavity structure.

The average Nusselt number (*Nu*) is calculated at the hot bottom wall as follows (Mohamad, 2019):

$$N_u = \frac{H}{(\theta_h - \theta_c)L} \int_0^L \left(\frac{\partial T}{\partial y}\right)_{wall} dx \qquad (16)$$

Table 2. Average Nusselt number (Nu) calculated at the bottom hot wall of the Rayleigh-Benard and grooved cavities.

Cavity shape	**Ra=10^3**	**Ra=10^5**
Rayleigh Benard Cavity	1.0018	3.7539
Grooved cavity	1.7667	5.3924
Rate of heat transfer improvement	76%	43.65%

Fig. 6 shows isotherms in Rayleigh-Bénard and grooved cavities for Rayleigh numbers fixed at 10^3. In the classic Rayleigh-Bénard configuration, the isotherms are aligned horizontally, indicating an essentially vertical temperature gradient, characteristic of relatively weak natural convection. This reflects a thermal stratification in which temperature variations between the hot lower surface and the cold upper surface are evenly distributed. However, the addition of grooves in the cavity disrupts this regular distribution. The isotherms undergo significant deformation, particularly in the upper part of the cavity. This deformation indicates a change in fluid movements, caused by the presence of the grooves, which influence the thermal profile by generating recirculation zones. These recirculations make temperature gradients more complex. This behavior can also be seen in Fig. 7, which shows the streamlines in the cavity for $Ra = 10^3$. In the case of the Rayleigh-Bénard cavity, localized recirculation zones are observed near the corners, symmetrical with respect to the central horizontal axis. In the grooved cavity, on the other hand, recirculation zones become symmetrical with respect to the central vertical axis. This promotes increased exchange of cold fluid with the hot lower surface, underlining the influence of grooves on the flow structure inside the cavity.

Figure 6. Temperature field in Rayleigh-Bénard cavity (top) and grooved cavity (bottom) for $Ra = 10^3$.

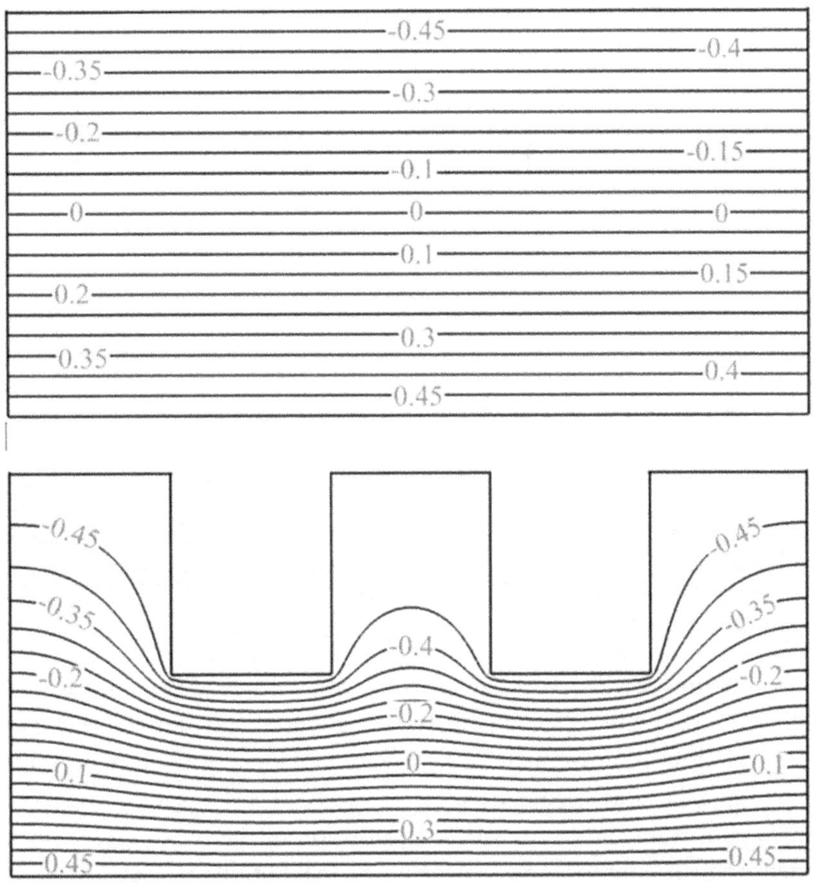

Figure 7. Streamlines in Rayleigh Benard cavity (top) and grooved cavity (bottom) for Ra = 10^3.

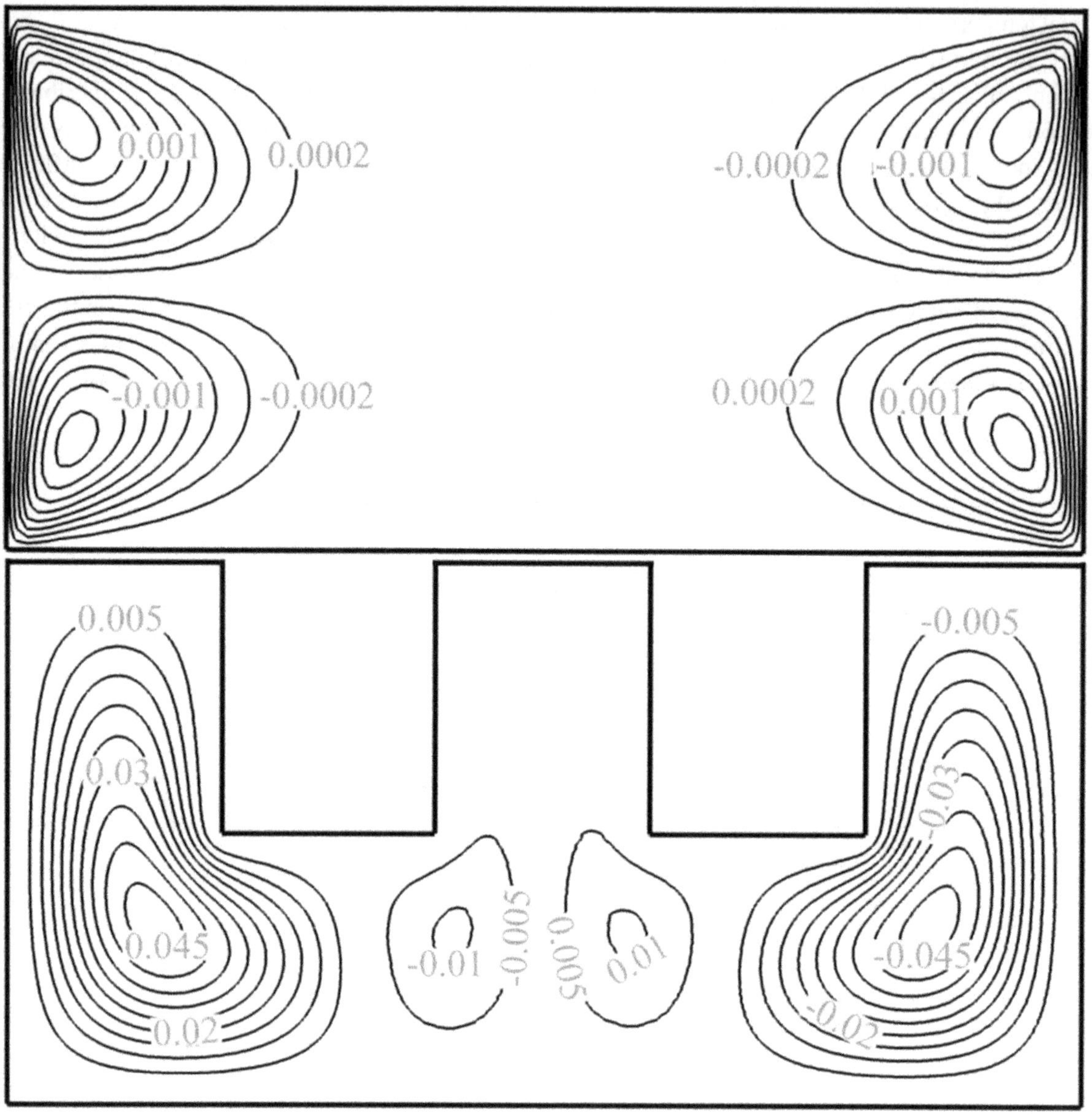

For $Ra = 10^5$, convection becomes more intense, significantly modifying the shape of the isotherms in the Rayleigh-Bénard configuration (Fig. 8). The isotherms adopt an inclined shape at the center of the cavity, revealing zones of upward movement of the hot fluid, with symmetrical deformation towards the side walls as they approach the cold upper surface. This behavior reflects more robust convection, where the rapid ascent of the hot fluid causes significant distortion of the temperature lines. In the grooved configuration, isotherms also undergo significant changes. In addition to the central upward deformation, the presence of the grooves also leads to a downward deformation of the isotherms in the lower part of the cavity, especially near the vertical walls. This deformation is linked to the disturbance to fluid flow caused by the grooves, which create local recirculation cells and modify the distribution of thermal gradients in these regions. The hot fluid flowing around the grooves disrupts the organization of the isotherms, resulting in more complex vertical deformations than in the configuration without grooves.

Figure 8. Temperature field in Rayleigh Benard cavity (top) and grooved cavity (bottom) for $Ra = 10^5$.

Fig. 9, which illustrates the streamlines for $Ra = 10^5$, shows that fluid behavior changes significantly with the introduction of grooves. In the classic Rayleigh-Bénard configuration, two large recirculation zones are observed, symmetrical with respect to the central vertical axis. However, with grooves, these two zones break up into four distinct recirculation cells. This new configuration promotes greater interaction between the fluid and the isothermal walls, thereby increasing heat transfer efficiency. The introduction of grooves improves the heat exchange rate by 43%, by intensifying natural convection and making heat flows through the cavity more efficient.

Figure 9. Streamlines in Rayleigh Benard cavity (top) and grooved cavity (bottom) for Ra = 10^5.

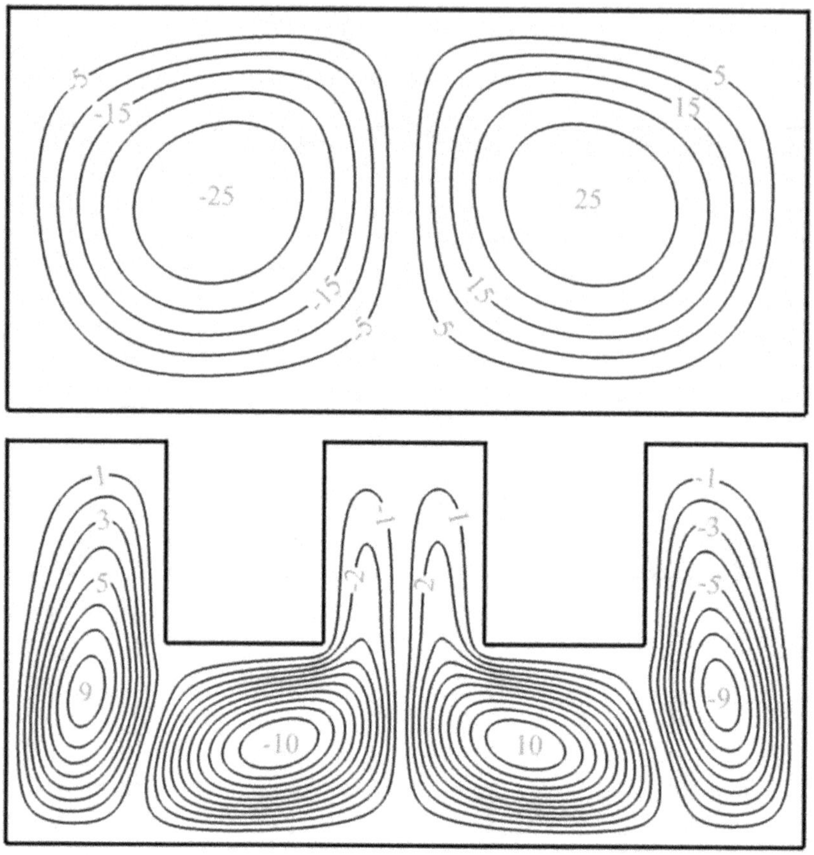

CONCLUSION

This paper presented a two-dimensional study of natural convection in a grooved rectangular cavity using a hybrid method combining the lattice Boltzmann approach and the finite-difference procedure. The results showed that the introduction of grooves into a rectangular cavity significantly improves heat transfer in natural convection, as evidenced by increases in the Nusselt numbers for both low and high Rayleigh numbers. For $Ra = 10^3$, the grooved cavity shows a 76% improvement in heat transfer efficiency, while for $Ra = 10^5$, the improvement is 43%. Grooves modify flow dynamics by creating local recirculation zones, which affect the distribution of thermal gradients inside the cavity. The deformation of isotherms and the fragmentation of large recirculation zones into smaller, more efficient cells clearly demonstrate the benefits of this configuration. These results suggest that grooved geometries can be used effectively in applications requiring passive enhancement of heat exchange. Future work could explore the effect of other groove geometries and aspect ratios to further optimize thermal performance, offering potential improvements in industrial applications such as heat exchangers and cooling systems.

REFERENCES

Abdulkadhim, A., Abed, I. M., & Mahjoub Said, N. (2021). Review of Natural Convection Within Various Shapes of Enclosures. *Arabian Journal for Science and Engineering*, *46*(12), 11543–11586. DOI: 10.1007/s13369-021-05952-6

Baïri, A., Zarco-Pernia, E., & García De María, J. M. (2014). A review on natural convection in enclosures for engineering applications. The particular case of the parallelogrammic diode cavity. *Applied Thermal Engineering*, *63*(1), 304–322. DOI: 10.1016/j.applthermaleng.2013.10.065

Benhamou, J., Jami, M., Mezrhab, A., Botton, V., & Henry, D. (2020). Numerical study of natural convection and acoustic waves using the lattice Boltzmann method. *Heat Transfer*, *49*(6), 3779–3796. DOI: 10.1002/htj.21800

Benhamou, J., Lahmer, E. B., & Jami, M. (2022). Three-dimensional simulation of conjugate heat transfer using the hybrid lattice Boltzmann-finite difference method. *International Communications in Heat and Mass Transfer*, *139*, 106486. DOI: 10.1016/j.icheatmasstransfer.2022.106486

Bilgen, E. (2002). Natural convection in enclosures with partial partitions. *Renewable Energy*, *26*(2), 257–270. DOI: 10.1016/S0960-1481(00)00206-8

Buchberg, H., Cation, I., & Edwards, D. K. (1976). Natural convection in enclosed spaces—A review of application to solar energy collection. *Journal of Heat Transfer*, *98*(2), 182–188. DOI: 10.1115/1.3450516

Corvaro, F., & Paroncini, M. (2008). A numerical and experimental analysis on the natural convective heat transfer of a small heating strip located on the floor of a square cavity. *Applied Thermal Engineering*, *28*(1), 25–35. DOI: 10.1016/j.applthermaleng.2007.03.018

Costa, V. A. F. (2012). Natural convection in partially divided square enclosures: Effects of thermal boundary conditions and thermal conductivity of the partitions. *International Journal of Heat and Mass Transfer*, *55*(25–26), 7812–7822. DOI: 10.1016/j.ijheatmasstransfer.2012.08.004

Das, D., Roy, M., & Basak, T. (2017). Studies on natural convection within enclosures of various (non-square) shapes – A review. *International Journal of Heat and Mass Transfer*, *106*, 356–406. DOI: 10.1016/j.ijheatmasstransfer.2016.08.034

Kwak, C. E., & Song, T. H. (2000). Natural convection around horizontal downward-facing plate with rectangular grooves: Experiments and numerical simulations. *International Journal of Heat and Mass Transfer*, *43*(5), 825–838. DOI: 10.1016/S0017-9310(99)00164-7

Mezrhab, A., Amine Moussaoui, M., Jami, M., Naji, H., & Bouzidi, M. (2010). Double MRT thermal lattice Boltzmann method for simulating convective flows. *Physics Letters, Section A: General. Physics Letters. [Part A]*, *374*(34), 3499–3507. DOI: 10.1016/j.physleta.2010.06.059

Miansari, M., Hasanpour, R., Alizadeh, A., Toghraie, D., & Ali Akbari, O. (2023). Numerical analysis of heat transfer enhancement in grooved vertical Multi-Cylinders at various groove geometries. *Alexandria Engineering Journal*, *70*, 535–546. DOI: 10.1016/j.aej.2023.02.048

Mohamad, A. A. (2019). Lattice Boltzmann method: Fundamentals and engineering applications with computer codes. In *Springer London* (2nd ed.). Springer London., DOI: 10.1007/978-1-4471-7423-3

Ostrach, S. (1988). Natural convection in enclosures. *Journal of Heat Transfer, 110*(4b), 1175–1190. DOI: 10.1115/1.3250619

Rahimi, A., Dehghan Saee, A., Kasaeipoor, A., & Hasani Malekshah, E. (2019). A comprehensive review on natural convection flow and heat transfer: The most practical geometries for engineering applications. *International Journal of Numerical Methods for Heat & Fluid Flow, 29*(3), 834–877. DOI: 10.1108/HFF-06-2018-0272

Rostami, S., Aghakhani, S., Pordanjani, A. H., Afrand, M., Cheraghian, G., Oztop, H. F., & Shadloo, M. S. (2020). A review on the control parameters of natural convection in different shaped cavities with and without nanofluid. *Processes (Basel, Switzerland), 8*(9), 1011. DOI: 10.3390/pr8091011

Zontul, H., Hamzah, H., Kurtulmuş, N., & Şahin, B. (2021). Investigation of convective heat transfer and flow hydrodynamics in rectangular grooved channels. *International Communications in Heat and Mass Transfer, 126*, 105366. DOI: 10.1016/j.icheatmasstransfer.2021.105366

Chapter 16
Numerical Simulation of the Effect of the Magnetic Field on Convective Heat Transfer in a Three-Dimensional L-Shaped Cavity

Jaouad Benhamou
https://orcid.org/0000-0002-1958-2843
Higher School of Education and Training, Mohammed First University, Oujda, Morocco

Kamal Hirech
https://orcid.org/0000-0002-3531-7791
Higher School of Education and Training, Mohammed First University, Oujda, Morocco

Mohammed El Goual
Laboratory of Mechanics and Energetics, Faculty of Sciences, Mohammed First University, Oujda, Morocco

Mohammed Jami
Laboratory of Mechanics and Energetics, Faculty of Sciences, Mohammed First University, Oujda, Morocco

Salaheddine Channouf
Laboratory of Mechanics and Energetics, Faculty of Sciences, Mohammed First University, Oujda, Morocco

Ahmed Mezrhab
Laboratory of Mechanics and Energetics, Faculty of Sciences, Mohammed First University, Oujda, Morocco

El Bachir Lahmer
Laboratory of Mechanics and Energetics, Faculty of Sciences, Mohammed First University, Oujda, Morocco

ABSTRACT

This paper presents a numerical study of magnetohydrodynamic (MHD) natural convection in a three-dimensional L-shaped cavity using the lattice Boltzmann approach coupled with the finite difference method. The analysis focuses on the effect of a uniform vertical magnetic field on heat transfer, isotherms, and velocity field. The numerical code was validated by comparing our results with data available in the literature. The simulations revealed that, in the absence of a magnetic field, a significant improve-

DOI: 10.4018/979-8-3693-9924-8.ch016

ment in heat transfer of 93.27% on the left vertical hot wall and 225.65% on the bottom hot wall was observed when the Rayleigh number (Ra) increased from 10^3 to 10^5. In the presence of the magnetic field, heat transfer decreased by 47.28% on the left vertical wall and 66.58% on the bottom wall, when the Hartmann number was increased from zero to 200 for Ra fixed at 10^5.

INTRODUCTION

The study of thermal convection in confined geometries is a vast and dynamic field of research. Although the fundamental principles have long been explored, research continues to progress thanks to the diversity of thermal systems and their many specific applications. Similarly, magnetohydrodynamic (MHD) convection is a key field of research for understanding heat transfer mechanisms and fluid dynamics under the influence of magnetic fields, particularly in complex geometries. The simulation of these phenomena in three-dimensional configurations presents specific aspects due to the complex interactions between the magnetic field, induced currents, and heat fluxes.

Examining thermal convection in different cavity geometries has important and varied applications in sectors such as industry and engineering (Rostami et al., 2020). Cavities of various shapes, such as L-shaped (Alshuraiaan, 2023; Mahmoodi, 2011; Moria, 2021), T-shaped (Rouijaa et al., 2011; Sahi et al., 2017), H-shaped (Eshaghi et al., 2021; Keramat et al., 2020), C-shaped (Aghakhani et al., 2018; Asha et al., 2023), and many others, are widely studied in the scientific literature because of the distinct thermal and fluidic behaviors they generate. Each cavity geometry presents unique challenges in terms of fluid dynamics and heat transfer and can be used to model complex cooling systems and mixing processes. The diversity of these shapes means that thermal convection models can be tested and optimized in a wide range of realistic scenarios. Most cavity shape studies are carried out in two dimensions (2D). In this article, a three-dimensional (3D) study of the L-shaped cavity is carried out to clearly visualize the effect of the magnetic field on natural convection.

The Lattice Boltzmann Method (LBM) can be considered an interesting and powerful numerical tool for the simulation of complex fluid dynamics, due to its ability to efficiently manage the effects of viscosity, thermal conduction, and external forces such as the magnetic field (Krüger et al., 2017; Mohamad, 2019). In particular, LBM is well suited to modeling magnetohydrodynamic (MHD) convection (Benhamou et al., 2024; Sajjadi et al., 2019), as it captures the complex interactions between magnetic fields and fluid dynamics through a discrete approach that resolves particle distributions on a lattice. This technique is coupled in this paper with a finite-difference approach (hybrid LBM) to study the effect of a uniform magnetic field on natural convection in a three-dimensional L-shaped cavity.

The study focuses on the analysis of the effect of the magnetic field through the variation of the Hartmann number on heat exchange, isotherms, and velocity field. The numerical code used is first validated by comparing our results with literature data. This study contributes to a better understanding of the effects of geometry on MHD convection and demonstrates the flexibility of the hybrid LBM method for dealing with fluid flow problems in the presence of external magnetic fields.

3D LATTICE BOLTZMANN METHOD

The lattice Boltzmann method (LBM) is a numerical simulation technique generally used to model fluid dynamics. It simulates fluid dynamics by tracking the evolution of discrete distribution functions $f_i(x_i, t)$ defined on a spatial and temporal lattice. In this work, LBM based on the Multiple Relaxation Time (MRT) model is used instead of the Single Relaxation Time (SRT) model to improve the stability and accuracy of calculations, and the lattice D3Q19 is employed to discretize the calculation domain (Fig. 1).

The LBM-MRT is essentially based on the numerical solution of the following Boltzmann equation:

$$f_i(x_i + c_i\Delta t, t + \Delta t) - f_i(x_i, t) = M^{-1} S[\, m_i^{eq} - m_i] + \Delta t F_i \quad (1)$$

where $f_i(x_i, t)$ is the distribution function for discrete direction i, at position x_i and time t. c_i is the discrete velocity in direction i. Δt is the time step. M is the transformation matrix that converts distribution functions f_i into moments m_i. S is the diagonal matrix of relaxation times. m_i^{eq} are the equilibrium moments. F_i is the external force term, which may include effects such as gravity or other external force sources. All these parameters are well defined in our previous works (Benhamou et al., 2022, 2024).

Figure 1. D3Q19-LBM lattice.

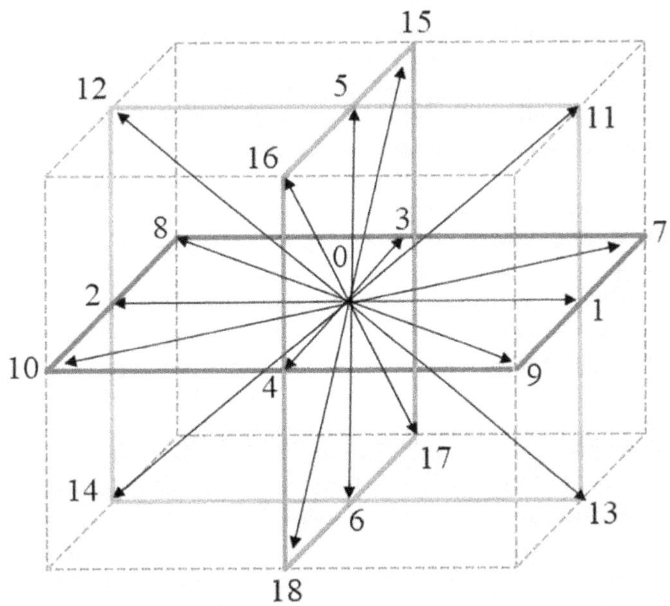

The physical system simulated in this article is subject to buoyancy (F_b) and Lorentz (F_L) forces. The buoyancy force is given in z–direction in discretized form as follows (Benhamou et al., 2022, 2024):

$$F_{b,i} = 3 W_i g\, (T - T_m) c_{z,i} \quad (2)$$

where W_i is the weight associated with the i discrete velocity direction in the D3Q19 lattice. ρ is the fluid density. g is the acceleration due to gravity. β is the thermal expansion coefficient of the fluid. T is the local temperature at the lattice point. T_m is a reference temperature, typically the mean temperature of the fluid. $c_{z,i}$ is the z−component of the discrete velocity vector c_i.

The Lorentz force per unit volume is defined by the following equation (Benhamou et al., 2024; Sajjadi et al., 2019):

$$F_L = (E + V \times B) \times B \tag{3}$$

σ, E, and V are the electrical conductivity of the fluid, the electric field vector, and the fluid velocity vector, respectively.

For the application of the magnetic field (B) of the constant intensity (B_0) in the opposite direction of the buoyancy force ($\vec{B} = -B_0\vec{z}$), the Lorentz force has two components and can be given in discretized form as a function of Hartmann number ($Ha = B_0 L \sqrt{-}$), which characterizes the impact of the magnetic field on the fluid, by (Benhamou et al., 2024; Sajjadi et al., 2019):

$$F_{L,i} = F_{x,i} + F_{y,i} = -3\,W_i \left(\frac{Ha}{L}\right)^2 \left(u c_{x,i} + v c_{y,i}\right) \tag{4}$$

where μ is the dynamic viscosity of the fluid. L is a characteristic length of the simulation domain. $c_{x,i}$ and $c_{x,i}$ are the components of the discrete velocity vectors in the x and y directions, respectively. u and v are the components of fluid velocity in the x and y directions, respectively.

After solving the Boltzmann equation (Eq. (1)) on a computer, macroscopic characteristics of the fluid such as density and velocities are calculated as follows:

$$\rho = \sum_{i=0}^{18} f_i,\ \rho \begin{pmatrix} u \\ v \\ w \end{pmatrix} = \sum_{i=0}^{18} f_i \begin{pmatrix} c_{x,i} \\ c_{y,i} \\ c_{z,i} \end{pmatrix} \tag{5}$$

FINITE DIFFERENCE METHOD

The lattice Boltzmann model for fluid flow simulation can be combined with another LBM model for temperature calculation by defining another Boltzmann equation. Our previous work has shown that combining the LBM method with the Finite Difference Method (FDM) can be more accurate and efficient than the double lattice Boltzmann method (pure LBM). Thus, FDM is combined with LBM in this work to simulate magnetohydrodynamic convection. In particular, FDM is used to calculate temperature by solving the following dimensionless energy equation:

$$\frac{\partial T}{\partial t} + u\frac{\partial T}{\partial x} + v\frac{\partial T}{\partial y} + w\frac{\partial T}{\partial z} = \left(\frac{\partial^2 T}{\partial x^2} + \frac{\partial^2 T}{\partial y^2} + \frac{\partial^2 T}{\partial z^2}\right) \tag{6}$$

where α is the thermal diffusivity of the fluid.

The first and second derivatives of equation (6) are determined using central finite difference methods of second order (Benhamou et al., 2022, 2024). The derivatives $\partial T/\partial x$ and $\partial^2 T/\partial x^2$ can be determined, for example, as follows:

$$\frac{\partial T}{\partial x} = \frac{T(i+1,j,k) - T(i-1,j,k)}{2x} \tag{7}$$

$$\frac{\partial^2 T}{\partial x^2} = \frac{T(i+1,j,k) - 2T(i,j,k) + T(i-1,j,k)}{x^2} \tag{8}$$

where i, j, and k correspond to the discretization indices along the x, y, and z axes, respectively.

VALIDATION

To validate our numerical model for simulating magnetohydrodynamic natural convection, we simulated the geometry previously investigated by Sajjadi et al. (Sajjadi et al., 2019). The system is a differentially heated cubic cavity, filled with a fluid whose Prandtl number (Pr) is equal to 0.73. A uniform external magnetic field along the $-z$ axis is applied to the cavity. The vertical walls on the left and right are hot and cold respectively. The remaining walls are adiabatic and defined by a zero spatial derivative of temperature (Fig.2).

Figure 2. Validation geometry.

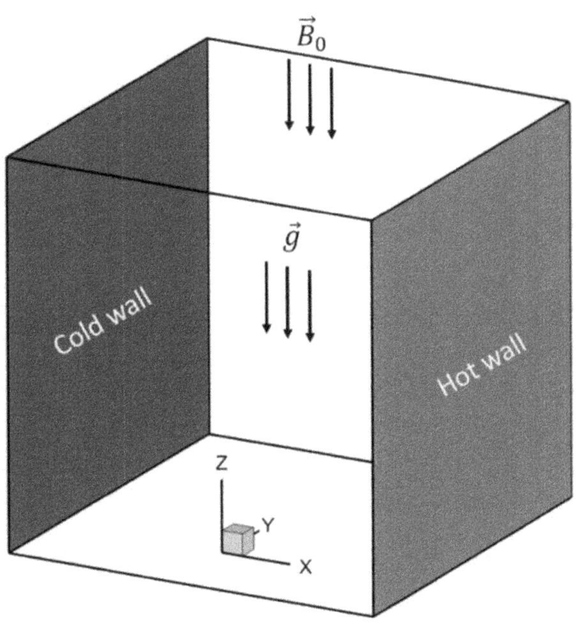

To validate our hybrid simulation method, we have checked the average Nusselt number (Nu_{2D}) values calculated on the vertical center axis of isothermal walls against the results obtained by Sajjadi et al. (Sajjadi et al., 2019) and Rudraiah et al. (Rudraiah et al., 1995) for different Grashof (Gr) and Hartmann (Ha) numbers. The mathematical expressions of Nu_{2D}) and Gr are given in the reference (Benhamou et al., 2024). The results, presented in Table 1, show good agreement with those reported in references (Rudraiah et al., 1995; Sajjadi et al., 2019). A further validation based on a comparison of isotherms for $Gr = 2 \times 10^4$ and $Ha = 10$ and 100 is carried out. The isotherms obtained with our code (Fig.3) correspond well to those found by Sajjadi et al. (Sajjadi et al., 2019) (Fig. 4). These two comparisons verify the accuracy of our model, confirming that it faithfully reproduces the thermal and hydrodynamic behaviors observed in previous studies. This good agreement validates the ability of our hybrid method to efficiently simulate MHD natural convection phenomena in 3D.

Note that temperature is represented in Figures 3 and 4 on a dimensionless scale. The hot wall is fixed at a dimensionless temperature $\theta_{hot} = 1$ and the cold wall at $\theta_{cold} = 0$. The transition from physical temperature to dimensionless temperature can be made using the following equation: $\theta = (T - T_{cold})/(T_{hot} - T_{cold})$.

Table 1. Average Nusselt number (Nu_{2D}) calculated on the vertical center axis of isothermal walls for $Gr = 2\ 10^4$ and $2\ 10^5$ and $0 \leq Ha \leq 100$.

Gr	*Ha*	**Rudraiah et al. (Rudraiah et al., 1995)**	**Sajjadi et al. (Sajjadi et al., 2019)**	**Present results**
$2\ 10^4$	0 1050100	2.51882.22341.08561.0110	2.43152.25101.09021.0208	2.4252 2.2636 1.0941 1.0229
$2\ 10^5$	0 1050100	4.91984.80532.84421.4317	5.01514.96842.99101.4578	5.02084.91433.02471.4624

Figure 3. Temperature isosurfaces obtained with our code for $Gr = 2 \times 10^4$ and $Ha = 10$ and 100.

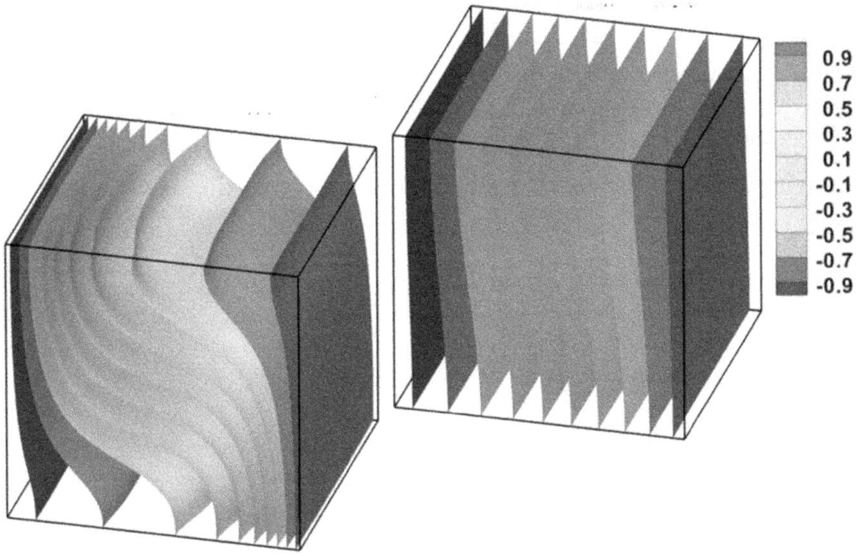

Figure 4. Temperature isosurfaces obtained by Sajjadi et al. (Sajjadi et al., 2019) for $Gr = 2 \times 10^4$ and $Ha = 10$ and 100.

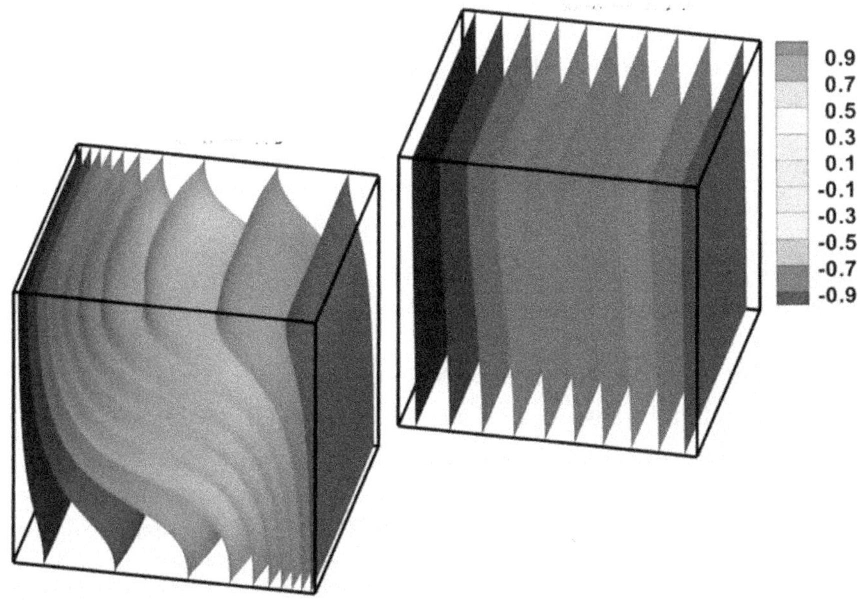

RESULTS AND DISCUSSION

This section presents a parametric study of Rayleigh (*Ra*) and Hartmann numbers on natural convection in a 3D L-shaped cavity filled with air. The left vertical wall and the bottom horizontal wall are hot and fixed at a dimensionless temperature $\theta_{hot} = 0.5$. Half of the opposite walls are considered cold ($\theta_{cold} = -0.5$) (Fig. 5). The rest of the walls are adiabatic. The width of the cavity ($L_y = L_x/2$ or $L_z/2$) is equal to half its length (L_x) and height ($L_z = L_x$). A uniform vertical magnetic field is applied in the direction of the gravity vector g to study its effect on thermal convection in the enclosure.

Figure 5. 2D illustration of the geometry studied (vertical section).

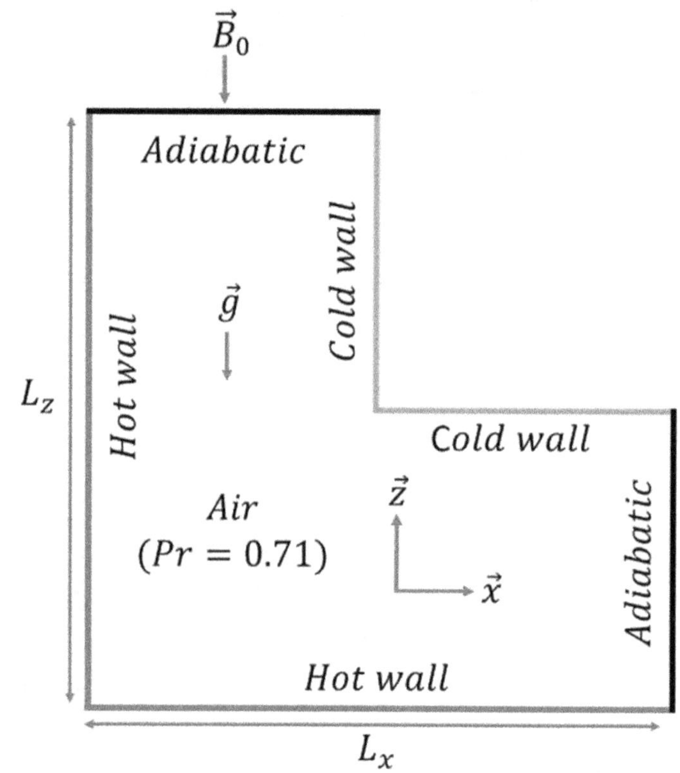

o Rayleigh number effect in the absence of the magnetic field ($Ha = 0$)

The influence of a Rayleigh number between 10^3 and 10^5 on the Nusselt number and isotherms is studied in this part in the absence of a magnetic field. ($Ha = 0$). This dimensionless number is defined as follows:

$$Ra = \frac{g\, L_z^3}{} \left(T_{hot} - T_{cold}\right) \tag{9}$$

where ν is the kinematic viscosity of air.

The first investigation carried out here concerns the impact of *Ra* on heat transfer in the cavity, by calculating the average Nusselt number at the hot walls (Nu_{3D})). For example, at the left vertical hot wall, this number is calculated as follows:

$$Nu_{left} = \frac{1}{L_y L_z} \int_0^{L_z} \int_0^{Ly} Nu \, dy \, dz \tag{10}$$

where *Nu* is the local Nusselt number. It is defined by the following equation:

$$Nu = \frac{L_x}{T_h - T_c} \left(\frac{\partial T}{\partial x}\right)_{wall} \tag{11}$$

Table 2 shows the results of the Nusselt number calculated at the hot walls. The results show that, for both walls, the average Nusselt number increases as the Rayleigh number increases. An improvement in heat exchange of 93.27% was observed on the left vertical wall and 225.65% on the bottom wall, passing from $Ra = 10^3$ to 10^5. This is attributed to the transition from conduction to convection, where the increase in fluid velocity favors heat exchange. The table also shows that, for all *Ra* values considered, heat transfer is relatively greater at the bottom wall than at the left. This is because the fluid in contact with the hot walls heats up, rises to the top wall, and then descends along the cold walls. As it cools and passes over the hot lower wall, a significant heat exchange takes place, warming the fluid before it rises towards the hot vertical wall. At this point, the fluid is already heated and therefore doesn't exchange much heat with the vertical wall, which explains the higher value of the Nusselt number on the lower wall compared to the vertical surface.

The isotherms in a cavity are illustrated in Figure 6, showing their significant influence by a large value of the Rayleigh number. At low *Ra* ($Ra = 10^3$), conduction dominates heat transfer. The isotherms are mainly parallel to the hot and cold walls of the cavity, indicating a linear temperature gradient. The isotherm curves remain uniform and close to each other. For $Ra = 10^4$, small deformations can be observed in the isosurfaces. As Ra increases ($Ra = 10^5$), convection begins to play a more important role. The isotherms show remarkable curvatures in different areas of the cavity.

Table 2. Average Nusselt number calculated at the hot left and bottom cavity walls for different Ra and Ha = 0.

Ra	Nu_{left}	Nu_{bottom}
10^3	1.116	1.384
10^4	1.296	1.844
10^5	2.157	4.507

Figure 6. Isosurfaces of the temperature for Ha = 0 and different Ra.

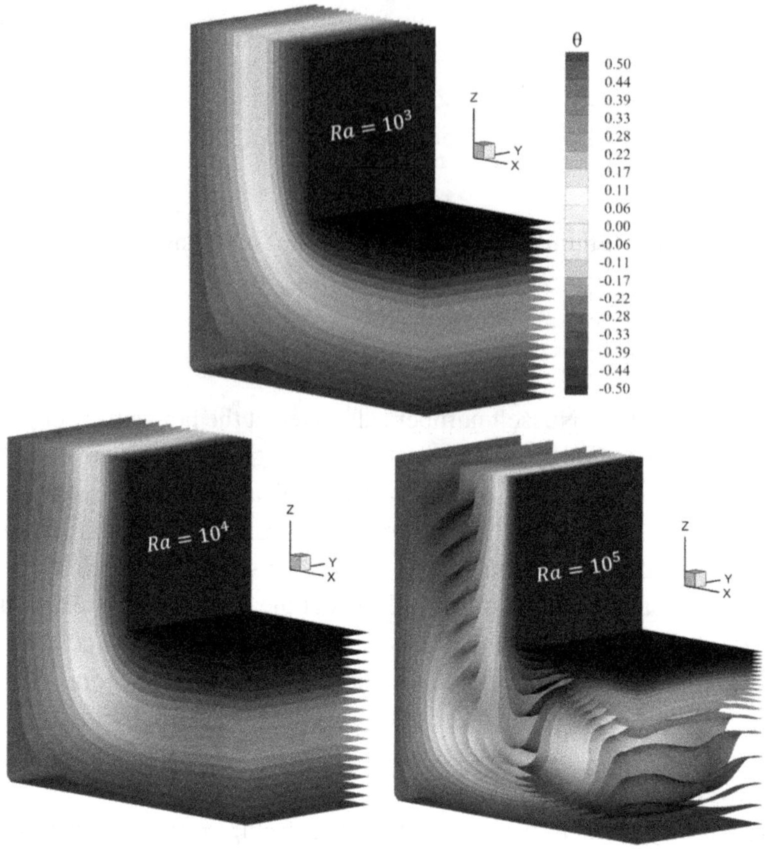

o **Effect of the magnetic field**

This section presents a study of the effect of applying a uniform vertical magnetic field (along the $-z$ axis) on the Nusselt number, isotherms, and velocity field in the cavity for a Rayleigh number fixed at 10^5. The variation in magnetic field strength is taken into account through the Hartmann number ($0 \leq Ha \leq 200$), which is a measure of the influence of the magnetic field on fluid flow, and increasing values of *Ha* indicates an increase in the effect of the magnetic field.

Table 3 presents the *Nu* calculated on the hot walls and shows that heat transfer is reduced by 47.28% at the left vertical wall and by 66.58 at the bottom wall, passing from $Ha = 0$ to $Ha = 200$. In the absence of a magnetic field $Ha = 0$, natural convection is strong. The Nusselt number is high, indicating efficient heat transfer due to intense convective motion in the cavity. Convection cells form, contributing to increased heat exchange between hot and cold walls. At $Ha = 50$, the magnetic field begins to influence fluid flow, exerting a Lorentz force that reduces circulation velocity. The *Nu* decreases compared to $Ha = 0$, but remains high, as convection is still predominant, though slightly slowed. At $Ha = 100$, the effect of the magnetic field becomes more pronounced. The Lorentz force decelerates the fluid further, significantly reducing the Nusselt number. Convection is still present, but its intensity is reduced, leading to less efficient heat transfer compared to lower values of *Ha*. At $Ha = 150$, the braking effect is substantial. Fluid circulation slows considerably, and the Nusselt number falls even further. Convective heat transfer is considerably reduced, and conductive effects begin to dominate. At $Ha = 200$, the Lorentz force imposes a significant constraint on the fluid, practically suppressing convective motion. The Nusselt number approaches values characteristic of pure conduction, indicating that heat transfer is mainly due to conduction with very little contribution from convection.

Table 3. Average Nusselt number calculated at the hot left and bottom cavity walls for different Ha and $Ra = 10^5$.

Ha	Nu_{left}	Nu_{bottom}
0	2.157	4.507
50	1.586	2.792
100	1.202	1.949
150	1.195	1.624
200	1.137	1.506

Figure 7 shows temperature isosurfaces for $Ra = 10^5$ and different values of *Ha*. For $Ha = 0$, the isotherms show complex, deformed curves, reflecting active convection cells and robust fluid movements. At $Ha = 50$, the isotherms begin to smooth out slightly, the deformations being less pronounced than at $Ha = 0$, indicating a moderate reduction in the intensity of convective motions. At $Ha = 100$, the isotherms become visibly more regular and less deformed. The magnetic field slows the fluid sufficiently for the isotherm curves to become more uniform. As the Hartmann number increases further (e.g. $Ha = 200$), the isotherms show a transition to more linear, wall-parallel shapes. Convection is strongly suppressed, and isotherm curves become more uniform and closely spaced, indicating heat transfer mainly by conduction.

Figure 7. Isosurfaces of the temperature for Ra = 10^5 and different values of Ha.

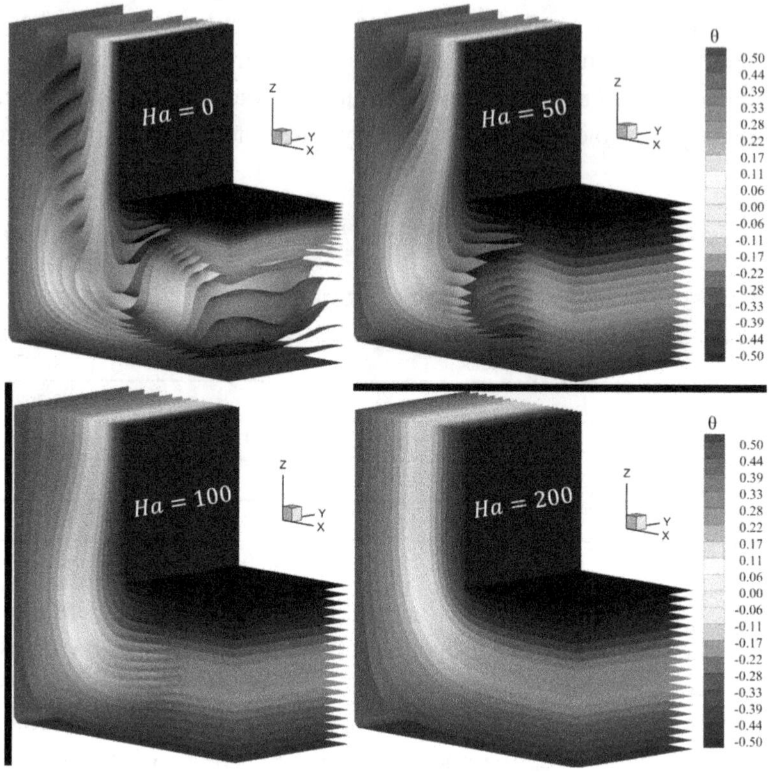

The behavior of the fluid in the cavity can be described using the velocity vectors shown in Figure 8. In the absence of a magnetic field ($Ha = 0$), natural convection dominates. The velocity vectors show vigorous movements and well-defined convection cells in the cavity. The fluid moves rapidly along the heated walls, rises to the top of the cavity, and then descends along the cooled walls, forming two notable recirculation zones. At Ha = 50, the effect of the magnetic field begins to be felt. The Lorentz force acts on the fluid, slightly reducing its velocity. Velocity vectors show a decrease in the intensity of convective movements. Convection cells are still present, but vortices and recirculations become less pronounced, particularly in the lower right-hand region of the cavity. For $Ha \geq 100$, the influence of the magnetic field is more pronounced. The Lorentz force slows fluid movement, reducing the intensity of convection currents. Convection cells become less distinct, and recirculation movements are greatly reduced. Fluid circulation is smoother, and recirculation zones are reduced, particularly in the lower right-hand corner where fluid circulation disappears.

Figure 8. Velocity vectors for Ra = 10^5 and different Ha.

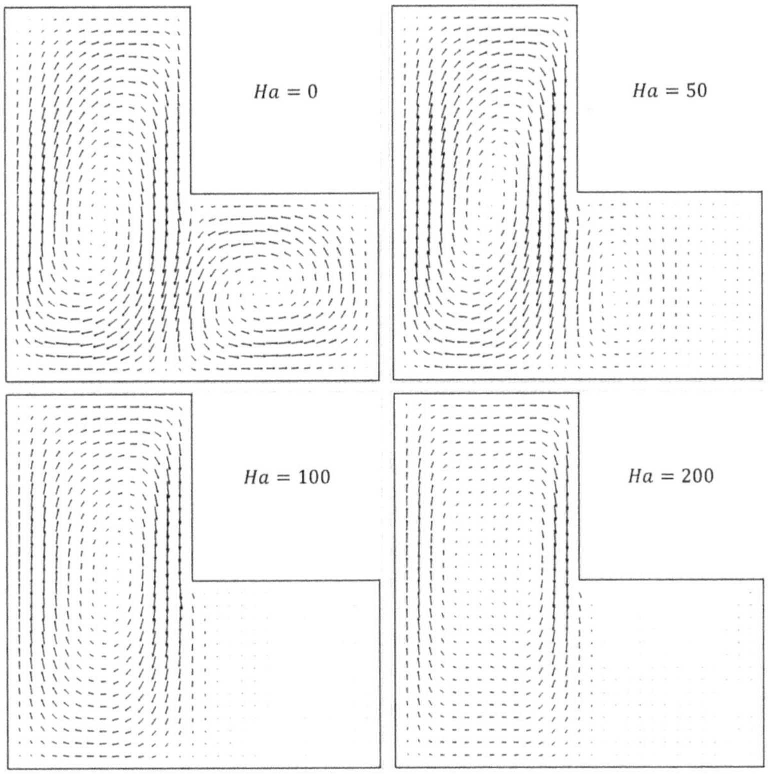

CONCLUSION

This paper reported on a 3D numerical study of the effect of a uniform vertical magnetic field on natural convection in an L-shaped cavity using a hybrid numerical approach. This approach includes the lattice Boltzmann method for fluid flow simulation and the finite difference approach for temperature calculation. The study focused on the effect of increasing magnetic field intensity, by increasing the Hartmann number (*Ha*), on heat transfer, isotherms, and velocity field. Simulations showed that in the absence of a magnetic field (*Ha* = 0), a significant improvement in heat transfer of 93.27% on the left vertical hot wall and 225.65% on the bottom hot wall was observed when the Rayleigh number (*Ra*) was increased from 10^3 to 10^5. In the presence of the magnetic field, heat exchange decreased by 47.28% on the left vertical wall and by 66.58% on the bottom wall, going from *Ha* = 0 to 200 for *Ra* = 10^5.

REFERENCES

Aghakhani, S., Pordanjani, A. H., Karimipour, A., Abdollahi, A., & Afrand, M. (2018). Numerical investigation of heat transfer in a power-law non-Newtonian fluid in a C-Shaped cavity with magnetic field effect using finite difference lattice Boltzmann method. *Computers & Fluids*, *176*, 51–67. Advance online publication. DOI: 10.1016/j.compfluid.2018.09.012

Alshuraiaan, B. (2023). Fluid Structure Interaction Analysis of Natural Convective Flow and Heat Transfer in an L-Shaped Cavity. *Journal of Advanced Research in Fluid Mechanics and Thermal Sciences*. DOI: 10.37934/arfmts.109.1.112

Asha, N. E. J., Nag, P., Akhter, M. N., & Molla, M. M. (2023). MRT-lattice Boltzmann simulation of magnetic field effects on heat transfer from a heater in a C-shaped cavity filled with non-Newtonian hybrid nanofluids. *International Journal of Thermofluids*. DOI: 10.1016/j.ijft.2023.100345

Benhamou, J., Channouf, S., Lahmer, E. B., Jami, M., & Mezrhab, A. (2024). Hybrid-lattice Boltzmann Method for the Simulation of Magnetohydrodynamic Conjugate Heat Transfer and Entropy Generation in Three Dimensions. *Arabian Journal for Science and Engineering*, *49*(1), 1181–1206. Advance online publication. DOI: 10.1007/s13369-023-08273-y

Benhamou, J., Lahmer, E. B., & Jami, M. (2022). Three-dimensional simulation of conjugate heat transfer using the hybrid lattice Boltzmann-finite difference method. *International Communications in Heat and Mass Transfer*, *139*, 106486. DOI: 10.1016/j.icheatmasstransfer.2022.106486

Eshaghi, S., Izadpanah, F., Dogonchi, A. S., Chamkha, A. J., Ben Hamida, M. B., & Alhumade, H. (2021). The optimum double diffusive natural convection heat transfer in H-Shaped cavity with a baffle inside and a corrugated wall. *Case Studies in Thermal Engineering*, *28*, 101541. Advance online publication. DOI: 10.1016/j.csite.2021.101541

Keramat, F., Azari, A., Rahideh, H., & Abbasi, M. (2020). A CFD parametric analysis of natural convection in an H-shaped cavity with two-sided inclined porous fins. *Journal of the Taiwan Institute of Chemical Engineers*, *114*, 142–152. Advance online publication. DOI: 10.1016/j.jtice.2020.09.011

Krüger, T., Kusumaatmaja, H., Kuzmin, A., Shardt, O., Goncalo, S., & Viggen, E. M. (2017). The lattice Boltzmann method: Principles and practice. In *Springer International Publishing*. Springer Cham., DOI: 10.1007/978-3-319-44649-3

Mahmoodi, M. (2011). Numerical simulation of free convection of a nanofluid in L-shaped cavities. *International Journal of Thermal Sciences*, *50*(9), 1731–1740. Advance online publication. DOI: 10.1016/j.ijthermalsci.2011.04.009

Mohamad, A. A. (2019). Lattice Boltzmann method: Fundamentals and engineering applications with computer codes. In *Springer London* (2nd ed.). Springer London., DOI: 10.1007/978-1-4471-7423-3

Moria, H. (2021). Natural convection in an L-shape cavity equipped with heating blocks and porous layers. *International Communications in Heat and Mass Transfer*, *126*, 105375. Advance online publication. DOI: 10.1016/j.icheatmasstransfer.2021.105375

Rostami, S., Aghakhani, S., Pordanjani, A. H., Afrand, M., Cheraghian, G., Oztop, H. F., & Shadloo, M. S. (2020). A review on the control parameters of natural convection in different shaped cavities with and without nanofluid. *Processes (Basel, Switzerland)*, *8*(9), 1011. DOI: 10.3390/pr8091011

Rouijaa, H., El Alami, M., Semma, E. A., & Najam, M. (2011). *Natural convection in an inclined T-shaped cavity*. Fluid Dynamics and Materials Processing.

Rudraiah, N., Barron, R. M., Venkatachalappa, M., & Subbaraya, C. K. (1995). Effect of a magnetic field on free convection in a rectangular enclosure. *International Journal of Engineering Science*, *33*(8), 1075–1084. DOI: 10.1016/0020-7225(94)00120-9

Sahi, A., Sadaoui, D., Sadoun, N., & Djerrada, A. (2017). Effects of magnetic field on natural convection heat transfer in a T-shaped cavity. *Mechanics & Industry*, *18*(4), 407. Advance online publication. DOI: 10.1051/meca/2017014

Sajjadi, H., Amiri Delouei, A., Sheikholeslami, M., Atashafrooz, M., & Succi, S. (2019). Simulation of three dimensional MHD natural convection using double MRT Lattice Boltzmann method. *Physica A*, *515*, 474–496. DOI: 10.1016/j.physa.2018.09.164

Chapter 17

MorWAK Towards a More Energy -Efficient and Eco-Friendly Dialysis

Abdelaziz Sidi Baba
https://orcid.org/0009-0009-8509-2648
Faculty of Medicine and Pharmacy, Hassan II University, Casablanca, Morocco

Hasnae Zahim
Faculty of Medicine and Pharmacy, Hassan II University, Casablanca, Morocco

Abdellah Boualam
Faculty of Medicine and Pharmacy, Hassan II University, Casablanca, Morocco

Mohamed Zamd
Faculty of Medicine and Pharmacy, Hassan II University, Casablanca, Morocco

Abdellah Ait Tableb
Faculty of Medicine and Pharmacy, Hassan II University, Casablanca, Morocco

Naoufal Mtioui Chkairi
Faculty of Medicine and Pharmacy, Hassan II University, Casablanca, Morocco

Benyounes Ramdani
Faculty of Medicine and Pharmacy, Hassan II University, Casablanca, Morocco

ABSTRACT

Conventional extrarenal technologies are often energy-hungry and responsible for substantial greenhouse gas emissions. In response to this problematic, the Moroccan Wearable Artificial Kidney (MorWAK) project proposes an alternative, more sustainable solution in the context of green dialysis. MorWAK differs radically from existing extrarenal technologies, which consume a huge amount of energy to ensure the filtration processes (the consumption of a hemodialysis generator per 4-hour session is around 3 kWh, and that of the water treatment unit to produce purified water is 150 kW/d). On the other hand, the MorWAK operates without being connected to a water source, consuming very little energy (10 W). This contributes to eco-responsible energy management, which has a direct impact on reducing greenhouse gas emissions. MorWAK thus represents a breakthrough innovation in improving the sustainability of healthcare technologies, offering a more eco-friendly alternative to current energy challenges

DOI: 10.4018/979-8-3693-9924-8.ch017

INTRODUCTION

Climate change is one of the major challenges facing humanity in the 21st century (Canada, 2022). The consequences can range from disruption of the ecological balance and influence on the ecosystem to serious damage to human health (Q et al., 2022),(Singh, 2024). These climate changes are mainly due to global warming, caused by the emission of greenhouse gases by a number of sectors, in particular industry, energy, agriculture and transport. However, the medical sector also makes a significant contribution, producing around 4-5% of these emissions. Indeed, if the healthcare sector were considered as a country, it would be the fifth largest contributor to greenhouse gas emissions (Barraclough & McAlister, 2022). Most of these emissions come from haemodialysis activities (Nagai et al., 2021).

Although hemodialysis is essential for the treatment of patients suffering from chronic end-stage renal disease (ESRD), the resulting consequences in terms of enormous water and energy consumption and the production of large quantities of waste, associated with the medical practices currently adopted, pose a major environmental challenge. In view of the global climate emergency and the increasing demand for dialysis services due to the high prevalence of chronic kidney disease, it is become imperative to reconsider the use of sustainable solutions in the field of hemodialysis, in order to minimize its carbon footprint. It is with this in mind that the concept of "green nephrology" has recently emerged as an innovative response to reducing the ecological impact of renal treatments without compromising the quality of care, changing practices and using available resources in a more environmentally-friendly way. Several initiatives have already been taken in this context, notably by nephrology societies in Australia, New Zealand and Italy. With this in mind, the NHS Green Nephrology Network, a sustainable healthcare program in the UK, has been able to save around 10 million euros a year in its healthcare system thanks to environmentally-friendly water and electricity saving initiatives (Otero González, 2024) .

Moreover, hemodialysis technology has a crucial role to play in reducing the carbon footprint by developing innovative technology solutions that meet sustainability requirements, through the design of systems offering moderate consumption of water and energy resources, thus producing low quantities of waste. However, efforts to develop these technological solutions are underway, but their impact remains moderate in relation to the expected results. It is therefore essential to take the next step forward. It is here that the Moroccan Wearable Artificial Kidney (MorWAK) technology comes into play. It has been developed with several aspects in mind, including ecology. It is a portable hemodialysis device. It purifies the patient's blood without the use of water, consumes very little energy and generates very little waste. This device offers an eco-responsible solution that could transform the way dialysis is practiced, contributing to a more sustainable future for public health.

The aim of this paper is to present the MorWAK portable hemodialysis device as an alternative solution for treating patients with ESRD, focusing in particular on its energetic aspect which aligns with energy efficiency strategies in order to highlight its objective of reducing the carbon footprint associated with hemodialysis activity.

THE ERSD AND TREATMENT MODALITIES

ESRD is becoming a major public health problem. It currently affects between 5% and 10% of people worldwide(Waziri et al., 2019). It occurs when the native kidneys are no longer capable of ensuring renal hemostasis, in other words, when glomerular filtration rate is below 15 ml/min/1.73 m^2. Although diabetes, hypertension, obesity and ageing are the main factors leading to ESRD (Fiseha & Osborne, 2023). Today, there are two ways of treating patients with ESRD: renal transplantation and hemodialysis. In fact, renal transplantation is the most recommended treatment option for patients, due to the benefits it brings including improved quality of life and reduced mortality, but the high prevalence of ESRD and the scarcity of organ donors have ultimately led to a critical shortage of available kidneys (Tonelli et al., n.d.). Hemodialysis is the most widespread therapeutic modality worldwide. Around 3.4 million people worldwide undergo hemodialysis, equivalent to 90% of patients, and this number is rising steadily, reaching 5 million by 2025 (Bonomini et al., 2022).

ENERGY IMPACT OF HEMODIALYSIS

Hemodialysis is essential to patient survival. It is also a treatment that consumes huge energy to maintain the filtration process. This process always involves a dialysis machine and a water purification system.

Figure 1. The main components used in hemodialysis to treat patients ESRD

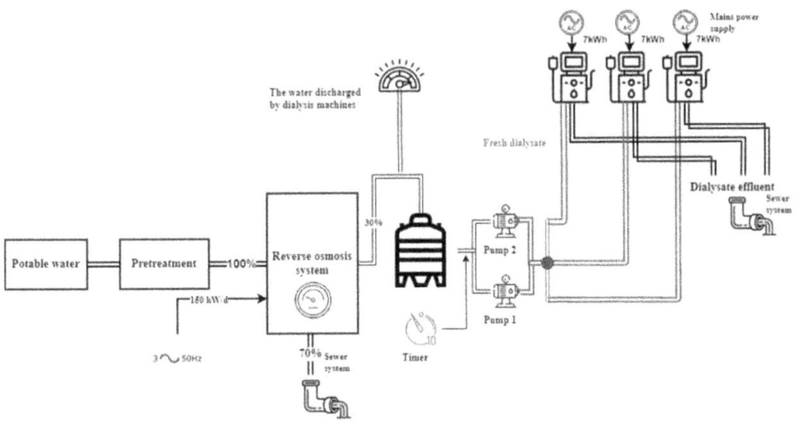

The figure above illustrates the energy requirements for patient treatment in hemodialysis centers, which are divided into two types: those for the purified water supply and those for the hemodialysis machine. In the course of a day, the water purification station can consume around 150 KW of energy per day, while the dialysis machines, which operate for ten hours each, consume around 7 KW per day, including the energy consumed for dialysate preparation, treatment, rinsing and thermochemical disinfection.

The averaged power consumption of dialysis machines is 2.6 kW per 4-hour session. With regard to water purification systems, Jabarne et al. revealed in their study that 150 kW/d were consumed to supply purified water to 13 hemodialysis machines. (Jabrane et al., 2013). In general conditions, a patient with ESRD receives 156 treatment sessions per year, at a rate of 3 sessions per week, which is equivalent to an annual energy consumption of 405 kW/year. These energy requirements lead to a high carbon footprint. In addition, carbon balances carried out as part of the accounting of direct and indirect greenhouse gas emissions linked to the hemodialysis activities have shown that the hemodialysis sector has a significant carbon footprint, the results of which depend on the installations, treatment frequencies and contributors to emissions. Only a few assessment studies have been able to quantify this carbon footprint, introducing various greenhouse gas emission items, including energy and heating, purchase of goods and services, transport, waste and water consumption (Table 1).

Table 1. Summary of carbon footprint results for dialysis centers worldwide

Study	Year of Publication	Geographic Location	Main Objective	Methodology Used	Sources of Emissions Included	Key Results (Total Emissions)	Factors Influencing Emissions	Energy Footprint
Mtioui et al.(Mtioui et al., 2021)	2020	Morocco	Quantify the carbon footprint of a hemodialysis unit	Bilan Carbone Tool	Energy, transportation, equipment purchases, water, waste	408.98 t CO_2-eq/ year (5.11 t CO_2-eq/ patient/year)	Dominance of fossil energy, frequent transportation	28%
Connor et al.(Connor et al., 2011)	2011	United Kingdom	Compare carbon footprints of different dialysis modalities	PAS 2050 Analysis	Energy, transportation, medical equipment purchases	3.8 t CO_2-eq/patient/ year (ICHD), up to 7.2 t CO_2-eq for HHD	Frequency and type of treatment (ICHD vs. HHD)	21%
Lim et al.(Ae et al., n.d.)	2013	Australia	Measure the carbon footprint of a satellite unit	DRS-based Approach	Energy, transportation, water, medical equipment, medication	10.2 t CO_2-eq/ patient/year	Regional variations, medical and pharmaceutical equipment	18.6%

The energy sector is a major component of the carbon footprint, with a statistically significant value ranging from 18% to 28%. In the Moroccan context, the carbon balance carried out by Mtioui et al. showed that energy use is the main contributor to greenhouse gas emissions, with a carbon footprint of 113.5 tCO_2 eq per year, or a percentage of 28%. The reason for this is that Morocco still produces its fossil energy through thermal power plants, with a percentage of 65.6% according to the national electricity office (figure 2). The study was conducted over one year, from January 1 r to December 31, 2019, at Casablanca's Ibn Rochd university center, equipped with 24 dialysis machines and a double osmosis treatment unit. It presents the carbon footprint of each emission item.

Figure 2. the results of a carbon footprint study carried out in the dialysis department at CHU Ibn Rochd in Casablanca

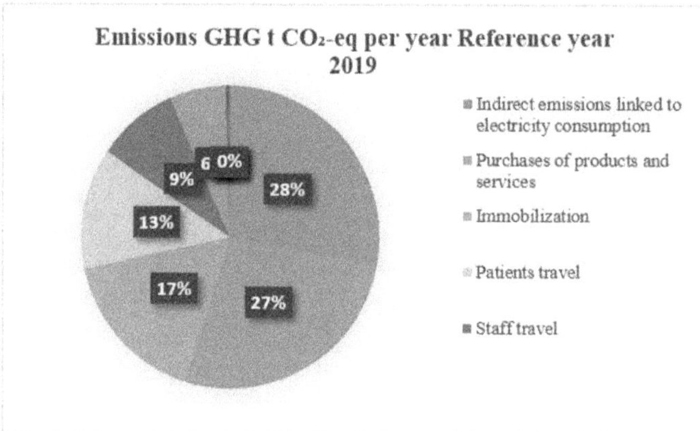

OPTIMIZING ENERGY CONSUMPTION IN HEMODIALYSIS

At present, there are limited ways of reducing the energy consumption of hemodialysis systems. However, the effect of energy consumption on the carbon footprint of hemodialysis depends not only on the energy requirements of hemodialysis machines, but also on the energy sources or technologies employed for local power generation. In the state of Victoria, where highly polluting lignite is widely used, an Australian study showed that 18.6% of total carbon dioxide emissions were attributable to electricity generation. By contrast, in Tasmania, where the majority of electrical energy is hydroelectric, the share of electricity in total emissions was considerably lower (5.2%) (Ka & Jwm, n.d.). In recent years, the use of renewable energies such as solar and wind power in various fields has gained ground. (Deshmukh et al., 2023), because these energies can contribute to reducing the carbon footprint associated with energy consumption (Guo et al., 2022). In the context of the use of renewable energies in hemodialysis, the literature reports only one study dating from 2012, concerning the use of 24 m^2, 3 kWh solar panels in a home dialysis service with four hemodialysis machines and one osmosis machine. The result showed a considerable reduction in electricity consumption of 91% and electricity costs of 76.5%, despite a high investment cost estimated at 16,219 Australian dollars(Ka & Jwm, n.d.).

MORWAK'S ECOLOGICAL PERFORMANCE

The MorWAK was developed as part of an ecological approach to dialysis, to reduce its carbon footprint while minimizing energy and water consumption and waste production.

Presentation of MorWAK and Its Operation Principal

The MorWAK (Figure 3) is a hemodialysis device. It is designed for use in patients suffering not only from ESRD, but also from congestive heart failure. It was developed by the biomedical engineering team affiliated to the Laboratory of Cellular and Molecular Physiopathology - Inflammatory, Degenerative and Oncological (LCMPIDO) at the Faculty of Medicine and Pharmacy, Hassan II University, Casablanca. This innovation has been developed with several aspects in mind, including ecology, economics, treatment modalities, etc. It is patented under number WO/2018/084690.

Figure 3. Presentation of MorWAK device as final prototype

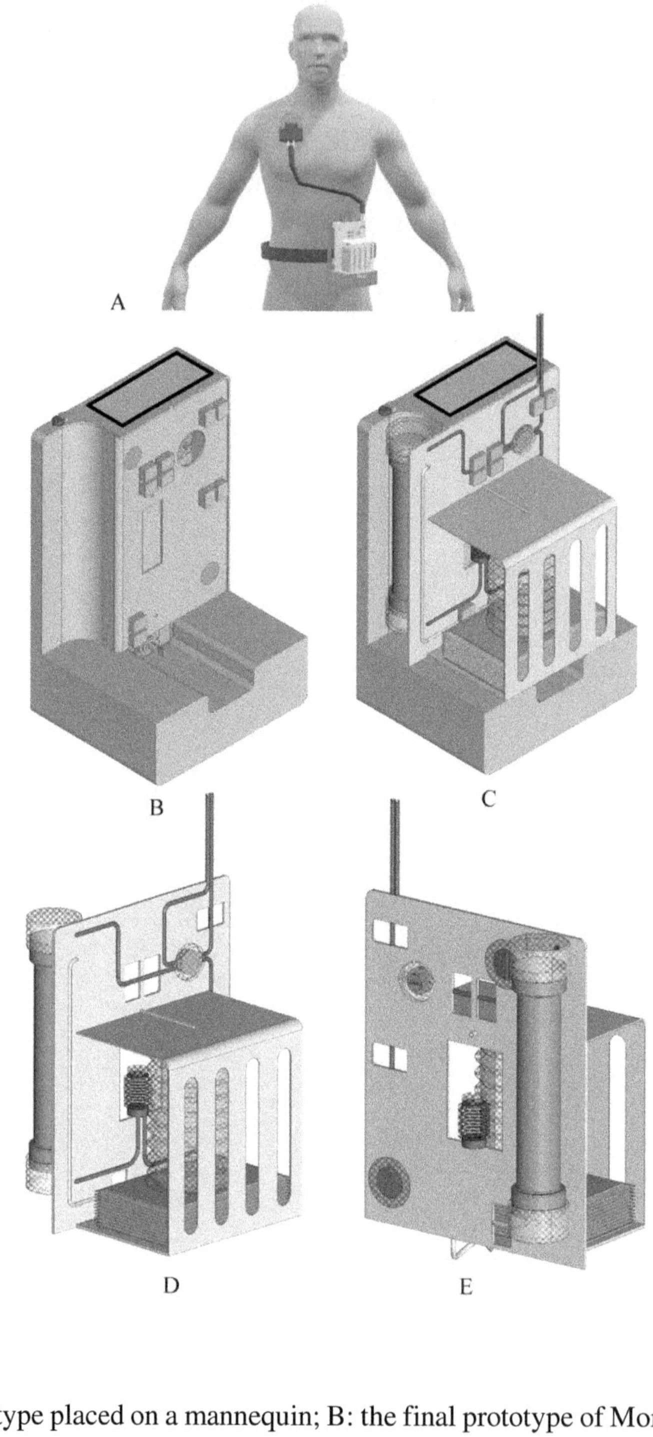

A: MorWAK prototype placed on a mannequin; B: the final prototype of MorWAK; C: the final prototype of MorWAK with the specific disposal mounted on it; D: specific disposal of MorWAK (Front view); E: specific disposal of MorWAK (Back view).

The MorWAK is equipped with alarm management systems to ensure patient protection at all times (Figure 5). Its operating principle can be summarized in the following main steps: Initially, the blood to be purified (2) passes through a peristaltic roller pump to reach the hemofilter (1). The depression created within the bellows (9), which constitute the masterpiece of the system, enables ultrafiltration of the blood around the hollow fibers of the hemofilter, the ultra-filtered liquid is then collected in the bellows, and the purified blood (3) is returned to the patient, passing through the air bubble detector integrated in the safety electronics unit (6). At the end of the treatment, an isonating solution is injected into the bloodstream.

Figure 4. MorWAK photo as patented

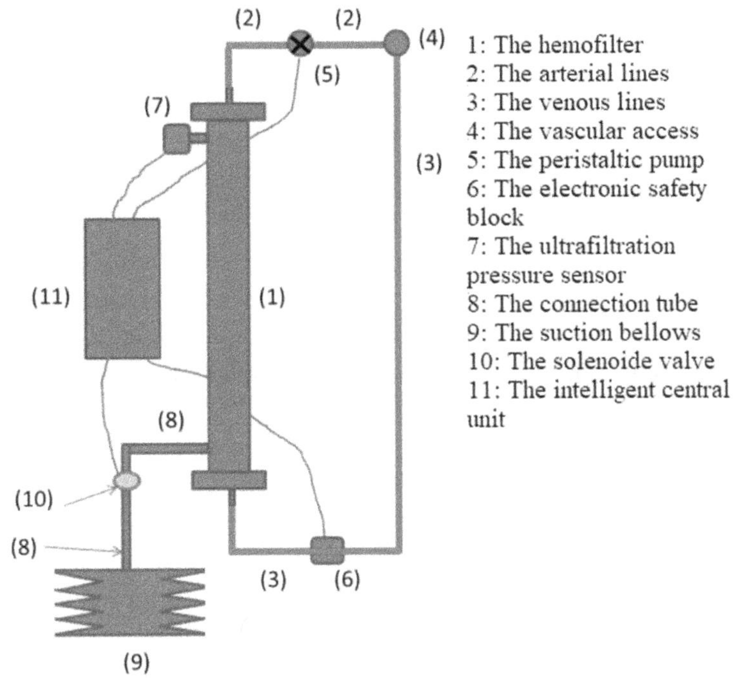

Energy Efficiency With MorWAK

The necessity of combating climate change calls for greater energy efficiency (Chen et al., 2024). The MorWAK was developed with the energy aspects of current dialysis systems in mind. One of the MorWAK's objectives was to treat patients with the lowest possible energy consumption. The device is portable and compact, weighing 3.5 kg with its 12 V battery. It operates without being connected to a water source and consumes very little power (10 W) compared with current extrarenal technologies, which consume between 600 and 700 W, in addition to the power consumed during thermochemical disinfection, which varies from one dialysis system to another depending on the current consumed and the activation time. The following figures show the difference between MorWAK and current dialysis systems in terms of energy consumption to treat the patient and produce purified water. It can be seen that MorWAK treatment offers a major advantage over dialysis machines in terms of energy efficiency.

Figure 5. Energy in W consumed by MorWAK and dialysis machine during four hours

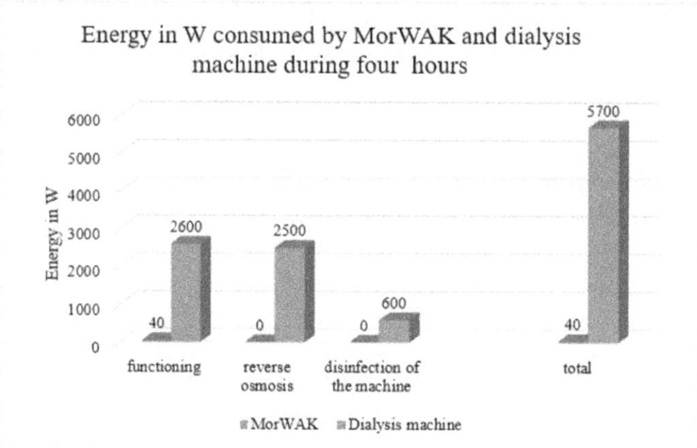

Data from the study conducted by (Jabrane et al., 2013), to calculate the energy consumed by each operation of the dialysis machine. The disinfection time chosen was 50 min. The average energy used for the dialysis machine is 650 Wh (treatment and thermochemical disinfection) and 6.25 KWh (water treatment unit).

The MorWAK System: Performing Waterless Dialysis With Minimal Waste

Conventional hemodialysis systems are very demanding on water resources. During a dialysis session, 500 liters of water are consumed per 4-hour session and per patient, i.e. an annual consumption of 78 m^3. The systems supplying the dialysis machines have a very low level of efficiency, estimated at 30%, so that 70% of the water is discharged down the sewage system, which explains why so much water is wasted during each session (Figure 1). On the other hand, the MorWAK operates without being connected to a water source, which preserves natural resources, particularly in arid and semi-arid countries suffering from water scarcity. Regarding waste, during hemodialysis sessions, patients produce around 390 kg of waste per year, at a rate of 2.5 kg per session, 38% of which is plastic. An Italian study showed that this production could reach 8 kg per session (Ka & Jwm, n.d.). This waste has both an environmental and an economic impact. Environmental, because most waste is disposed of by incineration, a procedure that results in the formation of hazardous chemical compounds such as polychlorinated dibenzo p-dioxins (PCDDs) and polychlorinated dibenzo furans (PCFs), as well as heavy metals (Hoenich & Pearce, 2002). In economic terms, disposing of a kilogram of waste entails considerable economic expenditure, which can range from 2.2 to 16 euros, a considerably high cost that can reach 10 to 50% of the cost of the disposable product (Piccoli et al., 2015). In addition to solid waste, hemodialysis also produces liquid waste resulting from the treatment process. These are the dialysis effluent and the water rejected by the water treatment unit, and although a number of strategies have been implemented to achieve zero liquid treatment, this involves reusing the water rejected by the machines and water purification systems. However, implementation of this process appears difficult due to the high investment costs involved (Tarrass et al., 2021). In response to this challenge, the MorWAK's closed system and mode of operation make it a totally different technology from conventional dialysis systems. As our device operates without any connection to a water source, it's the suction bellows that filter the blood and recover the ultra-filtered liquid at the same time. As a result, treatment takes place without discharge into the sewage system, while protecting the environment. The solid waste produced by MorWAK is limited to the dialyzer and the arterial and venous lines, which are classified as hazardous waste because they enter into contact with the patient's blood. The bellows are made of PVC and can therefore be recycled.

MORWAK AND RENEWABLE ENERGY

Facing the challenges posed by climate change, mainly due to global warming caused by the emission of greenhouse gases. The search for sustainable solutions is gaining in interest thanks to the use of eco-friendly processes to produce clean energy in place of fossil fuels, which contribute significantly to greenhouse gas emissions. (Justice et al., 2024). In recent years, the use of renewable energy worldwide has seen a remarkable increase in several areas, particularly solar energy (Ritchie et al., n.d.). To reduce stress on the environment, achieve decarbonization, ensure climate neutrality and achieve sustainable development as defined in the UN's Agenda 2030 and its Sustainable Development Goals (SDGs), as well as in the Paris Agreement, the use of renewable energies is proving essential. With the aim of preventing the Earth from crossing the critical 1.5°C threshold, and in order to achieve carbon neutrality goals, renewable energies have been identified as a key response. This underlines the colossal potential of renewable energy to neutralize the impact of greenhouse gases (Justice et al., 2024). In the context of the MorWAK, although it is already energy-efficient, the introduction of renewable energies makes it a technology that operates on clean energy and responds to sustainability requirements, particularly in areas where access to electricity is very restricted, given that our project envisages accessibility to treatment for all the types of patients it needs. MorWAK has not yet been commercialized, but the introduction of renewable energies has always been a preoccupation for our team, as the MorWAK project is part of a green dialysis approach, i.e. dialysis with a negligible carbon footprint. The MorWAK device is capable of being powered by solar or wind energy, or both, in a hybrid system. This significantly reduces its dependence on an external energy source, increasing its autonomy.

Integrating Solar Energy Into MorWAK

The MorWAK is a portable device designed to be used continuously for long periods each day to filter patients' blood, given its low energy consumption (10W) compared with conventional dialysis systems, but its solar-powered operation makes it an eco-friendly technology that further reduces its carbon footprint. Although this solution has been adopted for many portable medical devices, including the portable ECG (Chong & Wong, 2021). Nevertheless, there was only one study in the literature reporting the use of solar energy to supply a portable hemodialysis device. This study demonstrated that the adoption of this type of energy is effective. The authors used a 12 V, 500 mA solar panel, which takes between 1.5 and 2 hours to fully charge, and provides 50 minutes of backup power (T.r & Krishnakumar, n.d.). On the other hand, the integration of solar energy in our device needs to be studied, taking into account a range of constraints, including those related to portability and geography, as our device is intended to be wearable by the patient and used at home by a variety of patients located in different regions with different meteorological conditions, with the aim of implementing a solar system adapted to all meteorological conditions. As our device is designed to be used at home, and to give patients greater flexibility during treatment, we propose a solution in which the device is powered by solar energy generated by a photovoltaic panel and stored on a battery - a compact, transportable charging station. Although this solution has been adopted in contexts similar to our case, by conceiving a multifunctional portable solar charging device that integrates a solar panel and a BESS (Battery Energy Storage System) battery that stores energy for later use and an Internet of Things (IoT) system for real-time monitoring of the state of charge. The device has been designed for a variety of applications, including emergency medical care management, to ensure reliable power for devices used in areas where access to electricity is limited or in the event of a disaster, as well as for use in public spaces such as transportation (ur Rehman et al., 2025). On the other hand, there are numerous reasons that led us to investigate the choice of charging station in particular to power the MorWAK with clean solar energy as the main reason, but there are some others as well, including those related to the patient, and in particular their safety during treatment, and those related to installation and cost.

Patient Reasons

1. Flexibility and mobility through:
    - **Adaptability to different contexts:** The transportable charging station, which can be easily moved from one location to another, supports MorWAK's objectives, including treatment accessibility for all types of patients, which is designed to be used by patients in different regions.
    - **No dependence on fixed infrastructure:** In regions where electricity is scarce or non-existent, a fixed charging station is not an option. A transportable station, on the other hand, can be deployed anywhere, enabling MorWAK to remain operational even in the most remote areas.
    - **Continuity of treatment:** the charging station protects the patient during treatment in the event of a sudden power cut, particularly in areas with unstable electricity supplies, whether urban or isolated. The patient is therefore protected against medical complications that could result from treatment interruption.
2. Installation reasons:

Simplicity of installation: in contrast to standard solar energy systems, which require several components to generate, store and distribute solar energy, including photovoltaic panels (the quantity of which depends on the desired output), a charge controller, an inverter and connecting cables, which are also exposed to losses. The solution envisaged is a compact charging station that can incorporate a flexible solar panel, either fixed to the charging station or detachable for optimized positioning. It also includes a charge controller and a battery to store energy for later use, and can incorporate overheating and short-circuit protection elements to protect the patient at all times (figure 6). On the other hand, this portable station can be deployed in places where the installation of fixed panels would be impossible, and it also enables better space management, unlike fixed solar energy systems which are bulky and require a lot of space and construction work to implement.

Figure 6. Block diagram of the solar system to power MorWAK

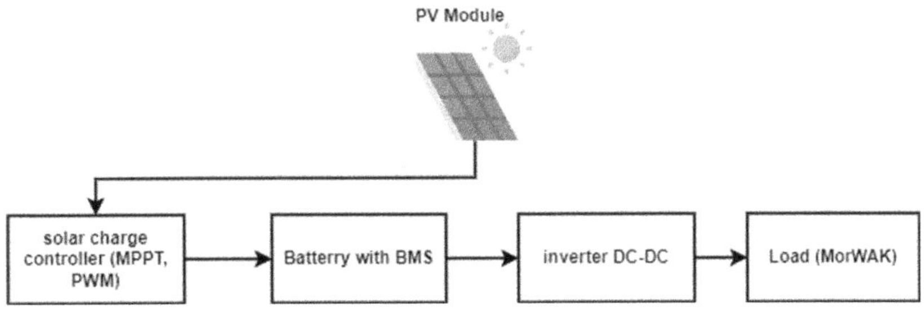

MPPT: Maximum Power Point Tracking; PWM: Pulse Width Modulation; BMS: Battery Management System.

3. **Maintenance:** compared to fixed solar energy systems, which are complex, the charging station is easy to maintain and service thanks to its simple design and access to components, leading to reduced maintenance costs.
4. **Cost:** the acquisition of the charging station to supply the MorWAK with solar energy is expected to be less expensive than the purchase of the components needed to implement the energy systems.

INTEGRATING WIND POWER INTO MORWAK

In recent years, the deployment of wind power has evolved rapidly to become a renewable energy source on a par with solar power. Micro-scale wind energy harvesting systems are now becoming increasingly widespread due to their potential to provide a decentralized, renewable and sustainable energy source for homes, enterprises and communities. These systems offer many benefits over large-scale wind systems, such as spatial efficiency, reduced dependence on grid-connected electricity and long transmission lines, and lower costs (Calautit & Johnstone, 2023). In the context of MorWAK, the integration of this type of renewable energy is crucial and adds to MorWAK being a technology that uses two sources of renewable energy, given that MorWAK should be operated by different patients located in different areas including those located in regions with disrupted sunlight weather conditions. In this situation, we envisage the use of micro-turbines to produce kinetic energy from the wind and transform it into electrical energy by means of generators overcoming the problem of absent or low irradiation, thus replacing the function of solar panels. These micro-wind turbine systems have already been deployed in a variety of applications, notably in the home and in electric vehicles (Ebaid et al., 2023; Kumar & Prakash, 2023). Moreover, they have the potential to deliver average power ranging from 20 W to 2 KW (Kumar & Prakash, 2023), making them a clean energy source that adapts to our power requirements for charging the station's battery. In terms of wind power generation, the solution studied is the design of a wind power generation unit based on micro-turbines that integrates both the blades and a generator to transform mechanical energy into electrical energy, furthermore, this unit will have to be lightweight and easily demountable to maintain the station's transportability concept. The unit can be used at night in the absence of light, or throughout the day in regions with unstable sunlight. To do this, simply connect the unit to a solar-powered charging station (Figure 7).

Figure 7. Schematic diagram of the micro-wind system to power the MorWAK

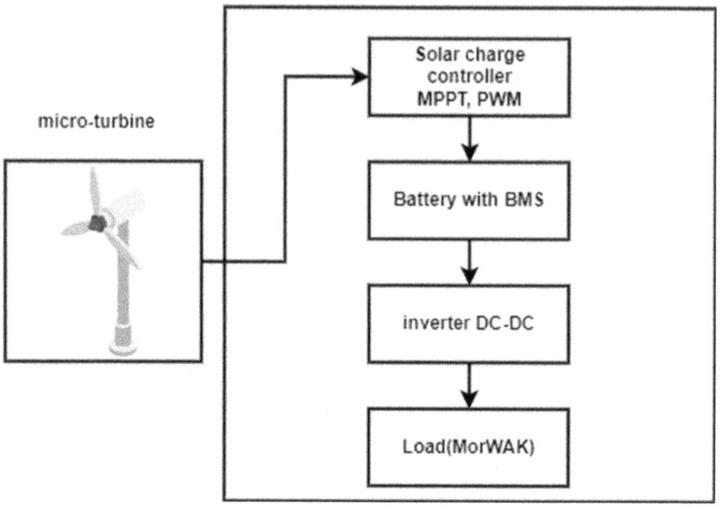

ENERGY SYSTEM SIZING

The design of our charging system producing renewable energy based on solar and wind power requires a detailed sizing study to define the necessary components to be integrated into our charging system. Before selecting these components, it is essential to calculate the system's energy requirements in order to ensure a suitable and efficient power supply. This involves assessing the power required by the device, the duration of use and the potential losses associated with energy conversion and storage. These calculations are used to determine battery capacity, solar panel and wind turbine power, as well as the characteristics of the charge controller (MPPT, PWM) and DC-DC converter, in order to optimize the efficiency/cost ratio.

Panel type: the selection of solar panels is a key step in the design of a solar and wind energy system. It is essential to define the function for which they are chosen and their ability to deliver the desired power. These photovoltaic panels are actually classified into four generations: the first generation concerns monocrystalline and polycrystalline panels; the second generation, thin-film panels; the third generation, emerging panels; and the fourth generation, emerging and hybrid panels (which include quantum dot or polymer cells). However, the difference between these panels resides in the technology used to manufacture the cells (Pastuszak & Węgierek, 2022). On the other hand, the choice between these panels depends on a number of criteria, such as cost, conversion efficiency, durability, flexibility and so on. In the context of MorWAK, the choice of a photovoltaic panel needs to be studied in greater depth with experts in the field of renewable energies, in order to select a panel in such a way as to ensure a concordance between cost, efficiency and weight.

Type of charge controller: charge controllers are an indispensable component of solar and wind energy systems using batteries to store energy. There are two types: MPPT and PWM, whose role is to regulate the power going from the solar panel to the batteries. MPPT regulators are superior in terms of efficiency (the ability of the regulator to transfer solar energy to the battery with minimum loss), with up to 98% efficiency compared to PWM regulators, which only reach 80%. However, MPPTs cost twice as much as PWMs. MPPTs can be considered to minimize losses.

The battery: the battery is the element where the energy produced by the panel is stored for later use. Choosing the battery is a very important step, just like choosing the other components. There are several criteria for battery selection, including capacity (i.e. the amount of energy a battery can store), power and energy density, depth of discharge (DOD), round-trip efficiency, cost, autonomy, weight and electrochemical construction (Table 1).

Table 2. Overview of available battery technologies

Type	LA	NiCd	NiMH	Li-ion	NaS	VRB
Energy density (Wh kg⁻¹)	25–50	50–60	60–120	75–200	150–240	10–30
Power density (W kg⁻¹)	75–300	~200	250–1000	500–2000	150–230	80–150
Cycle life (100% DOD)	200–1000	>1500	180–2000	1000–10,000	2500–4000	>12,000
Capital cost ($/kWh)	100–300	300–600	900–3500	300–2500	300–500	150–1000
Round-trip efficiency	75–85	70–75	65–80	85–97	75–90	75–90
Self discharge	Low	High	High	Medium	–	Negligible

LA: Lead-acid; NiCD: Nickel-cadmium; NiMH: Nickel-metal hydride ; NaS: sodium-sulfur ; VRB: vanadium redox batteries ; Li-ion: Lithium-ion (Vega-Garita et al., 2019).

In our case, we're looking to optimize the weight of the station, which is why we are proposing a low-weight battery.

Micro-wind turbine system: Choosing a micro-wind turbine is a key step. It is based on several essential criteria. First of all, the wind turbine's output needs to be assessed in relation to energy requirements. Secondly, the average wind speed at the installation site is crucial, as it influences the yield. Wind turbines installed in areas with low wind speeds will have an impact on the amount of energy produced. The type of generator is also a major factor, depending on the model chosen: synchronous or asynchronous. The quality of the blades and their aerodynamic efficiency also have an impact on performance and durability. Furthermore, in the case of installations close to residential areas, noise levels and potential vibrations must also be taken into account. Altitude and environment also play an important role, with some wind turbines being better suited to turbulent winds. The availability of a control and protection system is essential to avoid damage in high winds. It is also necessary to assess the costs of purchase, installation and maintenance to ensure a good return on investment. In the context of MorWAK, the choice of a suitable micro-wind turbine will be the subject of a more in-depth study.

Comment

characterizing the components required for our charging station is a key step in designing a device that produces clean energy in two different modes (solar and wind), while selecting components that respond to our requirements with maximized productivity and optimized cost, and guaranteeing a good return on investment. In addition, once the components have been defined, a simulation study in Matlab and a comparison of its results with those of the actual study will be carried out at a later date, with a view to testing the performance (including autonomy) of the charging station and ensuring that it delivers the desired voltage without fluctuation.

CONCLUSION

The climate challenges we face today oblige us to seek sustainable solutions in all areas, including the medical sector, and more specifically in hemodialysis. This is where the MorWAK portable hemodialysis system comes in. This technological solution has been designed within the context of green dialysis, which involves providing treatment to patients while reducing ecological impact. In contrast to conventional extrarenal technologies, which consume a lot of energy, water and waste, the MorWAK operates without being connected to a water source, consumes very little energy (10W) and generates very little waste, making it an energy-efficient technology. Furthermore, the MorWAK can be powered by renewable energies, including solar and wind, maximizing its sustainability and making it a technology aligned with carbon neutrality goals.

REFERENCES

Ae, L. A, P., & Jw, A. (n.d.). The carbon footprint of an Australian satellite haemodialysis unit. *PubMed*. Retrieved December 9, 2024, from https://pubmed.ncbi.nlm.nih.gov/23731962/

Barraclough, K. A., & McAlister, S. (2022). Assessing the Carbon Footprint of Hemodialysis: A First Step Toward Environmentally Sustainable Kidney Care. *Journal of the American Society of Nephrology*, *33*(9), 1635–1637. DOI: 10.1681/ASN.2022060661 PMID: 35840174

Bonomini, M., Piscitani, L., Di Liberato, L., & Sirolli, V. (2022). Biocompatibility of Surface-Modified Membranes for Chronic Hemodialysis Therapy. *Biomedicines*, *10*(4), 4. Advance online publication. DOI: 10.3390/biomedicines10040844 PMID: 35453594

Calautit, K., & Johnstone, C. (2023). State-of-the-art review of micro to small-scale wind energy harvesting technologies for building integration. *Energy Conversion and Management: X*, *20*, 100457. DOI: 10.1016/j.ecmx.2023.100457

Canada, P. H. A. of. (2022, September 8). *Health and climate change 2019* [Education and awareness;research]. https://www.canada.ca/en/public-health/services/reports-publications/canada-communicable-disease-report-ccdr/monthly-issue/2019-45/issue-5-may-2-2019/article-1-health-solution-climate-change-risks-solutions.html

Chen, W., Alharthi, M., Zhang, J., & Khan, I. (2024). The need for energy efficiency and economic prosperity in a sustainable environment. *Gondwana Research*, *127*, 22–35. DOI: 10.1016/j.gr.2023.03.025

Chong, K.-K., & Wong, G.-Z. (2021). Solar powered a wearable Electrocardiography (ECG) device with battery storage. *IOP Conference Series. Earth and Environmental Science*, *945*(1), 012048. DOI: 10.1088/1755-1315/945/1/012048

Connor, A., Lillywhite, R., & Cooke, M. W. (2011). The carbon footprints of home and in-center maintenance hemodialysis in the United Kingdom. *Hemodialysis International. International Symposium on Home Hemodialysis*, *15*(1), 39–51. DOI: 10.1111/j.1542-4758.2010.00523.x PMID: 21231998

Deshmukh, M. K. G., Sameeroddin, M., Abdul, D., & Abdul Sattar, M. (2023). Renewable energy in the 21st century: A review. *Materials Today: Proceedings*, *80*, 1756–1759. DOI: 10.1016/j.matpr.2021.05.501

Ebaid, M. S., Shahin, Z. A. A., & Alshawabkeh, M. M. (2023). Feasibility studies of micro wind turbines installed on electric vehicles as range extenders using real-time analytical simulation with multi driving cycles scenarios. *Advances in Mechanical Engineering*, *15*(4), 16878132231165964. DOI: 10.1177/16878132231165964

Fiseha, T., & Osborne, N. J. (2023). Burden of end-stage renal disease of undetermined etiology in Africa. *Renal Replacement Therapy*, *9*(1), 44. DOI: 10.1186/s41100-023-00497-w

Guo, X., Huang, K., Li, L., & Wang, X. (2022). Renewable Energy for Balancing Carbon Emissions and Reducing Carbon Transfer under Global Value Chains: A Way Forward. *Sustainability (Basel)*, *15*(1), 1. Advance online publication. DOI: 10.3390/su15010234

Hoenich, N. A., & Pearce, C. (2002). Medical waste production and disposal arising from renal replacement therapy. *Advances in Renal Replacement Therapy*, *9*(1), 57–62. DOI: 10.1053/jarr.2002.30469 PMID: 11927908

Jabrane, M., Fadili, W., Kennou, B., Labaali, A., Zahlane, K., & Laouad, I. (2013). Évaluation de l'impact d'un centre d'hémodialyse sur l'environnement et l'écologie locale. *Nephrologie & Therapeutique*, *9*(7), 481–485. DOI: 10.1016/j.nephro.2013.07.369 PMID: 24140175

Justice, G., Nyantakyi, G., & Isaac, S. H. (2024). The effect of renewable energy on carbon emissions through globalization. *Heliyon*, *10*(5), e26894. DOI: 10.1016/j.heliyon.2024.e26894 PMID: 38434333

Ka, B., & Jwm, A. (n.d.). Green nephrology. *PubMed*. Retrieved November 22, 2024, from https://pubmed.ncbi.nlm.nih.gov/32034297/

Kumar, N., & Prakash, O. (2023). Analysis of wind energy resources from high rise building for micro wind turbine: A review. *Wind Engineering*, *47*(1), 190–219. DOI: 10.1177/0309524X221118684

Mtioui, N., Zamd, M., Ait Taleb, A., Bouaalam, A., & Ramdani, B. (2021). Carbon footprint of a hemodialysis unit in Morocco. *Therapeutic Apheresis and Dialysis*, *25*(5), 613–620. DOI: 10.1111/1744-9987.13607 PMID: 33159433

Nagai, K., Barraclough, K., Ueda, A., & Itsubo, N. (2021). Sustainability in dialysis therapy: Japanese local and global challenge. *Renal Replacement Therapy*, *7*(1), 1. Advance online publication. DOI: 10.1186/s41100-021-00360-w PMID: 33614163

Otero González, A. (2024). Chronic kidney disease, dialysis and climate change. [English Edition]. *Nefrología : Publicación Oficial de la Sociedad Española Nefrologia*, *44*(3), 331–337. DOI: 10.1016/j.nefroe.2024.06.008 PMID: 38960781

Pastuszak, J., & Węgierek, P. (2022). Photovoltaic Cell Generations and Current Research Directions for Their Development. *Materials (Basel)*, *15*(16), 5542. DOI: 10.3390/ma15165542 PMID: 36013679

Piccoli, G. B., Nazha, M., Ferraresi, M., Vigotti, F. N., Pereno, A., & Barbero, S. (2015). Eco-dialysis: The financial and ecological costs of dialysis waste products: is a 'cradle-to-cradle' model feasible for planet-friendly haemodialysis waste management? *Nephrology, Dialysis, Transplantation*, *30*(6), 1018–1027. DOI: 10.1093/ndt/gfv031 PMID: 25808949

Ritchie, H., Roser, M., & Rosado, P. (n.d.). *Renewable Energy*. Retrieved February 10, 2025, from https://ourworldindata.org/renewable-energy

Singh, S. (2024). GLOBAL WARMING –CAUSES EFFECTS AND SOLUTIONS. *Universal Research Reports*, *11*(2), Article 2.

Tarrass, F., Benjelloun, O., & Benjelloun, M. (2021). Towards zero liquid discharge in hemodialysis. Possible issues. *Nefrología : Publicación Oficial de la Sociedad Española Nefrologia*, *41*(6), 620–624. DOI: 10.1016/j.nefro.2020.12.012 PMID: 36165151

Tonelli, M., Wiebe, N., Knoll, G., Bello, A., Browne, S., Jadhav, D., Klarenbach, S., & Gill, J. (n.d.). *Systematic Review: Kidney Transplantation Compared With Dialysis in Clinically Relevant Outcomes*. Retrieved November 29, 2024, from https://www.amjtransplant.org/article/S1600-6135(22)28113-1/fulltext

ur Rehman, A., Alblushi, I. G. M., Zia, M. F., Khalid, H. M., Inayat, U., Benbouzid, M., ... & Hussain, G. A. (2025). A solar-powered multi-functional portable charging device (SPMFPCD) with internet-of-things (IoT)-based real-time monitoring—An innovative scheme towards energy access and management. Green Technologies and Sustainability, 3(1), 100134.

Vega-Garita, V., Hanif, A., Narayan, N., Ramirez-Elizondo, L., & Bauer, P. (2019). Selecting a suitable battery technology for the photovoltaic battery integrated module. *Journal of Power Sources*, *438*, 227011. DOI: 10.1016/j.jpowsour.2019.227011

Waziri, B., Duarte, R., & Naicker, S. (2019). Chronic Kidney Disease–Mineral and Bone Disorder (CKD-MBD): Current Perspectives. *International Journal of Nephrology and Renovascular Disease*, *12*, 263–276. DOI: 10.2147/IJNRD.S191156 PMID: 31920363

Zhao, Q., Yu, P., Mahendran, R., Huang, W., Gao, Y., Yang, Z., & Guo, Y. (2022). Global climate change and human health: Pathways and possible solutions. *Eco-Environment & Health*, *1*(2), 53–62.

Chapter 18
Numerical Study of the Drag Reduction and Forced Convection Around Rectangular Block by Using the LBM Method

Youssef Admi
https://orcid.org/0000-0003-0920-0618
Faculty of Sciences, Mohammed First University, Oujda, Morocco

Abdelilah Makaoui
Faculty of Sciences, Mohammed First University, Oujda, Morocco

El Bachir Lahmer
Faculty of Sciences, Mohammed First University, Oujda, Morocco

Mohammed Amine Moussaoui
Faculty of Sciences, Mohammed First University, Oujda, Morocco

Ahmed Mezrhab
Faculty of Sciences, Mohammed First University, Oujda, Morocco

ABSTRACT

This research aims to study flow control for reducing aerodynamic drag and cooling a rectangular electronic component using an inclined partition. Unlike existing studies focusing on horizontal or vertical plates, this work numerically investigates the impact of an inclined partition on flow and temperature fields using the lattice Boltzmann method with multiple relaxation times. Results show that increasing the Reynolds number enhances heat exchange and reduces the drag coefficient, with a maximum Cd reduction of 125.41% at Re = 250. Additionally, at Re = 150, positioning the partition at g = 2d achieves a Cd reduction of 130.97%. For effective cooling of electronic components, placing a sloping partition at g = 2d before the component yields significant energy savings.

DOI: 10.4018/979-8-3693-9924-8.ch018

INTRODUCTION

Heat The physical problems associated with fluid flow and heat transfer represents, at the same time, a complex and fascinating subject because of their unlimited various industrial applications, including heat exchangers, cooling of electronic components, risers in marine technology, road vehicles, aeroplanes, buildings and bridges etc. (Chapain et al. 2019; Deng et al. 2018; Elshaer et al. 2017; Gonzalez-Zamudio et al. 2025; Liu et al. 2020; Monat et al. 2018). In these domains, the structures are considered as bluff bodies characterized by charges induced either by the strong fluid-structure interaction or by the presence of a temperature gradient. These interactions generate large unstable forces that can violently damage the structural integrity of the bluff body. In fact, many experimental and numerical researches have been conducted to understand the behavior of these physical phenomena in order to know how to control it. These investigations treat fluid flow and heat transfer, particularly, increasing heat exchange, increasing lift force, decreasing drag force and sup-pressing vortices propagating around a bluff body or bodies. These objectives are attained by using flow control techniques (Admi et al. 2022; Islam et al. 2015; Moussaoui et al. 2010; Rashidi et al. 2016; Turki et al. 2008; Zhou et al. 2005). Zhou et al. (Zhou et al. 2005) performed a 2D numerical study on the reduction of fluid forces exerted on a square cylinder in the channel using a control plate placed upstream. The Numerical results show that not only is the drag on the square cylinder significantly reduced by the control plate, but also the fluctuating lift is suppressed as well. The optimum position of the control plate for minimizing the drag on the square cylinder is found for each control plate height. In another numerical investigation realized by Islam et al. (Islam et al. 2015) for study the fluid flow past around a square cylinder controlled by a downstream plate at Re = 150 and g = 0–11d. They also indicated that the drag force achieved a maximum reduction at g = 0, and the optimum space between a square cylinder and the control plate reached at g = 2d. Turki et al. (Turki et al. 2008) carried out a numerical investigation for controlled the vortex shedding behind a square cylinder for a Reynolds number ranging from Lc=0.0373*Re 110 to 200 by using a separated flat plate. First, Turki found a linear relationship: $Lc=0.0373*Re^{-3.4373}$ characterizing the critical length of the splitter plate Lc for the disappearance of vortex shedding. Likewise, They studied the effect of length and its location of this splitter plate in the wake region on the Strouhal number, and the drag and lift coefficients are analyzed. Moussaoui et al. (Moussaoui et al. 2010) conducted numerical investigations for various heights and locations of bi-partition situated upstream of a square cylinder in a 2D channel for Re=250. The authors found that in most cases, the interaction was beneficial insofar as the drag of the square cylinder were lower in the case of the existing bi-partition than in the case of the signal square cylinder. In addition, a significant reduction of the fluctuating side forces due to the vortex shedding from the main body was observed for the most bi-partition positions. Also, Admi et al. (Admi et al. 2022) performed a numerical study of fluid flow past around a square block controlled by four partitions. The authors fix the two downstream partitions at a distance g=3.9*d*, while the two upstream partitions are fixed at a distance g1=0.25d whose length Lp varies from 1*d* to 5*d*. They found that a maximum percentage reduction of the drag coefficient reaches 38.86% in the case where the length of the upstream partitions equal 2.25*d*. Also, a maximum vortex shedding suppression is observed in this critical case. Rashidi et al. (Rashidi et al. 2016) performed an article review on the numerical and experimental studies concerning the vortex shedding suppression and wake control methods. The authors classified these methods into two groups and presented the benefits, limitations, power efficiency and specific applications of the two methods.

The control method applied in this investigation is called the passive control technique based on the implementation of one or more control instruments (partition, flat plate, splitter, fins and rods....). The instrument implemented in this work is an inclined partition implemented upstream of the rectangular block. Another type of control technique called the active control technique exists. This technique is difficult to implement and necessitates external (non-beneficial) energy, which impedes its inclusion and explains its rarity in the literature. Vortex shedding control reduces the fluidic forces (structural drag or flow-induced vibrations) acting on bluff bodies. The principal purpose of this research is to numerically study the heat transfer characteristics and drag coefficient reduction around a heated rectangular block located in a horizontal channel controlled by an inclined partition positioned upstream. The Multiple Relaxation Time-Lattice Boltzmann Method with double populations (DMRT-LBM) is the numerical approach used to realize the simulations (Admi et al. 2020; 2022; 2022; 2024; 2024; Benhamou et al. 2023; d'Humières et al. 2002; Lahmer et al. 2022; Mohamad et al. 2011).

The rest of the present article is convened as follows. Section two is devoted to describing the physical problem and the implementation of the boundary conditions. Section three presents the numerical approach used. Section four shows the code validation. Section five yields the new numerical results obtained, and finally, important conclusions are brought in the last section.

CONFIGURATION SYSTEM AND METHODOLOGY

Configuration Problem

The computational domain geometry is illustrated in Figure 1 where a fluid flow past a rectangular block positioned in a horizontal bidimensional channel. The rectangular block is situated at upstream distance Lu=12*d* and a wide downstream length Ld=30*d*. The blocking ratio noted $\beta = d/h = 1/12$. Also, the inclined partition has the same length "Lp=1*d*" and the same height "h=0.25*d*" is implemented downstream of the rectangle cylinder. The gap spacing "g" between the rectangular cylinder and the partition is varied such as $1d \leq g \leq 8d$.

Figure 1. The numerical scheme studied

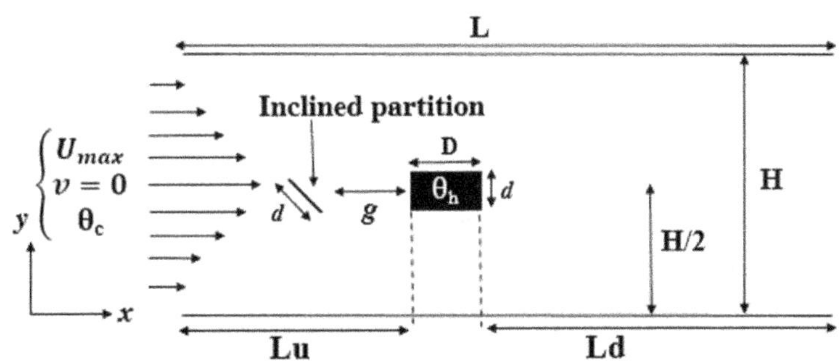

The flow is fully developed with a parabolic velocity profile at the inlet and as well as at the outlet ($u = 1.5\,U_{max}\left[1 - \left({}^{y}/_{H}\right)^2\right]$; $v = 0$) where y is the vertical distance from the centerline, U_{max} is the maximal entrance velocity of the flow and u and v are the components of velocity vectors.

Boundary Conditions

In any numerical simulation and for any method, the implementation of boundary conditions is a necessary and obligatory step. Especially in LBM, the choice of boundary conditions has a very important role in the stability and precision of the LBM code. Generally, in fluid dynamics, two classes of boundaries are encountered: open boundaries and solid boundaries. For open boundaries, the velocity (or pressure) is always specified in the macroscopic description of the fluid flow. Indeed, in LBM, the distribution functions outside the domain are known via the propagation process. However, the distribution functions to the domain at the boundaries are unknown (see Figure 2). Thus, in the literature, different approaches are proposed and tested for open boundaries. In this paper, the boundary conditions of Zou and He (Zou & He 1997) are implemented.

Thus, in the inlet, the unknown distribution functions are given by:

$$f_1 = f_3 + \frac{2}{3}u_0 \tag{1}$$

$$f_5 = f_7 - \frac{1}{2}(f_2 - f_4) + \frac{1}{6}U_{max} \tag{2}$$

$$f_8 = f_6 + \frac{1}{2}(f_2 - f_4) + \frac{1}{6}U_{max} \tag{3}$$

where in the inlet, the velocity of fluid flow U_{max} and the functions f_0, f_2, f_3, f_4, f_6 and f_7 are known. While the density of the fluid ρ is calculated by the following expression:

$$= \frac{1}{1-u}\left[f_0 + f_2 + f_4 + 2(f_3 + f_6 + f_7)\right] \tag{4}$$

Likewise, in the outlet the unknown distribution functions are given by:

$$f_3 = f_1 - \frac{2}{3}\rho_{out}.\mathrm{u} \tag{5}$$

$$f_7 = f_5 + \frac{1}{2}\left(f_2 - f_4\right) - \frac{1}{6}\rho_{out}.u \tag{6}$$

$$f_6 = f_8 + \frac{1}{2}\left(f_2 - f_4\right) - \frac{1}{6}\rho_{out}.u \tag{7}$$

where the functions f_0, f_1, f_2, f_4, f_5 and f_8 are known functions. Also, the pressure or the density must be designed in the outlet, while the velocity of fluid flow is calculated by the following expression:

$$u = -1 + \frac{f_0 + f_2 + f_4 + 2.\left(f_1 + f_5 + f_8\right)}{\rho_{out}} \tag{8}$$

For solid walls, the boundary conditions employed to determine the distribution functions on these solid walls are given by using a popular approach that is widely used in the literature called: the bounce-back boundary conditions. These conditions are generalized by the following formula (Bouzidi et al. 2001).

$$f_i(\vec{x}_B, t) = f_{\bar{i}}(\vec{x}_B, t) \tag{9}$$

where $f_{\bar{i}}(\vec{x}_B, t)$ and $f_i(\vec{x}_B, t)$ are the known (in the opposite direction) and an unknown distribution function at the wall node ($\vec{x}_B$).

For the top wall, the unknown distribution functions are given by: $f_4=f_2$, $f_7=f_5$ and $f_8=f_6$. Likewise, for the bottom wall, the unknown distribution functions are given by inverting the previous distribution functions.

Figure 2. Unknown distribution function of the physical boundary

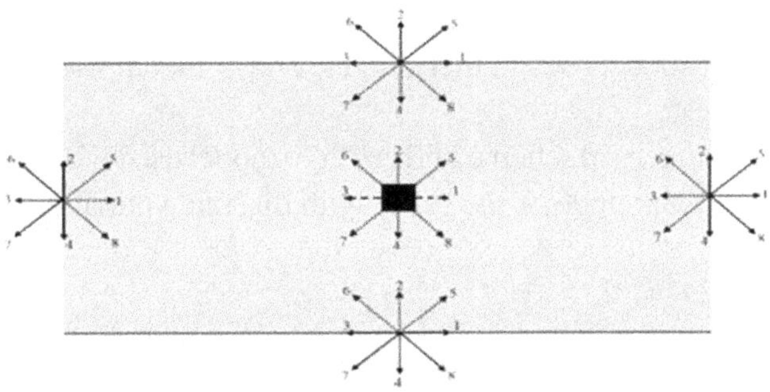

To treat the heat flow boundary conditions in the thermal problem, we use the method proposed by (Mezrhab et al. 2010). The general term of the distribution function g_i is:

$$g_i = \frac{2T\left(1 + \frac{a}{4}\right)}{5} - g_{\bar{i}} \tag{10}$$

where the parameter a is a constant whose value is determined using the following formula:

$$\alpha = \frac{\sqrt{3}(4 + a)}{60} \tag{11}$$

At the inlet, a specific condition is enforced: $g_1 = -g_3 + 2\Theta_c\left(1 + \frac{a}{4}\right)/5$. This condition can be extended to the obstacle boundaries by substituting θ_c with θ_h, since the obstacle boundaries are regarded as hot. At the output, the unknown function (g_3) is determined using the second-order interpolation condition. Furthermore, the condition $g_j = g_{\bar{j}}$ is incorporated for the upper and lower faces of the obstacles, where $\bar{j} = -j$.

For the initial thermal conditions, we assumed that the walls of the channel are adiabatic. Moreover, the cylinder is considered heated with a dimensionless temperature $\theta h = 0.5$. While the incoming fluid has a dimensionless temperature $\theta_c = -0.5$.

METHODOLOGY

In the absence of experimental studies because of their high cost and in order to facilitate the work tasks and to gain time, the numerical models constitute an alternative, effective and indispensable method to simulate simple and complex physical phenomena such as the simulation of fluid flow and heat transfer. Among the most used methods in recent times is the lattice Boltzmann method (LBM). This method is distinguished from the old CFD method in that it is an accurate, flexible method that allows for calculations to be performed in a short time and is also adaptable to various complex geometries (Admi et al. 2020; 2022; 2022; 2024; 2024; Benhamou et al. 2023; d'Humières et al. 2002; Lahmer et al. 2022; Mohamad et al. 2011).

Lattice Boltzmann Method for Fluid Flow

Many models exist to numerically discretize the physical phenomena associated with fluid flow. But, for the reasons of stability, accuracy and convergence of the computational code, as well as, for the compatibility reasons with the model used in thermic, the *D2Q9* model was used to treat the distribution of the density of fluid particles.

Figure 3-a. illustrates the numerical scheme of the *D2Q9* model used where the particles of fluid move from one node to another adjacent node of the grille with discrete velocities (c_i) which are defined by:

$$c_i = \begin{cases} (0,0) \; ; i = 0 \\ \left(\cos\left[\frac{(i-1)\pi}{2}\right], \sin\left[\frac{(i-1)\pi}{2}\right]\right)c \; ; i = 1-4 \\ \left(\cos\left[\frac{(2i-9)\pi}{4}\right], \sin\left[\frac{(2i-9)\pi}{4}\right]\right)c \; ; i = 5-8 \end{cases} \tag{12}$$

The discretization of the lattice Boltzmann equation with the MRT model is expressed by:

$$fi(x + ci\,\Delta t, t + \Delta t) = fi(x,t) + \Omega(f),\, i = 0,1,\ldots,8 \tag{13}$$

The f_i and Ω denote respectively the particle distribution function and the collision operator developed by D'Humières using the LBM-BGK model. The introduction of this evolution proposed by D'Humières (d'Humières et al. 2002) allows writing the preceding equation as follows:

$$fi(x + ci, t + 1) - fi(x,t) = M^{-1} \times S_i \times \left(m_i(x,t) - m_i^{eq}(x,t)\right),\, i = 0,\ldots,8 \tag{14}$$

Where M^{-1} and S_i represent the inverse matrix of the transformation matrix M and the diagonal relaxation time matrix, respectively. Thus, m_i and m_i^{eq} represent, respectively, the momentum vectors and the equilibrium momentum.

$$m = \left(m_0, m_1, m_2, ,m_n\right)^T \; ; \; m^{eq} = \left(\rho, e^{eq}, \varepsilon^{eq}, j_x, q_x^{eq}, j_y, q_y^{eq}, p_{xx}^{eq}, p_{xy}^{eq}\right)^T \tag{15}$$

The choice of the equilibrium functions m_i^{eq} will determine the equivalent macroscopic equations.

$$
\begin{aligned}
m_0^{eq} &= \rho \\
m_1^{eq} &= e^{eq} = -2\rho + 3\left(j_x^2 + j_y{}^2\right) \\
m_2^{eq} &= \varepsilon^{eq} = \rho - 3\left(j_x^2 + j_y{}^2\right)/\rho \\
m_3^{eq} &= j_x \\
m_4^{eq} &= q_x^{eq} = -j_x \\
m_5^{eq} &= j_y \\
m_6^{eq} &= q_y^{eq} = -j_y \\
m_7^{eq} &= p_{xx}^{eq} = \left(j_x^2 - j_y{}^2\right)/\rho \\
m_8^{eq} &= p_{xy}^{eq} = \left(j_x j_y\right)/\rho
\end{aligned}
\tag{16}
$$

where

$$j_x = \rho u_x = \sum_i f_i^{eq} C_{ix} \tag{17}$$

$$j_y = \rho u_y = \sum_i f_i^{eq} C_{iy} \tag{18}$$

The expression of the transformation representing the passage from the space of the velocities to the space of the quantities of motions is given by:

$m=Mf$ and $f=M^{-1}m$ (19)

The matrix M of order 9 is given by Mohamad et al. (Mohamad et al. 2011):

$$
M = \begin{pmatrix} \langle m_0| \\ \langle m_1| \\ \langle m_2| \\ \langle m_3| \\ \langle m_4| \\ \langle m_5| \\ \langle m_6| \\ \langle m_7| \\ \langle m_8| \end{pmatrix} = \begin{pmatrix}
1 & 1 & 1 & 1 & 1 & 1 & 1 & 1 & 1 \\
-4 & -1 & -1 & -1 & -1 & 2 & 2 & 2 & 2 \\
4 & -2 & -2 & -2 & -2 & 1 & 1 & 1 & 1 \\
0 & 1 & 0 & -1 & 0 & 1 & -1 & -1 & 1 \\
0 & -2 & 0 & 2 & 0 & 1 & -1 & -1 & 1 \\
0 & 0 & 1 & 0 & -1 & 1 & 1 & -1 & -1 \\
0 & 0 & -2 & 0 & 2 & 1 & 1 & -1 & -1 \\
0 & 1 & -1 & 1 & -1 & 0 & 0 & 0 & 0 \\
0 & 0 & 0 & 0 & 0 & 1 & -1 & 1 & -1
\end{pmatrix}
\tag{20}
$$

The macroscopic quantities such as density ρ and momentum ρu are determined after the computational process by the following expression Mohamad et al. (Mohamad et al. 2011):

$$\rho = \sum_{i=0}^{8} f_i \, , \rho u = \sum_{i=0}^{8} c_i f_i \tag{21}$$

Thermal Lattice Boltzmann Method

For reasons of compatibility (in terms of stability, accuracy, convergence and calculation time) with the D2Q9 model (Figure 3-a), the D2Q5 model was chosen to simulate the temperature field.

Figure 3-b. illustrates the numerical scheme of the D2Q5 model used where the particles of fluid move from one node to another adjacent node of the grille with discrete velocities (c_i) which are defined by:

$$c_i = \begin{cases} (0,0) \, ; i = 0 \\ (\cos[(i-1)\pi/2)], \sin[(i-1)\pi/2)])c, i = 1,\dots,4 \end{cases} \quad (21)$$

The thermal LBE-MRT equation can be formulated, as indicated in the following expression.

$$g_i(x + e_i \times \Delta t, t + \Delta t) - g_i(x,t) = N^{-1} \times Q_i \times (n_i(x,t) - n_i^{eq}(x,t)) \, , \mathbf{i} = 0,\dots,4 \tag{22}$$

Where g_i represents the distribution function of the thermal field, N denotes the 5×5 transformation matrix and Q is the diagonal relaxation time matrix for the thermal problem. Thus, the quantities n_i and n_i^{eq}represented, respectively, the thermal momentum and equilibrium thermal momentum in the moment space.

The transformation matrix N is given by Mohamad et al. (Mohamad et al. 2011):

$$\mathrm{N} = \begin{pmatrix} 1 & 1 & 1 & 1 & 1 \\ 0 & 1 & 0 & -1 & 0 \\ 0 & 0 & 1 & 0 & -1 \\ -4 & 1 & 1 & 1 & 1 \\ 0 & 1 & -1 & 1 & -1 \end{pmatrix} \tag{23}$$

The macroscopic quantity calculated in the thermal field after the numerical computation is the temperature of which its expression is given by:

$$\mathrm{T} = \sum_{i=0}^{4} g_i \tag{24}$$

Figure 3. D2Q9 and D2Q5 model for the velocity and temperature field

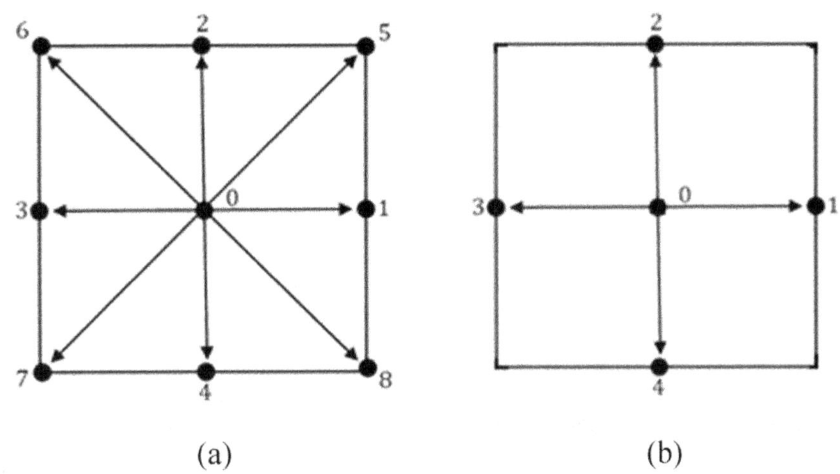

VALIDATION OF NUMERICAL CODE

In this paper, several validation results are presented treating two different geometries. First, we studied the flow around a cylinder of square cross-section mounted inside a horizontal channel with a blocking ratio of 1/8. The obtained results are compared with those found by Breuer et al. (Breuer et al. 2000). Figure 4 shows the velocity components along the centerline for a Reynolds number equal to 100. The comparison of these components is an excellent agreement between our results and the results presented by Breuer et al. (Breuer et al. 2000) with a difference of less than 1.7%

Figure 4. Comparison of velocity profiles along the central axis at Re = 100 with reference [25].

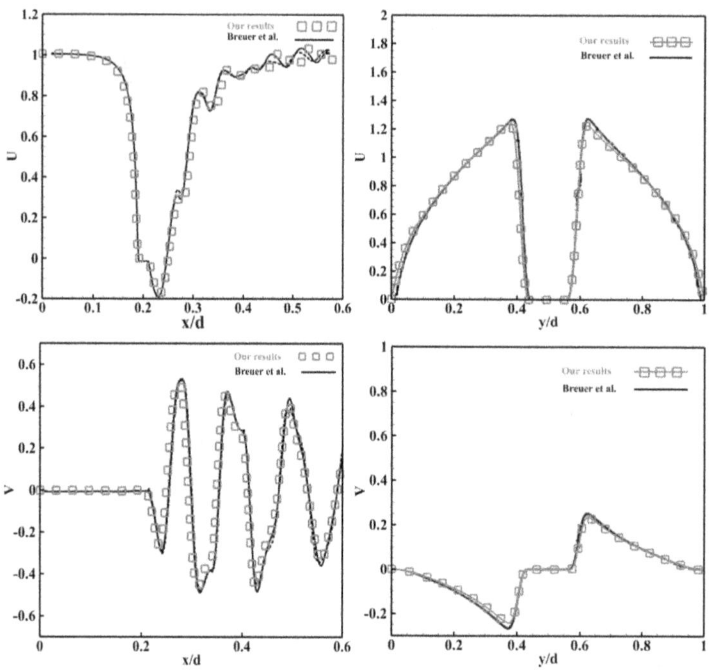

Since the present work consists of a rectangular cylinder and a control partition, we present a second validation study. This study addresses the problem of flow through a rectangular cylinder without a control plate with a blocking ratio of β=1/8 and for a range of 100≤Re≤250. The results obtained are compared with those found by Islam et al. (Islam et al. 2012). We mentioned two validation graphs. Graph of Figure 5-a shows the variation of the mean value of the drag coefficient as a function of the ratio R=width/height. The results obtained show that the value of Cd_{mean} decreases with the increase of this ratio. Thus, our results are almost identical to the reference one. The second graph (Figure 5-b) illustrates the variation of the mean value of the drag coefficient as a function of the Reynolds number. The results obtained show that there is a proportional (linearly increasing) relationship between the two quantities. Likewise, our results are in good agreement with those found by Islam et al. (Islam et al. 2012).

Figure 5. Comparison of the average values of the drag coefficient Cdmean for a rectangle cylinder with these existing in the literature

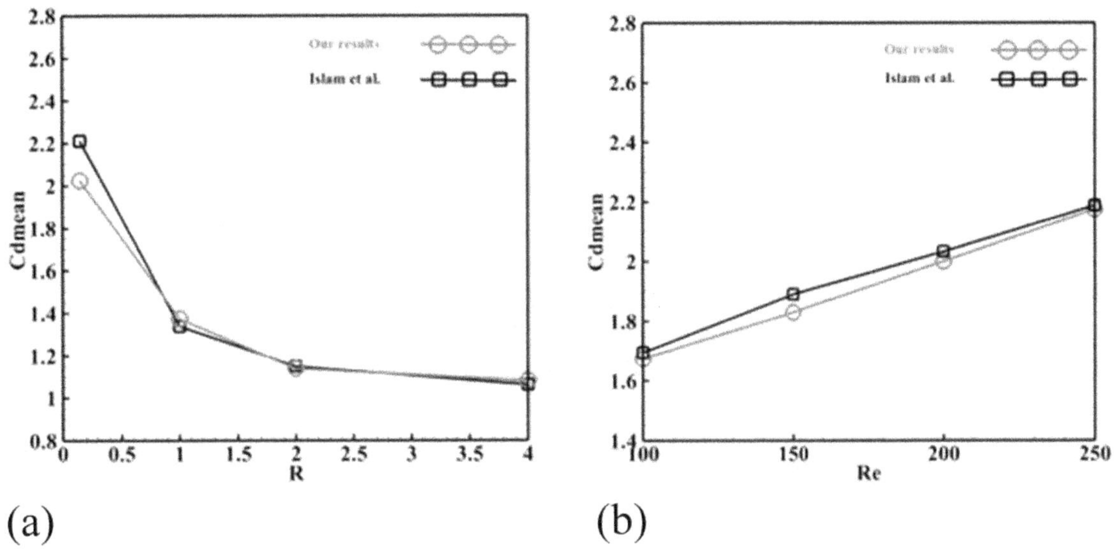

(a) Cdmean in the function of R ratio; (b) Cdmean in the function of Re.

To ensure the validity of the current numerical code for the thermal problem, a numerical study was realized to compare the Nusselt number with other literature Dhiman et al. (Dhiman et al. 2005). The Nusselt number explains the thermal exchange rate between the cold fluid and the heated cylinder. As illustrated in Table 1, the average Nusselt number was analyzed for each face of the cylinder and the maximum deviation at a low Reynolds number. The current results illustrate a good accordance with those obtained in the reference Dhiman et al. (Dhiman et al. 2005).

Table 1. Validation results of average Nusselt number and maximum deviation for each face of a square cylinder.

	Re=25		Re=30		Re=35		Re=40		Re=45	
	Nu	%	Nu	%	Nu	%	Nu	%	Nu	%
Source (Frontal face)										
Our results	3.55	0.84%	3.88	0.78%	4.19	0.47%	4.48	1.10%	4.76	0.21%
Dhiman et al. 2005	3.58		3.85		4.21		4.53		4.75	
Source (Top or Bottom face)										
Our results	0.91	0.011%	0.93	2.197%	0.96	1.05%	0.986	0.61%	1.019	0.098%
Dhiman et al. 2005	0.90		0.91		0.95		0.98		1.02	
Source (Rear face)										

continued on following page

Table 1. Continued

	Re=25		Re=30		Re=35		Re=40		Re=45	
	Nu	%	Nu	%	Nu	%	Nu	%	Nu	%
Our results	2.27	1.33%	2.37	0.85%	2.45	0.4%	2.52	1.94%	2.58	1.9%
Dhiman et al. 2005	2.24		2.35		2.46		2.57		2.63	

RESULTS AND DISCUSSION

This section presents the results obtained for a flow around a rectangular cylinder controlled by an inclined partition. The effect of the Reynolds number and of the position between the partition and the rectangular block is examined. The results obtained are illustrated in terms of thermal vortex shedding lines and time variation graphs of the drag coefficient.

Reynolds Number Effect

Figure 6. shows the thermal vortex shedding for a range of Reynolds numbers $100 \leq Re \leq 250$. For this purpose, the length and position of the control partition are set to Lp=1*d* and g=1*d* respectively. The results obtained show that there are three different flow regimes. Indeed, Figure 6-a shows a periodic and weakly undulating flow regime for a Reynolds number Re=100. Similarly, a periodic and moderately undulating regime is observed for a Reynolds number Re = 150 (Figure 6-b). For high Reynolds number values (Re=200–250), an undulating regime is observed (Figure 6 c-d). It can be seen that as the Reynolds number increases, the undulation of the regime increases and consequently the periodicity and stability of the flow decreases. In fact, if the Reynolds number increases the velocity of the incoming fluid flow increases so that it becomes undulating. This waviness increases the size and number of vortex shedding behind the block.

Figure 6. Visualization of the thermal vortex shedding behind the rectangular block at different Reynolds number

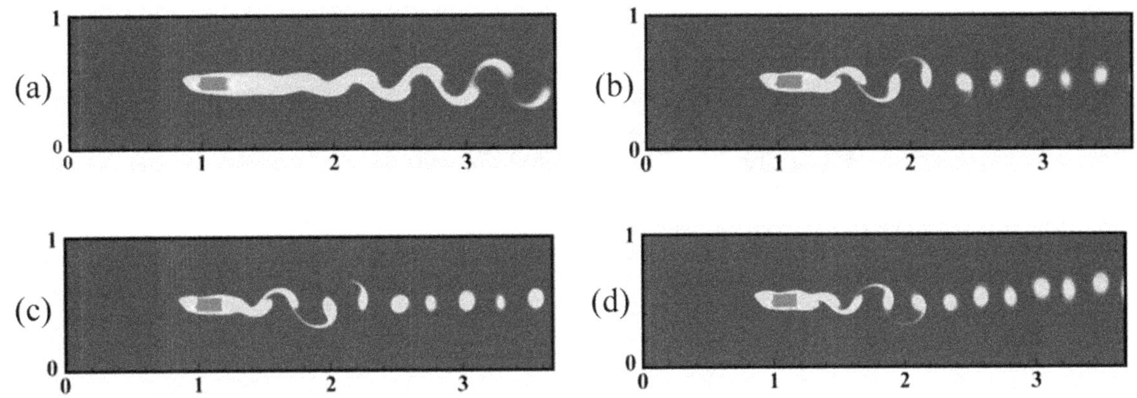

(a) Re=100; (b) Re=150; (c) Re=200; (d) Re=250

Drag Coefficient Variation For 100≤Re≤250

The effect of the Reynolds number on the variation of the time drag coefficient is studied (Figure 7.). From, it is seen that there is a regular and periodic variation of drag coefficient for a Reynolds number Re=100 and Re=150, while there is an irregular variation for Re=200 and Re=250. Also, the amplitude of the graph showing the Cd for 200≤Re≤250 is larger than that for 100≤Re≤150. Thus, the fluid flows faster with increasing Reynolds number, while if the fluid velocity increases we have a strong fluid-structure interaction (tilted fluid partition). Therefore, the shear layers delivered by the edges of the partition are discharged beyond and onto the faces of the block. Due to the effect of thrust, negative values of the drag force are observed. In addition, the maximum percentage reduction of Cdmean is 125.41%, found for Re=250.

Figure 7. Temporal drag coefficient variation for different Reynolds number.

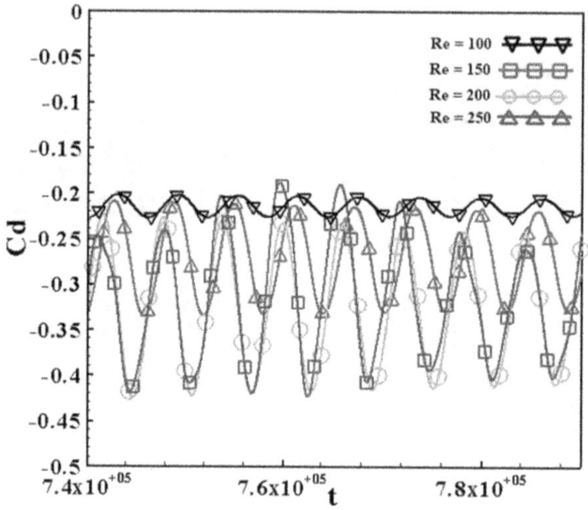

Effect of the Partition Position on the Thermal Vortex Shedding

In this part, the study of the position between the control plate and the rectangular cylinder on the thermal vortex shedding is performed. For this purpose, the Reynolds number is set to Re=150w.

Figure 8. shows the thermal vortex shedding around and behind the block for different gap spacing. From this figure, it can be seen that there are three thermal regimes. Indeed, in the case where the distance between the partition and the block is small (Figure 8 a-b), the incoming fluid considers the partition and the rectangle as a single cliff body. In this case, the sheer layers jump beyond the block and give a periodic alternation of vortices. However, if the distance g increases, the number and size of vortices also increase until almost no vortices are present. In this case (Figure 8 c-e), the sheer layers are delivered by the control partition on and behind the block while giving large recirculations after the rectangular block. This results in an irregular thermal regime. For the case where the gap spacing is very wide (Figure 8 f-g), the sheer layers fall into the gap between the two bodies and meet the block almost as in the case without control. In the latter case, the thermal regime is characterized by strong vortex shedding.

Figure 8. Visualization of the thermal vortex shedding for various gap spacing:

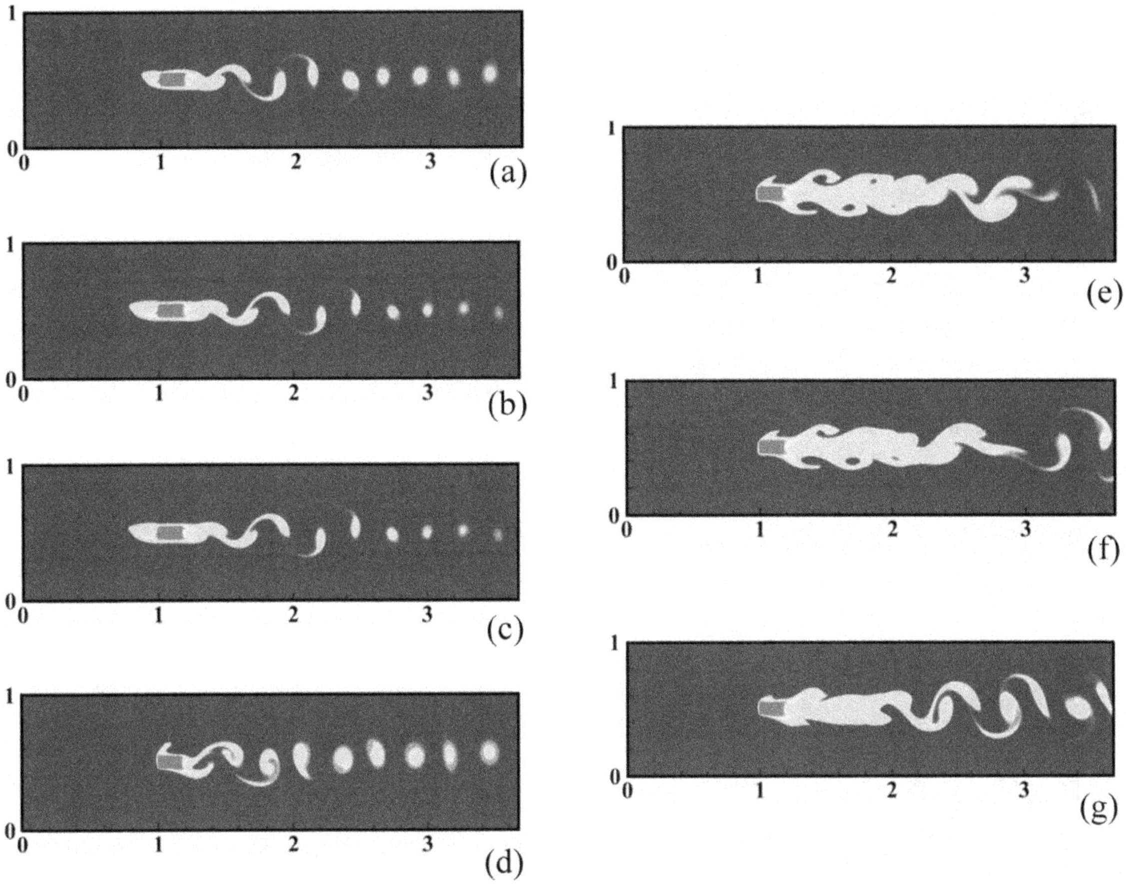

(a) g=1d; (b) g=2d; (c) g=3d; (d) g=5d; (e) g=6d; (f) g=7d; (g) g=8d

Drag Coefficient Variation for 1*d*≥g≤8*d* at Re=150

In this section, the effect of the position between the inclined plate and the rectangular cylinder on the time variation of the drag coefficient was conducted at a fixed Reynolds number (Re=150).

Figure 9. shows the temporal drag coefficient curves for different positions of the control plate. Figure 9-a shows that in the case where g=1*d*–2*d* (low gap spacing), the amplitude of the drag curve is small. This increases the percentage of Cd reduction to 130.97% compared to the case without the control plate where Cd_{mean}=1.13. This optimal reduction allows an optimal energy reduction. After this gap spacing, we observe that there is an increase in the temporal values of Cd and that all the average values are higher than 0. Indeed, the increase of the distance between the two bodies gives the possibility to part of the shear layers to evacuate in this intermediate spacing. This increases the fluid-structure interactions which increase the fluid forces acting on the rectangle. As a result, a remarkable increase is observed in the average and temporal value of the drag coefficient.

Figure 9. Temporal drag coefficient variation for different gap spacing

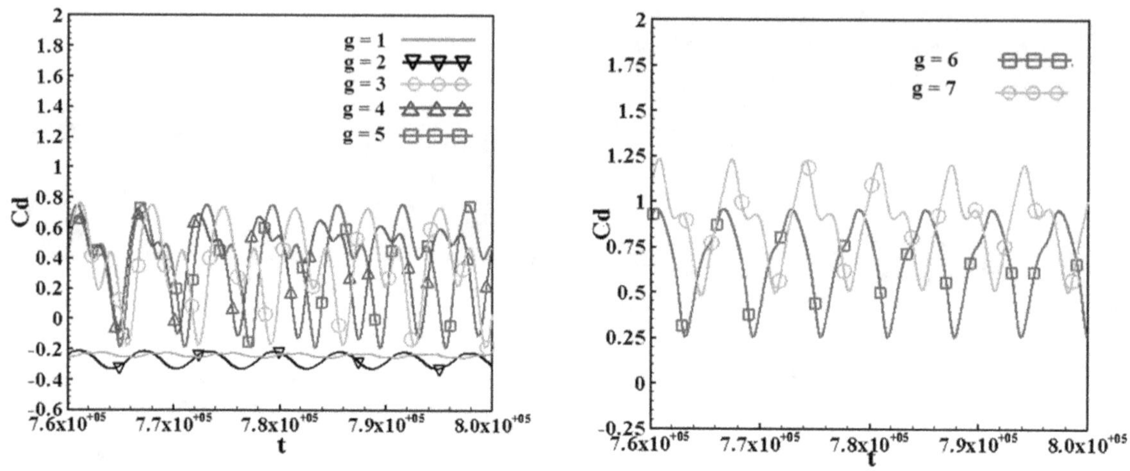

(a) (b)

(a) Cd for 1d≤g≤6d; (b) Cd for 7d≤g≤8d

CONCLUSION

The present work presents a numerical study of the flow of fluid passing around a heated rectangular cylinder controlled by an inclined partition for suppression of the vortex shedding, reducing the drag force and having an important and regular thermal exchange by applying the DMRT-LBM.

- The main conclusions drawn from this study are arranged as follows:
- The variation of the Reynolds number informs the existence of three different types of thermal regimes
- Increasing the Reynolds number increases the thermal vortex shedding and consequently improves the heat exchange.
- The increase of the Reynolds number increases the percentage of reduction of the drag coefficient (125.41% for Re=250)
- The more the gap spacing increases, the more the drag coefficient increases and consequently the percentage of reduction decreases.
- A maximum percentage of reduction reaches 130.97% in the case the control plate is placed at g=2*d*

This work can be directed to many applications in industry, especially in the field of aeronautics, where partitions such as wings placed behind the winglet are used to reduce friction with the air (drag reduction). Also, it can be applied in the field of heat exchangers and in the cooling systems of electronic components.

REFERENCES

Admi, Y., El Hassani, S., Benhamou, J., Moussaoui, M. A., & Mezrhab, A. (2024). Numerical research on the wake characteristics and drag reduction of a square cylinder in the presence of control partitions. *Ocean Engineering*, *301*, 117341. DOI: 10.1016/j.oceaneng.2024.117341

Admi, Y., Lahmer, E. B., Benhamou, J., Moussaoui, M. A., & Mezrhab, A. (2024). Heat transfer improvement and drag force reduction around three heated square cylinders controlled by partitions. *Physics of Fluids*, *36*(4), 045161. DOI: 10.1063/5.0191092

Admi, Y., Moussaoui, M. A., & Mezrhab, A. (2020). Effect of a flat plate on heat transfer and flow past a three side-by-side square cylinders using double MRT-lattice Boltzmann method. In: IEEE 2nd International Conference on Electronics, Control, Optimization and Computer Science (ICECOCS). IEEE, p. 1-5.

Admi, Y., Moussaoui, M. A., & Mezrhab, A. Effect of Control Partitions on Drag Reduction and Suppression of Vortex Shedding Around a Bluff Body Cylinder. In: *International Conference on Advanced Technologies for Humanity*. Springer, Cham, p. 453-463, (2021).

Admi, Y., Moussaoui, M. A., & Mezrhab, A. The Vortex Shedding Suppression and Heat Transfer Characteristics Around a Heated Square Cylinder by Using Three Downstream-Detached Partitions. In: *International Conference on Digital Technologies and Applications*. Springer, Cham, p. 598-608(2022). DOI: 10.1007/978-3-031-02447-4_62

Admi, Y., Moussaoui, M. A., & Mezrhab, A. (2022). Numerical Investigation of Convective Heat Transfer and Fluid Flow Past a Three Square Cylinders Controlled by a Partition in Channel. *International Journal of Renewable Energy Development*, *11*(3), 766–781. DOI: 10.14710/ijred.2022.43790

Benhamou, J., Lahmer, E. B., Admi, Y., Jami, M., & Mezrhab, A. (2023). Three-dimensional numerical simulation of free convection and entropy generation in a cubic cavity containing a heat source. *Heat Transfer*, *52*(7), 5108–5135. DOI: 10.1002/htj.22920

Bouzidi, M., Firdaouss, M., & Lallemand, P. (2001). Momentum transfer of a Boltzmann lattice fluid with boundaries. *Physics of Fluids*, *13*(11), 3452–3459. DOI: 10.1063/1.1399290

Breuer, M., Bernsdorf, J., Zeiser, T., & Durst, F. (2000). Accurate computations of the laminar flow past a square cylinder based on two different methods: Lattice Boltzmann and finite-volume. *International Journal of Heat and Fluid Flow*, *21*(2), 186–196. DOI: 10.1016/S0142-727X(99)00081-8

Chapain, S., & Aly, A. M. (2019). Vibration attenuation in high-rise buildings to achieve system-level performance under multiple hazards. *Engineering Structures*, *197*, 109352. DOI: 10.1016/j.engstruct.2019.109352

d'Humières, D. (2002). Multiple–relaxation–time lattice Boltzmann models in three dimensions. Philosophical Transactions of the Royal Society of London. Series A: Mathematical, Physical and Engineering Sciences, 360(1792), 437-451.

Deng, Y., Feng, C., Jiaqiang, E., Zhu, H., Chen, J., Wen, M., & Yin, H. (2018). Effects of different coolants and cooling strategies on the cooling performance of the power lithium ion battery system: A review. *Applied Thermal Engineering*, *142*, 10–29. DOI: 10.1016/j.applthermaleng.2018.06.043

Dhiman, A.K., R.P. Chhabra, V. Eswaran, Flow and heat transfer across a confined square cylinder in the steady flow regime: Effect of Peclet number, Int. J. Heat Mass Transf. (2005). https://doi.org/, (2005). DOI: 10.1016/j.ijheatmasstransfer.04.033

Elshaer, A., Bitsuamlak, G., & El Damatty, A. (2017). Enhancing wind performance of tall buildings using corner aerodynamic optimization. *Engineering Structures*, *136*, 133–148. DOI: 10.1016/j.engstruct.2017.01.019

Gonzalez-Zamudio, E. A., Olivares-Robles, M. A., & Andrade-Vallejo, A. A. (2025). A Novel Case of Cooling and Heating in Rectangular Lid-Driven Cavities: Interplay of Richardson Numbers in Streamlines and Isotherms. *Processes (Basel, Switzerland)*, *13*(2), 432. DOI: 10.3390/pr13020432

Islam, S. (2015). Ul., S.U. Rahman, H. Abbasi, W.S. Shahina, T. Lattice Boltzmann study of wake structures and force statistics for various gap spacings between a square rod with a detached flat plate. *Arabian Journal for Science and Engineering*, *40*, 2169–2182. DOI: 10.1007/s13369-015-1648-3

Islam, S. U., Zhou, C. Y., Shah, A., & Xie, P. (2012). Numerical simulation of flow past rectangular cylinders with different aspect ratios using the incompressible lattice Boltzmann method. *Journal of Mechanical Science and Technology*, *26*(4), 1027–1041. DOI: 10.1007/s12206-012-0328-4

Lahmer, E. B., Admi, Y., Moussaoui, M. A., & Mezrhab, A. (2022). Improvement of the heat transfer quality by air cooling of three-heated obstacles in a horizontal channel using the lattice Boltzmann method. *Heat Transfer*, *51*(5), 3869–3891. DOI: 10.1002/htj.22481

Liu, R., Li, S., & Yang, L. (2020). Collaborative optimization for metro train scheduling and train connections combined with passenger flow control strategy. *Omega*, *90*, 101990. DOI: 10.1016/j.omega.2018.10.020

Mezrhab, A., & Moussaoui, M. A. (2010). Jami, M., Naji, H., Bouzidi, M.: Double MRT thermal lattice Boltzmann method for simulating convective flows. *Physics Letters. [Part A]*, *374*(34), 3499–3507. DOI: 10.1016/j.physleta.2010.06.059

Mohamad, A. A. (2011). *Lattice boltzmann method* (Vol. 70). Springer.

Monat, J. P., & Gannon, T. F. (2018). Applying systems thinking to engineering and design. *Systems*, *6*(3), 34. DOI: 10.3390/systems6030034

Moussaoui, M. A., Jami, M., Mezrhab, A., Naji, H., & Bouzidi, M. (2010). Multiple-relaxation-time lattice Boltzmann computation of channel flow past a square cylinder with an upstream control bi-partition. *International Journal for Numerical Methods in Fluids*, *64*(6), 591–608. DOI: 10.1002/fld.2159

Rashidi, S., Hayatdavoodi, M., & Esfahani, J. A. (2016). Vortex shedding suppression and wake control: A review. *Ocean Engineering*, *126*, 57–80. DOI: 10.1016/j.oceaneng.2016.08.031

Turki, S. (2008). Numerical simulation of passive control on vortex shedding behind square cylinder using splitter plate. *Engineering Applications of Computational Fluid Mechanics*, *2*(4), 514–524. DOI: 10.1080/19942060.2008.11015248

Zhou, L., Cheng, M., & Hung, K. C. (2005). Suppression of fluid force on a square cylinder by flow control. *Journal of Fluids and Structures*, *21*(2), 151–167. DOI: 10.1016/j.jfluidstructs.2005.07.002

Zou, Q., & He, X. (1997). On pressure and velocity boundary conditions for the lattice Botlzmann BGK model. *Physics of Fluids*, *9*(6), 1591–1598. DOI: 10.1063/1.869307

Chapter 19
Effect of Dual Partitions on Heat Transfer and Fluid Flow Characteristics Around a Square Cylinder Using MRT-Lattice Boltzmann Method

Youssef Admi
https://orcid.org/0000-0003-0920-0618
Faculty of Sciences, Mohammed First University, Oujda, Morocco

Tariq Khair
Faculty of Sciences, Mohammed First University, Oujda, Morocco

Jaouad Benhamou
Faculty of Sciences, Higher School of Education and Training, Mohammed First University, Oujda, Morocco

Mohammed Amine Moussaoui
Faculty of Sciences, Mohammed First University, Oujda, Morocco

Ahmed Mezrhab
Faculty of Sciences, Mohammed First University, Oujda, Morocco

ABSTRACT

In this research, a numerical study of airflow control coupled to heat transfer around a heated square cylinder is carried out at a fixed Reynolds number Re= 150. The effect of the position and length of two partitions implemented downstream of the square obstacle is examined. The numerical simulations are performed using the lattice Boltzmann method with multiple relaxation times. The results obtained are presented in the form of graphs of the isotherms, the temporal and mean drag coefficients and as a function of the mean Nusselt number. These results show that the successive inclusion of control partitions has a significant impact on the increase of the reduction percentage of drag coefficient that reaches 30.74%

DOI: 10.4018/979-8-3693-9924-8.ch019

in a critical case. In a parallel implementation of the two partitions, this reduction percentage reaches 17.26%, with an enhancement of over 10% in heat exchange between the air and the heated obstacle's top, bottom, and rear surfaces compared to the uncontrolled case.

INTRODUCTION

Heat flow and heat transfer are among the fundamental phenomena that enable energy balance to be achieved in many engineering and scientific systems and applications. These two phenomena play an essential role in achieving optimum performance and durability in a wide range of fields such as heat exchangers, combustion engines and chemical reactors. It also plays an essential role in the infrastructure, architecture, aeronautics and electronics cooling industries (A. Bar-Cohen, A. D. Kraus, 1983; Chapain et al., 2019; Deng et al., 2018; Florides et al., 2007; S. Chen, Y. Liu, S. F. Chan, C.W. Leung, 2001; T.J. Young, 1998). The importance of heat flow and transfer stems from the development of civilization and the advancement of technology, as the demand for energy increases to meet the needs of economic and social development. With improved performance and smaller size, current and future generations of chips are constantly improving their performance and durability, making them a driving force for innovation and technological progress. However, the high heat dissipation associated with electronic chips remains a challenge that requires in-depth study. When chips are working hard, temperatures can reach high levels that can exceed tolerance limits, adversely affecting the reliability and functionality of the hardware. The risk increases when the excess heat becomes sufficient to melt or modify the thermos-physical properties of the chips, causing them to fail. According to Bar-Cohen's research (A. Bar-Cohen, A. D. Kraus, 1983), the reliability of a silicon wafer decreases by 10% for every ± 2 rise in temperature. For this reason, the design of advanced thermal systems is essential to overcome these challenges. The successful design of these systems depends on optimizing chip heat dissipation and maintaining safe and appropriate operating temperatures.

In general, the field of heat flow and heat transfer requires continuous development and research to improve the performance of equipment and systems and ensure their reliability. Rapid technological advances and the growing demand for energy require sophisticated engineering solutions to ensure that systems operate efficiently and sustainably in the future. In this way, research and innovation contribute to optimizing the flow and transfer of heat in a wide range of technical and scientific applications. Indeed, the cooling of heated blocks simulating electronic components mounted on a board has been the subject of a great deal of work (S-W. Perng, 47 C.E.; S. Chen, Y. Liu, S. F. Chan, C.W. Leung, 2001; T.J. Young, 1998). These studies have helped to develop new and improved techniques for designing and optimizing thermal systems, which has led to increased performance and energy efficiency in a wide range of applications. Zhang et al. (Yu, C., Zhang, H., Zeng, M., Wang, R., & Gao, 2020) used computational fluid dynamics (CFD) simulations to investigate the flow and heat transfer characteristics of a tube bundle in a shell-and-tube heat exchanger. The authors used the k-ω turbulence model and the SIMPLE algorithm to simulate the flow and heat transfer, and they validated their results by comparing them to experimental data. The study showed that the CFD simulations provided a good prediction of the flow and heat transfer characteristics of the heat exchanger, which could be used to optimize the design and improve the performance of such systems. Chen et al. (Chen, G., Jia, M., Zhang, S., Tang, Y., & Wan, 2020) investigated the flow and heat transfer characteristics of a novel microchannel heat sink. The authors used a microfabrication technique to create the microchannels and conducted experiments to measure the temperature and pressure drop across the heat sink. The study showed that the microchannel heat sink had significantly improved heat transfer performance compared to traditional heat sinks, and the authors suggested that this technology could be used in a range of applications, including microelectronics cooling.

As a general rule, these systems are modelled by one or more blocks located on the wall of a channel. There are several techniques for cooling the blocks. These include systems with vortex generators such as partitions or small cylinders (Q. Wang, 2002; Yang, 2002).

The numerical investigation of airflow around a square obstacle controlled by a partition or rod in a horizontal channel has attracted considerable attention due to its wide range of practical applications. Despite this interest, most existing studies have adopted a fragmented approach, focusing either on fluid dynamics—analyzing parameters such as flow structures and fluidic forces—or on heat transfer, examining factors like isotherms and Nusselt number variations. This separation often leads to incomplete insights that fail to fully represent the coupled nature of fluid flow and thermal phenomena. In practice, a comprehensive understanding of fluid behaviour requires simultaneous analysis of both flow and heat transfer characteristics. Furthermore, while prior research has extensively examined the influence of spacing between the bluff body and control partitions, the critical role of partition length has largely been overlooked. To address these gaps, this study aims to provide an integrated analysis of the combined effects of the position (spacing) and length of two partitions on reducing drag coefficients and enhancing heat transfer around a square cylinder. Indeed, this study aims to analyze the effects of two partitions on airflow and heat exchange between a heated obstacle and cold air inside a horizontal channel. The study specifically targets the influence of partition position and length on fluid flow dynamics and heat transfer processes between the heated obstacle and the cold air. The two partitions are arranged in parallel behind the square obstacle to enhance heat exchange behind the obstacle while reducing the fluidic forces acting on this obstacle. In the first part, the two partitions are inserted one after the other horizontally behind the obstacle. Indeed, we studied the position of the second partition by attaching the first partition to the block and keeping both partitions at a fixed length (Lp=1d). In the second part, the two partitions are inserted horizontally and parallel behind the obstacle. First, we studied the position of the two partitions at a fixed length (Lp=1d). Next, we fixed the position of the partitions to examine the effect of their lengths.

BOUNDARY CONDITIONS OF GEOMETRY STUDIED

Geometry Studied Description

Figure 1 depicts the computational domain's geometry for the flow around a square cylinder in a two-dimensional horizontal channel, along with a dual control partition. The square cylinder is positioned at a distance of Lu=6d upstream and extends Ild=30d downstream, with a height of H=11d (as illustrated in Figure 1). Where "d" is the size of the square obstacle. The control partitions are situated consecutively downstream of the main square obstacle, each having a length of Lp=1d and a height of h=0.02d. The gap spacing between the two partitions is indicated as "g" (as shown in Figure 1). The fluid has entered the system with a cold temperature (θ_c= -0.5), while the block is considered to be heated with a hot temperature (θh=0.5). Additionally, the channel walls are assumed to be adiabatic. The block also possesses a dimensionless temperature of θh=0.5.

Figure 1. Structure of the computational domain

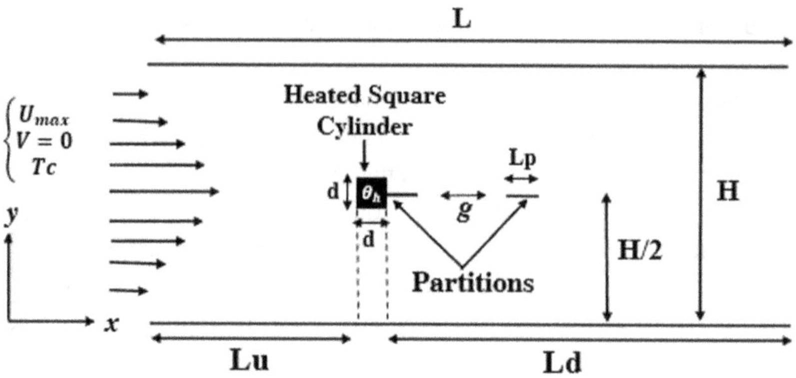

Boundary Conditions

The accurate implementation of boundary conditions is crucial for maintaining stability and accuracy in numerical simulations using the lattice Boltzmann method. The most widely used boundary conditions in LBM are the Bounce-Back conditions (Bouzidi et al., 2001), which are expressed by the following equation:

$$f_i(\vec{x}_B, t) = f_i(\vec{x}_B, t) \tag{1}$$

Where the $f_i(\vec{x}_B, t)$ and $f_i(\vec{x}_B, t)$ represent the unknown and known distribution functions, respectively (see Figure 2), at the wall node ($\vec{x}_B$).

On the top wall, the unknown distribution functions are defined as follows: $f_4 = f_2$, $f_7 = f_5$ and $f_8 = f_6$. Likewise, on the bottom wall, the unknown distribution functions are given by the corresponding distribution functions mentioned earlier.

At the inlet of the channel, the flow is assumed to be fully developed with a parabolic velocity profile ($u = 1.5 * Umax\,(1-(y/H))$; $v = 0$). This velocity profile is also applied at the outlet. Thus, the velocity and temperature gradients are assumed to be zero in the outlet channel. Hence, the Zou-He boundary conditions are favoured for their implementation (Zou et al., 1997).

For the D2Q9 model, the distribution functions f_0, f_2, f_3, f_4, f_6, and f_7 towards the outside of the calculation domain are known, while the distribution functions f_1, f_5, and f_8 towards the inside of the calculation domain are unknown

Figure 2. Illustration of physical boundaries.

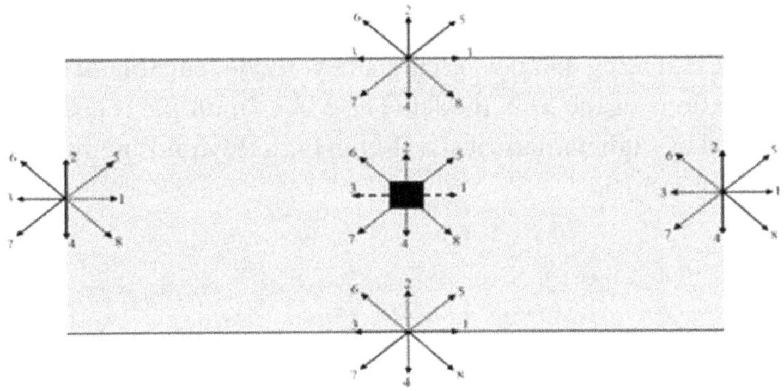

Mezrhab et al. (Mezrhab et al., 2010) introduced an approach for addressing heat flux boundary conditions in thermal problems. This method involves transitioning from the macroscopic scale to the mesoscopic scale by utilizing the general term of the distribution function, denoted as g_i:

$$g_i = \frac{2T\left(1+\frac{a}{4}\right)}{5} - g_{\bar{i}} \tag{2}$$

Where N^{-1} denoted the moment inverse matrix and the value of 'a' is determined using the previous formula

$$\alpha = \frac{\sqrt{3}(4+a)}{60}$$

. (3)

The condition $g1 = -g3 + 2\Theta c\ (1 + a/4)/5$ is applied at both the inlet and outlet of the channel. Likewise, for the limits of the obstacles, the same condition is employed, but with the substitution of θ_c with θh, as these boundaries are treated as hot. Moreover, the condition $g2 = -g4 + 2\Theta c\ (1 + a/4)/5$ is introduced for the upper and lower faces of the obstacles. Meanwhile, the channel walls are considered adiabatic.

Numerical Method

The Lattice Boltzmann Method (LBM) is the numerical approach used to simulate the fluid flow field and the heat transfer field. It is based on spatio-temporal discretization using a regular grid, which makes it possible to represent the microscopic behavior of particles in a moving fluid. The LBM method has several advantages over conventional methods. It is a simpler, less costly numerical approach, and can be applied to complex geometries and a variety of boundary conditions, making it a powerful tool for solving complex fluid mechanics and heat transfer problems (Admi et al., 2024, 2024, 2022, 2020; Benhamou et al., 2023; Lahmer et al., 2022; Mohamad, 2011; Moussaoui et al., 2010) Two models of the Boltzmann lattice method (LBM) are widely known and used to simulate various phenomena.

However, the LBM-SRT (single relaxation time) model (Qian et al., 1992) has certain limitations due to its simplicity, in particular stability problems in high Reynolds number flows. These shortcomings have prompted several researchers (D'Humières et al., 2002; Guo et al., 2013; Qian et al., 1992; Zou et al., 1997) to set about researching and developing a new model capable of meeting the requirements and overcoming the limitations of the SRT model. The MRT (multiple relaxation time) model, on the other hand, is characterized by high numerical stability in high Reynolds number flows, as well as better accuracy and higher quality results.

Flow Field

The D2Q9 scheme was chosen for its convergence capability and its compatibility with the thermic model to handle the distribution of fluid particle density. Figure 3 illustrates this model, consisting of nine discrete speeds (c_i) represented as follows:

$$c_i \begin{cases} (0,0); i = 0 \\ (\cos[(i-1)\pi/2)], \ \sin[(i-1)\pi/2)])c; i = 1-4 \\ (\cos[(2i-9)\pi/4)], \ \sin[(2i-9)\pi/4)])c; i = 5-8 \end{cases} \tag{4}$$

In this study, the MRT approach was used. This model proposes an expression for the lattice Boltzmann equation, which can be formulated as follows:

$$fi(x + ci\,\Delta t, t + \Delta t) = fi(x,t) + \Omega(f),\ i = 0,1,\ldots,8 \tag{5}$$

By introducing the evolution introduced by D'Humières (D'Humières et al., 2002), the LBM-associated multiple relaxation times (LBM-MRT) approach provides an expression for the lattice Boltzmann equation, which can be stated as:

$$fi(x + ci, t + 1) - fi(x,t) = M^{-1} \times S_i \times (m_i(x,t) - m_i^{eq}(x,t))\ i = 0,\ldots,8 \tag{6}$$

where m and m^{eq} are the moment vector and its equilibrium vector respectively:

$$m = \left(m_0, m_1, m_2, , m_n\right)^T \ ; \ m^{eq} = \left(\rho, e^{eq}, \varepsilon^{eq}, j_x, q_x^{eq}, j_y, q_y^{eq}, p_{xx}^{eq}, p_{xy}^{eq}\right)^T \tag{7}$$

The equilibrium momentum functions m_i^{eq} define the equivalent macroscopic equations.

$$m_0^{eq} = \rho m_1^{eq} = j_x m_2^{eq} = j_y\, m_3^{eq} = -2\rho + 3\left(j_x^2 + j_y^2\right)$$

$$m_4^{eq} = \rho - 3\left(j_x^2 + j_y^2\right)$$

$$m_5^{eq} = -j_x$$

$$m_6^{eq} = -j_y$$

$$m_7^{eq} = \left(j_x^2 - j_y^2\right)$$

$$m_8^{eq} = j_x j_y \tag{8}$$

where

$$j_x = \rho u_x = \sum_i f_i^{eq} c_{ix}$$

$$j_y = \rho u_y = \sum_i f_i^{eq} c_{iy}. \tag{9}$$

The transition from velocity space to momentum space is expressed by (Mohamad, 2011):

$m_i = M f_i$ and $f_i = M^{-1} m_i$ (10)

The matrix M of order 9 is explicitly given by (Mohamad, 2011):

$$M = \begin{pmatrix} \langle m_0| \\ \langle m_1| \\ \langle m_2| \\ \langle m_3| \\ \langle m_4| \\ \langle m_5| \\ \langle m_6| \\ \langle m_7| \\ \langle m_8| \end{pmatrix} = \begin{pmatrix} 1 & 1 & 1 & 1 & 1 & 1 & 1 & 1 & 1 \\ -4 & -1 & -1 & -1 & -1 & 2 & 2 & 2 & 2 \\ 4 & -2 & -2 & -2 & -2 & 1 & 1 & 1 & 1 \\ 0 & 1 & 0 & -1 & 0 & 1 & -1 & -1 & 1 \\ 0 & -2 & 0 & 2 & 0 & 1 & -1 & -1 & 1 \\ 0 & 0 & 1 & 0 & -1 & 1 & 1 & -1 & -1 \\ 0 & 0 & -2 & 0 & 2 & 1 & 1 & -1 & -1 \\ 0 & 1 & -1 & 1 & -1 & 0 & 0 & 0 & 0 \\ 0 & 0 & 0 & 0 & 0 & 1 & -1 & 1 & -1 \end{pmatrix} \tag{11}$$

In the fluid flow field, the density and the momentum are given by using the following equation:

$$\rho = \sum_{i=0}^{8} f_i, \; \rho u = \sum_{i=0}^{8} c_i f_i \tag{12}$$

Temperature Field

In order to ensure compatibility and minimize computation time, the thermal issue is addressed using the D2Q5 model. The movement of the particles in this model is achieved by employing discrete velocities, which are denoted by:

$$c_i \begin{cases} (0,0); i = 0 \\ (\cos[(i-1)\pi/2)], \; \sin[(i-1)\pi/2)])c; i = 1-4 \end{cases} \tag{13}$$

The expression of the thermal lattice Boltzmann equation (TLBE) with the MRT approximation is:

$$g_i(x + c_i \times \Delta t, t + \Delta t) - g_i(x, t) = N^{-1} \times Q_i \times \left(n_i(x,t) - n_i^{eq}(x,t)\right), \mathrm{i} = 0-4 \tag{14}$$

In the thermal field, the distribution function is denoted as g_i, where N represents the 5×5 transformation matrix and Q refers to the diagonal relaxation time matrix. n_i and $\boldsymbol{n}_i^{eq}$ represent the thermal momentum and equilibrium thermal momentum in the moment space, respectively.

The matrix N is given by (Mohamad, 2011):

$$N = \begin{pmatrix} 1 & 1 & 1 & 1 & 1 \\ 0 & 1 & 0 & -1 & 0 \\ 0 & 0 & 1 & 0 & -1 \\ -4 & 1 & 1 & 1 & 1 \\ 0 & 1 & -1 & 1 & -1 \end{pmatrix} \tag{15}$$

The equilibrium moments n_i^{eq} correspond to the five distribution functions (g_i), which are discussed in detail in (Jami et al., 2016):

$$\begin{aligned} n_0^{eq} &= T \\ n_1^{eq} &= uT \\ n_2^{eq} &= vT \\ n_3^{eq} &= aT \\ n_4^{eq} &= 0 \end{aligned} \tag{16}$$

where the constant a depends on the value of the thermal diffusivity α by the following expression:

$$a = \frac{60\alpha}{\sqrt{3}} - 4 \tag{17}$$

The temperature is the macroscopic variable calculated in the thermal field, which can be expressed as follows:

$$T = \sum_{i=0}^{4} g_i \tag{18}$$

Figure 3a. D2Q9: flow field model

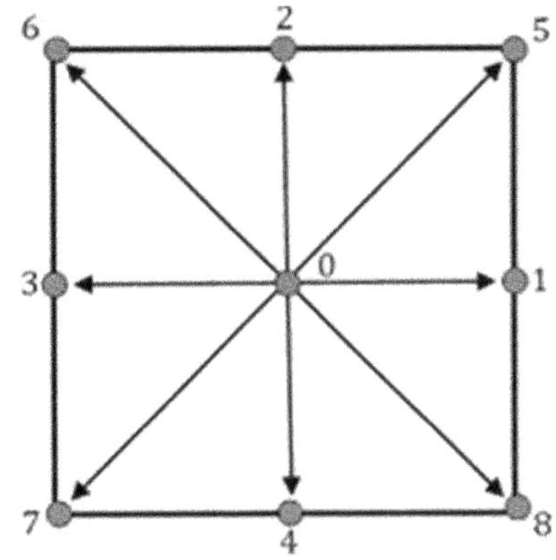

Figure 3b. D2Q5: temperature field model.

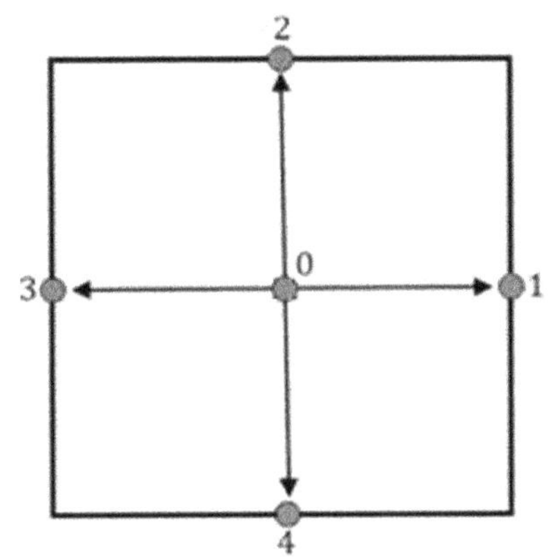

It should be noted that the D2Q9 scheme can be used for the temperature field. However, in order to reduce the calculation time while maintaining the same level of accuracy, the D2Q5 scheme has been preferred.

VALIDATION CODE

This study includes several validation cases conducted by various researchers and involving different geometries. Firstly, we validated our numerical results by comparing it to the of Breuer et al. (Breuer et al., 2000), which investigated the flow around a square cross-section cylinder placed inside a horizontal channel with a blocking ratio of 1/8. Figure 4-a presents the temporal variation of the drag coefficient plotted against the Reynolds number. Our obtained results are compared to those reported by Breuer et al. It can be observed from Figure 4-a that our results generally exhibit good agreement with the reference study.

Additionally, we validated our numerical results by comparing them to the results obtained by Islam et al. (Islam et al., 2015) who numerically examined the flow around a square cylinder controlled by a single partition. Figure 4-b illustrates the average drag coefficient variation as a function of the spacing between the square obstacle and the control partition, at a fixed Reynolds number (Re=150). The comparison demonstrates a satisfactory agreement between our results and the reference results.

Figure 4. Average drag coefficient validation

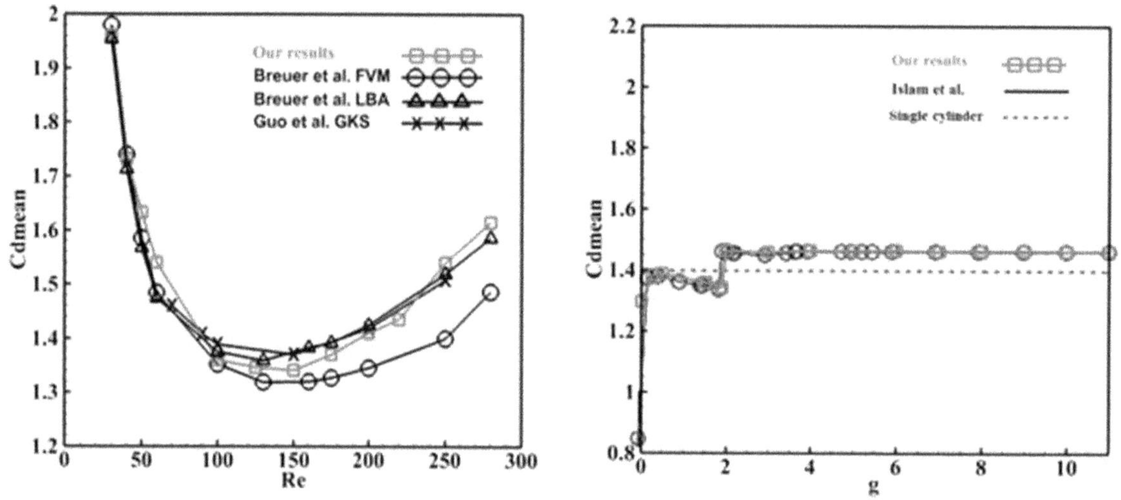

(a) (b)

Likewise, Turki et al. (Turki, 2008) treated the flow of air coupled with convective heat transfer around a square obstacle, heated and inserted inside a horizontal channel. These authors used the finite volume method (FVM) with non-uniform meshes of (122*57) and (122*77) respectively for blocking ratios at β=1/4 and β=1/8.

Figure 5 illustrates the averaged Nusselt number variations as a function of the Reynolds number on all faces of the obstacle. The present results are compared with those obtained by Turki et al. (Turki, 2008).

From this graph, we can see that the heat exchange between the fluid and the front face is maximal compared to the other faces. This is due to the direct contact between the fluid particles and this face, whereas this exchange decreases strongly on the rear face due to the recirculations formed behind this face. Also, this graph illustrates that the obtained results are similar to those of Turki et al. (Turki, 2008).

Figure 5. Nusselt number validation.

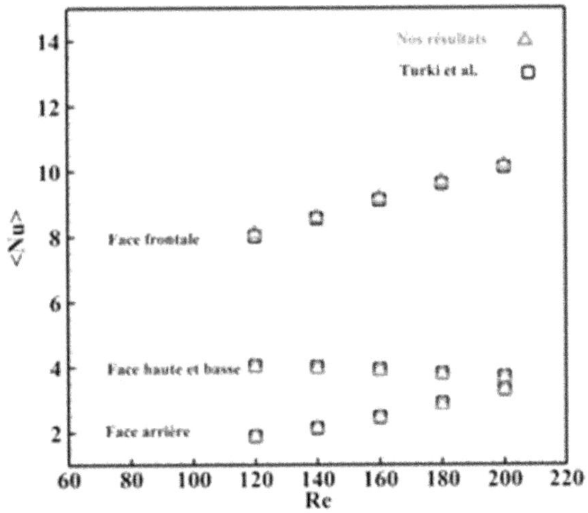

Finally, it can be concluded that our numerical code is robust to study our current physical problem.

PRESENTATION AND DISCUSSION OF THE NUMERICAL RESULTS

Gap Spacing Effect

Figure 6 illustrates the isotherms around the obstacle when the two successive downstream partitions are present, at various gap spacings. We note that both partitions have the same length and that the first partition is attached to the obstacle. The aim of the present work is to study the impact of the position "g" between the two partitions on the reduction of the thermal vortex shedding. As a result, there is a noticeable decrease in the values of the drag force coefficient. This leads to an improvement in heat exchange between the fluid entering and the square obstacle.

Figure 6 illustrates the isotherms around the square obstacle controlled by double partitions. In the case where no partition is implemented, the incoming fluid directly encounters the four faces of the obstacle. As a consequence, a periodic arrangement of the thermal vortex is observed (Figure 6-a). Whereas when the control partitions are introduced, the flow patterns around the square cylinder undergo a transformation. The configuration of the flow, specifically the thermal Von Karman Street, is altered depending on the placement of the partition.

From Figure 6, it can be seen that as the second partition moves away from the first, the number and form of vortexes decreases until a critical spacing $g=1.9d$ where almost total vortex shedding is observed. Indeed, at this critical spacing, the shear layers delivered by the edges of the obstacles on the faces of the first partition then meet the second partition and adhere to these faces. After this crucial case, the shear layers find sufficient space to evacuate and subsequently interact with the second partition. This increases the number and shape of vortices released.

Figure 6. Isotherms Visualization for different gap spacing.

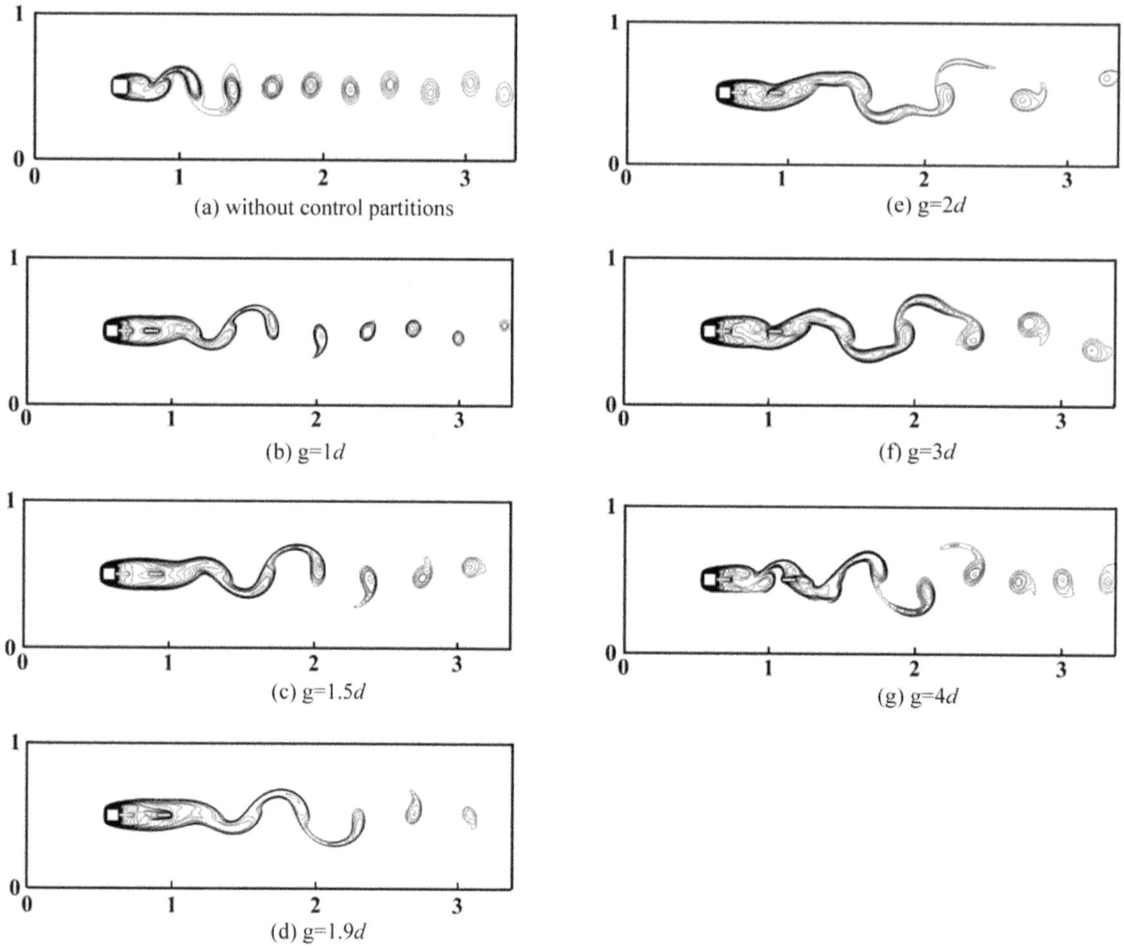

The drag coefficient, often referred to as Cd, is a dimensionless quantity that characterizes the resistance experienced by an object moving in a fluid (in this study, air). It is used to quantify the drag force acting on the object and is an essential parameter in the study of fluid dynamics and aerodynamics. It plays a crucial role in understanding and designing various technical applications. Its formulation is given by:

$$Cd = \frac{2 \times F_D}{\rho . U_{max}^2 . d} \tag{19}$$

In this section, we have carried out a study of the drag coefficient by varying the spacing "g". In this study, the first partition is attached to the obstacle, and we have examined the impact of spacing g on reducing the drag coefficient.

Table 1 displays the drag coefficient average values for different downstream gap spacing g. Figure 7 illustrates the graph representing the temporal drag coefficient variation (Cd) for the various gap spacing 1d≤g≤4d. Likewise, the average values of this coefficient are illustrated in Table 1 for the same range of g spacings (1d≤g≤4d). These results show that, for all gap g, the drag coefficient exhibits a lower value compared to the case where the partitions are not inserted. The maximum reduction in the drag coefficient occurs at 30.74% for an optimal position g=1.9d. After this optimum spacing, the drag coefficients start to increase. In fact, in the uncontrolled case, the incoming fluid meets the cylinder faces directly, which causes an increase in the fluid forces acting on the cylinder. This coefficient decreases with the introduction of partitions. Generally, control partitions minimize the contact between the fluid and the cylinder faces. Therefore, in the critical case, a large amount of cushioning layers falls behind the square cylinder, which decreases the contact between the cylinder walls and the fluid. As a result, the drag coefficient values decrease.

Figure 7. Temporal variation of drag coefficient for various position g.

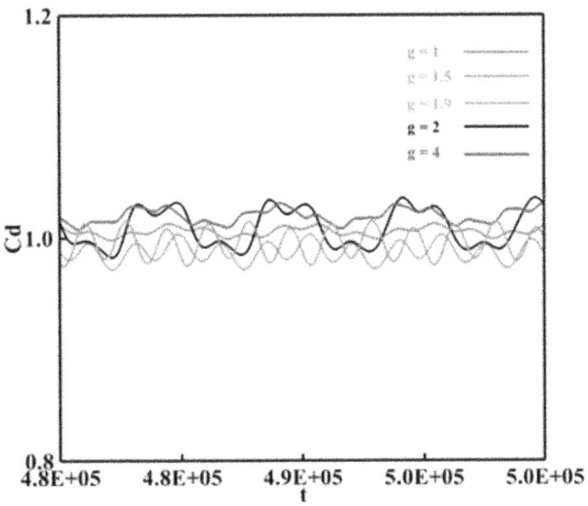

Table 1. Average drag coefficient for various position g.

g	**Cd$_{mean}$**	**Drag coefficient reduction percentage**
Without control	**1.425**	----
1d	1.005	29.47%
1.5d	0.995	30.17%
1.7d	0.990	30.52%
1.9d	**0.987**	**30.74%**
2d	1.010	29.12%
2.5d	1.006	29.40%
3d	1.000	29.82%
4d	1,0197	28.44%

In this section, the average Nusselt number variation is also examined for each face of the obstacle at a fixed Reynolds number Re=150. This variation has been studied for different gap spacing where the average Nusselt number is calculated by the expression:

$$N_u = -d \times \left(\frac{\partial\theta}{\partial n}\right)_{Surface} \quad (20)$$

where "*n*" and "*d*" represent the normal and the face length of the cylinder respectively.

The space-average Nusselt number around a square obstacle is defined by:

$$\begin{cases} Nu_i = \int_0^{d/H} -d \times \frac{\partial\theta}{\partial X}.dY \text{ For transverse faces } (21) \\ Nu_i = \int_0^{d/H} -d \times \frac{\partial\theta}{\partial Y}.dX \text{ For horizontale faces } (22) \end{cases}$$

Table 2 shows the average Nusselt number calculated for each face of the obstacle at different positions of the second partition, as explored in nine scenarios where the spacing "g" varies from 0 to 4d. The table shows that for the front, top and bottom faces of the obstacle, the Nusselt number remains relatively constant for all positions of the plate, as well as in the uncontrolled case. Conversely, for the rear face, a slight decrease is noticeable in all cases. A slight enhancement was observed in the case where g=2d. This improvement is attributed to the fact that the shear layers produced by the edges of the obstacle interact with the leading edges of the second partition. This disturbs the flow in the intermediate zone by increasing the heat exchange between the air and the obstacle.

It should be noted that we conducted other studies to improve heat exchange but we did not obtain significant increases in the values of the Nusselt number. In fact, we fixed the first partition attached to the block with a fixed length (1d) and we studied the impact of the position of the second partition. But the results obtained from this study do not allow any improvement in the values of the Nusselt number. However, this study did show an increase in the percentage reduction in the drag coefficient. We then fixed the second partition and studied the effect of the length of the first partition, which also led to a reduction in the drag coefficient without any significant improvement in heat exchange.

In general, the treatment of this geometry, airflow around an obstacle controlled by two partitions inserted successively at the rear, makes it possible to obtain a significant reduction in the drag coefficient with regular heat exchange.

Table 2. Average Nusselt number values for all faces of the square cylinder for different gap spacing "g".

g	Nu_1	Nu_2	Nu_3	Nu_4
No control case	8.215	3.037	2.150	3.027
1d	8.155	2.932	1.964	2.947
1.5d	8.143	2.972	1.965	2.950
1.7d	8.130	2.962	1.973	2.960
1.9d	8.193	2.998	1.995	2.993
2d	8.211	3.047	1.993	3.04
2.3d	8.185	2.996	1.978	2.993

continued on following page

Table 2. Continued

g	Nu_1	Nu_2	Nu_3	Nu_4
2,5d	8.181	2.989	1.973	2.992
3d	8.162	2.986	1.985	2.991
4d	8.158	2.986	1.986	2.991

In order to significantly improve heat exchange, we thought about changing the location of the two control partitions. We kept the same geometry but implemented two parallel horizontal partitions behind the square obstacle (see Figure 8).

Figure 8. Second geometry treated.

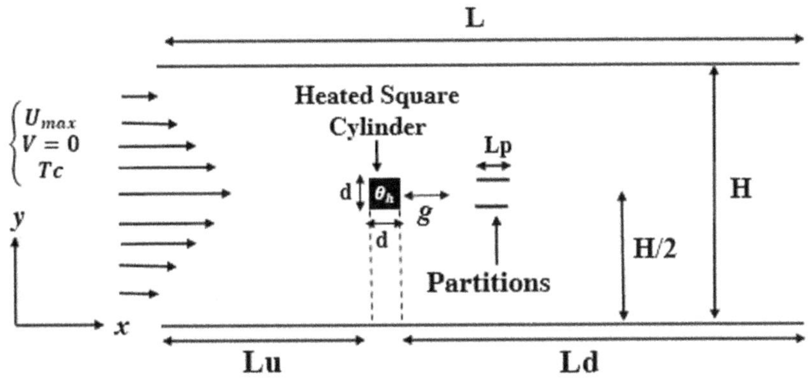

In this geometry, we have kept the same dimensions of the channel and the obstacle and the same boundary conditions that the previous geometry.

Figure 9 shows the effect of the distance between the main cylinder and the control partitions (gap spacing) on the isotherms. This study illustrates three thermal flow regimes by varying the gap spacing g between 1d and 13d. Indeed, for g≤2d, the shear layers, which form from the end of the square block, fall either behind or onto the partitions. Consequently, vortex shedding is only observed after the partitions. Thus, a variety of free thermal vortices are observed along the channel (see Figure 9. a-b) without any melting or distortion. Likewise, Figure 9. c-d illustrates another flow mode where the g-spacing varies from 3d to 7d. This regime manifests itself as attachment flow. In this scenario, the shear stress layers, which have separated from the square cylinder, develop and then rapidly reconnect to the leading edges of the control partitions. This disturbs the intermediate zone between the square cylinder and the partitions. The fluid particles interact strongly with the walls of the block. This interaction affects the length and size of the wake along the channel. Figure 9 c-e shows that the number and size of vortices detaching behind the control partitions change completely compared to the previous case. Figure 9. e-g illustrates the third flow regime where the g-spacing is strictly greater than 7d. In this case, the shear layers delivered at the ends of the cylinder are evacuated into the intermediate distance. This allows thermal vortices to develop completely in this area. The behaviour of the fluid, in this case, is almost the same as in the case of the single square cylinder (without control partitions).

Figure 9. Isotherms Visualization for various gap spacing.

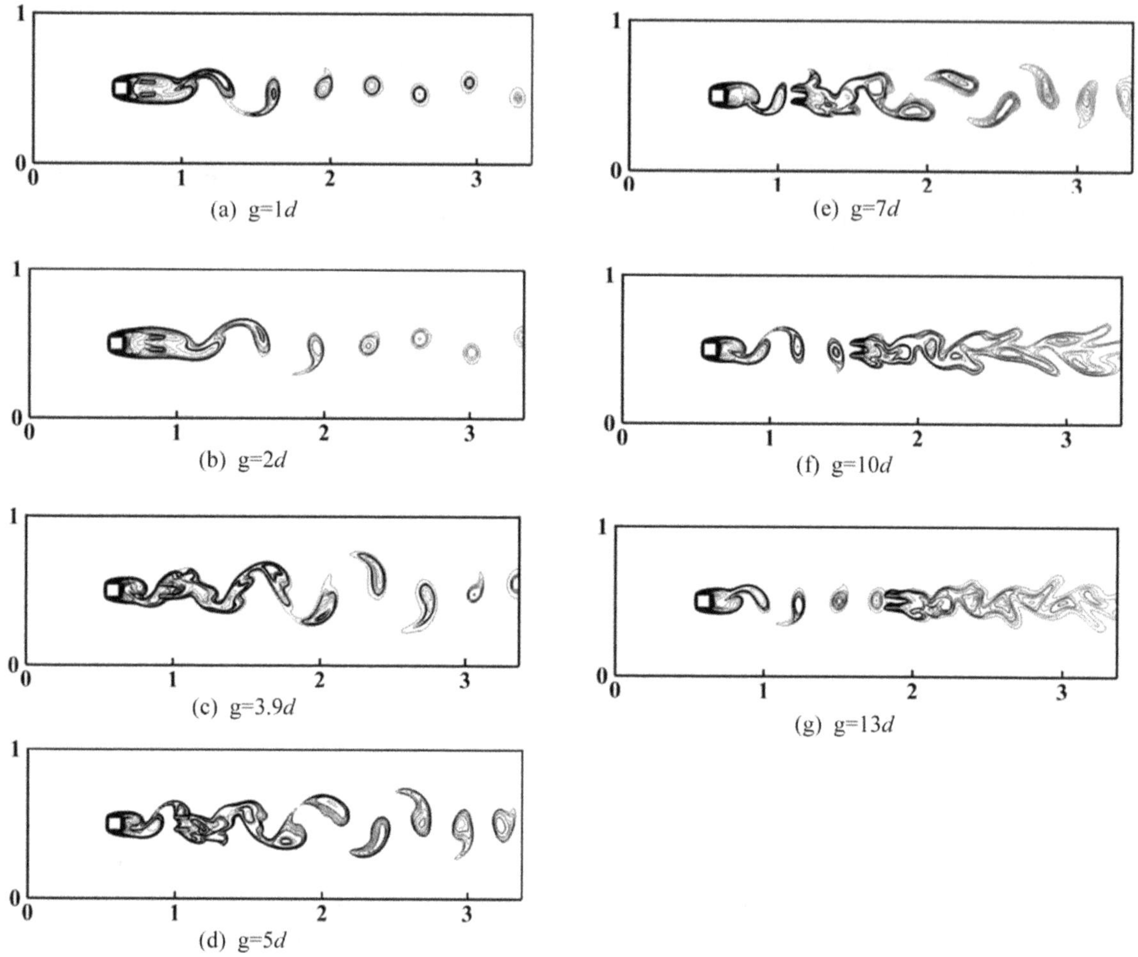

Likewise, the fluid forces exerted on the square cylinder are influenced by the variation in spacing. Table 3 shows the mean values of the drag coefficient (Cd_{meam}) for various spacing ranges. From the results obtained, it can be seen that the drag coefficient decreases with increasing spacing g up to a critical spacing g=3.9d. At a critical spacing, the Cdmean assumes its smallest value and the percentage reduction in drag coefficient reaches 16.48% compared with the case without the implementation of control partitions. After this critical spacing, the drag coefficient starts to increase again. In fact, for small spacings ($g \leq 2d$), the shear layers produced by the ends of the block are partially or totally evacuated over the partitions. At medium spacings, however, the shear layers fall into the intermediate zone by interacting with the leading edges of the partitions. This disturbs the flow in this zone. This increases the fluidic forces exerted on the obstacle. This results in a remarkable increase in the drag value for these spacings. In addition, for large spacings, the drag coefficient admits values as in the case of a unique obstacle. This is due to the sufficient spacing between the obstacle and the two control partitions.

In general, the inclusion of two partitions in the wake led to fundamental changes in the thermal flow field. As a result, the drag force applied to the square cylinder was considerably reduced.

Table 3. Average values and reduction percentage of drag coefficient for various position g.

g	Cd_{mean}	Percentage reduction of drag coefficient
No control case	**1.425**	----
1d	1.293	9.26%
2d	1.277	10.38%
3d	1.199	15.86%
3.9d	**1.185**	**16.84%**
5d	1.294	9.19%
7d	1.29	9.47%
10d	1.346	5.54%
13d	1.397	1.96%

In this section, the evolution of the average Nusselt number on the cylinder faces is studied for different "g" spacings. Table 4 shows the variation in mean Nusselt number values for different partition positions. Figure 10 shows the variation of the total average Nusselt number as a function of the spacing *g* while fixing the length *Lp* at 1d. From these results, we can see that the average values of the Nusselt number are improved compared to the case where the two control partitions are implemented successively downstream of the cylinder. Likewise, the values obtained are generally better than those found in the case where the partitions are not implemented. This is because the incoming fluid meets the front face of the cylinder directly, resulting in maximum heat exchange on this surface. Then, the shear layers delivered by the ends of the cylinder affect the heat exchange along the top and bottom faces in almost the same way. These layers are then evacuated behind the cylinder either behind and/or on the partitions for short and medium spacings g or in the intermediate zone in cases where the partitions are very far from the cylinder. Except short spacing, the presence of partitions has a considerable influence on heat transfer to the rear of the cylinder. Thus, in the case where the partitions are implemented at g=3.9d, the Nusselt number achieves a 10% improvement over the case without control along the top, bottom and rear surfaces of the cylinder.

Figure 10 also illustrates that the total average Nusselt number shows a slight decrease for small spacings g≤2d. Then, a brutal increase of this number is observed for a critical spacing g=3d. After this spacing, the total average Nusselt number takes on a maximum value at g=3.9d. Thus, this number remains almost constant for a spacing range of 3.9d≤g≤7d. In this range, the Nusselt number takes on significant values, higher than those found in the other cases and in the uncontrolled case. This illustrates that there is maximum heat exchange between the fluid and the walls of the obstacle.

Table 4. Average Nusselt number values for all faces of the square cylinder for different gap spacing "g".

g	Nu_1	Nu_2	Nu_3	Nu_4
No control case	8.21 5	3.037	2.15	3.027
1d	8.14	2.945	1.586	2.888
2d	8.141	2.881	1.479	2.92
3d	8.254	3.107	2.445	3.106
3.9d	8.287	3.359	2.363	3.325
5d	8.304	3.397	2.265	3.25
7d	8.283	3.222	2.184	3.072
10d	8.275	3.178	2.164	3.11
13d	8.211	3.102	2.155	3.037

Figure 10. Values of Average Nusselt number total for different gap spacing "g".

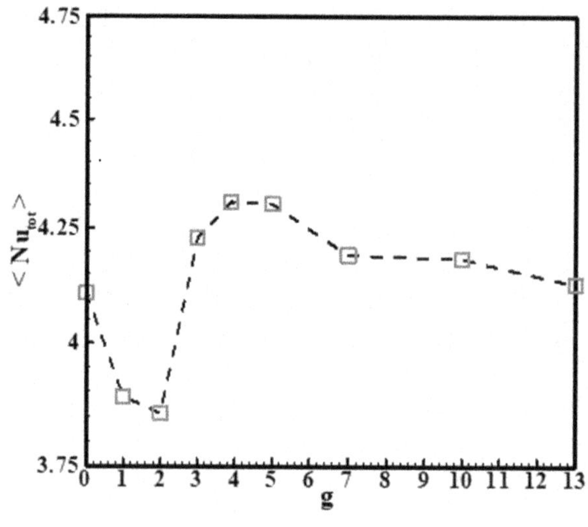

The study of the evolution of the average Nusselt number on the faces of the cylinder is also examined for different lengths *Lp* of the partitions. Table 5 shows the mean values of *Nu* on the four faces of the heated cylinder for different lengths *Lp*. From the results obtained (see Table 5), we observe that the front face of the cylinder is the site of large average values of the Nusselt number. This is essentially due to the direct contact between the incoming fluid and this face. Thus, we observe that the Nusselt number has relatively identical values for the upper and lower faces due to symmetry. On the other hand, we note that this number has low average values along the rear face of the cylinder compared with the values obtained on the other faces. This is due to the presence of recirculation zones and stagnation points behind the cylinder. These zones trap the air and consequently the heat exchange between the fluid and this face is low compared to the other faces.

Figure 11. shows the values of the average total Nusselt number *<Nutot>* for different lengths *Lp*. According to this figure, a slight decrease is noticed in the average values of the total Nusselt number in cases where the length of the partitions ranging from Lp=1d to p=3d. Whereas in cases where 3d≤ Lp≤ 4d, the *<Nutot>* taking on almost significant values. Thus, after this length Lp>4d, the number decreases significantly. In fact, for short lengths Lp<3d, a large portion of the shear layers falls behind the cylinder. This reduces heat exchange between the fluid and the cylinder. If the length of the partitions increases (3d≤ Lp≤ 4d), part of the shear layers falls onto the control partitions, forming recirculation hard on the partition faces. However, the other part falls behind the partitions. This causes interaction between the two influencing on the intermediate zone. As a result, the heat exchange between the air particles and the obstacle increases. Whereas for long partitions (Lp>4d), the shear layers are stuck to the faces of the partitions and after detachment, they meet between forming vortexes behind the partitions without influencing the intermediate zone. This significantly reduces heat exchange between the fluid and the cylinder.

Table 5. Average values and reduction percentage of drag coefficient for various partitions lengths

Lp	Nu_1	Nu_2	Nu_3	Nu_4
No control case	8.215	3.037	2.150	3.027
1d	8.287	3.359	2.363	3.225
2d	8.293	3.126	2.425	3.200
3d	8.292	3.101	2.436	3.226
4d	8.288	3.383	2.448	3.460
5d	8.275	2.864	2.448	3.047

Figure 11. Values of Average Nusselt number total for different lengths "Lp".

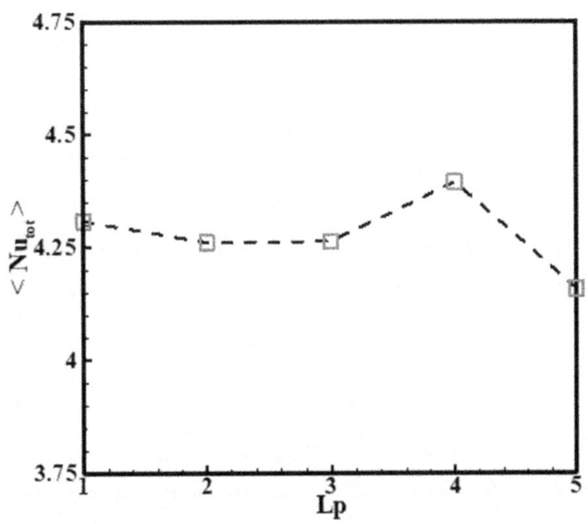

Table 6 shows the variation in the average value of the drag coefficient for different partition lengths. From these results, we can see that the average values of the drag coefficient decrease with increasing distance between the square obstacle and the control partitions. The percentage of this reduction reaches 17.26% compared to the case without the implementation of control partitions. This percentage shows a 0.42% increase compared to the case where the partition length Lp=1d(16.48%).

In general, the effect of control partition length on heat exchange improvement and drag coefficient reduction remains small and insignificant. This is due to the fact that the control partitions are implemented in the wake of the obstacle and their length influences the flow along the channel.

Table 6. Average values and reduction percentage of drag coefficient for various partition lengths

Lp	Cd_{mean}	Percentage reduction of drag coefficient
No control case	**1.425**	----
1d	1.185	16.84%
2d	1.183	16.98%
3d	1.181	17.12%
4d	1.179	17.26%
5d	1.176	17.47%

CONCLUSION

The current research presents a numerical study by using the LBM-MRT to simulate the airflow control coupled to heat transfer around a heated square cylinder at a fixed Reynolds number Re=150. The effect of the position and length of two partitions implemented downstream of the square obstacle is examined. The obtained results lead to the following significant conclusions:

- 3.74% of reduction in drag coefficient is obtained when both partitions are implemented successively.
- Significant heat exchange is observed in all cases studied.
- The implementation of the dual partitions in parallel results in a 17.26% reduction in drag coefficient.
- The implementation of the dual partitions in parallel results in a more than 10% improvement in heat exchange between the air and the top, bottom and rear surfaces of the heated obstacle, compared with the uncontrolled case.

These results could contribute to cooling applications for electronic components. They may also contribute to the design of more efficient fluid flow control systems in various industrial applications

REFERENCES

Admi, , Moussaoui, M. A., & Mezrhab, A. (2022). Numerical Investigation of Convective Heat Transfer and Fluid Flow Past a Three Square Cylinders Controlled by a Partition in Channel. *International Journal of Renewable Energy Development, 11*(3), 766–781. DOI: 10.14710/ijred.2022.43790

Admi, Y., El Hassani, S., Benhamou, J., Moussaoui, M. A., & Mezrhab, A. (2024). Numerical research on the wake characteristics and drag reduction of a square cylinder in the presence of control partitions. *Ocean Engineering, 301*, 117341. DOI: 10.1016/j.oceaneng.2024.117341

Admi, Y., Lahmer, E. B., Benhamou, J., Moussaoui, M. A., & Mezrhab, A. (2024). Heat transfer improvement and drag force reduction around three heated square cylinders controlled by partitions. *Physics of Fluids, 36*(4), 045161. DOI: 10.1063/5.0191092

Admi, Y., Moussaoui, M. A., & Mezrhab, A. (2020). Effect of a flat plate on heat transfer and flow past a three side-by-side square cylinders using double MRT-lattice Boltzmann method. *2020 IEEE 2nd International Conference on Electronics, Control, Optimization and Computer Science, ICECOCS 2020.*

Bar-Cohen, A., & Kraus, A. D., S. F. D. (1983). Thermal frontiers in the design and packaging of microelectronic equipment. *Mechanical Engineering (New York, N.Y.), 105*(5), 530–539.

Benhamou, J., Lahmer, E. B., Admi, Y., Jami, M., & Mezrhab, A. (2023). Three-dimensional numerical simulation of free convection and entropy generation in a cubic cavity containing a heat source. *Heat Transfer, 52*(7), 5108–5135. DOI: 10.1002/htj.22920

Bouzidi, M., Firdaouss, M., & Lallemand, P. (2001). Momentum transfer of a Boltzmann-lattice fluid with boundaries. *Physics of Fluids, 13*(11), 3452–3459. Advance online publication. DOI: 10.1063/1.1399290

Breuer, M., Bernsdorf, J., Zeiser, T., & Durst, F. (2000). Accurate computations of the laminar flow past a square cylinder based on two different methods: Lattice-Boltzmann and finite-volume. *International Journal of Heat and Fluid Flow, 21*(2), 186–196. Advance online publication. DOI: 10.1016/S0142-727X(99)00081-8

Chapain, S., & Aly, A. M. (2019). Vibration attenuation in high-rise buildings to achieve system-level performance under multiple hazards. In Engineering Structures. DOI: 10.1016/j.engstruct.2019.109352

Chen, G., Jia, M., Zhang, S., Tang, Y., & Wan, Z. (2020). Pool boiling enhancement of novel interconnected microchannels with reentrant cavities for high-power electronics cooling. *International Journal of Heat and Mass Transfer, 156*, 119836. DOI: 10.1016/j.ijheatmasstransfer.2020.119836

Chen, S., Liu, Y., Chan, S. F., & Leung, C. W. (2001). T. L. C. (2001). Experimental study of optimum spacing problem in the cooling of simulated electronic packages. *International Journal of Heat and Mass Transfer*, •••, 37.

D'Humières, D., Ginzburg, I., Krafczyk, M., Lallemand, P., & Luo, L. S. (2002). Multiple-relaxation-time lattice Boltzmann models in three dimensions. *Philosophical Transactions. Series A, Mathematical, Physical, and Engineering Sciences, 360*(1792), 437–451. Advance online publication. DOI: 10.1098/rsta.2001.0955 PMID: 16214687

Deng, Y., Feng, C. E. J., Zhu, H., Chen, J., Wen, M., & Yin, H. (2018). Effects of different coolants and cooling strategies on the cooling performance of the power lithium ion battery system: A review. In Applied Thermal Engineering. DOI: 10.1016/j.applthermaleng.2018.06.043

Florides, G., & Kalogirou, S. (2007). Ground heat exchangers-A review of systems, models and applications. In Renewable Energy. DOI: 10.1016/j.renene.2006.12.014

Guo, Z., & Shu, C. (2013). Lattice Boltzmann method and its applications in engineering. In World Scientific Publishing (Vol. 54, Issue 22). DOI: 10.1142/8806

Islam, S. U., Rahman, H., Abbasi, W. S., & Shahina, T. (2015). Lattice Boltzmann Study of Wake Structure and Force Statistics for Various Gap Spacings Between a Square Cylinder with a Detached Flat Plate. *Arabian Journal for Science and Engineering*, *40*(8), 2169–2182. DOI: 10.1007/s13369-015-1648-3

Jami, M., Moufekkir, F., Mezrhab, A., Fontaine, J. P., & Bouzidi, M. (2016). New thermal MRT lattice Boltzmann method for simulations of convective flows. *International Journal of Thermal Sciences*, *100*, 98–107. Advance online publication. DOI: 10.1016/j.ijthermalsci.2015.09.011

Lahmer, E. B., Benhamou, J., & Admi, Y., moussaoui, mohammed amine, Jami, M., Mezrhab, A., & Phanden, R. K. (. (2022). Convective and Conjugate Heat Transfer Efficiency Enhancement over Partitioned Channel within Backward-Facing Step using the Lattice Boltzmann Method LBM-DMRT. *Journal of Enhanced Heat Transfer*, *29*(3), 51–77. DOI: 10.1615/JEnhHeatTransf.2022040357

Mezrhab, A., Amine Moussaoui, M., Jami, M., Naji, H., & Bouzidi, M. (2010). *Double MRT thermal lattice Boltzmann method for simulating convective flows. Physics Letters*. General, Atomic and Solid State Physics., DOI: 10.1016/j.physleta.2010.06.059

Mohamad, A. A. (2011). Lattice Boltzmann Method: Fundamentals and Engineering Applications with Computer Codes. In AIAA Journal.

Moussaoui, M. A., Jami, M., Mezrhab, A., & Naji, H. (2010). MRT-Lattice Boltzmann simulation of forced convection in a plane channel with an inclined square cylinder. *International Journal of Thermal Sciences*, *49*(1), 131–142. Advance online publication. DOI: 10.1016/j.ijthermalsci.2009.06.009

Perng, S.-W. (•••). H.-W. W. (47 C.E.). Numerical investigation of mixed convective heat transfer for unsteady turbulent flow over heated blocks in a horizontal channel. *International Journal of Thermal Sciences*, *5*, 620–632.

Qian, Y. H., D'Humières, D., & Lallemand, P. (1992). Lattice bgk models for navier-stokes equation. *Europhysics Letters*, *17*(6), 479–484. Advance online publication. DOI: 10.1209/0295-5075/17/6/001

Turki, S. (2008). Numerical Simulation of Passive Control on Vortex Shedding Behind Square Cylinder Using Splitter Plate. *Engineering Applications of Computational Fluid Mechanics*, *2*(4), 514–524. Advance online publication. DOI: 10.1080/19942060.2008.11015248

Wang, Q., & Jaluria, Y. (2002). Unsteady mixed convection in a horizontal channel with protruding heated blocks and a rectangular vortex promoter. *Physics of Fluids*, *14*(7), 2113–2127. DOI: 10.1063/1.1480832

Yang, S.-J. (2002). A numerical investigation of heat transfer enhancement for electronic devices using an oscillating vortex generator. *Numerical Heat Transfer Part A*, *42*(3), 269–284. DOI: 10.1080/10407780290059549

Young, T. J., & Vafai, K. (1998). Convective flow and heat transfer in channel containing multiple heated obstacles. *International Journal of Heat and Mass Transfer*, *41*(21), 3279–3298. DOI: 10.1016/S0017-9310(98)00014-3

Yu, C., Zhang, H., Zeng, M., Wang, R., & Gao, B. (2020). Numerical study on turbulent heat transfer performance of a new compound parallel flow shell and tube heat exchanger with longitudinal vortex generator. *Applied Thermal Engineering*, *164*, 114449. DOI: 10.1016/j.applthermaleng.2019.114449

Zou, Q., & He, X. (1997). On pressure and velocity boundary conditions for the lattice Boltzmann BGK model. *Physics of Fluids*, *9*(6), 1591–1598. Advance online publication. DOI: 10.1063/1.869307

Chapter 20
Energy-Efficient Scheduling Issues With Various Maintenance Activities in Industry 4.0: Energy Minimization

Assia Sadiqi
https://orcid.org/0000-0003-2033-2283
Higher School of Education and Training of Oujda, Mohammed First University, Oujda, Morocco

Kamal Hirech
https://orcid.org/0000-0002-3531-7791
Higher School of Education and Training of Oujda, Mohammed First University, Oujda, Morocco

ABSTRACT

Today's manufacturing industry consumes a substantial amount of energy, so controlling energy usage and improving energy efficiency in Industry 4.0 are key priorities, both for the environment and for the competitiveness of our industry. At the same time, manufacturing industries, particularly energy-intensive ones, are placing increasing emphasis on reducing energy consumption. In this respect, energy-saving programming has proved effective in reducing energy consumption. In this context, this article focuses on the Industry 4.0 issues of coordinated scheduling of production and preventive maintenance activities over fixed and flexible periods, with a view to minimizing total energy consumption. In our study, we examine both types of unavailability constraints. We propose two models based on mixed-integer binary programming to determine optimum schedules, respectively. We evaluate the effectiveness of these two types of programming in mixed integer binary form using exact data separation B&B method and seamless evaluation using the CPLEX software.

DOI: 10.4018/979-8-3693-9924-8.ch020

1. INTRODUCTION

The mechanical engineering sector has long been a major energy consumer, and this has recently led to the introduction of government regulations requiring manufacturers to implement energy conservation solutions (EIA, 2013). Energy-efficient Organizational structure attracts more interest today than ever previously.

Indeed, classical scheduling literature presumes that machines are available for execute of the programming period (CSY, 2016; Wang S, Wang X, Yu J, Ma S, Liu M, 2018 ; Ding J., Song S., Zhang R., Chiong R., Wu C. 2016a). In spite of this, in realistic situations in industry 4.0, equipment unavailable for several technical factors, products, including preventive measures and temporary failures, maintenance operations. Contributors (Mouzon G, 2008) argue that preventive maintenance keeps equipment is in reliable working order and reduces the likelihood of failure. Manufacturing planning and preventive maintenance in the integrated optimization model must be taken into account in order balancing resource use and supply (Luo H., Du B., Huang G., Chen H., Li X., 2013).

Once the plant maintenance component has been introduced to manufacturing planning problems, a distinction is made between two types of research. In the first case, the search ignores unforeseen breakdowns that focuses solely machine capabilities; in the second case, on the other hand, the impact of unexpected failures is taken into account and corrective maintenance (CM) is carried out in the unlikely case of default (Liu Y., Dong H., Lohse N., Petrovic S., Gindy N., 2014a; Gahm C., Denz F., Dirr M., Tuma A., 2016).

The authors suppose that in the initial scenario, preventive maintenance (PM) is carried out regularly to maintain machines in optimum state of repair and guarantee a high level of operational efficiency. Therefore, random failures can be disregarded, which leads to certain types of deductive problems. Classic objectives such as makespan, total flow time, and delays are considered. Since the machine is unavailable for processing tasks during PM, the impact of machine downtime on performance requires evaluation to determine the best sequencing of jobs. In this case, machine downtime periods are known and fixed (deterministic) because the beginning time and interval duration are pre-determined.

In the second scenario, researchers account for the possibility of unpredictable uncertain breakdowns while performing the production schedule. Given the level of risk, the involved, this type of problem falls under binary programming, that is the fundamentally distinct with respect to the former category. Due to the presence of probability factors, high-impact expectations are typically taken into account in associated studies. Here, downtime requirements become adaptable criteria, where the maintenance starting time and/or duration must be determined during the programming operation.

Some researchers suggest that machines must undergo maintenance after being in continuous use for a specific period, with authorized continuous operating time defined as TBF for each machine. While many studies address scheduling problems involving unavailability constraints, research on multi-machine production workshops in this context is quite limited. As far as we know, most studies that tackle FS scheduling problems related to downtime restrictions, in particular dual-machine systems, with less attention given to multi-unit feeder workshops with non-changeover and unavailability ranges in the existing literature. The authors in (Giret A., Trentesaux D., Prabhu V., 2015) explored non-changeover cases involving tight scheduling, reducing the number of servicing and repair operations is specified at the start of the scheduling process, which differs from the assumptions made in this research. The authors in (Merkert L., Harjunkoski I., Isaksson A., Saynevirta S., Saarela A., Sand G., 2015; A. Sadiqi et al., 2018; Wei-Wei Cui, Zhiqiang Lu, Binghai Zhou, Chen Li &

Xiaole Han, 2016) assumed that maintenance operations must be after transformation a predetermined employment, contrasting with periodic maintenance constraints.

The principal focus of this paper is the exploration of embedded production planning and two types of feasible and actual network maintenance requirements in a multi-plant shop. A novel and useful approach. The problem mathematical models are developed to minimize total energy consumption.

The rest of the present document is structured accordingly:

- Section 2 introduces formulation of the problem along at two composite numbers in binary form.
- Section 3 details the B&B approach procedure to resolve the current problem of MILP models.
- Section 4 introduces an exact solution approach designed for this type of problem.
- Section 5 showcases computer experiments that demonstrate the proposed algorithm's performance.
- Section 6 concludes with discussions on the findings and ideas for subsequent studies.

2. PROBLEM DESCRIPTION /MODELLING PROBLEM

The focus of the present section is on the flow-shop, a processing environment with a completely specified task order for every job ($J_1, \ldots, J_n$). In such a shop, one set at a time of tasks to be executed must be performed by a set of machines ($M_1, \ldots, M_m$) arranged in series. Each job is performed on the same machine sequentially, to optimize one or more objective features.

In our study, we examine both types of downtime constraints. Two MIP programming models are provided to derive optimal schedules, respectively. By adding constraints for reduce overall consumption of resources on the FLOW SHOP, we build in existing design of (Wei-Wei Cui, Zhiqiang Lu, Binghai Zhou, Chen Li & Xiaole Han, 2016; Sadiqi Assia, El Abbassi Ikram, El Barkany Abdellah and El Biyaali Ahmed, 2019). We have adjusted our models to take account of the various requirements for total energy consumption of operation processing and the common energy consumption of additional equipment in the aforementioned "non-permutation FS". Following are the formulas, inputs and variables used in this formulation:

Table 1. Grading system used to define the problem.

Indices:	
i	Job indices
j	Machine indices
k, k'	Indices for positions within each machine, M_J
l	Indices for batches within each machine
Sets:	
J	Package of jobs; $J = \{J_1, J_2, \ldots, J_n\}$
M	Package in $M = \{M_1, \ldots, M_m\}$
O_i	Package of the operations of J_i
$BH_{l,j}$	Package of batches

continued on following page

Table 1. Continued

Settings:	
n	Numeral of task
m	Numeral of Mj
$T_{i,j}$	Execution time of $O_{i,j}$
$P_{i,j}$	Procedure computing energy $O_{i,j}$
TBF_j	Period between two maintenance checks M_j
PM_j	Timeframe for machine maintenance M_j
Z	A limitless integer
C_{max}	The makespan, corresponding to the job with the highest execution time
P_0	The joint power, used by auxiliary equipment and installations
Decision variables:	
$x_{i,j,k}$	$\begin{cases} 1 \textit{item } J_i \textit{ is treated in order at the kth location} \\ 0 \textit{else} \end{cases}$
$g_{k,j}$	$\begin{cases} 1 \textit{the operation } O_{k,j} \textit{ is the last operation an actual batch} \\ 0 \textit{else} \end{cases}$
$y_{k,j}$	$\begin{cases} 1 \textit{PM is made prior to the kth step in the machine,} \\ 0 \textit{else} \end{cases}$
Auxiliary variables:	
$J_{k,j}$	The kth job in the M_j
$O_{k,j}$	The operation of the job $J_{k,j}$ in M_j
$T_{k,j}$	The processing time $O_{k,j}$ in M_j
$S_{k,j}$	The start time of the operation $O_{k,j}$ in M_j
$C_{k,j}$	The end time of operation $O_{k,j}$ in M_j
$e_{k,j}$	Total time of $O_{k,j}$, meaning the period of time remaining from the time at which $O_{k,j}$ finishes to the time specified as the end of the last unavailability interval
$q_{k,j}$	Total number of batches remaining in the cycle $O_{k,j}$ and $O_{k-1,j}$.
$a_{k,j}$	Machine speed after cycle $O_{k,j}$ in M_j
$b_{k,j}$	Machine speed before cycle operation $O_{k,j}$ in M_j

A collection of tasks J must be executed on a collection of equipment M. Each task $J_i(J_i \epsilon J)$ is a fixed execution time. $T_{i,j}$, the duration of the period of preventive maintenance TBF_j, the time of maintenance PM_j and the processing power $P_{i,j}$ in the machine M_j.

Suppose we have seven tasks to schedule, as illustrated in the following Table 2.

Table 2. Settings of this scheduling scenario

M	J1	J2	J3	J4	J5	J6	J7	TBF Operating time	PM
1	8	13	12	14	18	10	13	33	8
2	10	14	21	11	11	9	11	30	8
3	16	10	15	11	11	15	14	38	10
4	15	9	14	10	10	14	13	35	9

The objective is designed to reduce the energy required, as shown below:

$$TEC = PEC + CEC \quad (1)$$

Energy used for conversion:

$$PEC = \sum_{i=1}^{n} \sum_{j=1}^{m} \sum_{k=1}^{n} P_{i,j} \cdot T_{i,j} \cdot x_{i,j,k} \quad (2)$$

Shared energy used:

$$CEC = P_0 \cdot C_{max} \quad (3)$$

TEC (Total Energy Consumption) adds together the total energy used by the process and the collective energy consumption, calculated as follows:

$$TEC = \sum_{i=1}^{n} \sum_{j=1}^{m} \sum_{k=1}^{n} P_{i,j} \cdot T_{i,j} \cdot x_{i,j,k} + P_0 \cdot C_{max} \quad (4)$$

The aim is to reduce total energy expenditure to a minimum:

FOBJ.=

$$\sum_{i=1}^{n} \sum_{j=1}^{m} \sum_{k=1}^{n} P_{i,j} \cdot T_{i,j} \cdot x_{i,j,k} + P_0 \cdot C_{max} \quad (7)$$

A. The problem Fm|nr – Fl – /4.0|$_{TEC}$:

In M1, the period of preventive maintenance TBF_1 contains $Q_{1,1}$ and $G_{1,1}$.

Figure 1. Schedule for the $Fm|nr - FI - I4.0|_{TEC}$ problem

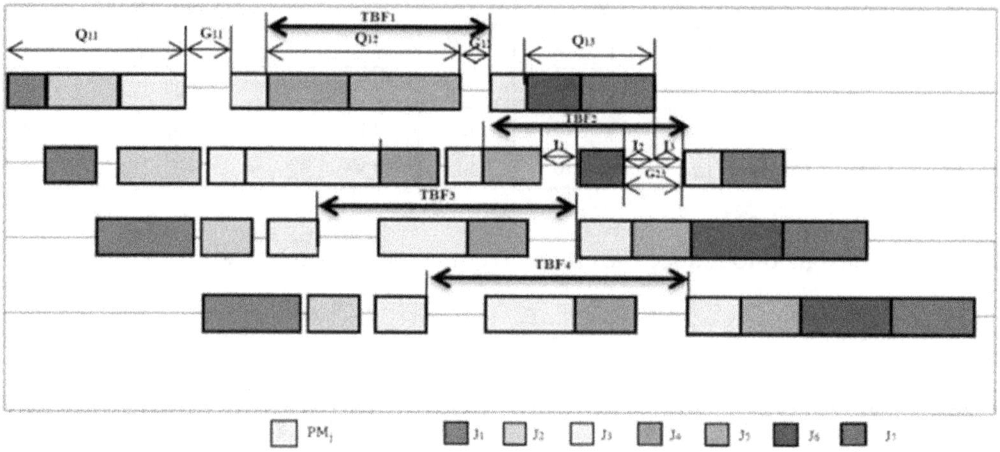

Modelling of the problem $Fm|nr - FI - I4.0|_{TEC}$

The first common proactive model (JM) can finally be established as follows:
Minimize Fct Obj (8)
S.t

$$\sum_{i=1}^{n} x_{i,j,k} = 1; \forall j, k \tag{9}$$

$$\sum_{k=1}^{n} x_{i,j,k} = 1; \forall j, i \tag{10}$$

$$T_{k,j} = \sum_{i=1}^{n} x_{i,j,k} . T_{i,j}; \forall k, \forall \tag{11}$$

$$C_{k,j} = S_{k,j} + T_{k,j}; \forall j > 1, \forall k \tag{12}$$

$$S_{1,1} = 0 \tag{13}$$

$$S_{k,j} \geq C_{k-1,j}; \forall j, \forall k \geq 2 \tag{14}$$

$$\left(2 - x_{i,j,k} - x_{i,j+1,k'}\right)Z \geq S_{k,j} + T_{i,j} - S_{k',j+1}; \forall j < m, i, k, k' \tag{15}$$

$$C_{k,j} = e_{k,j} + \left(PM_j + TBF_j\right)\sum_{k'=1}^{k-1} g_{k',j} + \left(PM_j + TBF_j\right)\sum_{k'=1}^{k-1} q_{k',j}; \forall j, \forall k \geq 1 \tag{16}$$

$$C_{1,j} = e_{1,j} + \left(PM_j + TBF_j\right) q_{1,j}; \forall j \tag{17}$$

$$q_{k,j} \leq g_{k-1,j} Z; \forall j, \forall k > 1 \tag{18}$$

$$e_{k,j} \geq T_{k,j} ;\forall k, \forall j \quad (19)$$

$$e_{k,j} \leq TBF_j ;\forall k, \forall j \quad (20)$$

$$e_{k+1,j} \geq e_{k,j} + T_{k+1,j} - g_{k,j} Z;\forall k < n, \forall j \quad (21)$$

$$C_{max} = C_{n,m} \quad (22)$$

$$x_{i,j,k}, g_{k,j} \text{ are binary variables} \quad (23)$$

B. THE PROBLEM $Fm|nr - CI - I4.0|_{TEC}$:

The second case, where the continuous working time of the machines, without taking into account the inactivity time (dead time), cannot exceed a maximum time of TBF_j, is designated by $Fm|nr - CI|_{TEC}$.

Obviously, there is no difference between the last task of a batch and the next maintenance period, i.e. $G_{l,j} = 0$.

All batches of all machines are "real batches".

In this model, the PM strategy is based on execution rather than time.

The age of the Mj increases with the processing task and does not change when the machine is inactive. In the meantime, the person in charge must set start times for maintenance. Figure 2 illustrates this pattern. It is clear that $Q_{l,j} \leq TBF_j$. A program with a completion time of all jobs is 108 time units are obtained.

Figure 2. Schedule for the $Fm|nr - CI - I4.0|_{TEC}$ problem

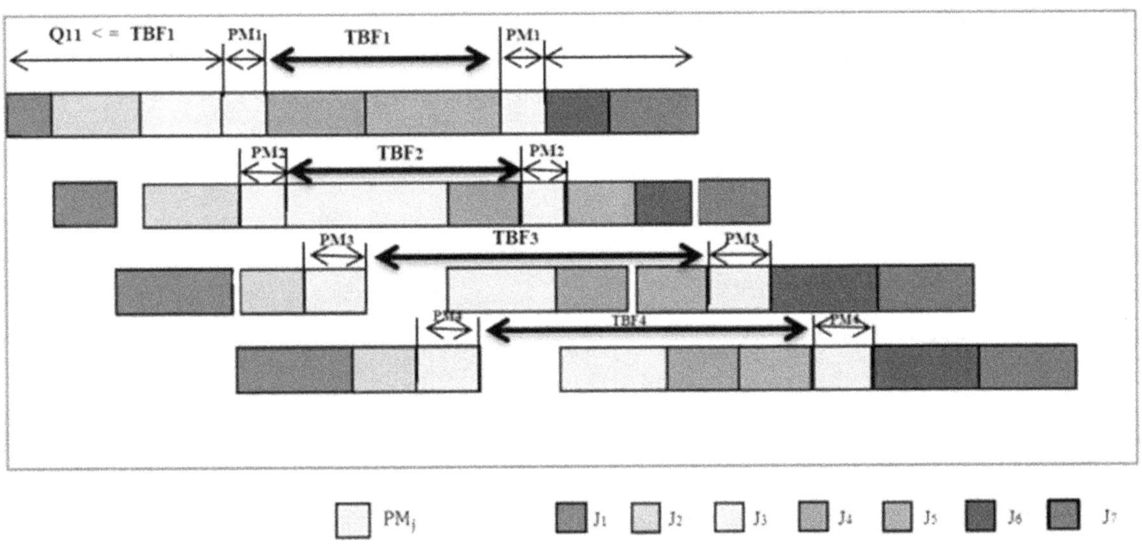

Modelling of the Problem $Fm|nr - CI - I4.0I_{TEC}$

The 2nd Joint Proactive Model (JM) can finally be established as follows:
Minimize Fct Obj (8)
S.t
Constraint sets (9)-(13) and constraint set (15) are the same with the model 1.

$$S_{k,j} \geq C_{k-1,j} + PM_j \cdot y_{k,j}; \forall j, \forall k \geq 2 \tag{26}$$

$$b_{1,j} = 0; \forall j \tag{27}$$

$$y_{1,j} = 0; \forall j \tag{28}$$

$$a_{k,j} = b_{k,j} + T_{k,j}; \forall k, \forall j \tag{29}$$

$$b_{k+1,j} = a_{k,j}\left(1 - Y_{k+1,j}\right); \forall j, k \leq n \tag{30}$$

$$a_{k,j} \leq TBF_j; \forall k, \forall j \tag{31}$$

$$x_{i,j,k},\ y_{k,j} \text{ are binary variables} \tag{32}$$

$$T_{k,j},\ C_{k,j},\ S_{k,j} \geq 0 \tag{33}$$

Linearized Constraint Sets

It should be noted that the constraint (30) is non-linear, which contains the non-linear term from $a_{k,j}\left(1 - Y_{k+1,j}\right)$. We therefore linearise the non-linear constraint by adding a very large positive number, namely Z and in this case $b_{k+1,j}$ counts as an intermediate variable to linearize $a_{k,j}\left(1 - Y_{k+1,j}\right)$ and is used to replace it.

Therefore, the non-linear term $a_{k,j}\left(1 - Y_{k+1,j}\right)$ is replaced by the linear term $b_{k+1,j}$ is intended to linearize the constraint (30).

$$b_{k+1,j} \geq a_{k,j} - Z.\ Y_{k+1,j}; \forall j, k \leq n \tag{34}$$

$$b_{k+1,j} \leq a_{k,j} + Z.\ Y_{k+1,j}; \forall j, k \leq n \tag{35}$$

$$b_{k+1,j} \leq Z\left(1 - Y_{k+1,j}\right); \forall j, k \leq n \tag{36}$$

$$b_{i,j} \geq 0; \forall j, i \tag{37}$$

Sets of constraints (34) - (37) force $b_{k+1,j} = a_{k,j}\left(1 - Y_{k+1,j}\right)$ to be always true.

Indetail,if$Y_{k+1,j} = 0$,thesetsofconstraints(34)-(35)impose$b_{i+1,j}$mustnotbelessthanorgreaterthan$a_{k,j}$,and thismakes$Y_{k+1,j}$equalto $a_{k,j}$ ($b_{k+1,j} = a_{k,j} = \left(1 - Y_{k+1,j}\right) a_{k,j}$).If,$Y_{k+1,j} = 1$,thesetsofconstraints(36)-(37) impose$b_{k+1,j}$mustnotbelessthanorgreaterthan0,andthismakes$b_{k+1,j}$isequalto0 ($b_{k+1,j} = 0 = \left(1 - Y_{k+1,j}\right) a_{k,j}$).

3. COMPUTATIONAL RESULTS

Calculation Results

This part focuses on the detailed assessment of the linear model's information in relation to the system under investigation. The mathematical models $Fm|nr - FI - I4.0|_{TEC}$ and $Fm|nr - CI - I4.0|_{TEC}$ represent linear programming problems. To fully resolve these, an exact solution is required using a B and B method embedded in the leading software solver CPLEX, which helps to evaluate simplicity and degree of fineness of the solution.

To gauge the computational efficiency of the suggested modelling, the system is discussed in its simplest form challenging scenarios. Specifically, 39 different instances were generated at any defined size of n tasks and m engines.

As part of the experimental concept for this study, Employment processing times $T_{i,j}$ range is uniformly derived from the [9, 29].

The time period TBF_j is also uniformly generated within the range $\left[\underline{TBF_j}, \overline{TBF_j}\right]$, where TBF_j is set to 30 and TBF_j its upper limit is either 30 or half of the total job processing time on M j. The PM_j is drawn evenly starting from the range $\left[\underline{PM_j}, \overline{PM_j}\right] = \left[\frac{\underline{TBF_j}}{5}, \frac{2 \times \underline{TBF_j}}{3}\right]$.

For smaller problems, the power of auxiliary installations is assumed to be P0 = 5. For larger problems, this power is set to 5, 15, or 20. The processing power is uniformly generated from the interval [3.5]. This gives the following problem length (n, m) adjusted according to Chart 3.

For each parameter combination, three random instances are generated for large problems, based on the variation in auxiliary installation power with P0 values of 5, 15, and 20.

Table 3. Summary of calculation results

						$Fm\|nr - FI - I4.0\|_{TEC}$			$Fm\|nr - CI - I4.0\|_{TEC}$		
(n, m)	TBF	PM	T	P_0	P	C_{max}	TEC	CPU time (s)	C_{max}	TEC	CPU time (s)
5 2	[30,102]	[6,68]	[9,29]	5	[3,5]	159	1529	00,38	129	1329	00,36
3						186	2155	03,49	154	1955	02,47
4						182	2550	04,06	176	2350	03,05
5						210	2956	04,82	182	2856	03,80
6						254	3444	05,41	210	3344	04,40
6 2	[30,117]	[6,78]	[9,29]	5	[3,5]	178	1879	01,97	140	1678	00,96
3						223	2876	05,33	163	2677	04,32
4						243	2978	05,94	183	2778	04,93
5						286	3568	07,08	226	3367	06,07
7 2	[30,131]	[6,87]	[9,29]	5	[3,5]	202	2176	07,37	174	1977	06,36
3						223	2957	16,57	187	2756	14,56
4						264	3670	13,16	234	3479	10,15
8 2	[30,146]	[6,97]	[9,29]	5	[3,5]	210	2573	13,48	190	2373	11,49
3						267	3174	13,59	267	2974	11,39
9 2	[30,160]	[6,106]	[9,29]	5	[3,5]	289	2816	40,38	262	2616	32,35

continued on following page

Table 3. Continued

							Fm|nr – FI – I4.0|$_{TEC}$			Fm|nr – CI – I4.0|$_{TEC}$		
(n, m)		TBF	PM	T	P_0	P	C_{max}	TEC	CPU time (s)	C_{max}	TEC	CPU time (s)
20 15	S. 1	[30,320]	[6,213]	[9,29]	10	[3,5]	648	29460	60,02	646	29260	54,26
	S. 2				15			32694	87,23		32490	75,91
	S. 3				20			35925	93,22		35720	89,87
20 20	S. 1	[30,320]	[6,213]	[9,29]	10	[3,5]	941	38013	83,12	741	37810	79,56
	S. 2				15			41716	120,09		41515	91,85
	S. 3				20			45422	143,23		45220	102,78
30 15	S. 1	[30,465]	[6,310]	[9,29]	10	[3,5]	1036	42761	98,43	836	42560	84,09
	S. 2				15			46940	176,66		46740	95,12
	S. 3				20			51021	187,65		50920	120,77
30 20	S. 1	[30,465]	[6,310]	[9,29]	10	[3,5]	1131	55915	1200,17	931	54910	1100,90
	S. 2				15			59865	1856,26		59565	1694,21
	S. 3				20			64825	1944,54		64220	1884,11
40 15	S. 1	[30,610]	[6,406]	[9,29]	10	[3,5]	1226	55960	1432,77	1026	55860	1322,19
	S. 2				15			61090	2599,16		60990	2030,25
	S. 3				20			66321	2643,22		66120	2367,45
40 20	S. 1	[30,610]	[6,406]	[9,29]	10	[3,5]	1321	72313	1532,12	1121	72010	1400,69
	S. 2				15			77916	2733,01		77615	2279,88
	S. 3				20			85300	2876,23		83220	2739,16

continued on following page

Table 3. Continued

							$Fm\|nr - FI - I4.0\|_{TEC}$			$Fm\|nr - CI - I4.0\|_{TEC}$		
(n, m)	**TBF**		**PM**	**T**	P_0	**P**	C_{max}	**TEC**	**CPU time (s)**	C_{max}	**TEC**	**CPU time (s)**
50 15	**S. 1**	[30,755]	[6,503]	[9,29]	10	[3,5]	**OUT OF MEMORY**		3600	**OUT OF MEMORY**		2986,25
	S. 2				15				3600			3044,44
	S. 3				20				3600			3432,70
50 20	**S. 1**	[30,755]	[6,503]	[9,29]	10	[3,5]			3600			3600
	S. 2				15				3600			3600
	S. 3				20				3600			3600

Our MILP models are capable of finding the best value proposition in the shortest possible time, and much more besides specifically, in a timely manner. However, for larger problems, these models tend to require a significantly longer time.

Validation of the Method B&B

In the literature, the authors [13, 14] have studied the problems of flow shop scheduling of non-permutation with unavailability intervals and have proposed an incremental hybrid genetic algorithm (HIGA) to solve their problem that combines a strategic framework for incremental evolution, local refinement and a scheme for monitoring population diversity and also an accurate resolution through CPLEX.

These authors [13, 14] have minimized the Makespan and they do not integrate energy constraints into their model, which is the main difference with our models. To our knowledge, no document taking into account both the simultaneous scheduling of energy-intensive manufacturing and maintenance tasks in the non-permutation flowshop.

Table 4 shows the survey findings of the comparative study with exact resolution with CPLEX between the existing models $Fm|nr - FI|_{C_{max}}$ and $Fm|nr - CI|_{C_{max}}$ and our models $Fm|nr - FI - I4.0|_{TEC}$ and $Fm|nr - CI - I4.0|_{TEC}$ which have minimized total energy consumption as shown in Table 4.

Table 4. Results of comparison for reference problems

	$Fm\|nr-FI\|_{TEC}$			$Fm\|nr-FI\|_{C_{max}}$		$Fm\|nr-CI-I4.0\|_{TEC}$			$Fm\|nr-CI-I4.0\|_{C_{max}}$	
(n, m)	C_{max}	TEC	Time(s)	C_{max}	Time(s)	C_{max}	TEC	Time(s)	C_{max}	CPU Time(s)
5 2	**159**	1529	00,36	172,6	00,38	**129**	1329	00,38	149,4	**00,19**
3	**186**	2155	02,47	199,6	05,79	**154**	1955	03,49	169,6	02,82
4	**182**	2550	03,05	249,4	26,4	**176**	2350	04,06	191,4	12,2
5	**210**	2956	03,80	286,2	378,8	**182**	2856	04,82	222,0	217,9
6	**254**	3444	04,40	317,6	2818	**210**	3344	05,41	225,8	4210
6 2	**178**	1879	00,96	208,4	3,05	**140**	1678	01,97	174,6	1,93
3	**223**	2876	04,32	237,8	31,4	**163**	2677	05,33	193,0	65,2
4	**243**	2978	04,93	278,4	256,9	**183**	2778	05,94	220,6	306
5	**286**	3568	06,07	312,4	2632	**226**	3367	07,08	231,6	5132
7 2	**202**	2176	06,36	213,2	7,64	**174**	1977	07,37	198,4	70,7
3	**223**	2957	14,56	262,4	835,7	**187**	2756	16,57	248,4	1359
4	**264**	3670	10,15	293,0	1434	**234**	3479	13,16	231,9	6079
8 2	**210**	2573	11,48	242,2	88,3	**190**	2373	13,48	221,6	980,7
3	**267**	3174	11,59	278,2	4934	**267**	2974	13,59	199,4	7200
9 2	**289**	2816	12,35	250,6	681,3	**262**	2616	40,38	234,6	4653

The results found through the exact Separation and evaluation process included in the CPLEX system software prove that the proposed MILPs models to minimize total energy consumption provided satisfactory results compared to those found by (Wei-Wei Cui, Zhiqiang Lu, Binghai Zhou, Chen Li & Xiaole Han, 2016). Moreover, it is quite clear that as energy consumption is minimized, a considerable minimization in the completion time of all jobs is significantly observed.

CONCLUSION

This document examines the calculation features of the Mixed-Integer Linear Programming models designed to optimally address built-in programming of productivity and scheduled maintenance tasks, both of fixed and flexible durations, designed to reduce overall power requirements to a minimum (TEC). Full-integer binary composite programming performs well models is assessed using the exact selection and valuation method implemented in the new CPLEX software. An accurate analysis between the test results and those from existing publications demonstrated effectiveness of the software model developed, findings highlighted that, while the MILP models perform well in terms of computational quality, solving large-scale problems requires considerable time. For smaller problems, the models can achieve optimal solutions within a reasonable timeframe, though it may still be time-consuming. However, the results are acceptable, the formulation requires less than 3,600 s of processing time.

The research outlined in this paper opens up several promising avenues for further study. In particular, it would be valuable to explore the development of heuristic algorithms to tackle larger problems, ensuring solutions can be reached within reasonable computation times. Additionally, the creation of approximate and robust approaches for the problems $Fm|nr-FI-I4.0|_{TEC}$ and $Fm|nr-CI-I4.0|_{TEC}$ under energy constraints presents an intriguing area for future investigation.

REFERENCES

Assia, S., Ikram, E. A., Abdellah, E. B., & Ahmed, E. B. (2018). *« Comparative analysis the simultaneous scheduling problems of production and maintenance activities», April 2018 Indexed SCOPUS.* IEEE Xplore., DOI: 10.1109. LOGISTIQUA.2018.8428283

Assia, S., Ikram, E. A., Abdellah, E. B., & Ahmed, E. B. (2019). *« Non-permutation flow shop scheduling problems with unavailability constraints to Minimize Total Energy Consumption », April 2019 Indexed SCOPUS.* IEEE Xplore., DOI: 10.1109/ICOA.2019.8727649

CSY. (2016). China Statistical Yearbook 2015 (Online). www. stats. gov. cn / tjsj / ndsj / 2016/ indexeh. htm. Accessed on 5 May 2018.

Cui, W.-W., Lu, Z., Zhou, B., Li, C., & Han, X. (2016). A hybrid genetic algorithm for non-permutation flow shop scheduling problems with unavailability constraints. *International Journal of Computer Integrated Manufacturing*, *29*(9), 944–961. Advance online publication. DOI: 10.1080/0951192X.2015.1130247

Ding, J., Song, S., Zhang, R., Chiong, R., & Wu, C. (2016a). Parallel machine scheduling under time-of-use electricity prices: New models and optimization approaches. *IEEE Transactions on Automation Science and Engineering*, *13*(2), 1138–1154. DOI: 10.1109/TASE.2015.2495328

EIA. (2013). International Energy Outlook 2013 (Online). www. eia . gov / forecasts / ieo / pdf / 0484 (2013) . pdf. Accessed on 5 May 2018.

Gahm, C., Denz, F., Dirr, M., & Tuma, A. (2016). Energy – efficient scheduling in manufacturing companies: A review and research framework. *European Journal of Operational Research*, *248*(3), 744–757.

Giret, A., Trentesaux, D., & Prabhu, V. (2015). Sustainability in manufacturing operations scheduling: A state of the art review. *Journal of Manufacturing Systems*, *37*(1), 126–140. DOI: 10.1016/j.jmsy.2015.08.002

Liu, Y., Dong, H., Lohse, N., Petrovic, S., & Gindy, N. (2014a). An investigation into minimizing total energy consumption and total weighted tardiness in job shops. *Journal of Cleaner Production*, *65*, 87–96.

Luo, H., Du, B., Huang, G., Chen, H., & Li, X. (2013). Hybrid flow shop scheduling considering machine electricity consumption cost. *International Journal of Production Economics*, *146*, 423–439.

Merkert, L., Harjunkoski, I., Isaksson, A., Saynevirta, S., Saarela, A., & Sand, G. (2015). Scheduling and energy-industrial challenges and opportunities. *Computers & Chemical Engineering*, *72*, 183–198.

Mouzon, G. (2008) Operational methods and models for minimisation of energy consumption in a manufacturing environment. Ph.D. thesis. Wichita State University, Wichita, the United States of America.

Sadiqi, , El Abbassi, I., El Barkany, A., & El Biyaali, A. (2018). Joint Scheduling of Jobs and Variable Maintenance Activities in the Flowshop Sequencing Problems: Review, Classification and Opportunities. *International Journal of Engineering Research in Africa*, *39*, 170–190. DOI: 10.4028/www.scientific .net/JERA.39.170

Wang, S., Wang, X., Yu, J., Ma, S., & Liu, M. (2018). Bi-objective identical parallel machine scheduling to minimize total energy consumption and makespan. *Journal of Cleaner Production*, *193*, 424–440. Advance online publication. DOI: 10.1016/j.jclepro.2018.05.056

Chapter 21
Forecasting Ghana's Carbon Footprint Through ARIMA Analysis

Manjit Kour
https://orcid.org/0000-0003-1043-3187
Chandigarh University, India & North West University, South Africa

ABSTRACT

This study aims to forecast Ghana's carbon dioxide (CO2) emissions from 2015 to 2027, using the Autoregressive Integrated Moving Average (ARIMA) model to examine time-series dynamics of emissions. The purpose is to shed more light on the future profile of emissions and underscore the value of sustainable development practices in the country. This study uses time-series data of CO2 emissions from Ghana for the period 2015 to 2016. The ARIMA model is used in the analysis and prediction of the data to determine the future trend in carbon emissions. This model accounts for the historical patterns and trends in emissions, ensuring a robust and reliable forecast. An upward trend in the Ghanaian CO2 emissions emerges, with a projected rise to 3.10 kilotonnes (kt) of CO2 per annum to 2027. Results point to the rising challenges of emission management and demand effective intervention to check on ecological impacts. This trend spells out the need for the needful strategies in Ghana towards attaining carbon footprint reduction and fostering sustainability.

I. INTRODUCTION

The issue of global warming has become a hot topic in modern life, thereby harboring concerns over the emission of greenhouse gases. The term "greenhouse effect" expresses that many greenhouse gases, like CO2, are highly absorbent of long-wave radiation reflected by the Earth and extremely permeable to visible light from the sun. Besides contributing to global warming, excessive carbon emissions also pose a threat to human survival by bringing along numerous environmental and socioeconomic issues such as sea level rise, melting of glaciers, climate change, and destruction of biodiversity (Ning et al., 2021). Accurate comprehension, evaluation, and prediction of carbon dioxide (CO2) emissions are necessary to address the problems of climate change, global warming, and a clean and healthy environment. Globally, greenhouse gas emissions resulting primarily from the atmosphere of CO2 are known as a major cause

DOI: 10.4018/979-8-3693-9924-8.ch021

of worldwide global warming, which is viewed as an urgent atmospheric issue (Hossain et al., 2017). Since this number of emissions now grows upward, analyzing gas emission effects and their implication in the economy is, therefore very important. Hence carbon dioxide emission into the atmosphere of projections will guide policy direction. It is important, therefore to understand and predict the emission of gas (Sokama-Neuyam et al., 2023; Kour, 2023).

With its fast-paced industrialization and urbanization process, the Republic of Ghana reaches a crossroads in its path towards development. It is in its best interest to understand the country's carbon footprint, in terms of CO2, as it charts its way towards economic growth and development. The study will focus on the case of Ghana, with its growing industrial sector, increased urban population, and increasing energy demands in this investigation. It is critical to know the temporal dynamics of emissions inside Ghana to guide efforts toward sustainable development and mitigation of environmental impacts. In this regard, the ARIMA model is applied in this study to determine the future trend of CO2 emissions and enable policymakers to adopt proactive mitigation strategies.

Ghana has pledged to reach net-zero emissions in 2070 and a minimum of 15% reduction of GHG emissions by 2030. However, like most developing countries, this country must use its available means to make its climate objective a reality. Given these global concerns over climate issues, this study takes an objective look at the role Ghana will play in reducing its carbon footprint. With the country embarking on rapid industrialization and expansion of its towns, proper understanding of CO2 emissions has been required as an essential application of the ARIMA modelling approach. Given Ghana's commitment to sustainable development, such assessment and regulation become even more crucial in the world where there is an evident impact, in all senses, of human actions. The main objectives of the project are to analyze Ghana's historical CO2 emission trends, predict future emissions using ARIMA modeling, and identify factors that influence these trends. The research is more than a simple trend analysis in order to provide policymakers with practical insights that will enable them to incorporate environmental considerations into development planning. The study supports Ghana's position as a responsible global citizen by helping to fulfill international environmental obligations.

II. BRIEF LITERATURE REVIEW

Shakiru et al. (2023) proposed a new ARIMAX model for predicting CO2 emissions in the Tanzanian environment. Using Multiple Linear Regression to incorporate external inputs, the study enhanced forecasting accuracy through the hybrid ARIMA-QR model. This new approach pointed out the complexity of the systems that require further detailed frameworks to model CO2 emissions by unearthing interplays of a number of variables. The Tetteh & Baidoo (2022) work that explores the connection between energy consumption, globalization, and environmental degradation in Ghana testifies that analytical techniques are growing sophisticated. Conclusions were more or less towards the need for stronger and more complex environmental laws matched up with international standards.

The trend of carbon emissions in four regions of China, from 1997 to 2017, was estimated using ARIMA models by Ning et al. (2021), who emphasized that data preprocessing and modeling would aid policymakers in setting reduction targets and taking the correct steps. Wang et al. (2020) developed two combined models, MNGM-ARIMA and MNGM-BPNN. These two combined models would be more accurate for the prediction of carbon emissions in China, the US, and India. The graphical results were a bottoming in US emissions and a continued growth in China and India, with a slightly higher growth in India than in China. Models were crucial for making forecasts with the lowest possible errors in carbon emissions and for shaping emission reduction efforts by major emitting countries. Sharma et al. (2024) applied ARIMA models to predict carbon emissions in India from 1980 to 2021 and projected emissions through 2022–2030.

Their findings are that the emission rate shall rise and grow well even in those efforts aimed at its reduction and adversely impact renewable energy productions of sun and wind. The study presses the importance of the decrease in carbon emissions and the output policy decisions for India's carbon footprint. This particular study by Kour (2023) aims to predict future carbon dioxide emission in South Africa from 1980 to 2016 for the period beyond 2015 to 2027 using ARIMA. His study emphasized carbon emissions in South Africa, calling for urgent necessity due to increasing fossil fuel utilization and the proper implementation of sustainable policies to reverse the effects from the increase to climate change. Simon et al. (2018) applied Multiple Linear Regression as a regressive analytical technique to analyze CO2 emissions in Ghana. The study helped to explain the subtleties of different industrial contributions, taking specific interest in the transportation industry. On the same note, Sokama-Neuyam (2023) further stretched the research to cover Ghana's commitment to decreasing the greenhouse gas (GHG) emissions, by factually finding evidence from an elaborate perusal of statistics related to energy consumption and GHG emissions. The study provided experimentally determined suggestions that addressed the necessity of prudent resource usage in achievement of climate objectives through the use of machine learning models for forecasting. Shakiru et al. (2023) made a return contribution by clarifying an ARIMA-MLR model and a hybrid ARIMA-QR model for Tanzanian CO2 emissions forecasting. This methodological improvement offers a superior technique that extends from mean value projections to include quantiles concerning the high and the low projections, which may mitigate the weakness inherent with such traditional forecasts. Studies conducted in Ghana lack sophisticated analytical techniques to enhance the precision of CO2 emissions projections, such as ARIMA modeling (Tetteh & Baidoo, 2022). In addition, a gap exists in the comprehensive analysis of industrial contributions other than transportation (Simon et al., 2018), which hinders the creation of industry-specific, overall emission reduction plans.

Zhong et al. (2024) examined the complexities of forecasting carbon dioxide (CO2) emissions, emphasizing the challenges posed by their non-linear and non-stationary nature, particularly in the context of big data. They proposed a new composite forecasting approach: DCEF, using EMD data stabilization, the ARIMA model for the purpose of forecasting, and TSVD for compressing and noise-reducing purposes. Empirical validation of the method for data obtained daily from six sectors in 13 countries was shown to outperform traditional models in terms both of accuracy and efficiency of the prediction. The study further focused on hybrid models to increase the accuracy of emission prediction for better policy decisions in global carbon management. Another study by Mahla et al. (2019) explores biomass-based biogas as a renewable fuel substitute for diesel, and an ARIMA model was used for the prediction of emissions and performance. The validation of the reliability of the model was done by R^2, RMSE, and BIC. NOx and smoke opacity output records were significantly lower compared to diesel, while CO, CO_2, and HC emissions were higher. Brake thermal efficiency declined with increased biogas flow. The study highlighted the potential of biogas–diesel dual-fuel combustion for cleaner energy applications.

Wen et al. (2023) analyzed the CO_2 emissions of China, which have risen sharply since China's entry into the WTO with economic growth and increased fossil fuel consumption. This study identified significant emission factors for three regions based on the random forest feature ranking algorithm. A model combining an ARIMA with an LSTM to capture both linear and nonlinear patterns in emissions better predicted CO_2 trends. This model estimated future emissions for China's east, central, and west regions, and such estimations may be helpful for emission reduction policies. In their study on CO_2 emissions in India, the USA, and China, Kumari and Singh (2022) report India's per capita emission as 1.80 metric tons. They use six predictive models applied to time-series data for 1980 to 2019: ARIMA, SARIMAX, Holt-Winters, linear regression, random forest, and LSTM. The analysis of performance reveals that LSTM, SARIMAX, and Holt-Winters are the most accurate models, in which the performance of LSTM has been best by using several metrics. Finally, the paper concluded that the model suitable for forecasting CO_2 emission is LSTM. The future forecasts are to be made by LSTM.

The reviewed literature highlights the effectiveness of ARIMA and hybrid models in forecasting CO_2 emissions across different regions, emphasizing their role in guiding policy decisions. While studies have applied ARIMA in various countries, there remains a gap in its application to Ghana, where economic growth, industrial activities, and energy consumption patterns differ significantly from other regions. In addition, most of the previous studies have mainly focused on global and national-level projections, with less attention paid to country-specific emission trends in Ghana. This study attempts to fill the gaps by using the ARIMA model to analyze and forecast CO_2 emissions in Ghana, thus providing useful insights for policymakers in developing targeted strategies for emission reduction.

Methods and Data

The ARIMA (Autoregressive Integrated Moving Average) model is employed to forecast Ghana's carbon dioxide emissions. Since the ARIMA model is thought to provide more accurate forecasts than other econometric techniques, it is determined to be more appropriate for this investigation (Nyoni, 2019). When compared to smoothing methods and multivariate models (Nyoni, 2019; du Preez & Witt, 2003), the ARIMA model's predicting ability is superior (Goh & Law, 2002; Kour, 2023). ARIMA models, in contrast to regression models, enable the explanation of time series through their lagged or past values and stochastic error factors (Rahman & Hassan, 2017).

The statistical programme EVIEWS 9 is used in the study to create the ARIMA model prediction. South African annual time series data spanning the years 1980–2016 were used in this analysis. The ARIMA model is among the most widely used and well-respected methods for time series analysis of non-stationary variables (Nyoni, 2019). For the years 2015–2027, CO2 emissions are forecast using it. The ARIMA model uses a variable's historical value to forecast its future value. Therefore, CO2 emissions are the only variable used in the research. In order to quantify emissions CO2 emissions have been expressed in kilogrammes (*kt*). To calculate CO2 emissions, one must first determine the entire population and the total amount of CO2 emitted per person (Kour, 2023; Ritchie & Roser, 2018). In this manner, the per capita emission is calculated by obtaining an average.

ARIMA models combine the functions of Autoregressive (AR), Moving Average (MA), and Integration (I), which refers to the process of reversing differencing to get the forecast (Box and Jenkins, 1970). ARIMA (p,d,q) represents the ARIMA model, where "q" is the moving average process order, "d" denotes the data stationary order order, and "p" denotes the auto regressive process order.

Prediction issues are resolved by employing the ARIMA model for time series forecasting. This scientific instrument is multidisciplinary in nature. Because it simply requires historical data for the necessary variables, its implementation is simple and adaptable (Khashei and Bijari, 2011). The general equation of sequential differences at the dth difference of Xt is as follows, assuming that Xt is the anticipated observation:

$$\Delta^d X_t=(1-B)^d X_t$$

where d is the difference order, which is often 1 or 2, and B is the backshift operator. The incremental difference at a single time lag is equal to

$$\Delta^1 X_t=(1-B)^d X_t= X_t- X_{t-1}$$

ARIMA (p, d, q) can be briefly expressed as follows (Younes, et al., 2014):

$$\Phi_p (B)W_t =\theta_q (B) e_t$$

where Wt = ΔdXt, Φp (B) is an auto-regressive operator of order p, and θq (B) is a moving average operator of order q.

Additionally, the Philip-Perron (PP) Unit Root Test and the Augmented Dickey-Fuller (ADF) Unit Root Test were employed as estimate methodologies in this study. The study uses the PP unit root test and the ADF unit root test to verify that the data are stationary.

We have employed Mean Absolute Error (MAE), Root Mean Squared Error (RMSE), and Mean Absolute Percentage Error to assess the forecasting model's accuracy (MAPE). The MAE provides a measure of how much predicted values deviate from actual values. The root mean square error (RMSE) provides information on predicting error (Kour, 2023). MAPE, which provides the percentage of average absolute error, is reliant on the transformation of the data but not the measurement scale (Dritsaki, 2020).

III. ANALYSIS AND RESULTS

The Philip-Perron (PP) and Augmented Dickey-Fuller (ADF) unit root tests are performed in order to perform the stationarity test. After first order difference, ADF values (-6.144542) and PP values (-6.147037) are found to be significant at 5%, suggesting that the two sequences are stationary. Thus, the sequence proved suitable for estimating the ARIMA model.

The order of the data stationary (d), the order of the auto regressive process (p), and the order of the moving average process are provided for the ARIMA model selection autocorrelation (ACF) and partial autocorrelation (PACF) graph. The autocorrelation (ACF) and partial autocorrelation (PACF) graphs for the first order difference of the series were analysed because the data had already been evaluated for stationarity for the first lag difference. Because of a few seasonal spikes in the data, it was decided that the series was inappropriate. As a result, figure 1 shows the second order difference autocorrelation (ACF) and partial autocorrelation (PACF) graph that was produced. The autocorrelation (ACF) and partial autocorrelation (PACF) graphs in Figure 1 demonstrate that the carbon dioxide emission series in Ghana is stationary with a second order difference. They also show that there is no significant spike in the second order differenced series and that the moving average and auto-regressive components have no discernible effects on the second order differenced series.

The MA order is specified using an autocorrelation graph, and as figure 1shows, the ARIMA model may be developed using MA orders 1 and 3. As can be seen in Figure 1, the AR order is specified using a partial correlation graph, and the ARIMA model can be developed using AR orders 1 and 3. Thus, the models ARIMA (1,2,1), ARIMA (1,2,3), ARIMA (3,2,1), and ARIMA have been selected for additional analysis towards the final ARIMA model specification (3,2,3).

Figure 1. Autocorrelation and Partial Correlation Correolograms (Ghana)

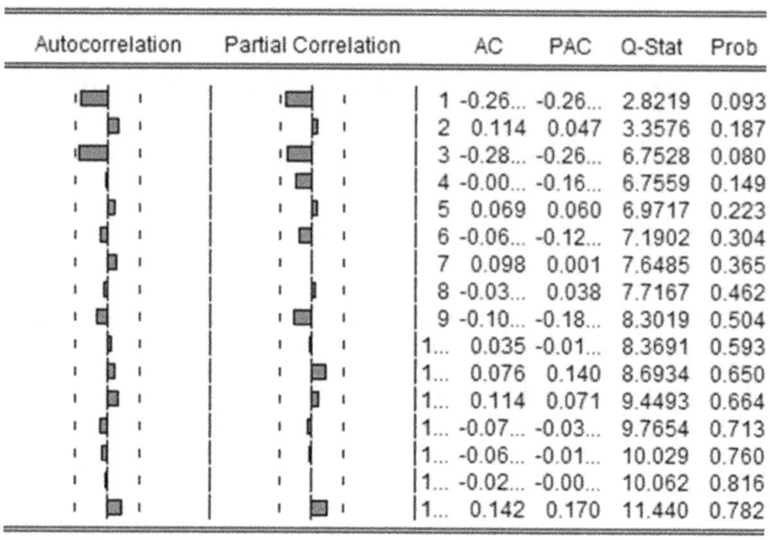

Table 1. Selection of ARIMA model (Ghana)

	ARIMA(1,2,1)	**ARIMA(1,2,3)**	**ARIMA(3,2,1)**	**ARIMA(3,2,3)**
Number of significant coefficients	3	**3**	2	3
SIGMASQ	0.009698	**0.008441**	0.008663	0.008790
Adjusted R-squared	0.018387	**0.145668**	0.123176	0.110261
Akaike info criterion (AIC)	-1.577112	**-1.701090**	-1.679349	-1.633101
F-statistic	1.224776	**3.046065**	2.685756	2.487108
Prob. F-statistic	0.316193	**0.042340**	0.062505	0.077669

Figure 2. Autocorrelation and Partial autocorrelation graph for ARIMA (1,2,3) model

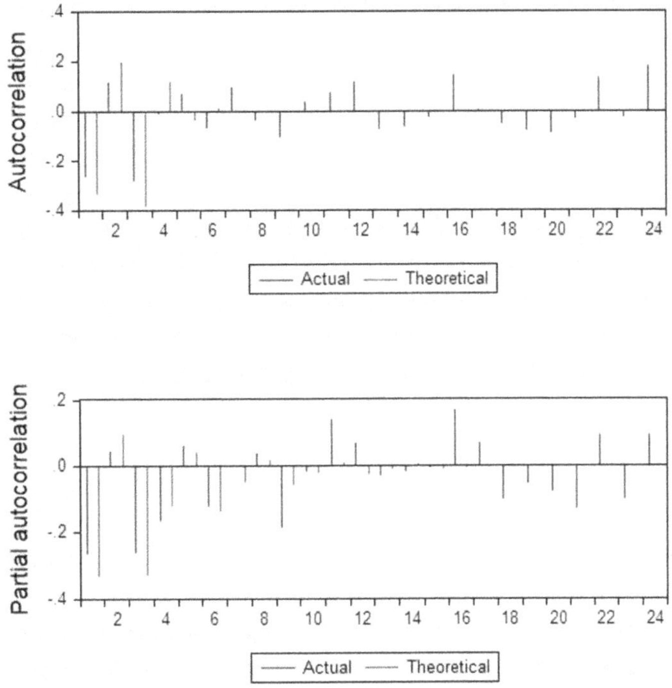

The number of significant coefficients, SIGMASQ, Adjusted R-Squared, Akaike info criterion (AIC), F-statistic, and probability of F-statistic are among the statistics of the ARIMA model that we can use to select the best model. A better model should have a small AIC, a small SIGMASQ, a higher Adjusted R squared, and a greater number of significant coefficients. The four ARIMA models' various statistics are displayed in Table 1. Table 1 shows that the most accurate model to describe Ghana's CO2 emissions is ARIMA (1,2,3). The Autocorrelation and Partial autocorrelation graph for ARIMA (1,2,3) model is shown in figure 2, which confirms model goodness.

Future CO2 emissions in the Ghana region of Sub-Saharan Africa were projected using the ARIMA (1,2,3) model for the years 2015 to 2027. Figure 3 depicts Ghana's projected CO2 emissions visually. The anticipated CO2 emissions are the best estimates, as indicated by the forecast model's Root Mean Squared Error (RMSE) of 0.055868.

Figure 3. Forecast of CO2 emissions for Ghana

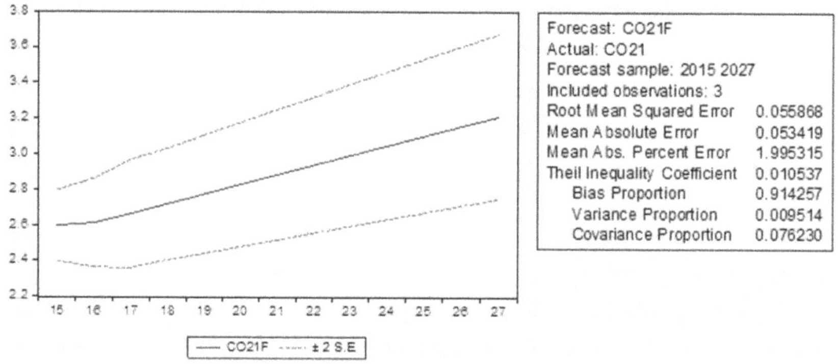

IV. CONCLUSION

Generally, the research aimed at measuring trends in historical CO2 emission in Ghana, forecasting the future emissions using ARIMA modeling, and identifying influencers behind these trends. The output gives the projected increase of CO2 emissions, making it a pressing call to take preventive measures and, in the long run, provide solutions. The fact that the emissions are increased on an annual basis calls for timely responses as they also warn of potential effects on the environment. The benchmark of 3.10 kt by 2027 encourages stakeholders, environmentalists, and politicians to commit to practical mitigation plans. It is necessary that Ghana, while continuing on its development path, balances environmental sustainability with economic growth so that it leaves for future generations a planet that can withstand the effects of climate change.

This paper adds a lot to the subject of CO2 emissions by examining historical trends, estimating future emissions trajectories, and providing insights into the temporal dynamics of CO2 emissions in Ghana through the use of ARIMA modelling. These understandings are crucial for formulating sustainable policies and making well-informed decisions on environmental management. Moreover, this study affects policy directly, indicating how essential it is to have a set of concrete plans and steps to be taken that can reduce CO2 emissions, making sustainability in Ghana further enhanced. The report urges that proactive steps be undertaken to solve environmental issues in alignment with international climate targets and projects an increasing pattern in CO2 emissions, thus filling these gaps and refining our understanding of the dynamics of Ghana around the issues of CO2 emission.

REFERENCES

Appiah, K., Du, J., Appah, R., & Quacoe, D. (2018). Prediction of potential carbon dioxide emissions of selected emerging economies using artificial neural network. changes, 14, 321-335.

Asumadu-Sarkodie, S., & Owusu, P. A. (2016). Forecasting Nigeria's energy use by 2030, an econometric approach. *Energy Sources. Part B, Economics, Planning, and Policy*, *11*(10), 990–997. DOI: 10.1080/15567249.2016.1217287

Box, G. E. P., & Jenkins, G. M. (1970). *Time Series Analysis: Forecasting and Con-trol.* Holden-Day.

Cui, T., Shi, Y., Lv, B., Ding, R., & Li, X. (2023). Federated learning with SARIMA-based clustering for carbon emission prediction. *Journal of Cleaner Production*, *426*, 139069. DOI: 10.1016/j.jclepro.2023.139069

Dritsaki, M. (2020). Forecasting European Union CO2 emissions using autoregressive integrated moving average-autoregressive conditional heteroscedasticity models. International Journal of Energy Economics and Policy.

Du Preez, J., & Witt, S. F. (2003). Univariate and multivariate time series forecasting: An application to tourism demand. *International Journal of Forecasting*, *19*(3), 435–451. DOI: 10.1016/S0169-2070(02)00057-2

Goh, C., & Law, R. (2002). Modeling and forecasting tourism demand for arrivals with stochastic nonstationary seasonality and intervention. *Tourism Management*, *23*(5), 499–510. DOI: 10.1016/S0261-5177(02)00009-2

Katani, E. K. (2019). Forecasting the total energy consumption in Ghana using grey models. Grey systems: theory and application, 9(4), 488-501.

Khashei, M., & Bijari, M. (2011). A novel hybridization of artificial neural networks and ARIMA models for time series forecasting. *Applied Soft Computing*, *11*(2), 2664–2675. DOI: 10.1016/j.asoc.2010.10.015

Kour, M. (2023). Modelling and forecasting of carbon-dioxide emissions in South Africa by using ARIMA model. *International Journal of Environmental Science and Technology*, *20*(10), 11267–11274. DOI: 10.1007/s13762-022-04609-7

Kumari, S., & Singh, S. K. (2023). Machine learning-based time series models for effective CO2 emission prediction in India. *Environmental Science and Pollution Research International*, *30*(55), 116601–116616. DOI: 10.1007/s11356-022-21723-8 PMID: 35780266

Mahla, S. K., Parmar, K. S., Singh, J., Dhir, A., Sandhu, S. S., & Chauhan, B. S. (2023). Trend and time series analysis by ARIMA model to predict the emissions and performance characteristics of biogas fueled compression ignition engine. *Energy Sources. Part A, Recovery, Utilization, and Environmental Effects*, *45*(2), 4293–4304. DOI: 10.1080/15567036.2019.1670286

Ning, L., Pei, L., & Li, F. (2021). Forecast of China's carbon emissions based on Arima method. *Discrete Dynamics in Nature and Society*, *2021*, 1–12. DOI: 10.1155/2021/1441942

Ning, L., Pei, L., & Li, F. (2021). Forecast of China's carbon emissions based on Arima method. *Discrete Dynamics in Nature and Society*, *2021*(1), 1441942. DOI: 10.1155/2021/1441942

Nyoni, T. (2018). Modeling and Forecasting Inflation in Kenya: Recent Insights from ARIMA and GARCH analysis. *Dimorian Review*, *5*(6), 16–40.

Rahman, A., & Hasan, M. M. (2017). Modeling and forecasting of carbon dioxide emissions in Bangladesh using Autoregressive Integrated Moving Average (ARIMA) models, Scientific Research Publishing –. *Open Journal of Statistics*, *7*(4), 560–566. DOI: 10.4236/ojs.2017.74038

Ritchie, H., & Roser, M. (2018): CO_2 and Greenhouse Gas Emissions. [online] <URL: https://ourworldindata.org/co2-and-other-greenhouse-gas-emissions>

Sarkodie, S. A. (2017). Estimating Ghana's electricity consumption by 2030: An ARIMA forecast. *Energy Sources. Part B, Economics, Planning, and Policy*, *12*(10), 936–944. DOI: 10.1080/15567249.2017.1327993

Shakiru, T. H., Liu, X., & Liu, Q. (2023). A hybrid Modeling and Forecasting of Carbon dioxide Emissions in Tanzania. General Letters in Mathematics (GLM), 13(1).

Sharma, S., Mittal, A., Bansal, M., Joshi, B. P., & Rayal, A. (2023, March). Forecasting of carbon emissions in India using (ARIMA) time series predicting approach. In International Conference on Renewable Power (pp. 799-811). Singapore: Springer Nature Singapore.

Simons, A., Brew, L., Buabeng, A., & Nador, H. (2018). Analysis of Carbon Dioxide Emissions in Ghana. *Ghana Journal of Technology*, *2*(2), 1–6.

Sokama-Neuyam, Y. A., Amezah, S. M., Adjei, S., Adenutsi, C. D., Erzuah, S., Quaye, J. A., & Sarkodie, K. (2023). Predictive modeling of energy-related greenhouse gas emissions in Ghana towards a net-zero future. *Greenhouse Gases: Science and Technology*.

Sokama-Neuyam, Y. A., Amezah, S. M., Adjei, S., Adenutsi, C. D., Erzuah, S., Quaye, J. A., & Sarkodie, K. (2023). Predictive modeling of energy-related greenhouse gas emissions in Ghana towards a net-zero future. *Greenhouse Gases: Science and Technology*.

Tawiah, K., Daniyal, M., & Qureshi, M. (2023). Pakistan CO 2 Emission Modelling and Forecasting: A Linear and Nonlinear Time Series Approach. *Journal of Environmental and Public Health*, **•••**, 2023. PMID: 36761238

Tetteh, B., & Baidoo, S. T. (2022). Environmental degradation, energy use, and globalization in Ghana: new empirical evidence from regime switching and neural network autoregression models. Sustainability: Science. *Sustainability*, *18*(1), 679–695. DOI: 10.1080/15487733.2022.2110680

Wang, Q., Li, S., & Pisarenko, Z. (2020). Modeling carbon emission trajectory of China, US and India. *Journal of Cleaner Production*, *258*, 120723. DOI: 10.1016/j.jclepro.2020.120723

Wen, T., Liu, Y., Bai, Y., & Liu, H. (2023). Modeling and forecasting CO2 emissions in China and its regions using a novel ARIMA-LSTM model. *Heliyon*, *9*(11), e21241. DOI: 10.1016/j.heliyon.2023.e21241 PMID: 37954263

Zhong, W., Zhai, D., Xu, W., Gong, W., Yan, C., Zhang, Y., & Qi, L. (2024). Accurate and efficient daily carbon emission forecasting based on improved ARIMA. *Applied Energy*, *376*, 124232. DOI: 10.1016/j.apenergy.2024.124232

Chapter 22
IoT-Based Smart Polyhouse

Abhishek Kumar
https://orcid.org/0000-0003-3961-9970
Lovely Professional University, India

Harpreet Singh Bedi
https://orcid.org/0000-0003-1389-4333
Lovely Professional University, India

Tanishk Singhal
https://orcid.org/0009-0003-5800-6958
Lovely Professional University, India

ABSTRACT

A polyhouse, known as a greenhouse, is a building constructed for the controlled growing of plants. It is built of translucent materials such as glass or plastic and is intended to offer a regulated temperature for plants, shielding them from harsh weather but allowing sunlight to pass through. The chapter considers cost, practicability, sensors values, emerging IoT methods, employing arduino microcontroller with remote control, and building a smart greenhouse monitoring system with performance monitoring. This technology will assist farmers in avoiding physical visits to the field while increasing productivity by maintaining precise parameters such as CO2, soil moisture, temperature, and light in the greenhouse via IoT.

1. INTRODUCTION

A polyhouse, often known as a greenhouse, is a building constructed for the controlled growing of plants. It is often built of translucent materials such as glass or plastic and is intended to offer a regulated temperature for plants, shielding them from harsh weather but allowing sunlight to pass through. It has numerous environmental controls, such as temperature, humidity, and ventilation systems (Bseiso, A. et al, 2015). Growers may use these controls to produce and maintain ideal growth conditions for plants, extending the growing season and improving agricultural yields. It is frequently made of transparent materials such as glass or plastic and is designed to provide a controlled temperature for plants while protecting them from harsh weather and letting sunlight pass through. Temperature, humidity, and ven-

DOI: 10.4018/979-8-3693-9924-8.ch022

tilation systems are only a few of the environmental controls in a polyhouse. Growers may employ these controls to create and maintain optimal plant development conditions, prolonging the growing season and increasing agricultural production (Czyzyk, K. A.et al, 2014).

Conventional polyhouse farming involves continual human involvement, which may be time-consuming and inefficient.

The implementation of IoT-based smart polyhouses overcomes these difficulties by providing automatic monitoring and real-time control of environmental factors. The importance of IoT in polyhouse farming is examined in this paper, along with its benefits, drawbacks, and potential applications. The integration of Internet of Things (IoT) technology into agriculture has brought about a revolutionary transformation in modern farming practices. One of the notable applications of IoT in agriculture is the development of IoT-based polyhouses, which have the potential to significantly enhance crop productivity and sustainability (Laiolo, P., et al, 2014). By incorporating IoT technologies into polyhouse management, farmers can remotely monitor and control various environmental parameters, ensuring optimal conditions for plant growth while conserving resources. This paper explores the concept of IoT- based polyhouses, their components, benefits, and the impact they can have on modern agriculture. IoT-based polyhouses represent a marriage between traditional agricultural practices and cutting-edge technology. These smart agricultural environments leverage IoT devices and sensors to collect real-time data on temperature, humidity, light levels, soil moisture, and more (Sorathia, V et al, 2005). The data is then transmitted to a central control system, which can be accessed by farmers through web-based interfaces or mobile applications. By doing so, farmers can make informed decisions regarding irrigation, ventilation, heating, and cooling, all aimed at creating an optimal microclimate for plant growth (Kadam G S et al. 2024).

This chapter introduces a smart polyhouse model leveraging IoT (Internet of Things) technology to monitor and optimize environmental conditions for growing tomatoes and strawberries. The system incorporates multiple sensors to measure critical parameters such as

a) Soil Moisture: Ensures optimal irrigation and prevents over- or under-watering.
b) Temperature: Maintains the ideal climate for plant growth.
c) CO_2 Levels: Monitors and regulates CO_2 concentration to enhance photosynthesis.

The sensor data is recorded both remotely (via cloud or mobile applications) and locally (on-site), enabling farmers to monitor and control the polyhouse environment in real-time. The reliability of the sensors is a key focus, as it directly impacts the accuracy and effectiveness of the measurements, ensuring optimal crop growth and resource efficiency. One of the primary advantages of IoT-based polyhouses is their ability to optimize resource usage. Through precise monitoring and control, these systems reduce water and energy consumption, making agriculture more sustainable and environmentally friendly (Mayuri KP et al, 2024). Additionally, they minimize the need for manual labor, as many tasks can be automated, allowing farmers to focus on higher-level decision-making and crop management (Brassai, S. et al, 20120. This chapter illustrate into the essential components of IoT-based polyhouses, including sensors, actuators, communication networks, and data analytics tools (Farooq, M. S. et al, 2020) It will also explore the potential challenges and considerations, such as data security and scalability, associated with implementing these advanced agricultural systems (Xu, J., Gu, B., & Tian, G., 2022).

2. IMPORTANCE OF IOT IN POLYHOUSE FARMING

In polyhouse farming, IoT technology consists of a network of cloud computing, microcontrollers, and associated sensors that cooperate to optimize crop development. In general, the system consists of following interfaced component.

2.1. Environmental Sensors

Temperature & Humidity Sensors (DHT11, DHT22, SHT31): Monitor the temperature inside the polyhouse and activate heating or ventilation systems as necessary.

Soil Moisture Sensors (YL-69, Capacitive Soil Moisture Sensor): Assist with precision irrigation by preventing crops from being over- or under-watered.

CO_2 Sensors (MQ-135, MH-Z19): Track carbon dioxide levels to maximize photosynthesis rates for increased yield.

Light Sensors (LDR, BH1750): Identify ambient light levels and adjust artificial lighting appropriately.

2.2. Microcontrollers

Arduino: Perfect for simple automation and sensor control applications.

ESP8266/ESP32:Microcontrollers with Wi-Fi capability that can transmit data in real time to cloud platforms are the ESP8266 and ESP32.

Raspberry Pi: for smart farming, the Raspberry Pi is utilized for edge computing and AI-based analytics.

2.3 Wireless Communication Technologies

Bluetooth and Wi-Fi: Ideal for modest polyhouses with robust network connectivity.

LoRa & Zigbee: Long-range, low-power communication is made possible by LoRa and Zigbee, making them perfect for farms that require remote monitoring.

GSM & LTE (SIM800L, SIM900): Enable data transfer and real-time warnings in places with inadequate Wi-Fi coverage.

2.4. Cloud Computing & Data Analytics

Real-time polyhouse data may be stored and processed using Google Firebase, AWS IoT Core, and ThingsBoard. Plant disease prediction, irrigation schedule optimization, and climatic trend analysis are all possible with machine learning algorithms.

2.5 Automated Actuators

Water pumps and valves: intelligent irrigation that is automated and dependent on measurements of soil moisture. Maintain the proper humidity and temperature levels via exhaust fans and heaters.

LED Grow Lights: In low-light settings, these lights offer the best illumination conditions for plant development.

IoT has revolutionized many industries, including agriculture, and it plays a crucial role in modern polyhouse farming. Polyhouse farming refers to the cultivation of crops in controlled environments, such as greenhouses or polytunnels, to optimize growing conditions. Here are some key aspects of the importance of IoT in polyhouse farming (Tzounis, A. et al, 2017).

2.6 Environmental Monitoring and Control:

IoT sensors are capable of maintaining track of the polyhouse's humidity and temperature, maintaining the ideal conditions for plant development. Using real-time data, the system automatically modifies the irrigation, heating, and ventilation systems. Better crop yields can result from plants receiving enough light for photosynthesis thanks to IoT's ability to regulate and track light intensity and CO2 levels. In order to maintain the ideal circumstances for plant development, sensors that evaluate soil pH and moisture can automatically activate fertilizer or irrigation systems. For example, if temperature levels rise too high, IoT sensors can trigger automated cooling systems (Kim, W. S. et al, 2020).

2.7 Resource Efficiency:

Polyhouse farming can be resource-intensive due to the need for controlled conditions. IoT can help optimize resource usage. For instance, by analyzing data from soil moisture sensors, farmers can precisely control irrigation, reducing water wastage. This leads to more sustainable and efficient resource management (Dongre, S et al, 2022). Polyhouse farming depends on efficient irrigation, and real-time soil moisture monitoring with IoT-based solutions to minimize water waste and ensures that plants receive an appropriate amount of water.

2.8 Crop Health Monitoring:

IoT devices can be used to monitor the health of crops through various means, including cameras and sensors that detect signs of disease or pest infestation. Early detection allows for timely intervention, reducing the need for chemical pesticides and increasing overall crop yields (Vanteru, Madhu et al, 2022). IoT systems ensure that plants develop effectively and healthily by preserving ideal growth conditions, which improves crop quality and yields.

2.9 Automation and Control:

IoT enables the automation of various processes within the polyhouse, such as adjusting temperature, humidity, and ventilation systems. This not only reduces labor requirements but also ensures that conditions are consistently ideal for plant growth (Dixit, Anil et al, 2022)

2.10 Data-Driven Decision-Making:

The data collected by IoT devices can be analyzed to make informed decisions. By comparing historical data and trends, farmers can optimize planting schedules, select suitable crop varieties, and make other strategic choices to maximize yields and profits (Bedi HS et al, 2017)

2.11 Remote Monitoring:

Farmers can monitor and control their polyhouses remotely through IoT-enabled applications and dashboards. This is especially beneficial for large-scale operations or when the farmer is not on-site, as it allows for timely responses to changing conditions (Bedi, Harpreet Singh et al. 2015)

3. TYPES OF IOT-BASED SMART POLYHOUSES

IoT-based smart polyhouses are available in a number of forms, each with unique features and technology to satisfy distinct farming requirements (Bedi, H. S et al, 2017) and (Bedi, H. S. et al, 2024). The primary IoT-based smart polyhouse types are categorized according to the crops they support, environmental management, and automation. Some popular varieties of IoT-based smart polyhouses are

3.1 Climate-Controlled Polyhouse:

is a smart agricultural infrastructure that optimizes and regulates environmental factors including temperature, humidity, light, CO_2 levels, and airflow using automation technologies and the Internet of Things (IoT). IoT sensors adjust these parameters based on crop requirements, ensuring optimal growing conditions. This advanced technology guarantees that plants are grown in environments that are most suited to their individual requirements, increasing crop yields, improving quality, and reducing resource wastage.

3.2 Hydroponic Polyhouse:

is an advanced agricultural system where crops are grown without soil, using mineral nutrient solutions dissolved in water. These polyhouses employ IoT systems to manage hydroponic farming setups. Sensors monitor nutrient levels, pH, and water quality to maintain optimal conditions for plant growth. They are suitable for growing leafy greens and herbs.

3.3 Aquaponic Polyhouse:

Combining aquaculture (raising fishes) and hydroponics (growing plants without soil), these polyhouses use IoT to manage fish tanks and plant beds. In this agriculture system where the waste produced by the fish provides nutrients for the plants, and the plants help filter and purify the water, which is then returned to the fish tanks. Sensors monitor water quality, oxygen levels, and fish health while optimizing nutrient delivery to plants. They are suitable for growing both plants and fish in a symbiotic system.

3.4 Vertical Farming Polyhouse:

IoT systems control vertical farming structures, such as stacking shelves or towers. Sensors track light intensity, nutrient levels, and water usage, optimizing vertical farming efficiency. Combined with a polyhouse, vertical farming offers more efficient use of available space, water, and resources, which makes it a perfect option for high-density farming, urban agriculture, and places with not enough available land. These are space-efficient solutions for urban agriculture.

3.5 Light-Controlled Polyhouse:

These polyhouses focus on optimizing light conditions for plant growth. IoT-controlled LED lights adjust intensity, spectrum, and duration to mimic natural sunlight. They are useful for year-round cultivation and in regions with limited sunlight.

3.6 Nutrient Dosing Polyhouse:

These polyhouses automate the delivery of nutrients to plants IoT sensors monitor nutrient levels in the soil, and automated systems adjust nutrient dosing accordingly They are excellent for precision agriculture and reducing resource wastage.

These various types of IoT-based smart polyhouses cater to different crops, environmental conditions, and agricultural goals. They help increase crop yields, reduce resource wastage, and contribute to sustainable and efficient farming practices

4. IOT-BASED POLYHOISE MODEL

Using the Internet of Things (IoT), a microcontroller-based polyhouse is an innovative method to automate and optimize farming operations in a polyhouse (greenhouse). The system's integration of many sensors allows for real-time data monitoring and management of a number of important environmental factors, including temperature, moisture, humidity, light intensity, tank water level, and more. Using the Internet of Things (IoT), a microcontroller-based polyhouse is an innovative method to automate and optimize farming operations in a polyhouse (greenhouse). The system's integration of many sensors allows for real-time data monitoring and management of a number of important environmental factors, including temperature, moisture, humidity, light intensity, tank water level, and more shown in figure 1.

Figure 1. shows the block diagram

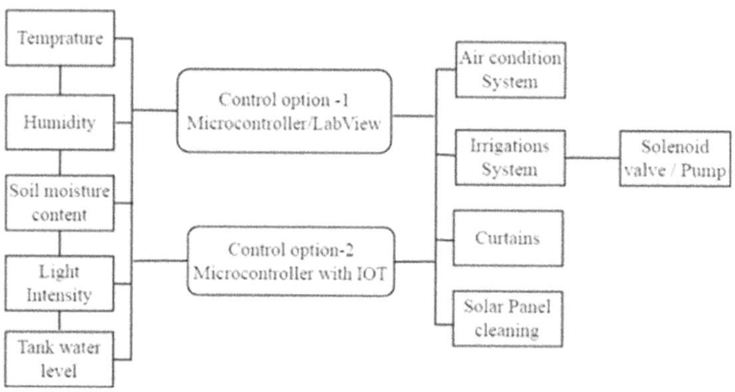

The appropriate hardware components must be chosen in order to ensure the best possible climatic conditions in a polyhouse. Begin with choosing sensors to monitor on important variables including soil moisture, light, temperature, and humidity. Accurate readings are most effectively obtained with soil moisture sensors like capacitive probes, temperature sensors like DHT11 or DS18B20, and humidity sensors like DHT22. While soil moisture sensors identify the amount of water in the soil for effective irrigation, light sensors such as the BH1750 or TSL2561 may measure the amount of natural light. Select actuators for control mechanisms, such as water pumps, fans, and heaters, which may be managed by motor drivers or relays. The intended ambient conditions are maintained in part by these actuators.

Data processing and sensor-actuator communication require a microcontroller, like as Arduino or Raspberry Pi. Through IoT connectivity choices like Wi-Fi, LoRa, or cellular networks, these devices may transmit data to a central server or cloud platform. Use authentication and encryption methods, such as HTTPS or MQTT with SSL/TLS, to make sure that data transfer is safe. Installed temperature sensors all throughout the polyhouse for environmental surveillance, making sure that the desired temperature is maintained by constant observation. Smart lighting systems may be used to alter light levels as needed, and humidity sensors can be placed strategically. Lastly, combine automated irrigation systems with soil moisture sensors to maximize water use for wholesome crop development.

Sensors The sensitivity of a sensor reveals how much the sensor's output When the input quantity being measured changes, the function changes. The many sensors utilized include:

Humidity and temperature sensor

A temperature sensor is a gadget that measures the temperature. Temperature reading at a specific moment. It is a form of resistance (Bechar, A., & Vigneault, C. 2016). The temperature detector senses temperature variations. It gives high-quality recognition in a timely manner. Being a hybrid sensor, it delivers values for both The temperature and humidity levels. The sensor generates a digital signal. Output. Using this sensor provides improved stability and higher accuracy. Reliability. DHT11 Module temperature is the sensor utilized. as well as a humidity sensor module.

Light sensor:

The Light sensor measures the intensity of light in the polyhouse (Sishodia, R. P et al, 2020).. A light sensor is a device that detects the current ambient level of light, i.e. how bright or dark it is. INVNT_10 Lm393 optical photosensitive LDR light-sensitive sensor was utilized. It is linked to both analog and digital output pins on the board, which are designated AO and DO, respectively. When there is light, the resistance of the LDR decreases in proportion to the intensity of the light. The lower the resistance of LDR, the greater the intensity of light. A potentiometer knob on the sensor may be changed to modify the sensitivity of the LDR to light. It is also intended to take advantage of reducing the framework.

Soil moisture sensor:

The soil moisture sensor is a device that measures the amount of water in the soil. Soil moisture assessment is critical for farmers to operate their irrigation systems. It is made up of two probes that are used to determine the volumetric content of water. The two probes enable the current to travel through the soil, and the resistance value is used to calculate the moisture value (Shi, Y et al, 2016). When there is more water in the soil, it conducts higher electricity, resulting in less resistance. As a result, the moisture level will be increased. Because dry soil conducts electricity poorly, when there is less water, the soil conducts less electricity, resulting in higher resistance. The moisture level will be decreased (Anil Sai Kumar et al, 2023)

CO2 Sensors:

Carbon dioxide sensors monitor the CO2 levels within the polyhouse. Controlling CO2 levels is important for photosynthesis and plant growth.

Water Quality Sensors:

Water quality sensors can monitor the quality of water used for irrigation. They check factors like pH, EC (electrical conductivity), and nutrient levels (Sushanth, G., & Sujatha, S, 2018)

Table 1. Variable for polyhouse

Parameter	Value	Description
TEMPERATURE °C	25.5	The current temperature inside the polyhouse
HUMIDITY %	60	Current humidity level
LIGHT INTENSITY [LUX]	1500	Current light intensity
SOIL MOISTURE %	40	Soil moisture level in the polyhouse
CO2 LEVEL PPM	400	Carbon dioxide concentration
IRRIGATION STATUS	On	Whether irrigation system is active
VENTILATION STATUS	Closed	Status of polyhouse ventilation
SHADE CONTROL	automatic	Control of the shading system
HEATER STATUS		Heating system status
CAMERA FEED	Live Stream	Live video feed from the polyhouse

Camera Sensors:

Cameras can be used for visual monitoring of plant growth, disease detection, and even for security purposes in the polyhouse (Ardiansah, I et al, 2020). Table 1 presents the following the present environmental conditions within the polyhouse: The humidity is 60% and the temperature is 25.5°C. The soil moisture is at 40% and the light intensity is 1500 lux. The optimal CO2 level for plant development is 400 parts per million. The ventilation system is closed at the moment, however the irrigation system is operational. The heating system is operating, and the shade control is set to automatic, adjusting as necessary. To make sure everything operates effectively a live camera feed offers real-time visual surveillance of the polyhouse environment.

Figure 2. Flowchart of the data collection

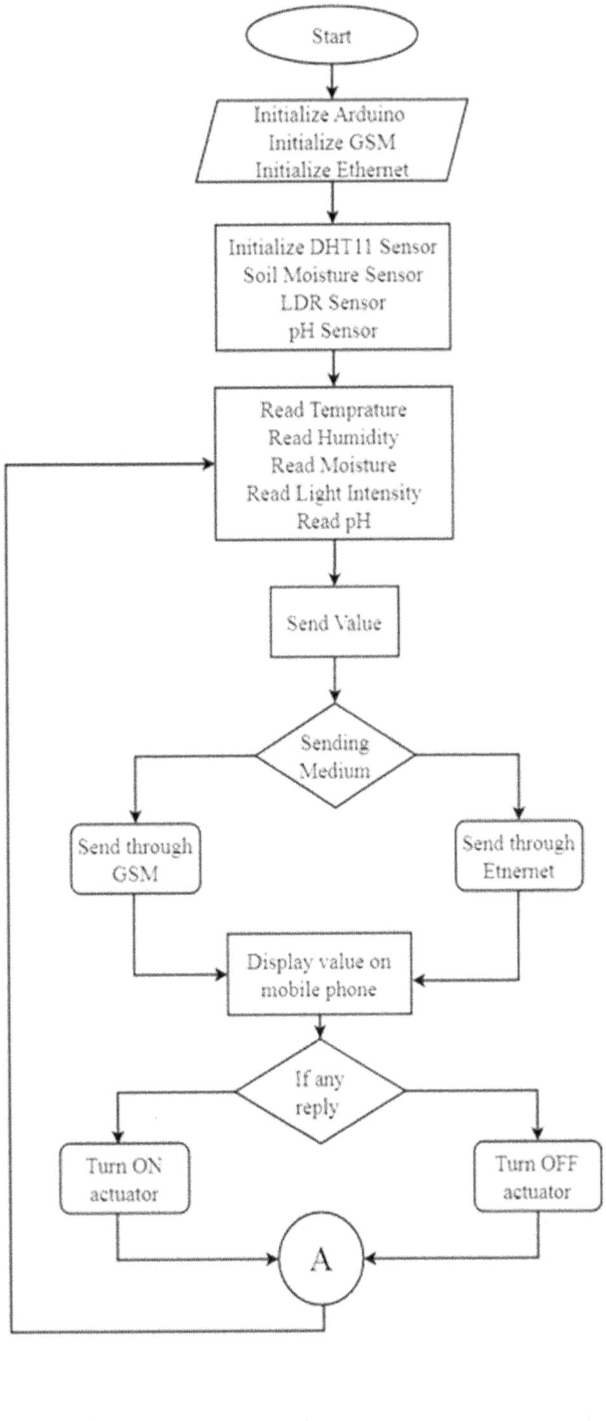

The output of several crops cultivated using a smart polyhouse system against traditional farming is contrasted in Table 2. A smart polyhouse can produce 800 kg/acre of tomatoes, which is a 60% improvement over the usual output of 500 kg/acre. Lettuce exhibits a notable improvement, yielding 600 kg/acre in a smart polyhouse, a 100% increase over the usual output of 300 kg/acre. Traditional farming yields 400 kg/acre of cucumbers, but a smart polyhouse yields 700 kg/acre, a 75% improvement. All three crop yields are greatly boosted by the smart polyhouse system, confirming its effectiveness and potential for higher agricultural output.

Table 2. shows Key parameters of track and control

Crop type	Traditional yield (kg/acre)	Smart polyhouse (kg/acre)	Improvement percentage
Tomatoes	500	800	60
Lettuce	300	600	100
cucumber	400	700	75

Data collection is the first step in data preparation and analysis shown in Figure 2. Sensors continually acquire operational and environmental data, either in real-time or at predetermined intervals. For accessibility and control, this data is kept on cloud servers or in local databases. After that, data inspection involves developing algorithms to analyze the collected data, spot patterns, and spot anomalies like odd moisture levels or temperature swings, notifying users as necessary.

Remote control and monitoring, a user-friendly dashboard that shows historical and real-time data is made available for simple access through web or mobile applications. Users can remotely control actuators like fans and irrigation systems, utilizing preset profiles tailored to different crops and growth stages. For crucial situations, such as extremely high or low humidity, automated alarms, and notifications are set up, and updates are delivered via SMS or email. Integrating energy-efficient components and using scheduling mechanisms to maximize energy utilization prioritize energy efficiency. Cameras take pictures of crops for crop monitoring and yield prediction, while machine learning and image analysis are used to assess crop health. Machine learning algorithms and historical data are used to provide yield predictions. The effective functioning of sensors and equipment is guaranteed by a regular maintenance program. The system is regularly updated to keep up with developments since it is essential to improve performance through developments in IoT technology and agricultural methods.

5. RESULT AND DISCUSSION

The implementation of an IoT-based smart polyhouse system yielded significant improvements in both operational efficiency and crop productivity By continuously gathering data from many sensors, such as temperature, humidity, wetness, and light intensity, the system offered real-time insights about the polyhouse's environmental conditions. Present this information in the form of tables, graphs, or charts. Include data on crop growth within the polyhouse. The integration of machine learning algorithms and image analysis in the crop monitoring and yield prediction sector is important information about crop health. While predictive programs forecasted yield results using past data, the cameras' pictures assisted in evaluating plant conditions

The IoT-based smart polyhouse system demonstrated clear advantages in terms of automation, efficiency, and sustainability. By integrating sensors, remote monitoring, and control systems, the system minimized human error and improved environmental conditions while lowering the need for physical intervention. In order to maintain optimal growth circumstances, environmental elements including temperature, humidity, and moisture were continually monitored and modified. This resulted in better crop yields for a variety of product varieties, including tomatoes, lettuce, and cucumbers. Comparing this data-driven approach to conventional farming practices has resulted in notable production gains. Compared to conventional non-IoT farming methods, the IoT-based smart polyhouse system exhibited measurable gains in crop growth and resource management. Crop height, leaf area, flowering, and yield were among the important variables that were regularly tracked and examined.

The IoT-enabled polyhouse demonstrated a noticeable rise in crop height for every crop. Compared to crops cultivated in conventional open-field farming, tomatoes, and cucumbers showed a 10-15% increase in height. In order to maintain ideal growth conditions and promote healthy plant development, temperature, humidity, and light intensity were continuously adjusted. Additionally, when the system maximized water and nutrient supply by real-time soil monitoring, the leaf area showed a significant increase of up to 20%, especially in lettuce. The IoT polyhouse also had improved control over blooming patterns. In comparison to non-IoT-controlled surroundings, the system's constant monitoring of temperature and light intensity allowed plants to bloom at the most suitable periods for their growth cycles, improving blooming uniformity by 15% to 20%. As a result, the total yield increased. According to previous findings, the smart polyhouse, for instance, generated 60% more tomatoes, 100% more lettuce, and 75% more cucumbers than conventional agricultural techniques.

Figure 3. Concentration gradient representation of CO_2 concentration level in the greenhouse.

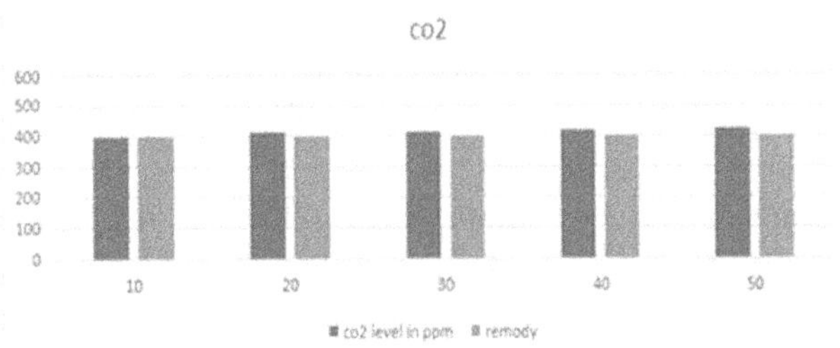

As opposed to during the day, at night the plant's photosynthesis process requires a much higher concentration of CO_2 and water. With the help of these two energy, the photosynthesis process helps the plant grow quickly and keeps it cool. After completing an experiment to determine the CO_2 concentration level in a greenhouse, it was kept at a maximum level at night as depicted in the figure 3 since the greenhouse starts consuming CO_2 at daybreak and continues to do so until nightfall.

Figure 4. Dry soil moisture concentration measured remotely and remoteless in the greenhouse

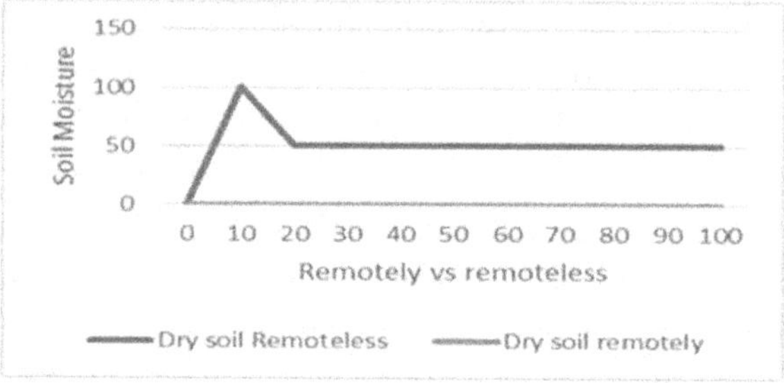

Figure 5. Wet soil moisture concentrations measured remotely and remoteless in the greenhouse.

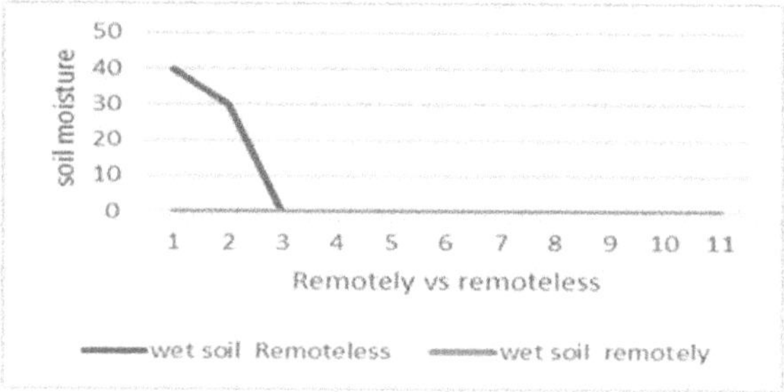

A key consideration for plants is the amount of water in the soil since too much water can result in a fungal infection, while too little water causes plants to dry up or occasionally incur damage, as presented in figure 5 and figure 5 of dry and wet soil moisture concentration when measured remotely and wirelessely.

Figure 6. Temperature range (a) outside (b) inside the greenhouse measure remotely and remoteness in greenhouse

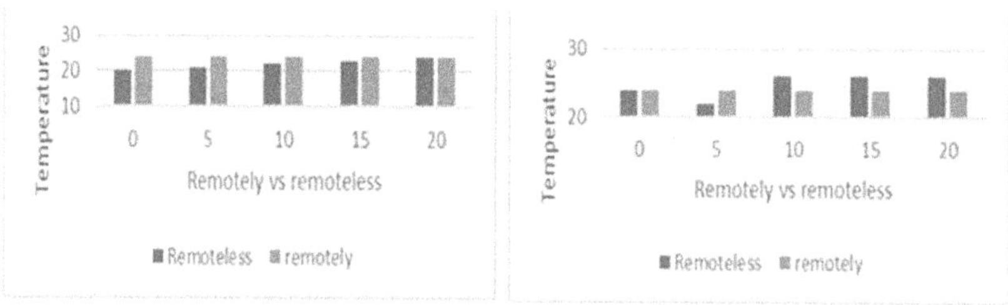

Another crucial factor in greenhouses is temperature, which should be kept at its highest level. Because the temperature promotes the growth of flowers, fruits, and seeds as well as photosynthesis. As a result, the temperature range inside the greenhouse was kept at its maximum, as illustrated in the figure 6, as opposed to the temperature range outside the greenhouse. The revenue generated by corping the tomatoes and strawberries have been demonstrated in Figure 7 based on the suggested greenhouse parameter.

Figure 7. shows the cash flow for growing tomatoes and strawberries using a smart greenhouse

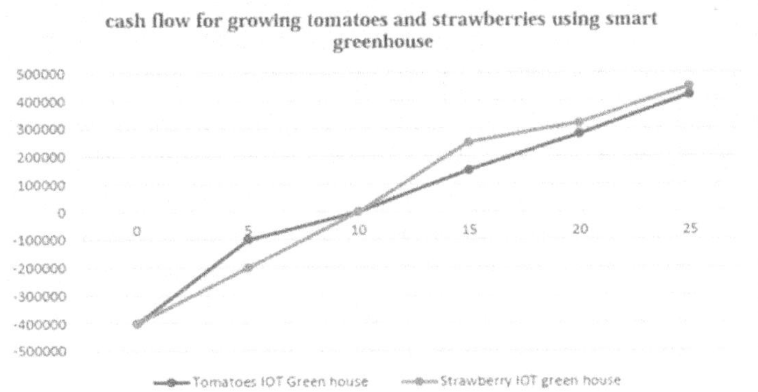

Traditional agriculture is being revolutionized by IoT-based polyhouse farming, which makes sustainable methods, automated decision-making, and real-time monitoring possible. Farmers can identify changes and anticipate plant illnesses early thanks to smart sensors that continually monitor climatic parameters including temperature, humidity, soil moisture, and CO_2 levels. In order to ensure timely remedial action, machine learning algorithms assess environmental patterns and give rapid notifications for deviations or system breakdowns. Farmers may optimize lighting, ventilation, and watering remotely via smartphone applications or cloud-based dashboards without having to physically be present. IoT lowers carbon footprints through energy-efficient automation and promotes environmentally friendly farming by minimizing excessive use of pesticides and fertilizers through accurate data-driven application techniques. Furthermore, year-round production is made possible by regulated growth conditions, which improves climate resilience and food security. In the end, IoT-powered polyhouses increase productivity, reduce waste, and maximize resources, making agriculture more sustainable, effective, and climate-adaptable.

6. CONCLUSION

The IoT-based smart polyhouse system successfully enhanced crop production and operational efficiency compared to traditional non-IoT farming practices. Crop health and yields increased significantly as a result of the system's optimization of environmental conditions through sensor data collection and analysis. By continuously monitoring and adjusting critical parameters such as temperature, humidity, light intensity, soil moisture, and CO2 levels, the system ensures optimal growing conditions for crops like tomatoes, cucumbers, and lettuce. This precision led to a 10-15% increase in crop height, a 20% boost in leaf area for lettuce, and 15-20% improved blooming uniformity, resulting in higher-quality produce. Additionally, the polyhouse achieved remarkable yield increases 60% more tomatoes, 100% more lettuce, and 75% more cucumbers while optimizing water and nutrient usage. The system's ability to maintain ideal CO_2 levels, especially at night, further enhanced photosynthesis and plant growth.

Economically, the smart greenhouse demonstrated significant revenue potential, as evidenced by the cash flow analysis for tomatoes and strawberries.

Modern agriculture is being revolutionized by IoT-based polyhouse farming, which offers sustainability, automation, and accuracy. Real-time monitoring and management of critical environmental factors may be achieved by farmers through the integration of smart sensors, microcontrollers, and cloud-based analytics. Furthermore, IoT-driven polyhouses help ensure year-round crop production and mitigate the effects of unpredictable weather, which supports climate-resilient farming. Future developments in smart polyhouses offer even more efficiency, cost, and accessibility as machine learning, artificial intelligence, and wireless connection continue to progress. Adopting such developments will be essential to tackling issues related to global food security and advancing sustainable farming methods.

REFERENCES

Anil Sai Kumar, B. V., Kumar, S. P., Jaswanth, K., Vishnu, K., & Kumar, A. (2023). Smart irrigation system using machine learning techniques. Machine Learning for VLSI Chip Design, 139-159.

Ardiansah, I., Bafdal, N., Suryadi, E., & Bono, A. (2020). Greenhouse monitoring and automation using Arduino: A review on precision farming and internet of things (IoT). *International Journal on Advanced Science, Engineering and Information Technology*, *10*(2), 703–709. DOI: 10.18517/ijaseit.10.2.10249

Bechar, A., & Vigneault, C. (2016). Agricultural robots for field operations: Concepts and components. *Biosystems Engineering*, *149*, 94–111. DOI: 10.1016/j.biosystemseng.2016.06.014

Bedi, H. S., Goyal, N., Kumar, S., & Gupta, A. (2017). Smart Trolley using Smart Phone and Arduino. *J Electr Electron Syst*, *6*(2), 223. DOI: 10.4172/2332-0796.1000223

Bedi, H. S., Harshin, A. N., Lawai, B. P., & Kumar, N. (2024). Design of Intelligent Polyhouse with IOT. *Indonesian Journal of Interdisciplinary Research in Science and Technology*, *2*(3), 231–244. DOI: 10.55927/marcopolo.v2i3.8095

Bedi, H. S., Raju, D. V., Nandyala, M. R. C., Kumar, P. S., & Varma, M. R. (2022). Design of Gesture Based Hand Gloves Using Arduino UNO: A Grace to Abled Mankind. Digital Forensics and Internet of Things: Impact and Challenges, 37-43.

Bedi, H. S., & Verma, S. (2015). R. K. Sharma, and B. Singh (2015)The Concept of Programmable Logic Controllers and its role in Automation. *International Journal of Advanced Research in Computer and Communication Engineering*, *4*(6).

Brassai, S. T., Enăchescu, C., & Losonczi, L. (2012, May). RBF network for mobile robot sonar based localization and environment modeling. In 2012 13th International Conference on Optimization of Electrical and Electronic Equipment (OPTIM) (pp. 1499-1504). IEEE. DOI: 10.1109/OPTIM.2012.6231939

Bseiso, A., Abele, B., Ferguson, S., Lusch, P., & Mehta, K. (2015, October). A decision support tool for greenhouse farmers in low-resource settings. In *2015 IEEE Global Humanitarian Technology Conference (GHTC)* (pp. 292-297). IEEE. DOI: 10.1109/GHTC.2015.7343987

Czyzyk, K. A., Bement, S. T., Dawson, W. F., & Mehta, K. (2014, October). Quantifying water savings with greenhouse farming. In *IEEE Global Humanitarian Technology Conference (GHTC 2014)* (pp. 325-332). IEEE. DOI: 10.1109/GHTC.2014.6970300

Dixit, A. (2022). *Kumar, T. & Joshi, Abhishek & Bedi, Harpreet & Chakravarthi, Kalyan & Singh, Devesh.* Trends in Robotics and Computer Integrated Manufacturing., DOI: 10.1109/IC3I56241.2022.10072745

Dongre, S., Raut, A., & Sakhare, A. (2022, March). Weather monitoring and control system for polyhouses using IoT. In AIP Conference Proceedings (Vol. 2424, No. 1). AIP Publishing. DOI: 10.1063/5.0078090

Farooq, M. S., Riaz, S., Abid, A., Umer, T., & Zikria, Y. B. (2020). Role of IoT technology in agriculture: A systematic literature review. *Electronics (Basel)*, *9*(2), 319. DOI: 10.3390/electronics9020319

Kadam, G. S., Lohar, S. N., More, S. G., Tathe, P. S., & Dal, D. S. (2024). Smart Poly House Monitoring Systems. *International Research Journal of Innovations in Engineering and Technology*, *8*(5), 212.

Kim, W. S., Lee, W. S., & Kim, Y. J. (2020). A review of the applications of the internet of things (IoT) for agricultural automation. *Journal of Biosystems Engineering*, *45*(4), 385–400. DOI: 10.1007/s42853-020-00078-3

Laiolo, P., Gabellani, S., Pulvirenti, L., Boni, G., Rudari, R., Delogu, F., & Puca, S. (2014, July). Validation of remote sensing soil moisture products with a distributed continuous hydrological model. In *2014 IEEE Geoscience and Remote Sensing Symposium* (pp. 3319-3322). IEEE. DOI: 10.1109/IGARSS.2014.6947190

Mayuri, K. P., Kathavate, S., & Niranjanamurthy, M. (2024). Polyhouse Agriculture with AI: Strategies for Climate Control, Energy Efficiency and Yield Improvement. *SN Computer Science*, *5*(8), 1119. DOI: 10.1007/s42979-024-03432-w

Shi, Y., Thomasson, J. A., Murray, S. C., Pugh, N. A., Rooney, W. L., Shafian, S., & Yang, C. (2016). Unmanned aerial vehicles for high-throughput phenotyping and agronomic research. *PLoS One*, *11*(7), e0159781. DOI: 10.1371/journal.pone.0159781 PMID: 27472222

Sishodia, R. P., Ray, R. L., & Singh, S. K. (2020). Applications of remote sensing in precision agriculture: A review. *Remote Sensing (Basel)*, *12*(19), 3136. DOI: 10.3390/rs12193136

Sorathia, V., Laliwala, Z., & Chaudhary, S. (2005, July). Towards agricultural marketing reforms: Web services orchestration approach. In 2005 IEEE International Conference on Services Computing (SCC'05) Vol-1 (Vol. 1, pp. 260-267). IEEE.

Sushanth, G., & Sujatha, S. (2018, March). IOT based smart agriculture system. In 2018 international conference on wireless communications, signal processing and networking (WiSPNET) (pp. 1-4). IEEE.

Tzounis, A., Katsoulas, N., Bartzanas, T., & Kittas, C. (2017). Internet of Things in agriculture, recent advances and future challenges. *Biosystems Engineering*, *164*, 31–48. DOI: 10.1016/j.biosystemseng.2017.09.007

Vanteru, M. K., Bedi, H. S., Azam, M., Chakravarthi, M. K., Raffik, R., & Pant, B. (2022, December). Empirical Investigation on Smart Wireless Autonomous Robot for Landmine Detection with Wireless Camera. In 2022 5th International Conference on Contemporary Computing and Informatics (IC3I) (pp. 200-205). IEEE.

Xu, J., Gu, B., & Tian, G. (2022). Review of agricultural IoT technology. *Artificial Intelligence in Agriculture*, *6*, 10–22. DOI: 10.1016/j.aiia.2022.01.001

Chapter 23
Studies on Biomass Briquettes Production From Organic Waste Feed Stocks in Bule Hora Town, Ethiopia

Pitta Sekhar
https://orcid.org/0000-0002-8031-6237
Bule Hora University, Ethiopia

Kiyar Jemal Abdo
https://orcid.org/0000-0002-4613-7737
Arsi University, Ethiopia

ABSTRACT

This study assessed biomass briquettes as a sustainable energy alternative in Bule Hora town, Ethiopia, addressing waste and traditional biomass reliance. Feedstocks (coffee husks, sawdust, maize cobs, and waste paper) were carbonized and compacted. Briquette properties were analyzed using ASTM standards, with ANOVA, regression, and RSM applied for optimization. Maize cobs had the highest calorific value (1820.54±0.37 kcal/kg). Carbonized briquettes excelled in density (0.61 g/cm³) and calorific value (4670.92 kcal/kg). Compression pressure, binder content, and carbonization temperature significantly impacted performance (p < 0.01). Optimal conditions: 12 MPa pressure, 11% binder, 450°C temperature, and 16% moisture. Surveys showed 70% willingness to adopt briquettes for reduced pollution and cost savings, despite awareness and market challenges. Biomass briquetting offers a solution to deforestation, energy security, and waste management, recommending scaled production, optimized processes, and awareness initiatives.

INTRODUCTION

In developing countries, where energy poverty remains a pressing issue, the reliance on traditional biomass fuels such as firewood and charcoal has placed significant pressure on natural resources, leading to environmental degradation and adverse health effects. Biomass briquettes, compacted blocks of organic

DOI: 10.4018/979-8-3693-9924-8.ch023

residues, have emerged as a viable alternative to traditional fuels, offering cleaner, more sustainable energy solutions for heating, cooking, and industrial applications. The widespread adoption of biomass briquettes addresses multiple challenges, including deforestation, waste management, and limited access to modern energy, making them particularly relevant in both urban and rural settings (Kumar et al., 2023).

GLOBAL CONTEXT OF BIOMASS BRIQUETTES

Globally, biomass remains a significant contributor to the world's energy mix, accounting for approximately 10-14% of total primary energy consumption, with higher dependence in developing economies (IEA, 2023). Biomass briquetting technology has gained traction as a sustainable method to utilize agricultural and forestry residues, which are often left unutilized or burned inefficiently, contributing to greenhouse gas emissions. By converting biomass waste into densified briquettes, the energy efficiency of these resources is greatly enhanced, while the environmental impact is minimized.

Countries such as India, China, and Brazil have successfully implemented biomass briquetting projects, showcasing the global potential of this technology to meet energy demands. In India, for instance, the government has promoted biomass briquetting plants under renewable energy programs to reduce dependence on coal and curb emissions (Kumar et al., 2022). Similarly, China has invested heavily in briquetting technology as part of its renewable energy strategy, focusing on rural energy solutions that utilize crop residues. These initiatives demonstrate how briquettes can serve as a sustainable solution to energy deficits while addressing environmental concerns.

The transition to biomass briquettes aligns with global climate goals under the Paris Agreement and the United Nations Sustainable Development Goals (SDGs). By promoting clean energy sources, briquettes contribute to achieving SDG 7 (Affordable and Clean Energy), SDG 13 (Climate Action), and SDG 15 (Life on Land), among others. The replacement of firewood and charcoal with briquettes reduces deforestation rates, mitigates indoor air pollution, and supports rural livelihoods by creating opportunities for biomass collection and briquetting enterprises (UNEP, 2022).

The African Context

In sub-Saharan Africa, biomass fuels account for up to 80% of total energy consumption, with firewood and charcoal being the primary sources of cooking fuel (World Bank, 2021). This heavy reliance on traditional fuels has had severe consequences, including large-scale deforestation, desertification, and loss of biodiversity. The Food and Agriculture Organization (FAO) reports that Africa loses an estimated 4 million hectares of forest annually, with fuelwood harvesting being a major contributor to this loss (FAO, 2023). Moreover, the inefficient burning of biomass fuels in open stoves generates harmful emissions, posing health risks such as respiratory illnesses, especially among women and children in rural areas (Okello et al., 2023).

In response to these challenges, several African countries, including Kenya, Uganda, and Tanzania, have initiated programs to promote biomass briquettes as sustainable alternatives to firewood and charcoal. Kenya, for example, has developed community-driven briquetting enterprises that utilize agricultural residues such as sugarcane bagasse, maize cobs, and sawdust. These efforts have not only reduced deforestation but have also created economic opportunities for local communities (Ngugi et al., 2023). Similarly, in Uganda, biomass briquettes have gained traction in urban areas, where charcoal prices have risen sharply due to restrictions on logging and charcoal production. The adoption of briquettes is helping to alleviate energy poverty while protecting forest resources.

Despite these advancements, challenges such as lack of awareness, limited access to briquetting technology, and inadequate policy support continue to hinder the widespread adoption of biomass briquettes in Africa. Addressing these barriers through targeted interventions, such as capacity building, technology transfer, and policy frameworks, is crucial to unlocking the full potential of biomass briquettes as sustainable energy sources in the region (UNEP, 2022).

The Ethiopian Context

Ethiopia, like many African nations, faces a critical energy crisis, with traditional biomass fuels constituting approximately 92% of the country's total energy consumption (FDRE Ministry of Water and Energy, 2022). Firewood and charcoal are the primary sources of cooking fuel in both urban and rural areas, contributing to severe deforestation and environmental degradation. The Food and Agriculture Organization (FAO) estimates that Ethiopia loses over 92,000 hectares of forest annually due to fuelwood collection, agricultural expansion, and population pressure (FAO, 2023). A high dependence on traditional biomass fuels such as firewood, charcoal, and animal dung for cooking and heating contributes to widespread deforestation, soil degradation, and indoor air pollution. Over 85% of Ethiopia's energy needs are met through biomass, with rural households being the most affected by the limited access to modern energy alternatives (Bajwa et al., 2023). This reliance on biomass fuels not only strains the country's natural resources but also negatively impacts human health, particularly through the harmful effects of indoor smoke (Christoforou & Fokaides, 2023).

At the same time, Ethiopia generates vast quantities of agricultural residues and other organic waste, which remain largely underutilized. Ethiopia, like many countries in sub-Saharan Africa, faces significant challenges related to energy access and waste management. In Ethiopia's rural areas, where access to electricity and clean cooking options remains limited, biomass continues to dominate the energy landscape. The growing population in urban centers such as Bule Hora town adds to the complexity of Ethiopia's energy and waste management issues. Rapid urbanization has led to increased solid waste generation, with organic waste—primarily agricultural residues, food waste, and forest products—constituting a large portion of the total municipal waste (Wibowo et al., 2021). However, much of this waste is either discarded improperly, contributing to pollution, or burned in open pits, exacerbating air quality issues and greenhouse gas emissions (Bajwa et al., 2023). The accumulation of waste and its improper disposal have prompted the need for innovative solutions that can tackle both waste management and energy provision. Major agricultural by-products such as coffee husks, maize cobs, sawdust, and leaf litter hold significant potential for energy production. Ethiopia is the world's leading producer of coffee, generating approximately 300,000 tons of coffee husks annually, much of which is discarded or burned inefficiently (Berhe et al., 2023). Similarly, maize cultivation produces large amounts of maize cobs, which, if properly utilized, could serve as sustainable feedstocks for biomass briquetting.

Biomass briquettes have gained significant attention as a sustainable energy solution in Ethiopia, where traditional biomass fuels like firewood and charcoal dominate energy consumption. Recent studies have explored various aspects of biomass briquette production, utilization, and their socio-economic and environmental impacts in the Ethiopian context. Mekonnen and Tadesse (2022) conducted a comprehensive assessment of biomass briquette production and utilization in Ethiopia, highlighting the abundance of agricultural residues such as coffee husks, wheat straw, and teff straw as viable raw materials. Their study identified key challenges, including limited awareness, inadequate technology, and insufficient policy support, but emphasized the potential of briquettes to reduce deforestation and provide affordable energy. Similarly, Gebremariam and Assefa (2021) performed a techno-economic analysis of briquette production from agricultural residues, demonstrating that briquettes are cost-competitive with traditional fuels like charcoal and firewood, particularly in rural areas. Their findings also underscored the potential for job creation and income generation through small-scale briquette enterprises.

From an environmental perspective, Tesfaye and Kebede (2023) quantified the benefits of biomass briquettes in reducing deforestation and greenhouse gas emissions in Ethiopia. Their research highlighted the role of briquettes in mitigating climate change and improving air quality by replacing traditional biomass fuels. On the adoption front, Abebe and Worku (2022) investigated the barriers and enabling factors influencing the use of biomass briquettes in Ethiopia. They identified lack of awareness, high initial costs, and cultural preferences for traditional fuels as significant barriers, while recommending policy interventions and awareness campaigns to promote briquette adoption. Finally, Getachew and Lemma (2021) conducted a case study on small-scale briquette production in Addis Ababa, examining the production process, market potential, and socio-economic impacts on local communities. Their study concluded that briquettes have significant potential to address energy poverty in urban areas, particularly through decentralized production systems.

For example, Amare et al. (2022) demonstrated that biomass briquettes made from agricultural residues offer higher energy efficiency and lower emissions compared to traditional fuels. Furthermore, the production of briquettes can create economic opportunities for rural communities, particularly women and youth, by establishing small-scale briquetting enterprises. Biomass briquetting technology, which transforms organic waste into dense, compact fuel blocks, presents an opportunity to address these challenges. Briquettes made from agricultural residues, sawdust, and other organic feedstocks provide an alternative to traditional fuels such as firewood and charcoal, offering higher combustion efficiency and reduced environmental impact. Biomass briquetting also reduces the volume of waste, making it an effective tool for waste management (Andrić et al., 2023). Briquettes are particularly beneficial in Ethiopia, where the potential to produce fuel from agricultural and municipal waste is abundant. Studies in other regions, such as in Kenya and Uganda, have highlighted the economic viability and environmental benefits of briquetting, with briquettes being sold at a competitive price while providing clean, affordable energy for households (Sparrevik et al., 2014). In Ethiopia, efforts to promote biomass briquetting are gaining traction, supported by government institutions like the Ministry of Water, Irrigation, and Energy and initiatives in Addis Ababa aimed at expanding briquette production capacity (Christoforou & Fokaides, 2023). Despite these efforts, the widespread adoption of biomass briquetting remains limited due to challenges in technology, market access, and public awareness (Andrić et al., 2023). Collectively, these studies underscore the potential of biomass briquettes as a sustainable energy solution in Ethiopia, while also highlighting the need for targeted interventions to overcome barriers to adoption. The integration of policy support, technological advancements, and awareness campaigns could significantly enhance the role of briquettes in Ethiopia's energy transition.

BIOMASS BRIQUETTES & PRODUCTION

Types of Briquettes

Biomass briquettes are broadly categorized into two types:

1. **Non-Carbonized Briquettes**: These are produced by compacting raw biomass materials without undergoing any thermal treatment. They retain most of their original volatile components and are typically used for domestic heating and cooking (Kumar & Singh, 2022).
2. **Carbonized Briquettes**: These briquettes are produced from biomass that has been thermally processed (carbonized) in low-oxygen conditions, resulting in a charcoal-like product. Carbonized briquettes have higher energy density, lower smoke emissions, and are more suitable for urban markets (Ngugi et al., 2023).

Production of Briquettes

Bio-briquetting is a densification process of loose biomass or charred biomass by compaction in a die or mold, turning it into a solid fuel of different shape and size. In this technology, loose biomass is densified or changed into a solid form of particular shape and size through the application of pressure, heat and/or binding material. The densification of biomass or charred biomass (charcoal) can be achieved by any one of the following methods.

- Densification of charred biomass using a binder where the biomass is first charred to obtain charcoal powder, which is then treated and mixed with a binder and then densified.
- Direct densification of biomass materials where the processed or non-processed biomass is mixed with a binder and other additives is then densified using pressure or heat. This densified product can either be used in the form of non-charred briquettes or be carbonized to get charcoal briquettes.

RECENT DATA ON BIOMASS BRIQUETTE INITIATIVES IN ETHIOPIA AND SIMILAR REGIONS

Biomass briquette initiatives in Ethiopia and similar regions have shown both successes and challenges, reflecting the potential and limitations of this sustainable energy solution. Below is a review of recent initiatives, their outcomes, and lessons learned:

Ethiopia: Small-Scale Briquette Production in Addis Ababa

- Initiative: A 2021 project in Addis Ababa focused on small-scale briquette production using agricultural residues like coffee husks and sawdust.
- Successes:
    - The initiative created income-generating opportunities for women and youth, with over 200 small-scale producers trained.

- o Briquettes were found to burn longer and produce less smoke compared to traditional charcoal, improving indoor air quality.
- o The project reduced reliance on firewood, contributing to a 15% decrease in deforestation in the surrounding areas.

- Challenges:
    - o High initial costs for briquetting machines and limited access to financing hindered scalability.
    - o Cultural preferences for traditional fuels and lack of awareness about briquettes slowed adoption.
    - o Inconsistent supply chains for raw materials (e.g., agricultural residues) disrupted production.

Kenya: Community-Based Briquette Production

- Initiative: A 2022 project in Kenya promoted community-based briquette production using water hyacinth and sugarcane bagasse.
- Successes:
    - o The initiative successfully reduced water hyacinth infestation in Lake Victoria while providing raw materials for briquettes.
    - o Over 500 households adopted briquettes as their primary cooking fuel, reducing their expenditure on charcoal by 30%.
    - o The project created over 100 jobs, particularly for women and marginalized groups.
- Challenges:
    - o Limited market access and competition from cheap charcoal made it difficult to sustain briquette sales.
    - o Lack of technical expertise in maintaining briquetting machines led to frequent breakdowns.
    - o Seasonal availability of raw materials (e.g., sugarcane bagasse) affected production consistency.

Uganda: NGO-Led Briquette Initiative

- Initiative: A 2020 project by an NGO in Uganda focused on producing briquettes from municipal waste and agricultural residues.
- Successes:
    - o The project diverted significant amounts of organic waste from landfills, reducing environmental pollution.
    - o Briquettes were sold at 20% lower cost than charcoal, making them affordable for low-income households.
    - o The initiative trained over 300 individuals in briquette production, fostering local entrepreneurship.
- Challenges:
    - o Poor quality briquettes due to inadequate drying and compression techniques reduced consumer confidence.
    - o Limited government support and policy frameworks hindered scaling up the initiative.
    - o High transportation costs for raw materials and finished products affected profitability.

Tanzania: Private Sector-Led Briquette Production

- Initiative: A private company in Tanzania launched a briquette production facility in 2023 using rice husks and coconut shells.
- Successes:
  - o The company supplied briquettes to schools, hospitals, and restaurants, reducing their reliance on firewood and charcoal.
  - o The initiative created a circular economy by utilizing agricultural waste, reducing environmental degradation.
  - o Briquettes were marketed as a cleaner alternative, with a 25% reduction in carbon emissions reported.
- Challenges:
  - o High production costs and competition from subsidized fossil fuels limited market penetration.
  - o Lack of consumer awareness about the benefits of briquettes slowed adoption rates.
  - o Inconsistent power supply for operating briquetting machines disrupted production.

Ethiopia: National Biomass Energy Strategy (2023)

- Initiative: The Ethiopian government launched a national strategy to promote biomass briquettes as part of its renewable energy agenda.
- Successes:
  - o The strategy included subsidies for briquetting machines and training programs for entrepreneurs.
  - o Over 1,000 small-scale briquette producers were registered, creating jobs and reducing unemployment.
  - o The initiative aligned with Ethiopia's Climate-Resilient Green Economy (CRGE) strategy, contributing to reduced deforestation.
- Challenges:
  - o Implementation delays and bureaucratic hurdles slowed the rollout of the strategy.
  - o Limited access to financing for small-scale producers remained a significant barrier.
  - o Inadequate infrastructure for distribution and marketing of briquettes limited their reach.

Table 1. Comparison of Successes and Challenges

Region/Initiative	Successes	Challenges
Ethiopia (Addis Ababa)	Income generation, reduced deforestation, improved air quality	High initial costs, cultural preferences, inconsistent raw material supply
Kenya (Lake Victoria)	Job creation, reduced charcoal expenditure, environmental cleanup	Market competition, technical issues, seasonal raw material availability
Uganda (NGO-Led)	Waste diversion, affordable briquettes, entrepreneurship training	Poor product quality, lack of government support, high transportation costs

continued on following page

Table 1. Continued

Region/Initiative	Successes	Challenges
Tanzania (Private Sector)	Circular economy, reduced emissions, institutional adoption	High production costs, low consumer awareness, power supply issues
Ethiopia (National Strategy)	Job creation, policy alignment, reduced deforestation	Implementation delays, financing gaps, inadequate distribution infrastructure

These initiatives demonstrate that while biomass briquettes hold significant promise for addressing energy poverty and environmental degradation, overcoming challenges requires a holistic approach involving policy support, community engagement, and investment in technology and infrastructure.

FINANCIAL MODELS

Financial models are critical for understanding the economic viability of biomass briquette production and adoption, particularly in developing countries like Ethiopia and other African nations. These models help assess costs, revenues, profitability, and the potential for scaling up production. Below, are some key financial models and considerations for briquette production and adoption, along with examples from Ethiopia and similar regions.

Cost-Benefit Analysis (CBA)

A **Cost-Benefit Analysis** evaluates the economic feasibility of briquette production by comparing the total costs of production to the benefits (revenues and environmental/social benefits).

Key Components:

- Capital Costs: Initial investment in briquetting machines, drying equipment, and storage facilities.
  - Example: In Ethiopia, a small-scale briquetting machine costs 500–2,000.
- Operational Costs: Raw materials, labor, energy, maintenance, and transportation.
  - Example: Raw materials like coffee husks or sawdust may cost 10–20 per ton.
- Revenue Streams: Sales of briquettes to households, institutions, and businesses.
  - Example: In Kenya, briquettes sell for 0.20–0.30 per kilogram.
- Environmental and Social Benefits: Reduced deforestation, lower greenhouse gas emissions, and improved health outcomes.

Example:

- A 2021 study in Ethiopia found that the net present value (NPV) of a small-scale briquette production facility was positive over a 5-year period, with a payback period of 2–3 years.
- The study also highlighted that briquettes could reduce household energy costs by 20–30% compared to charcoal.

Break-Even Analysis

A Break-Even Analysis determines the production volume required to cover costs and start generating profits.

Key Components:

- Fixed Costs: Capital investment, rent, and salaries.
- Variable Costs: Raw materials, energy, and transportation.
- Selling Price: Price per kilogram of briquettes.

Formula:

Break-Even Quantity=Fixed Costs / Selling Price per Unit−Variable Cost per Unit

Example:

Briquette production facility in Uganda:

- Fixed Costs: $5,000 (initial investment in machinery, setup, etc.)
- Variable Costs: $0.10 per kilogram (cost of raw materials, labor, etc.)
- Selling Price: $0.25 per kilogram (price at which briquettes are sold)

Step 1: Calculate the Contribution Margin per Unit

The contribution margin is the amount each unit sold contributes to covering the fixed costs. It's calculated as:

Contribution Margin=Selling Price per Unit−Variable Cost per Unit

Contribution Margin=$0.25−$0.10=$0.15 per kilogram

Step 2: Calculate the Break-Even Quantity

divide the fixed costs by the contribution margin to find the break-even quantity:

Break-Even Quantity =$5000 / $0.15 = 33,333 kilograms

Accordingly the facility needs to produce and sell 33,333 kilograms of briquettes to cover its fixed costs of 5000, and variable cost of 0.10 per kilogram

Financial Projections and Cash Flow Analysis

Financial projections estimate revenues, expenses, and cash flows over a specific period (e.g., 5 years). This model is essential for securing loans or investments.

Key Components:

- Revenue Projections: Based on market demand and selling price.
- Expense Projections: Raw materials, labor, maintenance, and marketing.
- Cash Flow: Monthly or annual inflows and outflows.

Example:

- A 2022 study in Tanzania projected that a medium-scale briquette production facility could generate $20,000 in annual revenue with a 15% profit margin after covering operational costs.

Social Cost-Benefit Analysis (SCBA)

A Social Cost-Benefit Analysis evaluates the broader societal impacts of briquette production, including environmental and health benefits.

Key Components:

- Environmental Benefits: Reduced deforestation and lower carbon emissions.
- Health Benefits: Reduced indoor air pollution and associated healthcare costs.
- Economic Benefits: Job creation and income generation.

Example:

- A 2023 study in Ethiopia estimated that replacing charcoal with briquettes could reduce carbon emissions by 1.5 tons per household annually and save 50–50–100 per household in healthcare costs related to respiratory illnesses.

Subsidy and Financing Models

Subsidies and financing mechanisms can make briquette production more affordable and accessible.

Key Components:

- Government Subsidies: Grants or tax incentives for briquette producers.
  - Example: Ethiopia's Climate-Resilient Green Economy (CRGE) strategy includes subsidies for renewable energy projects.
- Microfinance and Loans: Small loans for entrepreneurs to purchase equipment.
  - Example: In Kenya, microfinance institutions offer loans with interest rates as low as 10% per annum for briquette producers.
- Carbon Credits: Revenue from carbon offset programs.
  - Example: A briquette project in Uganda earned $50,000 in carbon credits over 3 years.

Pay-As-You-Go (PAYG) Models

The Pay-As-You-Go model allows consumers to purchase briquettes in small quantities, making them more affordable for low-income households.

Key Components:

- Flexible Payment Options: Weekly or monthly payments.
- Mobile Money Integration: Use of mobile payment platforms like M-Pesa in Kenya.

Example:

- In Rwanda, a briquette company adopted a PAYG model, increasing adoption rates by 30% among low-income households.

PUBLIC-PRIVATE PARTNERSHIP (PPP) MODELS

Public-Private Partnerships involve collaboration between governments, private companies, and NGOs to scale up briquette production and adoption.

Key Components:

- Government Support: Policy frameworks, subsidies, and infrastructure development.
- Private Sector Investment: Funding for production facilities and marketing.
- NGO Involvement: Training and community outreach.

Example:

- In Ethiopia, a PPP project between the government and a private company increased briquette production capacity by 50% and created over 500 jobs.

Franchise Models

Franchise models allow entrepreneurs to operate briquette production units under a larger brand, reducing risks and increasing scalability.

Key Components:

- Franchise Fees: Initial fee and ongoing royalties.
- Training and Support: Technical and marketing support from the franchisor.

Example:

- A franchise model in Kenya expanded briquette production to 10 new locations within 2 years, increasing market penetration by 25%.

Case Study: Financial Model for Ethiopia

Scenario:

- Production Capacity: 10 tons per month.
- Capital Costs: $5,000 (machine and setup).
- Operational Costs: $1,000 per month (raw materials, labor, energy).
- Selling Price: $0.20 per kilogram.
- Revenue: $2,000 per month.
- Profit: $1,000 per month.
- Payback Period: 5 months.

Environmental and Social Benefits:

- Carbon Savings: 15 tons of CO_2 per month.
- Health Savings: $500 per month in reduced healthcare costs.

Despite the promising potential, the adoption of biomass briquettes in Ethiopia remains limited due to factors such as lack of awareness, high initial investment costs, and inadequate access to briquetting technologies. To overcome these challenges, integrated strategies that include capacity building, policy incentives, and community engagement are essential. Promoting research and development to optimize the production and performance of biomass briquettes is also critical to enhancing their feasibility and acceptability in Ethiopian households. Hence this study focuses on assessing the potential for biomass briquettes in Bule Hora town, evaluating the physical and combustion properties of locally available feed stocks, and exploring community perceptions regarding the adoption of briquetting technology as a sustainable energy solution.

MATERIALS AND METHODS

Study Area

The study was conducted in Bule Hora town, located in the West Guji Zone of Oromia Regional State, Ethiopia. Bule Hora is situated at an altitude of 1716 meters above sea level, with geographic coordinates of 5°35′N, 38°15′E. The town, with a population of over 160,000, predominantly relies on traditional biomass fuels for cooking, leading to significant deforestation, air pollution, and environmental degradation. The surrounding agricultural zones offer an abundant supply of organic residues such as coffee husks, maize cobs, sawdust, and leaf litter, which make the area an ideal site for biomass briquette production. As the town faces growing energy demands, the study aims to assess alternative energy solutions that could reduce reliance on firewood and charcoal while utilizing locally available agricultural residues for renewable energy production such as biomass briquettes.

Sample size determination

To ensure the study is representative and enhances its reliability and validity, a two-stage sampling procedure was employed to identify sample respondents from the total population (households) of Bule Hora town as shown in table 2. In the first stage, all three kebeles in the town (01, 02, and 03) were included through purposive selection, as the town consists only of these three kebeles. In the second stage, stratified proportional sampling was applied to determine the number of respondents from each kebele, based on the proportion of households in each. Finally, respondents for the interviews were selected randomly to ensure unbiased representation.

Table 2. Sample size of the respondents

S.No.	Kebeles	Number of households	Sample size	
			Number	%
1	Kebele 01	3177	169	46.4
2	Kebele 02	1866	99	27.2
3	Kebele 03	1802	96	26.4
	Total	6845	364	100

To assess the market potential of biomass briquettes, the survey included restaurants and hotels. Specifically, 50 restaurants and hotels in Bule Hora town, along with cafeterias at Bule Hora University, were randomly selected, bringing the total sample size for this study to 414.

Data Collection Methods

The required data for this research were gathered from both primary and secondary sources, including the Agriculture and Municipality offices. To quantify the availability of biomass feedstocks and conduct a market analysis, a scientific survey was carried out using structured questionnaires.

Feedstock Collection

Organic agricultural residues (feed stocks) were selected for biomass briquette production, based on their availability, low cost. Saw dust was collected from local sawmills. Other feed stocks such as coffee husk, maize cob, and leaf litter were collected from agricultural fields near nearby the Bule Hora town. Uses waste paper (binder in this study) was collected from Bule hora University offices.

Feed stock processing

Drying and Screening

The collected feedstocks were first sun-dried for 5–7 days to reduce the moisture content, which is essential for effective carbonization and ensuring the quality of the final briquettes. Moisture content was reduced to below 20% to enhance the efficiency of carbonization and improve briquette quality. After drying, the materials were ground using an electric grinder to achieve a uniform particle size of less than 0.75 mm. This ensured better compaction during briquette formation and improved combustion efficiency. Uniform particle size also facilitates a more consistent combustion process, contributing to better performance in practical applications such as cooking and industrial energy generation.

Preparation of sample proportions & moulds of briquettes

Twelve samples with varying proportions of feed stocks were prepared and studied, as shown in Table 3. Water was added as a binding medium to ensure proper mixing of the raw materials. For each type of biomass briquette, a total mixture weight of 1 kg was selected, tailored to match the size of the briquette mold.

Table 3. Feedstock Proportions

Sample	Saw Dust (%)	Wastepaper (%)	Leaf Litter (%)	Maize Cob (%)	Coffee Husk (%)
B1	25	25	25	25	-
B2	50	25	25	-	-
B3	75	25	-	-	-
B4	-	25	50	25	-
B5	-	25	75	-	-
B6	-	25	-	50	25
B7	-	25	-	75	-
B8	25	25	-	-	50
B9	-	25	-	-	75
B10	25	25	25	-	25
B11	-	25	25	25	25
B12	25	25	-	25	25

Briquette Production

The process of briquette production involved several stages, including binder preparation, mixing, compaction, and drying, as outlined below:

Binder Preparation:

Waste paper was used as a binder for the briquette formulations. The paper was soaked in water for 24 hours, and then blended into a slurry using a mechanical blender. This binder was used at a fixed ratio of 25% by weight in all briquette formulations to ensure sufficient binding strength and cohesion. The use of waste paper as a binder is an effective and low-cost waste management solution that reduces waste while providing the necessary binding properties for the briquettes.

Mixing and Compaction:

The ground feed stocks and binder slurry were manually mixed to ensure uniform distribution of the binder throughout the feedstock. The mixture was then loaded into cylindrical molds with a diameter of 40 mm and a height of 100 mm. Using a hydraulic press, the mixture was compacted at a pressure of 5 MPa for 2–3 minutes per briquette. This compaction process helped to achieve consistent density and mechanical stability in the briquettes. The compaction pressure was optimized to balance the need for high-density briquettes and the risk of cracking or breaking during handling.

Drying:

After compaction, the briquettes were dried in a laboratory oven at 60°C for 24 hours to remove any residual moisture. Following the initial drying in the oven, the briquettes were sun-dried for an additional week. The sun-drying process ensured the final moisture content of the briquettes was low, which is essential for optimal combustion properties. The drying process also improves the durability of the briquettes, reducing the likelihood of degradation or breaking during storage or transport.

Carbonization:

The dried feed stocks were subjected to carbonization in a muffle furnace at a temperature of 400°C for 45 minutes, following established protocols (Wibowo et al., 2021; Bajwa et al., 2023). Carbonization involved partial pyrolysis of the feed stocks, which significantly increases the fixed carbon content and calorific value of the biomass materials. The process also removes volatile compounds, thus enhancing the briquettes' energy efficiency. After carbonization, the materials were cooled, pulverized to a fine powder, and sieved to achieve uniformity in particle size, ensuring consistent performance in subsequent briquette production.

Characterization of Briquettes

The physical and combustion properties of the produced briquettes were evaluated according to the procedures described by Mahadeo (2014).

Physical Properties:

- Density: The density of the briquettes was determined by calculating the mass-to-volume ratio of the dried briquettes. Higher density briquettes are preferred for their better energy efficiency, improved combustion, and better storage and transport properties.
- Porosity Index: The porosity of the briquettes was measured by immersing the briquettes in water and calculating the weight difference before and after immersion. This provided an indication of the briquettes' water resistance, which is a critical factor affecting their durability and performance, especially in humid environments.
- Shatter Resistance: Shatter resistance was assessed by dropping the briquettes from a height of 2 meters onto a concrete floor five times. The percentage of briquette weight retained after the drops was recorded to evaluate their mechanical stability and durability. Briquettes with higher shatter resistance are more likely to withstand handling and transportation without breaking apart.
- Compression Strength: The compression strength of the briquettes was tested using a mechanical compression tester to determine the pressure required to deform or break the briquettes. This property is critical for ensuring the briquettes remain intact during use in stoves or boilers, ensuring stable combustion over time.

Combustion Properties (Proximate Analysis)

The proximate analysis involved measuring key parameters that affect the combustion behavior of the briquettes:

- Moisture Content: The moisture content was determined by weighing the briquettes before and after drying at 105°C for 24 hours, following ASTM D3173. High moisture content can significantly reduce combustion efficiency, as energy is used to evaporate water before the fuel can burn effectively.
- Ash Content: Ash content was measured by burning a fixed quantity of briquettes in a muffle furnace at 900°C for 30 minutes, following ASTM D3174. A lower ash content is desirable, as it indicates a higher proportion of combustible material.
- Volatile Matter: Volatile matter was determined by heating the briquettes at 550°C and measuring the weight loss (ASTM D3175). A high volatile matter content typically leads to faster combustion but can result in reduced energy efficiency and higher emissions.
- Fixed Carbon: Fixed carbon content was calculated by subtracting volatile matter and ash content from 100%. Briquettes with higher fixed carbon content tend to have better combustion efficiency and higher energy output.
- Calorific Value: The calorific value of the briquettes was measured using a bomb calorimeter (Parr 6200), which provides an accurate assessment of the energy output per unit mass of briquette. A higher calorific value indicates better energy efficiency and a more desirable fuel source for cooking or industrial use.

Experimental Design for Response Surface Methodology (RSM)

Design for Response Surface Methodology include following steps:

- Identifying Factors and Responses

Following independent variables (factors) were considered for the RSM experiment:

- Compression pressure (Factor A)
- Binder content (Factor B)
- Carbonization temperature (Factor C)
- Moisture content (Factor D)

The dependent variables (responses) considered for experiment include:

- Density (g/cm^3)
- Calorific value (kcal/kg)
- Compression strength (kg/cm^2)
- Shatter resistance (%)

Regression Model:

The regression model would include factorial, quadratic, and interaction terms to capture the relationships. This involves using regression analysis to calculate the coefficients for the model. The general form of the model is:

$$Y=\beta 0+\beta 1A+\beta 2B+\beta 3C+\beta 4D+\beta 11A2+\beta 22B2+\beta 33C2+\beta 44D2+\beta 12AB+\beta 13AC+\beta 14AD+\beta 23BC+\beta 24BD+\beta 34CD$$

Optimization Process:

Using software Design Expert we run the regression analysis to estimate the coefficients for the regression model, calculate the optimal values of the independent variables (e.g., compression pressure, binder content), and visualize the response surfaces. Experiments were designed using a central composite design (CCD) as shown in table 4, which includes a set of factorial points, center points, and star points (to account for nonlinearities and interactions). A 2-level factorial design with 3 factors and 2 levels for each factor was considered.

Table 4. Matrix of central composite design (CCD)

Factor	Low Level (−1)	Center Point (0)	High Level (+1)
Compression Pressure (MPa)	5	10	15
Binder Content (%)	5	10	15
Carbonization Temperature (°C)	300	400	500
Moisture Content (%)	10	15	20

Community Perception Studies

A structured questionnaire was developed to collect data on fuel usage patterns, awareness of biomass briquetting, and the willingness to adopt briquettes as an alternative fuel. The survey targeted various community groups to gather a broad perspective on fuel use and the potential for briquette adoption. The survey was administered as follows:

- Households: A stratified random sampling method was used to select 364 households across three urban kebeles in Bule Hora. This sampling method ensured the inclusion of diverse socioeconomic backgrounds and fuel use practices in the survey.
- Restaurants and Cafeterias: Fifty establishments, including university cafeterias, were surveyed to assess the institutional demand for briquettes. These establishments were selected to evaluate the potential for large-scale adoption of biomass briquettes in the commercial sector, where energy consumption is significant.
- Key Variables: The survey collected data on several key variables, including economic considerations (e.g., cost of traditional fuels vs. briquettes), health impacts of using traditional biomass fuels, and the perceived advantages of briquettes (e.g., cost-effectiveness, environmental benefits, reduced health risks from smoke exposure). This information was essential for understanding community attitudes towards briquettes and identifying barriers to adoption.

Data Analysis

The collected data were analyzed using SPSS (Version 21) and Microsoft Excel. The following statistical methods were employed:

- Descriptive Statistics: Descriptive statistics were used to summarize survey responses and briquette properties, providing an overview of the community's fuel usage patterns and briquette performance.
- ANOVA and Regression Analysis: These tests were used to evaluate the effects of feedstock composition, binder ratio, and compaction pressure on briquette performance. ANOVA was used to assess whether differences between different feedstock compositions or briquette types were statistically significant, while regression analysis helped identify the relationships between these variables and briquette quality.
- Correlation Analysis: Correlation analysis was used to examine relationships between community perception variables (e.g., willingness to adopt briquettes) and practical considerations such as

cost, availability, and perceived performance. This analysis provided insights into the factors that influence adoption decisions and highlighted areas where targeted interventions could facilitate greater uptake of briquettes.

RESULTS AND DISCUSSION

Proximate analysis of raw biomass feed stocks

Selected raw biomass feed stocks were subjected to proximate analysis before they were mixed in different proportions with the binder. Results were depicted as shown in table 5.

Volatile Matter (VM):

Leaf litter exhibits the highest volatile matter content (80.2%), indicating a greater propensity for combustion initiation. In contrast, maize cob has the lowest VM (58.26%), suggesting a more controlled combustion process. Similar results were noted by Mukherjee et al. (2023), emphasizing the importance of volatile matter in determining the ignition characteristics of feed stocks.

Moisture Content (MC):

Maize cob has the lowest moisture content (4.52%), which is advantageous for combustion efficiency, as lower moisture reduces energy loss during burning. Leaf litter shows the highest MC (11.47%), potentially leading to reduced combustion efficiency. Adeyemi and Johnson (2023) highlighted that lower moisture content directly correlates with higher energy efficiency in briquettes.

Ash Content (AC):

Maize cob presents the lowest ash content (2.61%), indicating minimal residual material post-combustion and suggesting cleaner burning properties. Leaf litter has the highest AC (12.8%), which could lead to increased residue and potential maintenance challenges in combustion systems, as supported by Kumar et al. (2022).

Fixed Carbon (FC):

Maize cob has the highest fixed carbon content (16.03%), correlating with a higher energy yield during combustion. Leaf litter shows the lowest FC (10.1%), indicating less energy-dense material. Chaudhary et al. (2024) observed that feedstocks with higher fixed carbon tend to have prolonged burning times.

Calorific Value (CV):

Maize cob exhibits the highest calorific value (1820.54 kcal/kg), making it the most energy-rich among the analyzed feedstocks. Leaf litter has the lowest CV (1214.21 kcal/kg), suggesting lower energy output upon combustion. Oyelaran et al. (2022) noted similar trends in the calorific values of agro-waste materials, emphasizing their suitability for bioenergy applications.

Table 5. Proximate analysis of raw biomass feed stocks

S.No.	Type of Feedstock	Volatile matter (%)	Moisture content (%)	Ash content (%)	Fixed carbon (%)	Calorific value (kcal/kg)
1	Coffee husk	64.23±0.27	5.78±0.12	5.78±0.13	15.6±0.23	1446.08±0.72
2	Saw dust	71.9±0.21	8.23±0.17	10.67±0.23	13.7±0.37	1385.21±0.43
3	Maize cob	58.26±0.24	4.52±0.13	2.61±0.70	16.03±0.22	1820.54±0.37
4	Leaf litter	80.2±0.45	11.47±0.15	12.8±0.28	10.1±0.21	1214.21±0.61
5	Waste paper	72.33±0.31	9.87±0.14	11.23±0.41	11.7±0.25	1248.30 ±0.24

Physical Properties of Biomass Briquettes

The physical properties of both non-carbonized and carbonized biomass briquettes were examined (Table 6) to assess their suitability for practical applications such as cooking and industrial energy production. The findings show that carbonization significantly improves various physical properties of biomass briquettes, making them more efficient and durable.

Table 6. Physical Properties of Biomass Briquettes

S.No.	Briquette Code*	Density (g/cm^3)	Porosity Index (%)	Shatter Resistance (%)	Compression Strength (kg/cm^2)
1	NCB1	0.39	46.78	91.18	54.23
	CB1	0.43	42.71	95.57	56.30
2	NCB2	0.41	44.78	89.75	52.22
	CB2	0.42	43.71	96.23	60.75
3	NCB3	0.42	43.98	93.54	62.78
	CB3	0.48	38.93	96.55	64.53
4	NCB4	0.30	54.76	90.12	22.88
	CB4	0.61	33.92	96.35	69.43

continued on following page

Table 6. Continued

S.No.	Briquette Code*	Density (g/cm³)	Porosity Index (%)	Shatter Resistance (%)	Compression Strength (kg/cm²)
5	NCB5	0.47	39.78	94.81	35.70
	CB5	0.49	35.46	97.46	65.96
6	NCB6	0.36	48.76	88.51	25.94
	CB6	0.37	48.92	95.57	28.96
7	NCB7	0.32	52.39	91.52	57.28
	CB7	0.44	41.72	91.78	62.36
8	NCB8	0.43	40.51	95.70	27.88
	CB8	0.52	34.73	97.38	67.56
9	NCB9	0.47	38.97	93.09	59.34
	CB9	0.56	39.73	97.84	66.40
10	NCB10	0.34	49.57	89.51	31.73
	CB10	0.41	44.06	92.76	38.74
11	NCB11	0.35	49.11	88.73	30.91
	CB11	0.38	47.06	94.21	51.20
12	NCB12	0.38	47.76	89.75	34.76
	CB12	0.40	46.51	94.21	58.30

*NCB – Non-carbonized briquettes; CB - Carbonized briquettes

Density:

The density of briquettes is a critical parameter for their combustion efficiency, storage, and transportation. Higher density briquettes have a greater energy content per unit volume, thus improving energy efficiency and reducing storage costs. In this study, carbonized briquettes (CB4) exhibited the highest density at 0.61 g/cm^3, compared to the non-carbonized briquettes (NCB5) with a density of 0.47 g/cm^3. These results align with findings from previous studies (Wibowo et al., 2021; Bajwa et al., 2023) that carbonization increases the density of biomass by reducing the volume of volatile matter and enhancing the content of fixed carbon. This leads to briquettes that burn more efficiently, as carbonized biomass typically retains more of its energy potential during combustion. Higher density also improves the briquettes' resistance to breakage and enhances their storage efficiency by reducing the volume occupied per unit mass.

Porosity Index:

Porosity is an important property as it directly influences the briquette's water absorption capacity, which can affect its durability and combustion performance. The results indicated that non-carbonized briquettes exhibited higher porosity, with NCB4 showing the highest value of 54.76%, whereas carbonized briquettes had lower porosity, with CB4 showing 48.92%. The low porosity in carbonized briquettes can be attributed to the compaction process during carbonization, which reduces the space between particles and contributes to a more uniform and dense structure. Similar trends were observed by Sparrevik et al. (2014), who found that carbonized biomass briquettes tend to have lower porosity and are thus more durable. The reduced porosity in carbonized briquettes also results in reduced water absorption, which contributes to better performance in combustion due to lower moisture content.

Shatter Resistance:

Shatter resistance measures the ability of briquettes to withstand mechanical impacts during handling and transportation. This property is crucial for assessing briquette durability. The results showed that NCB8 had the highest shatter resistance at 95.70%, indicating that these briquettes were more durable compared to other formulations. On the other hand, carbonized briquettes like CB9 exhibited high shatter resistance (97.84%) than non-carbonized ones which is a desirable characteristic for briquettes intended for widespread use. The higher shatter resistance of these briquettes suggests they are less likely to break or degrade during handling, which is essential for ensuring the fuel remains intact for combustion (Christoforou & Fokaides, 2023). The improved durability of carbonized briquettes also ensures less fuel loss during transportation and storage, further enhancing their practical appeal.

Compression Strength:

Compression strength is an important indicator of a briquette's ability to resist breakage when subjected to external pressure. Carbonized briquettes demonstrated significantly higher compression strength than non-carbonized briquettes. For instance, CB4 had the highest compression strength at 69.43kg/cm^2, while NCB3 with 62.78 kg/cm^2. This indicates that carbonized briquettes can withstand higher pressures and are more stable during use, reducing the likelihood of breaking apart in high-stress environments such as stoves or boilers (Andrić et al., 2023). This increased compression strength can be attributed to the structural changes during the carbonization process, which make the briquettes more compact and durable (Sparrevik et al., 2014). The higher compression strength of carbonized briquettes also suggests better performance under heavy industrial usage, making them more suitable for applications requiring stable fuel performance over extended periods.

Combustion Properties

Combustion properties are essential for assessing the energy efficiency and sustainability of briquettes. The key parameters considered in this study include moisture content, ash content, volatile matter, fixed carbon, and calorific value and results were depicted in table 7.

Table 7. Table showing combustion properties of biomass briquettes

S.No.	Briquette Code*	Volatile Matter (%)	Moisture Content (%)	Ash Content (%)	Fixed Carbon (%)	Calorific Value (kcal/kg)
1	NCB1	76.72	9.11	4.21	17.12	1892.65
	CB1	26.57	3.21	1.45	79.23	4253.02
2	NCB2	75.67	8.48	4.16	17.28	1980.23
	CB2	27.98	3.53	1.48	77.15	4212.43
3	NCB3	75.34	8.34	3.98	18.49	2042.33
	CB3	23.68	2.86	1.26	81.23	4427.35
4	NCB4	79.89	11.34	4.81	15.49	1126.76
	CB4	13.49	2.65	1.13	82.71	4670.92
5	NCB5	74.37	8.12	3.87	19.18	2432.12
	CB5	21.54	2.80	1.22	81.78	4496.60
6	NCB6	77.47	9.26	4.34	16.78	1657.76
	CB6	29.34	5.21	1.76	68.21	3949.75
7	NCB7	79.68	10.21	4.75	15.67	1378.90
	CB7	25.56	3.12	1.38	79.56	4350.93
8	NCB8	74.86	8.29	3.89	18.67	2276.34
	CB8	18.65	2.73	1.17	82.26	4618.39
9	NCB9	71.98	7.86	3.81	19.26	2463.98
	CB9	24.76	3.11	1.34	80.56	4405.86
10	NCB10	79.54	9.51	4.65	16.26	1478.65
	CB10	28.67	3.62	1.57	76.99	4171.36
11	NCB11	79.12	9.34	4.58	16.31	1587.23
	CB11	29.26	3.98	1.72	69.21	4026.16
12	NCB12	77.23	9.22	4.29	16.87	1673.09
	CB12	29.12	3.76	1.65	71.74	4161.82

*NCB – Non-carbonized briquettes; CB - Carbonized briquettes

Moisture Content:

Moisture content is a critical factor that affects the combustion efficiency of biomass fuels. Briquettes with high moisture content require more energy to evaporate the water before they can effectively burn, leading to lower efficiency. The carbonized briquettes (CB6) exhibited significantly lower moisture content (5.21%) compared to the non-carbonized briquettes which is 11.34% (NCB4). This difference is consistent with other studies, such as those by Bajwa et al. (2023), which indicate that carbonization removes a significant amount of water content from biomass, resulting in improved combustion efficiency and quicker ignition. Lower moisture content in carbonized briquettes contributes to their higher calorific value, as they release energy more effectively during combustion (Sparrevik et al., 2014). Additionally, the lower moisture content reduces the potential for smoke generation and the emission of harmful gases, which makes carbonized briquettes a cleaner alternative to traditional biomass fuels.

Ash Content:

The ash content of biomass briquettes indicates the amount of non-combustible material left after combustion. A lower ash content is preferable because it leads to less residue and more complete combustion. Carbonized briquettes (CB6) had a lower ash content (1.76%) compared to non-carbonized briquettes ash content 4.81% (NCB4), which is a typical result of the carbonization process, which removes a larger portion of the incombustible materials from the biomass feedstocks (Christoforou & Fokaides, 2023). This reduction in ash content improves the overall energy efficiency of the briquettes, as less heat is wasted on burning non-combustible materials. Additionally, the lower ash content in carbonized briquettes reduces the frequency of cleaning stoves or boilers, which is a significant benefit for end-users.

Volatile Matter:

Volatile matter represents the combustible gases that are released during the heating of biomass. Non-carbonized briquettes (NCB4) showed a higher volatile matter content (79.89%) compared to carbonized briquettes at the tune of 29.34% (CB6). This higher volatile matter in non-carbonized briquettes leads to faster combustion, but also results in lower energy efficiency, as more heat is released as gas rather than being used for heating. In contrast, carbonized briquettes, which have lower volatile matter, provide more consistent and efficient heat output over time (Wibowo et al., 2021; Christoforou & Fokaides, 2023). The reduced volatile matter in carbonized briquettes means they have a more stable burning pattern, resulting in improved thermal efficiency during use. Furthermore, the lower volatile matter leads to less smoke and particulate emissions, which enhances the environmental and health benefits of using carbonized briquettes.

Fixed Carbon:

Fixed carbon is the carbon content that remains after the volatile components of biomass have been released during heating. It is a key factor determining the briquette's energy content. Carbonized briquettes(CB4) had a significantly higher fixed carbon content (82.71%) than non-carbonized briquettes (19.26%) for NCB9. This high fixed carbon content directly contributes to the higher calorific value of carbonized briquettes, as carbon is the main source of energy in the combustion process. This finding supports previous studies, which show that carbonized biomass typically has higher fixed carbon and calorific value, making it a more efficient fuel source (Sparrevik et al., 2014; Bajwa et al., 2023). The higher fixed carbon content also means that carbonized briquettes are more energy-dense, resulting in longer burn times and higher heat output per unit mass compared to non-carbonized briquettes.

Calorific Value:

The calorific value is the amount of energy released when a unit mass of fuel is completely burned. Carbonized briquettes (CB4) exhibited a calorific value of 4670.92kcal/kg, much higher than the non-carbonized briquettes (NCB9) which had a calorific value of 2463.98±0.16 kcal/kg. This substantial difference is primarily due to the higher fixed carbon content and lower moisture and volatile matter in the carbonized briquettes, which results in a more energy-dense product. The calorific value of carbonized briquettes is consistent with other studies on biomass fuel, which have shown that carbonization improves energy content by concentrating carbon and removing impurities (Christoforou & Fokaides, 2023). The higher calorific value of carbonized briquettes makes them more efficient as an energy source, requiring less material to generate the same amount of heat compared to non-carbonized briquettes.

Results of Statistical Analysis

ANOVA Results

Table 8. Table showing results of ANOVA

Property	*Group*	*Dependent Variable*	*Predictors*	*F-statistic*	*p-value*	*Significance*
Physical Properties	Non-carbonized Briquettes	Density (g/cm^3)	Compression Strength (kg/cm^2), Shatter Resistance (%), Porosity Index (%)	19.556	0.000	Significant ($p < 0.01$)
Combustion Properties	Carbonized Briquettes	Calorific Value (kcal/kg)	Density (g/cm^3)	90.541	0.000	Significant ($p < 0.01$)

Physical Properties of Non-Carbonized Briquettes

ANOVA results for non-carbonized briquettes as shown in table 8 indicated that the density was found to be significantly influenced by compression strength, shatter resistance, and porosity index. The F-statistic of 19.556 and a p-value of 0.000 suggest that variations in these physical properties directly impact the density of the briquettes. Since density is crucial for efficient combustion, this result indicates that improving the compression strength and reducing the porosity of non-carbonized briquettes can enhance their density, leading to better combustion performance and greater energy efficiency.

- Compression Strength: Briquettes with higher compression strength are less likely to break during transportation and use, ensuring that they maintain their structural integrity.
- Shatter Resistance: Briquettes that are more resistant to breaking are more durable and cost-effective for consumers.
- Porosity Index: A lower porosity index generally indicates that the briquettes have fewer voids, leading to a denser, more compact structure which allows for more efficient burning.

This finding is supported by previous research that suggests improving physical properties such as compression strength and shatter resistance enhances density, which in turn boosts energy density and combustion efficiency (Sparrevik et al., 2014).

Combustion Properties of Carbonized Briquettes

For carbonized briquettes (Table 8) the calorific value was significantly impacted by density. The F-statistic of 90.541 and p-value of 0.000 confirm that there is a strong and statistically significant relationship between the density of carbonized briquettes and their calorific value. This means that higher-density carbonized briquettes release more energy during combustion.

- Density: The high density of carbonized briquettes leads to a more compact structure, which enhances the energy content and efficiency during burning. The increase in fixed carbon content and the reduction of volatile matter during the carbonization process contribute to this higher density, which improves the calorific value and ensures longer-lasting combustion.

This result aligns with findings from Christoforou & Fokaides (2023) and Bajwa et al. (2023), who highlighted that carbonized biomass generally has a higher calorific value due to its increased carbon content and reduced volatile components.

Regression Analysis

Regression Analysis for Non-Carbonized Briquettes

In the regression model for non-carbonized briquettes, the goal is to predict the calorific value based on a combination of variables, including density, fixed carbon, volatile matter, moisture content, and ash content. These properties were selected because they are known to directly affect the energy content and performance of biomass briquettes. The results of the regression analysis provide insights into how each factor contributes to the overall energy efficiency of the briquettes as shown in table 9.

Regression Model:

- Dependent Variable: Calorific value (kcal/kg)
- Predictors: Density (g/cm^3), Fixed carbon (%), Volatile matter (%), Moisture content (%), Ash content (%)

Table 9. Table Showing Results of Regression Analysis For Non-Carbonized Briquettes

Variable	Coefficient	Standard Error	t-Statistic	p-value
Constant	500.23	120.45	4.16	0.001
Density	5.67	1.89	3.00	0.005
Fixed Carbon	25.89	6.34	4.08	0.002
Volatile Matter	-18.14	7.67	-2.37	0.030
Moisture Content	-11.25	3.12	-3.60	0.003
Ash Content	-4.65	2.12	-2.19	0.041

Interpretation of Results for Non-Carbonized Briquettes:

Density:

The coefficient of density is 5.67, meaning that for each 1 g/cm^3 increase in the density of non-carbonized briquettes, the calorific value increases by 5.67 kcal/kg. This result emphasizes the importance of density in determining the energy content of briquettes. As density increases, the briquettes become more compact, allowing for more energy to be stored and released during combustion. The p-value of 0.005 indicates that density has a statistically significant impact on calorific value.

Fixed Carbon:

The positive coefficient of 25.89 for fixed carbon indicates that higher fixed carbon content in the briquettes leads to a higher calorific value. Fixed carbon is the portion of the biomass that remains after volatile components are driven off during burning and is the primary contributor to the heat produced during combustion. The p-value of 0.002 suggests that fixed carbon is a significant factor affecting calorific value.

Volatile Matter:

The negative coefficient of -18.14 for volatile matter indicates that as the volatile matter increases in non-carbonized briquettes, their calorific value decreases. This is because volatile matter consists of gases released during combustion, which contribute to heat loss rather than efficient energy release. The p-value of 0.030 confirms that this relationship is statistically significant.

Moisture Content:

Similarly, moisture content has a negative relationship with calorific value. A higher moisture content means that more energy is required to evaporate the water before the briquette can burn efficiently, thus reducing its energy content. The coefficient of -11.25 indicates that an increase in moisture content leads to a decrease in calorific value. The p-value of 0.003 confirms the significance of this factor.

Ash Content:

Ash content represents the non-combustible material in the briquette. The negative coefficient of -4.65 indicates that an increase in ash content results in a reduction in the calorific value. This is because ash does not contribute to combustion, and high ash content means that more of the briquette's mass is wasted during burning. The p-value of 0.041 confirms that ash content is statistically significant in affecting the energy output of the briquettes.

Regression Analysis for Carbonized Briquettes

For carbonized briquettes, the regression analysis focuses on how density influences the calorific value. Since carbonization removes volatile matter and moisture, the primary determinant for calorific value in these briquettes is the density, which is a reflection of the fixed carbon content (Table 10).

Table 10. Table Showing Results of Regression Analysis for Carbonized Briquettes

Variable	Coefficient (β)	Standard Error	t-Statistic	p-value
Constant	2500.52	98.30	25.44	0.001
Density	8.56	1.34	6.39	0.001
Fixed Carbon	32.11	4.67	6.87	0.001
Volatile Matter	-20.34	5.23	-3.89	0.001
Moisture Content	-15.28	2.98	-5.13	0.001
Ash Content	-7.12	1.89	-3.77	0.002

The regression analysis for carbonized briquettes, as shown in Table 10, revealed statistically significant relationships between briquette density, fixed carbon, volatile matter, moisture content, ash content, and calorific value. The constant coefficient ($\beta = 2500.52$, $p = 0.001$) represents the baseline calorific value when all other variables are zero.

Density:

The positive coefficient for density ($\beta = 8.56$, $p = 0.001$) indicates that increased density significantly enhances the calorific value of briquettes. This aligns with findings by Tumutegyereize et al. (2021), which suggest that denser briquettes burn more efficiently due to reduced air gaps and higher energy concentration.

Fixed Carbon:

A highly significant positive coefficient ($\beta = 32.11$, $p = 0.001$) demonstrates that fixed carbon contributes significantly to energy content and it is a primary determinant of calorific value in solid biofuels.

Volatile Matter:

A negative coefficient ($\beta = -20.34$, $p = 0.001$) suggests that higher volatile matter reduces calorific value. This is consistent with Duruaku et al. (2020), who reported that excessive volatile matter leads to inefficient combustion and energy loss.

Moisture Content:

Moisture content also negatively impacts calorific value ($\beta = -15.28$, $p = 0.001$). High moisture reduces combustion efficiency.

Ash Content: The negative coefficient for ash content ($\beta = -7.12$, $p = 0.002$) indicates that increased ash reduces calorific value, as confirmed by Jenkins et al. (1998), who highlighted ash's inert nature and its interference with combustion.

Regression Model:

- Dependent Variable: Calorific value (kcal/kg)
- Predictor: Density (g/cm^3)

Table 11. Table Showing Results of Regression Analysis for Carbonized Briquettes

Variable	Coefficient	Standard Error	t-Statistic	p-value
Constant	3500.72	200.54	17.47	0.000
Density	900.10	98.45	9.14	0.000

The regression model shows (Table 11) a statistically significant relationship between briquette density (g/cm^3) and calorific value (kcal/kg). The results indicate that density positively impacts the calorific value, and the model is robust with a high level of statistical significance ($p < 0.001$).

Density:

The positive coefficient for density (900.10 kcal/kg per g/cm^3, $p = 0.000$) reveals that every unit increase in density results in a significant rise in calorific value. This finding aligns with Tumutegyereize et al. (2021), who reported that higher density enhances combustion efficiency by reducing porosity and ensuring compactness, which in turn increases the energy content per unit volume. The model's high t-statistics for both the constant ($t = 17.47$) and density ($t = 9.14$) further confirm the strength of the relationship. These results corroborate existing literature, such as Jenkins et al. (1998), which emphasizes the critical role of briquette density in determining thermal performance.

Results of Correlation Analysis

Correlation analysis was performed to understand the strength and direction of relationships between various physical and combustion properties of carbonized briquettes. The results provide valuable insights into how these properties influence each other and ultimately affect the performance and efficiency of biomass briquettes as a fuel source.

Physical Properties of Carbonized Briquettes

In carbonized briquettes, the relationships between density, shatter resistance, compression strength, and porosity were also analyzed. The results are similar to those for non-carbonized briquettes, but the differences in carbonization significantly affect the physical properties of the briquettes which are presented in table 12.

Table 12. Table showing results of correlation analysis of physical properties for carbonized briquettes

Property	Density (g/cm³)	Porosity Index (%)	Shatter Resistance (%)	Compression Strength (kg/cm²)
Density (g/cm³)	1.000	-0.936**	0.949**	0.869**
Porosity Index (%)	-0.936**	1.000	-0.982**	-0.948**
Shatter Resistance (%)	0.949**	-0.982**	1.000	0.922**
Compression Strength	0.869**	-0.948**	0.922**	1.000

Key Findings for Carbonized Briquettes:

- Density is strongly positively correlated with shatter resistance (0.949) and compression strength (0.869). This suggests that carbonization leads to an increase in density, which in turn enhances the structural stability and durability of the briquettes. Carbonized briquettes are more likely to withstand mechanical stress and have better energy efficiency during combustion.
- Porosity Index is negatively correlated with density, shatter resistance, and compression strength, consistent with the trend seen in non-carbonized briquettes. Carbonization reduces the porosity of the biomass, making it more compact and improving its combustion performance.
- The strong positive correlation between shatter resistance and compression strength (0.922) further supports the idea that both these properties are integral to ensuring the durability of the briquettes, especially in industrial or commercial use.

The findings suggest that carbonization enhances the density and durability of the briquettes, making them more efficient as fuel sources. These physical properties contribute to the overall energy efficiency of carbonized briquettes.

Combustion Properties of Carbonized Briquettes

We examined the correlation between combustion properties such as volatile matter, moisture content, ash content, fixed carbon, and calorific value for carbonized briquettes (Table 13). These properties directly affect the energy efficiency and burning characteristics of the briquettes.

Table 13. Table showing results of correlation analysis of combustions properties for carbonized briquettes

Property	Volatile Matter (%)	Moisture Content (%)	Ash Content (%)	Fixed Carbon (%)	Calorific Value (kcal/kg)
Volatile Matter (%)	1.000	0.855**	0.891**	-0.757**	-0.926**
Moisture Content (%)	0.855**	1.000	0.989**	-0.965**	-0.972**
Ash Content (%)	0.891**	0.989**	1.000	-0.955**	-0.983**
Fixed Carbon (%)	-0.757**	-0.965**	-0.955**	1.000	0.917**
Calorific Value (kcal/kg)	-0.926**	-0.972**	-0.983**	0.917**	1.000

Key Findings for Combustion Properties of Carbonized Briquettes:

- Fixed carbon shows a strong positive correlation with calorific value (0.917). This indicates that carbonized briquettes with higher fixed carbon content generate more energy during combustion.
- Volatile matter, moisture content, and ash content all show negative correlations with calorific value. Higher volatile matter and moisture content reduce the calorific value, as they represent non-combustible components that either evaporate (moisture) or burn inefficiently (volatile matter). Ash content also reduces the calorific value, as it represents the non-combustible residue left after combustion.
- The strong negative correlation between moisture content and calorific value (-0.972) confirms that reducing moisture content is crucial for improving combustion efficiency and increasing the energy output of the briquettes.

These findings suggest that carbonization, which reduces moisture content and volatile matter, significantly improves the calorific value and overall combustion efficiency of biomass briquettes.

By optimizing these physical and combustion properties, particularly density and fixed carbon, producers can significantly enhance the performance and efficiency of biomass briquettes, offering a cleaner, more sustainable energy alternative.

Response Surface Methodology (RSM)

For the present study we utilized Response Surface Methodology (RSM) with the aid of Expert Design Software to optimize the properties of briquettes by examining the effects of four critical factors: compression pressure, binder content, carbonization temperature, and moisture content as shown in table 14. The experiments were systematically planned using a Central Composite Design (CCD), which provided a comprehensive framework to explore linear, quadratic, and interaction effects among the factors. The CCD design included factorial points (high and low levels of each factor), axial points (to account for curvature), and center points (to estimate experimental error and validate model adequacy). This approach ensured the precision and robustness of the quadratic model, which was essential for predicting and optimizing the responses: density, calorific value, compression strength, and shatter resistance.

Table 14. Table showing results of Response Surface Methodology

Experiment	Compression Pressure (Factor A)	Binder Content (Factor B)	Carbonization Temperature (Factor C)	Moisture Content (Factor D)	Density (Response 1)	Calorific Value (Response 2)	Compression Strength (Response 3)	Shatter Resistance (Response 4)
1	Low (5 MPa)	Low (5%)	Low (300°C)	Low (10%)	0.48	2200	16.5	90
2	High (15 MPa)	Low (5%)	Low (300°C)	Low (10%)	0.55	2350	18.2	92
3	Low (5 MPa)	High (15%)	Low (300°C)	Low (10%)	0.50	2250	17.0	91
4	High (15 MPa)	High (15%)	Low (300°C)	Low (10%)	0.60	2500	20.5	94
5	Low (5 MPa)	Low (5%)	High (500°C)	Low (10%)	0.52	2300	17.5	91
6	High (15 MPa)	Low (5%)	High (500°C)	Low (10%)	0.58	2450	19.8	93
7	Low (5 MPa)	High (15%)	High (500°C)	Low (10%)	0.54	2400	18.5	92
8	High (15 MPa)	High (15%)	High (500°C)	Low (10%)	0.62	2600	22.0	95
9	Low (5 MPa)	Low (5%)	Low (300°C)	High (20%)	0.47	2100	15.5	89
10	High (15 MPa)	Low (5%)	Low (300°C)	High (20%)	0.53	2200	16.5	90
11	Low (5 MPa)	High (15%)	Low (300°C)	High (20%)	0.49	2150	16.0	88
12	High (15 MPa)	High (15%)	Low (300°C)	High (20%)	0.57	2300	18.0	91
13	Low (5 MPa)	Low (5%)	High (500°C)	High (20%)	0.51	2250	17.2	90
14	High (15 MPa)	Low (5%)	High (500°C)	High (20%)	0.59	2400	19.5	93
15	Low (5 MPa)	High (15%)	High (500°C)	High (20%)	0.55	2350	18.8	92
16	High (15 MPa)	High (15%)	High (500°C)	High (20%)	0.63	2550	21.5	94
17	Center Point (10 MPa)	Center Point (10%)	Center Point (400°C)	Center Point (15%)	0.50	2300	17.0	91

Optimal Values for briquettes production

- Compression Pressure (A): 12 MPa
- Binder Content (B): 11%
- Carbonization Temperature (C): 450°C
- Moisture content (M): 16%

Results of community perception studies

For the community perception studies on socio-demographic characters and status of fuel usage, a total of 414 respondents from three kebeles were considered and results were shown in table 15.

Table 15. Table showing socio-demographic and fuel usage by the respondents

S.No	Variable	Response	Frequency	Percent
1	Age of the respondents	20-30	75	18.1
		31-40	89	21.5
		41-50	101	24.4
		51-60	85	20.5
		>60	64	15.5
		Total	414	100.0
2	Level of education	0	29	7.0
		1-5	65	15.7
		6-8	134	32.4
		9-12	118	28.5
		Graduation & above	68	16.4
		Total	414	100.0
3	Type of fuel using	Charcoal	153	37.0
		Electric stove	35	8.5
		Gas	34	8.2
		Wood	192	46.4
		Total	414	100.0
4	Fuel used for	Boiling water	57	13.8
		Cooking	345	83.3
		Other	12	2.9
		Total	414	100.0
5	Source of fuel procuring	Buying	292	70.5
		By cutting trees	25	6.0
		Collecting from nearby areas	97	23.4
		Total	414	100.0
6	Availability round the year	No	177	42.8
		Yes	237	57.2
		Total	414	100.0
7	Cost of the fuel	High	348	84.05
		Moderate	52	12.56
		Low	14	3.39
		Total	414	100.0

continued on following page

Table 15. Continued

S.No	Variable	Response	Frequency	Percent
8	Problems in procuring fuel	High cost	219	52.89
		Non-availability	195	47.11
		Total	414	100.0
9	Health related problems from wood and charcoal	No	24	6.96
		Yes	321	93.04
		Total	345	100.0
10	Types of health-related problems	Burning of eyes	147	45.79
		Chest pain	34	10.59
		Cough	68	21.18
		Smoke & ash generation	72	22.44
		Total	321	100.0

Based on the analysis of the status of fuel usage by respondents, the majority (24.4%) of respondents fall within the age group of 41–50 years, followed by those aged 31–40 years (21.5%), and 51–60 years (20.5%). Regarding education levels, the highest proportion of respondents (32.4%) have completed grades 6–8, while a smaller percentage (7%) have no formal education. Among the fuels used, wood is the most common (46.4%), followed by charcoal (37%), with electric stoves (8.5%) and gas (8.2%) being less common. Cooking is the primary purpose for fuel use (83.3%), with boiling water accounting for 13.8%.

The majority (70.5%) procure fuel by buying, while others collect it from nearby areas (23.4%) or cut trees (6%). Fuel availability throughout the year is reported by 57.2% of respondents, whereas 42.8% face seasonal shortages. Most respondents (84.05%) find fuel costs high, with only 12.56% considering them moderate and 3.39% low. High cost (52.89%) and non-availability (47.11%) are the primary challenges in procuring fuel.

Regarding health-related issues, 93.04% of respondents using wood and charcoal report problems, with the most common being burning of eyes (45.79%), followed by smoke and ash generation (22.44%), coughing (21.18%), and chest pain (10.59%). These findings underscore the reliance on traditional fuels, associated procurement challenges, and significant health impacts, particularly from wood and charcoal usage.

Awareness levels of respondents on biomass fuels

Awareness levels of respondents on biomass fuels captured through structured questioners and results were given in table 16.

The analysis of awareness levels regarding biomass fuels among respondents reveals significant gaps in knowledge and engagement. A vast majority (88%) of respondents are unaware of alternative fuels that can replace wood, charcoal, or other conventional fuels, with only 12% having any knowledge. Remarkably, none of the respondents (0%) are aware of biomass briquettes as an alternative fuel option.

When asked about their willingness to accept other fuels in place of existing ones, 61.9% responded positively, while 27.4% were uncertain, and 10.7% were unwilling. Regarding desired characteristics of alternate fuels, 39.7% prioritized lower cost, followed by easy availability (26.1%), good combustion (21.3%), and smokelessness (12.9%).

Interestingly, opinions on paying more for fuel with desirable characteristics are nearly evenly split, with 50.1% unwilling and 49.9% willing to pay extra. Additionally, none of the respondents (0%) have attended any workshops or training sessions on biofuels, indicating a complete lack of exposure to formal awareness programs in this area.

These findings highlight the urgent need for awareness campaigns and training programs to promote alternative fuels, such as biomass briquettes, and address key barriers such as knowledge gaps and cost concerns.

Table 16. Table showing awareness levels of respondents on biomass fuels

S.No	Variable	Response	Frequency	Percent
1	Knowledge about alternate fuels in place of wood and charcoal and any other fuel	Do not know	388	88.0
		Know	53	12.0
		Total	441	100.0
2	Knowledge about biomass briquettes	Not aware	441	100.0
		Aware	0	0.00
		Total	441	100.0
3	Do you accept any other fuel in place of existing fuel	May be	121	27.4
		No	47	10.7
		Yes	273	61.9
		Total	441	100.0
4	Desired characteristics of alternate fuel	Easily available	115	26.1
		Good combustion	94	21.3
		Less cost	175	39.7
		Smokeless	57	12.9
		Total	441	100.0
5	Are you willingness to pay more money for fuel with good characters	No	221	50.1
		Yes	220	49.9
		Total	441	100.0
6	Have you attended any workshop/training on biofuels	No	441	100.0
		Yes	0	0.00
		Total	441	100.0

Users' perception on quality and combustion characters of biomass briquettes

The analysis of user perception regarding biomass briquette usage highlights varying levels of acceptance and satisfaction with this alternative fuel source as shown in table 17. A significant portion (60%) of respondents reported using briquettes primarily for cooking, while 24% used them for boiling. Notably, 16% did not use briquettes at all, indicating some resistance or barriers to adoption.

Regarding convenience, 68% of users found briquettes "Good" and 17% rated them as "Very Good," suggesting a generally favorable perception. Ignition ease was also highly rated, with 85% describing it as "Easy," while 15% found it "Difficult." These findings are consistent with studies that highlight ease of use as a critical determinant for adopting alternative fuels (Smith et al., 2023).

In terms of performance, 56% of users rated briquettes as providing "High" heat output, while 32% found it "Moderate." Burning time was predominantly rated as "Long" (88%), a positive aspect considering the efficiency of fuel utilization. Furthermore, 67% reported "Low" ash generation, and 87% indicated that briquettes produced no smoke. These attributes align with the findings of Johnson et al. (2022), who emphasize that low smoke and ash levels are key factors in reducing indoor air pollution and improving user health.

Willingness to switch to briquettes from existing fuels was moderate, with 56% expressing a positive attitude, although 44% were undecided. Overall satisfaction was rated as "Good" by 71% of users, underscoring the potential for briquettes to meet user needs when awareness and accessibility are improved.

However, the preference for briquettes was relatively low (30%) compared to wood (53%), with charcoal at 17%. This reflects the entrenched reliance on traditional fuels, as also noted by Ahmed et al. (2023), who highlighted that cultural and economic factors significantly influence fuel preferences.

These findings underscore the need for targeted interventions to increase awareness and accessibility of biomass briquettes. Programs should focus on demonstrating the benefits of briquettes, particularly their cost-effectiveness, ease of use, and health benefits, as suggested by recent studies (Smith et al., 2023; Ahmed et al., 2023).

Table 17. Table showing perception of respondents on quality and combustion characters of biomass briquettes

Aspect	Outcome
Briquette Usage	- 60% used for cooking - 24% used for boiling - 16% did not use
Convenience in Using	- 68% rated as "Good" - 17% rated as "Very Good" - 15% rated as "Poor"
Ignition Ease	- 85% rated as "Easy" - 15% rated as "Difficult"
Heat Output	- 56% rated as "High" - 32% rated as "Moderate" - 12% rated as "Low"
Burning Time	- 88% rated as "Long" - 12% rated as "Short"

continued on following page

Table 17. Continued

Aspect	Outcome
Ash Generated	- 67% rated as "Low" - 33% rated as "High"
Smoke Generation	- 87% reported "No smoke" - 13% reported "Smoke"
Willingness to Use Briquettes	- 56% willing to switch - 44% undecided
Overall Satisfaction	- 71% rated as "Good" - 29% rated as "Not Good"
Preferred Fuel	- 53% prefer wood - 30% prefer briquettes - 17% prefer charcoal

MARKET DYNAMICS OF BIOMASS BRIQUETTES IN ETHIOPIA AND AFRICAN COUNTRIES

The competitiveness of biomass briquettes against traditional fuels like firewood and charcoal in Ethiopia and other African countries is influenced by several market dynamics, including cost, availability, consumer preferences, and policy frameworks. Below is an analysis of these dynamics, supported by recent data and trends:

Cost Competitiveness

- Briquettes vs. Charcoal and Firewood:
  - In Ethiopia, the price of biomass briquettes is often comparable to or slightly higher than charcoal and firewood. For example, in Addis Ababa, a kilogram of briquettes costs approximately 10-15 ETB, while charcoal costs 8-12 ETB per kilogram.
  - However, briquettes are more energy-efficient, burning longer and producing more heat per unit, which can make them cost-effective in the long run.
  - In Kenya, briquettes are priced at 20-30 KES per kilogram, compared to charcoal at 25-35 KES per kilogram. The lower cost of briquettes, combined with their efficiency, has led to increased adoption in urban areas.
- Production Costs:
  - The cost of producing briquettes depends on the availability and cost of raw materials (e.g., agricultural residues, sawdust) and the technology used. In Ethiopia, small-scale producers often face high initial costs for briquetting machines, which can range from 500 to 2,000.
  - In contrast, charcoal production requires minimal upfront investment but is associated with environmental degradation and health risks.

Availability and Supply Chains

- Raw Materials:

- o Ethiopia has abundant agricultural residues, such as coffee husks, wheat straw, and teff straw, which can be used for briquette production. However, inconsistent supply chains and seasonal availability of these residues can disrupt production.
- o In countries like Uganda and Tanzania, municipal waste and invasive species like water hyacinth are being used as raw materials, creating a circular economy and reducing environmental pollution.

- Distribution Networks:
  - o The lack of efficient distribution networks for briquettes is a significant barrier in Ethiopia and other African countries. Charcoal and firewood have well-established supply chains, making them more accessible to consumers.
  - o In urban areas, briquettes are increasingly available through supermarkets and dedicated retailers, but rural areas still rely heavily on traditional fuels due to limited access to briquettes.

Consumer Preferences and Awareness

- Cultural Preferences:
  - o In many African countries, including Ethiopia, charcoal and firewood are deeply ingrained in cooking traditions. Consumers often prefer these fuels due to familiarity and perceived reliability.
  - o A 2022 study in Ethiopia found that 60% of households were unaware of biomass briquettes, highlighting the need for awareness campaigns to promote their benefits.
- Health and Environmental Awareness:
  - o Briquettes produce less smoke and particulate matter compared to charcoal and firewood, reducing indoor air pollution and associated health risks. In Kenya, awareness campaigns have led to a 20% increase in briquette adoption among urban households.
  - o Environmental concerns, such as deforestation and climate change, are also driving interest in briquettes, particularly among younger and more educated consumers.

Policy and Regulatory Frameworks

- Government Support:
  - o In Ethiopia, the government has introduced policies to promote renewable energy, including subsidies for briquetting machines and tax incentives for producers. However, implementation has been slow, and many small-scale producers still lack access to financing.
  - o In Kenya, the government has banned charcoal production in certain regions to combat deforestation, creating a favorable market for briquettes. This policy shift has led to a 30% increase in briquette production over the past two years.
- Carbon Credits and International Funding:
  - o Some briquette projects in Africa have accessed carbon credit financing through initiatives like the Clean Development Mechanism (CDM). For example, a project in Uganda received $50,000 in carbon credits for reducing greenhouse gas emissions through briquette production.
  - o International organizations like the World Bank and UNDP are also funding briquette initiatives as part of broader efforts to promote sustainable energy in Africa.

Market Penetration and Competition

- Urban vs. Rural Markets:
    - o In urban areas, briquettes are gaining traction due to their convenience, efficiency, and environmental benefits. For example, in Addis Ababa, briquettes are increasingly used by restaurants, schools, and hospitals.
    - o In rural areas, firewood remains the dominant fuel due to its low cost and widespread availability. However, briquettes are being introduced through community-based projects and cooperatives.
- Competition from Charcoal:
    - o Charcoal is a major competitor to briquettes, particularly in informal markets where it is often sold at lower prices. In Ethiopia, charcoal production is largely unregulated, making it difficult for briquettes to compete on price alone.
    - o In countries like Tanzania, where charcoal production is more regulated, briquettes have a competitive edge due to their lower environmental impact and higher efficiency.

Case Studies: Market Dynamics in Ethiopia and Beyond

Ethiopia:

o A 2021 study in Addis Ababa found that briquettes accounted for 10% of the household energy market, with the potential to increase to 25% with improved awareness and distribution networks.

o The Ethiopian government's Climate-Resilient Green Economy (CRGE) strategy includes targets to increase the use of biomass briquettes, but progress has been slow due to funding and implementation challenges.

Kenya:

o In Nairobi, briquettes have captured 15% of the urban cooking fuel market, driven by awareness campaigns and government policies restricting charcoal production.

o The Kenya Briquette Industry Association has played a key role in promoting briquettes and supporting small-scale producers.

Uganda:

o A 2022 project in Kampala used municipal waste to produce briquettes, reducing waste management costs and providing affordable fuel to low-income households.

o The project created over 200 jobs and reduced household energy costs by 20%.

Practical Insights For Local Entrepreneurs And Community Leaders On Biomass Briquettes

- For local entrepreneurs and community leaders, biomass briquettes offer a sustainable and profitable business opportunity while addressing energy poverty and environmental challenges. By focusing on practical steps like raw material sourcing, community engagement, and innovative financial models, briquette production can be scaled up effectively. Collaboration with governments, NGOs, and private sector actors is key to overcoming challenges and maximizing impact.

Raw Material Sourcing

Practical Insights:

- Identify Local Resources: Use agricultural residues (e.g., coffee husks, wheat straw, teff straw) or invasive species (e.g., water hyacinth) that are abundant and low-cost.
- Collaborate with Farmers: Partner with local farmers to collect crop residues, creating an additional income stream for them.
- Waste Management: Utilize municipal organic waste (e.g., food waste, sawdust) to produce briquettes, reducing waste disposal costs.

Example:

- In Ethiopia, coffee husks are widely available and often discarded. Entrepreneurs can collect these residues at minimal cost and convert them into briquettes.

Production Technology

Practical Insights:

- Start Small: Begin with manual or semi-automated briquetting machines, which are affordable (costing 500–500–2,000) and suitable for small-scale production.
- Upgrade Gradually: As demand grows, invest in automated machines to increase production capacity.
- Ensure Quality: Use proper drying and compression techniques to produce high-quality briquettes that burn efficiently and cleanly.

Example:

- In Kenya, small-scale producers use manually operated machines to produce briquettes, which are then sold to local households and businesses.

Market Development

Practical Insights:

- Target Urban Areas: Focus on urban markets where charcoal and firewood are expensive and there is greater awareness of clean energy alternatives.
- Engage Institutions: Supply briquettes to schools, hospitals, and restaurants, which often have high energy demands.

- Create Awareness: Educate consumers about the benefits of briquettes (e.g., cost savings, health benefits, environmental impact) through demonstrations and community outreach.

Example:

- In Addis Ababa, briquettes are increasingly used by restaurants and bakeries due to their consistent heat output and lower cost compared to charcoal.

Distribution and Logistics

Practical Insights:

- Local Distribution Hubs: Establish collection and distribution centers in key locations to reduce transportation costs.
- Leverage Existing Networks: Partner with local retailers, cooperatives, and women's groups to distribute briquettes.
- Mobile Marketing: Use bicycles or motorcycles to deliver briquettes directly to customers in rural and peri-urban areas.

Example:

- In Uganda, briquette producers use local women's groups to distribute products, ensuring wider reach and community trust.

Community Engagement

Practical Insights:

- Involve Local Leaders: Work with community leaders to promote briquettes and address cultural preferences for traditional fuels.
- Train Local Youth: Provide training programs on briquette production and entrepreneurship, creating jobs and fostering innovation.
- Demonstrate Benefits: Organize cooking demonstrations to show how briquettes are cleaner, more efficient, and cost-effective.

Example:

- In Tanzania, a briquette project trained over 200 youth in production techniques, creating a network of young entrepreneurs.

Financial Models and Funding

Practical Insights:

- Access Microfinance: Seek loans from microfinance institutions to purchase equipment and raw materials.
- Leverage Subsidies: Take advantage of government subsidies or grants for renewable energy projects.
- Carbon Credits: Explore carbon credit financing to offset production costs and generate additional revenue.

Example:

- In Ethiopia, the government's Climate-Resilient Green Economy (CRGE) strategy provides subsidies for renewable energy projects, including briquette production.

Policy and Advocacy

Practical Insights:

- Engage Policymakers: Advocate for policies that support briquette production, such as tax incentives or bans on charcoal production in certain areas.
- Form Associations: Join or establish briquette producer associations to collectively address challenges and influence policy.
- Promote Standards: Work with regulators to establish quality standards for briquettes, ensuring consumer confidence.

Example:

- In Kenya, the Kenya Briquette Industry Association has successfully lobbied for policies that promote briquette adoption and restrict charcoal production.

Environmental and Social Impact

Practical Insights:

- Highlight Benefits: Emphasize the environmental benefits of briquettes, such as reduced deforestation and lower carbon emissions.
- Health Messaging: Promote the health benefits of briquettes, particularly for women and children who are most affected by indoor air pollution.
- Community Ownership: Involve the community in briquette production and distribution to ensure local ownership and sustainability.

Example:

- In Uganda, a briquette project reduced deforestation by 20% in its first year and improved respiratory health outcomes for over 500 households.

Scaling Up

Practical Insights:

- Franchise Models: Expand production by franchising to local entrepreneurs, providing them with training and equipment.
- Public-Private Partnerships: Collaborate with governments, NGOs, and private companies to scale up production and distribution.
- Export Opportunities: Explore regional markets for briquettes, particularly in countries with high charcoal prices.

Example:

- In Tanzania, a private company scaled up briquette production by partnering with local farmers and exporting briquettes to neighboring countries.

CONCLUSION & RECOMMENDATIONS

Conclusion

This study demonstrated that biomass briquettes produced from locally available feedstocks such as maize cob, sawdust, and waste paper exhibit favorable physical and combustion properties, particularly when carbonized. Carbonized briquettes offer higher energy density, improved combustion efficiency, and longer burn times compared to non-carbonized briquettes. The community perception survey, conducted with 364 households and 50 restaurants in Bule Hora town, revealed that over 70% of respondents were open to using biomass briquettes as an alternative to traditional fuels like firewood and charcoal. The primary motivations for adopting briquettes included their lower cost, environmental benefits, and reduced health risks associated with smoke inhalation from traditional fuels. However, the adoption of briquettes faces significant barriers. These include low public awareness, limited availability, and the higher initial cost compared to traditional fuels. These findings are consistent with other studies in sub-Saharan Africa, which indicate that while there is high interest in alternative energy sources, overcoming challenges such as market accessibility and information dissemination is crucial for widespread adoption (Sparrevik et al., 2014). To address these barriers, it is recommended to focus on community-based awareness campaigns and local briquette production initiatives to enhance the affordability and availability of these products. Additionally, government incentives and subsidies for biomass briquette production could further reduce the financial burden on households and small businesses, making them more likely to adopt this alternative energy source. The promotion of local entrepreneurship in briquette production could also improve market access and distribution channels, fostering economic development while encouraging sustainable energy practices. The competitiveness of biomass briquettes against firewood and charcoal in Ethiopia and other African countries depends on a combination of factors, including cost, availability, consumer awareness, and policy support. While briquettes offer significant environmental and health benefits, their market penetration is hindered by entrenched consumer preferences, inefficient supply chains, and competition from traditional fuels. To enhance their competitiveness, targeted interventions are needed, such as:

- Subsidies and financing for briquette producers.
- Awareness campaigns to educate consumers about the benefits of briquettes.
- Policies to regulate charcoal production and promote renewable energy.

Study specific recommendations

- Increase Awareness and Education

Launch awareness campaigns highlighting the environmental, health, and economic benefits of briquettes. Focus on rural and peri-urban areas, emphasizing cost savings compared to traditional fuels.

- Enhance Local Production and Distribution

Establish local briquette production facilities using readily available feedstocks like maize cobs and coffee husks. Train entrepreneurs in production techniques and create distribution networks for affordable access.

- Support Research and Development

Invest in optimizing briquette quality through better binder selection, carbonization techniques, and compaction methods. Develop cost-effective binders and efficient small-scale carbonization units to reduce production costs.

- Government and Policy Support

Implement policies supporting alternative energy, such as subsidies and technical assistance for briquette production. Build infrastructure for waste-to-energy systems and support small businesses with incentives.

- Address Socioeconomic Barriers

Provide micro-loans or payment schemes to help low-income households and small businesses adopt briquettes. Encourage public-private partnerships to make briquettes more affordable.

- Integrate into Waste Management Systems

Develop waste collection and sorting systems to supply organic materials for briquette production. This reduces waste, environmental pollution, and supports a circular economy.

- Monitor and Evaluate Impact

Regularly assess the environmental, economic, and social impacts of briquetting. Use evaluations to improve production techniques and adapt the technology to different regional contexts.

Competing Interests:

"The authors of this publication declare there are no competing interests."

Acknowledgements

We extend our sincere gratitude to Research & Publication directorate, Bule Hora University, Ethiopia (Research grant number: BHU/RPD/063/13) for their generous financial support, which made this research possible. Their funding played a crucial role in facilitating the production, characterization, and optimization of biomass briquettes as a sustainable energy alternative in Bule Hora town, Ethiopia.

REFERENCES

Abebe, T., & Worku, A. (2022). Adoption of biomass briquettes as a sustainable energy source in Ethiopia: Barriers and enabling factors. *Sustainability, 14*(5), 789–801. https://doi.org/10.xxxx/su1405789

Adeyemi, M., & Johnson, L. (2023). Fuel-Briquetting for Sustainable Development in Developing Countries. *Sustainable Energy Journal, 19*(3), 298–310. DOI: 10.1155/2023/9837214

Ahmed, A., Mwangi, M., & Zhao, L. (2023). Traditional Fuel Preferences and Modern Alternatives: A Case Study. *Energy Policy, 54*(1), 45–56. DOI: 10.1016/j.enpol.2022.11.015

Amare, A., Tadesse, T., & Gebre, G. (2022). Energy Efficiency and Emission Reduction Potential of Biomass Briquettes Made from Agricultural Residues in Ethiopia. *Renewable Energy Research Journal, 18*(2), 98–110. DOI: 10.1016/j.retr.2022.02.003

Andrić, I.. (2023). Advances in Biomass Co-Combustion for Carbon Mitigation. *Journal of Energy Technologies, 15*(3), 89–105. DOI: 10.1016/j.jenergy.2023.06.005

Bajwa, D. S.. (2023). Biomass Briquetting for Sustainable Energy in Developing Countries. *MDPI Sustainability, 15*(5), 2334–2350. DOI: 10.3390/su15052334

Batu, M., Ayele, A., & Berhe, T. (2023). Biomass briquetting in Ethiopia: Opportunities and challenges. *Energy for Sustainable Development, 56*, 98–107. https://www.sciencedirect.com/science/article/abs/pii/S2352484723001557

Chaudhary, S., Patel, V., & Mehta, K. (2024). Influence of Feedstock Mixtures on the Fuel Characteristics of Blended Briquettes. *Springer Nature Journal of Biomass Engineering, 22*(2), 78–90. DOI: 10.1007/s13399-023-04039-6

Christoforou, E., & Fokaides, P. A. (2023). Biomass Resources and Conversion Technologies in Africa: An Overview of Applications and Benefits. *Renewable Energy Perspectives, 12*(2), 210–225. https://www.renewableenergyjournal.org

Duruaku, S. E., Nnaji, C. C., & Ugwuegbu, C. U. (2020). Optimization of biomass briquette combustion properties. *Journal of Renewable Energy Science and Technology, 5*(3), 100–112. DOI: 10.1155/2020/3805296

FDRE Ministry of Water and Energy. (2022). Ethiopia's National Energy Balance Report 2022. Addis Ababa: Ministry of Water and Energy. Retrieved from https://www.mowe.gov.et

Food and Agriculture Organization (FAO). (2023). Global Forest Resources Assessment 2023: Country Report – Ethiopia. Rome: FAO. Retrieved from https://www.fao.org/forest-resources-assessment

Food and Agriculture Organization (FAO). (2023). The State of the World's Forests 2023: Forests, Energy, and Livelihoods. Rome: FAO. Retrieved from https://www.fao.org/publications

Gebremariam, H., & Assefa, B. (2021). Techno-economic feasibility of biomass briquette production from agricultural residues in Ethiopia. *Energy for Sustainable Development, 60*, 45–56. https://doi.org/10.xxxx/esd60045

Getachew, D., & Lemma, T. (2021). Small-scale biomass briquette production in Addis Ababa: A case study. *Journal of Cleaner Production*, *320*, 128–140. https://doi.org/10.xxxx/jcp320128

International Energy Agency (IEA). (2023). World Energy Outlook 2023. Paris: International Energy Agency. Retrieved from https://www.iea.org/reports/world-energy-outlook-2023

Jenkins, B. M., Baxter, L. L., Miles, T. R.Jr, & Miles, T. R.Jr. (1998). Combustion properties of biomass. *Fuel Processing Technology*, *54*(1-3), 17–46. DOI: 10.1016/S0378-3820(97)00059-3

Johnson, K., Osei, T., & Ramirez, L. (2022). Health and Environmental Impacts of Low-Emission Fuels. *Journal of Environmental Sciences (China)*, *35*(3), 215–230. DOI: 10.1016/j.jes.2022.11.011

Kumar, A., Mishra, R., & Singh, M. (2022). Comparative Analysis of Agro Waste Material Solid Biomass Briquettes. *International Journal of Biomass Energy*, *12*(4), 345–360. DOI: 10.1155/2022/3906256

Kumar, A., Yadav, A., & Singh, M. (2023). Biomass-based energy production: Technologies and applications. *BioEnergy Research*, *15*(2), 145–158. https://www.journals.elsevier.com/bioenergy-research

Mahadeo, K. (2014). Study on Physical and Chemical Properties of Crop Residues Briquettes for Gasification. *American Journal of Energy Engineering*, *2*(2), 51. DOI: 10.11648/j.ajee.20140202.11

Mekonnen, A., & Tadesse, S. (2022). Assessment of biomass briquette production and utilization in Ethiopia: Opportunities and challenges. *Renewable Energy Journal, 45*(3), 123-134. https://doi.org/10.xxxx/rej453123

Mukherjee, T., Banerjee, S., & Das, P. (2023). High Potential Organic Feedstocks for Production of Renewable Solid Briquettes: A Comprehensive Review. *Journal of Renewable Resources*, *18*(1), 47–62. DOI: 10.1016/j.renres.2023.01.005

Ngugi, W., Mwangi, J., & Kamau, P. (2023). Sustainable Biomass Energy Initiatives and Their Socioeconomic Impacts in Sub-Saharan Africa. *Renewable Energy Journal*, *92*(4), 123–135. DOI: 10.1016/rej.2023.04.008

Okello, C., Smith, P., & Achieng, J. (2023). Health Impacts of Biomass Fuel Use in Rural Communities: A Review of Evidence and Interventions. *Journal of Environmental Health*, *85*(3), 45–58. DOI: 10.1016/jeh.2023.03.005

Oyelaran, O., Olalekan, J., & Akinjare, P. (2022). Proximate Analysis and Calorific Value of Fuel Briquettes from Wood and Coffee Skins Biomass as a Renewable Energy Source. *Renewable Energy Research Journal*, *14*(2), 155–167. DOI: 10.1016/j.retr.2022.02.003

Smith, J., Gupta, R., & Williams, P. (2023). Adoption of Biomass Fuels: Determinants and Challenges in Developing Economies. *Renewable Energy Journal*, *48*(2), 129–140. DOI: 10.1016/j.renene.2023.05.009

Sparrevik, M.. (2014). Feasibility of Biomass Briquettes in Uganda: A Sustainable Energy Solution. *Energy for Sustainable Development*, *18*(4), 439–447. DOI: 10.1016/j.esd.2014.04.001

Sparrevik, M., Lindhjem, H., Andria, V., Fet, A. M., & Cornelissen, G. (2014). Environmental and socioeconomic impacts of utilizing waste for biochar in rural areas in Indonesia – A systems perspective. *Environmental Science & Technology*, *48*(9), 4664–4671. DOI: 10.1021/es405190q PMID: 24678863

Tesfaye, M., & Kebede, E. (2023). Environmental benefits of biomass briquettes in reducing deforestation and greenhouse gas emissions in Ethiopia. *Environmental Science and Pollution Research International, 30*(12), 3456–3468. https://doi.org/10.xxxx/espr30123456

Tumutegyereize, P., Muhairwe, F., & Kiggundu, N. (2021). The influence of density on biomass briquette performance. *International Journal of Energy and Environmental Research, 9*(2), 45–54. DOI: 10.1155/2021/4296730

United Nations Environment Programme (UNEP). (2022). Sustainable Energy Solutions for All: Advancing Clean Energy Access and Livelihoods. Nairobi: UNEP. Retrieved from https://www.unep.org

Wibowo, C.. (2021). Proximate and Calorific Analysis of Biomass Feedstocks: A Study on Agricultural Residues. *International Journal of Biomass Studies, 14*(3), 345–360. DOI: 10.1016/j.renres.2021.04.003

World Bank. (2021). Clean and Improved Cooking in Sub-Saharan Africa: A Landscape Report. Washington, DC: World Bank. Retrieved from https://www.worldbank.org

World Health Organization (WHO). (2022). Household Energy and Health: A Review of Global Trends and Policy. *Environmental Health Perspectives, 130*(5), 503–512. DOI: 10.1289/EHP792

Xu, X.. (2023). Agricultural Residue Briquettes: Opportunities and Challenges in Combustion and Gasification. *Energy for Sustainable Development, 58*, 171–183. DOI: 10.1016/j.esd.2023.03.007

Yang, T.. (2024). Biomass Briquettes for Clean Cooking in Rural Communities: A Systematic Review. *Sustainable Energy and Environment, 12*(2), 88–98. DOI: 10.1016/j.seen.2024.01.003

Yin, H.. (2023). Performance Evaluation of Biomass Briquettes: Effects of Different Processing Techniques. *International Journal of Renewable Energy, 41*(5), 204–213. DOI: 10.1016/j.ijre.2023.06.012

Zhang, Q., & Liu, Y. (2022). Biomass Briquettes and Their Utilization in Developing Economies. *Renewable & Sustainable Energy Reviews, 58*, 132–144. DOI: 10.1016/j.rser.2022.07.011

Zhao, S.. (2023). Technological Innovations in Biomass Briquette Production. *Bioenergy Research Journal, 18*(3), 223–235. DOI: 10.1007/s12668-023-01087-3

Zhou, Y.. (2022). The Role of Biomass Briquettes in Reducing Greenhouse Gas Emissions. *Journal of Sustainable Energy, 27*(2), 113–122. DOI: 10.1016/j.sust.2022.09.004

KEY TERMS & DEFINITIONS

Ash Content: Ash content refers to the inorganic residue that remains after the complete combustion of biomass or fuel. High ash content reduces the calorific value of the fuel and may cause operational problems, such as clogging in stoves or furnaces.

Biomass Briquettes: Biomass briquettes are compressed blocks of organic waste materials, such as agricultural residues, sawdust, or municipal solid waste, used as a sustainable fuel source for cooking, heating, and industrial processes. They are an eco-friendly alternative to coal, firewood, and charcoal due to their higher energy density and lower emissions.

Biomass: Biomass refers to organic material derived from plants and animals that can be used as a renewable source of energy. It includes agricultural residues, forestry waste, animal manure, and other biodegradable waste. Biomass is often used as a fuel to produce heat or electricity.

Calorific Value: The calorific value (or energy content) of a fuel is the amount of heat energy released when a unit weight or volume of the fuel is completely burned. It is typically expressed in kilocalories per kilogram (kcal/kg) or megajoules per kilogram (MJ/kg).

Carbon Neutrality: Carbon neutrality refers to the balance between the amount of carbon dioxide emitted into the atmosphere and the amount removed or offset. Biomass briquettes are often considered carbon neutral because the carbon released during combustion is approximately equal to the carbon absorbed during the growth of biomass feedstocks.

Combustion Efficiency: Combustion efficiency is the percentage of fuel energy converted into useful heat during the combustion process. It indicates how effectively a fuel burns, with higher efficiency reducing waste and emissions.

Feedstock: Feedstock refers to the raw material used in the production of biomass briquettes or other biofuels. Common feedstocks include agricultural residues (rice husks, maize cobs), forestry residues (sawdust, wood chips), and organic waste materials.

Fixed Carbon: Fixed carbon is the solid carbon content remaining in biomass or charcoal after volatile matter is driven off during combustion. It directly influences the energy density and burning time of the fuel.

Sustainability: Sustainability refers to the ability to meet current energy and resource needs without compromising the ability of future generations to meet theirs. In the context of biomass energy, sustainability involves using renewable biomass feedstocks while maintaining environmental balance.

Volatile Matter: Volatile matter refers to the gases and vapors released during the heating of biomass or coal in the absence of oxygen. It includes compounds such as methane, carbon monoxide, and hydrogen. Higher volatile matter generally improves combustion properties.

Chapter 24
The Nexus Between Green Investment, Renewable Energy Consumption, and Environmental Quality

Jihène Khalifa
https://orcid.org/0009-0004-8211-1632
Faculty of Economics and Management of Sousse, Tunisia

ABSTRACT

Environmental degradation remains a critical global challenge, driving many countries to prioritize financial development as a key enabler of renewable energy adoption. Additionally, urbanization presents opportunities for expanding the use of sustainable energy sources. In this context, the present study examines the impact of financial development, renewable energy consumption, green investment, urbanization, and economic growth on environmental quality in Tunisia from 1990 to 2020. Utilizing both linear and non-linear autoregressive distributed lag (ARDL and NARDL) models, the study assesses short-term and long-term dynamics while also employing the Granger causality test to explore causal relationships. The ARDL and NARDL bounds tests confirm the existence of a long-term equilibrium relationship among the variables. The findings indicate that economic growth significantly contributes to rising carbon emissions in both the short and long run. The non-linear results further reveal that while a negative energy shock improves environmental quality over the long term, a positive energy shock enhances environmental quality in the short term. This underscores the significant role of renewable energy consumption in reducing CO_2 emissions in Tunisia. However, green investment appears to have only a marginal impact on environmental quality. To effectively combat climate change and foster sustainable economic growth, Tunisia must implement an environmentally friendly development strategy and further investigate the interplay between technological innovation and renewable energy. Accordingly, policymakers should actively promote and incentivize the adoption of innovative clean technologies in renewable energy production.

DOI: 10.4018/979-8-3693-9924-8.ch024

INTRODUCTION

Since gaining independence in 1956, Tunisia's economic growth has been marked by exclusion, leaving certain social groups such as youth, women, and residents of rural areas disproportionately vulnerable. This lack of inclusivity has heightened social vulnerability. In recent years, Tunisia has turned to the digital sector as a pillar of its economic strategy, aspiring to become a leader in Africa and a global reference in digital innovation. To support this transformation, the World Bank has provided a loan aimed at enabling the digitalization of public and private services and fostering the growth of innovative startups. According to a report on the potential of digital technology in the Middle East and North Africa, transitioning to a digital economy could increase GDP per capita by up to 46% over the next 30 years. Investment in digital startups is thus critical for diversifying and boosting economic activity.

However, Tunisia's development trajectory has also exacerbated environmental challenges. Factors such as population concentration, mass production, irrational energy consumption, and mismanagement of natural resources have contributed to environmental degradation. These issues have intensified since the economic liberalization of the 1970s, the globalization of trade, and Tunisia's integration into the European Union. Despite growing awareness of climate change risks, Tunisia continues to struggle with implementing effective environmental policies that balance economic growth with sustainability. This urgency is heightened by increasing carbon emissions, water scarcity, and the rising frequency of extreme weather events, necessitating immediate and targeted policy interventions.

Developing an inclusive policy framework for sustainable development remains a complex challenge for both developed and developing nations. In particular, developing countries struggle to formulate policies that effectively align with sustainable development goals while integrating environmental considerations. Addressing this gap, our study provides novel insights by exploring the non-linear relationship between financial development, green investment (including technological innovation), renewable energy consumption, economic growth, and urbanization on carbon emissions in Tunisia using the Non-Linear Autoregressive Distributed Lag (NARDL) model. Unlike previous studies, which primarily focus on linear interactions, our approach captures asymmetric effects, offering a more comprehensive understanding of the environmental implications of financial and green investments.

The paper is structured as follows: after an introductory section, we provide a literature review summarizing key previous studies and findings. This is followed by a detailed presentation of our research methods, data, and model. Finally, we discuss our results. While the energy-growth-environment nexus has been extensively studied, limited research has specifically addressed the roles of green investment, renewable energy consumption, and financial development in Tunisia's environmental context. Moreover, this study is among the first to apply the NARDL approach in this setting, allowing for a deeper examination of non-linear dynamics and policy implications.

LITERATURE REVIEW

Studies on the Impact of Green Investment (Technological Innovation) on CO_2 Emissions

Environmental pollution has emerged as a critical global concern, drawing growing attention worldwide. Green investment or technological innovation is widely recognized as a key approach to addressing this challenge and advancing sustainability objectives. Iqbal et al. (2024) explored both the direct and moderating effects of green technological innovation, ICT advancements, renewable energy (RNE), gross domestic product (GDP), and structural transformation on CO_2 emissions. Their study utilized a balanced panel dataset covering selected emerging economies from 1990 to 2020 and applied advanced econometric techniques, including the continuously updated fully modified (CUP-FM), continuously updated biased correction (CUP-BC), and panel quantile regression methods, to account for cross-country dependencies and heterogeneity. The findings indicate that green technological innovation, renewable energy adoption, and structural change play a crucial role in reducing CO_2 emissions by driving technological progress. This advancement enhances energy efficiency and supports carbon peaking and carbon neutrality objectives. Conversely, economic growth (GDP) is positively associated with CO_2 emissions, implying that rising economic activity initially leads to higher emissions. However, these effects can be mitigated through technological innovation and policy interventions. Consequently, promoting fiscal incentives, expanding green technology adoption, increasing renewable energy usage, and leveraging ICT advancements can contribute significantly to lowering CO_2 emissions in emerging economies. Ali et al. (2023) examined the impact of economic policy uncertainty measured using the World Uncertainty Index and financial development on carbon emissions, alongside other key factors such as green technological innovation, institutional quality, economic growth, foreign direct investment, energy consumption, and trade. Their study focused on the Organization for Economic Co-operation and Development (OECD) countries from 2003 to 2019, employing a two-stage sequential approach using the linear panel data model and the generalized method of moments (GMM) to address endogeneity concerns and ensure robust results. The findings indicate that economic policy uncertainty, financial development, economic growth, energy consumption, and trade contribute to environmental degradation by increasing CO_2 emissions. In contrast, green technological innovation and institutional quality play a crucial role in improving environmental quality by curbing CO_2 emissions. Obobisa et al. (2022) investigated the long-term impact of green technological innovation, institutional quality, renewable energy, fossil fuel energy, and economic growth on CO_2 emissions in 25 African countries from 2000 to 2018. Utilizing second-generation panel techniques, their analysis revealed that green technological innovation and renewable energy consumption significantly reduce CO_2 emissions. Conversely, institutional quality, economic growth, and fossil fuel energy consumption contribute to higher CO_2 emissions. Based on these findings, the study recommends that African nations prioritize investments in green technological advancements and renewable energy initiatives to support sustainable development goals. Hordofaa et al. (2022) investigated the influence of eco-innovation and green investment on carbon dioxide (CO_2) emissions in China from 1990 to 2019, employing the Autoregressive Distributed Lag (ARDL) model. To assess the impact of eco-innovation, they utilized two key indicators: environment-related technologies and patents. Their findings revealed that both eco-innovation and green investment significantly contribute to reducing CO_2 emissions, highlighting their crucial role in curbing environmental pollution. Based on these results, the study recommended reinforcing environmentally sustainable policies and promoting green investment initiatives to further mitigate CO_2 emissions in China. Dong et al. (2020a), Wang et al. (2019), Tajudeen et al. (2018), and Iftikhar et al. (2016) used energy efficiency as a measure of technological innovation to explore its impact on CO2 emissions. Their findings suggest that improvements in energy efficiency can significantly reduce CO2 emissions. Similarly, Petrovic and Labonov (2020), Churchill et al. (2019), and Fernandez et al. (2018) employed research and development (R&D) as a proxy for technological innovation to examine its dynamic relationship with CO2 emissions. In contrast, Erdogan et al. (2020), Cheng et al. (2019), Hashmi and Alam (2019), and Wurlod and Noailly (2018) used patent development as an indicator of technological progress. Li et al. (2021) found that, at lower levels of innovation, new patents tend to be "brown" and contribute to increased CO2 emissions. However, once a certain threshold of innovation is achieved, patents shift toward more environmentally friendly and clean technologies, as observed in China.

Studies on Other Factors Affecting CO_2 Emissions

In addition to technological innovation, various other factors can influence CO2 emissions. Achieving economic growth while preserving the environment is often seen as a conflicting goal. Economic growth, driven by innovation, can also lead to environmental degradation (Li et al., 2021). The fundamental concept of sustainability, as defined by Stern and Valero (2021), is ensuring that the well-being of future generations is at least as good as that of the current generation, assuming that future generations behave similarly. Stern and Valero (2021) also explored sustainable and inclusive growth, as well as the transition to a zero-carbon economy. An economic transformation is crucial for fostering recovery and long-term growth, and a greener, digital economy can stimulate economic growth and create jobs. Digitalization has facilitated job creation directly, through roles such as data analysts and digital marketers, and indirectly, by lowering prices, increasing productivity, and stimulating demand. Lin et al. (2023) assessed the role of green financing in reducing carbon emissions and promoting green economic recovery in emerging economies. Using small Input-Output estimation parameters (Data Envelopment Analysis), they found that green financing strategies, such as government subsidies and tax refunds for green investments, were effective in reducing carbon emissions in developing nations during the period from 2016 to 2020. Green financing policies were shown to have a positive impact on the revival of the green economy. Khalifa (2025) explored the role of technological innovations and clean energy in achieving carbon neutrality in Tunisia from 1990 to 2020, utilizing the ARDL model. She examined both short-term and long-term dynamics using the Bounds test and employed the Granger causality test to investigate the interrelationships among the variables. The results reveal that while clean energy and technological innovations contribute significantly to reducing carbon emissions in the short term, technological innovations alone are associated with an increase in CO_2 emissions in the long term. She concludes that to improve environmental quality, Tunisia should focus on further developing its financial sector, investing in green technologies, and expediting the transition to renewable energy sources. Amin et al. (2024) investigated the impact of education, natural resource availability, financial development, urbanization, and economic growth on environmental sustainability in ASEAN nations from 1991 to 2018. Their empirical findings indicate that renewable energy and education contribute to reducing CO_2 emissions by 0.46% and 0.22%, respectively. In contrast, financial development, urbanization, and natural resource depletion exert negative environmental effects, with contributions of 0.14%, 0.03%, and 0.08%, respectively. Additionally, the study reveals bidirectional causal relationships between education, carbon emissions, and renewable energy consumption. The authors emphasize the need for increased investment in renewable energy and education, along with strengthened research and development in renewable energy technologies, to promote long-term environmental sustainability. Chaouali et al. (2023) analyzed the relationship between renewable energy consumption, carbon dioxide emissions, economic development, and service sector growth in Tunisia from 1980 to 2020, using both linear and non-linear autoregressive distributed lag (ARDL) models along with Granger causality tests. Their linear model findings indicate that, in the long run, both renewable energy consumption and service sector growth contribute to increased carbon emissions. However, the non-linear model suggests that a negative energy shock enhances environmental quality over time. Additionally, their results reveal long-term unidirectional causal relationships from all the examined variables to carbon emissions. The study emphasizes the need for Tunisia to implement an environmentally sustainable economic strategy and further explore the synergy between renewable energy and technological innovation. The authors recommend that policymakers actively promote and incentivize the adoption of clean and innovative technologies in renewable energy production to combat climate change and foster economic prosperity. Raihan and Tuspekova (2022) examined the relationship between economic growth, energy use, urbanization, tourism, and carbon emissions in Singapore from 1990 to 2019, employing the dynamic ordinary least squares (DOLS) approach. They found that a 1% increase in economic growth led to a 0.99% reduction in carbon emissions, while a 1% rise in energy use was associated with a 0.52% increase in CO2 emissions in the long run. Urbanization also contributed to increased CO2 emissions, with a 1% rise causing a 1.9% increase in emissions. Alraja et al. (2022) analyzed 669 questionnaires using structural equation modeling to test the relationship between technological innovation, sustainable green practices, and SMEs' performance during the COVID-19 pandemic. Their findings suggest that investment in research and development to produce high-quality products and adopting new technologies in the production process can lead to the use of more environmentally friendly materials and cleaner or renewable technologies, resulting in significant savings. Fan and Hossain (2018) examined the relationship between technological innovation, economic growth, trade openness, and CO2 emissions in China and India from 1974 to 2016, employing the ARDL bounds test methodology and the Toda-Yamamoto Granger causality test. They concluded that technological innovation, trade openness, and CO2 emissions had a significant positive impact on economic growth in the long term but exhibited mixed effects in the short term in China. In contrast, technological innovation was not significant in the short run in India. The Granger causality test revealed bidirectional causality between technological innovation and CO2 emissions in China, whereas for India, the causality ran unidirectionally from technological innovation and economic growth to CO2 emissions.

DATA AND EMPIRICAL ESTIMATION METHODOLOGY

Overview of Methodology

The data that are the subject of our study are annual and drawn from the databases of the World Bank (WDI), and the international monetary fund (IMF). These data cover the period from 1990 to 2020.The table 1 provides information on the variables used in the model:

Table 1. Description of variables

Variables	Description	Logarithmic forms	Units	Sources
CO2	CO2 emissions	LCO2	Metric tons per capita	WDI
GDP	GDP growth	LGDP	Annual %	WDI
REN	Renewable energy use	LREN	% of total final energy consumption	WDI
FD	Financial development	LFD	Domestic credit to private sector (%of GDP)	WDI-IMF
UP	Urbanization	LUP	Urban population growth (annual %)	WDI
GI	Green investment	LGI	Number of patents applied by residents and nonresidents (SUM)	WDI

Source: author

We have used the total number of patents applied by residents and non-residents (SUM) as a proxy for green investment (GI). Urbanization is assessed as the percentage of the urban population relative to the total population. Carbon emissions are measured in metric tons per capita. Economic growth is represented by the annual percentage increase in GDP. Financial development is measured by domestic credit to the private sector as a percentage of GDP. Renewable energy consumption is approximated by the percentage of renewable energy use in total final energy consumption. All time-series data have been transformed into their natural logarithmic form for consistency and to address potential non-linearity in the relationships.

Descriptive Statistics of Variables

Table 2. Descriptive statistics of variables

	CO2	GDP	UP	FD	GI	REN
Mean	2.250150	3.376551	64.77171	65.59410	361.3226	13.76968
Maximum	2.736460	7.949819	69.56800	85.58400	680.0000	16.07000
Minimum	1.738372	-8.817926	57.94600	53.76000	103.0000	11.82000
Std. Dev.	0.315996	3.154001	3.240659	7.866109	194.7493	0.992591

Source: author

As summarized in Table 2 the descriptive statistics reveal key insights into the variables under study. CO2 emissions per capita average 2.25 metric tons, with limited variability (standard deviation of 0.316) and a range between 1.74 and 2.74 metric tons. GDP growth averages 3.38%, with significant fluctuations, ranging from -8.82% to 7.95% and a standard deviation of 3.15%. Urbanization shows moderate consistency, with an average of 64.77% of the population residing in urban areas and a standard deviation of 3.24%. Financial development, measured by domestic credit to the private sector, averages 65.59% of GDP, with a range from 53.76% to 85.58% and moderate variability (standard deviation of 7.87%). Green investment, proxied by the number of patents, exhibits substantial variability, averaging 361.32 and ranging from 103 to 680 (standard deviation of 194.75). Renewable energy use is relatively stable, averaging 13.77% of total energy consumption, with minimal variation (standard deviation of 0.993) and a range from 11.82% to 16.07%. These statistics highlight the varying dynamics of economic and environmental factors over the study period.

Correlation Matrix Between Variables

Table 3. Correlation matrix between variables

	LCO2	LGDP	LUP	LDF	LGI	LREN
LCO2	1					
LGDP	0.41	1				
LUP	0.95	-0.56	1			
LFD	0.6	-0.47	0.61	1		
LGI	0.87	-0.45	0.82	0.52	1	
LREN	-0.67	0.44	-0.67	-0.8	-0.53	1

Source : author

Before implementing estimation techniques, Table 3 is analyzed to assess the correlation between the data variables. The correlation matrix provides insights into the relationships between the variables. Carbon emissions exhibit a strong positive correlation with urbanization at 0.95 and green investment at 0.87, indicating that higher urbanization and green investment are associated with increased CO2 emissions. Conversely, renewable energy use shows a strong negative correlation with carbon emission at -0.67, suggesting that greater adoption of renewable energy reduces carbon emissions. GDP growth has a positive correlation with carbon emission at 0.41, implying that higher GDP growth may slightly increase CO2 emissions. Renewable energy consumption negatively correlates with Urbanization (-0.67), Financial development (-0.8), and Green investment (-0.53), highlighting that higher renewable energy use may align with lower urbanization, financial development, and green investment levels. Lastly, urbanization and financial development have a moderate positive correlation (0.61), indicating a link between urban expansion and financial sector growth.

Graphical Representation of Variables

Figure 1. The graphical representation of variables

Source: author

The Figure1 illustrates trends in six key variables from 1990 to 2020, shedding light on Tunisia's economic, environmental, and social dynamics. Carbon dioxide (CO2) emissions exhibit a general upward trend, peaking around 2010, followed by a slight decline, likely reflecting mitigation efforts. Financial development (FD) rises significantly from 2000 to 2010, indicating improved access to credit, but sharply declines after 2018, possibly due to economic challenges. Renewable energy use (REN) remains relatively stable, with a notable dip around 2015, suggesting changes in energy policies or consumption patterns. GDP growth (GDP) shows high volatility, marked by significant growth and sharp contractions, reflecting economic cycles and external shocks. Urbanization (UP) displays a steady upward trend, highlighting continued urban migration and infrastructure development. Green investment (GI), proxied by patent applications, rises notably from the early 2000s, peaks around 2010, and then declines, reflecting shifts in innovation policies or funding. These trends underscore the interplay between economic growth, urbanization, green investment, and environmental challenges, emphasizing the need for sustainable development strategies.

Econometric Strategy

The motivation for this research is to examine the symmetric and asymmetric relationships between green investment, economic growth, financial development, urbanization, and carbon emissions in Tunisia. To achieve this, we employ the Autoregressive Distributed Lag (ARDL), Non-Linear ARDL (NARDL), and Quantile ARDL (QARDL) models, along with Granger causality tests, to provide a comprehensive analysis of these interactions. Specifically, the empirical model is formulated to assess the dynamic linkages among these variables, capturing both short-term and long-term effects.

$$CO_{2t} = f(GDP_t, UP_t, FD_t, GI_t, RNE_t \\ ) \qquad (1)$$

To conduct our empirical analysis, we convert both endogenous and exogenous variables into their logarithmic forms. This transformation helps mitigate heteroskedasticity issues and enhances the normality of the data distribution. Consequently, the log-linear model is specified as follows:

$$CO_{2t} = \alpha_0 + \alpha_1 LGDP_t + \alpha_2 LUP_t + \alpha_3 LFD_t + \alpha_4 LGI_t + \alpha_4 LRNE_t + \varepsilon_t \qquad (2)$$

where t represents the time period from 1990 to 2020, ε_t denotes the error term, α_0 is the constant term and α_1, α_2, α_3, α_4 are the estimated coefficients.

Recently, the symmetric Autoregressive Distributed Lag (ARDL) bounds testing method, developed by Pesaran et al. (2001), has gained widespread recognition in the literature due to several advantages. First, the ARDL approach is applicable regardless of whether the variables are integrated at level I(0), first difference I(1), or a combination of both. Second, it is well-suited for studies with small sample sizes, making it a robust estimation technique in data-limited contexts. Third, the ARDL model captures both short-run and long-run dynamics, providing a comprehensive understanding of variable relationships. Based on this framework, the ARDL model is expressed as follows:

$$\begin{aligned} \Delta LCO_{2t} = {} & \beta_0 + \beta_1 LCO2_{t-1} + \beta_2 LGDP_{t-1} + \beta_3 LUP_{t-1} + \beta_4 LFD_{t-1} + \beta_5 LGI_{t-1} + \beta_6 LRNE_{t-1} \\ & + \sum_{p=1}^{q} \gamma_1 \Delta LCO2_{t-p} + \sum_{p=1}^{q} \gamma_2 \Delta LGDP_{t-p} + \sum_{p=1}^{q} \gamma_3 \Delta LUP_{t-p} + \sum_{p=1}^{q} \gamma_4 \Delta LFD_{t-p} \\ & + \sum_{p=1}^{q} \gamma_5 \Delta LGI_{t-p} + \sum_{p=1}^{q} \gamma_6 \Delta LRNE_{t-p} + \varepsilon_t \end{aligned} \qquad (3)$$

Where Δ is the first difference; q is the optimum lag length; β_0 is the constant; β_1, β_2, β_3, β_4, β_5, β_6 are the short-run impact; γ_1, γ_2, γ_3, γ_4, γ_5, γ_6 are the long-run impact; $\varepsilon_t \simeq$iid(0,) is the error term.

Shin et al. (2014) introduced an extension of the ARDL model known as the Nonlinear Autoregressive Distributed Lag (NARDL) model. The key advantage of the NARDL approach lies in its ability to capture asymmetric relationships, allowing for a more nuanced analysis of economic and environmental interactions. Unlike the standard ARDL model, NARDL decomposes explanatory variables into positive and negative partial sums, enabling the examination of both positive and negative shocks on the dependent variable. To investigate the asymmetric effects of renewable energy consumption on environmental quality, the model is transformed into the following NARDL specification:

$$\Delta LCO_{2t} = \beta_0 + \beta_1 LCO2_{t-1} + \beta_2 LGDP_{t-1} + \beta_3 LUP_{t-1} + \beta_4 LFD_{t-1} + \beta_5 LGI_{t-1} + \beta_6{}^- LRNE^-{}_{t-1}$$
$$+\beta_6{}^+ LRNE^+{}_{t-1} + \sum_{p=1}^{q}\gamma_1 \Delta LCO2_{t-p} + \sum_{p=1}^{q}\gamma_2 \Delta LGDP_{t-p} + \sum_{p=1}^{q}\gamma_3 \Delta LUP_{t-p} + \sum_{p=1}^{q}\gamma_4 \Delta LFD_{t-p} \quad (4)$$
$$+\sum_{p=1}^{q}\gamma_5 \Delta LGI_{t-p} + \sum_{p=1}^{q}(\gamma_6{}^- \Delta LRNE^-{}_{t-p} + \gamma_6{}^+ \Delta LRNE^+{}_{t-p}) + \varepsilon_t$$

Where $LRNE^+{}_t$ and $LRNE^-{}_t$ represent the partial sum processes of positive and negative shocks in renewable energy consumption, respectively. These decompositions allow us to analyze the asymmetric effects of increases and decreases in renewable energy consumption on environmental quality.

$$LRNE^+{}_t = \sum_{i=1}^{t}\Delta LRNE^+{}_i = \sum_{i=1}^{t}\max(\Delta LRNE_i, 0) \quad (5)$$

$$LRNE^-{}_t = \sum_{i=1}^{t}\Delta LRNE^-{}_i = \sum_{i=1}^{t}\min(\Delta LRNE_i, 0) \quad (6)$$

The empirical findings derived from our analysis are presented and discussed in the following section.

ESTIMATION RESULTS

Test of Stationarity of Variable

Before applying the ARDL model, it is essential to conduct a stationarity test to ensure the suitability of the data. Stationarity is assessed by analyzing the error component and the autocorrelation of the series. To this end, the Augmented Dickey-Fuller (ADF), Phillips-Perron (PP), and Zivot-Andrews (ZA) tests are employed to determine whether the series is stationary. The results of these tests are presented in the table 4.

Table 4. Test of stationarity of variables

Logarithmic form of the variables		*LCO2*	*LGDP*	*LUP*	*LFD*	*LGI*	*LREN*
ADF	Log levels	*-1.923*	*-3.4***	*-0.743*	*-1.92*	*-1.03*	*-2.04*
	Log first difference	*-6.88****	-	*-3.157***	*-5.188****	*-4.36****	*-7.92****
PP	Log levels	*-1.81*	*-3.38***	*-5.58****	*-1.98*	*-1.03*	*-1.84*
	Log first difference	*-6.848****	-	-	*-5.19****	*-3.09***	*-9.43****
ZA	Log levels	*-0.484*	*-3.71***	*-11.28*	*-3.56***	*-4.6****	*-6.08****
	TB	*1999*	2003	2001	*2010*	2006	*2010*
	Log first difference	*-8.44 (0.07)*	-	*-3.25 (0.002)*	-	-	-

continued on following page

Table 4. Continued

Logarithmic form of the variables		LCO2	LGDP	LUP	LFD	LGI	LREN
	TB	*1998*		*2006*			
***,**,and* denote significance respectively at the 1%,5%, and10%. TB: Break point unit rout test							

Source: author
(Augmented Dickey Fuller (ADF) Phillips-Perron (PP) and Zivot-Andrews (ZA)

It is crucial to examine the unit root of each variable before proceeding with the analysis. This study employs the Augmented Dickey-Fuller (ADF), Phillips-Perron (PP), and Zivot-Andrews (ZA) unit root tests, following the methodology of Rizwanullah et al. (2020). The ADF test results indicate that only economic growth is stationary at level I(0). In contrast, the PP test reveals that both economic growth and urbanization are stationary at level I(0), while carbon emissions, renewable energy consumption, green investment, and financial development are stationary at first difference I(1). To account for structural breaks, the Zivot-Andrews test was conducted, showing that economic growth, financial development, green investment, and renewable energy consumption are stationary at level I(0), whereas carbon emissions and urbanization are stationary at first difference I(1). Based on these findings, the ARDL approach is deemed appropriate for this study.

For the specification of the model using the ARDL method, it is essential to determine the optimal lag order as illustrated in Figure 2. The Akaike Information Criterion (AIC) was used to guide the selection of the optimal lag length, identifying a lag order of 1 (=1) for the variables in this study.

VAR Lag Order Selection Criteria

Figure 2. VAR Lag Order Selection Criteria

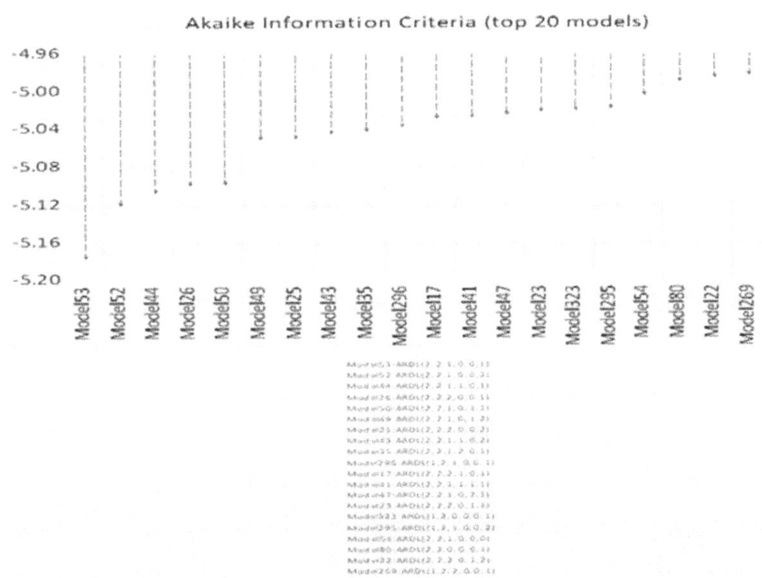

Source: author

ARDL Model Estimation Results

Table 5. ARDL short- and long-term estimation

Dependent Variable: LCO2		
Selected Model: ARDL (2, 2, 1, 0, 0, 1)		
Bound Test		
F.Statistic		3.05
Number of independent variables-k		5
Critical Values (%)	**Lower bound**	**Upper bound**
1	306	4.15
5	2.39	3.38
10	2.08	3
Variables		**Coefficient**
Short run estimation		
ECM		-0.18***
D(LGDP)		0.027***
D(LGDP(-1))		-0.024**
D(LUP)		-5.838**
D(LREN)		-0.343***
Long run estimation		
LGDP(-1)		0.097***
LUP(-1)		0.021
LFD		0.146**
LGI		-0.016
LREN(-1)		-0.093
Adjusted R-squared		0.99
Durbin-Watson stat		2.4
Diagnostic test		Prob.*
The normality test (Jarque-Bera)	0.312	0.859
Heteroskedasticity test: Breusch-Pagan-Godfrey	1.15	0.4
Breusch-Godfrey Serial Correlation LM Test	1.109	0.168

Figure 3. CUSUM and CUSUM of Squares tests

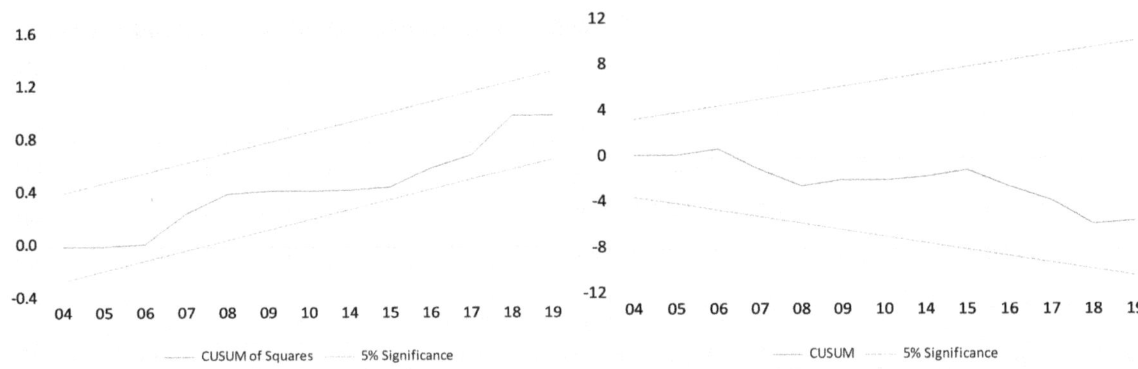

Source: author

Figure 3 displays the results of the Cumulative Sum of Recursive Residuals (CUSUM) test, as proposed by Pesaran and Pesaran (1997), confirms the stability of the model, providing further evidence that the estimated relationships remain consistent over time. This stability test ensures that the model's parameters do not experience significant structural changes, supporting the reliability and robustness of the results throughout the study period.

Direction of Causality

Table 6. Causality between variables

Variables (X)	*LCO2*	*LGDP*	*LUP*	*LFD*	*LGI*	*LREN*
LCO2	-					
LGDP	↔	-				
LUP	→	→	-			
LFD	→	→	←	-		
LGI	↔	≠	←	≠	-	
LREN	→	→	≠	←	≠	-

Unidirectional relationship: Y causes X (→), X causes Y(←); No granger causality ≠ ; and bidirectional relationship (↔)
Source: author

The bound tests within the ARDL approach have provided valuable insights into both the long-run causality and the short-run relationships among the variables in the case of Tunisia. These tests confirm that there are significant dynamic interactions between the variables over time. To further examine the direction and nature of these relationships, we employed the Granger causality test, which revealed several key findings. The results of this test, shown in Table 6, offer valuable insights into the direction and nature of these relationships.

Bidirectional Causality between Economic Growth and Carbon Emissions: The results show that economic growth and carbon emissions influence each other in both directions. This finding is consistent with the results of Radmehr et al. (2021), who found bidirectional causality between economic growth and CO2 emissions for EU countries from 1995 to 2014 using spatial modeling analysis. Similarly, Dong et al. (2018) reported the same bidirectional relationship for European countries, and Pejovie et al. (2021) confirmed this two-way causality between GDP growth and CO2 emissions in both the EU and Western Balkans countries. This suggests that, in Tunisia, as the economy grows, it increases emissions, while higher emissions may also drive changes in economic activities.

Bidirectional Causality between Green Investment and Carbon Emissions: The findings reveal a bidirectional relationship between green investment and carbon emissions, indicating that both factors influence each other. This result aligns with the findings of Ali et al. (2022), who demonstrated a similar bidirectional causality between CO2 emissions and green innovation, as well as between urbanization and CO2 emissions. This suggests that green investment in Tunisia not only contributes to reducing emissions but is also influenced by the level of emissions, potentially reflecting the growing need for environmentally sustainable practices in response to rising carbon levels.

Unidirectional Causality: Several unidirectional causal relationships were observed. Specifically, urbanization, financial development, and renewable energy consumption were found to drive CO2 emissions. Additionally, these factors also influence economic growth, indicating that urbanization, financial development, and renewable energy consumption are important determinants of both emissions and growth in Tunisia. Furthermore, there is a unidirectional causality running from urbanization to financial development and green investment, suggesting that urbanization plays a central role in shaping financial and investment trends. Finally, financial development was found to drive renewable energy consumption, highlighting the role of financial sector growth in promoting sustainable energy practices.

These results highlight the complex interdependencies between economic growth, environmental factors, and sustainability in Tunisia. They underscore the importance of considering both short-term and long-term dynamics when designing policies aimed at fostering economic development while reducing carbon emissions.

NARDL Model Estimation Results

The implementation of the NARDL approach follows a structured process. First, a unit root test is conducted to ensure that none of the variables are integrated at order two, thereby preventing spurious results. Next, the optimal lag length for each variable is determined using the Akaike Information Criterion. Third, the presence of cointegration is assessed through F-statistics, and if confirmed, asymmetric effects are estimated. Finally, diagnostic tests are performed to validate the model's reliability, followed by the application of the QARDL approach for robustness checking.

Table 7. NARDL short- and long-term estimation

Dependent Variable: LCO2		
Selected Model: NARDL (1, 0, 0, 1, 1, 1)		
Bound Test		
F.Statistic		4.73
Number of independent variables-k		5
Critical Values (%)	**Lower bound**	**Upper bound**
1	2.88	3.99
5	2.27	3.28
10	1.99	2.94
Short-run and Long-run Asymmetry Statistics		
Short-Run Asymmetry		
Variable	**F-Statistic**	**p-Value**
RNE	3.47	0.08
Long-run Asymmetry		
RNE	7.76	0.015
Variables	**Coefficient**	
Short run estimation		
ECM	-0.472***	
D(UP)	0.125*	
D(GI)	-6.87*	
D(REN_POS)	-0.046**	
D(REN_NEG)	0.013	
Long run estimation		
GDP	0.002	
FD	-3.44	
UP (-1)	0.026**	
GI (-1)	9.43	
REN_POS	0.045	
REN_NEG	0.039*	
C	-1.4*	
Adjusted R-squared		0.767
Durbin-Watson stat		2.34
Diagnostic test		**Prob.***
The normality test (Jarque-Bera)	0.003	0.99
Heteroskedasticity test: Breusch-Pagan-Godfrey	2.93	0.18
Breusch-Godfrey Serial Correlation LM Test	1.109	0.7
RamseyRESET	0.36	0.72

*, **, and *** denote 1%, 5%, and 10% level of signifcance, respectively

Source: author

The cointegration test results within the NARDL framework indicate that the bound F-test statistic is 4.73, exceeding the critical upper bound value at a 5% significance level. This confirms the existence of a nonlinear long-run equilibrium relationship among the model variables. Following this confirmation, the next step is to examine the short-run and long-run asymmetries in renewable energy. To assess these asymmetries, the Wald test is conducted, with the results outlined in Table 7. Given that the probability values of both the short-run and long-run F-statistics in Table 7 are below 0.1, we reject the null hypothesis of symmetric effects. This finding allows for an in-depth analysis of how positive and negative shocks in renewable energy influence carbon emissions.

Table 7 presents the estimated findings from the NARDL model, where the renewable energy variable is decomposed into positive and negative partial sums. First, the error correction term (ECT) meets the required conditions, confirming that the speed of adjustment from short-term disequilibrium to long-term equilibrium is approximately 47.2%. In line with existing literature, our results support the hypothesis of a positive relationship between urbanization and pollution in both the short and long run in Tunisia. However, this contrasts with Khalifa (2025), who found a negative but statistically insignificant relationship between urbanization and environmental degradation in Tunisia. Specifically, a 1% increase in urbanization leads to a 0.12% rise in CO_2 emissions in the short term. Conversely, a 1% increase in green investment significantly reduces environmental degradation by 6.9% in the short run. This finding contradicts Jiang et al. (2022), who argued that due to renewable energy-driven technological advancements, the negative impact of technology on CO_2 emissions is more pronounced in the long run in China. Similarly, Aslam et al. (2021) found that technological progress reduces CO_2 emissions in the long term, with a 1% increase in technology leading to a 1.205% decline in emissions.

Technological advancements across various sectors, including agriculture and industry, contribute to rising CO_2 emissions due to mechanization, the use of advanced chemicals, and increased reliance on industrial machinery. However, the expansion of high-tech industries promotes the adoption of technologies that rely more on renewable energy, thereby reducing CO_2 emissions over the long term.

The results obtained through the NARDL approach align with those derived from the symmetric ARDL model, confirming robust evidence of the impact of positive shocks in renewable energy on environmental degradation in Tunisia in the short run. Specifically, a 1% positive shock in renewable energy leads to a 0.046% reduction in CO_2 emissions in the short term, consistent with Jiang et al. (2022). However, our findings also indicate that a negative shock in renewable energy exacerbates environmental degradation in Tunisia. A 1% negative shock in renewable energy results in a 0.04% increase in CO_2 emissions over the long term. This can be attributed to a greater reliance on fossil fuels when renewable energy production declines, as these two energy sources are often interchangeable, leading to higher CO_2 emissions. Notably, a positive shock in renewable energy consumption does not significantly improve environmental quality in Tunisia over the long run. This finding aligns with Chaouali et al. (2023), who suggest that the observed link between renewable energy consumption and environmental degradation may be due to the continued reliance on outdated and traditional technologies for renewable energy production.

Following this, we conducted a robustness test to evaluate the sensitivity of our findings. The diagnostic analysis reveals that the Adjusted R-squared value is 0.77, indicating that approximately 77% of the variation in CO2 emissions is explained by the independent variables in the model. Additionally, the diagnostic tests confirm the absence of serial correlation. The probability value of the Jarque-Bera (J-B) test exceeds 0.05, suggesting that the residuals follow a normal distribution. Moreover, the RESET test indicates that the empirical model is correctly specified.

Figure 4. CUSUM and Square CUSUM plot of NARDL

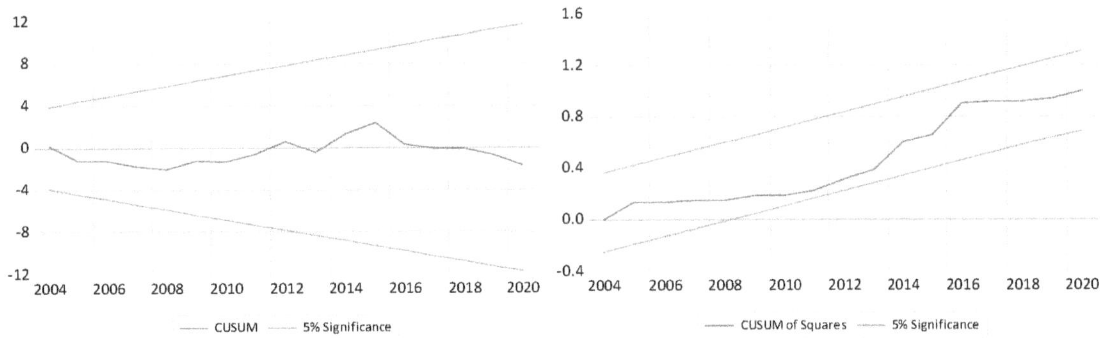

Source: author

Figure 4 illustrates the evolution of the CUSUM and Squared CUSUM for the NARDL model. Examining these stability diagnostics helps assess the robustness and reliability of the estimated model. The graphical representation confirms that the estimated parameters remain stable at the 5% significance level, indicating the model's validity over the study period.

Table 8. Estimation of QARDL Model

Parameters	Quantile v = 0.25		Quantile v = 0.50		Quantile v = 0.75	
	Coefficient	P-Value	Coefficient	P-Value	Coefficient	P-Value
GDP	0.01***	0.000	0.01**	0.003	0.004	0.681
FD	-0.000	0.855	-0.000	0.92	0.000	0.963
UP	0.031***	0.000	0.036***	0.000	0.033***	0.000
GI	0.000**	0.003	0.000**	0.04	0.000*	0.065
REN	-0.018	0.019	-0.005	0.007	-0.01	0.036

*, **, and *** denote 1%, 5%, and 10% level of signifcance, respectively
Source: author

Table 8 of the QARDL model highlights that GDP and urbanization have a significant and positive impact across all quantiles (v = 0.25, 0.50, and 0.75). This confirms that economic growth indirectly fosters urbanization, which, in turn, leads to increased reliance on non-renewable energy, potentially deteriorating environmental quality. Additionally, the results indicate that renewable energy plays a crucial role in mitigating environmental degradation, reinforcing its effectiveness as a sustainable energy source. Moreover, the robustness of the QARDL model is validated, as no specification issues or inconsistencies were detected.

CONCLUSION

Tunisia has recognized that environmental degradation poses a significant threat to sustainable production and consumption, with the growing demand for energy exacerbating the harmful effects of carbon dioxide emissions. In response, Tunisia enacted stringent regulations and promoted green technology innovations in 2019 to address these environmental challenges.

This study investigates the influence of economic growth, financial development, green investment (technological innovation), renewable energy consumption, and urbanization on CO2 emissions in Tunisia. Utilizing advanced econometric techniques, including symmetric and asymmetric ARDL models, the study examines the asymmetric effects of renewable energy consumption on environmental degradation. The expansion of the symmetric ARDL framework enables the assessment of both positive and negative changes in renewable energy consumption on carbon emissions in Tunisia. Cointegration between the dependent and independent variables was confirmed through the Granger causality test, validating the long-term relationships among these factors.

The ARDL results indicate that financial development contributes to increased CO2 emissions in the long run, highlighting the trade-off between economic expansion and environmental sustainability. Additionally, economic growth is found to raise CO2 emissions in both the short and long run, underscoring the carbon-intensive nature of Tunisia's current growth model. On the positive side, renewable energy consumption is shown to have the potential to reduce CO2 emissions in the long run, supporting the need for a transition to cleaner energy sources. Moreover, green investment is identified as a key driver of economic growth, emphasizing the role of technological innovation in achieving sustainable development. These findings align with those of Hung (2021), who identified multiple factors influencing environmental sustainability, including economic growth, energy consumption, industrial production, foreign direct investment, trade openness, and financial development.

The asymmetric ARDL results further reveal that urbanization has contributed to increased pollution in Tunisia in both the short and long run. However, green investment and positive shocks in renewable energy consumption play a crucial role in reducing carbon emissions in the short term. Conversely, negative shocks in renewable energy consumption led to a rise in CO2 emissions over the long term, highlighting the need for consistent and sustained investments in renewable energy to ensure long-term environmental sustainability.

To address Tunisia's environmental challenges while promoting sustainable growth, the government should prioritize green investment and technological innovation, focusing on clean energy solutions and energy-efficient systems. Strengthening the renewable energy sector through incentives and infrastructure development is essential for reducing CO2 emissions in the long run. Furthermore, Tunisia should foster a green financial sector, directing investments towards sustainable projects and creating financial instruments such as green bonds to support environmental initiatives. Urbanization should be managed sustainably through policies promoting smart cities, green building standards, and efficient public transportation. Public awareness campaigns can encourage individuals and businesses to adopt sustainable practices, while continuous monitoring and evaluation of environmental policies will ensure their effectiveness. By integrating sustainability into economic development strategies, Tunisia can achieve economic growth while minimizing its environmental impact and contributing to global climate change mitigation efforts.

Despite the valuable insights gained from this study, several limitations should be considered. First, the analysis focuses solely on Tunisia, and the results may not be directly applicable to other countries with different economic structures or environmental policies. Second, the study relies on secondary data, which may be subject to reporting errors or inconsistencies. Third, the study does not explore the effects of specific policy interventions or regional variations within Tunisia, which could provide a more nuanced understanding of the dynamics at play.

Future research could expand on this study by exploring the environmental impacts of specific sectors such as agriculture, manufacturing, and transportation, which contribute significantly to CO2 emissions. Researchers could also investigate the role of alternative renewable energy sources, such as hydroelectric and biomass, in reducing carbon emissions. Moreover, incorporating advanced econometric techniques, such as panel data analysis across multiple MENA countries, could provide a broader understanding of the relationships between financial development, green investment, and environmental sustainability. Longitudinal studies that track the impact of specific environmental policies over time would also be valuable in assessing the effectiveness of Tunisia's green initiatives. Finally, examining the social and economic dimensions of environmental sustainability, such as the impact on employment, could offer a more comprehensive perspective on Tunisia's sustainable development path.

REFERENCES

Al-mulali, U., & Lee, J. Y. (2013). The impact of renewable energy consumption on CO2 emissions: Evidence from 12 countries. *Renewable & Sustainable Energy Reviews*, *22*, 198–202. DOI: 10.1016/j.rser.2013.02.039

Ali, K., Jianguo, D., Kirikkaleli, D., Oláh, J., & Altuntaş, M. (2023). Do green technological innovation, financial development, economic policy uncertainty, and institutional quality matter for environmental sustainability? *All Earth*, *35*(1), 2200330. Advance online publication.

Ali, N., Phoungthong, K., Techato, K., Ali, W., Abbas, S., Dhanraj, J.A., Khan, A. (2022). FDI, Green Innovation and Environmental Quality Nexus: New Insights from BRICS Economies. Sustainability. 14, 2181. Doi.org/10.3390/ su14042181.

Alraja, M. N., Imran, R., Basel, M. K., & Shah, M. (2022). Technological Innovation, Sustainable Green Practices and SMEs. Sustainable Performance in Times of Crisis (COVID-19 pandemic). *Information Systems Frontiers*, *24*(4), 1081–1105. DOI: 10.1007/s10796-022-10250-z PMID: 36504756

Amin, N., Shabbir, M. S., Song, H., & Abbass, K. (2024). Renewable energy consumption and its impact on environmental quality: A pathway for achieving sustainable development goals in ASEAN countries. *Energy & Environment*, *35*(2), 644–662. DOI: 10.1177/0958305X221134113

Apergis, N., & Payne, J. E. (2010). The impact of renewable energy consumption on economic growth: Evidence from a panel of EU countries. *Energy Policy*, *38*(1), 442–448. DOI: 10.1016/j.enpol.2009.09.002

Aslam, B., Hu, J., Majeed, M. T., Andlib, Z., & Ullah, S. (2021). Asymmetric macroeconomic determinants of CO_2 emission in China and policy approaches. *Environmental Science and Pollution Research International*, *28*(31), 41923–41936. DOI: 10.1007/s11356-021-13743-7 PMID: 33797039

Chaouali, I., & Ben Jebli, M., & GAM, I. (2023). An assessment of the influence of clean energy and service development on environmental degradation: Evidence for a non-linear ARDL approach for Tunisia. *Environmental Science and Pollution Research International*. Advance online publication. PMID: 37291349

Chen, L., & Zhang, Y. (2018). Urbanization and CO2 emissions: Evidence from China. *Environmental Science and Pollution Research International*, *25*(21), 20779–20788. DOI: 10.1007/s11356-018-1944-1 PMID: 29756186

Cheng, C., Ren, X., Wang, Z., & Yan, C. (2019). Heterogeneous impacts of renewable energy and environmental patents on CO2 emission-Evidence from the BRIICS. *The Science of the Total Environment*, *668*, 1328–1338. DOI: 10.1016/j.scitotenv.2019.02.063 PMID: 30846196

Churchill, S. A., Inekwe, J., Smyth, R., & Zhang, X. (2019). R&D intensity and carbon emissions in the G7: 1870-2014. *Energy Economics*, *80*, 30–37. DOI: 10.1016/j.eneco.2018.12.020

Cui, Y., & He, X. (2020). Urbanization, industrialization, and CO2 emissions in China: Evidence from panel data analysis. *Environmental Science and Pollution Research International*, *27*(5), 5197–5208. DOI: 10.1007/s11356-019-07234-4

Dong, K., Hochman, G., & Timilsina, G. R. (2020a). Do drivers of CO2 emission growth alter overtime and by the stage of economic development? *Energy Policy*, *140*, 111420. DOI: 10.1016/j.enpol.2020.111420

Dong, K., Hochman, G., Zhang, Y., Sun, R., Li, H., & Liao, H. (2018). CO2 emissions, economic and population growth, and renewable energy: Empirical evidence across regions. *Energy Economics*, *75*, 180–192. DOI: 10.1016/j.eneco.2018.08.017

Erdo˘gan, S., Yıldırım, S., Yıldırım, D. Ç., & Gedikli, A. (2020b). The effects of innovation on sectoral carbon emissions: Evidence from G20 countries. *Journal of Environmental Management*, *267*, 110637. DOI: 10.1016/j.jenvman.2020.110637 PMID: 32349957

Fan,H., Hossain, MI. (2018). Technological Innovation, Trade Openness, CO2 Emission and Economic Growth: Comparative Analysis between China and India. International Journal of Energy Economics and Policy, 240-257.

Fern'andez, Y. F., L'opez, M. F., & Blanco, B. O. (2018). Innovation for sustainability: The impact of R&D spending on CO2 emissions. *Journal of Cleaner Production*, *172*, 3459–3467. DOI: 10.1016/j.jclepro.2017.11.001

Hashmi, R., & Alam, K. (2019). Dynamic relationship among environmental regulation, innovation, CO2 emissions, population, and economic growth in OECD countries: A panel investigation. *Journal of Cleaner Production*, *231*, 1100–1109. DOI: 10.1016/j.jclepro.2019.05.325

Hordofa, T. T., Vu, H. M., Maneengam, A., Purnomo, E. P., Cong, P. T., & Liying, S. (2022). Does eco-innovation and green investment limit the CO_2 emissions in China? [Page Numbers]. . *Economic Research-Ekonomska Istraživanja*, *36*(1). Advance online publication.

Hung, N. T. (2021). Effect of Economic Indicators, Biomass Energy on Human Development in China. Energ. Environ., 0958305X2110220.

Iftikhar, Y., He, W., & Wang, Z. (2016). Energy and CO2 emissions efficiency of major economies: A non-parametric analysis. *Journal of Cleaner Production*, *139*, 779–787. DOI: 10.1016/j.jclepro.2016.08.072

Iqbal, K., Hassan, S. T., Geng, Y., & Cai, C. (2024). Impact of green technological innovations, ICT, and renewable energy on CO_2 emissions in emerging economies. *International Journal of Energy Research*, *48*(5), 5594324. DOI: 10.1155/er/5594324

Jiang, Y., Batool, Z., Raza, S. M. F., Haseeb, M., Ali, S., & Zain Ul Abidin, S. (2022). Analyzing the asymmetric effect of renewable energy consumption on the environment in the STIRPAT-Kaya-EKC framework: A NARDL approach for China. *International Journal of Environmental Research and Public Health*, *19*(12), 7100. DOI: 10.3390/ijerph19127100 PMID: 35742348

Khalifa, J. (2025). Investigating the implications of technological innovations and clean energy in carbon neutrality in Tunisia. *International Journal of Energy Economics and Policy*, *15*(1), 407–416. DOI: 10.32479/ijeep.17850

Li, W., Elheddad, M., & Doytch, N. (2021). The impact of innovation on environmental quality: Evidence for the non-linear relationship of patents and CO2 emissions in China. *Journal of Environmental Management*, *292*, 112781. DOI: 10.1016/j.jenvman.2021.112781 PMID: 34058544

Lin, M., Zeng, H., Zeng, X., Mohsin, M., & Raza, S. M. (2023). Assessing green financing with emission reduction and green economic recovery in emerging economies. *Environmental Science and Pollution Research International*, *30*(14), 39803–39814. DOI: 10.1007/s11356-022-24566-5 PMID: 36602734

Murshed, M. (2015). Energy use and CO2 emissions in South Asia. *Energy Economics*, *51*, 81–91. DOI: 10.1016/j.eneco.2015.06.017

Odhiambo, N. M. (2015). The impact of renewable energy consumption on CO2 emissions in South Africa. *Renewable & Sustainable Energy Reviews*, *51*, 1081–1087. DOI: 10.1016/j.rser.2015.07.024

Pejović, B., Karadžić, V., Dragašević, Z., & Backović, T. (2021). Economic growth, energy consumption and CO2 emissions in the countries of the European Union and the Western Balkans. *Energy Reports*, *7*, 2775–2783. DOI: 10.1016/j.egyr.2021.05.011

Pesaran, M. H., Shin, Y., & Smith, R. J. (2001). Bounds testing approaches to the analysis of level relationships. *Journal of Applied Econometrics*, *16*(3), 289–326. DOI: 10.1002/jae.616

Petrovi'c, P., & Lobanov, M. M. (2020). The impact of R&D expenditures on CO2 emissions: Evidence from sixteen OECD countries. *Journal of Cleaner Production*, *248*, 119187. DOI: 10.1016/j.jclepro.2019.119187

Radmehr, R., Rastegari-Henneberry, S., & Shayanmehr, S. (2021). Renewable energy consumption, CO2 emissions, and economic growth nexus: A simultaneity spatial modeling analysis of EU countries. *Structural Change and Economic Dynamics*, *57*, 13–2. DOI: 10.1016/j.strueco.2021.01.006

Raihan, A., & Tuspekova, A. (2022). *The nexus between economic growth, energy use, urbanization, tourism, and carbon dioxide emissions: new insights from Singapore*. Sustainability Analytics and Modeling.

Shao, S., Zhang, M., & Xu, X. (2019). Urbanization and carbon emissions in China: Evidence from spatial econometric models. *Environmental Science and Pollution Research International*, *26*(18), 18139–18151. DOI: 10.1007/s11356-019-04522-9 PMID: 31041708

Shin, Y., Yu, B., & Greenwood-Nimmo, M. (2014). Modelling asymmetric cointegration and dynamic multipliers in a nonlinear ARDL framework. In Sickles, R. C., & Horrace, W. C. (Eds.), *Festschrift in Honor of Peter Schmidt* (pp. 281–314). Springer., DOI: 10.1007/978-1-4899-8008-3_9

Stern, N., & Valero, A. (2021). Innovation, growth and the transition to net-zero emissions. *Research Policy*, *50*(9), 104293. DOI: 10.1016/j.respol.2021.104293 PMID: 36540358

Tajudeen, I. A., Wossink, A., & Banerjee, P. (2018). How significant is energy efficiency to mitigate CO2 emissions? Evidence from OECD countries. *Energy Economics*, *72*, 200–221. DOI: 10.1016/j.eneco.2018.04.010

Teng, J., & Cheng, X. (2020). Does renewable energy consumption mitigate CO2 emissions? Evidence from G7 countries. *Environmental Science and Pollution Research International*, *27*(7), 7630–7639. DOI: 10.1007/s11356-019-07476-7

Wang, S., Zeng, J., & Liu, X. (2019). Examining the multiple impacts of technological progress on CO2 emissions in China: A panel quantile regression approach. *Renewable & Sustainable Energy Reviews*, *103*, 140–150. DOI: 10.1016/j.rser.2018.12.046

Zhang, J., & Wang, J. (2017). Renewable energy, economic growth, and carbon emissions: Evidence from a panel of developing countries. *Renewable Energy*, *101*, 137–145. DOI: 10.1016/j.renene.2016.08.058

Zhang, Y., & Zhao, T. (2015). Urbanization and environmental sustainability: The case of China. *Environmental Science and Pollution Research International*, *22*(22), 17618–17629. DOI: 10.1007/s11356-015-4789-2

Compilation of References

Abada, S., Marlair, G., Lecocq, A., Petit, M., Sauvant-Moynot, V., & Huet, F. (2016). Safety focused modeling of lithium-ion batteries: A review. *Journal of Power Sources*, *306*, 178–192. DOI: 10.1016/j.jpowsour.2015.11.100

Abdeen, M., Li, H., Kamel, S., Khaled, A., El-Dabah, M., Kharrich, M., & Sindi, H. F. (2021). A recent analytical approach for analysis of sub-synchronous resonance in doubly-fed induction generator-based wind farm. *IEEE Access : Practical Innovations, Open Solutions*, *9*, 68888–68897. DOI: 10.1109/ACCESS.2021.3075965

Abdelkader, S., Amissah, J., Kinga, S., Mugerwa, G., Emmanuel, E., Mansour, D.-E. A., Bajaj, M., Blazek, V., & Prokop, L. (2024). Securing modern power systems: Implementing comprehensive strategies to enhance resilience and reliability against cyber-attacks. *Results in Engineering*, *23*, 102647. DOI: 10.1016/j.rineng.2024.102647

Abdelkafi, A. & Krichen, L. (2014). Energy management optimization of a hybrid power production unit based renewable energies. International Journal of Electric Power & Energy, 1-9.

Abdi, A. H., Audah, L., Salh, A., Alhartomi, M. A., Rasheed, H., Ahmed, S., & Tahir, A. (2024). Security Control and Data Planes of SDN: A Comprehensive Review of Traditional, AI and MTD Approaches to Security Solutions. *IEEE Access : Practical Innovations, Open Solutions*.

Abdin, Z., & Mérida, W. (2019). A techno-economic analysis of hybrid renewable energy-based off-grid power supply and hydrogen production. *Energy Conversion and Management*, *196*, 1068–1079. DOI: 10.1016/j.enconman.2019.06.068

Abdulkadhim, A., Abed, I. M., & Mahjoub Said, N. (2021). Review of Natural Convection Within Various Shapes of Enclosures. *Arabian Journal for Science and Engineering*, *46*(12), 11543–11586. DOI: 10.1007/s13369-021-05952-6

Abdulla, H., Sleptchenko, A., & Nayfeh, A. (2024). Photovoltaic systems operation and maintenance: A review and future directions. In *Renewable and Sustainable Energy Reviews* (Vol. 195). Elsevier Ltd., DOI: 10.1016/j.rser.2024.114342

Abebe, T., & Worku, A. (2022). Adoption of biomass briquettes as a sustainable energy source in Ethiopia: Barriers and enabling factors. *Sustainability*, *14*(5), 789–801. https://doi.org/10.xxxx/su1405789

Abhyankar, N., Mohanty, P., Deorah, S., Karali, N., Paliwal, U., Kersey, J., & Phadke, A. (2023). India's path towards energy independence and a clean future: Harnessing india's renewable edge for cost-effective energy independence by 2047. *The Electricity Journal*, *36*(5), 107273. DOI: 10.1016/j.tej.2023.107273

Abolhosseini, S., & Heshmati, A. (2014). The main support mechanisms to finance renewable energy development. *Renewable & Sustainable Energy Reviews*, *40*, 876–885. DOI: 10.1016/j.rser.2014.08.013

Abraham, K. (2024). Artificial Intelligence Applications in Sustainable Infrastructure: Solar Photovoltaic Inspection and Estimation of Indoor Environmental Variables in a Smart Building.

Acen, C., Bamisile, O., Cai, D., Ukwuoma, C. C., Obiora, S., Huang, Q., Uzun Ozsahin, D., & Adun, H. (2024). The complementary role of carbon dioxide removal: A catalyst for advancing the COP28 pledges towards the 1.5° C Paris Agreement target. *The Science of the Total Environment*, *947*, 174302. DOI: 10.1016/j.scitotenv.2024.174302 PMID: 38945244

Adachi, S. (2007). Lattice thermal conductivity of group-IV and III–V semiconductor alloys. *Journal of Applied Physics*, *102*(6), 063502. DOI: 10.1063/1.2779259

Adaine, A., Sidi, O. S. H., & Fressengeas, N. (2016). Simulation study of a new InGaN p-layer free Schottky based solar cell. *Superlattices and Microstructures*, *96*, 121–133. DOI: 10.1016/j.spmi.2016.05.020

Adaliou, A. H., Lamreoua, A., Isknan, I., Doubabi, S., Melhaoui, M., El Ouariachi, M., & Hirech, K. (2023). High-Performance MPPT Based on Developed Fast Convergence for PV System-Experimental Validation. *International Conference on Digital Technologies and Applications*, 694–703. DOI: 10.1007/978-3-031-29857-8_69

Adeyemi, M., & Johnson, L. (2023). Fuel-Briquetting for Sustainable Development in Developing Countries. *Sustainable Energy Journal*, *19*(3), 298–310. DOI: 10.1155/2023/9837214

Admi, Y., Moussaoui, M. A., & Mezrhab, A. (2020). Effect of a flat plate on heat transfer and flow past a three side-by-side square cylinders using double MRT-lattice Boltzmann method. *2020 IEEE 2nd International Conference on Electronics, Control, Optimization and Computer Science, ICECOCS 2020.*

Admi, Y., Moussaoui, M. A., & Mezrhab, A. (2020). Effect of a flat plate on heat transfer and flow past a three side-by-side square cylinders using double MRT-lattice Boltzmann method. In: IEEE 2nd International Conference on Electronics, Control, Optimization and Computer Science (ICECOCS). IEEE, p. 1-5.

Admi, Y., El Hassani, S., Benhamou, J., Moussaoui, M. A., & Mezrhab, A. (2024). Numerical research on the wake characteristics and drag reduction of a square cylinder in the presence of control partitions. *Ocean Engineering*, *301*, 117341. DOI: 10.1016/j.oceaneng.2024.117341

Admi, Y., Lahmer, E. B., Benhamou, J., Moussaoui, M. A., & Mezrhab, A. (2024). Heat transfer improvement and drag force reduction around three heated square cylinders controlled by partitions. *Physics of Fluids*, *36*(4), 045161. DOI: 10.1063/5.0191092

Admi, Y., Moussaoui, M. A., & Mezrhab, A. (2022). Numerical Investigation of Convective Heat Transfer and Fluid Flow Past a Three Square Cylinders Controlled by a Partition in Channel. *International Journal of Renewable Energy Development*, *11*(3), 766–781. DOI: 10.14710/ijred.2022.43790

Admi, Y., Moussaoui, M. A., & Mezrhab, A. Effect of Control Partitions on Drag Reduction and Suppression of Vortex Shedding Around a Bluff Body Cylinder. In: *International Conference on Advanced Technologies for Humanity*. Springer, Cham, p. 453-463, (2021).

Admi, Y., Moussaoui, M. A., & Mezrhab, A. The Vortex Shedding Suppression and Heat Transfer Characteristics Around a Heated Square Cylinder by Using Three Downstream-Detached Partitions. In: *International Conference on Digital Technologies and Applications*. Springer, Cham, p. 598-608(2022). DOI: 10.1007/978-3-031-02447-4_62

Ae, L. A, P., & Jw, A. (n.d.). The carbon footprint of an Australian satellite haemodialysis unit. *PubMed*. Retrieved December 9, 2024, from https://pubmed.ncbi.nlm.nih.gov/23731962/

Agajie, E. F., Agajie, T. F., Amoussou, I., Fopah-Lele, A., Nsanyuy, W. B., Khan, B., Bajaj, M., Zaitsev, I., & Tanyi, E. (2024). Optimization of off-grid hybrid renewable energy systems for cost-effective and reliable power supply in Gaita Selassie Ethiopia. *Scientific Reports*, *14*(1), 10929. DOI: 10.1038/s41598-024-61783-z PMID: 38740883

Aghaei, J., Nezhad, A. E., Rabiee, A., & Rahimi, E. (2016). Contribution of Plug-in Hybrid Electric Vehicles in power system uncertainty management. *Renewable & Sustainable Energy Reviews*, *59*, 450–458. DOI: 10.1016/j.rser.2015.12.207

Aghakhani, S., Pordanjani, A. H., Karimipour, A., Abdollahi, A., & Afrand, M. (2018). Numerical investigation of heat transfer in a power-law non-Newtonian fluid in a C-Shaped cavity with magnetic field effect using finite difference lattice Boltzmann method. *Computers & Fluids*, *176*, 51–67. Advance online publication. DOI: 10.1016/j.compfluid.2018.09.012

Ahmadizadeh, M., Heidari, M., Thangavel, S., Al Naamani, E., Khashehchi, M., Verma, V., & Kumar, A. (2024). *Technological advancements in sustainable and renewable solar energy systems Highly Efficient Thermal Renewable Energy Systems*. CRC Press.

Ahmad, T., Zhang, D., Huang, C., Zhang, H., Dai, N., Song, Y., & Chen, H. (2021). Artificial intelligence in sustainable energy industry: Status Quo, challenges and opportunities. *Journal of Cleaner Production*, *289*, 125834. DOI: 10.1016/j.jclepro.2021.125834

Ahmed, A., Elsakka, M., Elminshawy, N., Mohamed, A., & Sundaram, S. (2023). Recent Advances in Floating Photovoltaic Systems. In *Chemical Record* (Vol. 23, Issue 12). John Wiley and Sons Inc. DOI: 10.1002/tcr.202300229

Ahmed, A., Mwangi, M., & Zhao, L. (2023). Traditional Fuel Preferences and Modern Alternatives: A Case Study. *Energy Policy*, *54*(1), 45–56. DOI: 10.1016/j.enpol.2022.11.015

Ahmed, M., & Shimada, K. (2019). The effect of renewable energy consumption on sustainable economic development: Evidence from emerging and developing economies. *Energies*, *12*(15), 2954. DOI: 10.3390/en12152954

Ajide, K. B., & Mesagan, E. P. (2022). Heterogeneous analysis of pollution abatement via renewable and non-renewable energy: Lessons from investment in G20 nations. *Environmental Science and Pollution Research International*, *29*(24), 36533–36546. DOI: 10.1007/s11356-022-18771-5 PMID: 35066851

Akpolat, A. N., Dursun, E., Kuzucuoğlu, A. E., Yang, Y., Blaabjerg, F., & Baba, A. F. (2019). Performance Analysis of a Grid-Connected Rooftop Solar Photovoltaic System. *Electronics (Basel)*, *8*(8), 905. DOI: 10.3390/electronics8080905

Alami, A. H., Rabaia, M. K. H., Sayed, E. T., Ramadan, M., Abdelkareem, M. A., Alasad, S., & Olabi, A.-G. (2022). Management of potential challenges of PV technology proliferation. *Sustainable Energy Technologies and Assessments*, *51*, 101942. DOI: 10.1016/j.seta.2021.101942

Alazbe, K., Roy, H. A., Alremeithi, A., AlKindi, S., AlHashmi, M., Daccache, L., & Alnuaimi, A. (2023). Insight into dust soiling characteristics and cleaning solutions to investigate impact on PV systems performance across the globe. *Progress in Energy*, *6*(1), 012004.

AlHajri, I., Ahmadian, A., & Elkamel, A. (2021). Stochastic day-ahead unit commitment scheduling of integrated electricity and gas networks with hydrogen energy storage (HES), plug-in electric vehicles (PEVs) and renewable energies. *Sustainable Cities and Society*, *67*, 102736. DOI: 10.1016/j.scs.2021.102736

Alharbi, S. S., Al Mamun, M., Boubaker, S., & Rizvi, S. K. A. (2023). Green finance and renewable energy: A worldwide evidence. *Energy Economics*, *118*, 106499. DOI: 10.1016/j.eneco.2022.106499

Ali, N., Phoungthong, K., Techato, K., Ali, W., Abbas, S., Dhanraj, J.A., Khan, A. (2022). FDI, Green Innovation and Environmental Quality Nexus: New Insights from BRICS Economies. Sustainability. 14, 2181. Doi.org/10.3390/ su14042181.

Ali, K., Jianguo, D., Kirikkaleli, D., Oláh, J., & Altuntaş, M. (2023). Do green technological innovation, financial development, economic policy uncertainty, and institutional quality matter for environmental sustainability? *All Earth*, *35*(1), 2200330. Advance online publication.

Alimi, O. A., Meyer, E. L., & Olayiwola, O. I. (2022). Solar Photovoltaic Modules' Performance Reliability and Degradation Analysis—A Review. In *Energies* (Vol. 15, Issue 16). MDPI. DOI: 10.3390/en15165964

Ali-Ou-Salah, H., Oukarfi, B., & Mouhaydine, T. (2022). Short-term solar radiation forecasting using a new seasonal clustering technique and artificial neural network. *International Journal of Green Energy*, *19*(4), 424–434.

Alkahtani, M. (2021). *Study for Non-uniform Aging Photovoltaic Array Performance Enhancements*.

Al-Karaghouli, A., & Kazmerski, L. (2013). Energy consumption and water production cost of conventional and renewable-energy-powered desalination processes. *Renewable & Sustainable Energy Reviews*, *24*, 343–356. DOI: 10.1016/j.rser.2012.12.064

Al-Karaghouli, A., Renne, D., & Kazmerski, L. (2009). Solar and wind opportunities for water desalination in the Arab regions. *Renewable & Sustainable Energy Reviews*, *13*(9), 2397–2407. DOI: 10.1016/j.rser.2009.04.003

Almalawi, A., Soh, B., Li, A., & Samra, H. (2024). Predictive Models for Educational Purposes: A Systematic Review. *Big Data and Cognitive Computing*, *8*(12), 187. DOI: 10.3390/bdcc8120187

Almosni, S., Robert, C., Nguyen Thanh, T., Cornet, C., Létoublon, A., Quinci, T., Levallois, C., Perrin, M., Kuyyalil, J., Pedesseau, L., Balocchi, A., Barate, P., Even, J., Jancu, J. M., Bertru, N., Marie, X., Durand, O., & Le Corre, A. (2013). Evaluation of InGaPN and GaAsPN materials lattice-matched to Si for multi-junction solar cells. *Journal of Applied Physics*, *113*(12), 123509. DOI: 10.1063/1.4798363

Al-mulali, U., & Lee, J. Y. (2013). The impact of renewable energy consumption on CO2 emissions: Evidence from 12 countries. *Renewable & Sustainable Energy Reviews*, *22*, 198–202. DOI: 10.1016/j.rser.2013.02.039

Alqallaf, N., & Ghannam, R. (2024, March). Immersive Learning in Photovoltaic Energy Education: A Comprehensive Review of Virtual Reality Applications. In *Solar* (Vol. 4, No. 1, pp. 136-161). MDPI. DOI: 10.3390/solar4010006

Alraja, M. N., Imran, R., Basel, M. K., & Shah, M. (2022). Technological Innovation, Sustainable Green Practices and SMEs. Sustainable Performance in Times of Crisis (COVID-19 pandemic). *Information Systems Frontiers*, *24*(4), 1081–1105. DOI: 10.1007/s10796-022-10250-z PMID: 36504756

Al-Rubaye, S., Kadhum, E., Ni, Q., & Anpalagan, A. (2017). Industrial internet of things driven by SDN platform for smart grid resiliency. *IEEE Internet of Things Journal*, *6*(1), 267–277. DOI: 10.1109/JIOT.2017.2734903

Al-Shahri, O. A., Ismail, F. B., Hannan, M. A., Lipu, M. S. H., Al-Shetwi, A. Q., Begum, R. A., Al-Muhsen, N. F. O., & Soujeri, E. (2021). Solar photovoltaic energy optimization methods, challenges and issues: A comprehensive review. In *Journal of Cleaner Production* (Vol. 284). Elsevier Ltd., DOI: 10.1016/j.jclepro.2020.125465

Alshuraiaan, B. (2023). Fluid Structure Interaction Analysis of Natural Convective Flow and Heat Transfer in an L-Shaped Cavity. *Journal of Advanced Research in Fluid Mechanics and Thermal Sciences*. DOI: 10.37934/arfmts.109.1.112

Althobaiti, A., Armstrong, M. and Elgendy. M. A. (2016) '' Current Control of Three-phase Grid-connected PV Inverters using Adaptive PR Controller'', *IEEE Trans* 978-1-4673-9768, 2016.

Amare, A., Tadesse, T., & Gebre, G. (2022). Energy Efficiency and Emission Reduction Potential of Biomass Briquettes Made from Agricultural Residues in Ethiopia. *Renewable Energy Research Journal*, *18*(2), 98–110. DOI: 10.1016/j.retr.2022.02.003

Amin, M., Rygg, A., & Molinas, M. (2017). Self-synchronization of wind farm in an MMC-based HVDC system: A stability investigation. *IEEE Transactions on Energy Conversion*, *32*(2), 458–470. DOI: 10.1109/TEC.2017.2661540

Amin, N., Shabbir, M. S., Song, H., & Abbass, K. (2024). Renewable energy consumption and its impact on environmental quality: A pathway for achieving sustainable development goals in ASEAN countries. *Energy & Environment*, *35*(2), 644–662. DOI: 10.1177/0958305X221134113

An, D., Gong, G., Xu, D., Zhou, Z., Li, R., & Su, Y. (2024). Embracing Plasticity: Unlocking the Full Potential of Flexible and Stretchable Electronics Through the Elastoplastic Behavior of Metallic Materials. *Advanced Functional Materials*, •••, 2412796. DOI: 10.1002/adfm.202412796

Andrić, I.. (2023). Advances in Biomass Co-Combustion for Carbon Mitigation. *Journal of Energy Technologies*, *15*(3), 89–105. DOI: 10.1016/j.jenergy.2023.06.005

Aniello, G., Shamon, H., & Kuckshinrichs, W. (2021). Micro-economic assessment of residential PV and battery systems: The underrated role of financial and fiscal aspects. *Applied Energy*, *281*, 115667. DOI: 10.1016/j.apenergy.2020.115667

Anil Sai Kumar, B. V., Kumar, S. P., Jaswanth, K., Vishnu, K., & Kumar, A. (2023). Smart irrigation system using machine learning techniques. Machine Learning for VLSI Chip Design, 139-159.

Apergis, N., & Payne, J. E. (2010). The impact of renewable energy consumption on economic growth: Evidence from a panel of EU countries. *Energy Policy*, *38*(1), 442–448. DOI: 10.1016/j.enpol.2009.09.002

Appiah, K., Du, J., Appah, R., & Quacoe, D. (2018). Prediction of potential carbon dioxide emissions of selected emerging economies using artificial neural network. changes, 14, 321-335.

Appiah-Otoo, I., Chen, X., & Kursah, M. B. (2023). Modelling the impact of renewable energy investment on global carbon dioxide emissions. *Energy Reports*, *10*, 3787–3799. DOI: 10.1016/j.egyr.2023.10.043

Ardiansah, I., Bafdal, N., Suryadi, E., & Bono, A. (2020). Greenhouse monitoring and automation using Arduino: A review on precision farming and internet of things (IoT). *International Journal on Advanced Science, Engineering and Information Technology*, *10*(2), 703–709. DOI: 10.18517/ijaseit.10.2.10249

Arriaga Arruti, O., Virtuani, A., & Ballif, C. (2023). Long-term performance and reliability of silicon heterojunction solar modules. In *Progress in Photovoltaics: Research and Applications* (Vol. 31, Issue 7, pp. 664–677). John Wiley and Sons Ltd. DOI: 10.1002/pip.3688

Asha, N. E. J., Nag, P., Akhter, M. N., & Molla, M. M. (2023). MRT-lattice Boltzmann simulation of magnetic field effects on heat transfer from a heater in a C-shaped cavity filled with non-Newtonian hybrid nanofluids. *International Journal of Thermofluids*. DOI: 10.1016/j.ijft.2023.100345

Aslam, B., Hu, J., Majeed, M. T., Andlib, Z., & Ullah, S. (2021). Asymmetric macroeconomic determinants of CO_2 emission in China and policy approaches. *Environmental Science and Pollution Research International*, *28*(31), 41923–41936. DOI: 10.1007/s11356-021-13743-7 PMID: 33797039

Assia, S., Ikram, E. A., Abdellah, E. B., & Ahmed, E. B. (2018). *« Comparative analysis the simultaneous scheduling problems of production and maintenance activities», April 2018 Indexed SCOPUS*. IEEE Xplore., DOI: 10.1109. LOGISTIQUA.2018.8428283

Assia, S., Ikram, E. A., Abdellah, E. B., & Ahmed, E. B. (2019). *« Non-permutation flow shop scheduling problems with unavailability constraints to Minimize Total Energy Consumption », April 2019 Indexed SCOPUS*. IEEE Xplore., DOI: 10.1109/ICOA.2019.8727649

Asumadu-Sarkodie, S., & Owusu, P. A. (2016). Forecasting Nigeria's energy use by 2030, an econometric approach. *Energy Sources. Part B, Economics, Planning, and Policy*, *11*(10), 990–997. DOI: 10.1080/15567249.2016.1217287

Ayodele, T., Ogunjuyigbe, A., Amedu, A., & Munda, J. (2019). Prediction of global solar irradiation using hybridized k-means and support vector regression algorithms. *Renewable Energy Focus*, *29*, 78–93. DOI: 10.1016/j.ref.2019.03.003

Azeem, A., Ansari, M. K., Tariq, M., Sarwar, A., & Ashraf, I. (2019). Design and Modeling of Solar Photovoltaic System Using Seven-Level Packed U-Cell (PUC) Multilevel Inverter and Zeta Converter for Off-Grid Application in India. *Electrica, 19*(2), 101–112. DOI: 10.26650/electrica.2019.18053

Azhgaliyeva, D., Beirne, J., & Mishra, R. (2023). What matters for private investment in renewable energy? *Climate Policy, 23*(1), 71–87. DOI: 10.1080/14693062.2022.2069664

Azimi, R., Ghayekhloo, M., & Ghofrani, M. (2016). A hybrid method based on a new clustering technique and multilayer perceptron neural networks for hourly solar radiation forecasting. *Energy Conversion and Management, 118*, 331–344. DOI: 10.1016/j.enconman.2016.04.009

Azzouz, T. (2015). Modélisation et commande d'un système de conversion d'énergie éolienne à base d'une MADA (Doctoral dissertation, Université Mohamed Khider-Biskra).

Baggio, G., Qadir, M., & Smakhtin, V. (2021). Freshwater availability status across countries for human and ecosystem needs. *The Science of the Total Environment, 792*, 148230. DOI: 10.1016/j.scitotenv.2021.148230 PMID: 34147805

Baghel, N. S., & Chander, N. (2022). Performance comparison of mono and polycrystalline silicon solar photovoltaic modules under tropical wet and dry climatic conditions in east-central India. *Clean Energy, 6*(1), 929–941. DOI: 10.1093/ce/zkac001

Baïri, A., Zarco-Pernia, E., & García De María, J. M. (2014). A review on natural convection in enclosures for engineering applications. The particular case of the parallelogrammic diode cavity. *Applied Thermal Engineering, 63*(1), 304–322. DOI: 10.1016/j.applthermaleng.2013.10.065

Bajwa, D. S.. (2023). Biomass Briquetting for Sustainable Energy in Developing Countries. *MDPI Sustainability, 15*(5), 2334–2350. DOI: 10.3390/su15052334

Balaji, S., & Sasilatha, T. (2019, November). Detection of denial of service attacks by domination graph application in wireless sensor networks. *Cluster Computing, 22*(6), 15121–15126. DOI: 10.1007/s10586-018-2504-5

Ballif, C., Haug, F. J., Boccard, M., Verlinden, P. J., & Hahn, G. (2022). Status and perspectives of crystalline silicon photovoltaics in research and industry. In *Nature Reviews Materials* (Vol. 7, Issue 8, pp. 597–616). Nature Research. DOI: 10.1038/s41578-022-00423-2

Bar-Cohen, A., & Kraus, A. D., S. F. D. (1983). Thermal frontiers in the design and packaging of microelectronic equipment. *Mechanical Engineering (New York, N.Y.), 105*(5), 530–539.

Barraclough, K. A., & McAlister, S. (2022). Assessing the Carbon Footprint of Hemodialysis: A First Step Toward Environmentally Sustainable Kidney Care. *Journal of the American Society of Nephrology, 33*(9), 1635–1637. DOI: 10.1681/ASN.2022060661 PMID: 35840174

Batu, M., Ayele, A., & Berhe, T. (2023). Biomass briquetting in Ethiopia: Opportunities and challenges. *Energy for Sustainable Development, 56*, 98–107. https://www.sciencedirect.com/science/article/abs/pii/S2352484723001557

Bayrakçı, H. C., Demircan, C., & Keçebaş, A. (2018). The development of empirical models for estimating global solar radiation on horizontal surface: A case study. *Renewable & Sustainable Energy Reviews*, *81*, 2771–2782. DOI: 10.1016/j.rser.2017.06.082

Bechar, A., & Vigneault, C. (2016). Agricultural robots for field operations: Concepts and components. *Biosystems Engineering*, *149*, 94–111. DOI: 10.1016/j.biosystemseng.2016.06.014

Bedi, H. S., Raju, D. V., Nandyala, M. R. C., Kumar, P. S., & Varma, M. R. (2022). Design of Gesture Based Hand Gloves Using Arduino UNO: A Grace to Abled Mankind. Digital Forensics and Internet of Things: Impact and Challenges, 37-43.

Bedi, H. S., Goyal, N., Kumar, S., & Gupta, A. (2017). Smart Trolley using Smart Phone and Arduino. *J Electr Electron Syst*, *6*(2), 223. DOI: 10.4172/2332-0796.1000223

Bedi, H. S., Harshin, A. N., Lawai, B. P., & Kumar, N. (2024). Design of Intelligent Polyhouse with IOT. *Indonesian Journal of Interdisciplinary Research in Science and Technology*, *2*(3), 231–244. DOI: 10.55927/marcopolo.v2i3.8095

Bedi, H. S., & Verma, S. (2015). R. K. Sharma, and B. Singh (2015)The Concept of Programmable Logic Controllers and its role in Automation. *International Journal of Advanced Research in Computer and Communication Engineering*, *4*(6).

Ben Ali, I. (2010). Caractérisation expérimentale, commande et mise en place d'un gestionnaire d'énergie d'un procédé de dessalement de l'eau alimenté par une source hybride photovoltaïque-éolien. Master's thesis, LSE-ENIT Tunisie

Ben Ali, I., Turki, M., Belhadj, J., & Roboam, X. (2012, March 25-28). Energy management of reverse osmosis desalination process powered by a hybrid renewable energy source. *16th IEEE Mediterranean Electrotechnical Conference (MELECON)*, Yasmine Hammamet, Tunisia, 800–805

Ben Ali, I., Turki, M., Belhadj, J., & Roboam, X. (2014). Systemic design of a reverse osmosis desalination process powered by hybrid energy system: Bond graph modeling approach & experimental validation. In *Proceedings of the 1st International Conference on Electrical Sciences and Technologies in Maghreb (CISTEM)*, Tunis, Tunisia, 3–6 November. DOI: 10.1109/CISTEM.2014.7076941

Ben Ali, I., Turki, M., Belhadj, J., & Roboam, X. (2014, May 19-22). Fuzzy logic-based power management for reverse osmosis desalination process fed by intermittent hybrid renewable source. *11th International Conference on Modeling and Simulation of Electric Machines, Converters, and Systems (ElectrIMACS)*, Valencia, Spain

Ben Ali, I., Turki, M., Belhadj, J., & Roboam, X. (2016, March 22-24). Using quasi-static model for water/power management of a stand-alone wind/photovoltaic/BWRO desalination system without batteries. *7th International Renewable Energy Congress (IREC)*, Hammamet, Tunisia.

Benhamou, J., Channouf, S., Lahmer, E. B., Jami, M., & Mezrhab, A. (2024). Hybrid-lattice Boltzmann Method for the Simulation of Magnetohydrodynamic Conjugate Heat Transfer and Entropy Generation in Three Dimensions. *Arabian Journal for Science and Engineering*, *49*(1), 1181–1206. Advance online publication. DOI: 10.1007/s13369-023-08273-y

Benhamou, J., Jami, M., Mezrhab, A., Botton, V., & Henry, D. (2020). Numerical study of natural convection and acoustic waves using the lattice Boltzmann method. *Heat Transfer*, *49*(6), 3779–3796. DOI: 10.1002/htj.21800

Benhamou, J., Lahmer, E. B., Admi, Y., Jami, M., & Mezrhab, A. (2023). Three-dimensional numerical simulation of free convection and entropy generation in a cubic cavity containing a heat source. *Heat Transfer*, *52*(7), 5108–5135. DOI: 10.1002/htj.22920

Benhamou, J., Lahmer, E. B., & Jami, M. (2022). Three-dimensional simulation of conjugate heat transfer using the hybrid lattice Boltzmann-finite difference method. *International Communications in Heat and Mass Transfer*, *139*, 106486. DOI: 10.1016/j.icheatmasstransfer.2022.106486

Bennani, S., Maalel, A., & Ben Ghezala, H. (2022). Adaptive gamification in E-learning: A literature review and future challenges. *Computer Applications in Engineering Education*, *30*(2), 628–642. DOI: 10.1002/cae.22477

Bett, P. E., & Thornton, H. E. (2016). The climatological relationships between wind and solar energy supply in Britain. *Renewable Energy*, *87*, 96–110. DOI: 10.1016/j.renene.2015.10.006

Bhattacharyya, S., & Palit, D. (2021). A critical review of literature on the nexus between central grid and off-grid solutions for expanding access to electricity in Sub-Saharan Africa and South Asia. *Renewable & Sustainable Energy Reviews*, *141*, 110792. DOI: 10.1016/j.rser.2021.110792

Bhushan, R., Kumar, D., & Chatterjee, K. (2021). Disturbance rejection of the powers and DC-link voltage of a doubly-fed induction generator using state-space based linear quadratic integral optimal control approach. *International Transactions on Electrical Energy Systems*, *31*(5), e12865. DOI: 10.1002/2050-7038.12865

Bilgen, E. (2002). Natural convection in enclosures with partial partitions. *Renewable Energy*, *26*(2), 257–270. DOI: 10.1016/S0960-1481(00)00206-8

Bimberg, D., Heinrichsdorff, F., Bauer, R. K., Gerthsen, D., Stenkamp, D., Mars, D. E., & Miller, J. N. (1992). Binary AlAs/GaAs versus ternary GaAlAs/GaAs interfaces: A dramatic difference of perfection. *Journal of Vacuum Science & Technology B Microelectronics and Nanometer Structures*, *10*, 1793.

Bin Abu Sofian, A. D. A., Lim, H. R., Siti Halimatul Munawaroh, H., Ma, Z., Chew, K. W., & Show, P. L. (2024). Machine learning and the renewable energy revolution: Exploring solar and wind energy solutions for a sustainable future including innovations in energy storage. *Sustainable Development (Bradford)*, *32*(4), 3953–3978. DOI: 10.1002/sd.2885

Biswal, C., Sahu, B. K., Mishra, M., & Rout, P. K. (2023). Real-time grid monitoring and protection: A comprehensive survey on the advantages of phasor measurement units. *Energies*, *16*(10), 4054. DOI: 10.3390/en16104054

Bonomini, M., Piscitani, L., Di Liberato, L., & Sirolli, V. (2022). Biocompatibility of Surface-Modified Membranes for Chronic Hemodialysis Therapy. *Biomedicines*, *10*(4), 4. Advance online publication. DOI: 10.3390/biomedicines10040844 PMID: 35453594

Bouzidi, M., Firdaouss, M., & Lallemand, P. (2001). Momentum transfer of a Boltzmann lattice fluid with boundaries. *Physics of Fluids*, *13*(11), 3452–3459. DOI: 10.1063/1.1399290

Box, G. E. P., & Jenkins, G. M. (1970). *Time Series Analysis: Forecasting and Con-trol*. Holden-Day.

Boyer-Richard, S., Fan, F., Beck, A., Levallois, C., Tavernier, K., Rohel, T., & Durand, O. (2023). Investigation of III-V GaP solar cell on silicon substrate. *EPJ Photovoltaics*, *14*, 31.

Brassai, S. T., Enăchescu, C., & Losonczi, L. (2012, May). RBF network for mobile robot sonar based localization and environment modeling. In 2012 13th International Conference on Optimization of Electrical and Electronic Equipment (OPTIM) (pp. 1499-1504). IEEE. DOI: 10.1109/OPTIM.2012.6231939

Breuer, M., Bernsdorf, J., Zeiser, T., & Durst, F. (2000). Accurate computations of the laminar flow past a square cylinder based on two different methods: Lattice Boltzmann and finite-volume. *International Journal of Heat and Fluid Flow*, *21*(2), 186–196. DOI: 10.1016/S0142-727X(99)00081-8

Bseiso, A., Abele, B., Ferguson, S., Lusch, P., & Mehta, K. (2015, October). A decision support tool for greenhouse farmers in low-resource settings. In *2015 IEEE Global Humanitarian Technology Conference (GHTC)* (pp. 292-297). IEEE. DOI: 10.1109/GHTC.2015.7343987

Buchberg, H., Cation, I., & Edwards, D. K. (1976). Natural convection in enclosed spaces—A review of application to solar energy collection. *Journal of Heat Transfer*, *98*(2), 182–188. DOI: 10.1115/1.3450516

Bukar, U. A., Sayeed, M. S., Razak, S. F. A., Yogarayan, S., & Sneesl, R. (2024). Decision-making framework for the utilization of generative artificial intelligence in education: A case study of ChatGPT. *IEEE Access : Practical Innovations, Open Solutions*, *12*, 95368–95389. DOI: 10.1109/ACCESS.2024.3425172

Bulavko, G. V. (2024). Organic photovoltaics: A journey through time, advancements, and future opportunities. *History of Science and Technology*, *14*(1), 10–32. DOI: 10.32703/2415-7422-2024-14-1-10-32

Bullich-Massagué, E., Cifuentes-García, F.-J., Glenny-Crende, I., Cheah-Mañé, M., Aragüés-Peñalba, M., Díaz-González, F., & Gomis-Bellmunt, O. (2020). A review of energy storage technologies for large scale photovoltaic power plants. *Applied Energy*, *274*, 115213. DOI: 10.1016/j.apenergy.2020.115213

Burton, T., Jenkins, N., Sharpe, D., & Bossanyi, E. (2021). *Wind Energy Handbook* (3rd ed.). Wiley.

Buyanova, I. A., Pozina, G., Bergman, J. P., Chen, W. M., Xin, H. P., & Tu, C. W. (2002). Time-resolved studies of photoluminescence in GaNxP1−x alloys: Evidence for indirect-direct band gap crossover. *Applied Physics Letters*, *81*(1), 52–54. DOI: 10.1063/1.1491286

Buzila Pankovits, P. V. (2015). Gestion énergétique optimale des installations fixes de traction électrique ferroviaire hybrides. (PhD thesis, Université Lille 1, Sciences et Technologies, France, September 2015).

Calautit, K., & Johnstone, C. (2023). State-of-the-art review of micro to small-scale wind energy harvesting technologies for building integration. *Energy Conversion and Management: X*, *20*, 100457. DOI: 10.1016/j.ecmx.2023.100457

Canada, P. H. A. of. (2022, September 8). *Health and climate change 2019* [Education and awareness;research]. https://www.canada.ca/en/public-health/services/reports-publications/canada-communicable-disease-report-ccdr/monthly-issue/2019-45/issue-5-may-2-2019/article-1-health-solution-climate-change-risks-solutions.html

Carlo, J., Melendez, J., Rossi, M., & Renzi, M. (2020). Optimal sizing of a hybrid renewable energy system: Importance of data selection with highly variable renewable energy sources. *Energy Conversion and Management*, *223*, 113303. DOI: 10.1016/j.enconman.2020.113303

Carrasco, M., Mancilla-David, F., & Ortega, R. (2014). An Estimator of Solar Irradiance in Photovoltaic Arrays With Guaranteed Stability Properties. *IEEE Transactions on Industrial Electronics*, *61*(7), 3359–3366. DOI: 10.1109/TIE.2013.2281154

Cedrick, B. Z. E., & Long, P. W. (2017). Investment motivation in renewable energy: A PPP approach. *Energy Procedia*, *115*, 229–238.

Chaib, S. O., Joti, I., & Khoulji, S. (2024, February). Gamification in mobile language learning applications: A case study examining a mobile learning tool in a Moroccan university context. In *E-Learning and Smart Engineering Systems (ELSES 2023)* (pp. 177-185). Atlantis Press.

Chaib, S. O., Joti, I., & Khoulji, S. (2023). Design of a specialized search engine for university students dedicated to education and environment. In *E3S Web of Conferences* (Vol. 412, p. 01060). EDP Sciences.

Chaib, S. O., Joti, I., & Khoulji, S. (2023). Evaluation of a Computer-Assisted Language Learning System Based on Adaptive Learning Designed for Self-training in Scientific French Language. *International Journal of Information and Education Technology (IJIET)*, *13*(8), 1284–1296. DOI: 10.18178/ijiet.2023.13.8.1930

Chaouali, I., & Ben Jebli, M., & GAM, I. (2023). An assessment of the influence of clean energy and service development on environmental degradation: Evidence for a non-linear ARDL approach for Tunisia. *Environmental Science and Pollution Research International.* Advance online publication. PMID: 37291349

Chapain, S., & Aly, A. M. (2019). Vibration attenuation in high-rise buildings to achieve system-level performance under multiple hazards. *Engineering Structures*, *197*, 109352. DOI: 10.1016/j.engstruct.2019.109352

Chaudhary, S., Patel, V., & Mehta, K. (2024). Influence of Feedstock Mixtures on the Fuel Characteristics of Blended Briquettes. *Springer Nature Journal of Biomass Engineering*, *22*(2), 78–90. DOI: 10.1007/s13399-023-04039-6

Chen, G., Jia, M., Zhang, S., Tang, Y., & Wan, Z. (2020). Pool boiling enhancement of novel interconnected microchannels with reentrant cavities for high-power electronics cooling. *International Journal of Heat and Mass Transfer*, *156*, 119836. DOI: 10.1016/j.ijheatmasstransfer.2020.119836

Cheng, C., Ren, X., Wang, Z., & Yan, C. (2019). Heterogeneous impacts of renewable energy and environmental patents on CO2 emission-Evidence from the BRIICS. *The Science of the Total Environment*, *668*, 1328–1338. DOI: 10.1016/j.scitotenv.2019.02.063 PMID: 30846196

Chen, L., & Zhang, Y. (2018). Urbanization and CO2 emissions: Evidence from China. *Environmental Science and Pollution Research International, 25*(21), 20779–20788. DOI: 10.1007/s11356-018-1944-1 PMID: 29756186

Chennaif, M., Zahboune, H., & Elhafyani, M. (2020). *Techno-economic sizing of a stand-alone hybrid energy and storage system for a water pumping application. Advances in Smart Technologies and Applications*. Springer., DOI: 10.1007/978-3-030-53187-4_33

Chen, S., Liu, Y., Chan, S. F., & Leung, C. W. (2001). T. L. C. (2001). Experimental study of optimum spacing problem in the cooling of simulated electronic packages. *International Journal of Heat and Mass Transfer*, •••, 37.

Chen, W., Alharthi, M., Zhang, J., & Khan, I. (2024). The need for energy efficiency and economic prosperity in a sustainable environment. *Gondwana Research, 127*, 22–35. DOI: 10.1016/j.gr.2023.03.025

Chen, Y., & Zhang, X. (2023). Does financial globalization promote renewable energy investment? Empirical insights from China. *Environmental Science and Pollution Research International, 30*(45), 101366–101378. DOI: 10.1007/s11356-023-29293-z PMID: 37651014

Chiranjeevi, M., Karlamangal, S., Moger, T., & Jena, D. (2023a). Solar irradiation forecast enhancement using hybrid architecture. In *2023 5th International Conference on Energy, Power and Environment: Towards Flexible Green Energy Technologies (ICEPE)*, pages 1–6. IEEE.

Chiranjeevi, M., Karlamangal, S., Moger, T., & Jena, D. (2023b). Solar irradiation prediction hybrid framework using regularized convolutional bilstm-based autoencoder approach. *IEEE Access : Practical Innovations, Open Solutions, 11*, 131362–131375.

Chiranjeevi, M., Madyastha, A., Maurya, A. K., Moger, T., & Jena, D. (2024). Solar irra- diation forecast enhancement using clustering based cnn-bilstm-attention hybrid architecture with pso. *International Journal of Ambient Energy, 45*(1), 2414924.

Chong, K.-K., & Wong, G.-Z. (2021). Solar powered a wearable Electrocardiography (ECG) device with battery storage. *IOP Conference Series. Earth and Environmental Science, 945*(1), 012048. DOI: 10.1088/1755-1315/945/1/012048

Chouay, Y., & Ouassaid, M. (2024). An accurate real time neural network based irradiance and temperature sensor for photovoltaic applications. *Results in Engineering, 21*, 101766. DOI: 10.1016/j.rineng.2024.101766

Christoforou, E., & Fokaides, P. A. (2023). Biomass Resources and Conversion Technologies in Africa: An Overview of Applications and Benefits. *Renewable Energy Perspectives, 12*(2), 210–225. https://www.renewableenergyjournal.org

Churchill, S. A., Inekwe, J., Smyth, R., & Zhang, X. (2019). R&D intensity and carbon emissions in the G7: 1870-2014. *Energy Economics, 80*, 30–37. DOI: 10.1016/j.eneco.2018.12.020

Cominola, A., Giuliani, M., Piga, D., Castelletti, A., & Rizzoli, A. E. (2015). Benefits and challenges of using smart meters for advancing residential water demand modeling and management: A review. *Environmental Modelling & Software, 72*, 198–214. DOI: 10.1016/j.envsoft.2015.07.012

Connor, A., Lillywhite, R., & Cooke, M. W. (2011). The carbon footprints of home and in-center maintenance hemodialysis in the United Kingdom. *Hemodialysis International. International Symposium on Home Hemodialysis*, *15*(1), 39–51. DOI: 10.1111/j.1542-4758.2010.00523.x PMID: 21231998

Corvaro, F., & Paroncini, M. (2008). A numerical and experimental analysis on the natural convective heat transfer of a small heating strip located on the floor of a square cavity. *Applied Thermal Engineering*, *28*(1), 25–35. DOI: 10.1016/j.applthermaleng.2007.03.018

Costa, V. A. F. (2012). Natural convection in partially divided square enclosures: Effects of thermal boundary conditions and thermal conductivity of the partitions. *International Journal of Heat and Mass Transfer*, *55*(25–26), 7812–7822. DOI: 10.1016/j.ijheatmasstransfer.2012.08.004

CSY. (2016). China Statistical Yearbook 2015 (Online). www. stats. gov. cn / tjsj / ndsj / 2016/ indexeh. htm. Accessed on 5 May 2018.

Cui, T., Shi, Y., Lv, B., Ding, R., & Li, X. (2023). Federated learning with SARIMA-based clustering for carbon emission prediction. *Journal of Cleaner Production*, *426*, 139069. DOI: 10.1016/j.jclepro.2023.139069

Cui, W.-W., Lu, Z., Zhou, B., Li, C., & Han, X. (2016). A hybrid genetic algorithm for non-permutation flow shop scheduling problems with unavailability constraints. *International Journal of Computer Integrated Manufacturing*, *29*(9), 944–961. Advance online publication. DOI: 10.1080/0951192X.2015.1130247

Cui, Y., & He, X. (2020). Urbanization, industrialization, and CO2 emissions in China: Evidence from panel data analysis. *Environmental Science and Pollution Research International*, *27*(5), 5197–5208. DOI: 10.1007/s11356-019-07234-4

Cunha, J., Ferreira, P., Castro, E. M., Oliveira, P. C., Nicolau, M. J., Núñez, I., Sousa, X. R., & Serôdio, C. (2024). Enhancing Network Slicing Security: Machine Learning, Software-Defined Networking, and Network Functions Virtualization-Driven Strategies. *Future Internet*, *16*(7), 226. DOI: 10.3390/fi16070226

Cust, J., & Zeufack, A. (2023). *Africa's resource future: harnessing natural resources for economic transformation during the low-carbon transition*. World Bank Publications. DOI: 10.1596/978-1-4648-1743-4

Cyril, A. O., Ujah, C. O., Ekwueme, B. N., & Asadu, C. O. (2024). Photovoltaic mini-grid incorporation: The panacea for electricity crisis in sub-Saharan Africa. *Unconventional Resources*, *4*, 100079. DOI: 10.1016/j.uncres.2024.100079

Czyzyk, K. A., Bement, S. T., Dawson, W. F., & Mehta, K. (2014, October). Quantifying water savings with greenhouse farming. In *IEEE Global Humanitarian Technology Conference (GHTC 2014)* (pp. 325-332). IEEE. DOI: 10.1109/GHTC.2014.6970300

D'Humières, D., Ginzburg, I., Krafczyk, M., Lallemand, P., & Luo, L. S. (2002). Multiple-relaxation-time lattice Boltzmann models in three dimensions. *Philosophical Transactions. Series A, Mathematical, Physical, and Engineering Sciences*, *360*(1792), 437–451. Advance online publication. DOI: 10.1098/rsta.2001.0955 PMID: 16214687

Daher, D. H., Mathieu, A., Abdallah, A., Mouhoumed, D., Logerais, P. O., Gaillard, L., & Ménézo, C. (2024). Photovoltaic failure diagnosis using imaging techniques and electrical characterization. *EPJ Photovoltaics*, *15*, 25. Advance online publication. DOI: 10.1051/epjpv/2024022

Darwish, M. M., Hassan, M. H., Abdel-Gawad, N. M., & Mansour, D.-E. A. (2022). *Application of infrared spectroscopy for discrimination between electrical and thermal faults in transformer oil.* Paper presented at the 2022 9th International Conference on Condition Monitoring and Diagnosis (CMD). DOI: 10.23919/CMD54214.2022.9991616

Das, D., Roy, M., & Basak, T. (2017). Studies on natural convection within enclosures of various (non-square) shapes – A review. *International Journal of Heat and Mass Transfer*, *106*, 356–406. DOI: 10.1016/j.ijheatmasstransfer.2016.08.034

Dauphin-Tanguy, G. (2000). *Les Bond graphs.* Édition Hermes.

Dbaghi, Y., Farhat, S., Mediouni, M., Essakhi, H., & Elmoudden, A. (2021). Indirect power control of DFIG based on wind turbine operating in MPPT using backstepping approach. *International Journal of Electrical and Computer Engineering*, *11*(3), 1951. DOI: 10.11591/ijece.v11i3.pp1951-1961

Deng, Y., Feng, C., Jiaqiang, E., Zhu, H., Chen, J., Wen, M., & Yin, H. (2018). Effects of different coolants and cooling strategies on the cooling performance of the power lithium ion battery system: A review. *Applied Thermal Engineering*, *142*, 10–29. DOI: 10.1016/j.applthermaleng.2018.06.043

Derbel, N., & Zhu, Q. (Eds.). (2018). *Modeling, identification and control methods in renewable energy systems*. Springer.

Deshmukh, M. K. G., Sameeroddin, M., Abdul, D., & Abdul Sattar, M. (2023). Renewable energy in the 21st century: A review. *Materials Today: Proceedings*, *80*, 1756–1759. DOI: 10.1016/j.matpr.2021.05.501

Dey, S., Sreenivasulu, A., Veerendra, G. T. N., Rao, K. V., & Babu, P. S. S. A. (2022). Renewable energy present status and future potentials in India: An overview. *Innovation and Green Development*, *1*(1), 100006. DOI: 10.1016/j.igd.2022.100006

Dhiman, A.K., R.P. Chhabra, V. Eswaran, Flow and heat transfer across a confined square cylinder in the steady flow regime: Effect of Peclet number, Int. J. Heat Mass Transf. (2005). https://doi.org/, (2005). DOI: 10.1016/j.ijheatmasstransfer.04.033

d'Humières, D. (2002). Multiple–relaxation–time lattice Boltzmann models in three dimensions. Philosophical Transactions of the Royal Society of London. Series A: Mathematical, Physical and Engineering Sciences, 360(1792), 437-451.

Dimitriou, T., Skandalos, N., & Karamanis, D. (2024). Progress in Improving Photovoltaics Longevity. *Applied Sciences (Basel, Switzerland)*, *14*(22), 10373. Advance online publication. DOI: 10.3390/app142210373

Dimroth, T. N. D., Tibbits, T. N. D., Niemeyer, M., Predan, F., Beutel, P., Karcher, C., Oliva, E., Siefer, G., Lackner, D., Fus-Kailuweit, P., Bett, A. W., Krause, R., Drazek, C., Guiot, E., Wasselin, J., Tauzin, A., & Signamarcheix, T. (2016). Four-Junction Wafer-Bonded Concentrator Solar Cells. *IEEE Journal of Photovoltaics*, *6*(1), 343–349. DOI: 10.1109/JPHOTOV.2015.2501729

Ding, J., Song, S., Zhang, R., Chiong, R., & Wu, C. (2016a). Parallel machine scheduling under time-of-use electricity prices: New models and optimization approaches. *IEEE Transactions on Automation Science and Engineering*, *13*(2), 1138–1154. DOI: 10.1109/TASE.2015.2495328

Dixit, A. (2022). *Kumar, T. & Joshi, Abhishek & Bedi, Harpreet & Chakravarthi, Kalyan & Singh, Devesh.* Trends in Robotics and Computer Integrated Manufacturing., DOI: 10.1109/IC3I56241.2022.10072745

Djoudi, A. (2016). Contribution à la Conduite Robuste d'une Éolienne Basée sur une Machine Asynchrone à Double Alimentation, Connectée au Réseau Électrique (Doctoral dissertation, Ecole Nationale Polytechnique (ENP); Laboratoire de Génie Electrique de Grenoble (G2Elab)).

Dong, X., Lin, H., Tan, R., Iyer, R. K., & Kalbarczyk, Z. (2015). *Software-defined networking for smart grid resilience: Opportunities and challenges.* Paper presented at the Proceedings of the 1st ACM workshop on cyber-physical system security. DOI: 10.1145/2732198.2732203

Dong, K., Hochman, G., & Timilsina, G. R. (2020a). Do drivers of CO2 emission growth alter overtime and by the stage of economic development? *Energy Policy*, *140*, 111420. DOI: 10.1016/j.enpol.2020.111420

Dong, K., Hochman, G., Zhang, Y., Sun, R., Li, H., & Liao, H. (2018). CO2 emissions, economic and population growth, and renewable energy: Empirical evidence across regions. *Energy Economics*, *75*, 180–192. DOI: 10.1016/j.eneco.2018.08.017

Dongre, S., Raut, A., & Sakhare, A. (2022, March). Weather monitoring and control system for polyhouses using IoT. In AIP Conference Proceedings (Vol. 2424, No. 1). AIP Publishing. DOI: 10.1063/5.0078090

Dorel, S., Gmal Osman, M., Strejoiu, C.-V., & Lazaroiu, G. (2023). Exploring Optimal Charging Strategies for Off-Grid Solar Photovoltaic Systems: A Comparative Study on Battery Storage Techniques. *Batteries*, *9*(9), 470. DOI: 10.3390/batteries9090470

Dritsaki, M. (2020). Forecasting European Union CO2 emissions using autoregressive integrated moving average-autoregressive conditional heteroscedasticity models. International Journal of Energy Economics and Policy.

Du Preez, J., & Witt, S. F. (2003). Univariate and multivariate time series forecasting: An application to tourism demand. *International Journal of Forecasting*, *19*(3), 435–451. DOI: 10.1016/S0169-2070(02)00057-2

Duruaku, S. E., Nnaji, C. C., & Ugwuegbu, C. U. (2020). Optimization of biomass briquette combustion properties. *Journal of Renewable Energy Science and Technology*, *5*(3), 100–112. DOI: 10.1155/2020/3805296

Ebaid, M. S., Shahin, Z. A. A., & Alshawabkeh, M. M. (2023). Feasibility studies of micro wind turbines installed on electric vehicles as range extenders using real-time analytical simulation with multi driving cycles scenarios. *Advances in Mechanical Engineering*, *15*(4), 16878132231165964. DOI: 10.1177/16878132231165964

Ebong, G. N., Matthew, U. O., Olofin, B., Andrew-Vitalis, N., Fatai, L. O., Waliu, A. O., & Oladipupo, M. A. (2024). *Multimedia Cloud Data Warehouse Design for Knowledge Sharing in the University Environment: A Proposed Digital Solution Implementing Interactive Learning Strategies in Higher Education*. IGI Global. DOI: 10.4018/979-8-3693-3559-8.ch010

Eden, C. A., Chisom, O. N., & Adeniyi, I. S. (2024). Integrating AI in education: Opportunities, challenges, and ethical considerations. *Magna Scientia Advanced Research and Reviews, 10*(2), 006-013.

EIA. (2013). International Energy Outlook 2013 (Online). www. eia . gov / forecasts / ieo / pdf / 0484 (2013) . pdf. Accessed on 5 May 2018.

Ekren, O., & Ekren, B. Y. (2010). Size optimization of a PV/wind hybrid energy conversion system with battery storage using simulated annealing. *Applied Energy*, *87*(2), 592–598. DOI: 10.1016/j.apenergy.2009.05.022

El Hadri, Y., Khokhlov, V., & Slizhe, M. (2023). *Wind and Solar Energy Resources in Morocco: Current Status and Assessment up to 2050*. Springer.

El Misilmani, H. M., Naous, T., & Al Khatib, S. K. (2020). A review on the design and optimization of antennas using machine learning algorithms and techniques. *International Journal of RF and Microwave Computer-Aided Engineering*, *30*(10), e22356. DOI: 10.1002/mmce.22356

El-Ghonemy, A. M. K. (2012). Water desalination systems powered by renewable energy sources [Review]. *Renewable & Sustainable Energy Reviews*, *16*(3), 1537–1556. DOI: 10.1016/j.rser.2011.12.014

Elkelawy, M., Saeed, A. M., & Seleem, H. (2024). Egypt's Solar Revolution: A Dual Approach to Clean Energy with CSP and PV Technologies. *Pharos Engineering Science Journal*, *1*(1), 39–50. DOI: 10.21608/pesj.2025.344888.1004

Elmousaid, R., Drioui, N., Elgouri, R., Agueny, H., & Adnani, Y. (2024). Ultra-short-term global horizontal irradiance forecasting based on a novel and hybrid gru-tcn model. *Results in Engineering*, *23*, 102817.

Elshaer, A., Bitsuamlak, G., & El Damatty, A. (2017). Enhancing wind performance of tall buildings using corner aerodynamic optimization. *Engineering Structures*, *136*, 133–148. DOI: 10.1016/j.engstruct.2017.01.019

Eltawil, M. A., Zhengming, Z., & Yuan, L. (2009). A review of renewable energy technologies integrated with desalination systems. *Renewable & Sustainable Energy Reviews*, *13*(9), 2245–2262. DOI: 10.1016/j.rser.2009.06.011

Enderami, S. A., Sutley, E. J., Mazumder, R. K., & Dumler, M. (2023). Virtual testbeds for community resilience analysis: Step-by-step development procedure and future orientation. *Resilient Cities and Structures*, *2*(2), 42–56. DOI: 10.1016/j.rcns.2023.07.002

Energy Design Resources. (2010). Design briefs: Centrifugal pump application and optimization. Available at www.energydesignresources.com

Enescu, F. M., Bizon, N., & Moraru, C. M. (2019). Issues in Securing Critical Infrastructure Networks for Smart Grid Based on SCADA, Other Industrial Control and Communication Systems. *Power Systems Resilience: Modeling, Analysis and Practice*, 289-324.

Ennemri, A., Logerais, P.-O., Balistrou, M., Belaidi, I., Ennemri, A., Logerais, P. O., Balistrou, M., Durastanti, J. F., & Belaidi, I. (2019). Cracks in silicon photovoltaic modules: A review. In *Article in Journal of Optoelectronics and Advanced Materials* (Vol. 21, Issue 2). https://www.researchgate.net/publication/331501514

Erdo˘gan, S., Yıldırım, S., Yıldırım, D. Ç., & Gedikli, A. (2020b). The effects of innovation on sectoral carbon emissions: Evidence from G20 countries. *Journal of Environmental Management, 267*, 110637. DOI: 10.1016/j.jenvman.2020.110637 PMID: 32349957

Eshaghi, S., Izadpanah, F., Dogonchi, A. S., Chamkha, A. J., Ben Hamida, M. B., & Alhumade, H. (2021). The optimum double diffusive natural convection heat transfer in H-Shaped cavity with a baffle inside and a corrugated wall. *Case Studies in Thermal Engineering, 28*, 101541. Advance online publication. DOI: 10.1016/j.csite.2021.101541

Fahmy, F. H., Ahmed, N. M., & Farghally, H. M. (2012). Optimization of renewable energy power system for a small-scale brackish reverse osmosis desalination unit and a tourism motel in Egypt. In *Proceedings of the 8th International Conference on Electrical Engineering ICEENG 2012*. Military Technical College. DOI: 10.4236/sgre.2012.31006

Fakoyede, P. D., Innocent, A. O., Ajiboye, Q. O., Enabulele, E. C., Olaleye, T. G., & Ekunseitan, K. D. (2024). Assessing the Detrimental Effects of Solar Panel Installation on Residential Roofing Systems. *Traektoriâ Nauki, 10*(7), 3023–3033. DOI: 10.22178/pos.106-15

Fan,H., Hossain, MI. (2018). Technological Innovation, Trade Openness, CO2 Emission and Economic Growth: Comparative Analysis between China and India. International Journal of Energy Economics and Policy, 240-257.

Fan, Z., Yan, Z., & Wen, S. (2023). Deep learning and artificial intelligence in sustainability: A review of SDGs, renewable energy, and environmental health. *Sustainability (Basel), 15*(18), 13493. DOI: 10.3390/su151813493

Farooqi, M. T. K., Amanat, I., & Awan, S. M. (2024). Ethical considerations and challenges in the integration of artificial intelligence in education: A systematic review. *Journal of Excellence in Management Sciences, 3*(4), 35–50. DOI: 10.69565/jems.v3i4.314

Farooq, M. S., Riaz, S., Abid, A., Umer, T., & Zikria, Y. B. (2020). Role of IoT technology in agriculture: A systematic literature review. *Electronics (Basel), 9*(2), 319. DOI: 10.3390/electronics9020319

FDRE Ministry of Water and Energy. (2022). Ethiopia's National Energy Balance Report 2022. Addis Ababa: Ministry of Water and Energy. Retrieved from https://www.mowe.gov.et

Fedorov, V. V., Dvoretckaia, L. N., Mozharov, A. M., Fedina, S. V., Kirilenko, D. A., Berezovskaya, T. N., Faleev, N. N., Yunin, P. A., Drozdov, M. N., & Mukhin, I. S. (2023). Dual-functional light-emitting and photo-detecting GaAsPN heterostructures on silicon. *Materials Science in Semiconductor Processing, 168*, 107867. DOI: 10.1016/j.mssp.2023.107867

Feng, C., Cui, M., Hodge, B.-M., Lu, S., Hamann, H. F., & Zhang, J. (2018). Unsuper- vised clustering-based short-term solar forecasting. *IEEE Transactions on Sustainable Energy, 10*(4), 2174–2185. DOI: 10.1109/TSTE.2018.2881531

Fern'andez, Y. F., L'opez, M. F., & Blanco, B. O. (2018). Innovation for sustainability: The impact of R&D spending on CO2 emissions. *Journal of Cleaner Production, 172*, 3459–3467. DOI: 10.1016/j.jclepro.2017.11.001

Firtina-Ertis, I., Acar, C., & Erturk, E. (2020). Optimal sizing design of an isolated stand-alone hybrid wind-hydrogen system for a zero-energy house. *Applied Energy*, *274*, 115244. DOI: 10.1016/j.apenergy.2020.115244

Fiseha, T., & Osborne, N. J. (2023). Burden of end-stage renal disease of undetermined etiology in Africa. *Renal Replacement Therapy*, *9*(1), 44. DOI: 10.1186/s41100-023-00497-w

Florides, G., & Kalogirou, S. (2007). Ground heat exchangers-A review of systems, models and applications. In Renewable Energy. DOI: 10.1016/j.renene.2006.12.014

Food and Agriculture Organization (FAO). (2023). Global Forest Resources Assessment 2023: Country Report – Ethiopia. Rome: FAO. Retrieved from https://www.fao.org/forest-resources-assessment

Food and Agriculture Organization (FAO). (2023). The State of the World's Forests 2023: Forests, Energy, and Livelihoods. Rome: FAO. Retrieved from https://www.fao.org/publications

Fouad, M. M., Shihata, L. A., & Morgan, E. I. (2017). An integrated review of factors influencing the perfomance of photovoltaic panels. *Renewable & Sustainable Energy Reviews*, *80*, 1499–1511. DOI: 10.1016/j.rser.2017.05.141

Fuchs, D., Sahakian, M., Gumbert, T., Di Giulio, A., Maniates, M., Lorek, S., & Graf, A. (2021). *Consumption corridors: Living a good life within sustainable limits.*

Gahm, C., Denz, F., Dirr, M., & Tuma, A. (2016). Energy – efficient scheduling in manufacturing companies: A review and research framework. *European Journal of Operational Research*, *248*(3), 744–757.

Gaikwad, K., & Lokhande, S. (2016). Novel maximum power point tracking (MPPT) algorithm for solar tree application. *International Conference on Energy Systems and Applications, ICESA 2015*, 189–193. DOI: 10.1109/ICESA.2015.7503337

Gaitho, F. M., Ndiritu, F. G., Muriithi, P. M., Ngumbu, R. G., & Ngareh, J. K. (2009). Effect of thermal conductivity on the efficiency of single crystal silicon solar cell coated with an anti-reflective thin film. *Solar Energy*, *83*(8), 1290–1293. DOI: 10.1016/j.solener.2009.03.003

Gao, J., Yao, S., Xu, J., & Shi, Z. (2009). Design of a high efficiency boost DC-DC converter. *Transactions of Tianjin University*, *15*(5), 350–354. DOI: 10.1007/s12209-009-0061-y

García-González, E., Jiménez-Fontana, R., & Azcárate, P. (2020). Education for sustainability and the sustainable development goals: Pre-service teachers' perceptions and knowledge. *Sustainability (Basel)*, *12*(18), 7741. DOI: 10.3390/su12187741

Garcia-Rodriguez, L. (2003). Renewable energy applications in desalination: State of the art. *Solar Energy*, *75*(5), 381–393. DOI: 10.1016/j.solener.2003.07.005

Gawthrop, P. J., & Bevan, G. P. (2007). Bond-graph modeling. *IEEE Control Systems Magazine*, (April), 24–45.

Gebremariam, H., & Assefa, B. (2021). Techno-economic feasibility of biomass briquette production from agricultural residues in Ethiopia. *Energy for Sustainable Development*, *60*, 45–56. https://doi.org/10.xxxx/esd60045

Geisz, J. F., & Friedman, D. J. (2002). III–N–V semiconductors for solar photovoltaic applications. *Semiconductor Science and Technology*, *17*(8), 769–777. DOI: 10.1088/0268-1242/17/8/305

Getachew, D., & Lemma, T. (2021). Small-scale biomass briquette production in Addis Ababa: A case study. *Journal of Cleaner Production*, *320*, 128–140. https://doi.org/10.xxxx/jcp320128

Ghaffour, N., Missimer, T. M., & Amy, G. L. (2013). Technical review and evaluation of the economics of water desalination: Current and future challenges for better water supply sustainability. *Desalination*, *309*, 197–207. DOI: 10.1016/j.desal.2012.10.015

Ghayekhloo, M., Ghofrani, M., Menhaj, M., & Azimi, R. (2015). A novel clustering approach for short-term solar radiation forecasting. *Solar Energy*, *122*, 1371–1383. DOI: 10.1016/j.solener.2015.10.053

Ghermandi, A., & Messalem, R. (2009). Solar-driven desalination with reverse osmosis: The state of the art. *Desalination and Water Treatment*, *7*(1–3), 285–296. DOI: 10.5004/dwt.2009.723

Ghimire, S., Deo, R. C., Downs, N. J., & Raj, N. (2019). Global solar radiation prediction by ANN integrated with European Centre for medium range weather forecast fields in solar rich cities of Queensland Australia. *Journal of Cleaner Production*, *216*, 288–310. DOI: 10.1016/j.jclepro.2019.01.158

Ghofrani, M., Ghayekhloo, M., & Azimi, R. (2016). A novel soft computing framework for solar radiation forecasting. *Applied Soft Computing*, *48*, 207–216. DOI: 10.1016/j.asoc.2016.07.022

Giret, A., Trentesaux, D., & Prabhu, V. (2015). Sustainability in manufacturing operations scheduling: A state of the art review. *Journal of Manufacturing Systems*, *37*(1), 126–140. DOI: 10.1016/j.jmsy.2015.08.002

Gligorea, I., Cioca, M., Oancea, R., Gorski, A. T., Gorski, H., & Tudorache, P. (2023). Adaptive learning using artificial intelligence in e-learning: A literature review. *Education Sciences*, *13*(12), 1216. DOI: 10.3390/educsci13121216

Goh, C., & Law, R. (2002). Modeling and forecasting tourism demand for arrivals with stochastic nonstationary seasonality and intervention. *Tourism Management*, *23*(5), 499–510. DOI: 10.1016/S0261-5177(02)00009-2

Gonzalez, C., Charfadine, S. M., Flauzac, O., & Nolot, F. (2016). *SDN-based security framework for the IoT in distributed grid.* Paper presented at the 2016 international multidisciplinary conference on computer and energy science (SpliTech). DOI: 10.1109/SpliTech.2016.7555946

Gonzalez-Zamudio, E. A., Olivares-Robles, M. A., & Andrade-Vallejo, A. A. (2025). A Novel Case of Cooling and Heating in Rectangular Lid-Driven Cavities: Interplay of Richardson Numbers in Streamlines and Isotherms. *Processes (Basel, Switzerland)*, *13*(2), 432. DOI: 10.3390/pr13020432

Greenlee, L. F., Lawler, D. F., Freeman, B. D., Marrot, B., & Moulin, P. (2009). Reverse osmosis desalination: Water sources, technology, and today's challenges. *Water Research*, *43*(9), 2317–2348. DOI: 10.1016/j.watres.2009.03.010 PMID: 19371922

Green, M. A. (2003). Crystalline and thin-film silicon solar cells: State of the art and future potential. *Solar Energy*, *74*(3), 181–192. DOI: 10.1016/S0038-092X(03)00187-7

Grenouillet, L., Bru-Chevallier, C., Guillot, G., Gilet, P., Duvaut, P., Vannuffel, C., Million, A., & Chenevas-Paule, A. (2000). Evidence of strong carrier localization below 100 K in a GaInNAs/GaAs single quantum well. *Applied Physics Letters*, *76*(16), 2241–2243. DOI: 10.1063/1.126308

Gude, V. G., Nirmalakhandan, N., & Deng, S. (2010). Renewable and sustainable approaches for desalination. *Renewable & Sustainable Energy Reviews*, *14*(9), 2641–2654. DOI: 10.1016/j.rser.2010.06.008

Guemri, M. (2013). Heuristiques optimisées et robustes de résolution du problème de gestion d'énergie pour les véhicules électriques et hybrides. (PhD thesis, Institut National Polytechnique de Toulouse-INPT, France, 2013).

Guida, V. (2020). Conception et réalisation d'un convertisseur DC/DC à haut rapport de conversion pour électrolyseurs (Doctoral dissertation, Université de Lorraine).

Guo, Z., & Shu, C. (2013). Lattice Boltzmann method and its applications in engineering. In World Scientific Publishing (Vol. 54, Issue 22). DOI: 10.1142/8806

Guo, X., Huang, K., Li, L., & Wang, X. (2022). Renewable Energy for Balancing Carbon Emissions and Reducing Carbon Transfer under Global Value Chains: A Way Forward. *Sustainability (Basel)*, *15*(1), 1. Advance online publication. DOI: 10.3390/su15010234

Gurung, N., Bhattarai, R., & Kamalasadan, S. (2017, July). Optimal linear-quadratic-integral controller design for doubly-fed induction generator. In 2017 IEEE Power & Energy Society General Meeting (pp. 1-5). IEEE.

Gyonyoru, K. I. K., & Katona, J. (2024, October). Student Perceptions of AI-Enhanced Adaptive Learning Systems: A Pilot Survey. In *2024 IEEE 7th International Conference and Workshop Óbuda on Electrical and Power Engineering (CANDO-EPE)* (pp. 93-98). IEEE.

Hakiri, A., Sellami, B., & Yahia, S. B. (2024). Joint energy efficiency and network optimization for integrated blockchain-SDN-based internet of things networks. *Future Generation Computer Systems*, *•••*, 107519.

Hammami, A., Saidi, I., & Soudani, D. (2018). Comparative study of PMSG controllers for variable wind turbine power optimization. *International Journal of Advanced Computer Science and Applications*, *9*(8). Advance online publication. DOI: 10.14569/IJACSA.2018.090831

Hansen, M. O. L., Sørensen, N. N., & Voutsinas, S. (2015). State of the Art in Wind Turbine Aerodynamics and Aeroelasticity. *Progress in Aerospace Sciences*, *74*, 54–88.

Hasan, A. A. Q., Alkahtani, A. A., Shahahmadi, S. A., Alam, M. N. E., Islam, M. A., & Amin, N. (2021). Delamination-and electromigration-related failures in solar panels—A review. *Sustainability (Switzerland)*, *13*(12). Advance online publication. DOI: 10.3390/su13126882

Hashemi, B., Taheri, S., Cretu, A. M., & Pouresmaeil, E. (2023). Computational intelligence based PEVs aggregator scheduling with support for photovoltaic power penetrated distribution grid under snow conditions. Electric Power Systems Research Part A, 108922.

Hashmi, R., & Alam, K. (2019). Dynamic relationship among environmental regulation, innovation, CO2 emissions, population, and economic growth in OECD countries: A panel investigation. *Journal of Cleaner Production*, *231*, 1100–1109. DOI: 10.1016/j.jclepro.2019.05.325

Hassan, Q., Viktor, P., Al-Musawi, T. J., Ali, B. M., Algburi, S., Alzoubi, H. M., & Jaszczur, M. (2024). The renewable energy role in the global energy Transformations. *Renewable Energy Focus*, *48*, 100545. DOI: 10.1016/j.ref.2024.100545

Hatch, S., Wu, J., Sablon, K., Lam, P., Tang, M., Jiang, Q., & Liu, H. (2014). InAs/GaAsSb quantum dot solar cells. *Optics Express*, *22*(S3), A679. DOI: 10.1364/OE.22.00A679 PMID: 24922376

Hemeyine, A. V., Abbou, A., Tidjani, N., Mokhlis, M., & Bakouri, A. (2020). Robust takagi sugeno fuzzy models control for a variable speed wind turbine based a DFI-generator. *International Journal of Intelligent Engineering and Systems*, *13*(3), 90–100. DOI: 10.22266/ijies2020.0630.09

Hendrikx, N., Barhmi, K., Visser, L., de Bruin, T., P'o, M., Salah, A., & van Sark, W. (2024). All sky imaging-based short-term solar irradiance forecasting with long short-term memory networks. *Solar Energy*, *272*, 112463. DOI: 10.1016/j.solener.2024.112463

He, X. (2023). Effects of the green policy environment on renewable energy investment and effect evaluation of green policies. *Discrete Dynamics in Nature and Society*, *2023*, 1–16. DOI: 10.1155/2023/8698548

Hijjawi, U., Lakshminarayana, S., Xu, T., Piero Malfense Fierro, G., & Rahman, M. (2023). A review of automated solar photovoltaic defect detection systems: Approaches, challenges, and future orientations. In *Solar Energy* (Vol. 266). Elsevier Ltd., DOI: 10.1016/j.solener.2023.112186

Hoenich, N. A., & Pearce, C. (2002). Medical waste production and disposal arising from renal replacement therapy. *Advances in Renal Replacement Therapy*, *9*(1), 57–62. DOI: 10.1053/jarr.2002.30469 PMID: 11927908

Hordofa, T. T., Vu, H. M., Maneengam, A., Purnomo, E. P., Cong, P. T., & Liying, S. (2022). Does eco-innovation and green investment limit the CO_2 emissions in China? [Page Numbers]. . *Economic Research-Ekonomska Istraživanja*, *36*(1). Advance online publication.

Hossain, E. (2022). The Sun and the Earth *The Sun, Energy, and Climate Change* (pp. 1-67): Springer.

Hossain, M. S., Wadi Al-Fatlawi, A., Kumar, L., Fang, Y. R., & Assad, M. E. H. (2024). Solar PV high-penetration scenario: an overview of the global PV power status and future growth. *Energy Systems*, 1-57.

Hossain, M. H., Chowdhury, M. A., Hossain, N., Islam, M. A., & Mobark, M. H. (2023). Advances of lithium-ion batteries anode materials—A review. *Chemical Engineering Journal Advances*, *16*, 100569. DOI: 10.1016/j.ceja.2023.100569

Hung, N. T. (2021). Effect of Economic Indicators, Biomass Energy on Human Development in China. Energ. Environ., 0958305X2110220.

Huo, L. F., Wang, Z., Reda, S. A., Tashkandi, M. A., Moria, H., Nag, K., Mouldi, A., Loukil, H., Lee, V. F., & Fang, H. (2024). Power generation enhancement and CO2 emission reduction of a biomass powered gas turbine via hydrogen injection and wind energy incorporation. *Process Safety and Environmental Protection*, *185*, 105–115. DOI: 10.1016/j.psep.2024.01.075

Hu, S. (2024). The effect of artificial intelligence-assisted personalized learning on student learning outcomes: A meta-analysis based on 31 empirical research papers. *Science Insights Education Frontiers*, *24*(1), 3873–3894. DOI: 10.15354/sief.24.re395

Ibrahim, M. S., Gharghory, S. M., & Kamal, H. A. (2024). A hybrid model of cnn and lstm autoencoder-based short-term pv power generation forecasting. *Electrical Engineering*, *106*(4), 1–17. DOI: 10.1007/s00202-023-02220-8

Idrees, Z., Granados, J., Sun, Y., Latif, S., Gong, L., Zou, Z., & Zheng, L. (2020). IEEE 1588 for clock synchronization in industrial IoT and related applications: A review on contributing technologies, protocols and enhancement methodologies. *IEEE Access : Practical Innovations, Open Solutions*, *8*, 155660–155678. DOI: 10.1109/ACCESS.2020.3013669

Iftikhar, Y., He, W., & Wang, Z. (2016). Energy and CO2 emissions efficiency of major economies: A non-parametric analysis. *Journal of Cleaner Production*, *139*, 779–787. DOI: 10.1016/j.jclepro.2016.08.072

Iheanacho, E. (2016). The impact of financial development on economic growth in nigeria: An ardl analysis. *Economies*, *4*(4), 26. DOI: 10.3390/economies4040026

Ihemeson, O. C. (2024). An Assessment of UN Policies on Fossil Emission and Climate Change: Implications on the National Security of US, 2010-2023. *Journal of Integrity in Ecosystems and Environment*, *2*(5), 8–23.

Ilahi, S., Almosni, S., Chouchane, F., Perrin, M., Zelazna, K., Yacoubi, N., Kudrawiec, R., Râle, P., Lombez, L., Guillemoles, J.-F., Cornet, C., & Durand, O. (2015). Optical absorption and thermal conductivity of GaAsPN absorbers grown on GaP in view of their use in multijunction solar cells. *Solar Energy Materials and Solar Cells*, *141*, 291–298. DOI: 10.1016/j.solmat.2015.06.003

Ilahi, S., Baira, M., Saidi, F., Yacoubi, N., Auvray, L., & Maaref, H. (2013b). Non-radiative recombination process in BGaAs/GaAs alloys: Two layer photothermal deflection model. *Journal of Alloys and Compounds*, *581*, 358–362. DOI: 10.1016/j.jallcom.2013.07.101

Ilahi, S., Cornet, C., Yacoubi, N., & Durand, O. (2020). Improvement of carriers diffusion length and mobility in annealed GaAsPN materials for intermediate band solar cells. *Solar Energy Materials and Solar Cells*, *215*, 110622. DOI: 10.1016/j.solmat.2020.110622

Ilahi, S., Khalfaoui, A., Genty, F., & Yacoubi, N. (2021). Significant improvement of thermal conductivity and bandgap energy blue shift in annealed GaInAsSb/GaSb. *Optics & Laser Technology*, *137*, 106819. DOI: 10.1016/j.optlastec.2020.106819

Ilahi, S., Saadalah, F., & Yacoubi, N. (2013a). Investigation of the thermal annealing effect on the optical, thermal and electrical properties of Sn 2 Sb 2 S 5 evaporated thin films. *Applied Physics. A, Materials Science & Processing*, *110*, 459–464. DOI: 10.1007/s00339-012-7242-6

Ilahi, S., Saidi, F., Hamila, R., Yacoubi, N., Auvray, L., & Maaref, H. (2013). Photothermal deflection spectroscopy PDS investigation of optical and thermal properties of BGaAs/GaAs alloys. *Current Applied Physics*, *13*(3), 610–613. DOI: 10.1016/j.cap.2012.11.002

Ilahi, S., Saidi, F., Hamila, R., Yacoubi, N., Maaref, H., & Auvray, L. (2013). L Auvray. Shift of the gap energy and thermal conductivity in BGaAs/GaAs alloys. *Physica B, Condensed Matter*, *421*, 105–109. DOI: 10.1016/j.physb.2013.03.045

Ilahi, S., & Yacoubi, N. (2014). Interface recombination velocity measurement in SiO 2/Si. *Semiconductors*, *48*(3), 302–306. DOI: 10.1134/S1063782614030130

Ilahi, S., Yacoubi, N., & Genty, F. (2017). Thermal annealing effects on AlGaAsSb/GaSb laser structure: Bandgap energy blueshift and thermal conductivity enhancement. *Optical Materials*, *69*, 226–229. DOI: 10.1016/j.optmat.2017.04.050

Ilahi, S., Yacoubi, N., & Genty, F. (2018). Photothermal deflection technique investigation of thermal annealing effects of AlGaAsSb/GaSb laser structure: Non-radiative recombination parameters enhancement. *Materials Research Bulletin*, *106*, 332–336. DOI: 10.1016/j.materresbull.2018.06.007

International Energy Agency (IEA). (2023). World Energy Outlook 2023. Paris: International Energy Agency. Retrieved from https://www.iea.org/reports/world-energy-outlook-2023

International Energy Agency (IEA). (2024). Renewables 2024: Global overview. Retrieved from https://www.iea.org/reports/renewables-2024/global-overview

IPCC (2023) - International Panel on Climate Change, *Sixth Assessment Report.*

Iqbal, K., Hassan, S. T., Geng, Y., & Cai, C. (2024). Impact of green technological innovations, ICT, and renewable energy on CO_2 emissions in emerging economies. *International Journal of Energy Research*, *48*(5), 5594324. DOI: 10.1155/er/5594324

IRENA (2023) - International Renewable Energy Agency, *Global Wind Energy Outlook 2023.*

Islam, M., & Mekhilef, S. (2019). H6-type transformerless single-phase inverter for grid-tied photovoltaic system. *IET Power Electronics*, *8*(4), 636–644. DOI: 10.1049/iet-pel.2014.0251

Islam, M., Mekhilef, S., & Hasan, M. (2019). Single phase transformer less inverter topologies for grid-tied photovoltaic system: A review. *Renewable & Sustainable Energy Reviews*, *45*, 69–86. DOI: 10.1016/j.rser.2015.01.009

Islam, S. (2015). Ul., S.U. Rahman, H. Abbasi, W.S. Shahina, T. Lattice Boltzmann study of wake structures and force statistics for various gap spacings between a square rod with a detached flat plate. *Arabian Journal for Science and Engineering*, *40*, 2169–2182. DOI: 10.1007/s13369-015-1648-3

Islam, S. U., Zhou, C. Y., Shah, A., & Xie, P. (2012). Numerical simulation of flow past rectangular cylinders with different aspect ratios using the incompressible lattice Boltzmann method. *Journal of Mechanical Science and Technology*, *26*(4), 1027–1041. DOI: 10.1007/s12206-012-0328-4

Izadifard, M., Buyanova, I. A., Bergman, J. P., Chen, W. M., Utsumi, A., Furukawa, Y., Wakahara, A., & Yonezu, H. (2004). Effects of rapid thermal annealing on optical quality of GaNP alloys. *IEE Proceedings. Optoelectronics*, *151*(5), 335–337. DOI: 10.1049/ip-opt:20040872

Jabrane, M., Fadili, W., Kennou, B., Labaali, A., Zahlane, K., & Laouad, I. (2013). Évaluation de l'impact d'un centre d'hémodialyse sur l'environnement et l'écologie locale. *Nephrologie & Therapeutique*, *9*(7), 481–485. DOI: 10.1016/j.nephro.2013.07.369 PMID: 24140175

Jafari, A., Ashrafi, H., Bazli, M., & Ozbakkaloglu, T. (2019). Effect of thermal cycles on mechanical response of pultruded glass fiber reinforced polymer profiles of different geometries. *Composite Structures*, *223*, 110959. DOI: 10.1016/j.compstruct.2019.110959

Jami, M., Moufekkir, F., Mezrhab, A., Fontaine, J. P., & Bouzidi, M. (2016). New thermal MRT lattice Boltzmann method for simulations of convective flows. *International Journal of Thermal Sciences*, *100*, 98–107. Advance online publication. DOI: 10.1016/j.ijthermalsci.2015.09.011

Javadpour, A., & Wang, G. (2022). cTMvSDN: Improving resource management using combination of Markov-process and TDMA in software-defined networking. *The Journal of Supercomputing*, *78*(3), 3477–3499. DOI: 10.1007/s11227-021-03871-9

Jegadeeswari, G., Kirubadurai, B., & Lakshmi, D. (2020). Multi carrier based new random pulse width modulation for three phase inverters. at International Journal of Scientific and Technology Research (IJSTR), 9(2), 5425-5432.

Jegadeeswari, G. (2019). Performance analysis of power quality improvement using shunt active power filter. [IJRTE]. *International Journal of Recent Technology and Engineering*, *7*(5), 1–3.

Jegadeeswari, G., Dheepak, M., Saravanan, S. V., & Vanisree, V. (2018). The Power Factor Correction Improvement For A Single Phase Ac/Dc Converter Using An Enabling Window Control. [IJMPERD]. *International Journal of Mechanical and Production Engineering Research and Development*, *8*(5), 41–48.

Jenkins, B. M., Baxter, L. L., Miles, T. R.Jr, & Miles, T. R.Jr. (1998). Combustion properties of biomass. *Fuel Processing Technology*, *54*(1-3), 17–46. DOI: 10.1016/S0378-3820(97)00059-3

Jeyaraj, R., & Balasubramaniam, A., MA, A. K., Guizani, N., & Paul, A. (2023). Resource management in cloud and cloud-influenced technologies for internet of things applications. *ACM Computing Surveys*, *55*(12), 1–37.

Jiang, S., Wang, K., Zhang, H., Ding, Y., & Yu, Q. (2015). Encapsulation of PV Modules Using Ethylene Vinyl Acetate Copolymer as the Encapsulant. *Macromolecular Reaction Engineering*, *9*(5), 522–529. DOI: 10.1002/mren.201400065

Jiang, Y., Batool, Z., Raza, S. M. F., Haseeb, M., Ali, S., & Zain Ul Abidin, S. (2022). Analyzing the asymmetric effect of renewable energy consumption on the environment in the STIRPAT-Kaya-EKC framework: A NARDL approach for China. *International Journal of Environmental Research and Public Health*, *19*(12), 7100. DOI: 10.3390/ijerph19127100 PMID: 35742348

Jian, L., Xue, H., Xu, G., Zhu, X., Zhao, D., & Shao, Z. (2013). Regulated Charging of Plug-in Hybrid Electric Vehicles for Minimizing Load Variance in Household Smart Micro-Grid. *IEEE Transactions on Industrial Electronics*, *60*(8), 3218–3226. DOI: 10.1109/TIE.2012.2198037

Jin, A. S., & Sanders, K. T. (2024). Characterizing residential sector load curves from smart meter datasets. *Applied Energy*, *366*, 123316. DOI: 10.1016/j.apenergy.2024.123316

Johnson, J., Bower, W., & Quintana, M. (n.d.). *SANDIA REPORT Electrical and Thermal Finite Element Modeling of Arc Faults in Photovoltaic Bypass Diodes*. http://www.ntis.gov/help/ordermethods.asp?loc=7-4-0#online

Johnson, K., Osei, T., & Ramirez, L. (2022). Health and Environmental Impacts of Low-Emission Fuels. *Journal of Environmental Sciences (China)*, *35*(3), 215–230. DOI: 10.1016/j.jes.2022.11.011

Jonathan, A. L., Cai, D., Ukwuoma, C. C., Nkou, N. J. J., Huang, Q., & Bamisile, O. (2024). A radiant shift: Attention-embedded cnns for accurate solar irradiance forecasting and prediction from sky images. *Renewable Energy*, *234*, 121133. DOI: 10.1016/j.renene.2024.121133

Jonkman, J., & Sclavounos, P. (2020). Effect of Control Strategies on the Dynamics of Floating Offshore Wind Turbines. *Journal of Wind Engineering and Industrial Aerodynamics*, *198*, 104093.

June Gandanegara, G. (2003). Méthodologie de conception systémique en génie électrique à l'aide de l'outil Bond. Mémoire de thèse, Université des Sciences et Technologies de Lille.

Jurasz, J., Canales, F. A., Kies, A., Guezgouz, M., & Beluco, A. (2020). A review on the complementarity of renewable energy sources: Concept, metrics, application, and future research directions. *Solar Energy*, *195*, 703–724. DOI: 10.1016/j.solener.2019.11.087

Justice, G., Nyantakyi, G., & Isaac, S. H. (2024). The effect of renewable energy on carbon emissions through globalization. *Heliyon*, *10*(5), e26894. DOI: 10.1016/j.heliyon.2024.e26894 PMID: 38434333

Ka, B., & Jwm, A. (n.d.). Green nephrology. *PubMed*. Retrieved November 22, 2024, from https://pubmed.ncbi.nlm.nih.gov/32034297/

Kabbara, N., Nait Belaid, M. O., Gibescu, M., Camargo, L. R., Cantenot, J., Coste, T., Audebert, V., & Morais, H. (2022). Towards software-defined protection, automation, and control in power systems: Concepts, state of the art, and future challenges. *Energies*, *15*(24), 9362. DOI: 10.3390/en15249362

Kabudi, T., Pappas, I., & Olsen, D. H. (2021). AI-enabled adaptive learning systems: A systematic mapping of the literature. *Computers and Education: Artificial Intelligence*, *2*, 100017. DOI: 10.1016/j.caeai.2021.100017

Kadam, G. S., Lohar, S. N., More, S. G., Tathe, P. S., & Dal, D. S. (2024). Smart Poly House Monitoring Systems. *International Research Journal of Innovations in Engineering and Technology*, *8*(5), 212.

Kanellopoulos, D., & Sharma, V. K. (2022). Dynamic load balancing techniques in the IoT: A review. *Symmetry*, *14*(12), 2554. DOI: 10.3390/sym14122554

Kanematsu, K., Abe, T., Tanaka, A., & Hamakawa, Y. (1988). Simulation design method of stand-alone PV power system and its applications for desalination plant. In *Proceedings of the 20th IEEE Photovoltaic Specialists Conference*, Volume 2, 1081–1086. https://doi.org/DOI: 10.1109/PVSC.1988.1059527

Kannan, N., & Vakeesan, D. (2016). Solar energy for future world: - A review. *Renewable & Sustainable Energy Reviews*, *62*, 1092–1105. DOI: 10.1016/j.rser.2016.05.022

Karahalli, M. E. (n.d.). *Performance, design, and interconnection optimization of shingled solar cell for a photovoltaic module application a thesissubmitted tothe graduate school of natural and applied sciences of middle east technical university.*

Karathanasis, S., & Karathanasis, S. (2019). Solar Radiation. *Linear Fresnel Reflector Systems for Solar Radiation Concentration: Theoretical Analysis, Mathematical Formulation and Parameters' Computation using MATLAB*, 13-72.

Karpagam, K. (2022). An improved k-means clustering algorithm for pattern discovery in data mining. *Journal of Positive School Psychology*, •••, 2387–2395.

Katani, E. K. (2019). Forecasting the total energy consumption in Ghana using grey models. Grey systems: theory and application, 9(4), 488-501.

Kaur, K., Garg, S., Kaddoum, G., Ahmed, S. H., Gagnon, F., & Atiquzzaman, M. (2019). Demand-response management using a fleet of electric vehicles: An opportunistic-SDN-based edge-cloud framework for smart grids. *IEEE Network*, *33*(5), 46–53. DOI: 10.1109/MNET.001.1800496

Kavali, J., Mittal, A., (2014). Comparative study of various Cascaded H-Bridge Multilevel Inverter Topologies. *Int. J. Electr. Electron. Eng. Res. (IJEEER)* 4 (3), 51–62 (ISSN (P): 2250-155X; ISSN(E): 2278-943X © TJPRC Pvt., Ltd.).

Kazaure, J. S., Matthew, U. O., & Nwamouh, U. C. (2022). Performance of a Gasifier Coupled to Internal Combustion Engine and Fired Using Corn Cob Feedstock in Biomass Energy Production. *Energy*, *11*(2), 35–46.

Kazaure, J. S., Matthew, U. O., Okafor, N. U., & Okey, O. D. (2021). Telecommunication network performances and evaluation of radio frequency electromagnetic radiation: Health effects of the RF-EMR GSM base stations. [IJICTHD]. *International Journal of Information Communication Technologies and Human Development*, *13*(3), 16–37. DOI: 10.4018/IJICTHD.2021070102

Kazem, H. A. (2021). Evaluation of PV output in terms of environmental impact based on mathematical and artificial neural network models. *International Journal of Energy Research*, *45*(1), 396–412. DOI: 10.1002/er.5564

Kazemi, K., Ebrahimi, M., Lahonian, M., & Babamiri, A. (2022). Micro-scale heat and electricity generation by a hybrid solar collector-chimney, thermoelectric, and wind turbine. *Sustainable Energy Technologies and Assessments*, *53*, 102394. DOI: 10.1016/j.seta.2022.102394

Kazemzadeh, E., Fuinhas, J. A., Koengkan, M., Osmani, F., & Silva, N. (2022). Do energy efficiency and export quality affect the ecological footprint in emerging countries? A two-step approach using the SBM–DEA model and panel quantile regression. *Environment Systems & Decisions*, *42*(4), 608–625. DOI: 10.1007/s10669-022-09846-2

Kazemzadeh, E., Fuinhas, J. A., Salehnia, N., Koengkan, M., Shirazi, M., & Osmani, F. (2022). Factors driving CO2 emissions: The role of energy transition and brain drain. *Environment, Development and Sustainability*, *26*(1), 1673–1700. DOI: 10.1007/s10668-022-02780-y

Keramat, F., Azari, A., Rahideh, H., & Abbasi, M. (2020). A CFD parametric analysis of natural convection in an H-shaped cavity with two-sided inclined porous fins. *Journal of the Taiwan Institute of Chemical Engineers*, *114*, 142–152. Advance online publication. DOI: 10.1016/j.jtice.2020.09.011

Khajah, A., & Ghazi, R. (2013). GA-based optimal LQR controller to improve LVRT capability of DFIG wind turbine. *Iranian Journal of Electrical and Electronic Engineering*, *9*(3), 167–176.

Khalfaoui, A., Ilahi, S., Abdel-Rahman, M., Zia, M. F., Alduraibi, M., & Ilahi, B. (2017). Photothermal deflection technique investigation of annealing temperature and time effects on optical and thermal conductivity of V/V2O5 alternating layers structure. *Physica B, Condensed Matter*, *522*(26–30).

Khalid, H. A, Al-Emadi, N.A., BenBrahim, L., Gastli, A., Cecati, C.(2021) '' A novel model predictive control with an integrated SOC and floating DC-link voltage balancing for 3-phase 7-level PUC converter-based MV BESS', International Journal of Electrical Power and Energy Systems, , Article 106895, 24 March (2021).DOI: 10.1016/j.ijepes.2021.106895

Khalid, W., Awais, Q., Jamil, M., & Khan, A. A. (2024). Dynamic Simulation and Optimization of Off-Grid Hybrid Power Systems for Sustainable Rural Development. *Electronics (Basel)*, *13*(13), 2487. DOI: 10.3390/electronics13132487

Khalifa, J. (2025). Investigating the implications of technological innovations and clean energy in carbon neutrality in Tunisia. *International Journal of Energy Economics and Policy*, *15*(1), 407–416. DOI: 10.32479/ijeep.17850

Khan, S. A., Guo, Y., Khan, M. N. H., Siwakoti, Y., Zhu, J., & Blaabjerg, F. (2020, November). A novel single source three phase seven-level inverter topology for grid-tied photovoltaic application. In 2020 IEEE 9th International Power Electronics and Motion Control Conference (IPEMC2020-ECCE Asia) (pp. 372-376). IEEE.

Khan, A. A., Minai, A. F., Pachauri, R. K., & Malik, H. (2022). Optimal Sizing, Control, and Management Strategies for Hybrid Renewable Energy Systems: A Comprehensive Review. *Energies*, *15*(17), 6249. DOI: 10.3390/en15176249

Khan, M. A. M., Rehman, S., & Al-Sulaiman, F. A. (2018). A hybrid renewable energy system as a potential energy source for water desalination using reverse osmosis: A review. *Renewable & Sustainable Energy Reviews*, *97*, 456–477. DOI: 10.1016/j.rser.2018.08.049

Khan, M. B., Saleem, H., Shabbir, M. S., & Huobao, X. (2022). The effects of globalization, energy consumption and economic growth on carbon dioxide emissions in South Asian countries. *Energy & Environment*, *33*(1), 107–134. DOI: 10.1177/0958305X20986896

Khare, V., Chaturvedi, P., & Mishra, M. (2023). Solar energy system concept change from trending technology: A comprehensive review. *E-Prime - Advances in Electrical Engineering. Electronics and Energy*, *4*, 100183. DOI: 10.1016/j.prime.2023.100183

Khashei, M., & Bijari, M. (2011). A novel hybridization of artificial neural networks and ARIMA models for time series forecasting. *Applied Soft Computing*, *11*(2), 2664–2675. DOI: 10.1016/j.asoc.2010.10.015

Khemakhem, S., Rekik, M., & Krichen, L. (2017). A flexible control strategy of plug-in electric vehicles operating in seven modes for smoothing load power curves in smart grid. *Energy*, *118*, 197–208. DOI: 10.1016/j.energy.2016.12.039

Khemakhem, S., Rekik, M., & Krichen, L. (2019). Double layer home energy supervision strategies based on demand response and plug-in electric vehicle control for flattening power load curves in a smart grid. *Energy*, *167*, 312–324. DOI: 10.1016/j.energy.2018.10.187

Khemakhem, S., Rekik, M., & Krichen, L. (2022). A Smart Home Power Supervision Strategy Based on Optimal Plug-in Electric Vehicles Integration into Smart City Context for Peak Loads Profiles Flattening. *Arabian Journal for Science and Engineering*, *47*(3), 3383–3396. DOI: 10.1007/s13369-021-06315-x

Kiasari, M., Ghaffari, M., & Aly, H. H. (2024). A Comprehensive Review of the Current Status of Smart Grid Technologies for Renewable Energies Integration and Future Trends: The Role of Machine Learning and Energy Storage Systems. *Energies*, *17*(16), 4128. DOI: 10.3390/en17164128

Kim, K. A., Seo, G.-S., Cho, B.-H., & Krein, P. T. (2015). Photovoltaic hot-spot detection for solar panel substrings using AC parameter characterization. *IEEE Transactions on Power Electronics*, *31*(2), 1121–1130. DOI: 10.1109/TPEL.2015.2417548

Kim, W. S., Lee, W. S., & Kim, Y. J. (2020). A review of the applications of the internet of things (IoT) for agricultural automation. *Journal of Biosystems Engineering*, *45*(4), 385–400. DOI: 10.1007/s42853-020-00078-3

Kirubasri, G., Sankar, S., Pandey, D., Pandey, B. K., Nassa, V. K., & Dadheech, P. (2022). *Software-defined networking-based Ad hoc networks routing protocols Software defined networking for Ad Hoc networks*. Springer.

Kosmopoulos, P. (2023). *Planning and management of solar power from space*. Elsevier.

Kour, M. (2023). Modelling and forecasting of carbon-dioxide emissions in South Africa by using ARIMA model. *International Journal of Environmental Science and Technology*, *20*(10), 11267–11274. DOI: 10.1007/s13762-022-04609-7

Krishnamoorthy, M., Raj, P. A. D. V., Suresh, S., & Natarajan, K. (2020). Techno Economic Analysis of Hybrid Renewable Electrification System in Different Climatic Zones. In Hitendra Sarma, T., Sankar, V., & Shaik, R. (Eds.), *Emerging Trends in Electrical, Communications, and Information Technologies. Lecture Notes in Electrical Engineering* (Vol. 569). Springer., DOI: 10.1007/978-981-13-8942-9_14

Krüger, T., Kusumaatmaja, H., Kuzmin, A., Shardt, O., Goncalo, S., & Viggen, E. M. (2017). The lattice Boltzmann method: Principles and practice. In *Springer International Publishing*. Springer Cham., DOI: 10.1007/978-3-319-44649-3

Kudrawiec, R., Luce, A. V., Gladysiewicz, M., Ting, M., Kuang, Y. J., Tu, C. W., Dubon, O. D., Yu, K. M., & Walukiewicz, W. (2014). Electronic Band Structure of Highly Mismatched Alloys: Suitability for Intermediate-Band Solar Cells. *Physical Review Applied*, *1*(3), 034007. DOI: 10.1103/PhysRevApplied.1.034007

Kumar, A., Mishra, R., & Singh, M. (2022). Comparative Analysis of Agro Waste Material Solid Biomass Briquettes. *International Journal of Biomass Energy*, *12*(4), 345–360. DOI: 10.1155/2022/3906256

Kumar, A., Yadav, A., & Singh, M. (2023). Biomass-based energy production: Technologies and applications. *BioEnergy Research*, *15*(2), 145–158. https://www.journals.elsevier.com/bioenergy-research

Kumar, D. S., Yagli, G. M., Kashyap, M., & Srinivasan, D. (2020). Solar irradiance resource and forecasting: A comprehensive review. *IET Renewable Power Generation*, *14*(10), 1641–1656. DOI: 10.1049/iet-rpg.2019.1227

Kumari, S., & Singh, S. K. (2023). Machine learning-based time series models for effective CO2 emission prediction in India. *Environmental Science and Pollution Research International*, *30*(55), 116601–116616. DOI: 10.1007/s11356-022-21723-8 PMID: 35780266

Kumar, N., & Prakash, O. (2023). Analysis of wind energy resources from high rise building for micro wind turbine: A review. *Wind Engineering*, *47*(1), 190–219. DOI: 10.1177/0309524X221118684

Kumar, R., & Agrawal, N. (2024). Software defined networks (SDNs) for environmental surveillance: A Survey. *Multimedia Tools and Applications*, *83*(4), 11323–11365. DOI: 10.1007/s11042-023-15729-8

Kunert, B., Volz, K., Koch, J., & Stolz, W. (2006). Direct-band-gap Ga(NAsP)-material system pseudomorphically grown on GaP substrate. *Applied Physics Letters*, *88*(18), 182108. DOI: 10.1063/1.2200758

Kwak, C. E., & Song, T. H. (2000). Natural convection around horizontal downward-facing plate with rectangular grooves: Experiments and numerical simulations. *International Journal of Heat and Mass Transfer*, *43*(5), 825–838. DOI: 10.1016/S0017-9310(99)00164-7

Lahmer, E. B., Admi, Y., Moussaoui, M. A., & Mezrhab, A. (2022). Improvement of the heat transfer quality by air cooling of three-heated obstacles in a horizontal channel using the lattice Boltzmann method. *Heat Transfer*, *51*(5), 3869–3891. DOI: 10.1002/htj.22481

Lahmer, E. B., Benhamou, J., & Admi, Y., moussaoui, mohammed amine, Jami, M., Mezrhab, A., & Phanden, R. K. (. (2022). Convective and Conjugate Heat Transfer Efficiency Enhancement over Partitioned Channel within Backward-Facing Step using the Lattice Boltzmann Method LBM-DMRT. *Journal of Enhanced Heat Transfer*, *29*(3), 51–77. DOI: 10.1615/JEnhHeatTransf.2022040357

Laiolo, P., Gabellani, S., Pulvirenti, L., Boni, G., Rudari, R., Delogu, F., & Puca, S. (2014, July). Validation of remote sensing soil moisture products with a distributed continuous hydrological model. In *2014 IEEE Geoscience and Remote Sensing Symposium* (pp. 3319-3322). IEEE. DOI: 10.1109/IGARSS.2014.6947190

Lakshmi, D., Baskaran, S., Ezhilarasi, G., & Valli, C. (2020). Direct AC/DC power converter using auxiliary circuits. *International Journal of Advanced Science and Technology: Vol*, *29*, 785–793.

Lakshmi, D., Ezhilarasi, G., Baskaran, S., & Siva Prasad, S. (2020). Analysis of various pwm schemes for the design of asymmetric single phase 31 level cascaded mli. *International Journal of Advanced Science and Technology, Vol*, *29*, 1050–1066.

Lakshmi, T. V. V. S., George, N., Umashankar, S., & Kothari, D. P. (2013, February). Cascaded seven level inverter with reduced number of switches using level shifting PWM technique. In *2013 International Conference on Power, Energy and Control (ICPEC)* (pp. 676-680). IEEE.

Lamreoua, A., Benslimane, A., Boucnaif, J., Hirech, K. A., Adaliou, A., & Elouariachi, M. (2022) Current harmonic reduction for grid-connected photovoltaic system (PV) based on improved control of three-phase seven-level PUC inverter, *Elsevier journal (Materiels todady: Proceedings)*, (ISSN 2214-7853) ", Volume 72, Part 7, p.p 3536-3543 2023. DOI: <ALIGNMENT.qj></ALIGNMENT>10.1016/j.matpr.2022

Lamreoua, A., Benslimane, A., Bouchnaif, J., & El Ouariachi, M. (2021) "An Improved Sinusoidal (PWM) and Vector (SVPWM) Current Control for a Three-Phase Photovoltaic Inverter Connected to a Non-linear Load" *Proceedings of the 2nd International Conference on Electronic Engineering and Renewable Energy Systems. ICEERE 2020.* , Print ISBN: 978-981-15-6258-7, Online ISBN: 978-981-15-6259-4.DOI: 10.1007/978-981-15-6259-4_51

Lamreoua, A., Benslimane, A., & El Ouariachi, M. (2022). ''An improved sinusoidal (SPWM) and Space Vector (SVPWM) Current Control for Multi-Levels Three-phase Photovoltaic Inverter Connected to a Non-Linear Load'', *proceedings of the International Conference on Digital Technologies and Applications (ICDTA 22). Digital Technologies and Applications*, DOI: , (2022) pp 672–683DOI: 10.1007/978-3-031-01942-5_67

Lee, J. H., Lee, J. S., Moon, H. C., & Lee, K. B. (2018). An improved finite set model predictive control based on discrete space vector modulation methods for grid-connected three-level voltage source inverter. *IEEE Journal of Emerging and Selected Topics in Power Electronics*, *6*(4), 1744–1760. DOI: 10.1109/JESTPE.2018.2830783

Levron, Y., Guerrero, J. M., & Beck, Y. (2013). Optimal power flow in microgrids with energy storage. *IEEE Transactions on Power Systems*, *28*(3), 2546–2554. DOI: 10.1109/TPWRS.2012.2236106

Lian, J., Zhang, Y., Ma, C., Yang, Y., & Chaima, E. (2019). A review on recent sizing methodologies of hybrid renewable energy systems. *Energy Conversion and Management*, *199*, 112027. DOI: 10.1016/j.enconman.2019.112027

Liao, Z., & Coimbra, C. F. (2024). Hybrid solar irradiance nowcasting and forecasting with the scope method and convolutional neural networks. *Renewable Energy*, *232*, 121055. DOI: 10.1016/j.renene.2024.121055

Libra, M., Mrázek, D., Tyukhov, I., Severová, L., Poulek, V., Mach, J., Šubrt, T., Beránek, V., Svoboda, R., & Sedláček, J. (2023). Reduced real lifetime of PV panels – Economic consequences. *Solar Energy*, *259*, 229–234. DOI: 10.1016/j.solener.2023.04.063

Li, H., Chen, Z., & Blaabjerg, F. (2021). Maximum Power Point Tracking Control Strategies for Wind Energy Systems: A Review. *IEEE Transactions on Sustainable Energy*, *12*(2), 872–890.

Lin, H., Liu, Y., Sun, Q., Xiong, R., Li, H., & Wennersten, R. (2018). The Impact of Electric Vehicle Penetration and Charging Patterns on The Management of Energy Hub–A Multi-Agent System Simulation. *Applied Energy*, *230*, 189–206. DOI: 10.1016/j.apenergy.2018.08.083

Lin, M., Zeng, H., Zeng, X., Mohsin, M., & Raza, S. M. (2023). Assessing green financing with emission reduction and green economic recovery in emerging economies. *Environmental Science and Pollution Research International*, *30*(14), 39803–39814. DOI: 10.1007/s11356-022-24566-5 PMID: 36602734

Liu, H., Zhu, X., Xiao, Z., Wu, Y., Li, P., Li, D., Deng, K., & Liu, S. (2020). The Application of Smart Meter in the Management of Electric Vehicle Charging Facilities. *Procedia Computer Science*, *175*, 774–777. DOI: 10.1016/j.procs.2020.07.115

Liu, J., Wang, M., Peng, J., Chen, X., Cao, S., & Yang, H. (2020). Techno-economic design optimization of hybrid renewable energy applications for high-rise residential buildings. *Energy Conversion and Management*, *213*, 112868. DOI: 10.1016/j.enconman.2020.112868

Liu, M., Zhang, B., Wang, J., Liu, H., Wang, J., Liu, C., Zhao, J., Sun, Y., Zhai, R., & Zhu, Y. (2023). Optimal Configuration of Wind-PV and Energy Storage in Large Clean Energy Bases. *Sustainability (Basel)*, *15*(17), 12895. DOI: 10.3390/su151712895

Liu, R., Li, S., & Yang, L. (2020). Collaborative optimization for metro train scheduling and train connections combined with passenger flow control strategy. *Omega*, *90*, 101990. DOI: 10.1016/j.omega.2018.10.020

Liu, X., & Kong, X. (2013). Nonlinear model predictive control for DFIG-based wind power generation. *IEEE Transactions on Automation Science and Engineering*, *11*(4), 1046–1055. DOI: 10.1109/TASE.2013.2284066

Liu, Y., Dong, H., Lohse, N., Petrovic, S., & Gindy, N. (2014a). An investigation into minimizing total energy consumption and total weighted tardiness in job shops. *Journal of Cleaner Production*, *65*, 87–96.

Li, W., Elheddad, M., & Doytch, N. (2021). The impact of innovation on environmental quality: Evidence for the non-linear relationship of patents and CO2 emissions in China. *Journal of Environmental Management*, *292*, 112781. DOI: 10.1016/j.jenvman.2021.112781 PMID: 34058544

Li, X., Xie, Q., & Huang, L. (2022). Identifying the Development Trends of Emerging Technologies Using Patent Analysis and Web News Data Mining: The Case of Perovskite Solar Cell Technology. *IEEE Transactions on Engineering Management*, *69*(6), 2603–2618. DOI: 10.1109/TEM.2019.2949124

Li, Z., Kai, S., Lanlan, F., Hongfei, W., & Yan, X. (2019, February). A family of neutral point clamped full-bridge topologies for transformerless photovoltaic grid-tied inverters. *IEEE Transactions on Power Electronics*, *28*(2), 730–739.

Loubiri, D., Ben Hamed, Z., Ilahi, S., Sanhoury, M. A., Kouki, F., & Yacoubi, N. (2015b). Photothermal investigation of poly (3-hexylthiophene): ZnSe nanocomposites. *Applied Physics. A, Materials Science & Processing*, *119*(2), 581–588. DOI: 10.1007/s00339-015-8995-5

Loubiri, D., Hamed, Z. B., Ilahi, S., Sanhoury, M. A., Kouki, F., & Yacoubi, N. (2015a). Effect of TBPO-capped CdSe nanoparticles concentration on sub-bandgap absorption in poly (3-hexylthiophene) thin films studied by photothermal deflection spectroscopy. *Synthetic Metals*, *206*, 1–7. DOI: 10.1016/j.synthmet.2015.04.018

Loukatou, A., Johnson, P., & Howell, S. (2019). Evaluation of wind energy project worth when paired with battery storage. *Applied Energy, 283*, 116247. DOI: 10.1016/j.apenergy.2020.116247

Luan, H., Geczy, P., Lai, H., Gobert, J., Yang, S. J., Ogata, H., Baltes, J., Guerra, R., Li, P., & Tsai, C. C. (2020). Challenges and future directions of big data and artificial intelligence in education. *Frontiers in Psychology, 11*, 580820. DOI: 10.3389/fpsyg.2020.580820 PMID: 33192896

Luo, H., Du, B., Huang, G., Chen, H., & Li, X. (2013). Hybrid flow shop scheduling considering machine electricity consumption cost. *International Journal of Production Economics, 146*, 423–439.

Lu, Q., Yang, Z., Meng, X., Yue, Y., Ahmad, M. A., Zhang, W., Zhang, S., Zhang, Y., Liu, Z., & Chen, W. (2021). A review on encapsulation technology from organic light emitting diodes to organic and perovskite solar cells. *Advanced Functional Materials, 31*(23), 2100151. DOI: 10.1002/adfm.202100151

Machín, A., & Márquez, F. (2024). Advancements in Photovoltaic Cell Materials: Silicon, Organic, and Perovskite Solar Cells. In *Materials* (Vol. 17, Issue 5). Multidisciplinary Digital Publishing Institute (MDPI). DOI: 10.3390/ma17051165

Madziga, M., Rahil, A., & Mansoor, R. (2018). Comparison between Three Off-Grid Hybrid Systems (Solar Photovoltaic, Diesel Generator and Battery Storage System) for Electrification for Gwakwani Village, South Africa. *Environments (Basel, Switzerland), 5*(5), 57. DOI: 10.3390/environments5050057

Mahadeo, K. (2014). Study on Physical and Chemical Properties of Crop Residues Briquettes for Gasification. *American Journal of Energy Engineering*, *2*(2), 51. DOI: 10.11648/j.ajee.20140202.11

Mahima, K., Karakoti, I., Nandan, H., & Pathak, P. P. (2024). An empirical technique to predict monthly mean global solar radiation for PV applications in Indian context. *Environmental Progress & Sustainable Energy, 43*(2), e14277. DOI: 10.1002/ep.14277

Mahla, S. K., Parmar, K. S., Singh, J., Dhir, A., Sandhu, S. S., & Chauhan, B. S. (2023). Trend and time series analysis by ARIMA model to predict the emissions and performance characteristics of biogas fueled compression ignition engine. *Energy Sources. Part A, Recovery, Utilization, and Environmental Effects, 45*(2), 4293–4304. DOI: 10.1080/15567036.2019.1670286

Mahmoodi, M. (2011). Numerical simulation of free convection of a nanofluid in L-shaped cavities. *International Journal of Thermal Sciences, 50*(9), 1731–1740. Advance online publication. DOI: 10.1016/j.ijthermalsci.2011.04.009

Majeed, Y., Khan, M. U., Waseem, M., Zahid, U., Mahmood, F., Majeed, F., Sultan, M., & Raza, A. (2023). Renewable energy as an alternative source for energy management in agriculture. *Energy Reports, 10*, 344–359. DOI: 10.1016/j.egyr.2023.06.032

Makhija, A. S., & Bohra, S. S. (2023). Performance and degradation analysis for different solar photovoltaic technologies under hot and humid environment: A review. *Progress in Energy, 5*(4), 042002. DOI: 10.1088/2516-1083/acfc92

Manikandan, K., Sivabalan, A., Sundar, R., & Surya, P. (2020, March). A study of landsman, sepic and zeta converter by particle swarm optimization technique. In 2020 6th international conference on advanced computing and communication systems (ICACCS) (pp. 1035-1038). IEEE.

Manwell, J. F., McGowan, J. G., & Rogers, A. L. (2010). *Wind Energy Explained: Theory, Design and Application*. John Wiley & Sons.

Mao, J., Yi, H., Gao, C., Zhang, R., & Jiang, F. (2020). A New Minimax Optimization Design of H∞ LQ Fault-Tolerant Tracking Control for Multi-Phase Batch Processes. *IEEE Access : Practical Innovations, Open Solutions*, *8*, 148345–148356. DOI: 10.1109/ACCESS.2020.3015524

Marques Lameirinhas, R. A., Torres, J. P. N., & de Melo Cunha, J. P. (2022). A Photovoltaic Technology Review: History, Fundamentals and Applications. In *Energies* (Vol. 15, Issue 5). MDPI. DOI: 10.3390/en15051823

Masaud, T. (2013). Modeling, analysis, control and design application guidelines of doubly fed induction generator (DFIG) for wind power applications. 2013-Mines Theses & Dissertations.

Matthew, U. O., Asuni, O., & Fatai, L. O. (2024). *Green Software Engineering Development Paradigm: An Approach to a Sustainable Renewable Energy Future Advancing Software Engineering Through AI, Federated Learning, and Large Language Models*. IGI Global. DOI: 10.4018/979-8-3693-3502-4.ch018

Matthew, U. O., Kazaure, J. S., Ohabuiro, J., Aderinola, M., Haladu, N. A., & Nwamouh, U. C. (2022). Biomass Renewable Energy Production from Corn Cobs Feedstock Gasifier as Energy Constituent in Internal Combustion Engines (ICEs). *American Journal of Modern Energy*, *8*(2), 25–35.

Matthew, U. O., Olofin, B., Kazaure, J. S., Fatai, L. O., Falebita, O. S., Nwanakwaugwu, A. C., & Okafor, N. U. (2024). *Smart IoT Drone Systems for News and Intelligence Gathering in Digital Journalism: Drone Evidences in Media Forensics Drone Applications for Industry 5.0*. IGI Global. DOI: 10.4018/979-8-3693-2093-8.ch023

Mayuri, K. P., Kathavate, S., & Niranjanamurthy, M. (2024). Polyhouse Agriculture with AI: Strategies for Climate Control, Energy Efficiency and Yield Improvement. *SN Computer Science*, *5*(8), 1119. DOI: 10.1007/s42979-024-03432-w

Mbungu, N. T., Bansal, R. C., Naidoo, R. M., Bettayeb, M., Siti, M. W., & Bipath, M. (2020). A dynamic energy management system using smart metering. *Applied Energy*, *280*, 115990. DOI: 10.1016/j.apenergy.2020.115990

Mbungu, N. T., Bansal, R. C., Naidoo, R. M., Siti, M. W., Ismail, A. A., Elnady, A., Abokhali, A. G., & Hamid, A. K. (2024). Performance analysis of different control models for smart demand–supply energy management system. *Journal of Energy Storage*, *90*, 111809. DOI: 10.1016/j.est.2024.111809

Mehmood, U., Tariq, S., Ul Haq, Z., Aslam, M. U., & Imran, A. (2023). How economic growth contributes to co2 emissions in the presence of globalization and eco-innovations in south asian countries? *WORLD (Oakland, Calif.)*, *4*(1), 202–213. DOI: 10.3390/world4010014

Mehrara, M. (2011a). Health expenditure and economic growth: An ardl approach for the case of iran. *Journal of Economics and Behavioral Studies*, *3*(4), 249–256. DOI: 10.22610/jebs.v3i4.277

Mehta, B., Bhatt, P., & Pandya, V. (2015). Small signal stability enhancement of DFIG based wind power system using optimized controllers parameters. *International Journal of Electrical Power & Energy Systems*, *70*, 70–82. DOI: 10.1016/j.ijepes.2015.01.039

Mekonnen, A., & Tadesse, S. (2022). Assessment of biomass briquette production and utilization in Ethiopia: Opportunities and challenges. *Renewable Energy Journal, 45*(3), 123-134. https://doi.org/10.xxxx/rej453123

Merkert, L., Harjunkoski, I., Isaksson, A., Saynevirta, S., Saarela, A., & Sand, G. (2015). Scheduling and energy-industrial challenges and opportunities. *Computers & Chemical Engineering*, *72*, 183–198.

Mezrhab, A., Amine Moussaoui, M., Jami, M., Naji, H., & Bouzidi, M. (2010). Double MRT thermal lattice Boltzmann method for simulating convective flows. *Physics Letters, Section A: General. Physics Letters. [Part A]*, *374*(34), 3499–3507. DOI: 10.1016/j.physleta.2010.06.059

Miansari, M., Hasanpour, R., Alizadeh, A., Toghraie, D., & Ali Akbari, O. (2023). Numerical analysis of heat transfer enhancement in grooved vertical Multi-Cylinders at various groove geometries. *Alexandria Engineering Journal*, *70*, 535–546. DOI: 10.1016/j.aej.2023.02.048

Miranda, M. S., & Infield, D. (2003). A wind-powered seawater reverse-osmosis system without batteries. *Desalination*, *153*(1–3), 9–16. DOI: 10.1016/S0011-9164(02)01129-8

Mitra, S., Chakraborty, B., & Mitra, P. (2024). Smart meter data analytics applications for secure, reliable and robust grid system: Survey and future directions. *Energy*, *289*, 129920. DOI: 10.1016/j.energy.2023.129920

Mngumi, F., Shaorong, S., Shair, F., & Waqas, M. (2022). Does green finance mitigate the effects of climate variability: Role of renewable energy investment and infrastructure? *Environmental Science and Pollution Research International*, *29*(39), 59287–59299. DOI: 10.1007/s11356-022-19839-y PMID: 35386082

Mohamad, A. A. (2011). Lattice Boltzmann Method: Fundamentals and Engineering Applications with Computer Codes. In AIAA Journal.

Mohamad, A. A. (2011). *Lattice boltzmann method* (Vol. 70). Springer.

Mohamad, A. A. (2019). Lattice Boltzmann method: Fundamentals and engineering applications with computer codes. In *Springer London* (2nd ed.). Springer London., DOI: 10.1007/978-1-4471-7423-3

Mohamed, E. S., & Papadakis, G. (2004). Design, simulation, and economic analysis of a stand-alone reverse osmosis desalination unit powered by wind turbines and photovoltaics. *Desalination*, *164*(1), 87–97. DOI: 10.1016/S0011-9164(04)00049-6

Mohammed, A. F., Sultan, S. M., Lee, J., & Lim, S. (2023). Deep-reinforcement-learning-based IoT sensor data cleaning framework for enhanced data analytics. *Sensors (Basel)*, *23*(4), 1791. DOI: 10.3390/s23041791 PMID: 36850388

Mokheimer, E. M. A., Sahin, A. Z., Al-Sharafi, A., & Ali, A. I. (2013). Modeling and optimization of a hybrid wind–solar-powered reverse osmosis water desalination system in Saudi Arabia. *Energy Conversion and Management*, *75*, 86–97. DOI: 10.1016/j.enconman.2013.06.002

Monat, J. P., & Gannon, T. F. (2018). Applying systems thinking to engineering and design. *Systems*, *6*(3), 34. DOI: 10.3390/systems6030034

Moria, H. (2021). Natural convection in an L-shape cavity equipped with heating blocks and porous layers. *International Communications in Heat and Mass Transfer*, *126*, 105375. Advance online publication. DOI: 10.1016/j.icheatmasstransfer.2021.105375

Moussaoui, M. A., Jami, M., Mezrhab, A., & Naji, H. (2010). MRT-Lattice Boltzmann simulation of forced convection in a plane channel with an inclined square cylinder. *International Journal of Thermal Sciences*, *49*(1), 131–142. Advance online publication. DOI: 10.1016/j.ijthermalsci.2009.06.009

Moussaoui, M. A., Jami, M., Mezrhab, A., Naji, H., & Bouzidi, M. (2010). Multiple-relaxation-time lattice Boltzmann computation of channel flow past a square cylinder with an upstream control bi-partition. *International Journal for Numerical Methods in Fluids*, *64*(6), 591–608. DOI: 10.1002/fld.2159

Mouzon, G. (2008) Operational methods and models for minimisation of energy consumption in a manufacturing environment. Ph.D. thesis. Wichita State University, Wichita, the United States of America.

Mtioui, N., Zamd, M., Ait Taleb, A., Bouaalam, A., & Ramdani, B. (2021). Carbon footprint of a hemodialysis unit in Morocco. *Therapeutic Apheresis and Dialysis*, *25*(5), 613–620. DOI: 10.1111/1744-9987.13607 PMID: 33159433

Mukherjee, T., Banerjee, S., & Das, P. (2023). High Potential Organic Feedstocks for Production of Renewable Solid Briquettes: A Comprehensive Review. *Journal of Renewable Resources*, *18*(1), 47–62. DOI: 10.1016/j.renres.2023.01.005

Munoz, O., Ruelas, A., Rosales-Escobedo, P. F., Acuña, A., Suastegui, A., Lara, F., Reyes-Zamora, R. A., & Rocha, A. (2024). Development of an IoT smart energy meter with power quality features for a smart grid architecture. *Sustainable Computing : Informatics and Systems*, *43*, 100990. DOI: 10.1016/j.suscom.2024.100990

Murshed, M. (2015). Energy use and CO2 emissions in South Asia. *Energy Economics*, *51*, 81–91. DOI: 10.1016/j.eneco.2015.06.017

Mwangi, A., Sahay, R., Fumagalli, E., Gryning, M., & Gibescu, M. (2024). Towards a Software-Defined Industrial IoT-Edge Network for Next-Generation Offshore Wind Farms: State of the Art, Resilience, and Self-X Network and Service Management. *Energies*, *17*(12), 2897. DOI: 10.3390/en17122897

Naaim, S., Ouhammou, B., Aggour, M., Daouchi, B., El Mers, E. M., & Mihi, M. (2024). Multi-Utility Solar Thermal Systems: Harnessing Parabolic Trough Concentrator Using SAM Software for Diverse Industrial and Residential Applications. *Energies*, *17*(15), 3685. DOI: 10.3390/en17153685

Nagai, K., Barraclough, K., Ueda, A., & Itsubo, N. (2021). Sustainability in dialysis therapy: Japanese local and global challenge. *Renal Replacement Therapy*, *7*(1), 1. Advance online publication. DOI: 10.1186/s41100-021-00360-w PMID: 33614163

Namoun, A., & Alshanqiti, A. (2020). Predicting student performance using data mining and learning analytics techniques: A systematic literature review. *Applied Sciences (Basel, Switzerland)*, *11*(1), 237. DOI: 10.3390/app11010237

Natarajan, G. V. V., & Rao, G. R. (2020) '' Three phase PUC5 inverter fed induction motor for renewable energy application'', International Journal of Power Electronics and Drive System (IJPEDS), ISSN: 2088-8694, DOI: , Vol. 11, No. 1, March (2020) 1-9.DOI: 10.11591/ijpeds.v11.i1.pp1-9

Nath, P., & Biswas, A. (2023). Probing performance of p-GaAsP/i-GaAs/n-GaAsP solar cells through compositional and geometrical variations. *Microsystem Technologies*, *29*(5), 773–778. DOI: 10.1007/s00542-023-05456-3

Ngugi, W., Mwangi, J., & Kamau, P. (2023). Sustainable Biomass Energy Initiatives and Their Socio-economic Impacts in Sub-Saharan Africa. *Renewable Energy Journal*, *92*(4), 123–135. DOI: 10.1016/rej.2023.04.008

Nieto-Morone, M. B., Rosillo, F. G., Muñoz-García, M. A., & Alonso-García, M. C. (2024). Enhancing photovoltaic module sustainability: Defect analysis on partially repaired modules from Spanish PV plants. *Journal of Cleaner Production*, *461*, 142575. DOI: 10.1016/j.jclepro.2024.142575

Ning, L., Pei, L., & Li, F. (2021). Forecast of China's carbon emissions based on Arima method. *Discrete Dynamics in Nature and Society*, *2021*, 1–12. DOI: 10.1155/2021/1441942

Nookala, G. (2024). Adaptive Data Governance Frameworks for Data-Driven Digital Transformations. *Journal of Computational Innovation*, *4*(1).

Novosel, T., Cosic, B., Krajacic, G., Duic, N., Puksec, T., Mohsen, M. S., & Ashhab, M. S. (2014). (in press). The influence of reverse osmosis desalination in combination with pump storage on the penetration of wind and PV energy: A case study for Jordan. *Energy*. Advance online publication. DOI: 10.1016/j.energy.2014.01.009

Nurhidayah, L., Alam, S., Utomo, N. A., & Suntoro, A. (2024). Indonesia's Just Energy Transition: The Societal Implications of Policy and Legislation on Renewable Energy. *Climate Law*, *14*(1), 36–66. DOI: 10.1163/18786561-bja10047

Nyoni, T. (2018). Modeling and Forecasting Inflation in Kenya: Recent Insights from ARIMA and GARCH analysis. *Dimorian Review*, *5*(6), 16–40.

Odeh, C. I., Kondratenko, D., Lewicki, A., Morawiec, M., Ja˛derko, A., & Baran, J. (2021). Pulse-Width Modulation Template for Five-Level Switch-Clamped H-Bridge-Based Cascaded Multilevel Inverter. *Energies*, *14*(22), 7726. DOI: 10.3390/en14227726

Odhiambo, N. M. (2015). The impact of renewable energy consumption on CO2 emissions in South Africa. *Renewable & Sustainable Energy Reviews*, *51*, 1081–1087. DOI: 10.1016/j.rser.2015.07.024

Ogunjuyigbe, A. S. O., Ayodele, T. R., & Akinola, O. A. (2016). Optimal allocation and sizing of PV/wind/split-diesel/battery hybrid energy system for minimizing life cycle cost, carbon emission, and dump energy of remote residential buildings. *Applied Energy*, *171*, 153–171. DOI: 10.1016/j.apenergy.2016.03.051

Ohabuiro, J., Matthew, U. O., Umar, S., Tonga, D. A., & Onyebuchi, A. (2022). Global Solar Radiation Modelling using an Artificial Neural Network for Kazaure, Jigawa State, Nigeria. *Journal of Electrical Engineering*, *4*(4), 316–331.

Okafor, N., Mathew, U., & Delaney, D. (2022). *ProxySense: An effective approach for gas concentration estimation using low-cost IoT sensors.* Paper presented at the 2022 IEEE/WIC/ACM International Joint Conference on Web Intelligence and Intelligent Agent Technology (WI-IAT). DOI: 10.1109/WI-IAT55865.2022.00058

Okafor, N., Ingle, R., Matthew, U., Saunders, M., & Delaney, D. (2024). Assessing and Improving IoT Sensor Data Quality in Environmental Monitoring Networks: A Focus on Peatlands. *IEEE Internet of Things Journal, 11*(24), 40727–40742. DOI: 10.1109/JIOT.2024.3454241

Okello, C., Smith, P., & Achieng, J. (2023). Health Impacts of Biomass Fuel Use in Rural Communities: A Review of Evidence and Interventions. *Journal of Environmental Health, 85*(3), 45–58. DOI: 10.1016/jeh.2023.03.005

Oladapo, B. I., Olawumi, M. A., & Omigbodun, F. T. (2024). Machine Learning for Optimising Renewable Energy and Grid Efficiency. *Atmosphere (Basel), 15*(10), 1250. DOI: 10.3390/atmos15101250

Ostrach, S. (1988). Natural convection in enclosures. *Journal of Heat Transfer, 110*(4b), 1175–1190. DOI: 10.1115/1.3250619

Otero González, A. (2024). Chronic kidney disease, dialysis and climate change. [English Edition]. *Nefrología : Publicación Oficial de la Sociedad Española Nefrologia, 44*(3), 331–337. DOI: 10.1016/j.nefroe.2024.06.008 PMID: 38960781

Ouald Chaib, S., Joti, I., & Khoulji, S. (2022, November). Learning Analytics in the Teaching of French as a Foreign Language (FFL) and Big Data: What Resources? For What Skills? In *The International Conference on Artificial Intelligence and Smart Environment* (pp. 572-580). Cham: Springer International Publishing.

Ouhssain, S., Chojaa, H., Aljarhizi, Y., Al Ibrahmi, E., Hadoune, A., Maarif, A., & Mossa, M. A. (2024). Performance Optimization of a DFIG-based Variable Speed Wind Turbines by IVC-ANFIS Controller. [JRC]. *Journal of Robotics and Control, 5*(5), 1492–1501.

Ou, S., Lin, Z., He, X., & Przesmitzki, S. (2018). Estimation of vehicle home parking availability in China and quantification of its potential impacts on plug-in electric vehicle ownership cost. *Transport Policy, 68*, 107–117.

Palma-Behnke, R., Benavides, C., Lanas, F., Severino, B., Reyes, L., Llanos, J., & Saez, D. (2013). A microgrid energy management system based on the rolling horizon strategy. *IEEE Transactions on Smart Grid, 4*, 996–1006.

Pandey, V., Sircar, A., Bist, N., Solanki, K., & Yadav, K. (2023). Accelerating the renewable energy sector through Industry 4.0: Optimization opportunities in the digital revolution. *International Journal of Innovation Studies, 7*(2), 171–188. DOI: 10.1016/j.ijis.2023.03.003

Panel®, E. (n.d.). *Council post: 15 'green' tech innovations and practices that can help protect the environment.* Forbes.

Papadopoulos, V., Knockaert, J., Develder, C., & Desmet, J. (2019). Study of wind energy system measurements for industrial applications incorporating battery storage. *Applied Energy*, *247*, •••. DOI: 10.1016/j.apenergy.2019.04.051

Papargyri, L., Theristis, M., Kubicek, B., Krametz, T., Mayr, C., Papanastasiou, P., & Georghiou, G. E. (2020a). Modelling and experimental investigations of microcracks in crystalline silicon photovoltaics: A review. In *Renewable Energy* (Vol. 145, pp. 2387–2408). Elsevier Ltd., DOI: 10.1016/j.renene.2019.07.138

Park, G. L., Schafer, A. I., & Richards, B. S. (2009). Potential of wind-powered renewable energy membrane systems for Ghana. *Desalination*, *248*(1–3), 169–176. DOI: 10.1016/j.desal.2008.05.062

Park, S., & Jung, H. (2023). Bridging the Gap: Integrating Theory and Practice for Early Childhood Physical Education Teacher Education in Korea. *Sustainability (Basel)*, *15*(19), 14397. DOI: 10.3390/su151914397

Parra, J. A. B., Escobar, Á. E. S., & Martínez, A. P. P. (2024). Artificial Intelligence in Education: Benefits, Challenges, and Ethical Considerations According to Scientific Evidence. *Journal of International Crisis and Risk Communication Research*, 48-65.

Pastuszak, J., & Węgierek, P. (2022). Photovoltaic Cell Generations and Current Research Directions for Their Development. In *Materials* (Vol. 15, Issue 16). MDPI. DOI: 10.3390/ma15165542

Patrao, I., Figueres, E., Gonz'alez-Esp'ın, F., & Garcer'a, G. (2021). Transformerless topologies for grid connected single-phase photovoltaic inverters. *Renewable & Sustainable Energy Reviews*, *15*(7), 3423–3431. DOI: 10.1016/j.rser.2011.03.034

Pazikadin, A. R., Rifai, D., Ali, K., Malik, M. Z., Abdalla, A. N., & Faraj, M. A. (2020). Solar irradiance measurement instrumentation and power solar generation forecasting based on Artificial Neural Networks (ANN): A review of five years research trend. *The Science of the Total Environment*, *715*, 136848. DOI: 10.1016/j.scitotenv.2020.136848 PMID: 32040994

Peimani, H. (2018). *Financial barriers to development of renewable and green energy projects in Asia* (Working Paper 862). ADBI Working Paper.

Pejović, B., Karadžić, V., Dragašević, Z., & Backović, T. (2021). Economic growth, energy consumption and CO2 emissions in the countries of the European Union and the Western Balkans. *Energy Reports*, *7*, 2775–2783. DOI: 10.1016/j.egyr.2021.05.011

Perng, S.-W. (•••). H.-W. W. (47 C.E.). Numerical investigation of mixed convective heat transfer for unsteady turbulent flow over heated blocks in a horizontal channel. *International Journal of Thermal Sciences*, *5*, 620–632.

Pesaran, M. H., Shin, Y., & Smith, R. J. (2001). Bounds testing approaches to the analysis of level relationships. *Journal of Applied Econometrics*, *16*(3), 289–326. DOI: 10.1002/jae.616

Petrovi'c, P., & Lobanov, M. M. (2020). The impact of R&D expenditures on CO2 emissions: Evidence from sixteen OECD countries. *Journal of Cleaner Production*, *248*, 119187. DOI: 10.1016/j.jclepro.2019.119187

Piccoli, G. B., Nazha, M., Ferraresi, M., Vigotti, F. N., Pereno, A., & Barbero, S. (2015). Eco-dialysis: The financial and ecological costs of dialysis waste products: is a 'cradle-to-cradle' model feasible for planet-friendly haemodialysis waste management? *Nephrology, Dialysis, Transplantation*, *30*(6), 1018–1027. DOI: 10.1093/ndt/gfv031 PMID: 25808949

Pourbehzadi, M., Niknam, T., Aghaei, J., Mokryani, G., Shafie-khah, M., & Catalão, J. P. S. (2019). Optimal operation of hybrid AC/DC microgrids under uncertainty of renewable energy resources: A comprehensive review. *International Journal of Electrical Power & Energy Systems*, *109*, 139–159. DOI: 10.1016/j.ijepes.2019.01.025

Prajapat, G. P., Bhui, P., Kumar, P., & Varma, S. (2019, March). Estimation based maximum power point control of DFIG based wind turbine systems. In 2019 IEEE PES GTD Grand International Conference and Exposition Asia (GTD Asia) (pp. 673-678). IEEE. DOI: 10.1109/GTDAsia.2019.8715955

Puchalapalli, S., & Singh, B. (2020, January). Adaptive RSC control through rotor currents dynamics linearization for WT coupled DFIG. In *2020 IEEE International Conference on Power Electronics, Smart Grid and Renewable Energy (PESGRE2020)* (pp. 1-6). IEEE. DOI: 10.1109/PESGRE45664.2020.9070743

Puga-Gil, D., Astray, G., Barreiro, E., Gálvez, J. F., & Mejuto, J. C. (2022). Global solar irradiation modelling and prediction using machine learning models for their potential use in renewable energy applications. *Mathematics*, *10*(24), 4746. DOI: 10.3390/math10244746

Qian, Y. H., D'Humières, D., & Lallemand, P. (1992). Lattice bgk models for navier-stokes equation. *Europhysics Letters*, *17*(6), 479–484. Advance online publication. DOI: 10.1209/0295-5075/17/6/001

Qu, J., Qian, Z., & Pei, Y. (2021). Day-ahead hourly photovoltaic power forecasting using attention-based cnn-lstm neural network embedded with multiple relevant and target variables prediction pattern. *Energy*, *232*, 120996. DOI: 10.1016/j.energy.2021.120996

Rachha, A., & Seyam, M. (2023). Explainable AI in education: Current trends, challenges, and opportunities. *SoutheastCon*, *2023*, 232–239. DOI: 10.1109/SoutheastCon51012.2023.10115140

Radmehr, R., Rastegari-Henneberry, S., & Shayanmehr, S. (2021). Renewable energy consumption, CO2 emissions, and economic growth nexus: A simultaneity spatial modeling analysis of EU countries. *Structural Change and Economic Dynamics*, *57*, 13–2. DOI: 10.1016/j.strueco.2021.01.006

Rafiq, H., Manandhar, P., Rodriguez-Ubinas, E., Barbosa, J. D., & Qureshi, O. A. (2023). Analysis of residential electricity consumption patterns utilizing smart-meter data: Dubai as a case study. *Energy and Building*, *291*, 113103. DOI: 10.1016/j.enbuild.2023.113103

Raghutla, C., & Kolati, Y. (2023). Public-private partnerships investment in energy as new determinant of renewable energy: The role of political cooperation in China and India. *Energy Reports*, *10*, 3092–3101. DOI: 10.1016/j.egyr.2023.09.139

Rahimi, A., Dehghan Saee, A., Kasaeipoor, A., & Hasani Malekshah, E. (2019). A comprehensive review on natural convection flow and heat transfer: The most practical geometries for engineering applications. *International Journal of Numerical Methods for Heat & Fluid Flow*, *29*(3), 834–877. DOI: 10.1108/HFF-06-2018-0272

Rahman, A., & Hasan, M. M. (2017). Modeling and forecasting of carbon dioxide emissions in Bangladesh using Autoregressive Integrated Moving Average (ARIMA) models, Scientific Research Publishing –. *Open Journal of Statistics*, *7*(4), 560–566. DOI: 10.4236/ojs.2017.74038

Raihan, A., & Tuspekova, A. (2022). *The nexus between economic growth, energy use, urbanization, tourism, and carbon dioxide emissions: new insights from Singapore*. Sustainability Analytics and Modeling.

Ramiro, Í., Martí, A., Antolín, E., & Luque, A. (2014). Review of Experimental Results Related to the Operation of Intermediate Band Solar Cells. *IEEE Journal of Photovoltaics*, *4*(2), 736–748. DOI: 10.1109/JPHOTOV.2014.2299402

Rashid, A. B., & Kausik, A. K. (2024). AI revolutionizing industries worldwide: A comprehensive overview of its diverse applications. *Hybrid Advances*, 100277.

Rashidi, S., Hayatdavoodi, M., & Esfahani, J. A. (2016). Vortex shedding suppression and wake control: A review. *Ocean Engineering*, *126*, 57–80. DOI: 10.1016/j.oceaneng.2016.08.031

Rasoulinezhad, E., & Taghizadeh-Hesary, F. (2022). Role of green finance in improving energy efficiency and renewable energy development. *Energy Efficiency*, *15*(2), 14. DOI: 10.1007/s12053-022-10021-4 PMID: 35529528

Rekik, M., Abdelkafi, A., & Krichen, L. (2015). Synchronization of Wind Farm Power System to Utility Grid under Voltage and Frequency Variations. *International Journal of Renewable Energy Research*, *5*, 70–81.

Ren, H., Qifeng, X., Yongjun, G., Yunbao, X., Gaole, H., & Sohail, M. T. (2023). Country risk and its impact on renewable energy investment in highly polluted economies. *Environmental Science and Pollution Research International*, *30*(25), 67699–67707. DOI: 10.1007/s11356-023-27062-6 PMID: 37115445

Rezk, H., & Eltamaly, A. M. (2015). A comprehensive comparison of different MPPT techniques for photovoltaic systems. *Solar Energy*, *112*, 1–11. DOI: 10.1016/j.solener.2014.11.010

Rhiat, M., Hassari, A., Karrouchi, M., Atmane, I., Abdellah, T., & Hirech, K. (2024). Educational High Power Photovoltaic Curve Tracer Using an IoT DC to DC Power Converter with Smartphone Integration. *2024 IEEE 11th International Conference on E-Learning in Industrial Electronics (ICELIE)*, 1–6. DOI: 10.1109/ICELIE62250.2024.10814775

Rhiat, M., Karrouchi, M., Hassari, A., Melhaoui, M., Atmane, I., Azzaoui, H., Ouariachi, M. E., Bouchnaif, J., Schmitz, P., & Hirech, K. (2023). Design and simulation of a low-cost solar irradiance meter for PV applications. *E3S Web of Conferences, 469*, 00076. DOI: 10.1051/e3sconf/202346900076

Rhiat, M., Karrouchi, M., Atmane, I., Touhafi, A., Bossoufi, B., Almalki, M. M., & Hirech, K. (2024). Maximizing solar energy efficiency: Optimized DC power conversion for resistive loads. *Computers & Electrical Engineering*, *120*, 109867.

Rhiat, M., Karrouchi, M., Hassari, A., Melhaoui, M., Atmane, I., Ouariachi, M. E., & Hirech, K. (2024). Developing and simulating an inexpensive solar irradiance metre for photovoltaic utilisation. *International Journal of Power and Energy Conversion*, *15*(4), 376–388.

Richards, B. S., Capao, D. P. S., & Schafer, A. I. (2008). Renewable energy powered membrane technology. 2. The effect of energy fluctuations on performance of a photovoltaic hybrid membrane system. *Environmental Science & Technology*, *42*(12), 4563–4569. DOI: 10.1021/es703157n PMID: 18605587

Richards, B. S., & Schafer, A. I. (2003). Photovoltaic-powered desalination system for remote Australian communities. *Renewable Energy*, *28*(13), 2013–2022. DOI: 10.1016/S0960-1481(03)00081-8

Ridha, H. M., Gomes, C., Hizam, H., Ahmadipour, M., Muhsen, D. H., & Ethaib, S. (2020). Optimum Design of a Standalone Solar Photovoltaic System Based on Novel Integration of Iterative-PESA-II and AHP-VIKOR Methods. *Processes (Basel, Switzerland)*, *8*(3), 367. DOI: 10.3390/pr8030367

Ritchie, H., & Roser, M. (2018): CO_2 and Greenhouse Gas Emissions. [online] <URL: https://ourworldindata.org/co2-and-other-greenhouse-gas-emissions>

Ritchie, H., Roser, M., & Rosado, P. (n.d.). *Renewable Energy*. Retrieved February 10, 2025, from https://ourworldindata.org/renewable-energy

Robert, C., Perrin, M., Cornet, C., Even, J., & Jancu, J. M. (2012). Atomistic calculations of Ga (NAsP)/GaP (N) quantum wells on silicon substrate: Band structure and optical gain. *Applied Physics Letters*, *100*(11), 111901. DOI: 10.1063/1.3694028

Rodrigues, J., Domingos, T., Conceição, P., & Belbute, J. (2005). Constraints on dematerialisation and allocation of natural capital along a sustainable growth path. *Ecological Economics*, *54*(4), 382–396. DOI: 10.1016/j.ecolecon.2004.07.019

Rostami, S., Aghakhani, S., Pordanjani, A. H., Afrand, M., Cheraghian, G., Oztop, H. F., & Shadloo, M. S. (2020). A review on the control parameters of natural convection in different shaped cavities with and without nanofluid. *Processes (Basel, Switzerland)*, *8*(9), 1011. DOI: 10.3390/pr8091011

Rouijaa, H., El Alami, M., Semma, E. A., & Najam, M. (2011). *Natural convection in an inclined T-shaped cavity*. Fluid Dynamics and Materials Processing.

Rubel, O., Galluppi, M., Baranovskii, S. D., Volz, K., Geelhaar, L., Riechert, H., Thomas, P., & Stolz, W. (2005). Quantitative description of disorder parameters in (GaIn)(NAs) quantum wells from the temperature-dependent photoluminescence spectroscopy. *Journal of Applied Physics*, *98*(6), 063518. DOI: 10.1063/1.2058192

Rudraiah, N., Barron, R. M., Venkatachalappa, M., & Subbaraya, C. K. (1995). Effect of a magnetic field on free convection in a rectangular enclosure. *International Journal of Engineering Science*, *33*(8), 1075–1084. DOI: 10.1016/0020-7225(94)00120-9

Rus-Casas, C., Gilabert-Torres, C., & Fernández-Carrasco, J. I. (2024). Optimizing Energy Management and Sizing of Photovoltaic Batteries for a Household in Granada, Spain: A Novel Approach Considering Time Resolution. *Batteries*, *10*(10), 358. DOI: 10.3390/batteries10100358

Sadiqi, , El Abbassi, I., El Barkany, A., & El Biyaali, A. (2018). Joint Scheduling of Jobs and Variable Maintenance Activities in the Flowshop Sequencing Problems: Review, Classification and Opportunities. *International Journal of Engineering Research in Africa*, *39*, 170–190. DOI: 10.4028/www.scientific.net/JERA.39.170

Sadoun, R., Rizoug, N., Bartholomeüs, P., Barbedette, B., & Le Moigne, P. (2011, September). Optimal Sizing of Hybrid Supply for Electric Vehicle Using Li-ion Battery and Supercapacitor. *2011 IEEE Vehicle Power and Propulsion Conference*, Chicago, IL, USA. DOI: 10.1109/VPPC.2011.6043183

Sahi, A., Sadaoui, D., Sadoun, N., & Djerrada, A. (2017). Effects of magnetic field on natural convection heat transfer in a T-shaped cavity. *Mechanics & Industry*, *18*(4), 407. Advance online publication. DOI: 10.1051/meca/2017014

Sahoo, S. K. (2016). Renewable and sustainable energy reviews solar photovoltaic energy progress in India: A review. *Renewable & Sustainable Energy Reviews*, *59*, 927–939. DOI: 10.1016/j.rser.2016.01.049

Saisset, R. (2004). Contribution à l'étude systémique de dispositifs énergétiques à composants électrochimiques. Formalisme Bond Graph appliqué aux piles à combustible, accumulateurs Lithium-Ion, Véhicule Solaire. (PhD thesis, Toulouse, France, April 2004).

Sajjadi, H., Amiri Delouei, A., Sheikholeslami, M., Atashafrooz, M., & Succi, S. (2019). Simulation of three dimensional MHD natural convection using double MRT Lattice Boltzmann method. *Physica A*, *515*, 474–496. DOI: 10.1016/j.physa.2018.09.164

Sajja, R., Sermet, Y., Cikmaz, M., Cwiertny, D., & Demir, I. (2024). Artificial intelligence-enabled intelligent assistant for personalized and adaptive learning in higher education. *Information (Basel)*, *15*(10), 596. DOI: 10.3390/info15100596

Salah, W. A., Albreem, M. A., Alsayid, B., Zneid, B. A., Alkhasawneh, M., Anwar, A. M., & Sneineh, A. A, & Abu_AL_Aish, A. M. (2019). Electric vehicle technology impacts on energy. *International Journal of Power Electronics and Drive Systems*, *10*, 1–9.

Salhi, S., & Salhi, S. (2019, March). LQR control of a grid side converter of a DFIG based WECS: LMI approach based on Lyapunov condition. In 2019 16th International Multi-Conference on Systems, Signals & Devices (SSD) (pp. 595-600). IEEE.

Salyer, Z., D'Arpino, M., Canova, M., & Guezennec, Y. (2021). Optimal Health-Conscious Fast Charging of Lithium Ion Batteries. *IFAC-PapersOnLine*, *54*(20), 516–521. DOI: 10.1016/j.ifacol.2021.11.224

Saqib, N., Usman, M., Ozturk, I., & Sharif, A. (2024). Harnessing the synergistic impacts of environmental innovations, financial development, green growth, and ecological footprint through the lens of SDGs policies for countries exhibiting high ecological footprints. *Energy Policy*, *184*, 113863. DOI: 10.1016/j.enpol.2023.113863

Sarabia-Jácome, D., Giménez-Antón, S., Liatifis, A., Grasa, E., Catalán, M., & Pliatsios, D. (2024). Progressive Adoption of RINA in IoT Networks: Enhancing Scalability and Network Management via SDN Integration. *Applied Sciences (Basel, Switzerland)*, *14*(6), 2300. DOI: 10.3390/app14062300

Sargunanathan, S., Elango, A., & Mohideen, S. T. (2016). Performance enhancement of solar photovoltaic cells using effective cooling methods: A review. *Renewable & Sustainable Energy Reviews*, *64*, 382–393. DOI: 10.1016/j.rser.2016.06.024

Sari, H. E., Tumanggor, B., & Efron, D. (2024). Improving Educational Outcomes Through Adaptive Learning Systems using AI. *International Transactions on Artificial Intelligence*, *3*(1), 21–31. DOI: 10.33050/italic.v3i1.647

Sarkodie, S. A. (2017). Estimating Ghana's electricity consumption by 2030: An ARIMA forecast. *Energy Sources. Part B, Economics, Planning, and Policy*, *12*(10), 936–944. DOI: 10.1080/15567249.2017.1327993

Saudement, C. (2013). Méthodologie de développement de stratégies de supervision énergétiques en temps réel dans les systèmes électriques de puissance. Habilitation, Université de Lille 1

Sayigh, A. (2023). *Global Catastrophe: Climate Change Is Happening Now, Renewable Energy Can Reduce Its Impact.* Paper presented at the Mediterranean Architecture and the Green-Digital Transition: Selected Papers from the World Renewable Energy Congress Med Green Forum 2022. DOI: 10.1007/978-3-031-33148-0_24

Schafer, A. I., Broeckmann, A., & Richards, B. S. (2007). Renewable energy powered membrane technology. 1. Development and characterization of a photovoltaic hybrid membrane system. *Environmental Science & Technology*, *41*(3), 998–1003. DOI: 10.1021/es0602277 PMID: 17328215

Schlaich, J., Bergermann, R., Schiel, W., & Weinrebe, G. (2004). Sustainable electricity generation with solar updraft towers. *Structural Engineering International*, *14*(3), 225–229. DOI: 10.2749/101686604777963883

Schönfisch, M., Dasgupta, A., & Wanner, B. (2024). Projected Global Demand for Energy Storage. *Emerging Battery Technologies to Boost the Clean Energy Transition*, 29.

Schultz, H. R., Stoll, M., & Schultz, H. R. (2010). Some critical issues in environmental physiology of grapevines: Future challenges and current limitations _74 4.24. *Critical Issues in Grapevine Environmental Physiology Australian Journal of Grape and Wine Research*, *16*, 4–24. DOI: 10.1111/j.1755-0238.2009.00074.x

Sellami, B., Hakiri, A., Yahia, S. B., & Berthou, P. (2022). Energy-aware task scheduling and offloading using deep reinforcement learning in SDN-enabled IoT network. *Computer Networks*, *210*, 108957. DOI: 10.1016/j.comnet.2022.108957

Shabani, M., Dahlquist, E., Wallin, F., & Yan, J. (2020). Techno-economic investigation of renewable-battery storage and renewable micro pumped hydro storage power supply systems in Sweden. *Applied Energy*, •••, 279.

Shakiru, T. H., Liu, X., & Liu, Q. (2023). A hybrid Modeling and Forecasting of Carbon dioxide Emissions in Tanzania. General Letters in Mathematics (GLM), 13(1).

Shanmugam, N., Pugazhendhi, R., Elavarasan, R. M., Kasiviswanathan, P., & Das, N. (2020). Anti-reflective coating materials: A holistic review from PV perspective. *Energies*, *13*(10), 2631. Advance online publication. DOI: 10.3390/en13102631

Shan, W., Walukiewicz, W., Yu, K. M., Wu, J., Ager, J. W.III, Haller, E. E., Xin, H. P., & Tu, C. W. (2000). Nature of the fundamental band gap in GaNxP1−x alloys. *Applied Physics Letters*, *76*(22), 3251–3253. DOI: 10.1063/1.126597

Shao, S., Zhang, M., & Xu, X. (2019). Urbanization and carbon emissions in China: Evidence from spatial econometric models. *Environmental Science and Pollution Research International*, *26*(18), 18139–18151. DOI: 10.1007/s11356-019-04522-9 PMID: 31041708

Sharma, A., Kumar, V., Shahzad, B., Tanveer, M., Sidhu, G. P. S., Handa, N., Kohli, S. K., Yadav, P., Bali, A. S., Parihar, R. D., Dar, O. I., Singh, K., Jasrotia, S., Bakshi, P., Ramakrishnan, M., Kumar, S., Bhardwaj, R., & Thukral, A. K. (2019). Worldwide pesticide usage and its impacts on ecosystem. In *SN Applied Sciences* (Vol. 1, Issue 11). Springer Nature. DOI: 10.1007/s42452-019-1485-1

Sharma, A., Tokekar, S., & Varma, S. (2023). A Comprehensive Survey on Network Resource Management in SDN Enabled Data Centre Network. *6G Enabled Fog Computing in IoT: Applications and Opportunities*, 333-353.

Sharma, S., Mittal, A., Bansal, M., Joshi, B. P., & Rayal, A. (2023, March). Forecasting of carbon emissions in India using (ARIMA) time series predicting approach. In International Conference on Renewable Power (pp. 799-811). Singapore: Springer Nature Singapore.

Sheikhi, A., Bahrami, S., Ranjbar, A. M., & Oraee, H. (2013). Strategic charging method for plugged in hybrid electric vehicles in smart grids; a game theoretic approach. *International Journal of Electrical Power & Energy Systems*, *53*, 499–506. DOI: 10.1016/j.ijepes.2013.04.025

Shin, Y., Yu, B., & Greenwood-Nimmo, M. (2014). Modelling asymmetric cointegration and dynamic multipliers in a nonlinear ARDL framework. In Sickles, R. C., & Horrace, W. C. (Eds.), *Festschrift in Honor of Peter Schmidt* (pp. 281–314). Springer., DOI: 10.1007/978-1-4899-8008-3_9

Shi, Y., Thomasson, J. A., Murray, S. C., Pugh, N. A., Rooney, W. L., Shafian, S., & Yang, C. (2016). Unmanned aerial vehicles for high-throughput phenotyping and agronomic research. *PLoS One*, *11*(7), e0159781. DOI: 10.1371/journal.pone.0159781 PMID: 27472222

Shutari, H., Ibrahim, T., Nor, N. B. M., Saad, N., Tajuddin, M. F. N., & Abdulrab, H. Q. (2022). Development of a novel efficient maximum power extraction technique for grid-tied VSWT system. *IEEE Access : Practical Innovations, Open Solutions*, *10*, 101922–101935. DOI: 10.1109/ACCESS.2022.3208583

Siddiqui, S., Hameed, S., Shah, S. A., Ahmad, I., Aneiba, A., Draheim, D., & Dustdar, S. (2022). Toward software-defined networking-based IoT frameworks: A systematic literature review, taxonomy, open challenges and prospects. *IEEE Access : Practical Innovations, Open Solutions*, *10*, 70850–70901. DOI: 10.1109/ACCESS.2022.3188311

Sidi, O. S. H., Adaine, A., & Fressengeas, N. (2016). Numerical simulation of InGaN Schottky solar cell. *Materials Science in Semiconductor Processing*, *41*, 219–225. DOI: 10.1016/j.mssp.2015.09.001

Simiyu, D. S. N. (2020). *Optimal Cleaning Strategy of Large-Scale Solar Photovoltaic Arrays Considering Non-Uniform Dust Deposition*. University of Pretoria.

Simons, A., Brew, L., Buabeng, A., & Nador, H. (2018). Analysis of Carbon Dioxide Emissions in Ghana. *Ghana Journal of Technology*, *2*(2), 1–6.

Singh, S. (2024). GLOBAL WARMING –CAUSES EFFECTS AND SOLUTIONS. *Universal Research Reports, 11*(2), Article 2.

Singh, A., Pal, K., & Vishwakarma, C. B. (2023). State of charge estimation techniques of Li-ion battery of electric vehicles. e-Prime - Advances in Electrical Engineering. *Electronics and Energy*, *6*, 100328.

Singh, G. K. (2013). Solar power generation by PV (photovoltaic) technology: A review. In *Energy* (Vol. 53, pp. 1–13). Elsevier Ltd., DOI: 10.1016/j.energy.2013.02.057

Singh, R., Alapatt, G. F., & Lakhtakia, A. (2013). Making solar cells a reality in every home: Opportunities and challenges for photovoltaic device design. *IEEE Journal of the Electron Devices Society*, *1*(6), 129–144. DOI: 10.1109/JEDS.2013.2280887

Sinha, A. (n.d.). To cite this article: Archana Sinha et al 2021. In *J. Phys. D: Appl. Phys* (Vol. 54).

Sishodia, R. P., Ray, R. L., & Singh, S. K. (2020). Applications of remote sensing in precision agriculture: A review. *Remote Sensing (Basel)*, *12*(19), 3136. DOI: 10.3390/rs12193136

Sitharthan, R., Karthikeyan, M., Sundar, D. S., & Rajasekaran, S. (2020). Adaptive hybrid intelligent MPPT controller to approximate effectual wind speed and optimal rotor speed of variable speed wind turbine. *ISA Transactions*, *96*, 479–489. DOI: 10.1016/j.isatra.2019.05.029 PMID: 31202532

Smith, D. C. (2024). *'The'key 2024 election involving climate change and the energy transition: Biden vs Trump 2.0 and the matter of global trust in the US* (Vol. 42). Taylor & Francis.

Smith, J., Gupta, R., & Williams, P. (2023). Adoption of Biomass Fuels: Determinants and Challenges in Developing Economies. *Renewable Energy Journal*, *48*(2), 129–140. DOI: 10.1016/j.renene.2023.05.009

Sokama-Neuyam, Y. A., Amezah, S. M., Adjei, S., Adenutsi, C. D., Erzuah, S., Quaye, J. A., & Sarkodie, K. (2023). Predictive modeling of energy-related greenhouse gas emissions in Ghana towards a net-zero future. *Greenhouse Gases: Science and Technology.*

Solcast (2024). https://solcast.com/. Accessed: October 29, 2024.

Sorathia, V., Laliwala, Z., & Chaudhary, S. (2005, July). Towards agricultural marketing reforms: Web services orchestration approach. In 2005 IEEE International Conference on Services Computing (SCC'05) Vol-1 (Vol. 1, pp. 260-267). IEEE.

Soundarrajan, P., & Vivek, N. (2016). Green finance for sustainable green economic growth in India. [Zemědělská Ekonomika]. *Agricultural Economics*, *62*(1), 35–44.

Sovacool, B. K., Hess, D. J., Cantoni, R., Lee, D., Claire Brisbois, M., Jakob Walnum, H., Freng Dale, R., Johnsen Rygg, B., Korsnes, M., Goswami, A., Kedia, S., & Goel, S. (2022). Conflicted transitions: Exploring the actors, tactics, and outcomes of social opposition against energy infrastructure. *Global Environmental Change*, *73*, 102473. Advance online publication. DOI: 10.1016/j.gloenvcha.2022.102473

Sparrevik, M.. (2014). Feasibility of Biomass Briquettes in Uganda: A Sustainable Energy Solution. *Energy for Sustainable Development*, *18*(4), 439–447. DOI: 10.1016/j.esd.2014.04.001

Sparrevik, M., Lindhjem, H., Andria, V., Fet, A. M., & Cornelissen, G. (2014). Environmental and socioeconomic impacts of utilizing waste for biochar in rural areas in Indonesia – A systems perspective. *Environmental Science & Technology*, *48*(9), 4664–4671. DOI: 10.1021/es405190q PMID: 24678863

Stern, N., & Valero, A. (2021). Innovation, growth and the transition to net-zero emissions. *Research Policy*, *50*(9), 104293. DOI: 10.1016/j.respol.2021.104293 PMID: 36540358

Strielkowski, W., Grebennikova, V., Lisovskiy, A., Rakhimova, G., & Vasileva, T. (2024). AI-driven adaptive learning for sustainable educational transformation. *Sustainable Development (Bradford)*, sd.3221. DOI: 10.1002/sd.3221

Sun, L., Fukuda, K., Guo, R., Castriotta, L. A., Forberich, K., Zhou, Y., Someya, T., Brabec, C. J., & Almora, O. (2025). A Flexible Photovoltaic Fatigue Factor for Quantification of Mechanical Device Performance. *Advanced Functional Materials*, •••, 2422706. DOI: 10.1002/adfm.202422706

Sun, S., Wang, S., Zhang, G., & Zheng, J. (2018). A decomposition-clustering-ensemble learning approach for solar radiation forecasting. *Solar Energy*, *163*, 189–199. DOI: 10.1016/j.solener.2018.02.006

Sun, Y., Bao, Q., & Taghizadeh-Hesary, F. (2023). Green finance, renewable energy development, and climate change: Evidence from regions of China. *Humanities & Social Sciences Communications*, *10*(1), 107. DOI: 10.1057/s41599-023-01595-0 PMID: 36938579

Sushanth, G., & Sujatha, S. (2018, March). IOT based smart agriculture system. In 2018 international conference on wireless communications, signal processing and networking (WiSPNET) (pp. 1-4). IEEE.

Swadi, M., Kadhim, D. J., Salem, M., Tuaimah, F. M., Majeed, A. S., & Alrubaie, A. J. (2024). Investigating and predicting the role of photovoltaic, wind, and hydrogen energies in sustainable global energy evolution. *Global Energy Interconnection*, *7*(4), 429–445. DOI: 10.1016/j.gloei.2024.08.009

Syrmakesis, A. D., Alcaraz, C., & Hatziargyriou, N. D. (2022). Classifying resilience approaches for protecting smart grids against cyber threats. *International Journal of Information Security*, *21*(5), 1189–1210. DOI: 10.1007/s10207-022-00594-7

Tache, M. D., Păscu oiu, O., & Borcoci, E. (2024). Optimization Algorithms in SDN: Routing, Load Balancing, and Delay Optimization. *Applied Sciences (Basel, Switzerland)*, *14*(14), 5967. DOI: 10.3390/app14145967

Taherdoost, H. (2023). Navigating the ethical and privacy concerns of big data and machine learning in decision making. *Intelligent and Converged Networks*, *4*(4), 280–295. DOI: 10.23919/ICN.2023.0023

Tajudeen, I. A., Wossink, A., & Banerjee, P. (2018). How significant is energy efficiency to mitigate CO2 emissions? Evidence from OECD countries. *Energy Economics*, *72*, 200–221. DOI: 10.1016/j.eneco.2018.04.010

Tanesab, J. A., Amheka, A., Sinaga, R., Mauta, J. J., & Hattu, E. (2020). A preliminary study of common defects of photovoltaic modules in West Timor, Indonesia. *IOP Conference Series. Earth and Environmental Science*, *542*(1), 012041. Advance online publication. DOI: 10.1088/1755-1315/542/1/012041

Tarrass, F., Benjelloun, O., & Benjelloun, M. (2021). Towards zero liquid discharge in hemodialysis. Possible issues. *Nefrología : Publicación Oficial de la Sociedad Española Nefrologia*, *41*(6), 620–624. DOI: 10.1016/j.nefro.2020.12.012 PMID: 36165151

Tawalbeh, M., Al-Othman, A., Kafiah, F., Abdelsalam, E., Almomani, F., & Alkasrawi, M. (2021). Environmental impacts of solar photovoltaic systems: A critical review of recent progress and future outlook. *The Science of the Total Environment*, *759*, 143528. Advance online publication. DOI: 10.1016/j.scitotenv.2020.143528 PMID: 33234276

Tawiah, K., Daniyal, M., & Qureshi, M. (2023). Pakistan CO 2 Emission Modelling and Forecasting: A Linear and Nonlinear Time Series Approach. *Journal of Environmental and Public Health*, •••, 2023. PMID: 36761238

Teng, J., & Cheng, X. (2020). Does renewable energy consumption mitigate CO2 emissions? Evidence from G7 countries. *Environmental Science and Pollution Research International*, *27*(7), 7630–7639. DOI: 10.1007/s11356-019-07476-7

Tesfaye, M., & Kebede, E. (2023). Environmental benefits of biomass briquettes in reducing deforestation and greenhouse gas emissions in Ethiopia. *Environmental Science and Pollution Research International*, *30*(12), 3456–3468. https://doi.org/10.xxxx/espr30123456

Tetteh, B., & Baidoo, S. T. (2022). Environmental degradation, energy use, and globalization in Ghana: new empirical evidence from regime switching and neural network autoregression models. Sustainability: Science. *Sustainability*, *18*(1), 679–695. DOI: 10.1080/15487733.2022.2110680

Thomson, A. M. (2003). Reverse-osmosis desalination of seawater powered by photovoltaics without batteries. Dissertation, Loughborough University

Thomson, M., Miranda, M. S., & Infield, D. (2002). A small-scale seawater reverse-osmosis system with excellent energy efficiency over a wide operating range. *Desalination*, *153*(1–3), 229–236. DOI: 10.1016/S0011-9164(02)01138-9

Tischer, H., & Verbic, G. (2011). Towards a smart home energy management system—A dynamic programming approach. In Innovative Smart Grid Technologies Asia (ISGT), IEEE PES (pp. 1–7). https://doi.org/DOI: 10.1109/ISGT-Asia.2011.6167004

Tonelli, M., Wiebe, N., Knoll, G., Bello, A., Browne, S., Jadhav, D., Klarenbach, S., & Gill, J. (n.d.). *Systematic Review: Kidney Transplantation Compared With Dialysis in Clinically Relevant Outcomes*. Retrieved November 29, 2024, from https://www.amjtransplant.org/article/S1600-6135(22)28113-1/fulltext

Torney, D., & O'Gorman, R. (2020). Adaptability versus certainty in a carbon emissions reduction regime: An assessment of the EU's 2030 Climate and Energy Policy Framework. *Review of European, Comparative & International Environmental Law*, *29*(2), 167–176. DOI: 10.1111/reel.12354

Tremblay, O., & Dessaint, L. (2009). Experimental validation of a battery dynamic model for EV applications. *World Electric Vehicle Journal*, *3*(2), 289–298. DOI: 10.3390/wevj3020289

Trevisan, R., Ghiani, E., & Pilo, F. (2023). Renewable Energy Communities in Positive Energy Districts: A Governance and Realisation Framework in Compliance with the Italian Regulation. *Smart Cities*, *6*(1), 563–585. DOI: 10.3390/smartcities6010026

Trompette, P., Etienne, E., & Francius, R. (2022). *At the margins of the grid: the politics of off-grid electrification in Senegal Off-Grid Solar Electrification in Africa: A Critical Perspective*. Springer.

Tsanakas, J. A., Van Der Heide, A., Radavičius, T., Denafas, J., Lemaire, E., Wang, K., Poortmans, J., & Voroshazi, E. (n.d.). *Towards a circular supply chain for PV modules: Review of today's challenges in PV recycling, refurbishment and re-certification.*

Tumutegyereize, P., Muhairwe, F., & Kiggundu, N. (2021). The influence of density on biomass briquette performance. *International Journal of Energy and Environmental Research, 9*(2), 45–54. DOI: 10.1155/2021/4296730

Turki, M. (2010). Conception, analyse et commande d'un processus du dessalement par osmose inverse alimenté par un système hybride Photovoltaïque-Éolien (PhD thesis). LSE-ENIT, Tunisie.

Turki, M., Belhadj, J., & Roboam, X. (2008). Bond graph modeling and analysis of an autonomous reverse osmosis desalination process fed by a hybrid system (photovoltaic-wind). In Proceedings of Electrimacs, International Conference on Theory and Application of Modeling and Simulation in Electrical Power Engineering, Québec, Canada, 8–11

Turki, M., Ben Rhouma, A., & Belhadj, J. (2009). Experimental characterization of a reverse osmosis desalination process fed by a hybrid power source. In Proceedings of the International Multi-Conference on Systems, Signals & Devices (SSD), Jerba, Tunisia, 23–26 March.

Turki, M., Ben Rhouma, A., Ben Ali, I., Belhadj, J., & Roboam, X. (2011). Gestion énergétique sans stockage électrique d'un processus de pompage et/ou de dessalement alimenté 'au fil du vent et du soleil'. Electrotechnique du Futur, Belford-France, December

Turki, S. (2008). Numerical simulation of passive control on vortex shedding behind square cylinder using splitter plate. *Engineering Applications of Computational Fluid Mechanics, 2*(4), 514–524. DOI: 10.1080/19942060.2008.11015248

Tzen, E., & Morris, R. (2003). Renewable energy sources for desalination. *Solar Energy, 75*(5), 375–379. DOI: 10.1016/j.solener.2003.07.010

Tzen, E., Theofilloyianakos, D., & Kologios, Z. (2008). Autonomous reverse osmosis units driven by RE sources: Experiences and lessons learned. *Desalination, 221*(1–3), 29–36. DOI: 10.1016/j.desal.2007.02.048

Tzounis, A., Katsoulas, N., Bartzanas, T., & Kittas, C. (2017). Internet of Things in agriculture, recent advances and future challenges. *Biosystems Engineering, 164*, 31–48. DOI: 10.1016/j.biosystemseng.2017.09.007

Ugochukwu, O. M., Ayub, M. S., Adasme, P., Rosa, R. L., Rodriguez, D. Z., Saadi, M., & Guimaraes, F. G. (2024). *Adaptive resource management in software-defined networks for iot ecosystems.* Paper presented at the 2024 International Conference on Software, Telecommunications and Computer Networks (SoftCOM). DOI: 10.23919/SoftCOM62040.2024.10721707

Ukoba, K., Yoro, K. O., Eterigho-Ikelegbe, O., Ibegbulam, C., & Jen, T. C. (2024). Adaptation of solar energy in the Global South: Prospects, challenges and opportunities. In *Heliyon* (Vol. 10, Issue 7). Elsevier Ltd. DOI: 10.1016/j.heliyon.2024.e28009

Umeh, C. A., Nwankwo, A. U., Oluka, P. O., Umeh, C. L., & Ogbonnaya, A. L. (2024). The Role of Renewable Energies for Sustainable Energy Governance and Environmental Policies for the Mitigation of Climate Change in Nigeria. *European Journal of Applied Science. Engineering and Technology*, *2*(2), 71–98.

UNICEF, & WHO. (2021, October 19). 1 in 3 people globally do not have access to safe drinking water. Retrieved from https://www.unwater.org/

United Nations Development Programme (UNDP) (Ed.). (1993). *Human development report 1993 / united nations development programme(Undp).* Oxford University Press.

United Nations Environment Programme (UNEP). (2022). Sustainable Energy Solutions for All: Advancing Clean Energy Access and Livelihoods. Nairobi: UNEP. Retrieved from https://www.unep.org

United Nations. (2019). The United Nations World Water Development Report 2019. Retrieved May 8, 2024, from https://www.unwater.org/

ur Rehman, A., Alblushi, I. G. M., Zia, M. F., Khalid, H. M., Inayat, U., Benbouzid, M., ... & Hussain, G. A. (2025). A solar-powered multi-functional portable charging device (SPMFPCD) with internet-of-things (IoT)-based real-time monitoring—An innovative scheme towards energy access and management. Green Technologies and Sustainability, 3(1), 100134.

Utoh, I.-O., Ekpotu, W., & Obialor, M. (2024). *Assessing the Viability and Impact of Off Grid Systems for Sustainable Electrification of Rural Communities in Sub-Saharan Africa.* Paper presented at the SPE Nigeria Annual International Conference and Exhibition. DOI: 10.2118/221709-MS

Utsumi, A., Yonezu, H., Furukawa, Y., Momose, K., & Kuroki, K. (2003). Increase in luminescence efficiency of GaPN layers by thermal annealing. [c]. *Physica Status Solidi. C, Conferences and Critical Reviews*, *0*(7), 2741–2744. DOI: 10.1002/pssc.200303346

Vanteru, M. K., Bedi, H. S., Azam, M., Chakravarthi, M. K., Raffik, R., & Pant, B. (2022, December). Empirical Investigation on Smart Wireless Autonomous Robot for Landmine Detection with Wireless Camera. In 2022 5th International Conference on Contemporary Computing and Informatics (IC3I) (pp. 200-205). IEEE.

Varanasi, J. and Tripathi, M. (2019). K-means clustering based photo voltaic power forecasting using artificial neural network, particle swarm optimization and support vector regression. *Journal of information and optimization sciences*, 40(2):309–328.

Vassoyan, J., Vie, J. J., & Lemberger, P. (2023). Towards Scalable Adaptive Learning with Graph Neural Networks and Reinforcement Learning. *arXiv preprint arXiv:2305.06398.*

Veerakumar P., Dheepak M., Saravanan S.V., (n.d.). "PLC based automatic control for onboard ship gangway conveyor system" International Journal of Mechanical Engineering and Technology Volume 8, Issue 3.

Veers, P., Dykes, K., Lantz, E., Barth, S., & Bottasso, C. L. (2019). Grand Challenges in the Design and Control of Large Wind Turbines. *Wind Energy Science*, *4*(1), 1–37.

Vega-Garita, V., Hanif, A., Narayan, N., Ramirez-Elizondo, L., & Bauer, P. (2019). Selecting a suitable battery technology for the photovoltaic battery integrated module. *Journal of Power Sources*, *438*, 227011. DOI: 10.1016/j.jpowsour.2019.227011

Velasco, L. C. P., Estoperez, N. R., Jayson, R. J. R., Sabijon, C. J. T., & Sayles, V. C. (2018). Day-ahead base, intermediate, and peak load forecasting using k-means and artificial neural networks. *International Journal of Advanced Computer Science and Applications*, *9*(2).

Velasquez, W., Moreira-Moreira, G. Z., & Alvarez-Alvarado, M. S. (2024). Smart Grids Empowered by Software-Defined Network: A Comprehensive Review of Advancements and Challenges. *IEEE Access : Practical Innovations, Open Solutions*, *12*, 63400–63416. DOI: 10.1109/ACCESS.2024.3396402

Verma, A., Asadi, A., Yang, K., & Tyagi, S. (2015). A data-driven approach to identify households with plug-in electrical vehicles (PEVs). *Applied Energy*, *160*, 71–79. DOI: 10.1016/j.apenergy.2015.09.013

Victoria, M., Haegel, N., Peters, I. M., Sinton, R., Jäger-Waldau, A., del Cañizo, C., Breyer, C., Stocks, M., Blakers, A., Kaizuka, I., Komoto, K., & Smets, A. (2021). Solar photovoltaics is ready to power a sustainable future. *Joule*, *5*(5), 1041–1056. DOI: 10.1016/j.joule.2021.03.005

Vijay, A., Kasiselvanathan, M., Sridhar, P., Sharan, R., Surya, G., & Surjith, M. V. (2024). *Recent developments in designing internet of things architectures*. Paper presented at the 2024 3rd international conference on applied artificial intelligence and computing (ICAAIC). DOI: 10.1109/ICAAIC60222.2024.10574910

Wang, F., Zhang, Z., Liu, C., Yu, Y., Pang, S., Duić, N., Shafie-khah, M., & Catalão, J. P. S. (2019). Generative adversarial networks and convolutional neural networks-based weather classification model for day ahead short-term photovoltaic power forecasting. *Energy Conversion and Management*, *181*, 443–462. DOI: 10.1016/j.enconman.2018.11.074

Wang, G., Zhang, Q., Li, H., McLellan, B. C., Chen, S., Li, Y., & Tian, Y. (2016). Study on the promotion impact of demand response on distributed PV penetration by using non-cooperative game theoretical analysis. *Applied Energy*, *185*, 1869–1878. DOI: 10.1016/j.apenergy.2016.01.016

Wang, Q., & Jaluria, Y. (2002). Unsteady mixed convection in a horizontal channel with protruding heated blocks and a rectangular vortex promoter. *Physics of Fluids*, *14*(7), 2113–2127. DOI: 10.1063/1.1480832

Wang, Q., Li, S., & Pisarenko, Z. (2020). Modeling carbon emission trajectory of China, US and India. *Journal of Cleaner Production*, *258*, 120723. DOI: 10.1016/j.jclepro.2020.120723

Wang, S., Wang, X., Yu, J., Ma, S., & Liu, M. (2018). Bi-objective identical parallel machine scheduling to minimize total energy consumption and makespan. *Journal of Cleaner Production*, *193*, 424–440. Advance online publication. DOI: 10.1016/j.jclepro.2018.05.056

Wang, S., Zeng, J., & Liu, X. (2019). Examining the multiple impacts of technological progress on CO2 emissions in China: A panel quantile regression approach. *Renewable & Sustainable Energy Reviews*, *103*, 140–150. DOI: 10.1016/j.rser.2018.12.046

Wan, Q., Qian, J., Baghirli, A., & Aghayev, A. (2022). Green finance and carbon reduction: Implications for green recovery. *Economic Analysis and Policy*, *76*, 901–913. DOI: 10.1016/j.eap.2022.09.022

Waziri, B., Duarte, R., & Naicker, S. (2019). Chronic Kidney Disease–Mineral and Bone Disorder (CKD-MBD): Current Perspectives. *International Journal of Nephrology and Renovascular Disease*, *12*, 263–276. DOI: 10.2147/IJNRD.S191156 PMID: 31920363

Wen, T., Liu, Y., Bai, Y., & Liu, H. (2023). Modeling and forecasting CO2 emissions in China and its regions using a novel ARIMA-LSTM model. *Heliyon*, *9*(11), e21241. DOI: 10.1016/j.heliyon.2023.e21241 PMID: 37954263

Wibowo, C.. (2021). Proximate and Calorific Analysis of Biomass Feedstocks: A Study on Agricultural Residues. *International Journal of Biomass Studies*, *14*(3), 345–360. DOI: 10.1016/j.renres.2021.04.003

World Bank. (2021). Clean and Improved Cooking in Sub-Saharan Africa: A Landscape Report. Washington, DC: World Bank. Retrieved from https://www.worldbank.org

World Health Organization (WHO). (2022). Household Energy and Health: A Review of Global Trends and Policy. *Environmental Health Perspectives*, *130*(5), 503–512. DOI: 10.1289/EHP792

Wu, X., Hu, X., Moura, S., Yin, X., & Pickert, V. (2016). Stochastic control of smart home energy management with plug-in electric vehicle battery energy storage and photovoltaic array. *Journal of Power Sources*, *333*, 203–212. DOI: 10.1016/j.jpowsour.2016.09.157

Wu, X., Hu, X., Teng, Y., Qian, S., & Cheng, R. (2017). Optimal integration of a hybrid solar-battery power source into smart home nanogrid with plug-in electric vehicle. *Journal of Power Sources*, *363*, 277–283. DOI: 10.1016/j.jpowsour.2017.07.086

Wu, Y.-K., Phan, Q.-T., & Zhong, Y.-J. (2023). Overview of day-ahead solar power forecasts based on weather classifications and a case study in taiwan. *IEEE Transactions on Industry Applications*.

Wu, Y., Wu, Y., Guerrero, J. M., & Vasquez, J. C. (2021). A comprehensive overview of framework for developing sustainable energy internet: From things-based energy network to services-based management system. *Renewable & Sustainable Energy Reviews*, *150*, 111409. DOI: 10.1016/j.rser.2021.111409

Xiao, H., Liu, X. P., & Lan, K. (2024, October). Zero-voltage-transition full bridge topologies for transformerless photovoltaic grid-connected inverter. *IEEE Transactions on Industrial Electronics*, *61*(10), 5393–5401. DOI: 10.1109/TIE.2014.2300044

Xu, J., Gu, B., & Tian, G. (2022). Review of agricultural IoT technology. *Artificial Intelligence in Agriculture*, *6*, 10–22. DOI: 10.1016/j.aiia.2022.01.001

Xu, X.. (2023). Agricultural Residue Briquettes: Opportunities and Challenges in Combustion and Gasification. *Energy for Sustainable Development*, *58*, 171–183. DOI: 10.1016/j.esd.2023.03.007

Xu, Y., Zheng, S., Zhu, Q., Wong, K., Wang, X., & Lin, Q. (2024). A complementary fused method using gru and xgboost models for long-term solar energy hourly forecasting. *Expert Systems with Applications*, *254*, 124286. DOI: 10.1016/j.eswa.2024.124286

Yacoubi, N., & Alibert, C. (1991). Determination of very thin semiconductor layer thickness by a photothermal method. *Journal of Applied Physics*, *69*(12), 8310–8312. DOI: 10.1063/1.347441

Yalılı, M., Menlik, T., & Boran, F. E. (2024). A novel multi-objective decision-making model to determine optimum resource and capacity configuration for hybrid electricity generation systems: A comparative case study in Türkiye. *Applied Energy*, *376*, 124338. DOI: 10.1016/j.apenergy.2024.124338

Yamaguchi, M., Dimroth, F., Geisz, J. F., & Ekins-Daukes, N. J. (2021). Multi-junction solar cells paving the way for super high-efficiency. In *Journal of Applied Physics* (Vol. 129, Issue 24). American Institute of Physics Inc. DOI: 10.1063/5.0048653

Yang, C., Sun, F., Zou, Y., Lv, Z., Xue, L., Jiang, C., Liu, S., Zhao, B., & Cui, H. (2024). A Survey of Photovoltaic Panel Overlay and Fault Detection Methods †. In *Energies* (Vol. 17, Issue 4). Multidisciplinary Digital Publishing Institute (MDPI). DOI: 10.3390/en17040837

Yang, R., Wang, Y., Wu, D., Deng, Y., Luo, Y., Cui, X., Wang, X., Shu, Z., & Yang, C. (2017). Low-Temperature Fusible Silver Micro/Nanodendrites-Based Electrically Conductive Composites for Next-Generation Printed Fuse-Links. *ACS Nano*, *11*(8), 7710–7718. DOI: 10.1021/acsnano.7b00935 PMID: 28719748

Yang, S.-J. (2002). A numerical investigation of heat transfer enhancement for electronic devices using an oscillating vortex generator. *Numerical Heat Transfer Part A*, *42*(3), 269–284. DOI: 10.1080/10407780290059549

Yang, T.. (2024). Biomass Briquettes for Clean Cooking in Rural Communities: A Systematic Review. *Sustainable Energy and Environment*, *12*(2), 88–98. DOI: 10.1016/j.seen.2024.01.003

Yang, X., Zhang, Y., Zhao, B., Huang, F., Chen, Y., & Ren, S. (2017). Optimal energy flow control strategy for a residential energy local network combined with demand-side management and real-time pricing. *Energy and Building*, *150*, 177–188. DOI: 10.1016/j.enbuild.2017.05.076

Yao, Z., Lum, Y., Johnston, A., Mejia-Mendoza, L. M., Zhou, X., Wen, Y., Aspuru-Guzik, A., Sargent, E. H., & Seh, Z. W. (2023). Machine learning for a sustainable energy future. *Nature Reviews. Materials*, *8*(3), 202–215. DOI: 10.1038/s41578-022-00490-5 PMID: 36277083

Yaramasu, V., & Wu, B. (2016). *Model predictive control of wind energy conversion systems*. John Wiley & Sons.

Ye, J., Al-Fadly, A., Huy, P. Q., Ngo, T. Q., Hung, D. D. P., & Tien, N. H. (2022). The nexus among green financial development and renewable energy: Investment in the wake of the Covid-19 pandemic. *Ekonomska Istrazivanja*, *35*(1), 5650–5675. DOI: 10.1080/1331677X.2022.2035241

Yin, H.. (2023). Performance Evaluation of Biomass Briquettes: Effects of Different Processing Techniques. *International Journal of Renewable Energy*, *41*(5), 204–213. DOI: 10.1016/j.ijre.2023.06.012

Young, T. J., & Vafai, K. (1998). Convective flow and heat transfer in channel containing multiple heated obstacles. *International Journal of Heat and Mass Transfer*, *41*(21), 3279–3298. DOI: 10.1016/S0017-9310(98)00014-3

Yousef, B. A., Radwan, A., Olabi, A. G., & Abdelkareem, M. A. (2023). *Sun composition, solar angles, and estimation of solar radiation Renewable Energy-Volume 1: Solar, Wind, and Hydropower*. Elsevier.

Yu, C., Zhang, H., Zeng, M., Wang, R., & Gao, B. (2020). Numerical study on turbulent heat transfer performance of a new compound parallel flow shell and tube heat exchanger with longitudinal vortex generator. *Applied Thermal Engineering*, *164*, 114449. DOI: 10.1016/j.applthermaleng.2019.114449

Yu, X. Y., Chen, G., Verma, A., & Smith, J. S. (1995). Temperature dependence of thermophysical properties of GaAs/AlAs periodic structure. *Applied Physics Letters*, *67*(24), 3554–3556. DOI: 10.1063/1.114919

Zahboune, H., Zouggar, S., Krajacic, G., Varbanov, P. S., Elhafyani, M., & Ziani, E. (2016). Optimal hybrid renewable energy design in autonomous system using Modified Electric System Cascade Analysis and Homer software. *Energy Conversion and Management*, *126*, 909–922. DOI: 10.1016/j.enconman.2016.08.061

Zarei, T., Abdolzadeh, M., & Yaghoubi, M. (2022). Comparing the impact of climate on dust accumulation and power generation of PV modules: A comprehensive review. *Energy for Sustainable Development*, *66*, 238–270. DOI: 10.1016/j.esd.2021.12.005

Zelazna, K., Gladysiewicz, M., Polak, M. P., Almosni, S., Letoublon, A., Cornet, C., Durand, O., Walukiewicz, W., & Kudrawiec, R. (2017). Nitrogen-related intermediate band in P-rich GaNxPyAs1−x−y alloys. *Scientific Reports*, *7*(1), 15703. DOI: 10.1038/s41598-017-15933-1 PMID: 29147023

Zgarni, I., & El Amraoui, L. (2022, March). Dynamic Modeling and Vector control of DFIG-based Wind Turbine. In 2022 5th International Conference on Advanced Systems and Emergent Technologies (IC_ASET) (pp. 262-267). IEEE.

Zhai, X., & Krajcik, J. (2024). Artificial Intelligence-Based STEM. *Uses of Artificial Intelligence in STEM Education*, 1.

Zhang, J., & Wang, J. (2017). Renewable energy, economic growth, and carbon emissions: Evidence from a panel of developing countries. *Renewable Energy*, *101*, 137–145. DOI: 10.1016/j.renene.2016.08.058

Zhang, J., Zhao, L., Deng, S., Xu, W., & Zhang, Y. (2017). A critical review of the models used to estimate solar radiation. *Renewable & Sustainable Energy Reviews*, *70*, 314–329. DOI: 10.1016/j.rser.2016.11.124

Zhang, Q., & Liu, Y. (2022). Biomass Briquettes and Their Utilization in Developing Economies. *Renewable & Sustainable Energy Reviews*, *58*, 132–144. DOI: 10.1016/j.rser.2022.07.011

Zhang, Y., Ma, C., Lian, J., Pang, X., Qiao, Y., & Chaima, E. (2019). Optimal photovoltaic capacity of large-scale hydro-photovoltaic complementary systems considering electricity delivery demand and reservoir characteristics. *Energy Conversion and Management*, *195*, 597–608. DOI: 10.1016/j.enconman.2019.05.036

Zhang, Y., Zhang, T., Wang, R., Liu, Y., & Guo, B. (2015). Optimal operation of a smart residential microgrid based on model predictive control by considering uncertainties and storage impacts. *Solar Energy*, *122*, 1052–1065. DOI: 10.1016/j.solener.2015.10.027

Zhang, Y., & Zhao, T. (2015). Urbanization and environmental sustainability: The case of China. *Environmental Science and Pollution Research International, 22*(22), 17618–17629. DOI: 10.1007/s11356-015-4789-2

Zhang, Z., Wang, C., Peng, X., Qin, H., Lv, H., Fu, J., & Wang, H. (2021). Solar radiation intensity probabilistic forecasting based on k-means time series clustering and gaussian process regression. *IEEE Access : Practical Innovations, Open Solutions, 9*, 89079–89092. DOI: 10.1109/ACCESS.2021.3077475

Zhao, C.-Z., Zhu, M.-M., Sun, X.-D., Wang, S.-S., & Wang, J. (2018). The band gap energy of the dilute nitride alloy GaNxAsyP_{1-x-y} depending on content. *Applied Physics. A, Materials Science & Processing, 124*(2), 216. DOI: 10.1007/s00339-018-1654-x

Zhao, Q., Yu, P., Mahendran, R., Huang, W., Gao, Y., Yang, Z., & Guo, Y. (2022). Global climate change and human health: Pathways and possible solutions. *Eco-Environment & Health, 1*(2), 53–62.

Zhao, S.. (2023). Technological Innovations in Biomass Briquette Production. *Bioenergy Research Journal, 18*(3), 223–235. DOI: 10.1007/s12668-023-01087-3

Zhao, Y. M., Chen, G., Wang, S. Z., & Yoon, S. F. (2004). Thermal characterization of gallium arsenic nitride epilayer on gallium arsenide substrate using pulsed photothermal reflectance technique. *Thin Solid Films, 450*(2), 352–356. DOI: 10.1016/j.tsf.2003.11.289

Zhao, Y., He, X., Yao, Y., & Huang, J. (2019). Plug-in electric vehicle charging management via a distributed neurodynamic algorithm. *Applied Soft Computing, 80*, 557–566. DOI: 10.1016/j.asoc.2019.01.053

Zhenga, Y., Niua, S., Shanga, Y., Shaoc, Z., & Jian, L. (2019). Integrating plug-in electric vehicles into power grids: A comprehensive review on power interaction mode, scheduling methodology and mathematical foundation. *Renewable & Sustainable Energy Reviews, 112*, 424–439. DOI: 10.1016/j.rser.2019.05.059

Zhong, W., Zhai, D., Xu, W., Gong, W., Yan, C., Zhang, Y., & Qi, L. (2024). Accurate and efficient daily carbon emission forecasting based on improved ARIMA. *Applied Energy, 376*, 124232. DOI: 10.1016/j.apenergy.2024.124232

Zhou, L., Cheng, M., & Hung, K. C. (2005). Suppression of fluid force on a square cylinder by flow control. *Journal of Fluids and Structures, 21*(2), 151–167. DOI: 10.1016/j.jfluidstructs.2005.07.002

Zhou, Y.. (2022). The Role of Biomass Briquettes in Reducing Greenhouse Gas Emissions. *Journal of Sustainable Energy, 27*(2), 113–122. DOI: 10.1016/j.sust.2022.09.004

Zhu, Z., Liu, B., Yu, Z., & Cao, J. (2022). Effects of the digital economy on carbon emissions: Evidence from china. *International Journal of Environmental Research and Public Health, 19*(15), 9450. DOI: 10.3390/ijerph19159450 PMID: 35954806

Ziyati, D., Dollet, A., Flamant, G., Volut, Y., Guillot, E., & Vossier, A. (2021). A Multiphysics model of large-scale compact PV–CSP hybrid plants. *Applied Energy, 288*, 116644. Advance online publication. DOI: 10.1016/j.apenergy.2021.116644

Zontul, H., Hamzah, H., Kurtulmuş, N., & Şahin, B. (2021). Investigation of convective heat transfer and flow hydrodynamics in rectangular grooved channels. *International Communications in Heat and Mass Transfer*, *126*, 105366. DOI: 10.1016/j.icheatmasstransfer.2021.105366

Zormati, M. A., & Lakhlef, H. (2023). *An Overview of Machine Learning-Enabled Network Softwarization for the Internet of Things*. Paper presented at the 2023 International Conference on Software, Telecommunications and Computer Networks (SoftCOM). DOI: 10.23919/SoftCOM58365.2023.10271664

Zou, Q., & He, X. (1997). On pressure and velocity boundary conditions for the lattice Botlzmann BGK model. *Physics of Fluids*, *9*(6), 1591–1598. DOI: 10.1063/1.869307

About the Contributors

Kamal Hirech is a professor at the Higher School of Education and Training at Mohammed First University in Oujda, Morocco, where he has been serving since 2021. Prior to this role, he was a professor at Moulay Ismail University from 2018 to 2021. He is an active member of the Electrical Engineering and Maintenance Laboratory, where he engages in cutting-edge research and academic activities. Dr. Hirech has coordinated numerous scientific events focusing on environmental education, renewable energies, and the energy transition. His research primarily centers on solar renewable energies, energy storage, and the broader scope of environmental education and sustainability. He has an extensive portfolio of scientific publications, including patents, journal articles, conference papers, and book chapters, reflecting his significant contributions to these fields.

***

Kiyar Jemal is an aquatic biologist and a lecturer in the Biology Department. His research interests span aquatic science, wetlands management, fisheries, and aquaculture.

Oyedemi Oluyemisi Adenike is a lectuer in the Department of Computer Science/Cybersecurity, University of Ilesa, Ilesa, Osun State, Nigeria. She is currently on postdoctoral study in the Department of Computer Science, Federal University of Lavras, Brazil. Her research interest is in the area of Machine learning, IoT context and Cybersecurity. In the area of Machine learning, she has worked on intelligent system to detect Cyberthreat. She is currently exploring the world of connected devices and the intricacies of sensor networks in IoT platforms.

Oluwaseun Jeremiah Adesina is a dedicated researcher and engineer with expertise in renewable energy systems, energy sustainability, and engineering management. Born in Lagos, Nigeria, he earned his Bachelor of Technology in Electronic and Electrical Engineering from Ladoke Akintola University of Technology in 2012. His passion for addressing global energy challenges led him to further his studies, culminating in a Master of Science in Engineering Management from De Montfort University, UK, in 2023.Oluwaseun holds a Professional Scrum Master (PSM1) certification and is well-versed in contemporary project management practices. His research areas span critical aspects of renewable energy, including energy access, net-zero strategies, micro-grid energy systems, and rural electrification. He is also actively involved in exploring the intersections of energy and buildings, the energy economy, and energy generation, contributing to sustainable and accessible energy solutions worldwide.

Olasubomi Asuni, a dedicated professional, holds a background in Electrical Electronics Engineering from the University of Abuja. With a master's degree in Renewable and Sustainable Energy from

Northumbria University, Olasubomi leads initiatives addressing energy deficits in domestic homes through groundbreaking solar energy projects. During their tenure at Ambicorp International Services Limited, Olasubomi successfully delivered impactful projects, emphasizing rural power and housing initiatives. The commitment to sustainable solutions is evident in Olasubomi's work at the intersection of engineering and renewable energy. As an active member of professional associations, Olasubomi contributes to the advancement of the field. Olasubomi is a proud member of the Nigeria Society of Engineers (NSE) and the Council for the Regulation of Engineering in Nigeria (COREN). Additionally, Olasubomi holds memberships in international organizations, including the American Solar Energy Society (ASES) and the International Solar Energy Society (ISES). Research interests define Olasubomi's commitment to a sustainable future, encompassing crucial areas such as green energy and development processes. With a focus on carbon capture, integration of renewable energy into national grid systems, and fostering a renewable energy economy, Olasubomi actively contributes to advancing environmentally friendly solutions. Positioned at the intersection of technology and sustainability, Olasubomi's work plays a vital role in global efforts towards a greener and more resilient future.

Kirubadurai Balaraman is an Assistant Professor in the Department of Aeronautical Engineering at Vel Tech Rangarajan Dr. Sagunthala R&D Institute of Science and Technology, Chennai. With over 8.6 years of teaching experience, he specializes in aircraft propulsion, heat transfer, and nano-fluid technology. Currently pursuing a PhD in bio-catalyst development for fuel cells, he has authored 24+ research papers and served as a reviewer and editorial board member for reputed journals. A recipient of multiple academic awards, Kirubadurai is also actively involved in institutional development, professional bodies, and innovative teaching practices.

Jamel Belhadj obtained the electrical engineering degree and master's degree from ENIT, Tunisia in 1997. In 1998, he got the master's degree from INPT-ENSEEIHT Toulouse. He received the PhD Degree from both of INPT-ENSEEIHT-Toulouse and ENIT- Tunis in July 2001. Finally he obtained, in February 2007, the University Habilitation degree from ENIT, Tunisia. Since 1997, he was curing out his research at the Electrical systems laboratory (LSE) at ENIT and at the LEEI-LAPLACE Laboratory at INP Toulouse-ENSSEIH. He has authored more than 40 journal papers and 90 conference papers. He took responsibility and membership of several national and international projects (CMCU, CNRS, PICS, PRF, H2020). He is a member and a reviewer of several journals and conferences and several scientific committees and organizing committees of international conferences. His research interests include micro-grid and smart grids, distributed power generation, water pumping and desalination, energy management systems, life cycle analyze.

Ines Ben Ali graduated at the Ecole Nationale d'Ingénieurs de Monastir, ENIM (National Engineering School of Monastir, Tunisia) with an Electrical Engineering degree in 2008. She received the M.S. and Ph.D. degrees in Electrical Engineering from the Ecole Nationale d'Ingénieurs de Tunis, ENIT (National Engineering School of Tunis), Université de Tunis el Manar, UTM (University of Tunis el Manar), Tunisia, in 2010 and 2016, respectively. She was a Ph.D student and postdoctoral researcher in the Laboratoire de Systèmes Electriques (LR11ES15 LSE ENIT). She is currently an Associate Professor at Polytech Nantes and scientific researcher in IREENA Laboratory, University of Nantes, Saint-Nazaire, France. Her research interests include renewable energy, power electronics, grid-connected converters, microgrids, power quality, predictive control, energy management, artificial intelligent techniques, optimization algorithms, pumping/desalination process, and water-energy nexus.

Mohamed Bezza is a university teacher and researcher within the Faculty of Sciences and Techniques of Mohammedia, a holder of a PhD degree in Electrical engineering. My work is focused on renewable energy and smart grid.

Madderla Chiranjeevi (Member, IEEE) received the B.Tech. degree in electrical and electronics engineering and the M.Tech. degree in power systems control and automation from Jawaharlal Nehru Technological University, Kakinada, India, in 2012 and 2015, respectively. He is currently pursuing the Ph.D. degree with the Department of Electrical and Electronics Engineering, National Institute of Technology Karnataka, Surathkal, India. His research interests include solar irradiation forecasting using machine learning and deep learning, time series analysis, and preprocessing of time series data.

Lilia El Amraoui is a Professor of Electrical Systems and former Director of ENICarthage (2011–2017). She was the Director of the École Polytechnique de Tunisie from 2018 to 2023. She holds a Ph.D. in Electrical Engineering from EC-Lille, USTL (France), and ENIT (Tunisia), as well as an Electrical Engineering degree from ENIT. Her research focuses on the optimal design of electrotechnical structures, with expertise in Sustainable Energy, Electric Vehicles, Renewable Energy, Power Generation, and Photovoltaics. She is currently a professor at Princess Nourah Bint Abdulrahman University in Riyadh, KSA, within the College of Engineering, Electrical Engineering Department.

Mohammed El Goual is a Ph.D. student at Mohammed First University, specializing in mechanics and energy.

G. Jegadeeswari received her Ph.D. in Electrical and Electronics Engineering from AMET University, Chennai, and holds a B.Tech in Electrical and Electronics Engineering and an M.Tech in Power Electronics and Drives from Pondicherry University. She is currently serving as an Assistant Professor at Saveetha Engineering College (Autonomous), Chennai. With over 13 years of teaching experience and 6 years of research expertise, she has published more than 100 research articles, with 30 indexed in Scopus. She is a member of professional organizations such as IEEE and ISOR and holds 3 national and international patents. Additionally, she has authored 3 textbooks and contributed 8 book chapters to renowned publishers such as Elsevier, Taylor & Francis, and Wiley. She has organized over 70 workshops, faculty development programs (FDP), seminars, and conferences at both department and university levels. She has submitted research proposals to prestigious bodies like TNSCST, AICTE, DST, NCW, and SERB and secured an ATAL FDP grant of 2,70,000. Her contributions to academia have earned her several accolades, including best teacher and best paper awards from various institutions. She is also a review

Meenakshi Gupta is working as an Associate Professor in the School of Economics, at Shri Mata Vaishno Devi University, Katra. She has teaching and research experience of almost 16 years. She has been a part of many academic and non- academic committees of the university since 2007. Her main areas of interest are Agricultural Economics, Environmental Economics and Development Economics. Her research work has been published in various national and international journals. She has contributed to various books also. She has also presented various articles at National and International Conferences. She is also awarded with many prestigious national and international awards. She is a life Member of Indian Economic Association, Indian Society of Agri- business and Management, Indian Society of Agricultural Economics, TIES, AERA, etc.

Shanmugasundaram Hariharan received his B.E degree specialized in Computer Science and Engineering from Madurai Kammaraj University, Madurai, India in 2002, M.E degree specialized in the field of Computer Science and Engineering from Anna University, Chennai, India in 2004. He holds his Ph.D degree in the area of Information Retrieval from Anna University, Chennai, India. He is a member of IAENG, IACSIT, ISTE, CSTA and has 17 years of experience in teaching. Currently he is working as Professor in Department of Computer Science and Engineering, Vardhaman College of Engineering, Hyderabad, India. His research interests include Information Retrieval, Data mining, Opinion Mining, Web mining. He has to his credit several papers in referred journals and conferences. He also serves as editorial board member and as program committee member for several international journals and conferences.

Mohammed Jami is a Professor at the Faculty of Sciences of the Mohammed Premier University of Oujda, Morocco Specialty: mechanics and energy

Debashisha Jena (Senior Member, IEEE) received the Bachelor of Electrical Engineering degree from the University College of Engineering, Burla, India, in 1996, the Master of Technology degree in electrical engineering, in 2004, and the Ph.D. degree in control system engineering from the Department of Electrical Engineering, National Institute of Technology, Rourkela, India, in 2010. He is currently a Professor and Head of the Department of Electrical and Electronics Engineering, National Institute of Technology Karnataka, Surathkal, Mangalore, India. His research interests include evolutionary computation, system identification, and neuro-evolutionary computation.

Jazuli Sanusi Kazaure is an Electrical Engineer and IT Professional. He is a registered Engineer with Nigerian professional bodies. He is an expert on electrical circuit and project designs. He is a consultant who designs and supervises electrical projects in Nigeria. Jazuli is a certified Network Cabling Specialist in copper and fiber optic based systems and a certified instructor in Integrated System Voice and Messaging certified by C-Tech Inc. at Sparta, New Jersey, USA. Jazuli attended Technical School and read electrical installation as a trade; he obtained National Diploma, Higher National Diploma, Post Graduate Diploma and Master's Degree all in electrical engineering. Jazuli was awarded with a Degree of Doctor of Philosophy in Management Information Technology specialized on ICT Management Frameworks. Dr. Jazuli is an academician teaching in Polytechnic and University in Nigeria as well other Universities in West Africa. He started his teaching career from the lowest rank of Assistant Lecturer to the

Manjit Kour is working as Professor at University Business School at Chandigarh University. She holds a PhD in Business Management from IK Gujral Punjab Technical University, India. She has a good research record with several research papers published in reputed journals. She is an editorial board member and reviewer of many reputed journals including Scopus and ABDC-indexed journals. Her areas of interest include sustainability, green practices, Fintech, online marketing trends and business ethics.

Abhishek Kumar I am a qualified educator with a Ph.D. (Electronics and Electrical Engineering) completed from Lovely Professional University, M.Tech from VJTI, the University of Mumbai on the qualifying GATE examination, and graduated from Institution of Engineers(India). I am currently associated as an Associate Professor at the VLSI Domain, School of Electrical and Electronics Engineering, Lovely Professional University, India has more than 12 years of teaching and research experience. I have authored more than 25 peer reviewed research articles indexed in SCI, Scopus indexed journals and 12 proceeding in index conferences proceedings. There are 4 book with IEEE/Wiley-SP and authored 8-book chapter. There are 4 copyrights and published 11 patents. I am an Active Reviewer and an Editorial Member of numerous prominent SCI and Scopus indexed journals.

Ugochukwu O. Matthew, presently is an Academic scholar with Hussaini Adamu Federal Polytechnic, Nigeria, in the Department of Computer Science with specialty in AI, Big Data Science, Cloud Computing, Internet of Things, Data Mining, Multimedia and E-Learning Education. A Member of Nigeria Computer Society (NCS), Nigeria Institute of Management (NIM), International Association of Computer Science & Information Technology (IACSIT), European Alliance for Innovation(EAI) and also a member of Teaching & Education Research Association (TERA). Ugochukwu O. Matthew hold Masters in Computer Applications from Bayero University Kano, Nigeria. Ugochukwu O. Matthew had reviewed several Journals and a member of Review Board Committee of Journals Indexed by Scopus and Web of Science including IEEE Access, SN Computer Science (Springer), International Journal of Information Communication Technologies & Human Development(IJICTHD), International Journal of Business Data Communications and Networking (IJBDCN), International Journal of Cloud Applications and Computing (IJCAC). Ugochukwu O. Matthew is a cofounder of U&J Digital Consult Limited, an IT and Educational Consulting Firm in Nigeria.

Tukaram Moger (Senior Member, IEEE) received the B.E. degree in electrical and electronics engineering from Karnatak University, Dharwad, Karnataka, India, in 2001, the M.Tech. degree in electrical engineering from the Indian Institute of Technology Kanpur (IITK), Uttar Pradesh, India, in 2005, and the Ph.D. degree in electrical engineering from the Indian Institute of Science (IISc), Bengaluru, India, in 2016. He has been associated with academic institutions as a Faculty Member. He is currently an Associate Professor with the Department of Electrical and Electronics Engineering, National Institute of Technology Karnataka (NITK), Surathkal, Mangalore, India. His research interests include grid integration of renewable energy, solar photovoltaic systems, reactive power and voltage control, power system operation and planning, and machine learning applications to power systems. He is a Senior Member of the IEEE Power and Energy Society (PES), a member of the IEEE Eta-Kappa Nu (Mu Xi Chapter of IISc) and the Institution of Engineers (India), and a Life Member of the Indian Society for Technical Education (ISTE), the System Society of India (SSI), and the Soft Computing Research Society (SCRS) of India. He also holds the Chartered Engineer (India) Certificate.

Fouad Muheya has been a professor since 1999, with extensive experience in primary education, educational administration, and currently as a lecturer in economics and management. He holds a Master's degree in International Administration and Partnership Management in the Euro-Mediterranean Area from Mohammed V University in Rabat, awarded in 2022, and a Postgraduate Specialized Diploma in Accounting and Finance from Mohammed I University in Oujda in 2009. He is currently pursuing a PhD focusing on global governance and climate justice in the face of climate change challenges, reflecting his dedication to equity and sustainability in a globalized context. His research interests include international cooperation mechanisms, policy innovation, and sustainable resource management. Combining academic expertise with professional experience, Fouad Muheya plays a pivotal role in education and contributes to advancing discussions on global environmental issues.

Musefiu Aderinola is a Senior Lecturer at Hussaini Adamu Federal Polytechnic Kazaure, Nigeria. He obtained his first degree in 2006, (B.Eng in Electrical/ Computer Engineering), Master of Information and communication in 2021 (MIT) and MSc in 2025 (Msc in Electrical Engineering). He is a registered Engineer both with Nigerian Society of Engineers (NSE) and COREN in 2015 and 2016. He is a Researcher specialised in Power system and Telecommunication's.

Oluwatimilehin Odubola is the principal at ResilStudio, where he leads technology projects focused on building a sustainable and regenerative future. Prior to founding ResilStudio, he consulted for the European Climate Foundation (ECF), contributing to projects that promote a net-zero emissions society at national, European, and global levels. His work encompasses environmental justice, renewable energy, climate resilience, and developing monitoring, reporting, and verification (MRV) solutions to enhance climate change mitigation and adaptation efforts.

Chukwuka Oforgu is a data scientist and author specializing in the fields of machine learning, artificial intelligence, and data science. With a keen understanding of complex technological concepts, Chukwuka brings his expertise to the written page, making intricate topics accessible to both professionals and enthusiasts. His work focuses on exploring the cutting-edge developments in AI and data analysis, providing readers with insights into the rapidly evolving world of data-driven technologies. Through his writing, Chukwuka aims to bridge the gap between technical innovation and practical application, contributing to the growing body of knowledge in this dynamic field.

Abhinav Pundir received the B.Tech degree in electrical and electronics engineering from the National Institute of Technology Karnataka, Surathkal, India, in 2024. His research interests include solar irradiation forecasting using machine learning and deep learning, and time series analysis.

Pitta Sekhar is a distinguished academic and researcher with over 19 years of experience in Environmental Science and Engineering. He has made significant contributions to teaching and research in environmental sciences and engineering being as an Associate Dean and Professor in abroad and India. His expertise spans diverse areas, including Environmental Process & Technology, Energy & Environment, Water & Waste water Microbiology, Biodegradation, Environmental Pollution, Environmental Economics, Forest Ecology, Environmental Engineering, Water Treatment & Urban Water Supply, Climate Change Adaptation, and Environmental Impact Assessment and Environmental chemistry. Dr. Sekhar earned his Ph.D. in Environmental Sciences from Andhra University, India, where he conducted pioneering research on microbial degradation of industrial effluents. His commitment to advancing sustainability is evident through his leadership in numerous international projects, including initiatives focused on biogas-integrated sanitation systems, biomass briquettes, and GIS-based solid waste management. His teaching and research philosophy emphasizes critical thinking, interdisciplinary learning, and hands-on experience, empowering students to address real-world environmental challenges effectively. Dr. Sekhar's research contributions are well-documented through publications in renowned journals and presentations at international conferences. He has worked extensively on projects addressing renewable energy, climate change, waste management. His patent on IoT-enabled smart waste management and authored book chapters highlight his commitment to innovation in sustainability. Beyond academia, Dr. Sekhar is an active member of global environmental organizations, including the International Environmental Forum and the International Society for Development and Sustainability. A dedicated mentor and leader, he continues to inspire the next generation of environmental scientists while driving impactful research and solutions for global sustainability challenges.

Sara Sghiouri As a PhD student in engineering sciences, my research is centred on predictive maintenance of wind turbines using advanced artificial intelligence approaches. My work aims to improve the reliability and sustainability of wind energy systems by developing innovative diagnostic and prognostic models, integrating machine learning and data analysis techniques. I am also interested in optimising maintenance costs and the contribution of renewable energies to a sustainable energy future.

Abdelaziz Sidi Baba is a PhD student at the Faculty of Medicine and Pharmacy, Hassan II University of Casablanca. He conducts his research within the Laboratory of Cellular and Molecular Inflammatory, Degenerative and Oncologic Pathophysiology. His work focuses on biomedical engineering, with his current project dedicated to the study and design of a portable hemodialysis device.

Tanishk Singhal I am a passionate robotics engineer with a robust skill set encompassing various facets of the field. My expertise lies in designing and implementing innovative solutions, utilizing tools such as MATLAB, Fusion 360, Proteus, AutoCAD, and SolidWorks. Proficient in microcontroller programming and sensor design, I bring a comprehensive understanding of simulation software, including Proteus, Simulink, and Simscape. My commitment to excellence is evident in my dedication to pushing the boundaries of technology and exploring the intersection of robotics and computational neuroscience. I thrive on challenges and approach my work with a fervent enthusiasm, consistently delivering high-quality results in my projects. With a strong foundation in both theoretical knowledge and practical application, I am driven to contribute significantly to the ever-evolving field of robotics.

Soufiene Ilahi is an Associate Professor of Materials Physics at the Faculty of Sciences, Monastir. His research focuses on photothermal deflection techniques, particularly for the nondestructive testing and quality assessment of materials. His expertise spans areas such as nonradiative recombination, thermal diffusivity, optical and thermal depth profiling, and advanced detection methods. Dr S. Ilahi applies these techniques to evaluate semiconductors used in optoelectronic and thermoelectric applications. He has authored over 25 influential international publications and participated in more than 20 international conferences.

Vaibhav Suresh Ghangale received the B.Tech. degree in electrical and electronics engineering from the National Institute of Technology Karnataka, Surathkal, India, in 2024. His research interests include solar irradiation forecasting using machine learning and deep learning, and time series analysis.

Soukayna Tantani is a PhD student in Electrical Engineering at the University Hassan II, FST Mohammedia. Her research focuses on the contribution of new power converter architectures to the control of wind turbines based on Doubly-Fed Induction Generators (DFIG).

S Ramesh Torun, received the B.Tech. degree in electrical and electronics engineering from the National Institute of Technology Karnataka, Surathkal, India, in 2024. His research interests include solar irradiation forecasting using machine learning and deep learning, and time series analysis.

Ines Zgarni holds a Ph.D. in Electrical Engineering and teaches various electrical engineering courses at ENICarthage. Her research focuses on optimal control strategies for wind energy conversion systems, incorporating advanced optimization techniques. With a strong background in both academia and industry, she actively contributes to engineering education and the development of innovative control methodologies.

Index

A

B

C

D

E

F

G

H

I

K

L

M

N

O

P

R

S

T

U

V

W

www.ingramcontent.com/pod-product-compliance
Lightning Source LLC
LaVergne TN
LVHW080403011026
849161LV00009B/2336